Third Edition

FORENSIC SCIENCE

An Introduction to Scientific and Investigative Techniques

Third Edition

FORENSIC SCIENCE

An Introduction to Scientific and Investigative Techniques

Edited by

Stuart H. James and Jon J. Nordby

CRC Press
Taylor & Francis Group
Boca Raton   London   New York

CRC Press is an imprint of the
Taylor & Francis Group, an **informa** business

CRC Press
Taylor & Francis Group
6000 Broken Sound Parkway NW, Suite 300
Boca Raton, FL 33487-2742

© 2009 by Taylor & Francis Group, LLC
CRC Press is an imprint of Taylor & Francis Group, an Informa business

---

Library of Congress Cataloging-in-Publication Data

Forensic science : an introduction to scientific and investigative techniques / editors, Stuart H. James, Jon J. Nordby. -- 3rd ed.
    p. cm.
  Includes bibliographical references and index.
  ISBN 978-1-4200-6493-3 (hardcover : alk. paper)
  1. Forensic sciences--Handbooks, manuals, etc. 2. Criminal investigation--Handbooks, manuals, etc. 3. Evidence, Criminal--Handbooks, manuals, etc. 4. Criminal laboratories--Handbooks, manuals, etc. I. James, Stuart H. II. Nordby, Jon J. III. Title.

HV8073.F5835 2009
363.25--dc22
                                                  2009000878

Visit the Taylor & Francis Web site at
http://www.taylorandfrancis.com

and the CRC Press Web site at
http://www.crcpress.com

Table of Contents

Jon J. Nordby

Law and Science
Lawyers and Scientists
Theoretical Natural Sciences and Practical Forensic Sciences
Forensic Experts
An Expert's Role
Scientific Method

SECTION I: Forensic Pathology and Related Specialties

Ronald K. Wright

Introduction
Coroner System
Medical Examiner System
Education and Training of Forensic Pathologists
Training Requirements for a Forensic Pathologist
Duties of a Forensic Pathologist
Obtaining Appropriate Specimens
Photography
Report Preparation
Testimony
Case Studies

Janet S. Barber Duval, Catherine M. Dougherty, and Mary K. Sullivan

Overview
Domain of Practice
Collaborative Relationships
Evolution of Forensic Nursing
International Frontiers of Forensic Nursing
What Is and Is Not Forensic Nursing
Career Opportunities
Responsibilities of Hospitals in Evidence Collection and Preservation
Status of Forensic Nurses in the Legal System
Educational Preparation of Forensic Nurses
Future of Forensic Nursing

Ronald K. Wright

Introduction
Cause and Mechanism of Death
Manner of Death
Time of Death
Classification of Traumatic Deaths
Case Study

SECTION II: Evaluation of the Crime Scene

SECTION III: Forensic Science in the Laboratory

PRECEPTS FOR ETHICAL CONDUCT BY EXPERTS IN THE JUSTICE SYSTEM[*]

Foreword

James E. Starrs

Professor of Law Emeritus &
Professor of Forensic Sciences
The George Washington University

Ethical improprieties of all kinds are rife within the community of forensic scientists and other experts in the justice system, both in the United States and internationally, No one, no agency and no crime laboratory, not even ASCLAD-Lab accredited laboratories, is exempt from involvement in this deepening and all pervasive contagion. Books have been written[1] pointedly exploring the deficiencies in the practice of courtroom experts and the laboratories supporting them. But the stain of unethical behavior has not been abated or curbed.

Entire laboratories, among them those in Detroit and Houston, have been closed when their work-product has been found to be substantially deficient. In Virginia, a DNA technician's errors led to countless reevaluations of her work. And the credentials, or alleged credentials, of persons tendered as experts have been a fertile source of critical commentary, more often in the media than in the halls of scientific academia.

Gary S. Stocco testified as an immersion burn expert in prosecutions for child abuse and worse in Massachusetts, Indiana, Ohio and New Jersey with the approval of the courts in those jurisdictions when in fact his asserted qualifications were bogus. His claims to have undergraduate and graduate academic credentials were spurious when it was proved that any degrees he claimed were from a diploma mill in Louisiana by the name of Columbia State University. That so-called university was investigated by the Louisiana attorney general and raided at its mail drop location in Orange County, California by a California postal inspector.[2]

A particularly odious case of puffery and more in the qualifications of an expert was that of firearms expert Joseph Kopera[3] who headed up the Maryland State Police firearms unit. After a career of some 37 years as a firearms expert for the prosecution in Maryland, Kopera took his own life when his lies concerning his credentials caught up with him. His one year of study at the University of Baltimore was falsely expanded by Kopera to include nonexistent degrees from the Rochester Institute of Technology and the University of Maryland. Not only did his claimed academic degrees not hold up to scrutiny but a transcript said by him to be from the University of Maryland was shown to be a forgery. The fall-out from these ethical and even illegal transgressions brought the possibility of false convictions based on Kopera's testimony under the searing light of law enforcement reevaluations.

Even fully qualified forensic experts have been found to have veered from the straight and narrow of their ethical obligations as imposed by the standards in place in their laboratories. In Scotland, police officer Shirley McKie was erroneously charged with a crime in giving a false statement as to her absence from the site of a home invasion based on a fingerprint reported to be hers found at that location. Three qualified fingerprint experts were in error in making that positive identification. As a result Ms. McKie has been awarded a significant monetary reparation.

In the United States, the F.B.I. fingerprint unit has been the scene of much distress after three of their highly placed and long-term fingerprint experts misidentified a fingerprint in the Madrid bombing investigation as belonging to Brandon Mayfield, an Islamic convert and lawyer living in Oregon. It has been said that "confirmation bias" was the cause of the misidentifications. The F.B.I., in a civil suit for damages filed on Mayfield's behalf, was the loser to the tune of a substantial sum.

Even in civil litigation experts have been seen to step beyond the pale. In the much-bruited damage action against Merck & Co in the federal court in Louisiana concerning its marketed product Vioxx,[4] an M.D. named Barry Rayburn appeared as a "central witness" supporting Merck's view that Vioxx did not cause the plaintiff's heart attack. When Rayburn was asked at trial "Are you board-certified" his response was "Yes, I passed boards in internal medicine and cardiovascular disease." Later, after the trial had ended with Merck having prevailed, it was learned that Rayburn's board certifications in internal medicine and cardiology had expired many years previously. When challenged as to the accuracy of his trial assertions, Rayburn said "I did not intend to represent that I wasn't lapsed or expired at all." And, in truth, his statement at trial was literally, if not forthrightly, true. However, New Orleans Federal District Judge Fallon was of a different mind altogether, finding that Rayburn had "misrepresented his status."

In federal court in Pennsylvania,[5] John B. Torkelsen, a securities expert, recently pleaded guilty to perjury for, among other charges, having submitted statements in which he falsely stated under oath that he had been retained to give his expert opinion on the valuations and appraisals of public and private corporations by the law firm of Milberg Weiss on a *non-contingent fee* basis. Experts who merit the status of experts in litigation are on notice that contingent fee arrangements are ethically improper.

The most palpably disturbing of all the most recent instances of an expert's egregious defalcations occurred not in the disciplines of criminalistics but in the field of paediatric forensic pathology. Ontario Canada's Dr. Charles Smith[6] was accused of using "the Sherlock Holmes approach to pathology," resulting in mistakes in autopsies of many children whose deaths were ascribed by Smith to homicidal foul play. Indeed, during a recent inquiry into this deplorable situation in Ontario, Smith himself admitted many of the mistakes charged to him, some resulting in the conviction of innocent people. Ontario's former chief coroner Dr. Jim Young, during whose tenure many of Smith's mistakes occurred, is reported to have said that it would be inapposite to point the finger of blame "at one person." To Young, Smith's wrongdoing involved "a whole justice system and the solution has to be a justice system solution or it just simply isn't going to work."

The "Inquiry into Pediatric Forensic Pathology in Ontario" released its report on its investigations into the trumpery of Dr. Smith in October 2008. The report pointed to the inadequate training of Dr. Smith and his "sloppy and inconsistent" documentation which passed muster in the Canadian courts. The problem, as Dr. Young had testified, was found not to be that of one expert but that of other pathologists who operated in a vacuum of "oversight and accountability." Furthermore Dr. Smith was not exposed by whistle-blowers but his flawed work was defended "and even covered up" by his colleagues and peers, so the Inquiry reported. Apparently the medical profession's maxim "primum non nocere" (first, do no harm) was placed way back in the dusty recesses of those pathologists' commitment to professional integrity.

One could go on endlessly back to West Virginia's Fred Zain[6] and even before the misconceived serology reports of Zain to others in the professional community of forensic experts.[7] In the face of this tsunami of unethical behavior, can an answer to it and its rectification be divined?

In the search for a remedy a first stop might encompass a perusal of the relevant codes of ethics of the applicable licensing, accrediting or certifying associations or organizations. Are they adequate to the task of holding the wave of expert malfeasance in check?

One must recognize at the start that contemplation bakes no ethical loaves, nor do pious platitudes, however well intended. The codes of ethics that I have canvassed can be fairly described as glorifying glittering generalities in the dictates they espouse. The Code of Ethics of the California Association of Criminalists is representative. Its member criminalists are adjured to "be guided by those practices and procedures which are generally recognized within the profession to be consistent with a high level of professional ethics." Immediately following this declaration appears the statement that the criminalist's actions "shall at all times be above reproach, in good taste and consistent with proper moral conduct." To be fed on such a skimpy diet of generalities is to offer the expert little but a starvation diet of unworkable platitudes.

Such ethical adjurations are so fabulous as to conjure up the work of the fabulist. And, to my thinking, the best of the lot among fabulists was Aesop. So let us turn to Aesop's Fables[8] for such instructions as they might offer for the ethical conundrums and precepts of the expert in forensics in today's juridical world.

The ingredients of Aesop's Fables are an animal as the modus operandi, a surprise ending, and a moral lesson, whether explicated or not in the fable. The lessons following each fable are not designed to be clichés like "two wrongs do not make a right" or "do good and avoid evil" but real world applications of an ethical mandate, as opposed to an ephemeral and glittering generality. To that extent the fables are akin to the case studies that appear and inform the chapters in this book.

Aesop's Fables were good enough for Aristotle and Socrates to hold them in high regard. That being so, most assuredly, they should have some special place in the pantheon of precepts for experts in the justice (civil and criminal) system. The moral strictures appended to each of Aesop's tales are not mere afterthoughts of limited importance but are, instead, meant to give the fables an active and permanent place as tutorials by which to live or even just to ruminate. Like the fables themselves these instructionals are of universal and timeless appeal and relevance. It became for me, therefore, an inviting task to reformulate those tales and their instructionals as precepts applicable to the work-a-day world of expert witnesses.

The selections from Aesop's Fables which are included in this foreword in reconstructed form are paraphrases of the Aesop translations of Olivia and Robert Temple. I have taken this occasion to substitute different titles for the fables, titles which are keyed to experts in forensics. The numbering of the fables in the Temple edition is retained and positioned next to the titles of the individual fables. These replacement titles are all one-worders, signifying in succinct compass the precept that each fable is intended to convey to the reader. Since this writer has selected fables that seem to him, without extreme warping, to fit the practices of expert witnesses, it did not seem inappropriate for new precepts to be substituted to the refashioned fables, more tailored to the activities of expert witnesses today.

The precepts which are postscripts to the tales as retold in this foreword are not designed, as in Aesop's Fables, to be moralistic in their thrust since morals are often considered to be a matter of personal conscience without any firm roots in objectivity. Nevertheless, the precepts that are appended are intentionally didactic and, from this writer's perspective, of wide, if not also universal, application to the diurnal activities of expert witnesses, whether scientists or technicians and whether laboratory based or not.

Although Aesop's fables are plainly "studies of human nature," they could be seen as lampooning and lambasting human nature in the guise of a jungle book. The motto of Aesop, gleaned from these 358 fables, might be the unwelcome one popularized by Randy Newman who tells us "it's a jungle out there." Even in such a jungle which may approximate the justice system as viewed from the perspective of the expert in forensics there is a law of the wild with its precepts to be learned or at least soberly contemplated. On then to the paraphrased fables and their retooled precepts.

Fable #246: Truth

Two teenaged boys were seen standing at the counter in a butcher shop waiting to be served. Their patience unnerved by the duration of their wait one of the boys (call him A) decides to help himself, surreptitiously, to a piece of meat in the tray fronting the counter. In no time at all the purloined meat went from A's larcenous hands to a secure and unseen location under the shirt of the other boy (call him B). So far so bad, with the boys confidently resting easy that they have pulled a fast one on the butcher.

But before they can depart the butcher approaches them, while observing the now obvious place on the meat counter where the stolen meat had previously been positioned. Looking directly at B (who has the meat concealed under his shirt) the butcher asks:

"Did you steal a piece of meat from the counter here?" B's reply is quick and direct. "No, I did not," he earnestly declares. Turning to A (the one who stole the meat) the butcher, skeptical of the truth of B's answer, inquires. "Do you have the meat from my counter's display case?" Once again he receives a negative reply, stated by A even more emphatically than did B.

Precept: The expert witness is sworn to tell the truth, the whole truth and nothing but the truth. It is not the literal truth that the expert swears to tell but the whole truth. The ineptitude of the one who questions the expert on direct or cross-examination does not excuse the expert from the sworn obligation to tell the whole truth.

Fable #105: Blind-Sided

A doe had been born with but one eye. Being by nature a browser she went to the seashore to survey the scene. She kept her good eye focused on the shore on the look out for hunters. Her blind eye she turned to the sea, not expecting any danger to appear from that direction. But, to the doe's surprise, she suffered a fatal wound when shot by a hunter-poacher who just happened to be boating by on the sea. The doe, with her death coming nigh, bemoaned her short-sighted failing to judge a danger being likely to appear from the sea.

Precept: So the expert witness, although well-prepared scientifically for a courtroom appearance in which his/her opinion will be solicited, must keep a weather eye out for the limitations the rules of evidence will impose on his/her direct and cross-examination, a failure of which can compromise even the soundest and surest opinion.

Fable #41: Self-Interest

A fox was distraught for his tail had been cut off in a trap. Seeking a way out of his embarrassment among his peers, he called a meeting of the foxes at which he urged them to shorten their tails mimicking his. He argued that their tails were just useless extra baggage and merely a rudimentary appendage. But one of the assembly of foxes spoke up saying: "You would not be so quick to have us follow your advice were it not for your own self-interest."

Precept: The expert witness who puts himself and his own interests ahead of the attainment of a just and true result is altogether likely to have his words and opinions come to nought.

Fable #39: Humility

A fox and a monkey were traveling together. Along the way they disputed who had more distinction and more accomplishments to his credit. When they came upon a cemetery the monkey broke down in heavy sobs. The fox inquired what troubled the monkey. The monkey pointed to the impressive monuments in the cemetery and said, "I weep for these my lordly ancestors long buried in this cemetery." "What?" replied the fax disdainfully. "You can lie to your heart's content when none of those you claim to be your forebears can arise to contradict you."

Precept: When an expert who appears at trial is not confronted by an opposing expert, the expert should not consider that fact to be an invitation to embellish his/her opinion or qualifications with exaggerations or untruths.

Fable #45: Multi-Faceted

A murderer was in flight from his murder victim's family. As he neared the banks of the Nile he came upon a menacing wolf. In his panic he climbed a tree growing by the riverside. But his refuge was not altogether safe for a savage-looking serpent was climbing up the tree towards him. In his terror he fell into the river where a crocodile found him to be a tasty meal.

Precept: The pursuit of criminals, especially violent ones, should never be declared to be fruitless for the unexpected has been known to trap the fugitive. Cold case investigations, resorting to the skills of experts and others, should never be abandoned. Criminals should not be given the comfort of feeling safe in their fugitivity.

Fable #34: Congruence

A fox was seeking to escape the hunters who were tracking him. Coming upon a woodcutter the fox asked for refuge. The woodcutter took the fox to his cottage where he let him remain unseen by the hunters.

Upon the arrival of the hunters the woodcutter boldly lied that he had not seen the fox. But in gesture, rather than words, the woodcutter made it clear to the hunters that the fox was hiding in his cottage. And the fox saw those gestures of the woodcutter from his hiding place from which he watched the hunters departing without realizing that they had been gestured to their prey.

Precept: Expert witnesses often render their opinions in the form of a written report with their oral testimony to follow in court or in a deposition. It is important that there be no unexplainable incongruities between that which is expressed in the courtroom or a deposition and that which appears in a written report. Major divergences between the two will lead the way to the quick-witted and well-prepared cross-examiner to suggest that such divergences are indicative of concealment or worse.

Fable #258: Appearances

There was once a group of travelers seeking a way to exit an island on which they were marooned. From a hilltop they observed what to them at that distance appeared to be a warship. And such they took it to be until, on hurrying down the hill and coming closer to the object, they realized they were mistaken, for the warship now appeared to be a cargo ship. Whereupon they rushed headlong to the beach from which their escape seemed assured. But, to their utter dismay, the cargo ship was in reality nothing but floating brushwood. Loudly did they bemoan their having been taken in by appearances.

Precept: An expert should not rely for a certain opinion upon the appearances derived from presumptive tests in contrast to more conclusive, confirmatory tests. A closer analysis with more definitive tests can change appearances to a more certain and accurate determination.

Fable #22: Chance

A skilled fisherman was disconsolate, for in many hours of fishing he had caught nothing. As he sat dejected in his boat he saw that the waters were being disturbed by a fish being chased by a school of larger fish. The fugitive fish, to save itself, jumped into the fisherman's boat. And the fisherman was glad for his good fortune, it being an accident of fate.

Precept: Experts cannot always rely only on their knowledge and their instruments to suffice to answer the problem before them. Sometimes chance will play a part in resolving the dilemma that besets them. Science and luck are not necessarily antipodes in competition with each other.

Fable #182: Ambiguity

A hound, while hunting, snared a hare. In between biting his prey, the hound licked it amicably. The hare, tiring of such divided attention, said to the hound either to bite or to kiss him as he wanted to know whether his opinion of him was as a friend or an enemy.

Precept: Experts, in their opinions, should studiously avoid ambiguous words or statements which may be construed, to their regret, to be deliberately deceptive or misleading.

Fable #38: Self-Importance

A monkey drew such applause from an assembly of animals for his skillful dancing that he was elected to be their king. But the fox was filled with envy. Later the fox, seeing a piece of meat in a trap, took the monkey to see what he had found.

Being enticed by the fox with claims of his royal prerogative, the monkey blindly took up the meat causing the trap to spring and catch him. When the monkey berated the fox for ensnaring him, the fox replied "if you expect to be king, you should take every precaution to ensure your royalty."

Precept: An expert should not accept the role of a witness in the courtroom with an inflated view of his own importance, for the snare awaits such unpreparedness and hauteur.

Fable #32: Defeatism

A hungry fox came in sight of grapes handing in bunches from a vine which had grown up a tree. He longed for and lunged for those grapes but to no avail for unassisted he could not reach them. Going on his way, disgruntled, he remarked "the grapes were not ripe anyway."

Precept: Experts in their laboratory testing are frequently unsuccessful, neither being able to say aye or nay but at best only to report an inconclusive result. Rather than fudging a more definite opinion the expert must recognize that the progress of scientific knowledge or the testing modalities in use may not yet have reached the unreal level portrayed in a CSI episode.

Other examples of situations which an expert might encounter with the imperative to toe the line of an ethical precept could be gleaned from the fables of Aesop but the above suffices as a representative sampling of the width and breath of an expert's professional responsibilities in the justice system. In short the expert should always be mindful of the necessity to keep his eyes on the prize which is the uncompromised commitment to professional integrity. And that commitment, when applied, as in Aesop, to the real world of the expert is no glittering generality.

Endnotes

[1] Fisher, J., *Forensics Under Fire*, Rutgers University Press, New Brunswick, NJ, 2008; Pyrek, K.M., *Forensic Science Under Siege*, Academic Press, Burlington, MA, 2007; Barnett, P., *Ethics in Forensic Science: Professional Standards for the Practice of Criminalistics*, CRC Press LLC, Boca Raton, FL, 2001; Candills, P.J., Weinstock, R. and Martinez, R., *Forensic Ethics and the Expert Witness*, Springer, 2007; Kitaeff, J., ed., *Malingering, Lies, and Junk Science in the Courtroom*, Cambria Press, Youngstown, NY, 2007, Koppl, R. and Balko, R., *Forensic Science Needs Checks and Balances*, *Engage*: 9 (2) June 2008, p. 43–47.

[2] Kitaeff, supra n. 1 at 358-364.

[3] McMenamin, J., Police expert lied about credentials, *The Baltimore Sun*, 9/3/07; Witte, B., *Md. Cases Reviewed After Suicide*, A.P., 3/12/07.

[4] *Plunkett v. Merck & Co.*, E.D.La., No. 05-4046, 5/29/07.

[5] *U.S. v. Torkelsen*, E.D. Pa., No. 2:08–cr - 00121, 2/28/08.

[6] Fisher, supra note 1, p. 236-240.

[7] Kohn, A., *False Prophets*, Basil Blackwell, Inc., New York, 1986.

[8] Temple, O. and R., *The Complete Fables: Aesop*, Penguin Books, New York, 1998.

Preface

Forensic Science: An Introduction to Scientific and Investigative Techniques, Third Edition provides the student as well as the forensic practitioner with a general overview, focused understanding, and appreciation of the wide scope of the forensic science disciplines. This textbook was developed primarily to provide a standard text for students of forensic science at the advanced high school and college undergraduate and graduate level. It serves well as a useful reference to those already involved in forensic science or the criminal justice system including investigators, forensic specialists, prosecutors, and defense attorneys.

Unlike this text, many existing texts offered as "introductions to forensic science" focus on only one discipline—the discipline of criminalistics and related laboratory subjects—to the exclusion of fundamental topics such as forensic pathology, forensic anthropology, forensic engineering, and bloodstain pattern analysis, to name a few. It is the intention of the editors to offer a much more comprehensive text in terms of general forensic science topics with technical and scientific detail that adequately introduces the breadth and richness of the forensic sciences to students and practitioners alike.

The editors, as well as CRC Press/Taylor & Francis Group, have assembled excellent contributing authors who are noted and highly respected forensic scientists and legal practitioners from throughout the United States as well Canada and the Netherlands. Many of the chapters that comprised the Second Edition have been revised and updated with the addition of a new chapter on Forensic Digital Photo Imaging. The book is divided into 7 sections with a total of 34 chapters representing a wide scope of forensic disciplines, as well as a broad set of issues concerning forensic science and the law. The chapters contain numerous references and suggested readings to provide further opportunity to explore forensic disciplines of interest to the reader in greater depth. Chapter 1, Here We Stand: What a Forensic Scientist Does, authored by Jon Nordby provides a glimpse of what it is like to be a forensic scientist. It presents what is involved when forensic scientists work together with law enforcement on emotionally and scientifically challenging cases.

Section 1: Forensic Pathology and Related Specialties introduces the reader to the forensic examination of the deceased. As the title indicates, death investigation involves a multifaceted approach of different forensic specialties. Issues include challenges encountered when identifying the deceased, determining the cause and manner of death, and establishing the postmortem interval. Chapter 2, The Role of the Forensic Pathologist, authored by Dr. Ronald K. Wright, M.D., focuses on issues involving differences between coroner and medical examiner systems, training requirements in forensic medicine, the role of forensic pathologists in both criminal and civil cases, and the autopsy procedure itself. Chapter 3, Forensic Nursing, authored by Janet S. Barber Duval, Catherine M. Dougherty, and Mary K. Sullivan, focuses on a discipline that plays an increasingly important role in forensic investigation with the collection of evidence from the deceased as well as from the living in cases of survivors from traumatic injuries and sexual assault. Chapter 4, Investigation of Traumatic Deaths, authored by Dr. Ronald K. Wright, M.D., emphasizes various types of traumatic injuries including blunt force, sharp force, and gunshot injuries. Other traumas presented include asphyxial, thermal, and electrical injuries. Chapter 5, Forensic Toxicology, authored by John Joseph Fenton, discusses areas including workplace drug testing and postmortem toxicology. Analytical methodologies are described as well as interpretation of toxicological findings as they relate to the workplace environment and to the cause and manner of death. Chapter 6, Forensic Odontology, authored by R. Tom Glass, is devoted to the role of the dentist in forensic science. Issues of dental identification of the deceased, bite mark recognition, personal injury analysis, and dental malpractice are presented. Chapter 7, Forensic Anthropology, authored by Marcella H. Sorg and William D. Haglund, provides an introduction to forensic anthropology and its role on the forensic team, involving examination and identification of decomposed and skeletonized human remains. Chapter 8, Forensic Taphonomy, authored by William D. Haglund and Marcella H. Sorg, is a relatively new collective forensic discipline basically defined as the study of postmortem debris. This chapter discusses the importance of anthropological, botanical, and entomological evidence associated with environmental and time of death issues. Chapter 9, Forensic Entomology, authored by Gail S. Anderson, is devoted to the study and forensic examination of the identification and life cycle of insects as they apply to the decomposition process and the estimation of the time of death.

Section 2: Evaluation of the Crime Scene focuses on the proper examination and reconstruction of the scene with emphasis on the recognition, documentation, collection, and interpretation of physical evidence, including bloodstain patterns. Chapter 10, Crime Scene Investigation, authored by Marilyn T. Miller, introduces the reader to the crime scene in terms of evidence search and recognition, procedural and photographic protocols, and the basic theory of crime scene reconstruction. Chapter 11, Forensic Digital Photo Imaging, authored by Patrick Jones discusses the principles of digital image photography and its application to the proper photography of crime scenes and physical evidence. Chapter 12, Recognition of Bloodstain Patterns, authored by Stuart H. James, Paul E. Kish and T. Paulette Sutton, discusses basic bloodstain pattern interpretation and its application to the crime scene, and the examination of bloodstained clothing. The third edition provides a new and interesting case presentation.

Section 3: Forensic Science in the Laboratory begins with Chapter 13, The Forensic Laboratory, authored by Linda R. Netzel and introducing the reader to the forensic laboratory. It discusses the multidisciplinary functions of the laboratory involving the analysis of physical evidence. Emphasis is placed on quality control and accreditation. Chapter 14, The Identification and Characterization of Blood and Bloodstains, authored by Robert P. Spalding, leads into more detailed chapters that discuss the analysis of specific types of physical evidence. Specific attention is given to the identification of blood utilizing chemical, crystal, and immunological tests as well as conventional serology. Chapter 15, Identification of Biological Fluids and Stains, authored by Andrew Greenfield and Monica M. Sloan, presents the methods for the identification of semen, saliva, and other biological fluids that are important types of physical evidence, especially in cases of sexual assault. Chapter 16, Forensic DNA: Technology, Applications and the Law authored by Susan Herrero, offers practical application and up-to-date information on DNA analysis. The role of DNA analysis assisting with missing body and cold cases is emphasized. Chapter 17, Microanalysis and Examination of Trace Evidence, authored by Thomas A. Kubic and Nicholas Petraco, presents the examination of trace evidence with the microscope and analytical instrumentation. Chapter 18, Fingerprints authored by R.E. Gaensslen discusses fingerprint classification and current methods of identification. Section 3 then concentrates on Chapters 19, Forensic Footwear Evidence, and Chapter 20, Forensic Tire Tread and Tire Track Evidence, both authored by William J. Bodziak. Chapter 21, Firearm and Tool Mark Examinations, authored by Walter F. Rowe discusses features of firearms and ammunition and the comparison and identification of projectiles and casings. Chapter 22, Questioned Documents, authored by Howard Seiden and Frank H. Norwitch, discusses the process of document analysis and ink identification. Chapter 23, Analysis of Controlled Substances, authored by Donnell R. Christian, Jr. discusses the methodologies of solid dose drug identification utilizing laboratory instrumentation and techniques.

Section 4: Forensic Engineering begins with the forensic investigation of structural failures in Chapter 24, Structural Failures, authored by Randall K. Noon, with the focus on the collapse of the World Trade Center buildings caused by the terrorist attack with hijacked commercial jetliners on September 11, 2001. This section is then complemented with a detailed discussion on fire and explosion investigation in Chapter 25, Basic Fire and Explosion Investigation, authored by David R. Redsicker, that explains the chemistry of fire, origin determination, and the investigation of accidental and intentionally set fires and explosions. Chapter 26, Vehicular Accident Reconstruction, authored by Randall K. Noon, discusses the principles of physics utilized in the reconstruction of vehicular accidents.

Section 5: Cybertechnology and Forensic Science concentrates on a relatively new forensic discipline concerning the use of computers in forensic science and the investigation of computer crime. Chapter 27, Informatics in Forensic Science, authored by Zeno Geradts, focuses on the extensive use of forensic databases and some of the modern computer image enhancement techniques. Chapter 28, Computer Crime and the Electronic Crime Scene, authored by Thomas A. Johnson, explores aspects of computer and Internet crime.

Section 6: Forensic Application of the Social Sciences introduces the reader to the applications of forensic psychology, forensic psychiatry, and criminal profiling. Chapter 29, Forensic Psychology, authored by Louis B. Schlesinger, discusses psychological research and the law, psychological testing, and applications of clinical psychology. Chapter 30, Forensic Psychiatry, authored by Robert A. Sadoff, introduces the reader to the role of the forensic psychiatrist, including issues of competency and legal insanity as well as those involving felony crimes, such as homicide, sexual crimes, and juvenile cases. Chapter 31, Serial Offenders: Linking Cases by Modus Operandi and Signature, authored by Robert D. Keppel, then discusses the linking of cases through modus operandi that leads into the subject of criminal profiling which is explained in Chapter 32, Criminal Personality Profiling, authored by Michael R. Napier and Kenneth P. Baker.

Section 7: Legal and Ethical Issues in Forensic Science complements the technical chapters of this book with detailed information concerning the relationship of the law and forensic science. Chapter 33, Forensic Evidence, authored by Terrence F. Kiely, presents the legal issues from the point of view of a forensic scientist and addresses the ethical challenges and application of scientific logic in the concept of the criminal justice system. Chapter 34, Countering Chaos: Logic, Ethics and the Criminal Justice System, authored by Jon J. Nordby, presents an overview of ethical and evidentiary issues from an attorney's point of view. The book also presents an extensive glossary of terms and a set of comprehensive appendices. The glossary provides the reader with a quick reference to forensic terminology gleaned from the chapters. (Words highlighted in **blue** are terms included in the Glossary.) The appendices have been designed to provide additional information to the reader. They include a section devoted to biohazard safety that is essential to those in forensic science who enter the crime scene, the autopsy room, and the laboratory, as well as anyone who must handle biological forms of physical evidence. A comprehensive listing of important and useful forensic-related websites has been included. These Web sites represent a broad spectrum of forensic topics and many contain extensive links to sites of forensic interest. Each of these Web site addresses has been verified to date. Finally, a trigonometric table of sine and tangent functions, as well as metric measurements and equivalents, have been included as a reference for scientific calculations.

This third edition text, in its attempt to bring the discipline of forensic science to the student, includes numerous photographs, which some may interpret as unpleasantly graphic. They are necessary to communicate the science and are offered in the spirit of education.

The editors and the contributing authors have combined time and talent to provide a solid, up-to-date, general forensic science text. The essential forensic disciplines have been addressed in these chapters. The field of forensic science constantly expands to include many additional areas of expertise. Many of these areas fall under the rubrics of developments in existing, established forensic sciences.

Acknowledgments

The editors extend their appreciation to the contributing authors for their outstanding chapters that form the body of this text. These authors are all working professionals in their discipline of forensic science who have given their time and effort to this third edition that will continue to educate the forensic scientists of the future. The staff at Taylor & Francis in Boca Raton, including Senior Acquisitions Editor Becky Masterman, Senior Project Coordinator Jill Jurgensen, and Project Editor Suzanne Lassandro, deserve special recognition for their guidance and assistance throughout the development of this text. A special thank you to Lisa DiMeo at Arcana Forensics, La Mesa, California for her assistance with proof-reading of chapters. Additionally, the continuing patience, support, and understanding of Victoria Wendy Hilt, R.N., of Fort Lauderdale and Kim Nordby of Tacoma, Washington, are truly appreciated.

Stuart H. James
Jon J. Nordby

The Editors

Stuart H. James is a forensic consultant with James and Associates Forensic Consultants, Inc., in Fort Lauderdale, Florida. He received a B.A. in biology and chemistry from Hobart College (Geneva, New York) and completed graduate courses in forensic science at Elmira College (Elmira, New York). He has been consulted on cases in 47 states as well as in Australia, Canada, England and the U.S. Virgin Islands. He also has been a consultant to the U.S. Army regarding cases in South Korea and Bosnia, and has given expert testimony in many of these jurisdictions in state, federal, and military courts. He has taught both basic and advanced bloodstain interpretation schools with Paul Kish throughout the United States, Canada, and The Netherlands, and has lectured extensively throughout the United States on these topics at forensic and law-related conferences and seminars. He coauthored the text, *Interpretation of Bloodstain Evidence at Crime Scenes* (CRC Press, 1999) with Dr. William G. Eckert, M.D., and is the editor of the text *Scientific and Legal Applications of Bloodstain Interpretation* (CRC Press, 1998). Mr. James is a fellow of the American Academy of Forensic Sciences, a charter and distinguished member of the International Association of Bloodstain Pattern Analysts, and a member of the FBI's Scientific Working Group on Bloodstain Pattern Analysis (SWGSTAIN).

Jon J. Nordby, Ph.D., D-ABMDI, received his advanced degrees from the University of Massachusetts–Amherst. He works as a forensic science consultant for Final Analysis, an independent consulting practice in death investigation, forensic science, and forensic medicine. He specializes in scene reconstruction, evidence recognition, collection, and analysis, as well as bloodstain pattern analysis and the investigation of police shootings. He serves the National Disaster Medical System as a medical investigator and forensic specialist with the Region X DMORT team. As such, he served in New York City after the World Trade Center terrorist attack. He is a professor emeritus and former department chair with Pacific Lutheran University, a consultant with the B.C. Coroner's Service Forensic Unit, the King County Medical Examiner's Office, the Pierce County Medical Examiner's Office, and the Puyallup Police Department's Homicide Investigation Section. An instructor at the Washington State Criminal Justice Training Commission (Police Academy), Dr. Nordby formerly worked as a medical investigator and training officer with both the Pierce and King County Medical Examiner's Offices. A Fellow of the American Academy of Forensic Sciences, he currently serves as General Section member of the Academy Board of Directors and member of the Academy's Ethics Committee. He belongs to the Association for Crime Scene Reconstruction, the International Association of Bloodstain Pattern Analysts, the Pacific Northwest Forensic Science Study Group, and the Philosophy of Science Association. He formerly served as chair of the International Law Enforcement Expert Systems Association. His many publications include the book *Dead Reckoning: The Art of Forensic Detection* (CRC Press, 1999).

Contributors

Gail S. Anderson
School of Criminology
Simon Fraser University
Burnaby, British Columbia, Canada

Kenneth P. Baker
The Academy Group, Inc.
Manassas, Virginia

William J. Bodziak
Bodziak Forensics
Jacksonville, Florida

Donnell R. Christian, Jr.
U.S. Department of Justice's International Criminal Investigative
    Training Assistance Program (ICITAP)
Washington, D.C.

Catherine M. Dougherty
Baylor Health Care System
Baylor, Texas

Janet S. Barber Duval
Colonel, retired, USAF Nurse Corps

John Joseph Fenton
Crozer-Keystone Health Systems
Media, Pennsylvania

Robert E. Gaensslen
University of Illinois at Chicago
Chicago, Illinois

Zeno Geradts
University of Utrecht
Utrecht, The Netherlands

R. Tom Glass
Oklahoma State University Center for Health Sciences
Tulsa, Oklahoma

Andrew Greenfield
Centre of Forensic Sciences
Toronto, Ontario, Canada

William D. Haglund
International Forensic Program for Physicians for Human Rights
Shoreline, Washington

Susan Herrero
Center of Equal Justice
New Orleans, Louisianna

Thomas A. Johnson
California Forensic Science Institute
Los Angeles, California

Patrick Jones
Forensic Lab Director
Purdue University
West Lafayette, Indiana

Robert D. Keppel
Sam Houston State University
Huntsville, Texas

Terrence F. Kiely
Center for Law and Science
DePaul University College of Law
Chicago, Illinois

Paul E. Kish
Forensic Consultant & Associates
Corning, New York

Thomas A. Kubic
John Jay College of Criminal Justice
City University of New York

Marilyn T. Miller
Forensic Science Program and the Wilder
    School of Government and Public Affairs
Virginia Commonwealth University
Richmond, Virginia

Michael R. Napier
The Academy Group
Manassas, Virginia

Linda R. Netzel
Kansas City Police Crime Laboratory
Kansas City, Missouri

Randall K. Noon
Noon Consulting
Hiawatha, Kansas

Frank H. Norwitch
Norwitch Document Laboratory (formerly
    Associated Forensic Services)
West Palm Beach, Florida

Nicholas Petraco
John Jay College of Criminal Justice
City University of New York
New York

David R. Redsicker
Peter Vallas Associates, Inc.
Endicott, New York

Walter F. Rowe
Department of Forensic Sciences
George Washington University
Washington, D.C.

Robert L. Sadoff
University of Pennsylvania
Philadelphia, Pennsylvania

Louis B. Schlesinger
John Jay College of Criminal Justice
City University of New York
New York

Howard Seiden
Crime Laboratory
Broward County Sheriff's Office
Fort Lauderdale, Florida

Monica M. Sloan
Biology Section of the Centre of Forensic
    Sciences
Toronto, Ontario, Canada

Marcella H. Sorg
Department of Anthropology
University of Maine
Orono, Maine

Robert P. Spalding
Spalding Forensics
Centreville, Virginia

James E. Starrs
George Washington University Law School
Washington, D.C.

Mary K. Sullivan
Carl T. Hayden VA Medical Center
Phoenix, Arizona

T. Paulette Sutton
 University of Tennessee Health Science
   Center
Memphis, Tennessee

Ronald K. Wright
Forensic Pathologist
Fort Lauderdale, Florida

Here We Stand

What a Forensic Scientist Does

Jon J. Nordby

"… When a man who is honestly mistaken hears the truth, he will either cease being mistaken, or cease being honest."

Thomas Paine (1737–1809)

The hard chair in the witness stand reminds me of the hard seat in the classroom—good for keeping one alert, bad for inducing anything remotely resembling comfort. But neither the courtroom nor the classroom is designed for comfort. Ideally, along with any discomfort, the occupants of both the witness chair and the classroom desk also share the quest for truth, or as near to truth as humanly possible. Unlike attorneys who become advocates for a position, right or wrong, the forensic scientist must remain an advocate only for the best that reliable science can bring to the data that we call "evidence." The scientist must remain true to this calling despite any pressures from either the prosecution or the defense intended to bend the scientific results in a self-interested direction.

To be a good forensic scientist is to recognize and resist these pressures to conform scientific truth to any advocate's position regardless of who pays the scientific bill—a police crime lab, a public defender, or a high-powered defense attorney. The fee covers only the scientific work: it must never be used to buy a favorable opinion that remains unsupported by the best science applied to all the available data.[1] Resisting these pressures is only part of what remains unseen in the forensic scientist's career and only part of what it means to be a good forensic scientist. Just appropriately bypassing these pressures isn't the only nonscientific skill required of the good forensic scientist: in short, there is much more to being a good forensic scientist than simply a keen understanding of natural science. Scientific detachment is not always possible.

In fact, it's impossible to come away from work on any event involving human suffering and death, especially a violent death or a murder, and retain the detachment that keeps us sane.[2] For example, feeling deeply sad for the victims of the World Trade Center attack and their families, while feeling morally outraged by their deaths, seems rational, normal, and common to all of us. The job of a forensic scientist can be exciting; but we must be aware that, often dealing with death, it is indeed a serious business that can, on occasion, be emotionally draining—as was my work for

1

Figure 1.1   Region 10 of the Disaster Mortuary Operational Response Team (DMORT) covers Washington, Oregon, and Alaska. (Photo courtesy of Jon J. Nordby.)

DMORT[3] at the World Trade Center in 2001 (Figure 1.1 and Figure 1.2).

To be a forensic scientist requires the appropriate integration of basic human emotions with basic rational enterprise, what the ancient Greeks called *reason*. Their term did not carry the unfortunate contemporary connotation of mere mechanical calculation. Forensic science, as science, demands cognitive skills but, more, it demands reason in the ancient Greek sense: it demands of us a rational soul, with emotive and cognitive elements operating in harmony.[4] Dealing with life's grimmest realities dispassionately while never losing sight of the feelings that keep us human—this is what a forensic scientist does.[5]

For ancient Greek thinkers, the highest human calling involved what they termed *politics*. In the ancient view, politics equated more or less with "ethics." Our use of the term today has obviously lost this connection. In less cynical times, contributing to the *polis*, to the rational order holding civil society together, represented the very highest human calling. Citizens were to be focused on maximizing justice, the moral cement of civil society, through each of their many individual endeavors.

Figure 1.2   Photo of the World Trade Center attack on America, September 11, 2001. (Photo courtesy of Michael Rieger, Federal Emergency Management Agency [FEMA], Washington, D.C.)

Without justice, society degenerates into a disharmonious mess that philosopher Thomas Hobbes called "a war of all against all, making life nasty, brutish, and short."

Almost every TV viewer knew early on how the victims of the World Trade Center attack died; their causes of death are obviously blunt-force injuries, thermal injuries, or some combination. Even the manners of their deaths pose no great mystery—obvious homicides. Even if heart attacks took some, they must have been induced by the attack. So, why do we bother trying to recover, identify, and document each individual's death through seemingly unending 12-hour shifts at such mass disasters as the collapse of the Twin Towers of the World Trade Center? The answer points to our society's moral foundation reflected through our system of law, and remains independent of our actual success or failure in the specific endeavor. We value human life as a matter of social order and, thereby, attach great significance to human death. Many questions must be asked and answered, not only for relatives and friends of the dead, but for the greater social good as well.

Certainly not all questions connected with the World Trade Center attack involve human death. Airline officials require answers to security questions, which in turn may depend on the results of a thorough forensic analysis and reconstruction of the hijacking. Airport security methods obviously failed; a thorough investigation sheds light on what went wrong and how we must fix problems revealed. Similarly, airplane manufacturers must incorporate improved security designs to prevent hijackers from gaining control of both cockpit communications and flight controls.

Addressing public safety questions about high-rise office buildings or issues of building design may teach us how to build safer tall structures in the future. Questions concerning the safety of the site itself and the extent of damage to surrounding structures also requires careful analysis from many different scientists working to address these and other questions of public significance. Perhaps the most obvious public safety issues concern identifying the perpetrators and preventing them from committing further acts of violence. The results of any investigation must also anticipate future questions by supplying facts relevant to many yet-to-be formulated inquiries. Clearly such enterprises require the expertise of scientists from many different backgrounds, and the knowledge gained through many specialties. Chemists, engineers, psychologists, and computer technicians are among those who will be asked to apply their knowledge in the forensic context.

While our U.S. courts became involved in the first terrorist attack on the World Trade Center in 1993, some legal issues arose from the September 11 attack that may block the court's involvement with "acts of war." Yet many investigative results will help point toward the identity, origin, and location of both conspirators and co-conspirators who may indeed fall under one or more court's jurisdictions.

When the results of such investigations establish criminal intent, justice demands punishment for the perpetrators of crime, as well as society's protection from further nefarious acts. But as former Attorney General of the United States Ramsey Clark put it, "There are few crueler injustices directly inflicted on an individual by government than conviction for a crime one did not commit."[6] Through their contribution of scientific reliability, the forensic sciences must help the court ensure that the guilty receive punishment and that the innocent remain free.[7] It is important to consider that, contrary to what movies portray, large numbers of the accused have been exonerated by forensic science.

Law and Science

The philosophical foundation of the criminal justice system remains to protect the innocent and to ensure that the truth emerges for any matter before the court, thereby ensuring that justice is done.[8] Given the number of cases to be heard, however, the criminal justice system has the potential to sacrifice values of truth and justice to organizational efficiency. While crime laboratory scientists may pride themselves as being "independent finders of fact," most operate under police jurisdiction or administration, and many scientists, perhaps unconsciously, develop the attitude that they work exclusively for the best interests of the police or the prosecutor.

When emotions overcome reason, a zealous forensic scientist may intentionally or inadvertently deny real justice. Results are misinterpreted, or worse, falsified. Such flawed science may not be easy to spot, since it can only appear through the results of the scientific investigation. While no one can ever attain anything close to a perfect harmony of reason with emotion, forensic scientists at least have a political duty to strike the best balance possible under life's most difficult circumstances. Of course, completely satisfying this duty remains both difficult and elusive. The commitment to ethics should be stressed in the education of a forensic scientist. The values inherent in "good science," including both these moral elements and the nonmoral elements distinguishing reliable from unreliable scientific practice, should be a part of official forensic scientific curricula.

In some tiny jurisdictions, **coroners** may also work as sheriffs, prosecutors, or funeral directors. Some **medical examiners** work under the administrative umbrella of prosecutors' offices, and, in some rare cases, the sheriff also moonlights as both coroner and district attorney. At the very least, such organizational structures risk potential conflicts of interest. The potential exists for overseers to influence reports, compromising appropriate objectivity. In the practical world, only the competence and rigorous honesty of the individuals holding such perilous positions preserve the philosophical basis of the criminal justice system designed to protect the innocent and expose the truth about complex actions. Under these organizational structures, the system works if, and only if, morally honest individuals hold key positions of power.[9]

Without the underpinnings of high ethical standards, forensic scientists may become what is known in the profession as "hired guns." The student considering this profession should resist the temptation of selling whatever opinion is needed by defense or prosecution. Not all hired guns become forensic frauds merely through nonexistent or meaningless credentials. Properly educated, experienced scientists may also act as gunslingers[10] through ignorance or misapplication of method. This might involve purposefully omitting relevant tests or suppressing relevant results. Many such experts may develop an entirely unjustified sense of their own scientific abilities and observational powers. Generally, such experts offer firm, certain, and conclusive opinions designed to fit the relevant courtroom advocate's agenda. Such a forensic expert may even resort to defining scientific error as any interpretation that disagrees with his or her own.

In the real forensic sciences, individual scientists always work as members of a larger team, perhaps with other specialized scientists, law enforcement investigators, prosecutors, defense attorneys, judges, juries and the media, each contributing his or her efforts toward the bigger picture of a public trial, or an investigation capturing the public interest. The job of a forensic scientist is not one of glamorous celebrity.

If Sherlock Holmes, the detective invented by Sir Arthur Conan Doyle, worked a shift with us at the Manhattan medical examiner's office in the aftermath of the 9/11 attack, he might be assigned to pick up trash in the parking lot at Memorial Park. He might have to check the fuel levels and the temperature gauges on the refrigerated trucks, or to water the potted plants decorating the entrance to each trailer. In this situation, it is not a waste of time or talent if those tasks need to be done, and Holmes has time and ability to do them. It's what a forensic scientist does. One important characteristic of forensic scientists is adaptability, and a willingness to advance the common good.

Of course, the extreme situation surrounding the World Trade Center attack does not illustrate the everyday nature of unique forensic sciences discussed in the following chapters of this book, nor does it describe the role of specialized scientists operating within their specific areas. It does, however, illustrate the unifying element of scientific work to be presented in courts of law to help resolve legal disputes. The forensic sciences uniquely share their applications to legal issues for resolution in a public forum. Without courts of law, there could be no forensic sciences; without the polis, there could be no law. Forensic sciences operate inextricably in the service of the public, represented through the rule of law by the courts. Different functions, but all necessary for the common good.

Lawyers and Scientists

All men are liable to error; and most men are, in many points, by passion or interest, under temptation to it.

John Locke (1632–1704)

Lawyers and forensic scientists enjoy a close, yet often uneasy, relationship. Forensic scientists must not forget that lawyers have moral and legal obligations that often generate conflict and misunderstanding among those with scientific minds. For example, defense lawyers have an obligation to conduct a spirited defense of the accused, especially if they are guilty. Like it or not, the fundamental purpose of the criminal justice system is to protect the rights of the accused.

Lawyers work in adversarial situations where the clear objective remains winning a favorable decision for one's client through knowledge of the law. The adversarial system depends for its success upon the vigilance of opposing counsel, who also works toward the same objective. In this sense, law is outcome based. In law, a judge or a jury determines the truth. What juries or judges say, through their verdicts, is what is so. This legal goal has nothing whatever to do with proper, logical, scientific practice.

In sharp contrast to the practice of law, science remains justification based. Reaching the truth, or as close as one can come to it, depends upon the available evidence combined with a reliable method and not upon the rhetoric of persuasion. Scientists remain dependent upon data and present their conclusions as tentative, conditional, or probable in nature where appropriate. Lawyers, however, represent one of two rival positions arguing for acceptance. They may be operating with a different set of facts. The scientist may present the data, but the lawyer may argue that the data is inadmissible and prevent the data from becoming evidence. Where a scientist may see a complex issue consisting of many related parts whose interactions may be unclear to varying degrees, a lawyer may see the issue simply as yes or no, black or white, on or off, true or false. In other cases, what the scientist sees as black and white data may become more complex in the law's view.

In this sense, at least, forensic scientists and lawyers speak different languages with different objectives, unfortunately using many of the same words. The words truth, fact, certainty, possible, and probable can mean very different things in law and in science. These points remain to be considered later in the book.

Theoretical Natural Sciences and Practical Forensic Sciences

It is one thing to show a man that he is in error, and another to put him in possession of truth.

John Locke (1632–1704)

Unlike theoretical natural scientists, forensic scientists have an obligation to become familiar with both lawyers and the law. And while all scientists are required to uphold a high ethical standard, as mentioned earlier, forensic scientists are particularly bound to combine scientific skills with a sworn duty to the public good. It is for this reason that forensic science has been called a public science. Forensic scientists must be prepared to battle dubious cultural expectations, either inappropriately elevating or denigrating the powers of science. Such expectations are usually generated through crime novels, popular theatre, movies, and television. These inappropriate expectations when found among jurors, lawyers, and even judges can negate conservative scientific testimony. From **crime scene** to conviction, a good forensic scientist will be teaching others, an ability that requires patience and the communication of complex principles in simple terms. This, too, is what a forensic scientist does.

Currently, legal challenges to many established forensic science techniques, such as fingerprint and hair comparisons, are being made. The law is questioning whether such evidence is truly scientific. The natural sciences from this adversarial position remain theoretical, while the forensic sciences remain pejoratively practical. The forensic scientist must work to counteract this misguided view without appearing defensive.[11] The following table summarizes the contrasts usually developed by applying such a view.

Misguided View

Natural Sciences are said to be:	Forensic Sciences are said to be:
Theoretical	Practical
Pure knowledge	Applied to problems

(continued on next page)

Misguided View (continued)

Natural Sciences are said to be:	Forensic Sciences are said to be:
Orderly	Disorderly
Pristine	Contaminated
Controlled	Chaotic
General	Specific
Covering Laws	Approximations
Predictions	Conjectures
Certain	Uncertain

Unlike the carefully controlled experiment set up in a laboratory, consider the slightly smudged half fingerprint on a glass. If forensic science is conjectural, operating in chaotic situations where data are likely to become contaminated, can we trust the fingerprint as evidence? The so-called covering law model of natural science accounts for expectations of scientific certainty, which no forensic science allegedly approximates: epistemically certain laws of nature cover and, thereby, through deduction, explain cases.

There are many examples of these certain deductions, such as Fick's law for diffusion, Fourier's law for heat flow, Newton's law for shearing force, and Ohm's law for electric current. But these laws assume that a single cause explains a single specific given effect. Laboratory conditions or observational situations artificially manipulate phenomena to fall within the parameters of the law under investigation. Hence, they are ceteris paribus laws; that is, they hold only with "other things being equal" or "other things being right," such as with situations in an artificially controlled laboratory environment. In contrast, the crime scene is anything but a controlled setting.

Of course, almost all cases requiring explanation in the forensic setting involve many combinations of so-called causes all mixed together in the world existing outside of the laboratory. Without the ceteris paribus clause, such laws become manifestly false; with the clause, they cover only artificially limited and trivially unrealistic cases. General laws, however, describing complex interactions of such ceteris paribus laws in concrete cases, are unavailable. There simply is no body of theory or law readily available to cover particular, unique, complex phenomena such as the World Trade Center attack and collapse on September 11, 2001.

Yet such events demand some kind of scientific explanation. Our ability to supply the best explanation of the World Trade Center attack and collapse precedes our knowledge of any scientific law that may, in fact, cover the unique situation.[12] At best, the existence of various scientific specialties helps us to break the vague request for an explanation into its various specific components. The search for some single covering law becomes sheer myth. Until we discover some such law, it is up to science to supply acceptable explanations in the absence of any so-called certain knowledge.[13] In practice, the forensic sciences have an important element in common with the natural sciences. While their scientific goals obviously differ, their scientific common ground rests within an identical method of inquiry.

The aims of the so-called scientific method remain solidly within a procedural scope, focusing on scientific reliability. Follow these steps and the results will be consistent. With this methodological focus, illusive certainty becomes attainable reliability; natural laws and causes disappear in favor of explanatory connections, and the quest for comprehensive theory is replaced by relevant experience.

Laws of thermodynamics aside, bodies tend to cool at generally predictable rates given ambient temperature and other environmental factors. Logical methods, rather than some unattained body of accepted laws and proposed theories, characterize reliable scientific explanations in either theoretical or forensic contexts.[14] This distinction can be summarized in the following table:

Reliable Method of Inquiry: The Common Ground of Theoretical and Forensic Science

Reliable Methods Possess Characteristics of	
Integrity	Defensible technique
Competence	Relevant experience

Forensic Experts

An expert is someone knowing more and more about less and less, eventually knowing everything about nothing.

Attributed to Sir Bernard Spillsbury, M.D.[15]

Neither natural scientists nor forensic scientists start from theories or laws when facing the need to explain some puzzling phenomenon. They start from data. And not from commonplace data, but from the surprising anomalies raising the puzzles requiring explanation. Unusual observations suggest explanatory connections to pursue and test. Such connections define evidence and distinguish data that are evidence from data that remain merely coincidental. In that effort, the natural scientist and the forensic scientist share a fundamental approach belying any simplistic distinction between real science and forensic science.

As a forensic scientist, whether working an average case or one as catastrophic as the World Trade Center investigation, it never becomes my job to convict or punish the perpetrators. The job description only includes ensuring the best data collection and control, or determining the clearest relevant scientific explanations supported by reliable methods, always limited to the available data. I aim for methodological reliability, even if that notion remains limited to matching up numbers on the inside and the outside of plastic pouches.

An Expert's Role

Regardless of one's role in an investigation, no one can accurately claim to be an **expert witness** by profession. Expert witnesses, by law, can only be declared such by a judge. There are experts who are not scientists; for example, experts in office design, river rafting, school bus driving, fashion design, art history, or scuba diving. Of course, there are also experts in the natural sciences and medicine, as well as those with forensic practices.

But only the court creates expert witnesses. Forensic scientists first and foremost must remain scientists. Those practicing forensic medicine remain, first and foremost, medical professionals. Forensic scientists and forensic pathologists may or may not be declared expert witnesses by the court.

Usually scientific or other experts offered by attorneys to the court as potential expert witnesses give opinions only within their areas of expertise. Sometimes, lawyers hire an expert simply because the other side hired one first. Usually, lawyers engage experts when the facts of a case remain unclear, when analytical procedures in some field might help clarify those facts, or when specialized training can help educate the jury in turn to help the jurors make better informed decisions. The goal remains to apply some reliable method to those facts to help the court render its decisions. For forensic scientists, it is all about reliable scientific methods.

Scientific Method

… thou shalt far more easily and happily attain to the knowledge of … Natural Philosophy … by long use and much exercise than by much reading of books or daily hearing of teachers.

Ambroise Paré (1510–1590)

Attempting to characterize reliable scientific methods, as if describing some lifeless nonexistent abstraction, remains doomed to failure. There simply is no such generalized abstraction available to describe.[16] Science must be data centered and data driven. With that in mind, at most, we can point out a simple list detailing some of the many features reliable methods implement, enabling the productive scientific investigation of facts before the court. These—

- Help distinguish evidence from coincidence without ambiguity.
- Allow alternative results to be ranked by some principle basic to the sciences applied.

- Allow for certainty considerations wherever appropriate through this ranking of relevant available alternatives.[17]
- Disallow hypotheses more extraordinary than the facts themselves.
- Pursue general impressions to the level of specific details.
- Pursue testing by breaking hypotheses (alternative explanations) into their smallest logical components, risking one part at a time.
- Allow tests either to prove or to disprove alternative explanations (hypotheses).

In the forensic sciences, we reason from a set of given results (a crime scene, for example) to their probable explanations (hopefully, a link to the perpetrator). The aims of forensic science and medicine rest with developing justified explanations. Obviously, not all forensic explanations are alike. Some involve entirely appropriate statistical assessments and degrees of error suitably dependent on accurate mathematical models and accurate population studies. The reader will meet such explanations, for example, when studying DNA and other population-based sciences presented in this text. However, not all forensic scientific explanations involve such statistical issues. Instead, individual, nonrepeatable events with no statistical characteristics may demand scientific explanation.

A medical diagnosis, for example, involves selecting the best explanation of abnormalities in the observed data from among the clinically available alternatives. Clinical diagnosis may involve prior probabilities in the Bayesian sense,[18] but ultimately the diagnosis concerns what is wrong with one individual, not just what affliction correlates to some population group. In forensic medicine, the diagnosis focuses on the cause and manner of an individual's death. Mathematics rightly plays no statistical role in these explanations or their possibilities for error. The accused may have known the victim, but cannot be logically convicted on the basis that statistically most murderers know their victim.

In either clinical medicine or in the forensic sciences, how one's opinion is constructed determines its certainty. The certainty of forensic explanations is measured by assessing their explanatory justifications. This, in turn, involves showing, first, that the explanation is justified, and second, that the explanation is better justified than any available alternative explanation.

In this forensic setting, certainty assessments address the scientific explanation's rational justification, leaving the question of the explanation's truth and role in legal deliberations to the court. This allows for a clearer understanding of requests for certainty assessments when scientists are asked by attorneys to attach some degree of certainty to their work product. First, omitting the larger issues of truth leaves out the difficult determination of how frequently justified opinions could be false. Second, it leaves the entirely inappropriate **precision** of mathematics and probability theory out of this sort of nonstatistical certainty assessment.

All reliably constructed scientific explanations are best viewed by their creators as works in progress. We could always learn additional facts that may alter our views. Sometimes, however, no additional information would be relevant. In either case, our opinions must be held with what American philosopher and scientist Charles Sanders Peirce called contrite fallibilism—an awareness of how much we do not know, and the humility to acknowledge the possibility of making mistakes. He describes this intellectual stance to a friend in personal correspondence.

The development of my ideas has been the industry of thirty years …. For years in the course of this ripening process, I used to, for myself, collect my ideas under the designation "fallibilism." Indeed, the first step toward finding out is to acknowledge you do not satisfactorily know already. No blight can so surely arrest all intellectual growth as the blight of cocksureness, and ninety-nine out of every hundred good heads are reduced to impotence by that malady—of whose inroads they are most strangely unaware!

Indeed, out of a contrite fallibilism, combined with a high faith in the reality of knowledge, and an intense desire to find things out, all my philosophy has always seemed to grow.[19]

This basic intellectual stance remains necessary both for essential humility and for the very possibility of scientific advance. Forensic scientists must develop an intellect not too sure of what must remain uncertain, not too uncertain about what must remain sure. In the spirit of intellectual honesty and judicial prudence, the best advice for the forensic scientist to carry from the scene to the lab and into court throughout a long career comes from a 20th-century Viennese philosopher, Ludwig Wittgenstein: "Whereof one can not speak," he said, "thereof one must remain silent."

Endnotes

1. Any practice of forensic science and forensic medicine must be independent. The work must present scientific results based upon the available evidence. It remains data driven. What may be "obviously so" may not be supported scientifically: that means that there simply may be no available data to support any interpretation, one way or another. The results of any forensic analysis must be developed through the application of sound scientific and medical methods applied to all the relevant data completely without regard for their potential adversarial consequences.

2. Of course, how we deal with those feelings takes on a rational component when considering the ethical dimensions of being a forensic scientist.

3. DMORT, the Disaster Mortuary Operational Response, falls under the National Disaster Medical System, which was in 2001 a part of the U.S. Public Health Service. Ten DMORT units make up 10 regions covering the United States. Each unit consists of medicolegal death investigators, forensic odontologists, forensic pathologists, other forensic scientists, and assorted specialists all suited to form a single team to help local officials continue to conduct their operations during a mass disaster involving a large number of casualties.

4. Aristotle formulates his entire moral theory on an understanding of this basic human harmony or balance. His student, Plato, also develops this view into a robust moral theory based on functions proper for humans.

5. Such views have been associated with Aristotle's notion of *eudaemonia*, loosely meaning "living well," "fulfillment of proper function," or "happiness." This notion remains at the heart of a U.S. Army recruiting slogan: "Be all that you can be." Aristotle, however, adds certain requirements, which eliminate obvious counterexamples.

6. Ramsey Clark, in *Medicolegal Investigation of Death: Guidelines for the Application of Pathology to Crime Investigation,* 2nd ed, Werner Spitz and Russell Fisher, Eds., Charles C Thomas, Springfield, IL, 1980.

7. It must never be the practical goal of the forensic sciences to prove guilt or to prove innocence. The goal becomes to present a reliable analysis of the evidence that withstands scientific assaults. Unfortunately, some forensic scientists recently exposed in the media have taken it upon themselves to prove defendants guilty in the absence of any reliable scientific analysis. Those working for defense attorneys who ignore science and advocate for innocence, stand on the same irrational, prejudicial ground.

8. These concepts of truth and justice have received great philosophical attention over the centuries. Social and political philosophers working in this area evaluate these questions as well as arguments about implementations from a theoretical point of view. In contrast, the areas known as "political science" and sociology provide a descriptive view of societies as they, in fact, are, rather than producing philosophical arguments about how societies ought to be.

9. Plato, the ancient Greek philosopher, when asked about the minimum requirements for a just society answered that "a just society exists only when that society is composed of morally just individuals properly positioned." Upon reflection, this answer betrays a fairly skeptical view of our chances for justice. For Plato's view of justice, see *The Republic of Plato*, translated with an Introduction by Francis MacDonald Cornford, Oxford University Press, New York and London, 1967.

10 In a sense, all forensic experts are hired guns. This is not necessarily a problem. To illustrate, consider Richard Boone's portrayal of a hired gun named Paladin in the classic TV western, "Have Gun, Will Travel." The classically educated Paladin travels to help settle disputes using a large tool bag, including his guns. He does not permit his clients to dictate which tool he is to use, even though they may pressure him to use his gun to "kill the problem." Resisting this pressure, he works to solve the problem applying (morally and practically) appropriate terms dictated by the problem itself. In turn, forensic scientists are hired by clients (including agencies) to apply our scientific knowledge and experience to a specific problem. As I use the term, a gunslinger simply kills the problem to please the client. A hired gun, like Paladin, chooses both the morally and practically appropriate approaches, and then accepts responsibility for his choice. I think that this is a useful distinction to bear in mind.

11 For a detailed discussion of this issue, see Nancy Cartwright and Jon J. Nordby, *Essay #6: How the Laws of Physics Lie,* Clarendon Press, Oxford University Press, New York, 1983, pp. 100–127. Also see Jon J. Nordby, Introduction, in *How the Laws of Physics Lie,* pp.1–20, 1983. See also Science is as Science Does: The Question of Reliable Methodologies in "Real Science," in *Shepard's Exp. Sci. Evid. Q.,* 2, 3, Winter 1995.

12 Such laws, even if they exist, would be so complex that they would be useless for any purpose whatever, even for developing some hypothetical set of robust laws of the universe. The point here is simply that all work to explain the complex phenomenon occurring September 11, 2001 must precede any invocation of law. Hence, such "laws" not only fail to supply the certainty of deductions from true premises; they remain extraneous artifacts of real scientific work.

13 In fact, even if we could devise such a law, it would be so specific that it would be useless as a law.

14 For an extended discussion of this point, see Jon J. Nordby, "Is Forensic Taphonomy Scientific?" in *Forensic Taphonomy*, Vol. II, W. D. Haglund and M. H. Sorg, Eds., CRC Press, 2001, pp. 31–42. Also see Jon J. Nordby, Science is as Science Does: The Question of Reliable Methodologies in "Real Science," *Shepard's Exp. Sci. Evid. Q.*, 2, 3, Winter 1995.

15 See my discussion of this quotation in Footnote 1 found in Jon J. Nordby, Can We Believe What We See If We See What We Believe? Expert Disagreement, *J. Forens. Sci.,* 37, 4, July 1992; reprinted in The International Society of Air Safety Investigators Forum, 26, 3, September 1993.

16 Whatever the scientific investigation at issue, how one's scientific opinion is constructed mirrors the certainty of the result. Certainty, in the medical and scientific sense, remains determined by the method of derivation applied in the investigation. Medical and scientific certainty remains distinctly independent from either absolute certainty or mere mathematical probability. That independence, of course, makes such certainty assessments challenging and difficult—no one has a decision procedure that establishes certainty assessments.

17 For example, if 300 patients present to a campus health center with fever, sore throat, cough, general body aches, and fatigue, the prior probability that the 301st student presenting the identical symptoms has the flu remain extraordinarily high. To venture the explanation that the 301st student has an exotic tropical virus is ceteris parabis, "unreasonable" in the Bayesian sense. Note, however, that it may still be true that the 301st student does indeed have an exotic tropical virus. That diagnosis would depend upon discovering new data such as a recent trip to an exotic tropical location. Verified, this may lead to epidemiological studies and even to revision of the original diagnosis of flu.

18 Charles Sanders Peirce to Cassius J. Keyser, in a letter dated April 10, 1908, found at Columbia University, Cassius Jackson Keyser Collected Papers, archived.

Credit

Photo on page 3 is courtesy of Michael Rieger, Federal Emergency Management Agency (FEMA), Washington, D.C.

SECTION I

Forensic Pathology and Related Specialties

The Role of the Forensic Pathologist

Ronald K. Wright

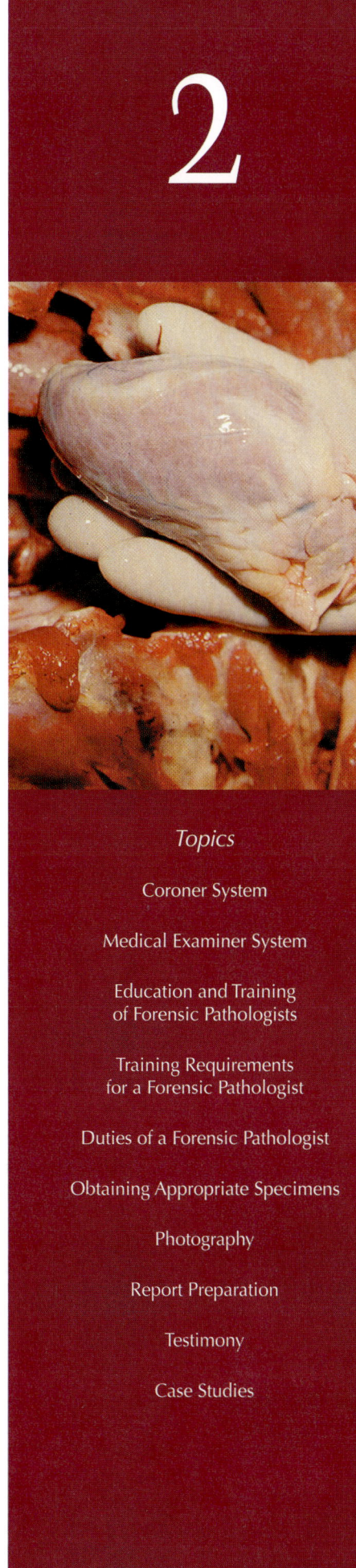

2

Introduction

A death that is unexpected or is thought to have been caused by injury or poison is always investigated for the purpose of determining whether it was a homicide—a death caused by an act of another, which was done with intent to produce bodily injury, or death, or done with disregard for the possibility that it could produce injury or death.

Coroner System

Forensic pathologists are physicians specializing in pathology, the diagnosis of disease, and who then subspecialize in the borderline area between law and medicine that emphasizes the determination of the cause of death. The role of the forensic pathologist can best be understood by examining the history of the coroner.

In English-speaking countries, the coroner is the government agent charged with responsibility for death investigations. The office of coroner has existed in England since before the 10th century. A wonderful series of articles about the history of the offices of the coroner and sheriff written by Professor Bernard Knight can be found at http://britannia.com/history/coroner1.html. The crowner of the king (the source of the word coroner) assumed a judicial function as early as the Norman invasion of England. The result was an office unique in modern English law: an inquisitional judge. Generally, English law employs noninquisitional judges who listen to the evidence brought by parties in litigation. The coroner and most judges who operate under the Roman system of law (Spain, France, Germany, Russia), are inquisitional judges. Their duty is to actually investigate the matters before them. Coroners had many duties in medieval times, including the investigation of the causes of deaths.

English law formed the basis for what became the initial law of the American colonies and subsequently the laws of the states.

Death investigation was considered a local, governmental or county function when the United States was created, and no provision was made in the Constitution for a coroner. The federal government had no death investigation operations, except in the District of Columbia, until the creation of the Federal Medical Examiner's Office in the Armed Forces Institute of **Pathology** in the 1990s. Each state enacted coroner's laws, and coroners are most often elected county officials.

Medical Examiner System

A movement beginning in the latter part of the 19th century established standards of education and training for certain professions. Despite the development of standards of professionalism in some areas, a significant problem arose related to the office of the county coroner. While the coroner was charged with a quasijudicial function and determined the causes of deaths, no particular education or training was required. Massachusetts was the first state to license nurses, physicians, and lawyers. In 1877, the Massachusetts legislature passed a statute that replaced coroners with medical examiners, and required medical examiners to be licensed to practice medicine.

As advances in industrialization in both manufacturing and agriculture caused the migration of huge numbers of people from farms to urban areas, big cities in the United States found that many institutions such as the county coroner did not transition well from rural areas. The medical examiner system of death investigation was adopted by cities such as Baltimore, Richmond, and New York around the time of World War I.

Generally, this change was set in motion by local scandals arising from deaths that were improperly investigated by coroners. Several New York City deaths were caused by incompetent administration of anesthesia during surgery. As a result, the law establishing the medical examiner in New York required an investigation into all deaths that occurred during surgery.

At the end of World War II, increasing trade and communication brought about by the automobile, train, and airplane made doing business in multiple jurisdictions increasingly complicated because of the nonuniformity of the laws from state to state and from county to county. The Commission on Uniform State Laws was created to develop model laws that could be adopted by every state and thus allow more efficient commerce. One such model uniform statute was the Medical Examiner's Act, which was passed by many states that already had coroners performing death investigations.

In recent years, more laws have established medical examiners, and this development eliminated or weakened the position of the coroner. However, rural areas are still generally served by elected coroners who are not required to have particular training or experience. Some states retained their coroners, but require training and continuing education. Interestingly, a few states have both coroners and medical examiners, and this often causes confusion.

The federal government was not involved in death investigation when the Constitution was written. The creation of the District of Columbia led to establishment of the first federal governmental coroner. The district abolished the coroner's office in favor of a medical examiner's office in 1970. No other federal death investigation program existed until the Federal Medical Examiner's Office was created in 1990. It serves the military and is administered from the U.S. Armed Forces Institute of Pathology.

Education and Training of Forensic Pathologists

Pathologists began to appear in hospitals in Europe and the United States in the middle of the 19th century after advances in the use of microscopes to examine tissues from patients led to the employment of physicians who used these new methods. These doctors came to be called pathologists from the

Greek *pathos* meaning "suffering" or disease and *logos* meaning "word" or "writing." Thus, a pathologist studies disease, its causes, and its diagnosis. Early pathologists examined tumors removed from patients to determine whether the tumors were cancerous. They also examined the bodies of deceased persons to determine the causes of death.

Pathologists later began to manage the laboratories where blood and urine were tested to determine the kinds and amounts of cells and the concentrations of chemicals they contained. By the middle of the 20th century, most pathologists specialized. Anatomic pathologists performed autopsies and examined tissues under microscopes. Clinical pathologists managed laboratories where body fluids were tested. Most of these physicians worked in hospitals.

The police and the coroners recognized that pathologists were needed to perform autopsies and determine the causes of deaths of people who died suddenly and unexpectedly. Thus, pathologists began doing autopsy examinations for the police, coroners, and medical examiners. By the end of World War II, the formal specialty of **forensic pathology** was recognized by the American Board of Pathologists. Today, in most large cities in the United States, the medical examiner is required to be a forensic pathologist. Forensic pathologists also handle autopsies for coroners in rural areas.

Training Requirements for a Forensic Pathologist

Medical School

The first requirement for a forensic pathologist is to graduate from a recognized allopathic (M.D.-granting) or osteopathic (D.O.-granting) medical school. Of course, entry into medical school requires a bachelor's degree or its equivalent. Medical school requires 4 years of study, and the curriculum generally includes very few elective courses. Graduates of medical schools located outside the United States or Canada must pass a test developed by the Education Council for Foreign Medical Graduates (ECFMG).

Postgraduate Training in Pathology

At least 4 additional years of postmedical school training are required. The training may be in anatomic pathology or a combination of anatomic and clinical pathology. Postgraduate pathology training takes place primarily in hospitals owned by or affiliated with medical schools. Postgraduate training was formerly called internship and residency. The training takes place on the job, and it is the equivalent of an apprenticeship in a trade. Unlike medical school, which costs about $25,000 per year in tuition, postgraduate training pays a salary of about $25,000 per year.

Additional Forensic Pathology Training

An additional year of training is required after completion of postgraduate training in anatomic or anatomic and clinical pathology. The training must be completed at a large coroner's or medical examiner's office. It constitutes a fifth year of postgraduate training and pays approximately $60,000 per year.

Training in Other Areas of Forensic Science

Many forensic pathologists undertake training in the broader areas of forensic science, such as toxicology, serology, tool mark examination, **firearms** examination, crime scene analysis, **forensic anthropology**, and **forensic odontology**.

After a candidate completes 5 years of postgraduate training, he or she must pass a 2- or 3-day examination to become a board-certified forensic pathologist. It is estimated that 500 forensic pathologists practice in the United States. Because forensic pathology deals with the intersection of law and medicine, an increasing number of forensic pathologists go to law school and obtain juris doctor (J.D.) degrees. About 35 forensic pathologists presently have medical as well as law degrees.

Duties of a Forensic Pathologist

Forensic pathologists are primarily employed by counties to investigate the deaths of persons who die suddenly and unexpectedly or as a result of injury. Civil and criminal litigation often arises from the work done by forensic pathologists. A few forensic pathologists work primarily as consultants in litigation.

Reviewing Medical History

Although forensic pathologists deal primarily with determining the causes of death, obtaining past medical history and understanding the issues raised by that history are important parts of the process of death investigation. Indeed, a medical history is generally the starting point of any investigation.

When a death is reported to the coroner or medical examiner, the first issue is to determine whether jurisdiction exists to investigate the death. As most reported deaths do not involve apparent injury, the issue in most jurisdictions is whether the death meets a two-pronged test. First, is the death sudden? The general definition of *sudden* is a death that occurs within a few hours of the onset of symptoms or death without any symptoms. Second, is the death unexpected? That determination requires a perusal of medical records. If the person has been diagnosed with a disease, the most common of which is cardiovascular disease, then death is somewhat expected, even if sudden, and the death does not fall within the jurisdiction of the coroner or medical examiner.

In addition to inquiring about sudden and unexpected deaths, the medical records must be examined for determination of jurisdiction based on delayed effects of injury. For certifying **cause of death**, forensic pathologists do not recognize a statute of limitations for fatal injuries. If a person who suffers a gunshot wound that renders him unconscious dies a few years later from pneumonia, the coroner or medical examiner has jurisdiction to determine whether the pneumonia was a consequence of the gunshot wound. Careful study of medical records is required to properly determine the causes and manners of death of persons with histories of trauma.

Finally, because of the efficiency of the rescue squads in the United States and Europe, most persons who die of injuries have been treated for the injuries even if the decedent showed no signs of life. Treatment includes the insertion of needles, creation of small or large **incised wounds**, and even fracture of bones. Although it is generally possible to discern between injuries produced after death from those produced before death, such distinction can be difficult when vigorous resuscitation takes place. Review of the medical history is extremely important in situations where people have been treated after injuries.

Reviewing Witness Statements

Knowing what witnesses recall of the activities of the deceased prior to death or injury is extremely important to a forensic pathologist. First, this information helps determine jurisdiction in cases where injury is not obvious. Also, since forensic pathology deals with recreating the circumstances of death, knowing what witnesses say happened is extremely valuable in developing questions to be answered. Forensic pathology is generally very effective for refuting witness statements. Understanding the contents of witness statements allows a forensic pathologist to know what questions will be asked. Access to such information can potentially create a problem by prejudicing the judgment of the forensic pathologist, but on the whole, the system works best when a hypothesis in a witness statement can be tested scientifically.

Scene Examination

In the best of situations, a forensic pathologist would examine the scene of death or the location where the body was found in every case he or she investigates. From a practical standpoint, this type of examination is impractical because of the cost. Many coroner and medical examiner offices in urban areas send forensic pathologists to the scenes of deaths that appear to be complicated or unusual. The information that can be gleaned from examining the scene in person is invaluable. The examination of scene photographs and reviewing the impressions of well trained and experienced crime scene personnel can compensate for much that is lost by not having a forensic pathologist examine the scene. However, the perspective of a forensic pathologist at a crime scene cannot be duplicated. Questions of postinjury movement, time between injury and death, time of injury, time of death, and questions that address exactly what happened to cause the death are reliably raised and sometimes answered by examination of the scene.

Autopsy Examination

The dissection of the human body to determine the cause of death has been practiced since the early middle ages. *Autopsy* means to look at oneself, so that term hardly seems appropriate. A more technically correct term for the dissection is *necropsy* or looking at the dead. Autopsy is more commonly used in the United States. In the premodern period, dissections of bodies were undertaken only as a means of limiting or stopping postmortem decomposition. Dissection was practiced by the early Egyptians.

Dissection of the bodies of deceased persons is forbidden by Middle Eastern religions. It also was forbidden by Egyptian polytheism, although it was required when bodies were prepared for mummification. The prohibitions imposed by Judaism, Islam, and Christianity vary. Such prohibitions occasionally make the performance of an autopsy difficult, if not impossible. Under English common law, the body of a deceased was treated as property for the purposes of burial, and burial was both a duty and a right of the surviving next of kin. Since dissection of a body is deemed to be interference with the duty and right of burial, it requires the permission of the surviving next of kin. One exception is the duty of the coroner or medical examiner to determine the cause of death in certain circumstances. Obviously, if the surviving next of kin is suspected of causing the death, he or she cannot be allowed to prevent the dissection.

Objections by the next of kin to the autopsy or to specific aspects of the autopsy are generally respected as far as possible. One requirement may be to carefully preserve any blood spilled and return it and all removed organs for burial with the body. This is an important issue for those who practice religions that teach that such preservation of the body is required for proper resurrection. Generally, in the United States, postmortem dissection is mandatory if a death is properly within the jurisdiction of the coroner or medical examiner, and the cause of death is not determinable without dissection.

Autopsy examinations generally entail the removal, through incisions, of the internal organs of the chest, abdomen, and head. The customary technique in the United States is the inframammary incision beginning at each shoulder, extending to the midline of the body in the lower chest, and extended to the top of the pubic bone. The T-shaped incision has been adopted because it facilitates the examination of the tongue and neck. Figure 2.1 shows a T-shaped incision that goes from shoulder to shoulder, then to the midline of the upper chest and down to the pubic bone.

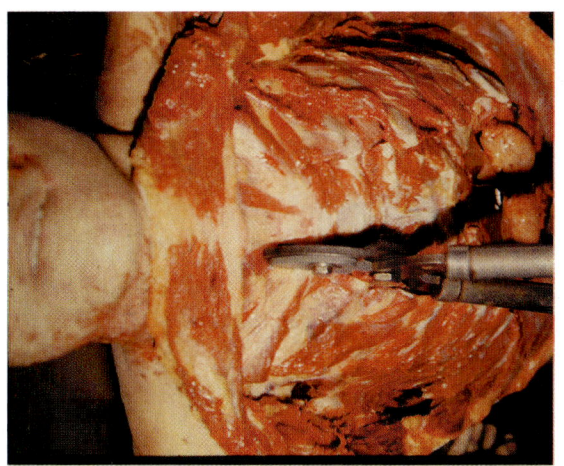

Figure 2.1 Typical United States technique for opening the chest and abdomen during autopsy.

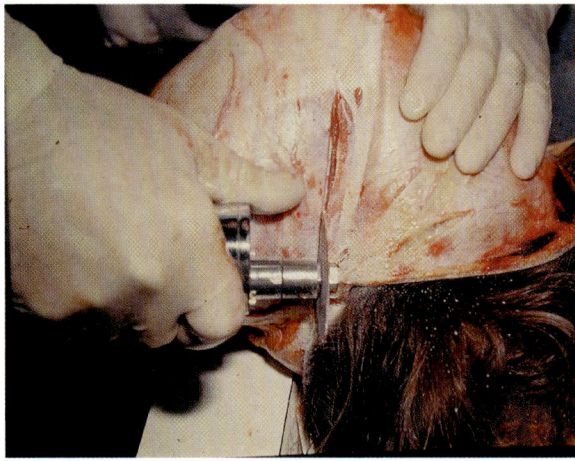

Figure 2.2 Typical United States technique for examining the brain.

Examination of the brain entails an incision from behind one ear to behind the other ear, reflection of the scalp by peeling it upward and backward, and then sawing of the skull in a circular or tonsorial cut, followed by removal of the resulting skull cap. Figure 2.2 shows such a dissection. The brain is sometimes dissected immediately or it may be put in a solution of formaldehyde for a week to "fix" the tissue for better dissection and examination. Fixation is a chemical process that causes proteins to harden; that preserves the tissue and prevents further decomposition.

After removal, organs are weighed and then dissected to determine disease or injury. Figure 2.3 shows a heart removed from a body in preparation for dissection. Additional

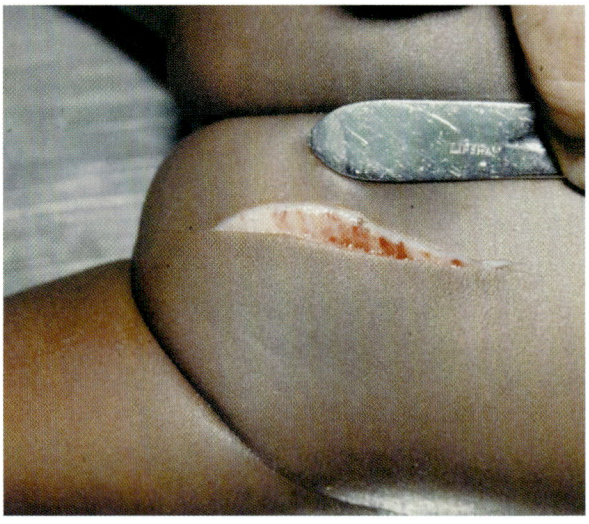

Figure 2.4 Bruised buttocks examined by autopsy incision.

dissections may be done, but are not generally considered part of a routine autopsy. For example, dissection of the spine and removal of the spinal cord may be required in a case involving possible spinal injury. Occasionally, especially in suspected child abuse cases, a posterior neck dissection is done, to show injury to the muscles, ligaments, and spinal cord. Also in children who are suspected of being abused, incisions are made to demonstrate bruises that do not show externally. Figure 2.4 and Figure 2.5 show incisions made in the buttocks of a child suspected to have died from abuse. Figure 2.4 illustrates bruising of the underlying tissue that was invisible

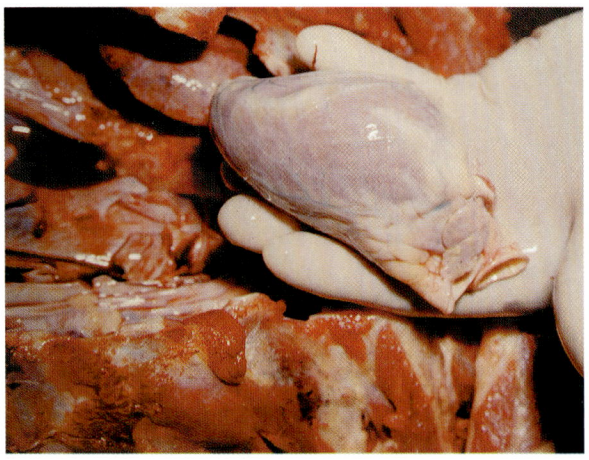

Figure 2.3 A normal heart as viewed at time of autopsy.

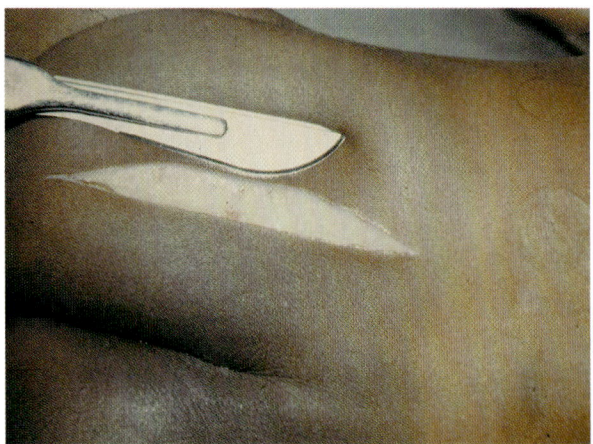

Figure 2.5 Normal color of subcutaneous tissue when incised at autopsy (same child shown in Figure 2.4).

without dissection. Figure 2.5 shows the other buttock that had no evidence of bruising.

Dissection of the legs is generally done if blood clots are found in the lung because clots often originate in the legs. In cases where death occurs in police custody, extensive dissections are required to rule out torture. This includes the examination and dissection of the soles of the feet, arms, and legs to look for evidence of subtle blunt trauma. Information derived from the medical history, witness statements, scene examination, and the autopsy guides the performance of additional dissections.

Obtaining Appropriate Specimens

Toxicology

In most forensic autopsies, specimens are removed for toxicology testing. The usual method of obtaining urine is shown in Figure 2.6. A syringe and needle are used to remove urine for testing.

Blood is usually taken from the **aorta** and sometimes from large veins. Some drugs redistribute in the postmortem (after death) period and venous blood is considered more reliable than heart or aorta blood for many drugs. Bile is taken from the gall bladder. Blood and urine are routinely used to determine the presence of common **drugs** of abuse. Alcohol is generally measured in the blood. **Opiates**, diazepines, and **cocaine** are measured in the urine. If positive results are found on urine screening, more extensive testing is generally performed on the blood, including quantitation of the drug in question. In addition, information received from the medical history, witness statements, scene examination, and the autopsy itself may trigger searches for other drugs and poisons.

Microscopic Examination

Small portions of the internal organs are put into a solution of formaldehyde to "fix" them and preserve them for further study. Figure 2.7 shows the usual way of preserving tissues. The forensic pathologist selects appropriate sections of diseased or injured and normal tissue. The tissue sections are usually prepared by histology technicians who encase the tissue sections in paraffin. Thinly sliced (5 microns thick) sections of the paraffin blocks containing tissues are mounted on

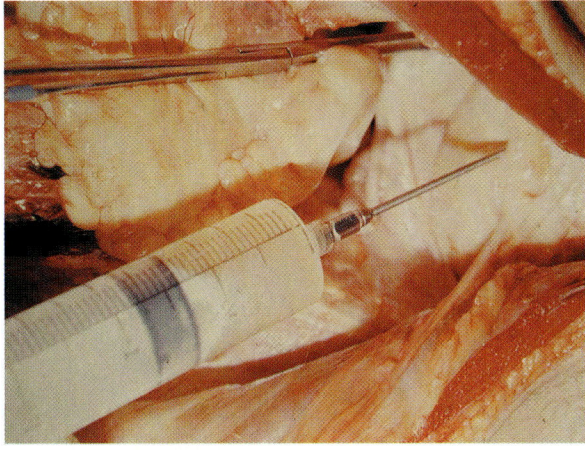

Figure 2.6 An 18-gauge needle and 60-cubic centimeter syringe removing urine from bladder during autopsy.

Figure 2.7 Formaldehyde solution containing sections of tissue removed at autopsy.

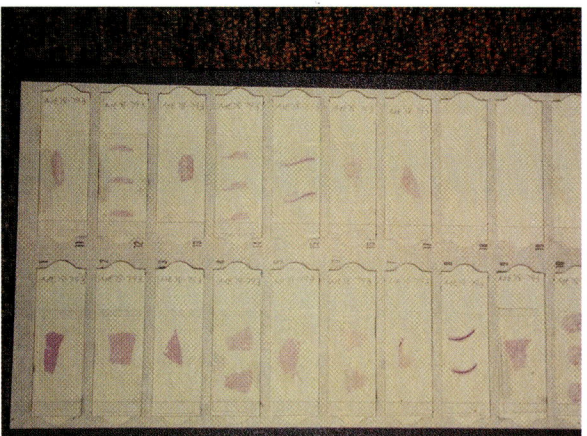

Figure 2.8 Prepared microscopic slides.

glass slides and stained with hematoxylin and eosin (H&E) dye for examination under a light microscope. Figure 2.8 shows a tray of slides prepared from an autopsy and ready for examination under a microscope.

DNA Analysis

Most coroners and medical examiners preserve one specimen from an autopsy that can be used for future DNA analysis. Two common methods are used. A spot of blood can be placed on absorbent paper, allowed to dry, and then stored in an envelope. A second common method is to pull head hairs and place them in an envelope. It is important when pulling hairs to remove the bulbs that contain nuclear DNA. Cut hair contains mitochondrial DNA. Both techniques, although presenting a slight biohazard, will provide reasonable **samples** for DNA studies. The potential biohazard arises from the possibility that the person had hepatitis B or C, or HIV infection. These diseases can be transmitted by exposure to the virus contained in hair or blood, even if dried.

In older cases and in situations where no DNA sample was preserved, the paraffin-fixed tissue saved after making microscopic slides or the slides themselves generally will provide adequate DNA samples. The only problem is that if the tissue sits for weeks in the formaldehyde solution before embedding in paraffin, the DNA may be hydrolyzed and unsuitable for study. DNA embedded in paraffin blocks or

cut into sections and made into slides will not further decompose.

Photography

Taking and preserving photographs of the scene and the autopsy are important duties of a forensic pathologist. In larger urban areas, coroners and medical examiners employ professional photographers to perform this function. However, photographs taken by a forensic pathologist are often preferable to those taken by a photographer because the years of training and experience required for a forensic pathologist ensure the taking of relevant and ultimately admissible photographs.

As higher resolution **digital** photographs are now routine, this photographic technique is used almost exclusively in forensic work. Careful use of a digital camera can provide documentation that is equal to or better than images on 35-millimeter film. Because the cost of digital photography does not increase with the number of photographs taken, multiple photographs can actually provide better detail that can be obtained with film.

Report Preparation

It is customary for a forensic pathologist to prepare a written report of each autopsy examination. Autopsy examinations are termed gross if they deal with what is seen by the unaided eye. An examination is microscopic if it involves examination of tissue under a microscope. Gross examination reports are generally dictated during or shortly after an autopsy. Most pathologists do not dictate during examinations because it is cumbersome to make a recording in an environment that has ambient noise and where bio-hazardous droplets and sprays are present. In addition, the order of the dissections does not lend itself to the usual organization of a report. For that reason, most pathologists dissect first and then dictate.

Most reports of a gross autopsy consist of discussions of external examination, medical treatment evidenced on the body, evidence of injuries, dissection technique, and diagnoses based on the gross autopsy. **Microscopic examination** results are dictated after the gross autopsy because the histology laboratory takes several days to prepare and deliver microscopic slides to the forensic pathologist. The toxicology report is usually prepared by the toxicologist, reviewed by the forensic pathologist, and appended to the autopsy report.

A forensic pathologist also may prepare a final summary of the external examination, internal dissections, microscopic examination, and toxicology report.

Testimony

Forensic pathologists spend vast amounts of time testifying about their findings and opinions. In criminal court, the testimony of a forensic pathologist is almost required to prosecute a **defendant** for manslaughter or murder.

In addition to testimony in criminal courts, forensic pathologists also testify in civil courts concerning torts (civil wrongs) alleged to have caused death. It is difficult for a **plaintiff** in a wrongful death suit to proceed without the testimony of the forensic pathologist who examined the body. It is far easier to prove that a death resulted from a tort by presenting the testimony of the forensic pathologist who performed the autopsy. In civil cases, testimony is obtained via deposition (sworn testimony compelled by subpoena and obtained out of court) more often than testimony obtained in court. In a few jurisdictions (Florida, for example), deposition testimony may be taken in criminal cases as well as civil cases.

Forensic pathologists often testify as expert witnesses in cases in which they did not actually examine the body of the deceased.

This testimony may be presented in cases where no one died or no one requested an autopsy. It may be presented when a litigant feels that the opinions and conclusions of the forensic pathologist who did the autopsy are not correct. He or she may have another forensic pathologist review the information or even perform a second autopsy.

A forensic pathologist is subject to *ad hominem* attack by the opposing parties whose interests are adverse to his or her findings and opinions. The job of the attorney for the litigants is to discredit any witness whose testimony is adverse to the interests of his or her client. The questioning by the opposition can include extensive inquiry into the professional and private life of the forensic pathologist for the purposes of finding anything that might make him or her less believable to the jury. A forensic pathologist should be aware of the probability that he or she will sometimes be subjected to embarrassing questioning.

Case Studies

To demonstrate the types of investigations medical examiners conduct, either as outside consultants or as primary investigators, I will present three cases from my files. Certain facts have been altered slightly to obscure the identities. All these cases were litigated in open court and, thus, constitute public record.

Case 1

In the morning following a thunderstorm, a prepubescent child was found lying between a metal street light pole and a metal bus bench shelter. The child had been reported missing by his mother the day before. Initially, the child was seen to have his leg lying on the exposed conduit that ran from the bus shelter to the pole. Figure 2.9 shows the child's leg on the conduit. The leg appeared to have an electrical burn. Figure 2.10 shows another view of the apparent electrical burn.

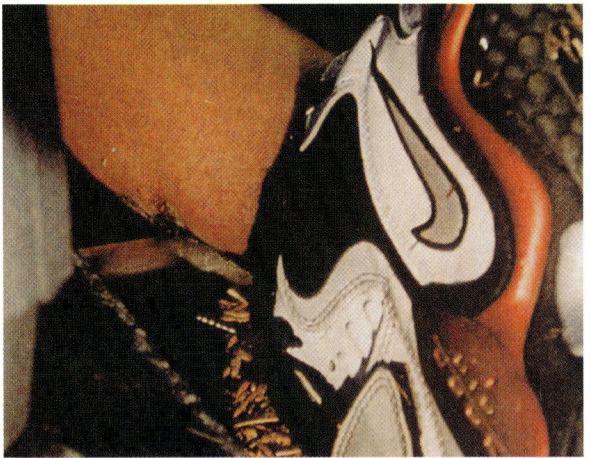

Figure 2.9 Scene photograph showing electrical burn on child's leg as found.

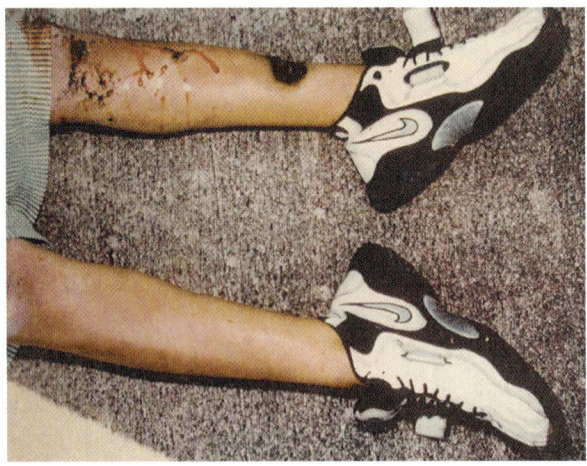

Figure 2.10 Scene photograph of electrical burn after child's body was turned.

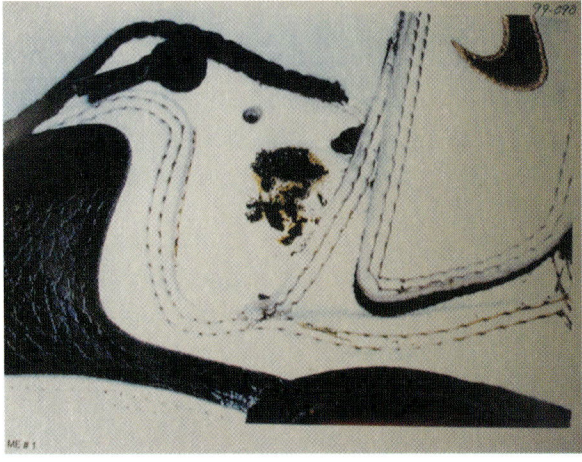

Figure 2.11 Arc electric burns on tennis shoe of child who sustained electrical burns.

The forensic pathologist on call went to the scene and examined the body, noting the presence of **rigor mortis** (stiffening of the body after death), the presence of an electrical burn on the leg, and multiple skin lesions from ants that attacked the body after death. Subsequent autopsy revealed an electrical burn as noted at the scene, along with burns of the foot (current penetrated the child's rubberized athletic shoe) and the knee. The hole in the shoe and the burned foot are shown in Figures 2.11 and 2.12. No preexisting disease was noted. The forensic pathologist determined that the cause of death was low voltage electrocution.

Back at the scene, building inspectors and electrical engineers determined that the transformer that converted the voltage of the street lights to 120 volts of alternating current for the bus shelter developed a short circuit through the insulation. Two ground rods that were supposedly 8 feet in length were only 4 feet long and were a few inches closer together than the building code required. Measurement of the exposed bus bench revealed a small amount of current (estimated to be about 40 volts). The source of the voltage was the short through the transformer. The building department had not issued a building permit for the shelter.

On the basis of these findings, the prosecuting attorney brought charges of manslaughter against the company that contracted

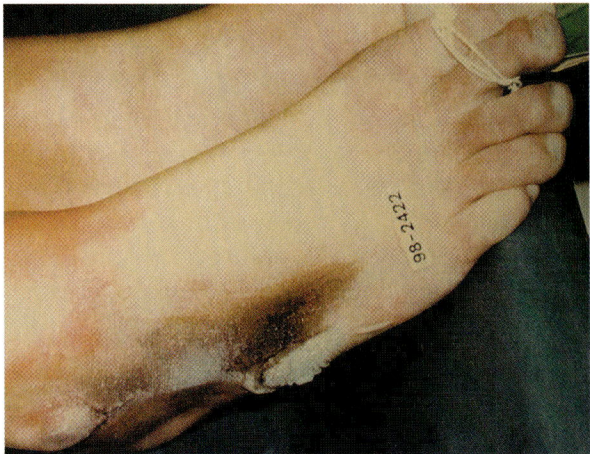

Figure 2.12 Electrical burns of foot noted during autopsy.

for construction of the shelter, against the subcontractor who actually built it, and against the electrician who did the wiring. I was asked by the defense attorneys to independently examine the evidence collected by the forensic pathologist and the various other investigating agencies. I concluded from my examination that the actual cause of death was **lightning**, not low voltage electrocution. Thousands of volts were required to arc through the rubber of the athletic shoe. The maximum voltage available from the wiring was 480. Further, thousands of volts were required to break down the insulation in the transformer. Both the shoe and the transformer were damaged at the same time by the same lightning bolt. The subsequent burn to the child's leg occurred postmortem as the body lay exposed to the leakage current that energized the conduit.

This case clearly demonstrates the requirement for scene and autopsy examination and preservation of evidence.

Case 2

A man was reported to have been involved in a chase by the police after he failed to stop at a stop sign. During the chase he was reported to have lost control of his motorcycle and was severely injured in an ensuing accident. He was transported from the scene by rescue personnel to a nearby emergency room where he was treated and admitted. He died 2 days later without regaining consciousness.

The medical examiner's office was notified of the death, and the body was transported for autopsy examination. I was able to secure the blood collected when the deceased was admitted to the hospital for the purposes of toxicology testing. A sample taken after death would be of little help as most drugs or alcohols present at the time of the crash would be gone. The sample proved negative.

Although some time elapsed between injury and death, examination of the head revealed impact injuries from the front, sides, and back. In addition, I noted wounds to the little finger side of the forearm. These are called defense wounds, and are rarely seen in motor vehicle crashes, particularly motorcycle crashes.

Although the scene no longer existed in its original state, I went to the location of the crash and reviewed the diagram made after the event. No photographs of the scene were available. I then examined the motorcycle for damage, and enlisted the aid of a mechanical engineer who specializes in reconstructing vehicular collisions. The engineer's findings and mine indicated impact from multiple directions to the motorcycle. These findings were inconsistent with the scene and the witness statements about the crash. The engineer's opinion was that the motorcycle had been beaten with flashlights to simulate a crash.

My opinion from the examination of the body was the same. The injuries were inconsistent with a motorcycle crash. They were consistent with being beaten with a flashlight about the head. Figure 2.13 shows the injuries on the front, right, and left sides of the head.

This case exemplifies the importance of obtaining witness statements and performing a scene examination even after the fact, even though the witness statements were not factually correct. Knowing what the witnesses said allowed the construction of a hypothesis that could be tested and found untrue.

These are examples of the types of cases in which forensic pathologists become involved. They clearly portray the vital role played by the forensic pathologist.

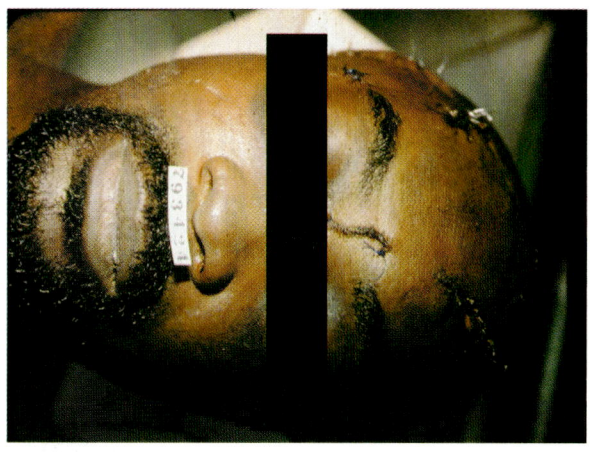

Figure 2.13 Autopsy photos of treated head wounds of person said to be in motorcycle collision.

Questions

1. The coroner is unique in English law because he is an _____ judge.

2. Forensic pathologists are employed primarily by _____.

3. What other forensic science disciplines are included in forensic pathology training?

4. What is the first determination that must be made when investigating a death?

5. The jurisdiction of the coroner/medical examiner to investigate deaths generally can be categorized as including deaths of what type?

6. Attending the scene of death by the forensic pathologist is most helpful in _____.

7. What is the medical term for the stiffening of the body due to postmortem depletion of glycogen?

8. Drug screens are usually performed on what specimen taken during an autopsy?

9. Dissection of the legs is done if what condition is found during the autopsy?

10. Microscopic sections are fixed in a solution of what chemical?

References and Suggested Readings

Baselt, R. C., *Disposition of Toxic Drugs and Chemicals in Man*, 5th ed., Chemical Toxicology Institute, Foster City, CA, 2000.

Knight, B., *Forensic Pathology*, 2nd ed., Oxford University Press, New York, 1996.

Knight, B., http://britannia.com/history/coroner1.html.

Moenssens, A. A. et al., *Scientific Evidence in Criminal Cases*, Foundation Press, Westbury, New York, 1995.

Moritz, A.R., The Pathology of Trauma, 2nd ed., Lea & Febiger, Philadelphia, 1954.

National Association of Medical Examiners, Writing Cause of Death Statements, http://www.thename. org/CauseDeath/main.htm, 2001.

Spitz, W. U., Ed., *Medicolegal Investigation of Death*, 4th ed., Charles C Thomas, Springfield, IL, 2006.

Forensic Nursing

*Janet S. Barber Duval, Catherine M. Dougherty,
and Mary K. Sullivan*

3

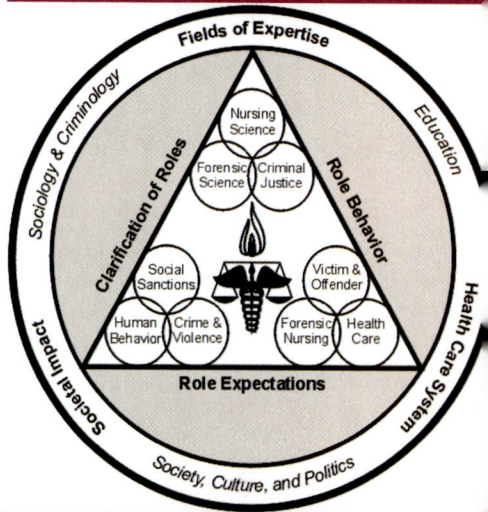

Overview

Forensic nursing is defined by the International Association of Forensic Nurses (IAFN) as the global practice of nursing when healthcare and legal systems interact. (IAFN, 2008). The forensic nurse functions as a staff nurse, nurse scientist, nurse investigator, or as an independent consulting nurse specialist to public and private operatives or to individuals in the medicolegal investigation of the injury or death of victims of violence, criminal activity, and traumatic accidents. The forensic nurse provides direct and indirect services to individual clients; provides consultation services to nursing-, medical-, and law-related agencies; and provides expert court testimony in areas encompassing (a) evidence collection, preservation, and analysis, (b) questioned death investigation processes, (c) adequacy of services delivered, and (d) specialized diagnosis of specific conditions related to nursing practice (International Association of Forensic Nurses, 1997).

Forensic nursing is a relatively new idea in the United States, but it has long been established in England, Canada, and Australia. The roots of forensic nursing go back over 200 years to clinical forensic medicine in England and other countries.

Clinical forensic practice is derived from the broad field of forensic medicine. This field focuses on the civil and criminal investigation of traumatic injury or patient treatment with law-related issues. It encompasses both victims and perpetrators who survive. The forensic pathologist, however, is solely concerned with the deceased and the scientific investigation of death.

However, the newest and perhaps most important domain of forensic experts (including nurses) is a field called **living forensics**, which relates to the identification and collection of evidence derived from the living, as opposed to the deceased. This shift in focus for forensic science (i.e., moving out of the morgue and into the emergency department, the clinics, physician offices, and other locations where there may be living victims of crime) paved the way for nursing's involvement. It permits law enforcement and judicial systems to join the efforts of health care to salvage victims of abuse, neglect, and other crimes of interpersonal violence before their bodies reach the morgue—an opportune time for strategic interventions to salvage the life of a victim and to curb activities of offenders.

The birth of living forensics is attributed to Harry C. McNamara, chief medical examiner for Ulster County, New York. In 1986 McNamara first defined clinical forensic medicine as "the application of clinical medicine to victims of trauma involving the proper processing of forensic evidence" (McNamara, 1986). This definition stresses the importance of health-care providers being aware of evidentiary materials and legal issues associated with their patients or clients.

There are many disciplines that participate in the processes of living forensics and still others that work in conjunction with law enforcement and death investigation. Among these are forensic pathologists, odontologists, document experts, photographers, entomologists, anthropologists, engineers, attorneys, psychiatrists, criminalists, and, of course, nurses.

The pictorial model of the theoretical foundation of forensic nursing (Figure 3.1) developed by Virginia A. Lynch (1990) portrays a nursing professional whose nursing science background incorporates forensic science and criminal justice principles. The model requires a forensic nurse to deliver

care within health care institutions to alleged victims, perpetrators, and significant others, diagnosing specific conditions related to nursing. Further, the forensic nurse must understand human behavior as it pertains to the acts of, and responses to, crime and violence in the light of social sanction. The model is composed of a triangle within a circle. At the top of the triangle are three areas from which the knowledge base of forensic nursing is composed: nursing science, forensic science, and criminal justice. The model assumes that the nurse embraces the philosophies and principles of these three disciplines and shares a responsibility with law enforcement agencies and courts in protecting the human rights of victims. In addition, the forensic nurse must recognize and protect the rights of the suspect or perpetrator of criminal acts. Represented on the lower left are social sanctions, human behavior, and crime and violence, recognizing that each society has its unique aggregate of crime and violence as dictated by its dynamic social constraints. The lower right corner portrays the emerging discipline of forensic nurses and the locus of their domain. In the center are the scales of justice with the caduceus superimposed. The bundle of public service is at the base. The flame of nursing is noted at the heart of the triangle, denoting enlightenment in a new field of practice. The triangle is enclosed by an interlocking circle that emphasizes the multidisciplinary coordination and cooperation among sociology, criminology, education, culture, politics, and health-care systems.

According to Lynch (1990), the model focuses on the necessity for society to respond to problems that develop between the related fields of nursing, forensic science, and the criminal justice system. Lynch feels that the effectiveness of the forensic nurse is based, in some part, on his or her ability to interact with other scientific, legal, medical, and social professionals as well as with victims, suspects, and perpetrators. Nurses must care for patients in their charge, whether they be victims of violent crimes, patients in legal custody, traumatic injury patients, or those with other liability-related injuries, and

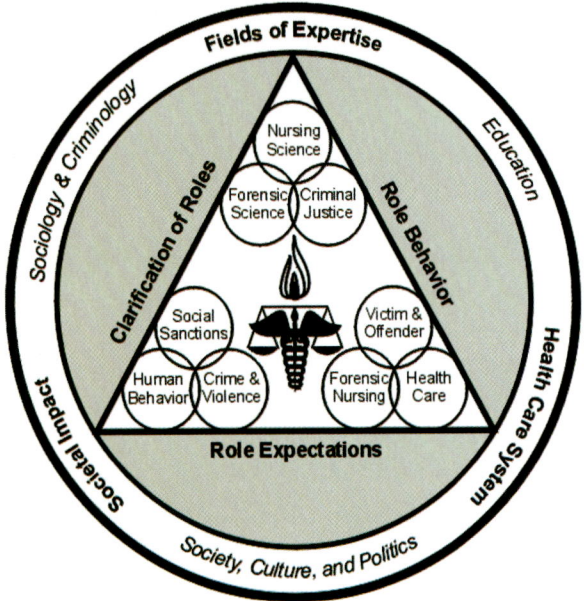

Figure 3.1 Pictorial model of the theoretical foundation of forensic nursing.

participate in the investigation of clinical and community-based deaths. However, it is clearly evident that the legal needs of these patients are not always met because of lack of education and training and, sometimes, ignorance on the part of a nurse or physician. With the acknowledgement of their comprehensive role in total care of these patients whose nursing problems bring them into actual or potential contact with the legal system, nurses accept the challenge to take responsibility for the legal issues concerning their participation in patient care.

Domain of Practice

The health care and social service burdens comprise much of the pain, suffering, and rehabilitation of victims who often carry life-long scars of interpersonal crime. Depression, fear, and guilt on the part of family members and the inability of animals to function harmlessly within a family unit or the larger society account for untold suffering and can result in the need for counseling, therapy, or rehabilitation that may extend throughout the individual's lifetime. Forensic nursing's earliest roots in the United States were related to caring for victims of sexual assault, human abuse, and domestic violence. As a result, many nurses assumed roles, by extension, of victim advocacy. The sensitivity, caring, protecting, and nurturing behaviors so inherent within nursing actually proved to be counterproductive, since a court of law seeks truth, based on scientific evidence, not emotion or empathy. Leaders within this new discipline of forensic science soon realized that if forensic nurses were to be taken seriously, they needed to align themselves with science. They would have to abandon identities that associated them exclusively with victims. It became apparent that alliances with groups such as Violence against Women and the Coalition for Sexual Assault Victims and other advocacy groups cast doubts about their value in court-room testimony. It was virtually impossible for these nurses to profess objectivity and impartiality when they were viewed as an arm of the prosecution.

Barber, a charter member of the International Association of Forensic Nurses, remarked that "one of the biggest stumbling blocks we've had is not deciding if we want to be nurses who dabble in forensic science or if we want to be forensic scientists who are also nurses. There's a big difference" (Pyrek, 2003). Today, sexual assault nurse examiners and other members of the forensic nursing community are struggling in their efforts to serve victims, suspects, and known perpetrators with equality. This indeed is a formidable hurdle for many within the forensic nursing ranks whose interest in the specialty originally stemmed from a desire to comfort, to heal, and to directly assist victims of violence. Forensic science is designed to determine the truth. It does not exist to gather facts for those who merely seek vengeance. Courts use objective data to ultimately decide who is to be punished, who is to be vindicated, and who should receive reparations for wrongdoing. They must ensure that the human rights of all citizens are protected, not merely the victim's.

Collaborative Relationships

Like traditional nurses in all patient-care settings, forensic nurses are trained to work within a multidisciplinary team. The traditional nurse works with other disciplines to provide patient care and is often the primary patient advocate, seeing to it that the most appropriate care is actually given. Depending on the setting and the task at hand, the forensic nurse will also interface with many different disciplines on any given case or situation, but the nurse is not an advocate for anything other than identifying the facts in the pursuit of truth and justice. The forensic nurse is not a witness for the prosecution in all cases, but must also be equally willing to serve the needs of the defense within the legal arena.

Forensic nurses may be a part of a crime scene response team in which they have specific responsibilities with regard to crime scene processing and evidence collection and management. The forensic nurse is the ideal person to provide the medicolegal information to the medical examiner as well as to identify possible sources of physical evidence that someone who is not trained in the medical field might overlook. This nurse is trained to work a crime scene alongside other specialists and within specific boundaries. Such crime scenes include not only indoor and outdoor homicides, suicides, traffic accidents, mass disasters, and other incidents of suspected foul play, but every area within the health-care arena.

Depending on the jurisdiction, a forensic nurse will need to establish collaborative relationships with a wide range of agencies and offices, including local law enforcement at the city, state, county, and federal levels; district and U.S. attorney's offices as appropriate; special investigative agencies such as the Office of the Inspector General, the Federal Bureau of Investigation, Department of Defense Offices of Special Investigations, and offices of the medical examiner or coroner. Within the health-care arena, forensic nurses will interface with quality assurance and risk management personnel, hospital attorneys, and security officers.

In other roles outside hospitals, there are many other potential collaborative relationships, depending on the practice setting.

Evolution of Forensic Nursing

In order to appreciate the development of forensic nursing in the 21st century, it is important to understand how forensic science has evolved over the previous centuries, eventually leading to the complex system we have today where law enforcement, health care, and the criminal justice system have joined forces to respond to interpersonal violence and other matters in which law and medicine are necessarily entwined.

European Influences

Forensic medicine is an old discipline, and its roots in the U.K. can be traced back for over 200 years. However, in 1248 a Chinese book, *Hsi Duan Yu*, literally, "the washing away of wrongs," described how to distinguish drowning from strangulation, becoming the first recorded application of medical knowledge to the solution of crime (Almirall and Furton, 1995).

U.S. Development

In the United States, the foundation for formalizing the discipline of forensic nursing was built on the initiatives of nurses involved in the examination of sexual assault victims. During the summer of 1992 in Minneapolis, Minnesota, 74 sexual assault nurse examiners convened and spearheaded the vision of an organization encompassing and defining the broad boundaries of forensic nursing. This endeavor crystallized the founding of the International Association of Forensic Nurses (IAFN). One year later, with over 200 members, the fledgling organization held its first scientific assembly in Sacramento, California. One of the earliest initiatives of the IAFN members was to formally petition the American Nurses Association (ANA) for the designation of forensic nursing as a specialty. In 1995 ANA's Congress of Nursing Practice formally recognized forensic nursing as a unique specialty by accepting its Scope and Standards of Forensic Nursing document. The IAFN also ratified its Code of Ethics in 1995. In 2008, the IAFN claimed nearly 3000 members, although the overwhelming majority of these are still aligned with specialization in sexual assault, many without an academic degree in nursing. In addition to an online membership newsletter, "On the Edge," the organization offers several networking opportunities on its Web site and is in the initial stages of defining a comprehensive core curriculum that will establish the groundwork for formal certification of its members in the specialty of forensic nursing. In the meantime, the rapid proliferation of sexual assault nurse examiners (SANEs) who must testify within the judicial system compelled the IAFN to establish a certification program for these nurses. In 2000 the IAFN offered its first practice validation and certification program for SANEs (Cammuso et al., 2001). The *Journal of Forensic Nursing*, a quarterly publication, was also recently launched. The dearth of advanced academic degrees among the IAFN membership and sparse legitimate research within the field of forensic nursing have been challenges for the advancement of the specialty and its peer-reviewed journal thus far. However, with growth of forensic nursing curricula at both the undergraduate and graduate levels, it is expected that more nurses with advanced-practice certification and master's and doctorate degrees will become the nucleus required to propel scholarly writing and research-based practice in forensic nursing.

As forensic nurses gained credibility and support for their nontraditional roles in traditional health-care settings, the way was paved for greater acknowledgement and respect from other disciplines of forensic sciences worldwide. In 1991 the American Academy of Forensic Sciences (AAFS) recognized forensic nursing as a unique discipline within its membership. Eleven nurses have achieved the status of "fellow" along with other eminent forensic scientists from a myriad of other disciplines.

In 1996 the University of Louisville's Clinical Forensic Medicine Program hired the first full-time clinical forensic nurse specialists in the United States. The university's forensic services program currently consists of an interdisciplinary team of experts, including five forensic pathologists, two clinical forensic emergency physicians, one forensic odontologist, and three clinical forensic nurses. The team uses state-of-the-art technologies coupled with traditional methodologies to assist in the medicolegal evaluation of victims and alleged perpetrators, and it testifies to its findings in relevant legal cases. Examinations are performed with the consent of the victim, a legal caregiver, a protective agency, or the courts. The team is called on to consult in, on average, 250 cases a year. The majority of cases have been pediatric victims.

Perhaps the most easily recognizable and currently accepted forensic nursing role is that of the sexual assault nurse examiner (SANE). Many states have SANE programs, and some can boast of 10 or more years of service to their communities. Many such programs were the direct result of victims' advocacy groups looking for a timely, and compassionate system in which to work with victims of sexual assault. Program sites, team compositions, and training requirements vary, but the mission of the SANE is to conduct a forensic examination of victims and, sometimes, the perpetrators.

In 2007, the Academy of Forensic Nursing Science was founded to offer online education, networking, and mentoring for both career development and practice. According to its Mission Statement, the Academy of Forensic Nursing Science (AFNS) is dedicated to meeting the professional developmental needs of forensic nurses and other interdisciplinary forensic specialists by providing a member-driven organizational structure for education, communication and networking. Its programming addresses the scientific basis of forensic nursing and the *objective* role it serves in society. Distinguishing itself from the IAFN, the AFNS endeavors to indoctrinate, educate, and support nurses as scientists, *not advocates*, in responding to interpersonal trauma or violent crime. AFNS emphasizes its ethical obligations to serve, with parity, both victims and offenders through the forensic nursing process, and pledges to work collaboratively and cooperatively with law enforcement officers, judicial authorities, and members of other forensic disciplines in search of truth and justice. The organization provides basic and continuing education courses to equip its members for roles in clinical practice as well as in assisting with public policy-making essential for healthy living and social order.

There are several additional professional forensic associations that have admitted nurses into membership and have opened their certification channels to nurse members. These include the Board of Forensic Examiners, Society of Medical Jurisprudence, the National Association of Medical Examiners, and the American Board of Forensic Death Investigators. Forensic nurses tend to be affiliated with two or more professional organizations to meet their personal and professional needs.

International Frontiers of Forensic Nursing

In addition to the United States, there are several countries with emerging initiatives in forensic nursing, including South Africa, India, China, Turkey, Pakistan, Italy, and Japan. With major issues of escalating crime and the need to ensure human rights for women and children where they have been limited by cultural norms, there are multiple driving forces for the development of forensic nursing within these areas. The minister of health in South Africa led the way for its becoming the first country to designate forensic nursing as a national priority program. Several countries have made significant strides in developing courses within nursing curricula to address forensic content. In 2007, the University of Bari, Italy, awarded master's degrees to 42 nurses who completed requirements of its rigorous curricula in forensic nursing. In 2009, this university also expects to award doctorates in the specialty. Forensic nurses from the United States in a visiting professor capacity have supplemented their faculty resources during the program's initial development. Students also had a unique opportunity to participate in extensive forensic nursing initiatives of the Harris County (Houston, Texas) Medical Examiner's Office. Such cooperative international endeavors are characteristic of the collaboration required to maximize educational resources required for basic and advanced nursing education.

What Is and Is Not Forensic Nursing

There are many career opportunities within the several subspecialties of forensic nursing, dependent upon educational background, forensic science training, and personal interests. Although some of these roles may be assumed after short courses, continuing education, and on-the-job-training, many require advanced degrees in nursing or forensic science.

Nurses who seek to work within the forensic arena must be careful to recognize or identify what constitutes a real forensic nursing role or contribution rather than seeing an opportunity to identify or affiliate with the popular culture of contemporary forensic science. Nurses must revert to the original and basic definition of forensic nursing, i.e., nursing science interfacing with the law or legal arena.

For example, a nurse who works in a correctional facility is not necessarily a forensic nurse. Providing health care to a prison population is not forensic nursing. However, assuming that correctional nurses have the appropriate background and training, they will have many opportunities to recognize forensic implications of any given situation and initiate the correct chain of events. For example, if a patient is found dead in a cell block from a series of stab wounds, the nurse indoctrinated in forensic science will know exactly what to do, and what not to do, to ensure that all potential evidence is recoverable, unaffected by mismanagement of first responders. However, the first and principal duties of the nurse in such an incident would be to ensure that death, in fact, had occurred and that proper notifications are made to appropriate supervisory and administrative personnel. The health care and welfare of others within the prison population would continue to be the focus for the nurse, not the crime scene and its investigation.

All psychiatric nurses are not necessarily forensic nurses. Some work in roles where they may encounter forensic incidents from time to time, but in their everyday activities, they utilize nursing processes, not forensic processes. Another example still being debated within the forensic nursing community is that of the forensic nurse involved in critical incident stress debriefing (CISD). It is extremely important that those who are first responders to critical incidents or anyone involved with working a traumatic event (from the emergency room [ER] to the homicide of a child, a hostage situation, a raging fire consuming a nursing home, or a mass disaster such as TWA 800 or September 11) be given immediate and appropriate stress debriefing by those not only trained to do so, but by those who fully understand the implications of how this event may have long-term consequences for the responders. To say that this is a forensic nursing role, however, is debatable. The debriefing sessions are extremely confidential and often fragile in nature. Unless there are extremely unusual circumstances, there should be no legal implications and no interfaces with forensic science or courts of law. This is a private, acutely focused, confidential, therapeutic session for those who may have performed specific forensic roles and who have just participated in an extremely traumatic event. However this very specific focused therapy is not, in the purest sense, a forensic function on the part of the debriefer.

Career Opportunities

There are many career opportunities (see Table 3.1) within the several subspecialties of forensic nursing, depending on level of

Table 3.1
Career Opportunities

Practice Area	Forensic Responsibilities
Medicolegal death investigator	Investigates deaths under the direction of the medical examiner; functions on a multidisciplinary team of other forensic specialists
Clinical forensic nurse (CFN)	Identifies forensic cases and medicolegal cases in all patient care settings; works with quality-assurance and risk-management in-house investigations as well as external investigative agencies when criminal activity has been identified (patient areas include OB/GYN, nursing home, operating room, inpatient care settings, primary care, etc.)
Clinical forensic nurse specialists	As in the previous job description, but with an advanced degree in forensic nursing/science; practice is broader in scope
Forensic psychiatric nurse	Assesses, evaluates, and plans treatment for those individuals who have court-ordered psychiatric evaluations or have been committed to forensic psychiatric facilities
Legal nurse consultant	Functions as a consultant on medicolegal issues, particularly as they relate to health-care and nursing practice
Emergency department nurse	Identifies forensic cases and medicolegal evidence, collects and preserves evidence, initiates chain of custody, and begins documentation and referrals; occurs in the course of providing emergent patient care of victim or perpetrator (this nurse could be a CFN)
Sexual-assault nurse examiner	Conducts a forensic examination of victims of sexual assault, including a physical examination and written, photographic, and colposcopic documentation, and makes appropriate referral for follow-up care; this examiner should not be involved in follow-up treatment or victim advocacy in any way
Forensic nurse educator or consultant	Specially trained in the practice of forensic nursing with the appropriate forensic nursing credentials
Forensic accident examiner/reconstructionist	Specializes in the reconstruction of accidents or crime scenes based on mechanisms of injury
Nurse attorney	Practices law within the traditional scope, often specializing in cases involving health care

education, forensic science training, and personal interests. Although some of these roles may be assumed after short courses, continuing education, and on-the-job training, most require advanced degrees in nursing or forensic science.

The Clinical Forensic Nurse Specialist

In her article in *Forensic Nurse*, "Clinical Forensic Nursing: a Higher Standard of Care," Sullivan makes a case for a clinical forensic nurse examiner (CFNE) role. She points out that, like traditional nursing, forensic nursing subspecialties may stratify, with the clinical forensic nurse being underrepresented. Sullivan describes a hospital staff member who could work in tandem with law enforcement agencies. One primary function of the clinical forensic nurse role would be to systematically assess trauma patients and ensure that all findings were precisely documented. Recognition, collection, and preservation of forensic evidence in suspicious patient events or unexpected patient deaths would be an additional responsibility.

Sullivan further envisions the CFNE as an essential part of the hospital team with the duty to evaluate and perform the root-cause analysis of adverse patient events. Adverse patient outcomes may cause only minor concern, or they may be extremely serious in nature, but most are not criminal in nature. Regardless of the impact, precise recognition, collection, preservation, and management of facts, data, and medical evidence are critical. Only in this way does Sullivan see that nursing can truly deliver a higher level of patient care. She suggests that it is the duty of health-care providers to be accountable in the fullest sense for their actions and to provide and ensure a safe environment for their patients. Sullivan also notes that it will require time, energy, and money to implement a CFNE position. She feels that the positive outcomes generated by a CFNE could easily be tracked and qualified. Further, this process may result in the discovery of additional areas that could be impacted by CFNEs.

Goll-McGee, a clinical forensic nurse consultant, agrees with Sullivan. Goll-McGee postulates that the living forensic patient in the critical care setting requires another dimension of nursing practice and care. She believes that this added dimension includes assessment of the living forensic patient with an index of suspicion to uncover the how and why of the mechanism of injury that placed that patient in the critical-care unit and the documentation of findings. As represented by the forensic nursing model, Goll-McGee believes that ideally all critical-care nurses would be cross-trained in the principles and philosophies of nursing science as it may relate to criminal justice in order to address the prevalence of patient-care cases involving accidents, homicide, assault, and abuse. She acknowledges that critical care nursing responsibilities have always included activities that may involve future court testimony, evidence collection, crisis intervention, death notification, and the pursuit of anatomical gifts, to name a few. The role of the CFN, she points out, has been supported in the trauma arena, although probably not formally identified as a CFN. Until certification and title clarifications can be achieved, the CFN's purpose and function may be ambiguous and uncertain.

To be most effective, a CFN would have advanced training in death investigation, forensic wound identification, critical incident stress management, domestic violence, sexual assault, and all types of abuse. Like Cammuso, Madden, and Wallen (2001), Goll-McGee feels that a CFNE should also have training in law, principles of criminal justice, and fundamental forensic science.

The index of suspicion required to determine the how and why of a patient's mechanism of injury is also addressed by Winfrey. Nurses have long employed the use of insight, intuition, suspicion, and "gut" to care for patients. Winfrey feels that the cultivation of suspiciousness is not paranoia but rather is motivated by the nurse's responsibility to perform a comprehensive assessment of the many and complex determinants of patient care. The nurse endeavors to uncover what is not, at first, completely evident. Incorporating the principles of forensic science into their knowledge base and developing the index of suspicion allow nurses to take action before they have an explanation of the problem. Winfrey proposes that if a nurse does not act intuitively, valuable evidence may be lost or destroyed. Suspicion coupled with forensic knowledge may make a critical care nurse unique in that he or she has a realistic set of responses that assumes the justice system is a logical part of any interdisciplinary response to patient-care needs.

Forensic Case Management

Many hospitals are now designating personnel to manage forensic cases and to ensure that a systematic approach is used to obtain, safeguard, and transmit evidentiary materials required by the coroner, medical examiner, or law enforcement officers. This individual is likely to be a nurse who has a broad role as a forensic point-of-contact within the facility. The important criterion is that the designee should have the education that prepares her/him for gathering facts and preserving evidence to reconstruct the circumstances of death or injury.

It is in the best interest of the public to provide forensic standards that guide the practice of health care professionals. A nurse who does not recognize forensic principles could miss crucial evidence that could later affect the outcome of a criminal case. Moreover, evidence gathered illegally or improperly cannot be admitted in a court of law, and the perpetrator of a crime could go unpunished or an innocent individual might not be exonerated.

Although it is a surprise to many health care personnel, they are regularly in contact with essential criminal evidence. Evidence from patients or their associates may be collected, and when used in legal proceedings, may be key to solving a crime. Conviction of an offender in a court of law is fundamental to interrupting the cycle of violence. Nurses who willingly accept their roles in forensic cases can be primary patient advocates, thus playing a vital role in the process of stemming intentional trauma, abuse, and neglect.

Responsibilities of Hospitals in Evidence Collection and Preservation

Nurses who provide immediate care to crime victims and victims of catastrophic accidents have an increasing responsibility to ensure that proven, reliable methods of evidence collection are used in a program sensitive to traumatized victims. A forensic protocol defining proper procedures is critical to the outcome of the case when a victim later seeks justice through the legal system. Evidence of many types provides legal proof in a court of law.

Evidence may be tangible, such as the written word or a photograph or drawing of the sustained wound or injury; it may also be intangible, such as "excited utterances" or odors noted or recalled by a witness interviewed during treatment and documented in writing.

Often the nurse is the only person in the right place at the right time to note and document what she/he sees, smells, hears, and touches when caring for a patient or interacting with his or her caretakers or "loved ones." In this era of violence, nurses are in a unique position to possibly interrupt the cycles of violence. To accomplish this, the nurse must recognize, collect, and preserve evidence in a legally acceptable manner.

When the victim of violent crime is treated by ambulance or emergency department personnel, valuable forensic evidence is often lost because these individuals may not be aware of what constitutes important evidence, or they may not perceive how important it is to safeguard it during medical treatment procedures.

Failure to recognize evidence is only one part of the problem. Evidence is often fragile or perishable and can be altered or lost during medical procedures. Therefore, it is very important to document and photograph physical injuries as soon as they are noted. Sometimes there is poor communication between medical and law enforcement personnel, which complicates evidence management. To that end, in

2005, during the 79th regular legislative session, Texas amended its Nurse Practice Act by adding 301.306, "Forensic Evidence Collection Component in Continuing Education." Under this new provision, a licensed nurse "employed to work in an emergency room setting" will be required to complete a minimum of 2 hours of targeted continuing education in forensic evidence collection no later than September 1, 2008, or by the second anniversary of the initial issuance of a license under the aforementioned chapter. Other states have proffered similar legislation. The Connecticut chapter of the International Association of Forensic Nurses began networking in 2007 with legislators and agencies including the Connecticut Sexual Assault Crisis Service to develop and sponsor a bill which would establish a statewide forensic nursing program. Senate bill 1013 would have been amended to convene a working group to develop recommendations for the establishment of such a system and was passed in the Senate but did not make it through the House legislative session.

The forensic nurse must remember that evidence is not always noted upon admission. Bruises may not be revealed for hours or even days after the injury. When something unusual or unexplained is noted, it must be photographed or documented in terms of location and appearance. Several days later, the significance of such observations and actions may become apparent.

Status of Forensic Nurses in the Legal System

Since forensic nursing is a relatively new discipline, the role of its members within the legal system has not been fully established in all states and jurisdictions. There are certain forensic nurse specialists who have qualified to testify within courts of law, both as experts and fact witnesses, and to serve as consultants for attorneys.

Forensic nurses who function in the roles of medicolegal death investigator and sexual assault nurse examiner have come a long way in establishing themselves as part of the multidisciplinary team of forensic scientists and the legal personnel involved in criminal and civil cases. State and national certification in areas such as sexual assault and medicolegal death investigation has been instrumental in conveying their qualifications.

The clinical forensic nurse will pioneer new ground with regard to how this role fits into the legal system and which boundaries or obstacles must be overcome before rules and regulations are firmly and clearly standardized in the world of health care. There are some specific questions that beg to be answered if the role of the clinical forensic nurse in health care is to be maximized. With the increase of serial murders committed by health-care providers, as well as other situations involving patients where foul play has occurred before or after the patient enters the hospital and the medical evidence is crucial to obtain and document, it is well for the forensic nursing profession to examine these questions carefully, recognizing that all health care systems are not created equally and that jurisdictional rules and regulations will vary (for example, federal health care facilities versus for-profit private-sector hospitals).

What is the legal position of a nurse working in a hospital in terms of evidentiary specimens? If a nurse suspects something suspicious or recognizes an inconsistency that is not appropriate in the course of a particular patient's care, can that nurse collect specimens from patients without a doctor's order, even if the patient requests and approves it? Does law enforcement override a doctor's orders? How much information should be entered into the medical record? Who must be notified, and how and where are these notifications documented? How soon does the hospital attorney need to be involved? At what point do nurses consult an attorney or law enforcement on their own? What guidelines do nurses have if they are discouraged or forbidden to gather evidence that they believe is in the patient's best interest to obtain? Photography continues to be a concern in many health-care settings as issues of confidentiality and patient consent are blurred with the identification and management of medical evidence, even when it is in the patient's best interest.

Educational Preparation of Forensic Nurses

The educational preparation of today's forensic nurse represents diverse pathways, since formal curricula in undergraduate and graduate programs have only been introduced within the past two decades within the United States. In 1986 the University of Texas at Arlington opened a graduate-level forensic nursing option for its master's degree candidates. Only a small group of students completed the curriculum over the next four years, and in 1990 the program was discontinued by the School of Nursing due to lack of human and fiscal resources to support the offering. In 1992 Beth El College in Colorado Springs offered forensic nursing courses at the undergraduate level. In 1996 Fitchburg State College in Massachusetts opened a graduate program designed to prepare the clinical nurse specialist. Other formal offerings have occurred at Gonzaga State University in Washington, the University of Virginia, the University of Texas at Austin, the University of Central Oklahoma, the University of Arkansas, Rutgers University, Johns Hopkins University, Hawaii Pacific University, Indiana University, the University of California (Riverside), Quinnipiac University, Louisiana State University, Xavier University, and Cleveland State University. Canada has forensic nursing educational offerings at Mt. Royal College (Calgary), the University of Calgary, and at the British Columbia Institute of Technology.

Outside of North America, there are forensic nursing programs at the University of Natal and the University of Zimbabwe in Africa, at Flinders University in Adelaide, Australia, and at Notre Dame University in Perth, Australia, at the Punjabi University in Punjab, India, and at Kobe University in Japan. As other states and countries continue to work for funding in order to establish forensic nurse examiner programs, the interest and demand for education and opportunities remain encouragingly high. The most recent countries waiting for funding and support include Turkey, Thailand, Mexico, Sweden, Italy, and Iran.

The barriers to introducing this specialty are largely attributed to the lack of understanding about the discipline itself. Many nurses cannot appreciate the relationships and close associations between the scientific methods of forensic science and the nursing process. Since nurse educators have traditionally lacked indoctrination into the forensic sciences, a shift of paradigm has been difficult. Furthermore, since forensic nursing is a new discipline, it lacks nursing educators with doctoral preparation who can champion the specialty within the faculty groups who determine the curricular structure of undergraduate and graduate study. At the dawn of this millennium, there are fewer than ten forensic nurses with doctoral preparation who have aligned themselves with forensic nursing. Although there is a larger pool of forensic nurses with a master's degree, they usually must work outside the specialty for their earning power, pursuing forensic nursing as a secondary field of interest, since there are currently few paid positions for forensic nursing specialists. Although the Catch-22 represented by this lack of qualified educators (and thus a lack of programs to prepare forensic nursing specialists) is frustrating, it is commonly experienced with emerging scientific disciplines. In the meantime, a small but highly motivated group of qualified educators has been carrying the banner.

Educational Models for Forensic Nursing

The most common model for today's forensic nurse is postlicensure education in a selected aspect of the specialty. For example, diploma nurses, as well as those with degrees (associate, baccalaureate, and master's), have obtained their forensic education by "piecing together" content from continuing education courses, workshops, professional meeting programs, and certificate offerings. Some have been fortunate enough to be able to obtain sufficient credentials for employment as a sexual assault nurse examiner, death investigator, or other forensic specialist. Most, however, still await the opportunity to enroll in degree-granting academic programs. In the meantime, due to the outcry for forensic educational courses, distance learning programs have become popular as a method of accessing forensic nursing information, since few nurses are lucky enough to live in proximity to the few programs now offering formal academic curricula. Although Internet and correspondence courses are effective tools for some content, they have limitations, of course. Many skills associated with forensic nursing require actual "real world" opportunities to collect physical and biological evidence and to interface with the human situations germane to the discipline. In addition, clinical experience with law enforcement, with members of the judiciary system, and with other forensic scientists is imperative for preparing the forensic nurse specialist. Well-designed simulations, supervised chat rooms, and written scenarios are indeed valuable learning tools, but they cannot fully replace what is gained from a well-qualified clinical mentor.

Undergraduate programs with required or elective forensic nursing courses are few and far between, even today. New strategies for nursing education must be forthcoming in order to meet the needs for those individuals who want to pursue this nursing specialty.

Continuing Education for Forensic Nursing

There are several schools of nursing and organizations offering singular programs and forensic series in forensic nursing. Due to lack of faculty support for establishing forensic nursing within the undergraduate or graduate programs as formal offerings, the University of Texas at Austin's School of Nursing offered a highly successful series of continuing education programs in forensic topics from 1993 to 1996. It was hoped that if the value of the offering could be demonstrated and the students were enthusiastic about participating, then the faculty might be leveraged to support formal courses when fiscal and human resources permitted. Topics in the CEU series included introductory concepts, education and career opportunities, forensic photography, crime scene investigation, environmental issues in forensic science, sexual assault, child and elder abuse, domestic violence, correctional and forensic psychiatric issues, and the role of the forensic nurse in legal proceedings. Guest experts in the various subjects provided lectures and discussion sessions and supervised clinical practice. However, in 1999 the Continuing Nursing Education Program division was closed, and forensic nursing educational opportunities once again were closed. In the meantime, the champions of forensic education have moved on, channeling their energies into other pursuits. In 1999 the School of Nursing at the University of New Mexico initiated a series of forensic programs to meet the needs of students clamoring for forensic education. Two types of students enrolled in the offerings: undergraduate for credit and graduate for CEUs only. In addition to these two examples, there are several other resources for continuing education in forensic nursing, but what is sorely needed is a widespread proliferation of undergraduate exposure to the subject.

Two approaches can be used effectively, namely, elective or required courses on certain forensic topics, or forensic principles or concepts woven throughout the curriculum as "threads." Examples of such threads include human abuse, forensic assessments and collection of evidence, documentation and reporting of crimes of abuse and negligence, social justice, jurisprudence, and prevention strategies for societal violence. The value of using curricular "threads" would be to effect a highly flexible way to ensure that students would be exposed to the core curriculum of forensic nursing. They could be introduced early in the curriculum and could be fleshed out as the student builds other components of a generic nursing education. There is not a singular clinical area to which the student is exposed that would be an inappropriate locus for testing forensic principles and concepts.

Future of Forensic Nursing

The Joint Commission on Accreditation of Healthcare Organizations (JCAHO) has laid the groundwork for the presence of forensic nurses in hospitals by its guidelines for patient assessment and management. These guidelines address the needs for all hospital personnel to be trained to identify victims of abuse and to take essential steps to preserve and safeguard evidentiary materials. JCAHO also requires that hospitals refer the subjects of forensic cases for follow-up care and treatment. These important elements of regulatory guidance now mandate that hospitals have specific educational and training programs for all personnel in basic forensic concepts and principles. It also requires that facilities enact policies and procedures regarding evidence collection and preservation, and align themselves with the needs of law enforcement and of the legal system within the community. Although most hospitals do not have a dedicated specialist in forensic nursing, several leading facilities have developed full-time or part-time nurse-specialist positions to ensure that JCAHO guidelines are properly addressed and that the forensic program is managed properly in all departments.

The most compelling role of forensic nursing seems to exist within the hospital and other patient-care settings where individuals receive physical and mental health care. The emergency department is particularly vital in the role of initial assessment and treatment of victims of trauma and interpersonal violence. Resuscitation attempts or patient deaths may also occur during this emergent phase of care, and important legal concerns accompany such events, including advanced directives, sustained life support, organ procurement, and autopsy requests. Deaths in the emergency department are invariably referred to the medical examiner. Mother–infant services, pediatric clinics, and psychiatric centers frequently require the knowledge and skills of the forensic nurse. Quality-assurance or risk-management programs as well as the department of patient safety may employ forensic nurses to assist in benchmarking the hospital's performance in regard to medical errors, adverse or sentinel patient events, health insurance or Workman's Compensation fraud, bomb threats, bioterrorism, theft of narcotics or controlled substances, and physical threats or assaults within the patient-care environment. An example of a position description for a forensic nurse who functions for the facility as a point-of-contact regarding any clinical forensic issues is presented on the following page.

Other forensic nursing roles may emerge in school systems, within the community-health arena, within law enforcement departments, and within the judicial system. Due to the international shortage of forensic pathologists, positions for forensic nurse death investigators will continue to increase. It seems apparent that at least for now, sexual assault nurse examiners will be in demand, especially in small communities and rural areas where physician resources are seriously constrained.

The real test of career opportunities will ultimately be based on the availability of resources to pay for the services of the forensic nurse. Up to now, many nurses have assumed forensic roles as a volunteer or as a minimally paid public servant, compelling these individuals to be gainfully employed in another demanding job as well. This situation impacts role commitment, since few nurses can afford to pay for education and training and then serve essentially for minimal wages, if paid at all.

The public currently views hospitals and health care with skepticism. Wrong-site surgeries, fatal medication errors, caregiver-perpetrated homicides, patient neglect and abuse, and other issues have caused patients and their families to demand answers. Nurses continue to rate very high in regard to public respect and trust, thus placing them in an ideal position to be the "watchdog" for adverse events that endanger patients during their hospital stay. Hospital administrators now perceive that they must take an active role in making the facility safer. The acquisition and marketing of a forensic nursing team may offer the public increased confidence that someone is looking out for their welfare.

Forensic nursing is a new nursing specialty. As with any scientific discipline, its future will depend on the ability of its members to respond to today's needs while concurrently preparing to fill an unknown niche that might emerge tomorrow. As forensic nurses evolve in education, practice, and research, they will be better prepared for independent and collaborative roles, and they will be ready to fill a niche in tomorrow's health-care and forensic arenas.

Resources for Forensic Nurses

Professional Associations

Academy of Forensic Nursing Science
American Institute of Forensic Education
Society of Medical Jurisprudence
International Association of Forensic Nurses
American Board of Forensic Death Examiners
American Academy of Forensic Sciences

Publications

Accreditation Manual for Hospitals, Core Standard and Guidelines Joint Commission on Accreditation of Healthcare Organizations (JCAHO), current edition.
Scope and Standards of Forensic Nursing, American Nurses Association and the International Association of Forensic Nurses, 1998.
Forensic Nurse, Virgo Publishing, Inc., Phoenix, AZ; www.forensic nursemag.com.
On the Edge, online newsletter of the International Association of Forensic Nurses, Pitman, NJ.

Internet Resources and Web Sites

Academy of Forensic Nursing Science www.tafns.org
American Academy of Forensic Sciences (AAFS) www.aafs.org
American Institute of Forensic Education www.taife.ord
American Professional Society on the Abuse of Children www.apsac.org
International Association of Forensic Nurses www.forensicnurse.org
International Society for Prevention of Child Abuse and Neglect
Web site: *www.ispcan.org*; Email: ispcan@ispcan.org
National Center for Victims of Crime www.ncvc.org
American Forensic Nurses www.forensictrak.com

Forensic Nurse Position Description

Personnel Requirements

- Exemplary clinician with high level of interest in medicolegal aspects of nursing; minimum of two years of professional nursing experience
- Completion of an introductory forensic nursing course
- Willingness to pursue further education and training in specific forensic topics pertinent to health-care settings

- Displays strong physical assessment, documentation, and communication skills pertinent to forensic patient care
- Impeccable integrity; uses good judgment and discretion when communicating sensitive patient information within the health-care facility and with other internal and external organizational contacts
- Thoroughly familiar with all regulatory guidance regarding management of forensic cases within the medical environment (i.e., sexual assault, child/elder abuse, neglect, trauma, domestic violence, unexplained death, etc.)

Primary Responsibilities

- Manages the forensic nursing program within the facility
- Provides direct services to individual patients and family and consultation services to nursing, medical, and law enforcement agencies both in the hospital and the local community
- Maintains current knowledge of case law, legal evidence collection procedures, and investigative responsibilities of medicine and law enforcement
- Applies forensic nursing knowledge and skills to the immediate interventions and ensures referrals for follow up care of victims and perpetrators of violence
- Establishes mechanisms for providing forensic education and training within the health-care facility as required by the Joint Commission on Accreditation of Healthcare Organizations (JCAHO)
- Ensures that forensic nursing assessment and referral services, mandated by JCAHO guidance, are offered within the medical treatment facility
- Provides case-centered forensic consultation services to hospital personnel
- Participates in staff orientation programs to convey critical information regarding forensic responsibilities of health-care personnel
- Serves as liaison with external agencies in regard to forensic-related medical issues and problems, including sexual assault examinations, domestic violence and human abuse cases, and investigations regarding unexplained deaths
- Participates in the review, revision, or updating of policies and procedures that include medicolegal content
- Performs periodic audits of medical records to ensure that forensic cases have been identified and managed in accordance with policies, procedures, and relevant laws
- Monitors and evaluates clinical forensic activities as an integral part of the medical facility's risk management program
- Represents the hospital in community endeavors associated with sexual assault, domestic violence, child and elder abuse, or other forensic-related topics
- Maintains the hospital's forensic database and resource materials
- Prepares reports/memoranda and composes correspondence that supports medicolegal investigative efforts
- Creates and directs the implementation of a forensic nursing care plan in response to episodic crisis situations (i.e., sudden or violent death, hostage negotiation, bioterrorism, internal disaster)
- Provides information and guidance for medical personnel relating to organ donation/procurement and interface with victims' family members in requesting anatomical gifts
- Reviews all memoranda of understanding (MOUs) pertinent to forensic issues within the facility
- Engages in problem-solving processes with agencies and services in order to improve/refine medicolegal management of victims of violence
- Maintains current with the emerging trends in forensic science and clinical practice that impact the medicolegal management of victims and perpetrators of violence

- Initiates/participates in research related to clinical forensics within the health care facility
- Represents the medical facility in community efforts to address domestic violence, human abuse, and sexual assault issues
- Maintains current knowledge of forensic nursing through literature, continuing education courses, workshop attendance, and the use of sound benchmark practices

Customer Service Responsibilities

- Endeavors to provide the optimum level of service to all customers, appreciating that they not only include alleged victims of assault, domestic violence, trauma, and neglect, but also the alleged perpetrators of violent acts
- Responsible for protecting forensic medical information from unauthorized release or loss, alteration, or unauthorized deletion; complies with applicable regulations regarding computerized files and release of access codes as set out by the information systems department

Organizational Relationships

- Reports directly to the hospital administrator for patient services and the risk manager in order to maintain patient confidentiality and shorten the chain of communication in the event of future legal proceedings
- Maintains functional relationships with:
  - Hospital security forces
  - Family advocacy
  - Hospital attorney
  - Risk management and quality assurance program managers
  - Medical examiner's office
  - County attorney's office
  - Local civilian law enforcement agencies
  - Community agencies and resources, as applicable

Questions

1. How is forensic nursing defined by the International Association of Forensic Nurses?

2. What year was forensic nursing formally accepted as a discipline by the American Academy of Forensic Sciences?

3. Describe the practice domain of the forensic pathologist.

4. How is "living forensics" distinguished from other branches of forensic science?

5. What are two barriers to the advancement of forensic nursing as a discipline and a science?

6. List three established roles for the clinical forensic nurse.

7. According to the model of forensic nursing, what are the three sources for the knowledge base of forensic nursing?

8. What forensic nurse specialist should assume responsibilities for investigating adverse clinical events in hospitals?

9. Describe the various educational options for preparing and credentialing the basic and advanced forensic nurse clinician.

10. List and briefly describe five current career opportunities for the forensic nurse.

References

Almirall, J. R. and Furton, K. G., The evolution, practice and future of the use of science in the administration of justice, *Standardization News (ASTM)*, 23(4), 1995.

Cammuso, B. S., Madden, B. P., and Wallen, A. J., in *Nursing Practice and the Law*, O'Keefe, M. E., Ed., F. A. Davis Co., Philadelphia, 2001.

Hoyt, C. A., personal correspondence with Catherine M. Dougherty and Janet Barber Duval, 1999.

International Association of Forensic Nurses (Membership Brochure), Arnold, MD, 2008.

International Association of Forensic Nurses and American Nurses Association, *Scope and Standards of Forensic Nursing Practice*, American Nurses Publishing, Washington, D.C., 1997.

Lynch, V., Clinical Forensic Nursing: A Descriptive Study in Role Development, thesis, School of Nursing, University of Texas at Arlington, 1990.

Lynch, V., Clinical forensic nursing: a new perspective in the management of crime victims from trauma to trial, *Crit. Care Clinics North Am.*, 7(3), 489–507, 1995.

McNamara, H., Living Forensics (seminar pamphlet), Office of the Medical Examiner, Ulster County, New York, 1986.

Pyrek, K., Forensic nursing's past, present and future, *Forensic Nurse* (May–June), 14–16, 2003.

Investigation of Traumatic Deaths

Ronald K. Wright

Introduction

A forensic pathologist investigates sudden and unexpected deaths of persons who are in apparent good health and deaths suspected to be traumatic. This chapter will examine some of the major issues related to death due to trauma, beginning with the concepts of cause and manner of death. The classification of trauma as mechanical, chemical, or thermal will be discussed. This chapter covers common traumatic deaths. Investigating traumatic death constitutes most of the life work of forensic pathologists, and it can only be covered in the broadest terms by a single chapter.

The purpose of investigating a death is to determine its cause and manner, and a forensic pathologist must be prepared to present physical evidence to support his or her conclusions about the cause and manner of death.

Cause and Mechanism of Death

The cause of death is the disease or injury that initiated the lethal chain of events, however brief or prolonged, that led to death. In other words, the cause of death is the underlying cause, even though a number of complications and contributing factors may have been involved. The **mechanism of death** is a biochemical or physiologic abnormality produced by the cause of death that is incompatible with life. A nonpathologist physician will ask a pathologist about the mechanism of death instead of the cause of death.

Consider the example of a middle-aged man who went to a hospital after having been shot multiple times during a robbery. He underwent emergency surgery during which the organs affected by bullets were repaired. The man's condition improved somewhat, but he developed pneumonia followed by renal failure, liver failure, and finally heart failure. Autopsy examination revealed that he had preexisting severe lung and heart diseases. The cause of death was multiple gunshot wounds because those injuries set in motion a lethal chain of events. However, the treating physician was interested in the mechanism of death, the multiorgan failure, and questioned why this patient developed multiorgan failure when all his injuries were repaired. The treating physician also wanted to know about the preexisting heart and lung diseases, without which the man probably would have survived. The preexisting diseases were not the cause of death because injury takes precedence over disease in determining cause of death. In other words, injury trumps disease.

The classic example of the precedence of injury over natural disease is a fable told in law school about a person with a paper-thin skull who was knocked down in an altercation. The paper-thin skull was a severe disease. The man probably would have survived the altercation and might not have even been injured if he did not have the severe disease. The cause of death was blunt trauma to the head. The paper-thin skull was a contributory factor.

Manner of Death

A forensic pathologist also is called upon to determine the **manner of death**. The manner of death is defined as the fashion in which the cause of death came to be. This explanation is not terribly clear without an explanation of the range of possible manners of death. There are hundreds or thousands of possible causes of death, but only four manners of death (or five, if you consider undetermined or unclassified death as a separate category). The four manners of death are *natural, accidental, homicidal,* and *suicidal.* Natural deaths are caused solely by disease, without the intervention of trauma.

The other manners of death all involve trauma. Accidental deaths are due to trauma occurring from acts no reasonable person would have felt had a high probability of producing bodily injury or death. Homicidal and suicidal deaths arise from acts a reasonably prudent person would have felt had a high probability of producing bodily injury or death. The difference between suicide and homicide is merely the person who acted. If the deceased took the action, the death is a suicide. If someone other than the deceased took the action, the death is a homicide.

Time of Death

When a person dies there are a number of changes that occur which can be used to estimate the time of death: **rigor mortis, livor mortis,** and **algor mortis**.

It is unknown when these changes were first noted, but certainly by the early 18th century there are writings that indicate that these changes were noted and utilized to determine the time of death.

Rigor mortis is the stiffening of the muscles that occurs following the death of a person. This is a chemical reaction that occurs when the glycogen normally found in muscles is used up following death and is not reformed. Glycogen is used to provide energy for the contraction of muscles, and it is depleted slowly after death. Generally, rigor mortis is seen about 4 hours after death. It can occur sooner if the glycogen has been depleted by the exercise of muscles just before death. This is called instant rigor mortis and is described in some war deaths. Electric shock can also lead to shorter periods between death and rigor's onset. Rigor generally disappears during the period from 24 to 36 hours after death as further decomposition of the muscles leads to their loss of ability to remain fixed in rigor.

Livor mortis is the discoloration of the body which occurs from the settling of red blood cells after the blood stops circulating. This can be seen within minutes of death where the blood cells have an increased sedimentation rate from infectious or other disease. Generally, in light skinned individuals livor mortis or lividity can be seen within an hour or so after death. In some dark skinned individuals it may not be possible to see lividity. If a person has died and lost most of their blood volume, then lividity may also not be able to be seen. Lividity becomes fixed, meaning that finger pressure will not blanch the lividity, about 12 hours after death. Lividity slowly disappears with decomposition after 36 hours.

Algor mortis is the cooling of the body that occurs after death, assuming the ambient temperature is lower than body temperature. The general rule of thumb for a nearly nude body exposed to 18° to 20°C is 1.5°C of temperature drop per hour for the first 8 hours. The normal body temperature is 37°. Thus, if a body has been dead for 4 hours the temperature will be 31°. Unfortunately, the temperature may go up or down from 37° right before death, which can cause the estimates to be wrong.

None of the changes after death result in accurate estimation of the time of death, but can be of great help in estimating the time of death.

Classification of Traumatic Deaths

Traumatic deaths may be classified as *mechanical, thermal, chemical,* or *electrical.* A mechanistic classification termed asphyxial death overlaps the other causes. Asphyxial death is caused by interference with the oxygenation of the brain. This asphyxia can occur from mechanical causes (strangulation), chemical causes (cyanide poisoning), and electrical causes (low-voltage electrocution).

Mechanical traumas are divided into sharp and blunt categories. Blunt traumas are further subdivided into nonfirearm and firearm groups. Firearm trauma can be divided into low velocity and high velocity. Trauma surgeons classify trauma as penetrating or nonpenetrating. Penetrating traumas include gunshot and **stab wounds**. Nonpenetrating traumas consist primarily of motor vehicle collisions and falls. The discussion below will cover a few of the high points of mechanical traumatic deaths. For further discussion of this topic and other types of traumatic deaths,

consult the forensic pathology textbooks listed in the references.

Mechanical Trauma

Sharp Force Injury

Mechanical trauma occurs when applied physical force exceeds the tensile strength of the tissue to which the force is applied. Sharp force refers to injuries received from sharp implements, such as knives, swords, and axes. The amount of force required for a sharpened instrument to exceed the tensile strength of tissue is significantly less than the force required with a blunt object.

Of great importance is the fact that blunt objects produce **lacerations** and sharp objects produce incised wounds. Examining a wound allows one to know whether a sharp or blunt object caused the wound. Unfortunately, many physicians call all relatively discrete injuries lacerations, thereby destroying the distinction between the two types of wounds. Figure 4.1 shows an incised wound, Figure 4.2 shows a laceration. Figure 4.3 shows representative stab wounds and incised wounds of the neck. A stab wound is produced by a sharp object whose longest dimension is depth as opposed to other surface dimensions. Figure 4.4 shows an incised artery. Note again the sharp edges of the wound that distinguish it from an injury produced by a blunt object.

It is difficult to precisely determine the size of a sharp object from examination of

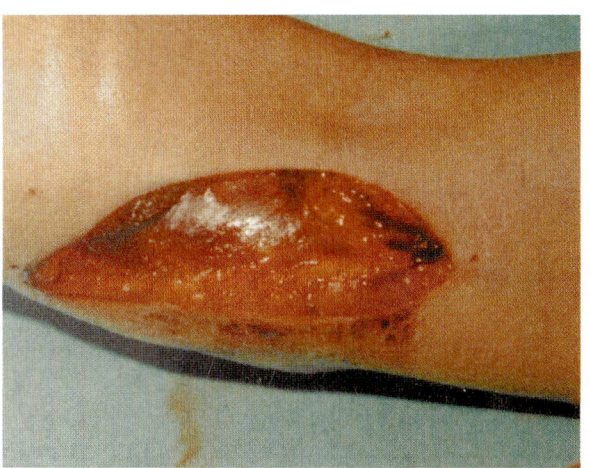

Figure 4.1 Incised wound.

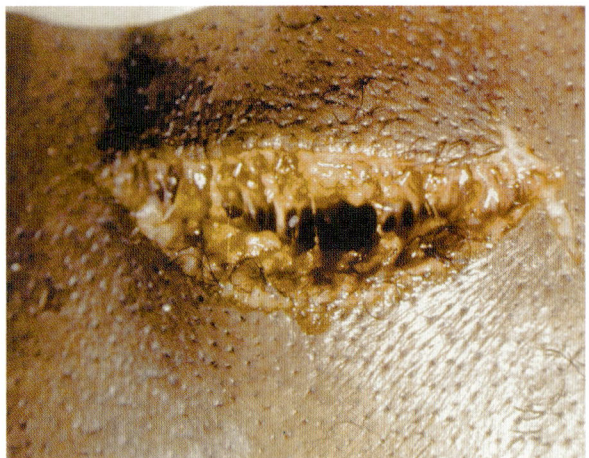

Figure 4.2 Typical laceration produced by blunt trauma.

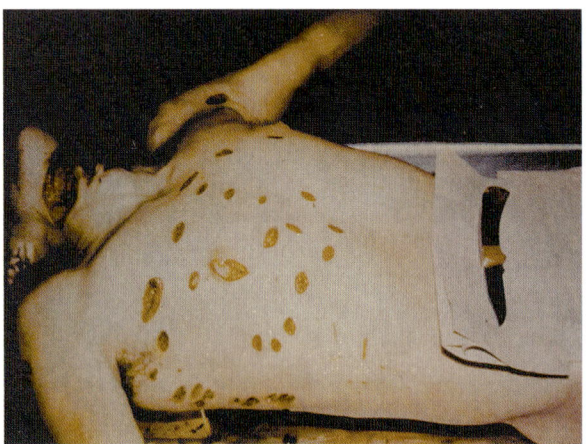

Figure 4.3 Stab wounds of the chest and incised wound of the neck.

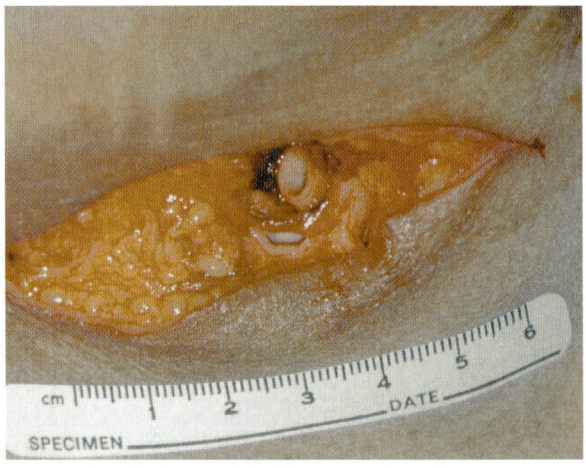

Figure 4.4 Incised wound of skin and artery.

the characteristics of the wound. A stab wound measuring 1½ inch on the surface and 4 inches deep could have been produced by a knife of the same dimensions. However, a ½–2-inch knife could produce the same wound if thrust in with great force and removed at a different angle from the angle of entry. A larger knife could have produced the wound if it was not completely inserted.

Death from blunt and sharp trauma arises from multiple mechanisms, but sharp trauma most commonly causes death by **exsanguination**. This means that a major artery or the heart must be damaged to produce death from sharp trauma.

Blunt Trauma

Blunt trauma causes death most commonly when the brain has been significantly damaged. However, blunt trauma can lacerate the heart or aorta, leading to exsanguination, or can produce many other complications.

Firearm Injury—A firearm causes a special kind of blunt trauma. Firearm injuries are the most common suicidal and homicidal wounds seen in the United States. That reflects the ready availability of firearms and the remarkable lethality of these weapons. Firearm injuries may be classified on the basis of the propellant used to accelerate the projectile. The most common propellants are gunpowder and smokeless powder (nitrocellulose). However, gunpowder is very rarely seen, so smokeless powder propellant is generally assumed in most classification schemes.

Another distinction is between rifled weapons and smooth bored weapons. Most firearm deaths are from rifled weapons—rifles or **handguns**. Antique weapons and **shotguns** are smooth bored weapons.

Wounds also may be classified on the basis of the diameter of the projectile or bullet. Currently, a combination of English and metric measures is used to classify firearms. See Chapter 20 on firearms examination for a further discussion.

Of more importance, from the standpoint of injuries produced, is the velocity of the projectile. The extent of injury produced by a firearm projectile increases as the square of

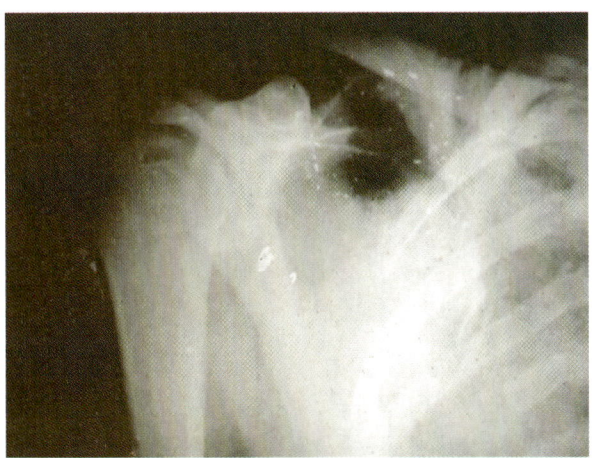

Figure 4.5 Massive destruction and lead "snowstorm" caused by high speed projectile.

the velocity increases, times the mass of the projectile. Thus, there is a near-quantitative difference between a high velocity projectile and a low velocity projectile. The cutoff point between high and low velocity is generally noted as 300 meters per second.

Figure 4.5 is an x-ray of a person who received a high speed projectile injury to the right upper chest. The dark area is missing tissue. Around the missing tissue are white fragments of lead, often called a lead snowstorm, that are diagnostic of a high speed projectile. High speed projectiles are only seen in wounds inflicted by high-powered hunting rifles and military rifles. The .44 magnum handgun may achieve the lower ranges of velocity of a high-speed projectile and is the only commonly encountered handgun that is capable of such speed and destruction.

Another classification scheme of gunshot wounds is whether they are penetrating or perforating. A penetrating gunshot wound has an entrance wound and no exit wound. A corollary of this is that a projectile must be recovered from the body for every penetrating gunshot wound. A perforating gunshot wound has an entrance wound and exit wound. A corollary of this is that generally no projectile will be recovered.

When a firearm is discharged, the force that propels the projectile is the gas produced by the rapid burning of the gunpowder or smokeless powder. We will discuss smokeless powder only, because of the rarity of gunpowder weapons. To ignite smokeless powder, it is necessary to first have a primer initiate the fire. In all cartridges except .22 **caliber** weapons (also called *rim fire weapons* as their primer is around the rim of the cartridge), the primer is in a small cup at the rear of the inside of the cartridge case. Striking (or heating) the primer causes it to ignite, and this in turn ignites the smokeless powder. The rapidly burning smokeless powder produces a large amount of **carbon monoxide**, nitrogen dioxide, carbon dioxide, and other gases. In addition, the components of the primer are heavy metals including lead, antimony, and bismuth that are admixed in the gases. All of these materials and the bullet are projected from the barrel of the weapon along with burning and unburned smokeless powder (no cartridge is 100 percent efficient at burning all of the powder).

How far each component travels is the basis for determining the distance of the barrel from the deceased at the time the weapon was discharged. The gases, including the heavy metals, and some smoke from unburned but gaseous carbon, are projected only a few inches. The effects of the gas produce what can be discerned as contact or near-**contact wounds**. What can be seen is the blackening of the skin. In addition, the skin will show variable amounts of laceration because the gas blown into the wound tears the skin apart. Finally, the carbon monoxide reacts with the hemoglobin and myoglobin in the wound to produce carboxyhemoglobin and carboxymyoglobin. These compounds are bright red, compared to the dull red color of normal hemoglobin and myoglobin. Figure 4.6 shows a typical contact gunshot wound. It shows several small lacerations, blackening around the edges of the wound, and red discoloration of hemoglobin in the wound.

Figure 4.7 shows a relatively ordinary contact gunshot wound of the head. Because of the tearing characteristics of the scalp and the reflection of gases by the skull, large lacerations are characteristic of head contact wounds. Indeed, **explosion** of the head and evacuation of the brain are common effects of gunshots that produce large amounts of gas.

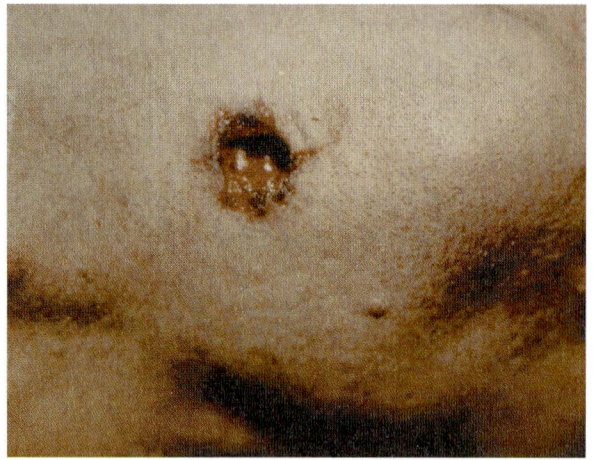

Figure 4.6 Contact gunshot wound of the chin. Note smoke, small laceration, and carboxyhemoglobin.

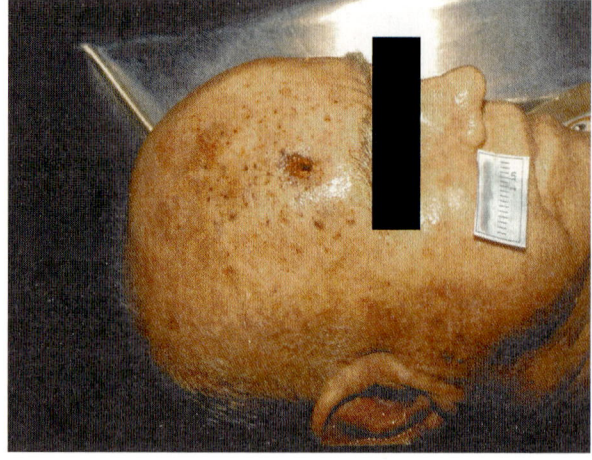

Figure 4.8 Intermediate range gunshot wound with 2 to 3 inches of stippling.

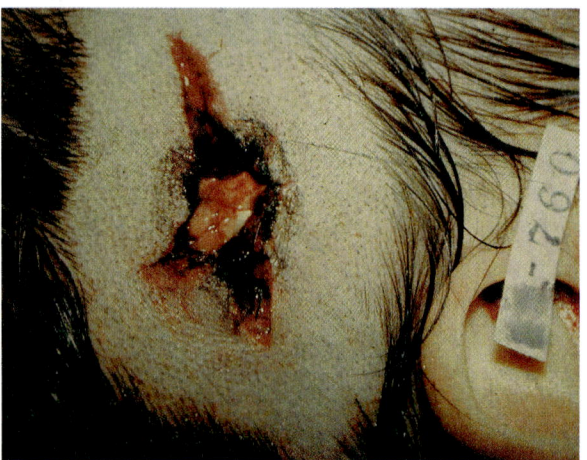

Figure 4.7 Contact gunshot wound of the head. Note large lacerations.

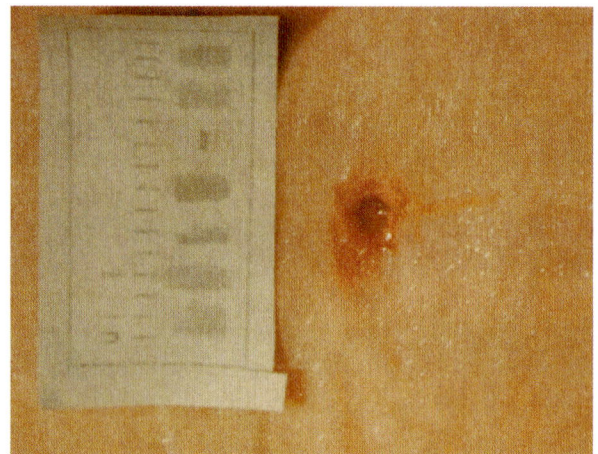

Figure 4.9 Typical distant gunshot wound.

As the distance from the barrel to the skin increases, the effect of the gas diminishes and only the unburned powder and bullet are capable of penetrating the skin. Unburned powder that penetrates the skin produces **stippling** or tattooing around the defect produced by the bullet. These wounds are called **intermediate range gunshot wounds** (Figure 4.8). Most handguns produce stippling when the skin-to-muzzle distance is 0.5 centimeter to 1 meter. The pattern enlarges as the distance increases. At 1 meter, the speed of the powder slows sufficiently so that it cannot penetrate the skin. A speed of 100 meters per second is generally considered necessary to produce penetration.

Distant gunshot wounds lack smoke and powder effects. Since distant gunshot wounds lack evidence of anything other than the effect of the bullet, the range is indeterminate because clothing and other objects can block the effects of gas and powder. Figure 4.9 shows a typical distant entrance gunshot wound. Distant gunshot wounds lack smoke, soot, and stippling. A typical **distant wound** has a circular skin defect and a rim of abraded skin around the edges. The diameter of the skin defect is some indication of the diameter of the bullet, but the estimate is not very reliable because of small differences in diameters of common bullets used by civilians. Bullets vary in size from a nominal 0.22 inches to

0.45 inches. The difference of 0.2 inches is not easily discernible by most observers.

The primary variable in the size of a distant entrance gunshot wound is the elasticity of the skin. The skin of younger people is much more elastic than the skin of older people. The elasticity means skin defects will be smaller. A nominal 0.38 caliber wound in a 20-year-old may appear more like a 0.22 or 0.25 caliber wound in a 50-year-old. Obviously, ascertaining the caliber of the weapon from a contact wound is not possible, as the wound bears little relationship to caliber due to the tearing of the skin.

Gunshot exit wounds typically are lacerated. Although the conventional wisdom is that gunshot wounds of exit are larger than gunshot wounds of entrance, this is not always the case, as can be seen with contact wounds that are much larger than corresponding exit wounds. Indeed, the error rate of emergency room physicians without forensic training in determining directionality of suicidal contact gunshot wounds to the head is almost 100 percent. These physicians usually rely upon the general rule, but suicidal gunshot wounds to the head are nearly always contact wounds.

Some estimation of the velocity of an exiting bullet can be discerned from the appearance of the exit wound. Exit wounds that are small and slit-shaped and have few small side lacerations are traveling at slow speeds and the bullet will generally be found near the body (or even in the clothing). Conversely, exit wounds with many side lacerations travel at a very high speed; this is seen best with high velocity weapons such as military and hunting long arms.

An exit wound will be supported or *shored* if a gunshot victim wears tight constrictive clothing such as a heavy leather coat or garments made of tight woven fabrics, or is against a material such as dry wall that can be penetrated by the exiting bullet and will support the skin. **Shored exit wounds** look remarkably like entrance wounds. See Figure 4.10 for a fairly representative example of this phenomenon. Often, the rim of abrasion is wider than is typically seen in an entrance wound. This may help in differentiating the two types of wounds. It is important to note that entrance

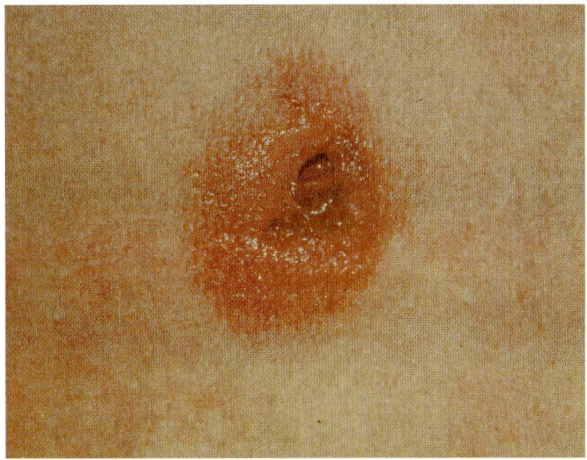

Figure 4.10 Shored exit wound. Note pattern of clothing that provided the shoring.

wounds have a unique appearance because they are all supported or shored. An entrance wound is shored by the underlying soft tissue and bone; that is why a rim of abrasion appears around entrance wounds. The skin is compressed for a time before the bullet penetrates the shoring material, then the nose of the bullet abrades the skin. If the skin is not supported, it tears and no abrasion occurs. This is the typical case with exit wounds.

Figure 4.11 shows a shored or supported exit wound that could very easily be an entrance wound with an intermediary target. Supported exits and entrance wounds with intermediary targets are probably not discernible. Of importance is the rectangular shape of the entrance wound. Entrance wounds are generally round when bullets are fired from

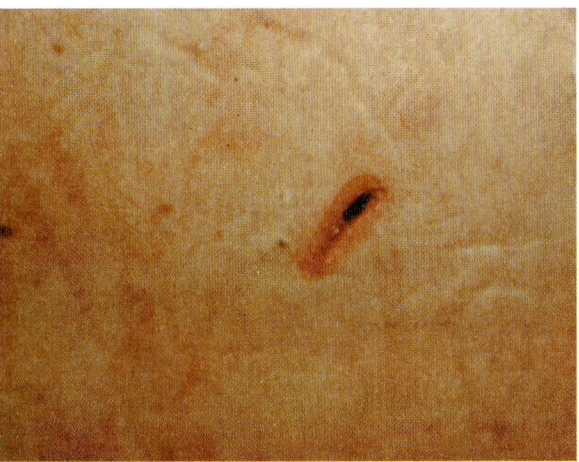

Figure 4.11 Shored exit wound or entrance with intermediary target.

rifled barrels because the bullet, spinning in an axis 90° off the direction it is traveling, moves through air with its nose pointed in the direction in which it travels. The rotation exerts a gyroscopic action on the bullet that keeps its nose pointed in the direction of travel, a consequence of Newton's first law of kinetics. The rotation causes a bullet fired from a rifled barrel to be much more accurate than alternative projectiles and is the reason firearms using other types of projectiles are generally antiques that predate the invention of the bullet.

The rotation causes the entrance wound from a bullet to be round or perhaps elliptical if the bullet impacts the skin at an angle other than 90°. If a bullet enters a body sideways, as it appears to have done in Figure 4.11, the bullet was **yawing**. Bullets do not yaw when fired from a properly designed gun with a rifled barrel. Bullets will yaw if they pass through a medium more viscous than air. Thus, a bullet that ricochets or passes through another person before hitting a second person will yaw. If at the time of entry the bullet is on its side as compared to its direction of flight, it will produce a bullet-shaped entrance wound. Supported exit wounds by definition are caused by bullets that have passed through a person.

The destruction produced by a bullet is proportional to the kinetic energy loss of the bullet in the body of a person. The kinetic energy is equal to the mass of the projectile multiplied by the velocity times the velocity, a consequence of Newton's second law of kinetics. Thus, the amount of energy and the amount of damage increase with the square of the velocity of the bullet. Increasing the speed and decreasing the diameter of bullets for military weapons has enhanced their effectiveness. During the Civil War, many of the bullets were larger than 0.5 inches in diameter. Most military ammunitions are near 0.2 inches in diameter.

Bullets that exit a body waste kinetic energy. It is difficult to design bullets that will not exit if fired from high powered, high velocity long arms. However, with handguns, alteration of the characteristics of the bullet will affect the probability of exiting. Handgun bullets that are designed to enlarge their diameter during passage through tissue are common. They are called *hollow points* when they are cast with defects in their noses. If jacketed, a common configuration is the half-jacketed hollow point that has a copper or aluminum cast around the lead core and an exposed nose with a hollow point. These bullets expand the most, but the difference in extent of injury is subtle at best. Thus, although bullets that expand should be more effective because they are less likely to exit and will expend maximal kinetic energy, in practice the nose characteristics make little difference in wounding power.

A bullet wound is caused by the creation of a temporary cavity as the bullet passes through a person, the collapse of the cavity, and the shock waves from the cavity's creation and subsequent collapse. When a bullet enters a person, it is traveling much faster than the velocity at which tissues tear, so it pushes tissue out of its way. This stretches tissue beyond its breaking point, but it does not break. It only breaks at a much slower speed than the bullet travels. In the case of high velocity long arms where velocities of 1000 meters per second are achieved, the bullet will cross the body entirely before tearing occurs.

The increased velocity can produce entrance wound soot and carbon monoxide effects on the exit wound side. Luckily for those determining directionality, these changes are to the inside of the exit wound side. When the tissue finally tears, it retracts back toward its prebullet passage location and beyond it due to the velocity the tissue picks up while rebounding. This retraction creates a temporary cavity that again is proportional to the kinetic energy of the bullet. The cavity subsequently collapses after rebounding a few more times. The passage of the shock wave and the subsequent collapse of the temporary cavity lacerate the tissues through which the bullet passes and the surrounding tissues. The extent of the damage depends on the organ, but even for relatively slow speed handgun bullets, the estimate, generally, is three times the diameter of the bullet. For high speed long arms, the extent of destruction may be

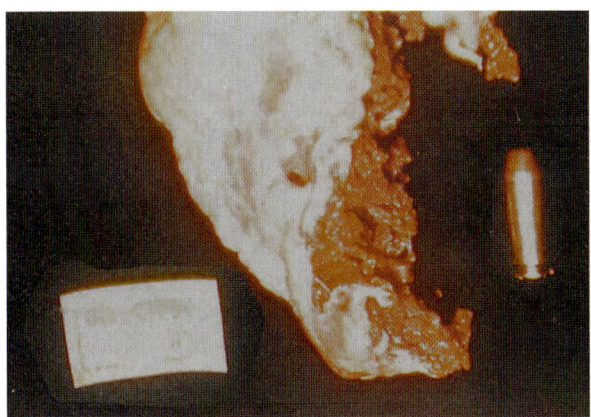

Figure 4.12 A heart with most of the left ventricle destroyed by a bullet cartridge case shown at right.

10 or more times the diameter of the bullet. Figure 4.12 shows a human heart with most of the left ventricle destroyed by a high speed projectile. The large amount of destruction is typical with high speed rounds.

One concept integral to the effects of firearms is "stopping power." Although it may have been discussed before the 1890s, it came to the forefront in the Spanish-American war. The story was told of soldiers attacking U.S. Army personnel. The enemy soldiers continued the attack after being shot in the heart. The outcome of the publicity was that the standard side arm of the American army officer became a 0.45 caliber weapon instead of a 0.38.

Newton's third law of kinetics says that for every action there is an equal and opposite reaction. Thus, if a handgun is fired, the backward force or recoil is equal to the force of the bullet. Obviously, the force of a fired handgun is not capable of knocking a person down. The person is stopped by a loss of brain function that is almost instantaneous after a firearm injury to the head that penetrates or perforates the brain. This is true for almost any modern firearm. A gunshot wound to an organ other than the brain stops the person through loss of perfusion of the brain.

Destruction of the heart as seen in Figure 4.12 will cause near-instantaneous loss of blood pressure and, thus, perfusion of the brain. However, the brain will function for 10 to 15 seconds after it loses perfusion. Thus, a charging person can successfully bayonet a victim within 10 to 15 seconds after being shot in the heart. A gunshot wound to a less vital organ will provide even more time. Therefore, the concept of stopping power is not terribly accurate. Any firearm has tremendous stopping power if used to shoot a person in the head. A firearm has no stopping power if used to shoot a person in a place other than the head.

Other Blunt Force Injury—Most common blunt force injuries in our society are from transportation collisions, usually motor vehicle collisions. Deaths resulting from such incidents are usually classified as accidents. Rarely are such collisions suicidal or homicidal.

Generally, with the exception of gunshot wounds, homicidal blunt trauma in an adult requires a lethal head injury. Injuries to other areas of the body rarely produce death. In children, lethal battery is most commonly due to head injury, but chest and abdominal traumas with laceration of internal organs, such as the spleen, liver, and heart are also seen.

Figure 4.13 shows a female prostitute. Such individuals have extremely high rates of on-the-job injury and death. The Department of Labor does not have reliable statistics on such deaths and injuries because of the illegal nature of the work. The woman was beaten repeatedly about the head with a blunt object that produced the lacerations on her scalp. Bleeding from such injuries is copious because of the vascularity of the scalp.

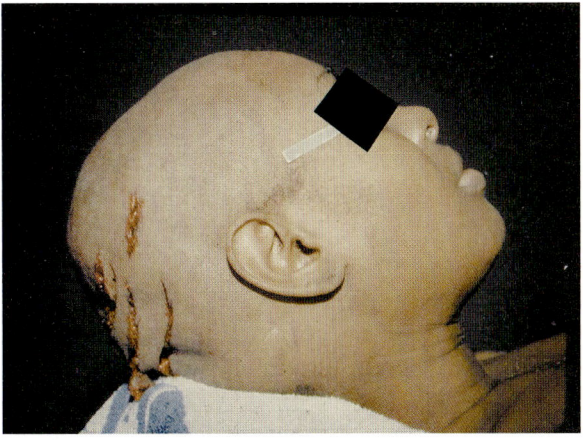

Figure 4.13 Typical lacerations of blunt trauma injury.

Figure 4.14 Weapon that caused lacerations shown in Figure 4.13.

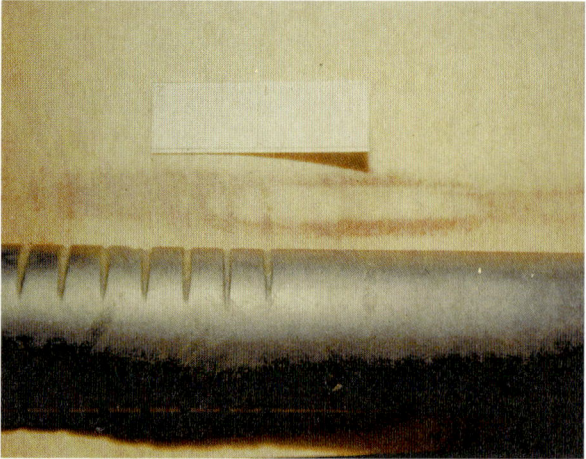

Figure 4.15 Patterned contusion of the skin.

Figure 4.14 shows the weapon that produced the lacerations. Note the considerable **spatter** visible on the weapon and the blood that flowed from the wound. The most common mechanism of death from blunt head trauma is drowning in the blood that has aspirated into the lungs. Blunt trauma causes fracturing of the bones at the base of the skull that form the roof of the nose and mouth. When these bones are fractured, copious blood exits from the veins at the base of the brain that are lacerated as the result of the fractures. The blood flows into the back of the throat and is inhaled, causing death by the mechanism of drowning.

Contrary to popular fiction, including movies and television, a single blow or even multiple blows to the head rarely produce loss of consciousness. Even if a blow is of a severity sufficient to fracture the skull and eventually produce death, it will probably not be sufficient to produce instant incapacitation. Persons who are beaten to death by blows to the head typically have defense wounds on the little finger side of the forearm and hand. If a person can see that he is about to be hit, he will instinctively attempt to ward off the blow by intercepting the trajectory of the weapon, usually with the little finger side of the forearm. If the victim is restrained by physical or chemical means, no defense wounds will occur. The most common chemical restraint is alcohol. Severe intoxication slows reaction time sufficiently to eliminate defense wounds.

Two other terms need to be examined. The first is **contusion**. A contusion is an accumulation of blood in the tissues outside the blood vessels. It is most commonly caused by blunt impact that distorts the tissues sufficiently to break small blood vessels that then leak blood. An important concept is that the pattern of the striking object may be transferred to the person who is struck. Figure 4.15 shows a patterned contusion caused by a striking object. Patterned injuries are important because of the possibility of determining what type of object caused the injury.

A second important term is **hematoma**. A hematoma is a blood tumor. *Hema* comes from **heme**, the Latin word for blood, and *toma* is Latin for tumor. Hematomas are contusions with more blood. Characteristically, blunt impact to the head will often produce a hematoma, referred to as a "goose egg."

Chemical Trauma

Deaths from trauma include deaths that result from the use of drugs and poisons. The most common drug seen in forensic practice rarely kills directly, but is a contributory factor in approximately 50 percent of traumatic deaths. That drug is ethyl alcohol, also called **ethanol**. It is the active ingredient in beers, wines, and distilled liquors. It is probably the drug with the longest history of abuse by humans, and certainly the most widely abused drug today. Alcohol is generally prohibited by

Islamic cultures and some Christian ones, but prohibition is not sufficient to eliminate alcohol as a causative agent in many if not most traumatic deaths.

Alcohol can also kill directly. It is a central nervous system **depressant**; it slows the reactions and communications from brain and spinal cord neurons. At low levels of intoxication, less than 0.03 gram percent blood alcohol concentration, the equivalent of perhaps a 330 milliliter bottle of 5 percent ethyl alcohol-containing beer, most people note a slight improvement in reaction time, probably because of slowing of inhibitory neurons. At levels of blood alcohol concentration above 0.03 gram percent, slowing of brain function occurs and reaction time slows. At a level of about 0.25 gram percent, a person who has not been exposed previously to ethyl alcohol will go into coma if not stimulated. Stimulation will cause a regaining of consciousness. At a level around 0.30 gram percent, the person will be in a deep coma. He or she cannot be roused and will breathe slowly enough to eventually die. The lack of oxygen causes death resulting from alcohol overdose. Such deaths are infrequent, primarily because a person not exposed to alcohol will start vomiting at a blood level of about 0.10 gram percent and further absorption will soon stop. Alcohol overdose deaths are generally seen during chug-a-lug contests in which participants bet on the person who can consume the most distilled spirits. With massive doses of alcohol, the vomiting reflex can be extinguished before it is initiated, resulting in death.

The numbers quoted above are for naïve consumers of alcohol. People who consume alcohol and most other drugs of abuse develop a tolerance that causes the effects of the alcohol or drug to diminish at a certain level. As an example, persons with blood alcohol concentrations above 0.30 gram percent occasionally are seen driving vehicles.

Drugs of abuse other than alcohol produce death generally from the same mechanism. These drugs include barbiturates, diazepams, and opiates. They all produce increasing degrees of coma followed by cessation of breathing and subsequent death. **Marijuana** is an exception to the general rule concerning drugs of abuse. Marijuana is not known to have ever produced an overdose death. Cocaine is another exception. Cocaine is a central nervous system **stimulant**. Deaths from cocaine are more unusual than deaths from depressant drugs. At high doses, resulting seizures, extremely high body temperatures, and uncontrolled quivering of the heart are all mechanisms that have been reported to cause death.

While not a drug of abuse, carbon monoxide (CO) is a common chemical that produces death. It is an odorless, colorless, explosive gas produced by the incomplete combustion of carbon-containing fuels. Deaths due to CO may be accidental, suicidal, or homicidal. CO is also produced in minute amounts by the body through a reaction that produces porphyrin, a component of hemoglobin.

The customary way to measure CO is as the percentage of hemoglobin that combines with it. CO kills by asphyxiation. It cuts off the oxygen to the brain because it binds to hemoglobin 300 times more strongly than oxygen does. Air at sea level contains about 20 percent oxygen. If one breathes air that is 1/300 of 20 percent or 0.06 percent, the result will be a 50 percent CO level. Because it is normally produced by the body, about 1 to 2 percent of hemoglobin is in the form of carboxyhemoglobin, a redder hemoglobin than the oxyhemoglobin that combines with oxygen.

Persons who smoke tobacco products as nicotine delivery vehicles commonly have CO levels above 2 percent and as high as 10 percent. Blood levels of CO as low as 20 percent may prove fatal. Levels in persons trapped in enclosed fires often reach 90 percent before the persons quit breathing and thus expose their blood to CO.

Historically, CO was widely used to commit suicide. One method was putting one's head in an unlit gas stove. During the late 19th and early 20th centuries, the most common source of domestic heating was city gas, a byproduct of the coke produced during the manufacture of steel. With the discovery of natural gas and the popularization of compressed propane, city gas is rarely used. City gas was a very efficient suicide method because it contained up to 5 percent CO.

The second CO source that became available in the late 19th century was exhaust from internal combustion engines. Although this method is still employed, it is more difficult because catalytic converters in modern automobiles efficiently reduce the CO concentrations in automobile exhaust, and lethal levels are difficult to attain.

Cyanide is similar to CO in that it interferes with the oxygenation of the brain, acting primarily on the enzymes in the mitochondria of the brain. Cyanide consists of carbon and nitrogen. Like CO, cyanide can be produced by burning, but its effect in producing or contributing to deaths in fires is less important. Cyanide is generally available as the sodium or potassium salt and is widely used in industry in electroplating and metal polishing applications. It gained some notoriety as a means of administering the death penalty in California, until it was held to be cruel and unusual punishment and abandoned.

Cyanide has a distinctive odor. It smells like almonds and can be detected in concentrations as little as 1 part per million or 0.00001 percent by those who can smell cyanide. Unfortunately, as much as 50 percent of the population cannot smell cyanide. Forensic pathologists should be able to smell cyanide or hire personnel who can smell it. A pathologist who opens the stomach of a person who committed suicide by swallowing potassium cyanide can be killed by the released gas.

The California authorities used potassium cyanide tablets that were dropped into hydrochloric acid to cause release of cyanide gas in the gas chamber. Swallowing cyanide tablets has exactly the same effect in the stomach because the stomach contains hydrochloric acid. Figure 4.16 shows the stomach of a person who died of a cyanide overdose by swallowing potassium cyanide tablets. The intense red color is a hint for the forensic pathologist who cannot smell cyanide, but compounds other than cyanide can also produce a red stomach.

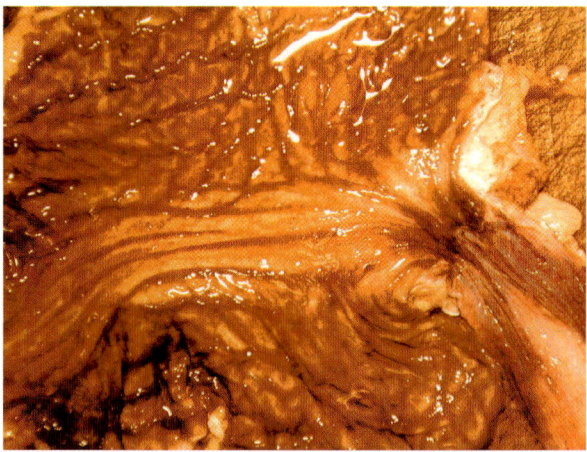

Figure 4.16 Stomach with intense gastritis caused by cyanide.

Thermal Trauma

Exposure to excessive heat or cold may produce death. Hypothermia is excessive cold; hyperthermia is excessive heat. Both conditions can cause death via a breakdown in the normal mechanisms that maintain body temperature around 37 degrees Celsius. In both types of deaths, few demonstrable signs can be found at autopsy to certainly diagnose either condition as the cause of death. Diagnosis requires the absence of other causes of death coupled with a history of exposure to an environment in which either hyperthermia or hypothermia could be expected.

Hypothermia deaths are common in individuals who are intoxicated with alcohol and exposed to cold temperatures. Air temperatures of only 5 degrees Celsius (41 degrees Fahrenheit) have been reported to cause hypothermia deaths. Alcohol intoxication reduces the appreciation of the cold while increasing the loss of body heat because of dilatation of the blood vessels on the surface of the body.

Hyperthermia deaths are common in elderly people in northern cities and in infants left in parked automobiles during heat waves. The ability to maintain homeostasis declines as people age. Dwelling units generally are heated and hypothermia deaths are often not seen in the elderly population, even though this group is susceptible. However, in the northern states, older dwelling units often lack air conditioning, and heat waves are often associated with large numbers of deaths of the elderly. Small children who are incapable of escaping the interiors of closed automobiles are very susceptible to hyperthermia. The inside temperature of an automobile in the sun can exceed 60 degrees Celsius (140 degrees Fahrenheit) and can be fatal in 10 minutes.

Thermal burns are localized wounds caused by hyperthermia. Generally, temperatures above 65 degrees Celsius (150 degrees Fahrenheit) will produce thermal burns upon direct contact with an object for a few minutes. Deaths from thermal burns occur under a wide range of circumstances, from exposure to hot liquids to burns from flaming hydrocarbons. Deaths from burns are usually delayed and arise from complications after medical treatment. The mechanism of death is generally multiple organ failure.

Persons who die at the scenes of fires most commonly succumb from the inhalation of products of combustion, the most common and deadly of which is CO. They usually die from inhaling the CO that accumulates in enclosed fires before they can die from burns. Most often, when a body is found in a burned structure, the level of CO can determine whether the person was alive or dead at the time of the fire. It is not uncommon for a person who commits a murder to attempt to camouflage the crime by burning the body of the victim. Finding a body with a 1 or 2 percent CO level in a burned structure is presumptive evidence that the person was dead, or at least not breathing, after the fire started.

A gasoline fire that traps a person inside a fireball can kill before the person has time to inhale CO. One example is a person who douses himself with gasoline as a means of suicide. Another is a person trapped inside a fireball that results from a gasoline transport trailer fire. In this situation, inhalation stops before the CO level is elevated. The reason for cessation of inhalation is not clear, but at least one mechanism is laryngeal edema which blocks the upper airway and produces asphyxial death.

Electrical Trauma

The passage of electricity through a person may produce death by a number of different mechanisms. If a circuit of alternating current (AC) at low voltages (below 1000 volts) crosses the heart, the heart will experience **ventricular fibrillation**, a nonpropulsive quivering that leads to nonresucitability within minutes. The heart fibrillates because the AC acts as a pacemaker. AC in the Americas alternates from positive to negative 3600 times per minute (2500 times per minute in Europe). Ventricular fibrillation produces about 300 quivers per minute, which is about as fast as the heart can beat. Low voltage may or may not produce electrical burns, depending on the length of exposure to the circuit. Many seconds of exposure are required to produce electrical burns.

Ventricular fibrillation is less likely with high voltage because the amount of current becomes defibrillatory instead of fibrillatory. High voltage current forces the heart into tetany, a sustained contraction that is broken when the circuit is broken. The heart generally will start again with a normal rhythm. However, electrical burn can occur within a fraction of a second with high voltage. In addition to burns, the flow of current through tissues creates holes in the membranes of cells. This is called *poration* and causes the devastating loss of limbs seen in persons exposed to high voltage circuits.

Asphyxias

We discussed above a few of the causes of chemical and thermal asphyxias. However, a number of other mechanisms that can cause asphyxias fall outside the chemical and thermal categories.

Drowning is death by asphyxiation from immersion in water or other liquid. Some deaths from immersion are not asphyxial and result from hypothermia. Exposure of a person to water temperatures below 20 degrees Celsius (68 degrees Fahrenheit) will result in death from hypothermia after exposure of many hours. Exposure to water temperatures near 0 degrees Celsius (32 degrees Fahrenheit) will produce death in a matter of a few minutes.

Drowning victims die as the result of asphyxia, the interruption of oxygenation of the brain. A person typically attempts to keep his head above water so he can continue to breathe air. When this becomes difficult, he struggles to maintain the airway, and this increases the need for oxygen. Inhalation of water adds to the excitement. Water that enters the back of the throat is reflexively swallowed. This transmits the negative pressure associated with trying to inhale water to the middle ear via the Eustachian tubes that open during swallowing. The swallowed water enters the stomach. Further efforts to breathe cause water to enter the upper air passages, triggering coughing and additional reflex inhalation. As the water enters the smaller air passages, the lining muscles go into spasm, thus protecting the alveoli or small air sacs from the entry of anything but air. The spasms create the equivalent of a severe acute asthma attack that traps air in the lungs. Loss of consciousness generally occurs within 1 to 2 minutes of the onset of the struggle, although consciousness may be prolonged if some air can be obtained. Loss of consciousness may be followed by involuntary inhalation attempts and vomiting. Heart cessation occurs a few minutes later. While the heart continues to beat, the pressure that the heart produces in the circulation of the lungs increases greatly, and the right side of the heart dilates from the increased pressure and perhaps from the increase in blood volume from absorbed water from the lungs.

The autopsy findings from drowning will vary depending on whether the drowning event followed the full sequence of events described above. If a person is unconscious when he enters the water, many effects of the excitation phase are not seen because an unconscious person cannot become excited.

The excitation results in transmittal of the negative pressure from the upper airway to the middle ears. The negative pressure along with other asphyxial changes in blood clotting factors results in hemorrhage into the mastoid air sinuses. In addition, water and materials in the water will be found in frontal and ethmoidal sinuses and in the stomach.

The lungs will show hyperinflation as a result of the spasm of the muscles protecting the alveoli. The lungs are generally heavier than normal because of the addition of aspirated water and the fluid that accumulates in the lung in all asphyxias.

Small unicellular organisms called diatoms are found in most fresh and salt waters in the world. These organisms have silica in their cell walls and, thus, they resist degradation by acids. During the late stages of drowning, aspirated water containing diatoms is circulated by the still-beating heart to all the organs. Diatoms are not ordinarily found in the bone marrow. Thus, removing bone marrow, digesting it with strong acids, and examining it under the microscope for diatoms can confirm a drowning. Since the types of diatoms present in water vary from place to place and time to time, it is possible to determine the time and place of drowning by identifying diatoms. This technique is especially helpful if a body is severely decomposed and skeletonized.

Asphyxia has many other possible causes including manual strangulation (with the hands) and strangulation by ligature. Manual strangulation constricts the airway by compressing the neck. Much has been written about the finding of fracture of the hyoid bone in manual strangulation. Actually, this is relatively infrequent and seen primarily in elderly women who have osteoporosis that makes fracturing the hyoid bone easier. Figure 4.17 shows a fractured hyoid bone. Note the hemorrhage around the fracture site. This is extremely important to document, as it is very easy to break the hyoid bone while removing it for examination. If a fracture is present and hemorrhage is absent, the fracture occurred after death.

The more common finding in manual strangulation is fracture of the cornu of the thyroid cartilage. The cornu is located in the larynx or voice box and rests against the front of the cervical spine. If the throat is squeezed to close the air passage, the cornu is forced backward against the front of the spine. An even more common finding is hemorrhage into

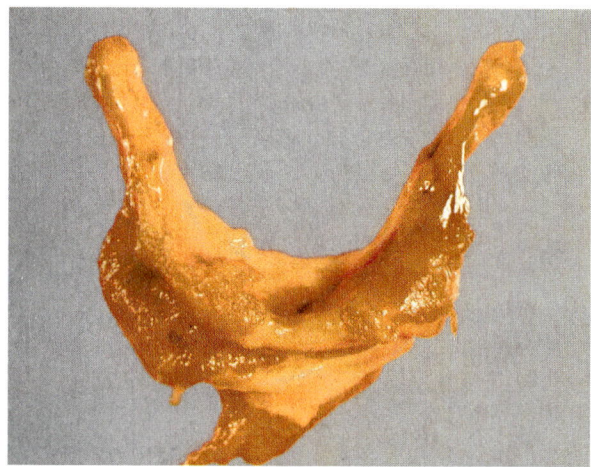

Figure 4.17 Fractured hyoid bone with hemorrhage.

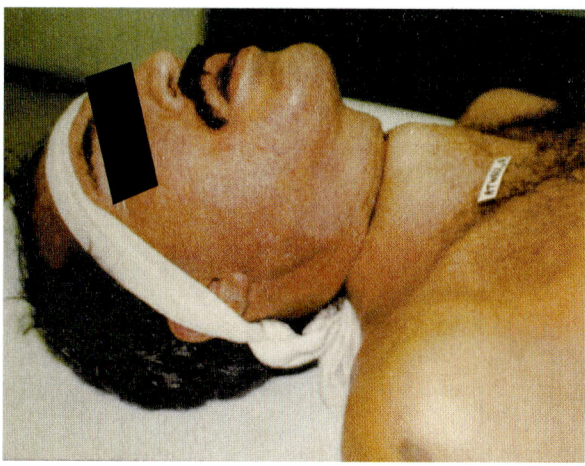

Figure 4.18 Furrow caused by ligature (hanging).

the muscles of the neck. They are collectively called *strap muscles* and are contused by manual strangulation.

Ligature strangulation, whether from hanging or garroting, characteristically does not involve fracture of the hyoid, fracture of the cornu of thyroid cartilage, or hemorrhage into the strap muscles. Generally, the only findings are asphyxial and the presence of a furrow in the neck (Figure 4.18).

Case Study

The police were called by a man who said he shot his neighbor. He told the police that his neighbor attacked him with a knife while he was holding his infant child. He said he feared for his own life and that of his child, obtained his firearm, and shot and killed his neighbor. A store clerk across the street who heard at least part of the altercation confirmed the man's account. The shooter's brother who arrived at the scene just as the altercation ended also confirmed the events.

The family of the deceased asked me to review the case to determine what happened. The family was unhappy with the prosecutor's decision not to prosecute the shooter. I reviewed the scene photographs, the autopsy photographs, and the autopsy report, and subsequently went to the scene of the shooting. While no blood was present, the bullet hole in the wall still existed. Figures 4.19 and 4.20 show the bullet hole in the hallway some months after the shooting. Figure 4.21 shows the body of the decedent as he lay when the police arrived.

The shooting was said to have occurred at the landing, but the bullet hole is at the top of the stairs. As will be discussed in a subsequent chapter, it is possible to ascertain the distance between a firearm and a person shot. The deceased had two gunshot wounds—one

Figure 4.19 Scene photo of shooting. Defect in wall above and to left of banister is from a projectile.

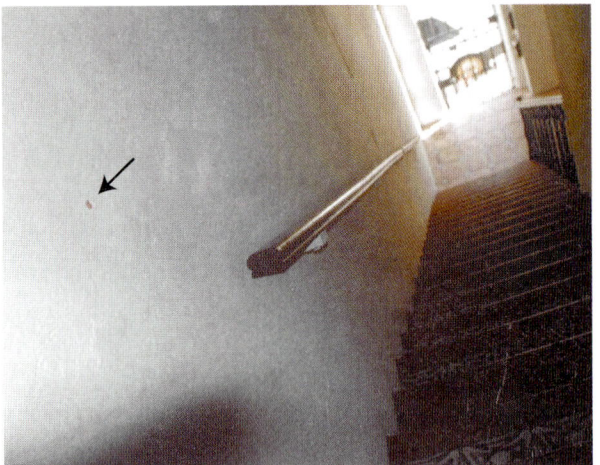

Figure 4.20 Scene photo of shooting, taken after the body was removed, showing projectile defect and stairs.

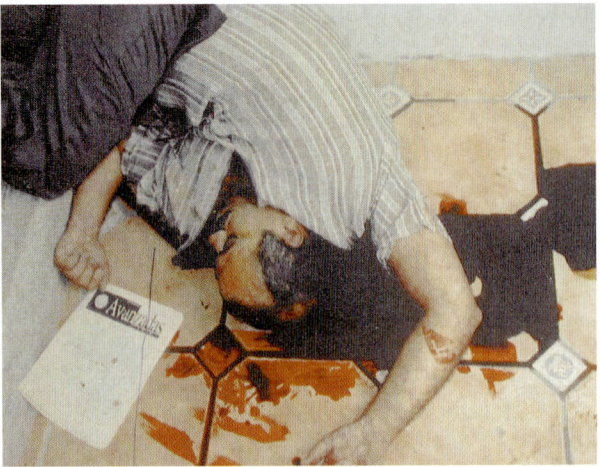

Figure 4.21 Scene photo of deceased at foot of stairs.

distant and the other close. Thus, the weapon was more than 3 feet away for one shot, and it was fired only a few inches away for the second shot.

Other issues that can be determined are the direction of the bullet through the body and the organs injured. One shot hit the side of the abdomen. It did not strike a major artery and exited the body on the other side. That bullet struck the wall and was the distant shot. The close shot was to the back of the head. It traversed the brain from front to back and slightly upward.

Another important aspect of gunshot wounds is the length of time between wound and collapse of the victim. A gunshot wound of the abdomen that does not hit any major vessels can exert effects for hours or even days or longer. A gunshot wound to the back of the head that traverses the brain will cause instant coma.

In this case, the physical evidence refuted the account of the shooter. The shot to the abdomen was the first. The shooter was standing when he fired the shot that produced the hole in the wall. The first shot was fired from more than 3 feet away, not close enough for the shooter to be threatened with a knife. The second shot in effect was a *coup de grace* to the back of the head as the victim fled down the stairs after being shot.

Questions

1. Describe four causes of traumatic death.

2. Which causes of traumatic death may be produced by asphyxia?

3. Describe the four manners of death.

4. How are gunshot wounds of entrance classified?

5. Which drug of abuse is most often encountered in the practice of forensic pathology?

6. What three features differentiate lacerations from cuts or incised wounds?

7. Differentiate between a perforating and a penetrating gunshot wound.

8. Describe the unicellular organisms that may prove helpful in diagnosing drowning.

9. Describe the common finding in manual strangulation.

10. Describe the common findings in ligature strangulation.

References and Suggested Readings

Baselt, R.C., *Disposition of Toxic Drugs and Chemicals in Man,* 5th ed., Chemical Toxicology Institute, Foster City, CA, 2000.

DiMaio, V.J.M., *Gunshot Wounds: Practical Aspects of Firearms, Ballistics, and Forensic Techniques,* 2nd ed., CRC Press, Boca Raton, FL, 1999.

DiMaio, V.J.M., and DiMaio, D.J., *Forensic Pathology*, 2nd ed., CRC Press, Boca Raton, FL, 2001.

Knight, B., *Forensic Pathology,* 2nd ed., Oxford University Press, New York, 1996.

Knight, B., http://britannia.com/history/coroner1.html.

Moenssens, A.A. et al., *Scientific Evidence in Criminal Cases*, Foundation Press, Westbury, New York, 1995.

Moritz, A.R., *The Pathology of Trauma*, 2nd ed., Lea & Febiger, Philadelphia, 1954.

National Association of Medical Examiners, Writing Cause of Death Statements, 2001, http://www.thename.org/CauseDeath/main.htm.

Patrick, U.W., Handgun Wounding Factors and Effectiveness, U.S. Department of Justice, 1989, available at http://www.firearmstactical.com/hwfe.htm.

Shuman, M. and Wright, R.K., Evaluation of clinician accuracy in evaluating gunshot wound injuries, *J. Forens. Sci.*, 44, 339, 1999.

Spitz, W.U., Ed., *Medicolegal Investigation of Death*, 4th ed., Charles C Thomas, Springfield, IL, 2006.

Forensic Toxicology

John Joseph Fenton

Introduction

Toxicology is the study of poisons. The term *forensic* means "pertaining to legal matters." From the simplest perspective, therefore, **forensic toxicology** is an examination of all aspects of toxicity that may have legal implications. This is a functional definition that allows us to distinguish forensic toxicology from other subspecializations of toxicology. For example, occupational toxicology deals with chemical hazards in the workplace. Environmental toxicology focuses on toxic chemicals often of human origin and present in the environment. In the forensic sphere, procedures and testing methods may be similar to those of other areas of toxicology, but the legal ramifications of such findings are what make forensic toxicology unique.

Applications of Forensic Toxicology

Forensic toxicology is employed today primarily in the three areas cited in Table 5.1. The first is **postmortem drug testing**. This consists of death investigation with a goal of establishing whether drugs were the cause or a contributing factor in death. There are many fatalities due to accidental or deliberate drug overdose. A small percent of such deaths are homicides. The task of the forensic pathologist is to explain what caused each death that comes under his or her jurisdiction and to determine the manner of death, i.e., whether it was accidental, suicidal, or homicidal. In the case of drug-related deaths, the forensic pathologist is immeasurably assisted by the forensic toxicologist who does comprehensive analyses of a wide variety of toxins from a large variety of tissue sources. The ability of forensic investigators to find poisons in human remains is a major factor in the dramatic decline in poisoning cases noted over the past 150 years.

A second major area of forensic toxicology is workplace drug testing. This consists of testing of biofluids, primarily urine and blood from employees or job applicants, for possible presence of drugs. The law usually allows random (unscheduled) drug testing for employees only if they have specific occupations, such as customs

Table 5.1
Divisions of Forensic Toxicology

Applications
Postmortem drug testing
Workplace drug testing
Investigation of contraband materials

agents, police officers, and some others. In these occupations, the law usually places the public safety above the employee's right to privacy. Employees may also be ordered to provide specimens for drug testing for cause, that is, if they appear to be impaired at work as a result of abusive use of alcohol or other drug.

Most employees are never subjected to forensic toxicology testing. Job applicants, on the other hand, are often required to be tested for drugs as a condition of hire. All federal government jobs in the United States require preemployment drug testing. Over 90 percent of the largest U.S. corporations also make drug testing a condition of employment. The rationale for this testing and the rejection of drug-positive individuals is that numerous studies have shown that persons who use drugs of abuse are often less reliable and less productive employees. This aspect of forensic toxicology, workplace drug testing, is a massive enterprise with several hundred thousand specimens examined each week in the United States. The nature of this testing is described below.

A third area of forensic toxicology is the evaluation of **contraband** materials to identify prohibited drugs. As part of the national effort to stop drug abuse, police agencies need laboratory support to prove that a seized material is, in fact, identical to a prohibited substance. If cocaine, for example, is found in a suspect's possession, a conviction for unlawful possession requires that the substance be shown unequivocally to be cocaine. Police agencies establish such presumptive identifications in the field with portable rapid testing kits. In the laboratory, with the application of sophisticated methods, field identifications can be confirmed.

The Testing Process

Specimen Collection

Blood—Because the concentration of toxin present in blood often correlates more closely with lethal outcome than concentrations in other specimens, blood is the most important specimen in postmortem toxicology. Blood is also preferred to establish driving under the influence (DUI). Two blood specimens are customarily collected in postmortem studies. One is taken from the heart and the second should be collected from a peripheral site. An appropriate volume is 50 to 100 milliliters.

Urine—In preemployment screening, urine is preferred over blood because large volumes can usually be collected, and the collection process does not involve venipuncture. A potential problem with urine is that correlation between drug concentration in urine and drug effects is usually poor. However, the purpose of preemployment screening is simply to determine whether an individual has been exposed to an abusable substance, and urine will provide an answer to that question. Urine should also be collected in postmortem investigation because certain toxins appear in urine in much larger quantities than amounts found in blood, and false negative findings could be obtained if blood alone were tested.

Gastric contents—Testing of gastric contents may be beneficial in the case of the sudden death of a person who has large quantities of a lethal agent in his stomach. For highly toxic substances, very low concentrations may be present in the blood in contrast with large amounts in the stomach. If the manner of death is suicide, large amounts of drugs in the stomach may help to establish this conclusion.

Vitreous humor—This specimen should be collected in postmortem investigation. Since the eye is an isolated bodily area, the **vitreous humor** is resistant to putrefaction. It is sometimes possible, therefore, to obtain a reliable measure of a biochemical or drug in vitreous humor at a time when the same biochemical has decomposed in the blood compartment. Vitreous humor may be the only fluid remaining in a decomposed cadaver, and assay of the vitreous may show chemical abnormalities that suggest the cause of death. Postmortem increases of certain substances such as potassium and hypoxanthine occur in the vitreous humor in a time-wise fashion, and the analysis of the vitreous humor may be helpful in establishing the time of death.

Bile and liver—The liver is the organ most heavily involved in drug metabolism. It is likely to contain significant quantities of most drugs and may, on occasion, permit identification of an agent that caused death even when that substance cannot be found in the blood. Bile drains from the liver and is very rich in certain types of drugs, such as opiates.

Hair—Hair testing is not common in forensic investigations for several reasons including the fact that drugs in the hair are present in extremely low concentrations, usually in the picogram (trillionth of a gram) range. Furthermore, controversies relating to active versus passive drug exposure continue (i.e., did the drug enter the hair from the body or via incidental external exposure?). Nevertheless, some companies prefer that individuals applying for work pass drug tests in which hair is tested. Advantages of hair testing for this purpose are the long lifetime of drugs in hair (drugs in hair are eliminated only when the hair is cut) and the ability to segment the hairs into small pieces that enable estimates of the approximate time when an individual was exposed to a drug.

Oral fluid—This specimen is derived from the oral cavity and is approximately equivalent to saliva. It has been studied recently because it may provide the advantages associated with urine without, however, raising issues of invasion of privacy in the collection process. Oral fluid is an ultrafiltrate of blood. Drugs within the blood that are extensively protein-bound will not diffuse into the oral fluid. Not surprisingly, oral fluid concentrations also appear to correlate with recent drug use. This specimen type may become popular for drug testing in the future.

Breath—An equilibrium exists between alcohol in the bloodstream and alcohol in the lung such that, on average, the concentration of blood alcohol is 2100 times greater than the concentration of breath alcohol. Since this equilibrium exists, one can measure the breath alcohol and infer the corresponding alcohol concentration in the bloodstream. The obvious advantage of breath testing is that it may be carried out without requiring blood collection with its hazards and associated difficulties. Further, unlike blood testing, the result is immediately available so that an impaired driver can be removed from the highway right away. In the past there were objections to the specificity limits of breath testing. Modern methods that are based on infrared analysis with narrow bandpass instruments provide specificity that approaches that of blood testing by sophisticated instruments.

Drugs of Abuse

Workplace drug testing is concerned primarily with drugs of abuse. Since the object of such testing is to evaluate the cause of erratic behavior on the job or to prevent the hiring of drug abusing persons, this testing focuses on drugs of abuse. Medical examiner laboratories have a wider mission. They try to establish the cause of death and, thus, typically examine blood and other tissues for drugs of abuse and also for other potentially lethal pharmaceuticals and environmental toxins.

It is important to recall that abusable substances often have legitimate medicinal applications. Consequently, the reader should bear in mind that many of the compounds cited below are used by persons who derive medical benefit from them. Despite their chronic use of such substances, many of these persons never use them in an abusive manner. From this perspective, it would be better to call these materials "drugs subject to abuse."

A program was initiated in the United States under the auspices of the National Institute on Drug Abuse (**NIDA**) to curb drug use by requiring drug tests as a condition of federal employment. This was a rigorous program that also aimed to assure the **accuracy** of laboratory findings. NIDA specified that screening could cover only the drugs listed in Table 5.2.

NIDA focused on drugs that had the highest abuse potentials. Certified laboratories could not test for any drug beyond those listed in the table. The amphetamines group included both **amphetamine** and methamphetamine. Opiates included morphine and codeine. Laboratories could also test for other substances that are sometimes abused, but this testing could not be done for specimens that were regulated, primarily those from federal employees and from the Department of Transportation. Thus, many laboratories provide testing for benzodiazepines, barbiturates, **hallucinogens** (e.g., lysergic acid diethylamide), and others.

Opiates

Opiates constitute a large class of drugs distinguished by their ability to cause profound euphoria. Many possess high potency as pain relievers. Best known among them are the natural members of the group, morphine and codeine. Both grow in copious quantities in Southeast Asia and several other areas of the world. Heroin is easily prepared from morphine and has the advantage of being less polar than morphine. As a result, heroin readily enters the central nervous system. It has 250 percent of the euphoric power of morphine and is more sought after as a drug of abuse.

The great pharmaceutical potential of morphine has resulted in a relentless research effort to generate even better analgesics (pain relievers) that are less addictive. Only modest success has been achieved, and this has led to production of semisynthetic and fully synthetic analogs of morphine (Figure 5.1).

Semisynthetic opiates are those made by a simple modification of the morphine or codeine molecule. These include hydromorphone, hydrocodone, oxymorphone, and oxycodone. OxyContin® is a relatively new sustained form of oxycodone that has caused many deaths. It was designed to be taken orally and allow

Table 5.2
Drugs Cited in the National Laboratory Certification Program

NIDA 5
Amphetamines
Opiates
Phencyclidine
Cocaine
Cannabinoids

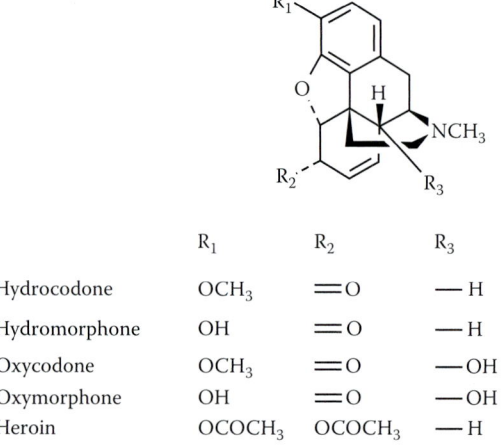

	R_1	R_2	R_3
Hydrocodone	OCH_3	$=O$	—H
Hydromorphone	OH	$=O$	—H
Oxycodone	OCH_3	$=O$	—OH
Oxymorphone	OH	$=O$	—OH
Heroin	$OCOCH_3$	$OCOCH_3$	—H

Figure 5.1 Molecular structures of morphine and codeine, natural opiates, and semisynthetic opiates.

patients who suffer from cancer and related pain to experience long-term relief. Drug abusers, however, discovered that these tablets could be dissolved and then administered by injection to produce extreme euphoria. Many fatal overdoses of opiates have resulted.

Synthetic opiates are not structurally similar to morphine and are manufactured by processes that do not begin with natural morphine or codeine. This group includes meperidine, pentazocine, and fentanyls, which are highly potent forms of **narcotic** analgesics.

Opiates are classified as depressants. Accordingly, they produce, in addition to the initial euphoria, reduced muscle activity, depressed respiration and heartbeat, and an inclination to sleep. In overdose, they cause death, usually, by paralysis of the respiratory center.

Amphetamines

In contrast to opiates, amphetamine and its analogs are stimulants that create an excitatory condition characterized by elevations of heart rate, blood pressure, and respiratory rate. They also provoke intense euphoria. Amphetamine and its close analog, methamphetamine, are occasionally prescribed for medical reasons. Most of the time, however, they are used in an abusive manner. Methamphetamine can be synthesized easily by clandestine laboratories starting with ephedrine. In an effort to prevent the illicit production of methamphetamine governments have recently passed legislation that limits the availability of ephedrine. Many compounds resemble amphetamine both structurally and pharmacologically. These compounds are sold by prescription or over the counter and have decongestant, antiinsomniac, and anorexic (appetite suppressant) actions. Since their abuse potential is less than that of amphetamine or methamphetamine and their medical benefits are significant, regulatory agencies permit their sale. These compounds are numerous and include ephedrine, phenylpropanolamine, phenylephrine, phenmetrazine, and others (Figure 5.2). Because of the molecular similarities of these legitimate pharmaceuticals to amphetamine

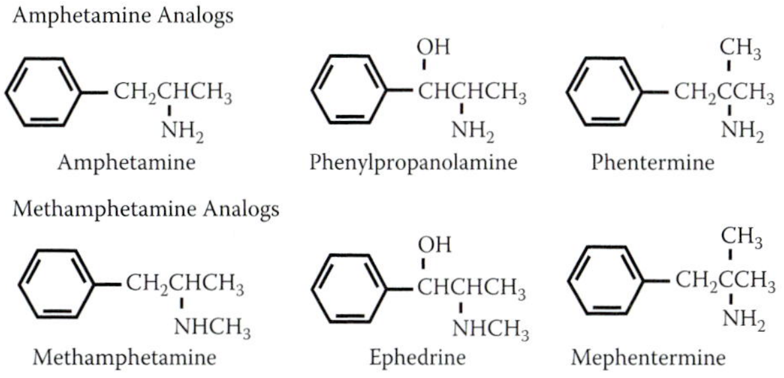

Figure 5.2 Amphetamine, methamphetamine, and similar medicinal agents.

and methamphetamine, drug testing methods must be rigorously specific in order to prevent confusion and false positives.

Cocaine

Cocaine is a stimulant that resembles amphetamine in its abuse potential and pharmacological responses. Indeed, many users of one of these substances report that a switch to the other is associated with the same euphoric and behavioral responses.

Unlike amphetamine, however, cocaine is a natural product found in the coca leaf. *Erythroxylon coca*, the natural source of cocaine, grows in damp, mountainous regions, especially the Andes range of South America. Cocaine is extracted by a simple process from the plant material. Since cocaine is alkaline in nature and is usually extracted with hydrochloric acid, the substance produced is cocaine hydrochloride in which the hydrochloric acid is bonded to the nitrogen atom of cocaine.

Cocaine hydrochloride may be treated with a base and extracted into an organic solvent such as ether. This additional treatment produces "free base" or "crack" cocaine. In actuality crack cocaine and free base are chemically the same. The two names refer to slight differences in the manner of preparation. Free base and crack cocaine have much lower boiling points than cocaine hydrochloride. This difference in physical properties is very important because it makes it possible to smoke cocaine. An attempt to smoke cocaine hydrochloride would simply burn the drug. When a drug is smoked, the large surface area of the lungs is available for drug absorption. The result is that larger amounts may

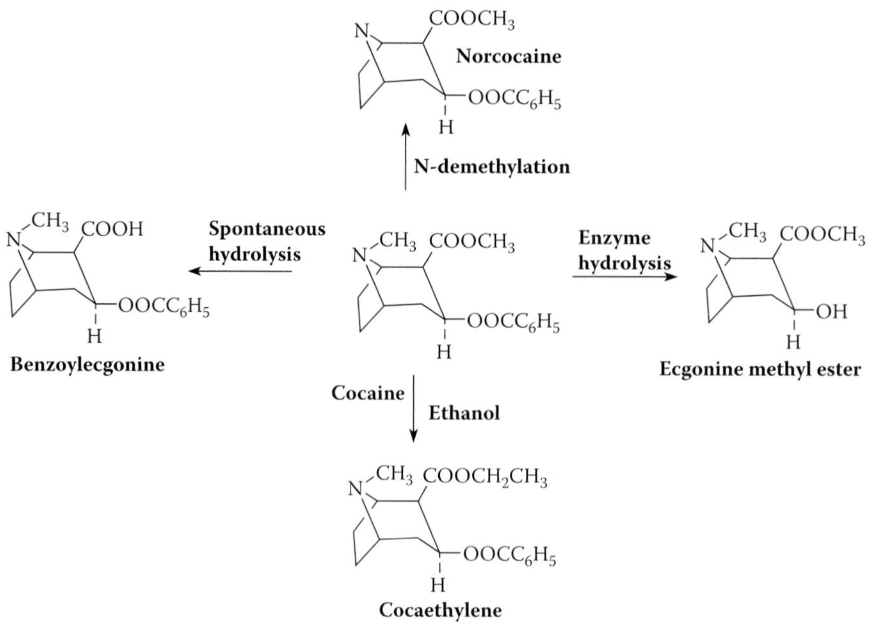

Figure 5.3 Cocaine and metabolites formed from it in blood.

be absorbed per unit of time and a greater drug effect is experienced. Free base and crack cocaine were introduced into the United States during the 1980s.

Cocaine as the free base (cocaine hydrochloride minus hydrochloric acid bonded to the nitrogen atom) engendered great increases in the popularity of cocaine and in drug-related morbidity and death.

Cocaine in the blood is metabolically converted to methylecgonine (Figure 5.3). Benzoylecgonine is also a product of the nonenzymatic hydrolysis of cocaine in the blood. For purposes of demonstrating use of cocaine, laboratories typically analyze urine for benzoylecgonine rather than cocaine because cocaine appears in only small amounts in urine and for a brief period. Benzoylecgonine, on the other hand, appears in large quantities and is usually present for approximately 3 days following cocaine use.

Cannabinoids

Marijuana is a name that applies to parts of the *Cannabis sativa* plant. Many related psychoactive compounds come from this plant and the collective term **cannabinoids** is often applied to them. Tetrahydrocannabinol (THC)

is the major active agent and is present to the extent of 2 to 6 percent by weight in cannabis (Figure 5.4). An oily extract of the plant, hashish, has much higher THC content (12 percent) and accordingly produces a greater psychoactive response when used.

Most marijuana users smoke hand-rolled cigarettes that contain about 75 milligrams of THC. This has a bioavailability of only 2 to 20 percent, but it is rapidly absorbed into the blood, reaching a peak concentration around 10 to 20 minutes later. This produces a drug state that typically lasts about 2 hours. The marijuana drug state is characterized by euphoria, perceptive alterations, and memory impairment. Mood swings and hallucinations are possible with moderate intoxication while heavy usage may provoke delusions and paranoia.

THC, the active agent in marijuana, is metabolized to 11-OH-THC (11-hydroxytetrahydrocannabinoic acid), also an active compound (Figure 5.4). Further metabolic transformation leads to 9-carboxy-THC, the major urinary metabolite and an inactive compound. Laboratories measure THC in blood as the best index of drug-related impairment. To demonstrate drug use, however, the best test is urinary measurement of 9-carboxy-THC

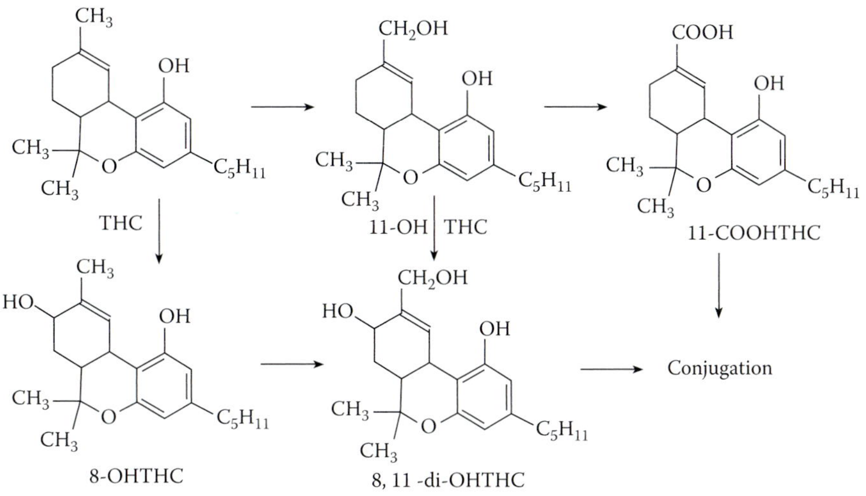

Figure 5.4 Metabolic conversion of THC, the major psychoactive component of marijuana, to inactive metabolites.

which has been shown to be present in urine as long as 2 months after the discontinuation of heavy usage.

Marijuana is the subject of intense controversy. Its proponents claim that legalization is reasonable because cannabinoids have no harmful effects. This is disputed by others who note that driving under the influence of marijuana is clearly a hazardous activity. Further, pulmonary damage appears evident since marijuana users often manifest bronchial inflammation and precancerous changes (marijuana has twice the carcinogen content of regular tobacco). Lastly, many marijuana users demonstrate amotivational syndrome, a personality trait characterized by lack of normal concern about life. Life-threatening effects are, however, clearly absent from the constellation of negative findings about cannabinoids.

Medicinal use of marijuana is also a topic for lively controversy. Marijuana is alleged to be beneficial to cancer patients and for the therapy of glaucoma. Drug companies sell medicinal marijuana, for example, as Dronabinol, structurally identical to THC. Users report that it is less pharmacologically effective than marijuana, possibly because of the multitude of other cannabinoids found in natural marijuana.

Phencyclidine

Phencyclidine is better known as PCP, a name that derives from its recreational use during the 1960s, when it was known as the "peace pill." PCP also stands for its chemical name, phenylcyclohexyl-piperidine (Figure 5.5).

PCP was developed by Parke, Davis & Co. and intended for use as a surgical anesthetic. It was found, however, to be unsatisfactory because some patients experienced severe toxicity, including manic behavior after its application. These patients had elevated blood levels either because of high dosage or slow elimination through metabolism. PCP was then relegated to veterinary use and eventually eliminated from all medicinal applications, human or veterinary.

PCP provides a feeling of euphoria, plus feelings of detachment from the world, strength, power, and even invulnerability.

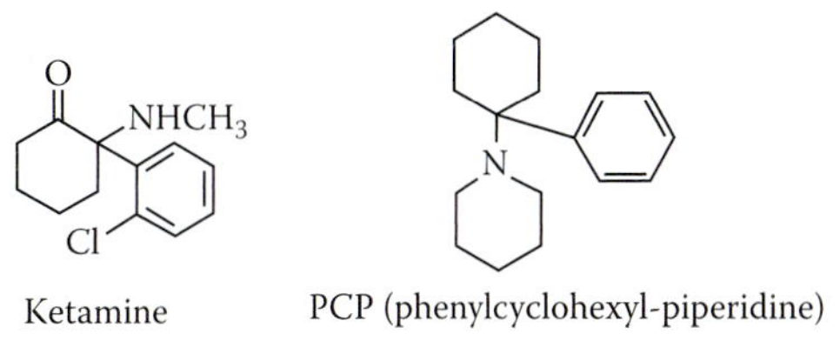

Figure 5.5 PCP and ketamine, a related psychoactive compound.

Table 5.3
Major Categories of Medicinal Agents Involved in Deaths

Category	Medicinal Purpose	Examples
Sedative hypnotics	Minor tranquilizers, sleep-inducing agents	Digoxin, procainamide, lidocaine, etc.
Cardioactive agents	Improve heart beat, regularize heart rhythm	Digoxin, procainamide, lidocaine, etc.
Antipsychotic agents	Reduce psychotic behavior	Phenothiazines, butyrophenones, etc.
Antiepileptic drugs	Reduce seizure frequency and severity	Phenobarbital, Dilantin, valproic acid, etc.
Antidepressants	Improve mood and outlook	Tricyclic antidepressants, selective serotonin uptake inhibitors

With higher doses, PCP usually causes severe perceptual distortions. PCP users who significantly overdose present with psychiatric manifestations that include violent behavior, psychosis, paranoia, and hallucinations.

PCP is measured in the blood or urine. Blood levels around 25 nanograms per milliliter are associated with excitation. Levels that are greater than 100 nanograms per milliliter have been related to seizures and some deaths. Urine is tested to demonstrate PCP use. After regular use of PCP, urine remains positive by the customary testing methods for about one week.

Pharmaceutical Materials

Forensic laboratories cannot limit their search for a cause of death to abused drugs only. Many individuals die from overdoses of medicinal agents. Such deaths are sometimes suicides but many are accidental. In the latter case they may occur because of medicinal errors such as taking the wrong drug or the wrong dose of the right drug. On other occasions, a patient may suffer some level of organ damage from a primary medical condition and the organ injury renders him or her incapable of metabolizing a drug in the normal manner. This latter type of problem often results in a buildup of the drug in the blood to a point where the drug's concentration is greater than the lethal concentration.

Most forensic toxicology laboratories are capable of postmortem identification of a wide variety of medicinal agents. They logically focus their efforts on drugs that are very dangerous in relatively small amounts (Table 5.3).

Nonmedicinal Agents

Many deaths are due to chemicals that are not medicinal but are encountered in the environment or in industrial activity. The major members of this category are alcohols, **cyanide**, carbon monoxide, and hydrocarbons.

Alcohols

Ethanol is beverage alcohol. Other low molecular weight alcohols such as methanol and isopropanol are also present in the environment or workplace and may cause human injury. Alcohols enter the membranes of nerve cells and disrupt their normal architecture. This disruption alters normal nerve-to-nerve signaling and is believed to be the major reason for the behavioral changes brought on by alcohol consumption. Alcohols sometimes are injurious because of the toxic properties of their metabolites. For example, methanol is converted into formaldehyde and formic acid principally by the liver. These two compounds are far more toxic than methanol and, indeed, treatment of methanol overdose is best accomplished by preventing the conversion of methanol into its metabolites.

Beverage alcohol enters the blood mainly from the small intestine. Within the liver, about 90 percent of the average dose of ethanol is converted into acetaldehyde and acetic acid. The remainder is eliminated via sweat or urine. Ninety minutes after ethanol ingestion is the approximate time that peak blood levels are reached. This figure is certainly highly variable and is based on the interaction of many factors. The volume of distribution equation predicts the relationship between blood concentration and alcohol dosage:

$$Cp\ (g/L) = D\ (g)/(Vd\ (L/kg) \times W(kg))$$

where Cp = blood concentration, D = dose, Vd = volume of distribution (0.70 in men and 0.60 in women), and W = body weight in kilograms.

One could use this equation to demonstrate the rule of thumb that one 12 ounce can of beer or one highball (1.5 ounces of 100 proof alcohol) raises the blood concentration of an average size individual by 0.02 percent. It is also true that the average rate of ethanol clearance from blood is one drink per hour. As a result, about 1.5 hours after drinking 4 drinks an individual is above the legal limit for driving in the United States (0.08 percent blood alcohol) and about 4 hours later the blood concentration of ethanol approaches zero. These approximations are very inexact because an individual's handling of alcohol depends on many factors, one of which is the person's drinking experience.

The toxicity of beverage alcohol is well known. Acute toxicity correlates fairly well with dose and blood level. Alcohol contributes to numerous disorders as a result of chronic abuse. The liver is the organ that is most vulnerable and it shows pathological response to alcohol ranging from fatty accumulation up to hepatoma (liver cancer). The brain may also be attacked with very significant injury including several psychosis-like syndromes.

Alcohol testing is very important in forensic toxicology. Since the law is explicit about permissible blood levels, it is difficult to obtain a conviction for driving under the influence without proof of elevated blood alcohol. Further, the test should be conducted by a respected method in a legally defensible manner. The latter requirement refers mainly to rigid record keeping (including specimen custody and **chain of custody** record keeping).

It is preferable to use **gas chromatography** to measure blood alcohol. It is superior to spectrophotometric methods since the latter sometimes are not of sufficient specificity and can exhibit cross-reactivity to other alcohols. Blood is the preferred specimen and it should be collected in a gray top tube that contains sodium fluoride as a preservative. Some states allow evidence from urine alcohol to be presented in court. It is very difficult, however, to infer that driving ability was impaired on the basis of a urine alcohol concentration.

What concentration of blood alcohol is sufficiently elevated as to constitute a reasonable cause of death? Concentrations greater than 350 milligrams per deciliter are listed as consistent with a cause of death. Although this is generally true, it is interesting that many individuals have survived much higher concentrations. One study also showed that the average blood alcohol concentration found in persons who appeared to have died with alcohol as the only apparent cause was just 290 milligrams per deciliter.

Cyanide

Cyanide is a highly toxic substance that is present in myriad forms in nature. The fastest acting form is the gas, hydrogen cyanide. Salts such as sodium cyanide are highly poisonous but their onset of action is somewhat slower than hydrogen cyanide gas. Many industrial chemicals, such as acetonitrile, are metabolized to cyanide and produce the same symptoms as cyanide, although to a lesser degree. Finally, many plant materials such as amygdalin and linamarin are poisonous to man because their ingestion leads to the production of cyanide in vivo.

Cyanide is dangerous because it binds to ferric ions in cytochrome oxidase, an enzyme in the electron transport system within the mitochondria of cells. It interrupts the electron transport cascade, the central pathway for energy generation in human biochemistry. Without the biochemical energy generated from electron transport, life is not possible. Death occurs quickly. Inhalation of large amounts of hydrogen cyanide is fatal in less than 1 minute.

An antidote for cyanide poisoning exists and would be fairly effective were it not for the very rapid action of cyanide. Those exposed to large doses are beyond treatment because death occurs so quickly. However, the antidote can save persons exposed to smaller amounts. Cyanide antidote contains nitrite which oxidizes hemoglobin to methemoglobin. The latter acts as a sink for cyanide, i.e., it forms cyanomethemoglobin, a much less toxic form of cyanide than cytochrome oxidase. Because the cyanide is bound up with methemoglobin it does not reach its customary target, cytochrome oxidase.

Forensic laboratories can test for cyanide in whole blood and its concentration correlates well with severity of poisoning. Normal level is less than 40 nanograms per milliliter. Levels greater than 1000 nanograms per milliliter are associated with stupor. Amounts above 2500 nanograms per milliliter are usually fatal.

Carbon Monoxide

Some studies suggest that carbon monoxide (CO) causes more deaths than any other toxic substance. This may be true since CO is present in fires and, together with other poison gases that result from combustion, causes more fire deaths than thermal injury. Most carbonaceous materials in the world are reduced forms of carbon (carbon bonded to hydrogen). Heat converts them to oxidized forms of carbon (carbon monoxide and carbon dioxide). In the presence of adequate heat and oxygen, reduced carbon is fully converted to relatively harmless carbon dioxide. In the presence of inadequate oxygen, however, the product of carbon oxidation is CO. Faulty heaters, indoor fires, and other situations in which carbon is not fully oxidized are the usual scenarios in which high amounts of CO are produced.

CO is toxic for a number of reasons. It binds hemoglobin much more tightly than oxygen so that hemoglobin is unable to fulfill its normal function of transporting oxygen to tissue. CO also causes a left shift in the hemoglobin dissociation curve. This means that hemoglobin binds oxygen more tightly at any given partial pressure of oxygen. This is dangerous because the oxygen, tightly bound to hemoglobin, cannot be transferred to its intended destination—cells in need of oxygen. Finally, CO binds to myoglobin and to cytochrome oxidase. These latter effects are detrimental to aerobic respiration. The major effect, however, is the binding of CO with hemoglobin in the bloodstream.

If equilibrium occurs, i.e., if hemoglobin spends enough time in the presence of CO, the resulting concentration of carboxyhemoglobin (CO-bound hemoglobin) may be found from the following equation:

$$pO_2/pCO = 240(HbO_2)/(HbCO)$$

where pO_2 = partial pressure of oxygen, pCO = partial pressure of CO, $HbCO$ = concentration of carboxyhemoglobin in blood, and HbO_2 = concentration of oxyhemoglobin in blood. From this equation we can show that only 1 millimeter CO pressure (about 1 part per thousand) in the air results in 62 percent saturation of hemoglobin with CO (partial pressure of O_2 = 151 torrs). A carboxyhemoglobin level greater than 50 percent is a cause of death after a short time.

CO testing is commonly conducted both in clinical laboratories and in postmortem evaluation. Persons whose blood carboxyhemoglobin levels exceed 60 percent are at great risk of death. Lesser degrees of hemoglobin–CO binding are associated with lesser morbidity, but it is important to be cautious in interpreting carboxyhemoglobin levels in the blood because removal from the source of CO immediately starts the process of unloading CO from hemoglobin. If testing is delayed, a carboxyhemoglobin level will underestimate the degree to which a patient was exposed to CO.

Other Gases

Hydrocarbons are commonly found in nature. They are widespread but of less inherent toxicity than cyanide or CO. Deaths have resulted from hydrocarbon exposure. Death is likely due to oxygen deprivation (if the victim is in a hydrocarbon-rich, air-deficient atmosphere) or accident following erratic behavior brought on by the effects of hydrocarbons on the brain.

Analytic Methods in Forensic Toxicology

The discussion which follows on the methods of testing in forensic toxicology concentrates on the chemistries of the available technologies. However, the tactical approach to testing must also be described. A sensible tactic is to employ a two-fold approach: screening and confirmation. In other words, we attempt to achieve maximum accuracy by demonstrating the presence of a toxin by two different methods. This double positive finding reinforces the credibility of the result. Moreover, laboratories often attempt to demonstrate a toxin not merely by two methods, but also in two locations, e.g., blood and urine, blood and liver, or another combination.

Using both a screen and a confirmatory method is not merely a device to guarantee accuracy. It makes sense for other reasons. One could immediately test specimens by the most sophisticated method available, an approach that might seem sensible, but sophisticated methods are labor-intensive. More commonly, a laboratory will screen by an automated method and confirm the presumptive positives (specimens found positive by the screening test—usually a small percent of the total workload) by a sophisticated method. This two-phase approach requires that the screening method be very sensitive although not necessarily specific. The second test, the confirmatory one, must be both sensitive and specific.

Screening Tests

Immunoassays

Immunoassays are tests in which antibodies are used. They are employed because they enable the reagents to react only with a substance that recognizes the **antibody**. An antibody is prepared against an analyte such as morphine or methamphetamine. The introduction of antibodies into forensic testing has replaced many other time- and labor-intensive preparatory steps. Immunoassays use antibodies to recognize the analyte of interest. Absorbance spectrophotometry, **fluorescence**, **chemiluminescence**, or some other technology is used to complete the measurement.

Immunoassays are objective, relatively specific, capable of high sensitivity, and compatible with automation. Their major drawback is lack of 100 percent specificity, but this is not an insurmountable problem because immunoassay positive test results are followed by confirmatory testing.

Thin Layer Chromatography

Chromatography is a means of separating chemicals. Many chromatographic methods exist. In **thin layer chromatography (TLC)**, the specimen is extracted into an organic solvent and spotted onto a glass plate coated with silica. The plate is placed into a tank that contains a mobile phase that migrates up the plate and separates whatever chemicals were originally present in the specimen. They can then be visualized by spraying with specific reagents. Toxins are identified on the basis of the distance they migrate up the plate and on the basis of the colors they produce with various identifying reagents.

TLC is an old technique that still has certain advantages. It is capable of identifying literally hundreds of compounds in a single run. It can also be inexpensive since it does not require major equipment. Disadvantages include relatively high detection limits for many compounds. The technique is labor-intensive, and significant operator experience is needed before a forensic scientist is fully competent with this method.

Ultraviolet-Visible Spectrophotometry (UV-VIS)

Many drugs absorb light. Each one produces a characteristic spectrum that allows the drug to be identified because it has peak absorbance at a certain wavelength and other specific spectral characteristics. This technology is limited by the fact that **ultraviolet**-visible spectra tend to be characterized by a small number of broad peaks and many spectra lack highly specific features. A further disadvantage is a high risk of misidentification when an additional drug or some other chemical is present that absorbs light in the region under analysis.

An automated analyzer, the Remedi (BioRad Laboratories, Hercules, California), has greatly improved the potential of UV-VIS identification of toxins. This instrument incorporates UV-VIS detection of drugs as they emerge from a high performance liquid chromatographic column. The Remedi has a photo diode array detector that enables rapid scanning of each fraction coming off the high performance liquid chromatography column. Computerized **enhancement**s provide reliable compound recognition.

Confirmatory Analyses

Confirmatory analyses, as mentioned above, are sophisticated analytical methods selected to provide answers that are essentially incontrovertible. They must be sensitive and specific. If they do not meet the specificity criterion, a positive finding could be due to something other than the reported substance. Confirmatory analyses are the last analytical step. They must, therefore, be extremely reliable.

Gas Chromatography (GC)

Gas chromatography is a method in which the substance to be tested is separated from other components of a mixture on a column. As it emerges from the column, a signal is generated by a physicochemical interaction with the emerging chemical, and the **retention time** (transit time on the column) is measured. The time may be compared to the known retention times of many substances. Under appropriate conditions, gas chromatography may meet the conditions required of a confirmatory method. However, one must keep in mind that several substances could conceivably emerge from the column at the same time and confuse the interpretation.

Gas Chromatography–Mass Spectrometry (GC-MS)

Clearly, for absolute specificity, something more than mere retention time is needed. That additional assurance is provided by gas chromatography–electron impact **mass spectrometry**. With this technology, the components of a mixture are also separated on a gas chromatographic column and retention time is determined. In addition, however, a mass spectrum of each component in the mixture is recorded (Figure 5.6). The mass spectrum is obtained by bombarding the substance under study with electrons. Short-lived molecular fragments are formed, and they are recorded and quantified. The resulting mass spectrum is often highly unique, a virtual "fingerprint" of the molecule. One can, therefore, base the identification of a substance on its retention time plus its unique mass spectrum. This combination is the most perfect that can be obtained in contemporary chemical analysis. Such identifications are usually beyond question.

In some cases, mass spectra are not unique. Some molecules are resistant to fragmentation and produce fragments in low abundance that strongly resemble mass spectra of other substances. One can still obtain positive results because of the additional factor of the retention time and other types of mass spectrometry can add additional options to this technology. **Chemical ionization** is a type of mass spectrometry in which a molecule reacts under relatively low energy with a reagent gas rather than fragmenting extensively. This provides a different spectrum from that found in mass fragmentation. A third approach is tandem mass spectrometry. In the tandem method a fragment produced in the first stage

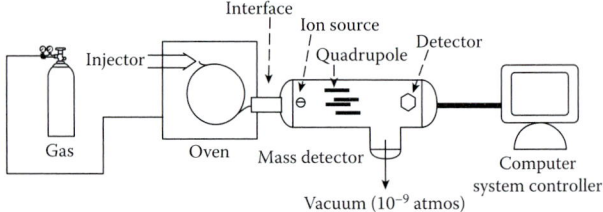

Figure 5.6 Schematic representation of a gas chromatography–mass spectrometry.

of electron impact is isolated and subjected to further bombardment. This produces a daughter spectrum. The daughter spectrum may be unique even if the first mass spectrum produced is similar to the spectrum of another substance. We see that mass spectrometry coupled to gas chromatography is an extremely powerful technology for recognition of specific substances.

How are substances emerging from chromatographic columns identified? This is usually left to a computer search, in which the unknown mass spectrum is matched with spectra of known toxins. The National Institute of Standards and Technology (**NIST**) has a library of over 140,000 compounds that can be searched for matches. Other libraries can also be employed, but most are smaller than the NIST library. In some circumstances, such as searching specific areas, they may have advantages over the NIST library.

GC-MS is a highly respected technology. Experts regard it as providing the highest level of certitude currently available for the identification of most chemical substances. In fact, national accrediting agencies mandate that laboratories must use GC-MS to confirm results or they will not be designated as certified laboratories.

Despite its great power, GC-MS has some disadvantages. Most analyses are time-consuming and they involve labor-intensive steps. Efforts to introduce automation into GC-MS testing have not, on the whole, been successful, mainly because the steps in the analysis involve a wide variety of manipulations. Lastly, the substances under study must be gases or capable of conversion to vapors in the analysis. Substances with very high boiling points are usually not amenable to assay by GC-MS.

Liquid Chromatography–Mass Spectrometry (LC-MS)

LC-MS is simply a technology in which a liquid chromatograph replaces the gas chromatograph used in GC-MS. The sample mixture under study is swept into a liquid solvent instead of a gaseous stream and is carried to a detector. Since the chemical of interest does not have to be heated for conversion into the gaseous state, this technology is compatible with virtually every known organic chemical. By contrast, it is estimated that GC-MS cannot be used with 80 percent of organic chemicals either because the substance being tested is thermolabile or hydrophilic. In either case, the substance cannot be vaporized, a necessary condition for analysis by gas chromatography.

LC-MS is expensive. The instrument costs are approximately three times the acquisition costs of comparably equipped GC-MS equipment. The high cost is partly due to highly complex interfaces between the liquid chromatograph and the mass detector. The interface must remove all carrier liquid from the mixture and send the sample of interest into the detector under conditions of high temperature and extreme vacuum. This is technically difficult and, not surprisingly, expensive. LC-MS also provides lower resolution than GC-MS, although the resolution it provides is adequate for most forensic applications. In the major technologies employed in LC-MS, electrospray ionization and atmospheric pressure chemical ionization, mass spectra are less complex than electron impact mass spectra, the kind that are typically collected in GC-MS. As a result, these less complex spectra are also less unique than electron impact spectra. It may not be possible, therefore, to make definitive identifications in some cases. Finally, LC-MS is a new technology for the forensic sciences, so a large body of applications and data are not yet available.

LC-MS is a supplement to GC-MS in most contemporary laboratories. Its most valued contribution appears to be for the analysis of more polar, unstable, and low concentration drugs, especially in blood plasma. For more widespread use some problems with this technology must be overcome. One problem is the lack of universality, such that spectra produced by one manufacturer's instruments are often not identical to spectra obtained from a different instrument. By contrast, GC-MS is very reproducible from one instrument to another. Similarly, much work remains to be done to develop universal work-up procedures that would isolate analytes of interest and reduce specific interferences that currently lead to ambiguous results.

It is a reasonable conjecture that it will be used more extensively as laboratories become more acquainted with its power, technological problems continue to be overcome, and acquisition costs drop.

Metal Analyses

Many metals are toxic to humans. Those that present the greatest danger include lead, mercury, arsenic, and cadmium. These are encountered often as environmental and occupational contaminants and occasionally as agents of homicide. Frequency of human poisonings is based on the inherent toxicity of the metal and, very importantly, on the fact that these substances are present extensively in the environment. Some of the most toxic metals (e.g., plutonium) are, fortunately rarely encountered and this rarity accounts for the equally rare incidence of poisonings. Other metals that are of importance to forensic toxicologists include iron, nickel, copper, zinc, bismuth, and thallium.

Colorimetric Assays

Any metal can be accurately measured by basic colorimetric testing. Only simple reagents and an inexpensive photometer are needed. Methods based on colorimetric endpoints were developed long ago—many in the 19th century. In particular, analyses for arsenic were discovered at that time in response to the widespread use of arsenic for homicide. The Marsh and Gutzeit tests were arsenic colorimetric methods introduced for forensic purposes during the 19th century.

Metal assays based on photometry have high detection limits. This disadvantage is partly circumvented by using large specimen size. For example, one procedure for arsenic testing calls for a 50-milliliter sample. This is feasible for testing urine but it is not readily compatible with other biological fluids.

Atomic Absorption Spectrophotometry (AAS)

The most popular instrumental method for metal analysis is **AAS**. In this technique the ion of a metal in solution is reduced to an atom, usually by a flame, and a specific wavelength of light is used to raise a valence electron of the metal to a high energy state. A detector measures the light impinging on it before and after the sample is introduced into the light path. Finally, the decrease in light striking the detector is proportional to the concentration of metal in the solution.

AAS is a common, inexpensive technology that has excellent detection limits. Some metals can be measured in the parts-per-billion range. Occasional interferences require specialized specimen preparation.

Neutron Activation Analysis (NAA)

NAA is a highly specialized method for metal testing. The sample is placed in the presence of low energy neutrons during which time it undergoes radioactive changes. The products of such neutron activation emit gamma rays or x-rays. The characteristics of the emitted radiation allow the analyst to identify the metals that are present and to measure their concentrations.

NAA is advantageous from the perspective that the atomic **nucleus** is the site of activation. Therefore, whether the metal is organic or inorganic makes no difference in the analysis. A disadvantage is the fact that the neutron source is an accelerator or a reactor, relatively unavailable items of equipment.

Inductively Coupled Plasma–Mass Spectrometry (ICP-MS)

ICP-MS is the best and most modern technique for metal analysis. Cost considerations and the newness of this technology prevent it from complete domination of the metal testing arena. Argon atoms in an ICP-MS torch are subjected to radio frequency energy that makes them collide. This drives the temperature of the torch to greater than 6000 degrees

Celsius. Atoms in the specimen are ionized and then directed into a mass detector where they can be separated on the basis of their masses and charges. Quantification of over 50 elements has been described and several metals may be analyzed at one time. Very low detection limits are achieved and the linear range of detection ranges over several orders of magnitude.

ICP-MS is a high quality methodology that will undoubtedly increase in applications within the forensic sphere. The technique has limitations relating to interference. For example, the masses and charges of **isotopes** of certain metals are very similar to those of other elements. When such ambiguities occur, specific techniques are available that allow the analyst to circumvent any possible misidentification.

Interpreting Drug Findings

Toxicogenomics

The term **toxicogenomics** refers to how genomes, the total amount of genetic material in an individual, respond to toxins. In a broad sense it is the applications of new knowledge about the human genome that will have applications in toxicology. Although this is a very new and largely theoretical field at the present time, it is generally believed that it will have many applications and great value in the future.

The human genome project produced a complete picture of the sequence of bases in human DNA, a study that was almost completed in 1999. At the present time, this knowledge base is being enhanced by studying specific genomes of persons with various diseases and specific genetic traits. The goal is to correlate physical and mental characteristics with unique genes and regions of the genome. Perhaps there are specific sequences of DNA that are related to some types of cancer and perhaps there are regions of the genome that relate to toxicology. The latter supposition is

almost certainly true although the extent of its value to science is essentially unknown.

Scientists are hopeful that in the future it might be possible to learn from the genome why some smokers get lung cancer and others do not. Susceptibility to many toxins must relate to genetic dispositions that ultimately correlate with the sequence of bases in DNA.

In the forensic area toxicogenomics will probably be very useful. For example, at this time DNA found at the scene of a crime is not immensely helpful unless a match with a suspect is possible. If the perpetrator's DNA has never been tested, then a match cannot be made. Imagine, however, the value of DNA if conclusions could be made about the perpetrator's race, physical features, ethnicity, and a host of other characteristics on the basis of matching that person's DNA to a massive database that correlates DNA sequences with human traits. This possibility will almost certainly be available in the future.

Interpreting Workplace Drug Tests

Workplace drug testing is usually conducted for the purpose of answering one of two questions. In the context of making a hiring decision, the question is whether or not the prospective employee is a drug addict. The other situation is when the behavior of an established employee in the workplace is erratic and suggestive of drug use. An employer will want to know if drug or alcohol use are causative. In collecting evidence of exposure to drugs of abuse by a potential employee, it is best to examine urine for the presence of drugs or drug metabolites. In the case of an established employee, demonstrating that erratic behavior was due to abusive drug use at the workplace requires a blood test.

Table 5.4 shows the cutoff levels for declaring a urine specimen positive for certain drugs. A person whose urine contains a drug in an amount greater than the screening and confirmation limits listed is presumed to have been using the drug in question. These limits were established by the U.S. government and are imposed on testing laboratories that have met the criteria of the government's rigid

Table 5.4
Cutoff Concentrations for Positive Specimens

Drug	Screening Threshold (Cutoff) (Nanograms per Milliliter)	Confirmation Threshold (Cutoff) (Nanograms per Milliliter)
Marijuana metabolite	50	15
Cocaine metabolite	300	150
Phencyclidine	25	25
Amphetamine, Methamphetamine	1000	500
Morphine, Codeine	2000	2000

certification program. The limits are much greater than zero so it appears that some degree of drug use is tolerated, but this is not the case. The nonzero limits were set because a finite quantity of a drug can result from passive exposure. For example, merely being in the presence of an individual who is smoking marijuana can cause some of the drug metabolite to be present in the urine of a nonusing person. Studies show that the amounts deposited in the urine in this manner are below the 15 nanogram per milliliter cutoff established by the accrediting agency.

Another factor that prevents us from setting zero as the concentration cutoff is the deteriorating precision and accuracy at extremely low concentrations of even the finest assay methods. It is reasonable, therefore, to set limits that are greater than zero despite the preference on the part of legislative agencies that no drug whatever be found in specimens.

Supporting an allegation that an employee is under the influence of a drug in the workplace is more difficult than merely demonstrating that his or her urine contains a drug of abuse. With the exception of alcohol, no specific threshold concentrations, above which everyone agrees that impairment is certain, have been determined. That is, there is no widespread consensus on what drug concentrations equal impairment. Nevertheless, finding substantial quantities of the parent drug in blood coupled with witnessed accounts of erratic behavior is often sufficient to conclude that impairment exists. Lacking a table with specific numbers, experts must testify that the amount found in blood by the laboratory is a probable explanation of the subject's behavior.

Interpreting Postmortem Test Results

The goal of the forensic toxicologist in postmortem investigation is to collaborate with the forensic pathologist in determining the cause and manner of death. In simple terms, we infer that a death is due to a specific toxin when appropriate quantities of that toxin are found, when other findings (e.g., congestion in the lungs) are consistent with that conclusion, and no other apparent cause of death is discovered. In addition to explaining the cause of death, the presence of a toxin may help to determine whether the death was an accident, a suicide, or a homicide.

This is a complex enterprise. It is important to recall, while keeping in mind the goal listed above, that many factors undermine the apparent straightforward character of a toxicologist's endeavors. For example, it is almost impossible to state that a specific death was not poison-related. Negative claims infer that all possible toxins were examined. This is not possible in the real world. Even the most comprehensive analytical scheme invariably omits many exotic (but potential) toxins.

A second problem is that, for obvious reasons, controlled experiments have not been done to determine the lethal doses or lethal blood concentrations of toxins in man. The best effort is made to get accurate values based on animal experiments and previous forensic experience. The lethal values that have been tabulated are usually reliable, but they have shortcomings. For example, some animals are much different from humans in susceptibility to the effects of specific toxins. Cats are far more sensitive than humans to benzoic acid. Rabbits are much less sensitive than humans to amanita mushrooms. Dioxin may be the strangest such example. Humans are little affected by dioxin whereas the **LD$_{50}$** (the quantity that kills 50 percent of a population) in guinea pigs is 0.6 micrograms per kilogram; in dogs it is 200-fold greater than that amount.

A third factor that complicates the interpretation of postmortem drug levels is postmortem redistribution. This term refers to concentration changes that occur after death as drugs move from one bodily region to another. For many years scientists believed that, since blood flow stops at the time of death, the postmortem concentration of a drug will remain constant until decomposition is in an advanced stage. This is not the case, and specimens collected at different times after death usually differ to some degree in drug concentration. A good example of this phenomenon of postmortem redistribution is found in the debate over the death penalty and lethal injection as a means of ending life. Currently, the death penalty is carried out by a series of injections. A common protocol involves the initial administration of sodium thiopental, then pancuronium bromide, and, finally, potassium chloride. Any one of these is lethal in the dosages given. The thiopental induces rapid unconsciousness. Pancuronium is a muscle relaxant that causes paralysis. Finally, potassium will stop the heart from beating. This protocol was developed as a means to rapidly and painlessly cause death and eliminate objections that are raised by some opponents of the death penalty on grounds that the method is "cruel and unusual" and, therefore, unconstitutional in the United States. Some of these opponents of the death penalty studied autopsy data from persons who were executed

by lethal injection. They pointed out that postmortem amounts of sodium thiopental were fairly low. It was possible, they concluded, that such persons were effectively suffocated. In other words, they were not entirely comatose because of low concentrations of thiopental and so died painfully from the next injection, the pancuronium bromide. This meant that the lethal injection protocol was a form of "cruel and unusual" punishment.

Supporters of the lethal injection method investigated these claims. They suspected that the finding of low post-mortem thiopental was an artifact because 10–20 times a lethal dose of this drug is given. One researcher noted that the postmortem results were determined 7–8 hours after death, a time by which significant postmortem redistribution could have occurred. He tested one executed person by drawing blood immediately after death and also 8 hours later. Results were noteworthy. Shortly after death blood thiopental was 29.6 milligrams per liter whereas 8 hours later it was 9.4 milligrams per liter. It is now recognized that sodium thiopental, being a lipophilic drug, gradually leaves the blood and enters the fat tissue after death. This finding, based on an understanding of postmortem redistribution, seems to explain the relatively low concentrations found in specimens collected in a delayed autopsy. It suggests that the drugs are having their intended effects, and the lethal injection method does not violate the constitutional protection against "cruel and unusual" punishment.

In arriving at the critical conclusion that the cause and manner of death is toxin-related, toxicologists must, therefore, carefully measure all toxins with the greatest care. They must consider all associated findings for each case. They must consult data collections of lethal blood concentrations and toxin amounts in other tissues (see Table 5.5) to arrive at the soundest conclusions. Finally, any effect due to postmortem redistribution must be considered. All variables must be accounted for before a final conclusion is drawn.

Table 5.5
Lethal Concentrations of Some Drugs

Drug	Blood Concentrations Found in Fatality Cases (Microgram per Milliliter)
Acetylsalicylic acid	3360
Amitriptyline	1.9
Amphetamine	0.8
Butabarbital	112
Cocaine	0.9
Digoxin	0.0033
Glutethimide	45
Meperidine	30
Methaqualone	11.3
Nicotine	29
Pentazocine	3.3
Propoxyphene	5.2
Thioridazine	4.5

Case Studies

Case 1: Wrongfully Accused?

A 41-year-old male applied for a position as an accountant with a large, multinational firm. He was provisionally hired, provided that he pass a preemployment drug screen. Three days later he was informed that he had failed the drug test because of the presence of methamphetamine in his urine. The offer of employment was withdrawn. The applicant vigorously denied use of any drug of abuse. Friends attested to his good character, but neither his denials nor the testimonials reversed the decision of the employer. Eventually, the applicant hired an attorney and brought suit against the laboratory and the company.

Is a laboratory error possible? If the methamphetamine finding was incorrect, how could such an error be avoided? What aspects of the subject's health history should be considered in interpreting the drug test results?

The man had a severe head cold 2 days prior to the drug test and treated his symptoms with Vick's Inhaler, which permitted him to breathe more easily. At the time of specimen collection for the drug test, he was asked whether he took any medications. He did not mention the inhaler because he assumed that the use of an over-the-counter product was irrelevant.

The laboratory found a positive result for amphetamine compounds in its screening test. Gas chromatography–mass spectrometry confirmed a positive result for methamphetamine. The active agent in Vick's Inhaler is L-methamphetamine, an optical isomer of D-methamphetamine. The inhaler ingredient is a decongestant, whereas D-methamphetamine is primarily used in an abusive manner. The compounds are so similar they are not differentiated by the customary method for analyzing methamphetamine.

During the discovery phase of the case, the litigant acknowledged that he failed to state that he used the inhaler. The laboratory admitted that its method of analysis could not distinguish between the two drugs. The case was dropped and the applicant was reoffered the position. This suit could have been avoided by a review of the entire case by a medical review officer or by testing of all positives by a more specific **GC-MS** method. These steps are currently employed by laboratories certified by the National Laboratory Certification Program.

Case 2: A Mysterious Death

A 32-year-old man went fishing with friends. He suddenly collapsed and complained of dizziness and nausea. He lapsed into a coma and medical assistance was summoned. His respirations became very weak, he became cyanotic, and eventually suffered a cardiac arrest from which he could not be resuscitated.

An autopsy revealed no signs of trauma or evidence of any disease condition. The case was referred to a laboratory that found fentanyl in the blood of the decedent at a concentration of 15 nanograms per milliliter. What were the cause and manner of death?

In the absence of any other findings and in view of the concentration of fentanyl found, fentanyl was ruled to be the cause of death. Concentrations greater than 12 nanograms per milliliter are regarded as sufficient to cause death. Investigation of the decedent's work history revealed that he was a known drug abuser who worked in a funeral home. He had recently handled the body of a cancer patient. Two fentanyl patches on her corpse at the time of her funeral were missing and presumed stolen by the funeral parlor worker. One known pattern of drug abuse is to extract the drug from the patches and then inject it to achieve an opiate high. The patches contain significant quantities of drugs and unintentional overdose is a very real possibility. The manner of death was ruled an accident.

Questions

1. What are the three areas covered by forensic toxicology?

2. Name six specimen types that are often tested in forensic toxicology. Under what circumstances is each specimen preferred?

3. Name the NIDA 5. Draw a table showing the following characteristics of each drug: structure of a representative molecule, drug group, symptoms of overdose, and drug source.

4. Name several groups of medicinal drugs often involved in fatalities. What characteristics render a drug most likely to be associated with overdose deaths?

5. A 210-pound male consumes three highballs each of which was made with 2 ounces of 80 proof whiskey. What is the expected peak in his blood alcohol concentration?

6. Name three methods for drug screening and describe the advantages and disadvantages of each.

7. Contrast gas chromatography with and without a mass spectrometer detector. Describe the advantages of the latter technology.

8. What are three methods of metal analysis? Which is the optimal method and why?

9. Describe the process of interpreting drug results in the context of preemployment drug testing. Why is drug testing for employed individuals more difficult?

10. In a published case, an elderly woman with cancer dies. Three fentanyl patches are found on her body. Discuss the investigation of her death with respect to factors that would be significant to the forensic toxicologist in arriving at the cause and manner of death.

References

Baselt, R. et al., Therapeutic and toxic concentrations of more than 100 toxicologically significant drugs in blood, plasma, or serum: a tabulation, *Clin. Chem.*, 21, 44, 1975.

Drummer, O. H. and Gerostamoulos, J., Postmortem drug analysis: Analytical and toxicological aspects, *Ther. Drug Monitoring*, 24(2): 210–221, 2002.

Ellerman, M. J., *Ellenhorn's Medical Toxicology*, 2nd ed., Williams & Wilkins, Baltimore, 1997.

Fenton, J. J., *Toxicology: A Case Oriented Approach*, CRC Press, Boca Raton, FL, 2002.

Fenton, J. J., *The Laboratory and the Poisoned Patient*, AACC Press, Washington, D.C., 1998.

Flanagan, L. et al., Fentanyl patches left on dead bodies: potential source of drug for abusers, *J. Forens. Sci.*, 41, 320, 1996.

Goeringer, K. E. et al., Postmortem forensic toxicology of selective serotonin reuptake inhibitors, *J. Forens. Sci.*, 45(3): 633–648, 2000.

Karch, S. B., *Drug Abuse Handbook*, CRC Press, Boca Raton, FL, 1998.

Leikin, J. B. and Paloucek, F. P., *Poisoning and Toxicology Handbook*, Lexicomp, Hudson, OH, 1995.

Levine, B., *Principles of Forensic Toxicology*, AACC Press, Washington, D.C., 1999.

Liu, R. H. and Goldberger, B. A., *Handbook of Workplace Drug Testing*, AACC Press, Washington, D.C., 1995.

Marquet, P., Progress of liquid chromatography-mass spectrometry in clinical and forensic toxicology, *Ther. Drug Monitoring*, 24 (2): 255–276, 2002.

Maurer, H. H., Screening procedures for simultaneous detection of several drug classes used for high throughput toxicological analyses and doping control. A review. *Comb. Chem. High Throughput Screen.*, 3(6): 467–480, 2000.

Moeller, M. R. and Kraemer, T., Drugs of abuse monitoring in blood for control of driving under the influence of drugs. *Ther. Drug Monitoring*, 24 (2): 199–209, 2002.

Moffat, A. C., *Clarke's Isolation and Identification of Drugs*, Pharmaceutical Press, London, 1986.

Van Boexlaer, J. F. et al., Liquid chromatography-mass spectrometry in forensic toxicology. *Mass Spectrum. Reviews*, 19(4): 165–214, 2000.

Forensic Odontology

R. Tom Glass

6

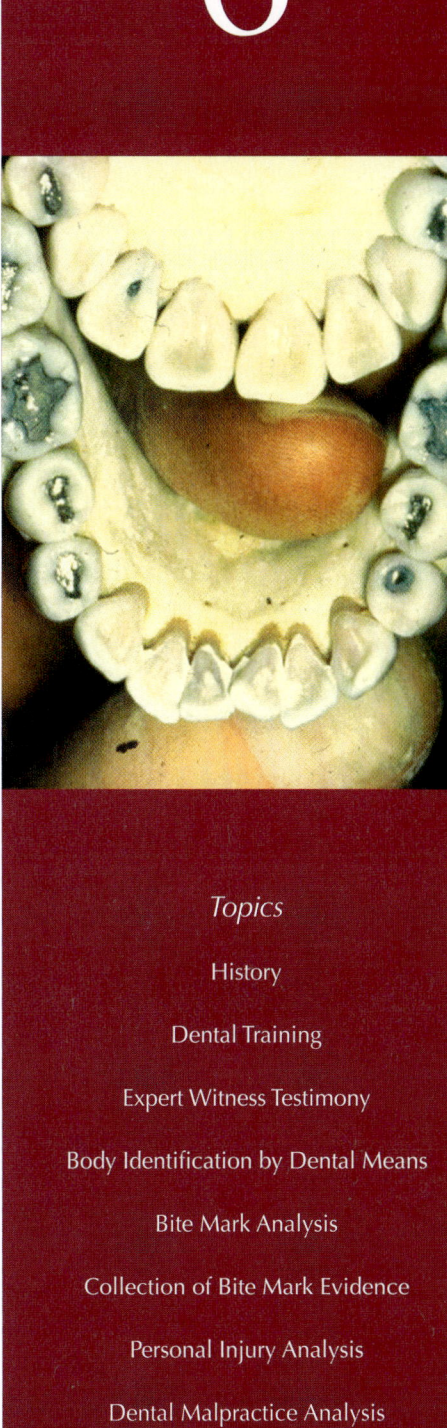

Introduction

Forensic **odontology** (forensic dentistry) is the application of the arts and sciences of dentistry to the legal system, and while some feel that the two terms should be separated, for this chapter they will be used interchangeably.[1] More specifically, forensic odontology is the:

1. Use of **dentition** in the identification of individuals by comparing the unique aspects of the victim's teeth with predeath dental records, and the use of teeth as an excellent source of DNA for identification purposes.
2. Comparison of dentition with pattern injuries in human tissue (**bite marks**) or comparison with the biting patterns in inanimate substances (e.g., foodstuffs, pencils, etc.) and the possible utilization of the same sources for obtaining DNA to assist in the identifying of suspects in crimes.
3. Analysis of personal injuries to the head and neck to determine the presence of and reasonable cause for the claimed injury.
4. Analysis of **negligence**, standard of care issues, and injury in dental malpractice cases.

The objective of this chapter is to provide only basic information in each of these areas. This chapter is not intended to make a student a forensic dentist. For a more thorough study of the subject, use the references and suggested readings at the end of the chapter.

History

Earlier uses of forensic dentistry have been described in many anecdotes. One of the oldest dates from the 1st century A.D. when the jealous wife of Roman Emperor Claudius demanded to see the decapitated head of his mistress, because the mistress was known to have a discolored anterior tooth.[2] Legend also holds that King William the Conqueror (circa 1066 A.D.) sealed his mail by biting into the soft sealing wax, leaving the outline of his misaligned teeth.[3] In 1776, Paul Revere was said to have used a denture he made to identify a deceased friend and patient at the battle of Bunker Hill.[4] Not until 1849 was dental evidence admitted into U.S. courts for identification purposes.

The first recorded use of dentitions in body identifications associated with mass disasters was at the Vienna Opera House fire in 1849.[4] Almost 100 years later, human bite mark evidence was allowed in court for the identification of a biting assailant (*Doyle v. Texas*, 1952).[5]

Clinical dentistry, like clinical medicine, has made great strides in the last 50 years. In the 1950s, body identification by dental means became more widely used. In the 1970s, national and international organizations formed to advance the science and technology of forensic dentistry. Dental offices began to use computers for patient records, x-rays, and photographs in the 1980s. Extraoral cameras were replaced by intraoral fiber-optic cameras that provided enormous amounts of detail about the teeth. Computer programs were developed to assist in the identification process and became especially useful in mass disasters involving large numbers of fatalities and a large number of fragmented body parts. Despite advances in forensic dentistry, dental schools often devote too little or no attention to this discipline.

Dental Training

One must first become a licensed dentist in order to be a forensic dentist. For most people, this includes at least 3 years of college and 4 years of dental school. A dentist graduates as a doctor of dental surgery (D.D.S.) or a doctor of dental medicine (D.M.D.). Basically, education for both degrees is the same; the only difference is the philosophy of the degree-granting institution.

The undergraduate or predental education is usually in a science-related field such as chemistry or biology. The first year of dental education is often devoted to the study of the normal human body in its entirety. The

second year focuses more directly on the structures and diseases that affect the oral cavity. During the second year, a dental student begins to deal with filling teeth and treating gum diseases. The third and fourth years are devoted to treatment of dental patients under the supervision of licensed practicing dentists.

After 4 years of training, dentists may take additional training (residency) in one of the nine areas of dental specialization recognized by the American Dental Association (ADA): oral and maxillofacial pathology, oral and maxillofacial surgery, oral and maxillofacial radiology, periodontics, pediatric dentistry, endodontics, prosthodontics, orthodontics and dentofacial orthopedics, and dental public health. These training programs usually span 1 to 4 years and lead to certificates. After earning a certificate and practicing 3 to 5 additional years, a dental specialist may elect to take a series of examinations that lead to specialty board certification that is recognized by the ADA. Once the specialist has been board-certified, they can then be listed as a specialist in their respective field. All dental graduates, whether they specialize or practice general dentistry, must pass federal and state competency examinations in order to practice dentistry. They must also maintain a specified number of continuing education hours each year in order to renew their licenses.

Any licensed dentist may elect to practice forensic dentistry. Because the area has some unique aspects that are often not well covered in formal dental education, additional training in forensic dentistry has been recommended. The author has long advocated that every licensed dentist practices forensic dentistry to a degree on a daily basis.[6]

Expert Witness Testimony

One of the objectives of forensic odontology is to provide witness testimony in courts of law.[7] If a forensic dentist is called to provide information on the dentition of a victim, he or she is considered a fact witness and certifies only predeath conditions and restorations of the victim's dentition (missing teeth and fillings). If, however, the forensic dentist is not the victim's treating dentist and is asked to compare predeath dental records with postmortem dental findings on a victim for purposes of identification, he or she is considered an expert witness. Expert testimony always uses the standard of reasonable medical/dental certainty because nothing is absolute in science.

A special use of the expert forensic dentist is in bite mark analyses. A forensic dentist is called to record and analyze pattern injuries on victims, assailants, or inanimate objects. These analyses allow the forensic dentist to assist the court in identifying the individual who did the biting. Of course, unless a life-and-death issue exists or the defense attorney stipulates to the identification by dental means, both fact and expert dental witnesses in body identification and bite mark analyses are required to appear in court in criminal cases.

In personal injury cases, a dentist who treats a patient can only be a fact witness and testify about the preinjury and postinjury conditions of the injured party. The expert dentist is given the task of reviewing the history of the patient, both with the patient and through the patient's medical/dental records, and usually examines the patient clinically. The expert is then asked to assess the cause of the injury, the extent of the injury, the disability (transient and permanent) caused by the injury, the treatments and their effectiveness, and a **prognosis** (forecast) for future improvement or disability. Again, this testimony uses the dictum of reasonable medical/dental certainty.

Finally, forensic dentists may be called as expert witnesses in cases of dental/medical malpractice. A forensic dentist will be asked to evaluate the patient and/or patient records to answer two questions:

1. Did negligence by the dentist/physician and/or his or her staff result in an injury to the patient?
2. Was the treatment of the patient by the dental/medical professional below the standard of care for the community?

If the expert finds that one or both questions are answered in the affirmative, juries will often find on behalf of the plaintiff (injured party) in the case. Once the verdict is rendered, the jury may assess monetary awards against the health professional for actual damages, punitive damages (punishment awards), or both.

Body Identification by Dental Means

Dentition Descriptions

The identity of a body is one of the most important answers to the questions arising in any medicolegal investigation. In order to explain the use of the dentition in body identification, it is important to review the anatomy of the oral cavity. Every individual has two dentitions during his or her life: the **primary dentition** (pediatric or childhood) and **permanent dentition** (adult). Primary tooth development begins around 4 months in utero, with tooth eruption starting at about 6 months of age. By 2 years of age, all 20 primary teeth usually have erupted. Adult tooth development begins soon after birth and usually takes 6 to 8 years to completely form and erupt. An additional 2 years are required for the roots to completely form. As the adult teeth in the incisor, cuspid, and bicuspid areas develop and begin their eruption process, the primary tooth root above the permanent tooth resorbs. When the root of a primary tooth is insufficient to hold the tooth in place, the tooth is exfoliated (lost). The permanent molar teeth have no preexisting primary tooth, so their eruption is without loss of a primary tooth. Humans have 20 primary teeth and 32 permanent teeth.

The teeth are equally divided between the two jaws: the **maxilla** (upper jaw) and the **mandible** (lower jaw). In the primary dentition from the midline to the posterior, the teeth on each side are the central incisor, lateral incisor, cuspid, first primary molar, and second primary molar. Each tooth is lettered consecutively, beginning with the upper right second primary molar (A) to the upper left second primary molar (J) to the lower left second primary molar (K) to the lower right second primary molar (T). While a child is using his or her primary teeth to chew, the adult dentition is forming in the bone beneath each primary tooth. The exceptions are the permanent molars that form without primary teeth above them.

In the adult dentition from the midline to the posterior, the teeth are the central incisor, lateral incisor, cuspid, first bicuspid, second bicuspid, first molar, second molar, and third molar. Like the primary dentition, a unique number identifies each tooth in the adult dentition. The numbering begins with the upper right third molar (#1) to the upper left third molar (#16) to the lower left third molar (#17) to the lower right third molar (#32).

All teeth (primary and permanent) have the same basic structures (Figure 6.1), but are anatomically unique to each individual. All teeth are divided into a crown (the structure seen in the mouth) and a root (the structure that anchors the tooth into the jaw bone).

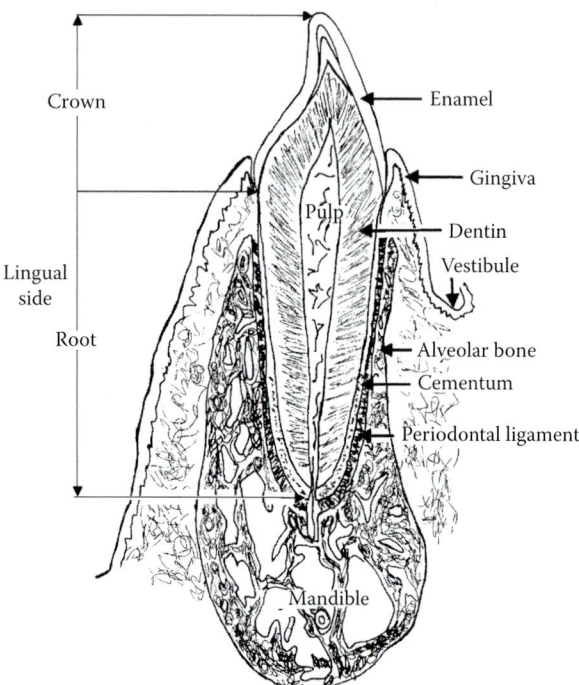

Figure 6.1 A cross-section of a lower (mandibular) incisor tooth, alveolus (tooth socket), and gingiva (gum tissue). Note that the tooth is held in the alveolus by a specialized structure known as the periodontal ligament.

While each tooth has only one crown, maxillary bicuspids may have two roots, mandibular molars have two roots, and maxillary molars have three roots. The crown is composed of an outer covering called **enamel**. The root is composed of an outer covering called **cementum**. The cementum has specialized fibers (**Sharpey's fibers**) that join ligaments (**periodontal ligaments**) in holding a tooth in the socket (**alveolus**) of the bone. The underlying structure of the tooth is the dentin. In the center of the tooth is the **pulp** or nervous tissue, which is well protected from the environment. Therefore, the pulp tissue is an excellent source of both nuclear and mitochondrial DNA that can be used in the identification process.[8,9] When a tooth is removed, the bone heals by eliminating the socket. The unique combinations of tooth and bone structures allow body identification by dental means.

Postmortem Examination and Record

Body identification by dental means begins with access to the dentition. If a forensic pathologist conducting an autopsy determines that the body can be reconstructed sufficiently to be viewed in a funeral, the forensic dentist is asked to examine the dentition without removing it (Figure 6.2). Because rigor mortis often makes it impossible to open the lower jaw for

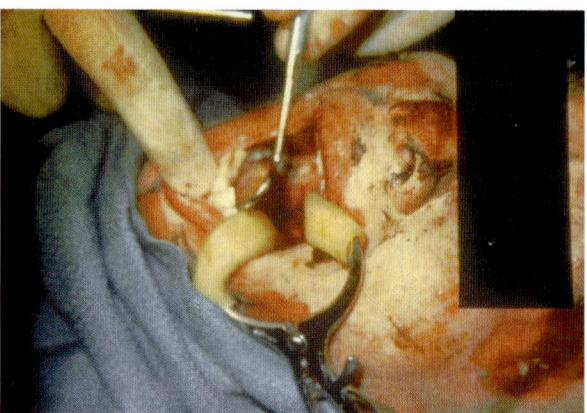

Figure 6.2 A typical unidentified victim. Despite areas of bruising and skin abrasions, a certified funeral director can restore the face so that it can be viewed at the funeral. For this reason, the dentition is examined in situ (in place) and the postmortem record is made from the information gathered.

this examination, a threaded conical-shaped device or screw is inserted between the first and second bicuspid teeth and twisted until the rigor is broken. More recently, facial myotomy (cutting of facial muscles) techniques have been described to release rigor.[10] Once the rigor is broken, the dentition is examined in a systematic manner from tooth A to tooth T in the primary dentition or tooth #1 to tooth #32 in the adult dentition.

If, because of dismemberment, fire, or decomposition, the victim cannot be viewed at the funeral, the jaws can be removed and examined outside the mouth. In many morgues, jaw removal is performed by the autopsy assistant. To accomplish the removal an incision is made at the corners of the mouth (**commissures**) and extended posteriorly to expose the **ramus of the mandible**. The ramus is cut with a bone saw and the floor of the mouth dissected along the lingual (tongue side) surface, completely freeing the lower jaw. An incision is then made in the upper **vestibule**, exposing the maxillary bone that covers the sinuses and floor of the nose. The bone saw cuts through the lateral and medial walls of the sinuses and the floor of the nose, completely freeing the maxilla (Figure 6.3). If soft tissue remains, the jaws can be placed in 30 percent hydrogen peroxide for 24 hours, washed in running water for 4 hours, and allowed to dry. The hydrogen peroxide will remove the residual tissue, but will not change the evidence value of the dentition. A standard caution after immersion in

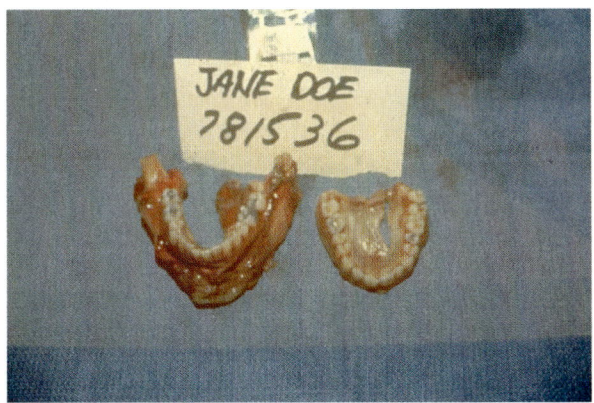

Figure 6.3 If a face is deformed or decomposed to the point where it cannot be restored for viewing, the jaws can be removed and the postmortem examination and record made more easily.

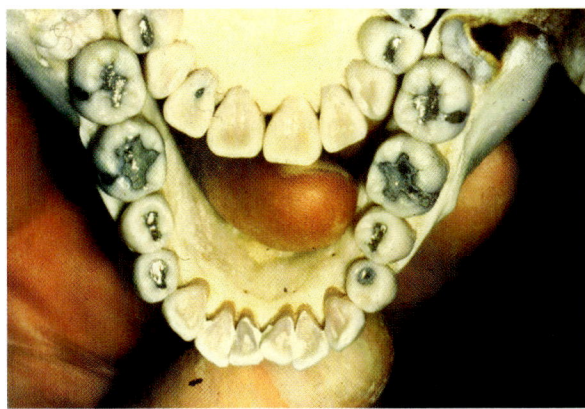

Figure 6.4 Due to substantial decomposition of the victim in this case, the jaws were placed in 30 percent hydrogen peroxide for 24 hours and then washed with running water for 3 hours prior to examination. Note that while some evidence of oxidation of fillings is present, their anatomies and structures have not changed.

hydrogen peroxide is to handle the jaws with heavy rubber gloves because the concentration of peroxide is very strong and will bleach the skin of the handler (Figure 6.4).

During dentition examination, a statement is made concerning the presence or absence of each tooth. If a tooth is missing, a statement is made as to whether the socket is present or healed. If the socket is present, the tooth may have been recently extracted, **avulsed** (expelled) in the terminal event, or removed by wild animals. If the tooth is present, a statement is made as to whether it is erupted or unerupted, is restored (filled), and, if restored, with what dental filling material. The basic materials that fill or restore a tooth after a cavity forms are silver amalgam, composite, glass compounds, and gold. Crowns are composed of gold, semiprecious metals, porcelain, or a combination. Any other changes in dentition (such as gum or periodontal disease) or the general anatomy of the oral cavity are noted as comments at the bottom of the postmortem recording form (Figure 6.5).

After the dentition is examined and findings are recorded on the appropriate form, the teeth are photographed. Whenever possible, the case number and date should appear in photographs. The dentition is then x-rayed using typical dental x-rays because they show the entire anatomy of each tooth and can be used to compare with **antemortem** (predeath)

POSTMORTEM FINDINGS

Date of Examination ___4-28-92/1530___   **Medico-Legal Number** ___9202806___

Forensic Dentist ___R.T. Glass___   Date of Recording ___4-28-92___   Recorder ___Martin Wilder___

Date of Examination ___Removed, decomposed maxilla and mandible, 30% H_2O_2 for 24 hours___

	Tooth #	M	O	D	B	L	Comments
Maxillary	X1						Missing with well healed socket
	2		X				O amalgam, crossing the transverse ridge
	3		X X			X	O, OL amalgam, not crossing the transverse ridge
	4		X				O amalgam
	5		X				O amalgam
	6						Present
	7						Present
	8						Present
	9						Present
	10					X	L amalgam
	11						Present
	12		X				O amalgam
	13		X				O amalgam
	14		X X			X X	O, OL amalgam, not crossing the transverse ridge; L cusp of Carabelli amalgam
	15		X				O amalgam, crossing the transverse ridge
	X16						Unerupted
Mandibular	X17						Unerupted with distal caries
	18		X		X		O amalgam B pit amalgam
	19		X		X		OB amalgam
	20		X				O amalgam
	21						O amalgam in distal pit
	22						Present
	23						Present
	24						Present
	25						Present
	26						Present
	27						Present
	28		X				O amalgam
	29		X				O amalgam
	30		X		X		O amalgam; B pit amalgam
	31		X		X		O amalgam; B pit amalgam
	X32						Unerupted

General Comments

There is generalized mild periodontal disease in the lower anterior region with calculus

Figure 6.5 Completed postmortem dental record (of the dentition seen in Figure 6.4). (From Dr. R. Tom Glass. With permission.)

x-rays obtained from the victim's dentist. Finally and most importantly, all data are checked one last time to assure accuracy and then signed by the forensic dentist who performed the examination and all of those involved with recording, photography, and x-ray imaging. To assure accuracy and completeness, the author finds it helpful to use a checklist for any forensic dentistry procedure. As a final step, the checklist is consulted to assure that each detail has been completed.

Antemortem Record and Examination

As important as the postmortem dental findings are in the identification process, the completeness and accuracy of the antemortem (predeath) dental records allow the forensic dentist to make an identification with reasonable medical/dental certainty.[11] Unfortunately, as variable as the dentition might be, antemortem dental records are often even more unpredictable. Antemortem dental records are usually found in the possession of the victim's dentist, but may also be found in military or prison archives or in hospital records. If a victim has personal effects such as a driver's license or credit cards in his or her possession, locating the dentist is made easier. The victim's next of kin can be consulted to determine who the victim's dentist is. This dentist can then be contacted and the victim's antemortem records can be requested.

If a victim has no reliable personal effects available, the process becomes more complicated. Once an individual is reported missing, the next of kin should be contacted concerning the missing person's dentist. The antemortem dental record and other physical details of the individual can then be placed on national and international missing persons registries (databases). The forensic authority responsible for the identification process will consult these missing person registries for possible victims. Regardless of the sources of the antemortem records, the dental findings are recorded on a form similar to the postmortem recording form so comparisons can be made more easily (Figure 6.5).

Record Comparison and Reporting

If the postmortem and antemortem data are complete, the comparison and identification process is straightforward. Each tooth of the postmortem record is compared to the same tooth in the antemortem record. Of course, because human error can be found in written records, the most reliable comparisons are made by using antemortem and postmortem dental x-rays (Figure 6.6). A statement is made regarding the comparison of each tooth. The possible statements are

1. The antemortem findings of tooth #X are consistent with the postmortem findings of tooth #X (no inconsistencies).
2. The differences between the antemortem findings of tooth #X and the postmortem findings of tooth #X can be explained by intervening cavities, fillings, or extractions (explainable inconsistencies).
3. The antemortem findings of tooth #X are inconsistent with the postmortem findings of tooth #X (unexplainable inconsistencies) and, therefore, eliminate [name of person] as the victim.
4. If the postmortem or antemortem findings are insufficient to make an accurate comparison, the term undeterminable is used.

After the analyses are made, a report is generated by the forensic dentist and addressed to the forensic pathologist responsible for the autopsy. The final report should list the time, date, specimen examined (e.g., removed maxilla and mandible; in situ or in place examination of the upper and lower dentition), and personnel involved. Similar information should be stated regarding the antemortem record, including the patient's name, their dentist's name, what records were examined, and what were the dates of treatment. A statement should be made concerning each of the 32 in both records, even if no information is available (e.g., tooth #12—no information). As with other parts of the process, the report is checked and rechecked before it is sent. Because nothing in biology is absolute, all findings in the

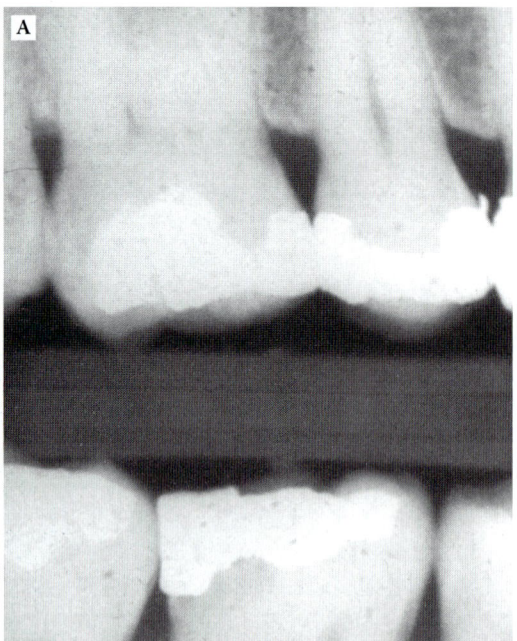
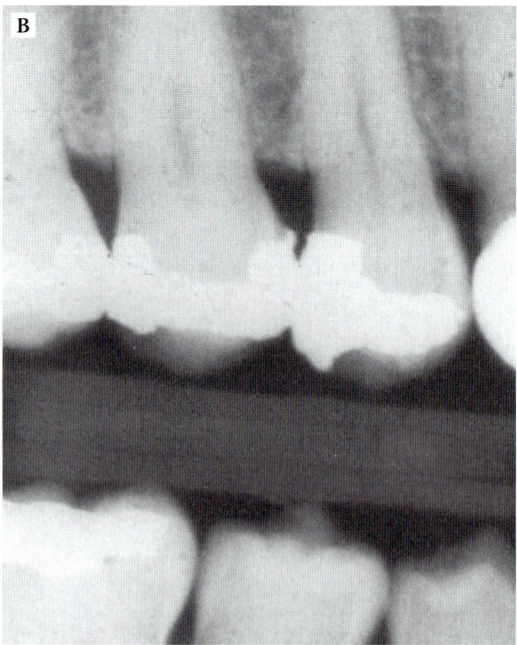

Figure 6.6 (A) Bite wing x-ray of the right posterior dentition of an unidentified victim. (B) Bite wing x-ray of the same area from the antemortem dental records received from the dentist of the possible victim. Note the consistencies between the fillings (white areas of teeth) in the two x-rays, leading the forensic odontologist to conclude that the person of the antemortem record cannot be ruled out as the victim in this case.

final report are made with reasonable medical/dental certainty.[12] Also, given the gravity of so many body identification cases, those involved in the identification process must maintain absolute confidentiality concerning the findings of the case.

While the previous discussion makes body identification by dental means seem straightforward, in many cases the process is complicated and demanding. Due to the limits of this chapter, the details of complicated cases can only be addressed superficially. In terms of postmortem difficulties, the major problems deal with dismemberment, incomplete dentition, complete tooth loss, and changes due to fire (Figure 6.7). The major antemortem difficulties deal with incomplete or inaccurate dental records. No regulations require uniformity of dental records and lack of uniformity often translates into major difficulties for the forensic dentist responsible for body identification. Finally, because the forensic dentist is asked to place one of his or her most intimate parts (their hands) in one of the victim's most intimate parts (their mouth), there is a high level of stress associated with the entire process. The stress becomes an important factor

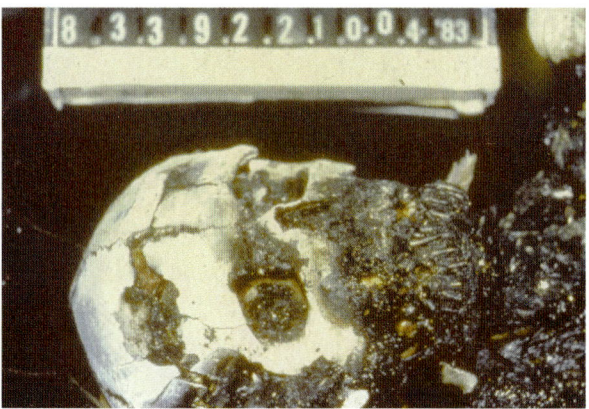

Figure 6.7 This incinerated body is that of a petroleum hauler involved in a motor vehicle accident. While the fire was intense enough to ash the skull, the teeth remained intact enough to allow identification of the victim within reasonable medical/dental certainty.

in the accuracy of the records. Thus, forensic dentists must be absolutely diligent to check each record carefully.

Forensic Dentistry in Mass Disasters

The forensic dentist probably faces no more difficult challenges than those that accompany mass disasters. Whether the disaster is of natural origin such as an earthquake, a tornado, a hurricane, or a forest fire, or of

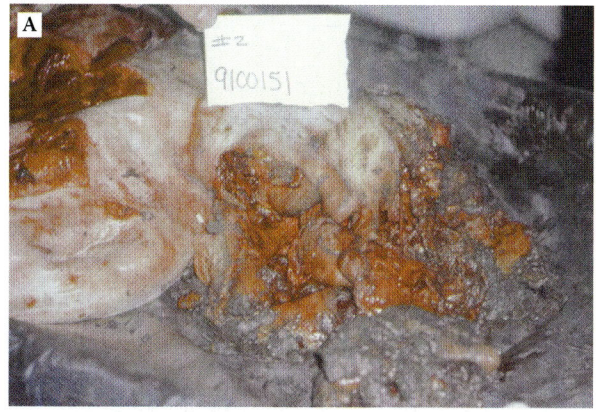

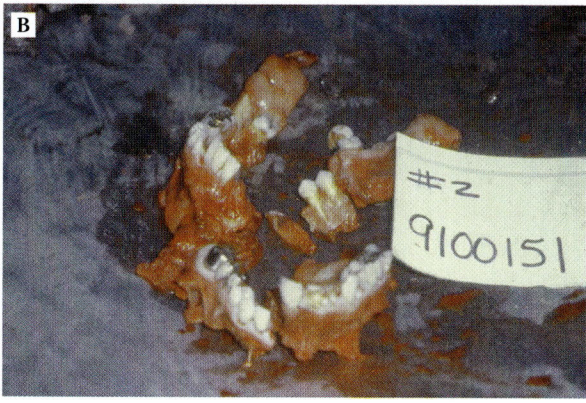

Figure 6.8 (A) The skull and upper torso of the victim of a small airplane crash. The face was deformed by the impact to the point where identity cannot be determined visually. (B) While the dentition was scattered over a 300-yard area, it was recovered and provided sufficient information for identity purposes.

man-made origin such as an explosion, a bomb, or an airplane crash, the challenge for the forensic dentist is often great. Because each of these events produces substantial dismemberment and alterations of the dentition, proactive planning and training are essential (Figure 6.8). Also, as demonstrated by several recent disasters, the victims often are from a number of countries and acquiring antemortem data is very difficult.[13]

Prior to a disaster, several dental teams should be formed and their roles well defined.[14] A chief forensic dentist must be identified as the person to make all dental identifications. Using the author's experience with a number of disasters (the largest was the bombing of the Murrah Federal Building in Oklahoma City), the teams and their roles must be well delineated and rehearsed prior to the event and no deviations made during the resolution.[15,16]

The first dental team is the one sent to the disaster scene. Prior to the arrival of the team, it is imperative that the disaster scene be secured from all elements that might put the scene dental team at risk. This dental team uses grids to systematically look for body parts, particularly jaw and tooth fragments throughout the entire scene. As each body part is discovered, a flag is placed and corresponding sector identification markings are made on the bagged body part so that the area can be reinvestigated if necessary. Because the structure of the bombed federal building in Oklahoma City was so unstable, no dental site teams were placed there. Instead, a member of the dental team was placed at the site where the debris was taken after removal from the building to provide consultation on body parts to investigating FBI agents.

The second dental team remains in the morgue and performs the postmortem dental examinations. Usually, two dental professionals (dentist and dental hygienist or dental assistant) are assigned to each victim. The most important thing for the members of this team to remember is that even though most disasters involve multiple bodies, team members must deal with each victim as an individual and not feel overwhelmed by the number of bodies. It is also imperative that those in the morgue constantly check and recheck each other. The objective of each dental examination is to provide the essential dental data that will allow the chief forensic dentist to make the identification. The Oklahoma City bombing experience taught us that due to the stress on the dental team assigned to the morgue, the team should work shifts no longer than 3 hours and no more than one shift every 2 days. The same holds true for those who perform the x-ray examinations and photography. All teams that come into direct contact with the victims also need debriefing with mental health experts after their shifts and after the disaster is resolved.

The third team is the antemortem record team, given the responsibility for contacting victims' dentists and recording the dental findings. Again, because of the nature of the disaster, great stress is associated with this process. The information must be received by a dental team member and then read back to the victim's dentist for accuracy. Another member of this team needs to function as a dental record librarian who maintains the postmortem and antemortem records. Victims' dental records or certified copies of the records must be sent to the morgue as rapidly as possible for use in the identification process and for final verification. Now that many dental records are computerized, they can be sent very rapidly using a digital format.[17]

The fourth team is responsible for inputting the postmortem findings and antemortem data into computers. Again, given the stress of the disaster, the work of even the most experienced computer person must be checked and rechecked for accuracy. Once the data are input, several programs can be used to compare postmortem dental findings with antemortem dental data. The military computer program (CAPMI) was used in Oklahoma City. It gave three to five possibilities for identifications, but the chief forensic dentist made the final identifications.

Finally, the role of the chief forensic dentist is very important. The chief forensic dentist in Oklahoma City elected to make all the dental identifications. This election allowed for continuity of the identification process because so many dental team members worked on the various teams. The chief forensic dentist also was able to watch various team members and relieve them of their duties when it appeared that stress caused them to make mistakes. Because the identifications were performed by only one person, there was no question about who would testify in court regarding the identity of the victims.

Bite Mark Analysis

Bite Mark Recognition

Bite marks have been used to identify victims and assailants from the beginning of recorded history. Bite marks most often are seen in cases of rape, murder, child abuse, and spousal abuse. While almost no one dies from a human or even animal bite, such injuries may lead to loss of function, infection, and gross disfigurement.[18] Bite marks, however, may place an assailant who performed rape, murder, child abuse, or spousal abuse in proximity of the victim. In order to consider the use of bite mark evidence in courts of law, it is imperative (1) to be able to recognize the bite mark pattern, both in animate and inanimate materials, and (2) to know what to do with the evidence once the pattern is determined to be a bite mark.

Probably the easiest way to understand the distinctive pattern of a bite mark is to make one. Go to the refrigerator; select an apple, a pear, or a block of cheese; bite into the food, then remove or retain the bitten plug in the original foodstuff. What can be seen is a typical opposing horseshoe bite mark pattern. Of course, the actual pattern will vary depending upon the material bitten, the presence or absence of your teeth, and the alignment of your teeth. Note that the mark in the foodstuff shows both patterns and indentations.

Analyze the bite mark in the foodstuff by looking at the outline of your teeth that did the biting. If there are no missing teeth and tooth alignment is good, two opposing well-defined horseshoes are seen. If, however, your dentition is malaligned or is missing anterior teeth, this pattern too should be reflected in the bite mark. Looking either at the initial entry point or the dragging pattern, the six upper anterior and six lower anterior teeth should be identifiable (central incisors, lateral incisors, cuspids). The first and second bicuspids are rarely involved in bite marks.

Bite marks in foodstuffs usually leave a well defined biting pattern. Bite marks in human tissue vary in definition, depending upon the tissue bitten, the biting force, and the resistance to biting by the victim.[19] Biting into tissues with a great deal of muscle or connective tissue tends to make a distinct bite mark outline. Biting into tissues with a great deal of fat leaves a less distinct pattern. A playful bite usually leaves a more distinct pattern (offensive bite mark) and a defensive or aggressive bite mark tends to be less distinct, have more bruising, and may show tearing or repeated biting. If the victim is resisting, the bite mark pattern is usually less defined. A bite mark is better defined if a victim is more compliant.

Most importantly, bite marks found on deceased victims are more useful in identifying the biter than bite marks made on living victims. The reason is that bite mark analysis relies, in part, upon dimension comparisons. When a living individual is bitten, the body responds by a complex process of inflammation that involves swelling and loss of the bite mark's dimensional integrity. Also, living individuals often manipulate their bite marks, again making the evidence less reliable. For bite marks on a living individual to be useful, the evidence must be collected within the first 8 hours after the bite is inflicted, and the bite mark cannot have been manipulated or washed. The bite mark also must be photographed multiple times over the next 7 days so that when the swelling subsides, the dimensional integrity of the bite can be recorded. The author has found it useful to use two descriptors related to bite mark patterns in deceased victims: the time of death relative to the bite mark and the response of the assailant or victim.[1]

Time of Death

1. Antemortem bite marks—because the injury is inflicted while the heart is still beating, a great deal of bruising will be found around the bite mark pattern, and this bruising is usually diffuse.
2. Agonal or **perimortem** bite marks (within 5 minutes of death)—since the injury is inflicted near the time the heart stops, a well-defined bruising pattern is associated with the bite mark.
3. Postmortem bite marks—because the heart is not beating and the blood has coagulated, the mark has well-defined indentions but no evidence of bruising.

Response of Assailant or Victim

1. Offensive bite marks—well defined biting pattern; usually a singular biting pattern; often associated with arousal and most often seen in postmortem bite marks.

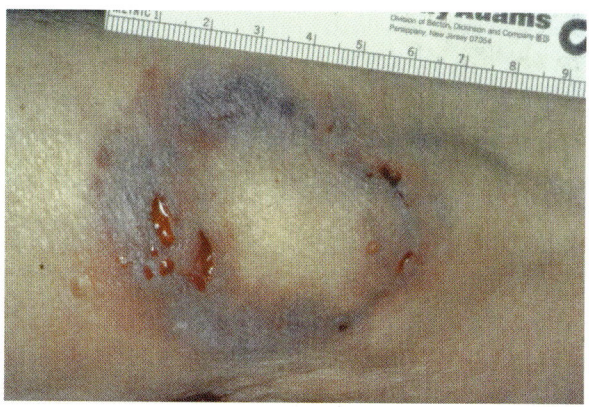

Figure 6.9 This figure represents defensive bite marks. Note that there is substantial bruising and tearing of the tissue, consistent with a struggle. The small, elevated mass in the middle of the bite mark area is herniated subcutaneous fat, demonstrating the power of a human bite.

2. Defensive bite marks—usually ill defined biting pattern with possible tearing; may have multiple bite marks; often used to defend against the assault by the assailant (Figure 6.9).

Finally, sometimes the assailant may leave traces of DNA in the bite mark (both in deceased and living victims), and retrieval of such specimens may also help in identifying the biter. Even with these considerations, the validity of bite marks as scientific evidence has recently been challenged.[20] These challenges make bite mark analysis both demanding and difficult. A forensic dentist should be extremely careful to evaluate all of the evidence before drawing conclusions.[21]

Collection of Bite Mark Evidence

Even if a pattern injury of a bite mark is recognized, it can only be useful as evidence if meticulous care is executed in gathering all the data. This discussion will focus on the function of the forensic dentist in gathering bite mark data in deceased individuals using the author's experience[1,22] and information from the American Board of Forensic Odontology.[23] The first thing to do when consulted regarding a bite mark is to start a log showing the date, the time, and the names of persons contacted regarding the pattern injury.

It is also imperative for a forensic dentist to have a bite mark kit that contains all the instruments and materials needed to collect the necessary evidence. Remember that the question that must be answered in bite mark analysis is whom the evidence excludes as the biter. If the analysis cannot exclude the suspect as the biter, a second question arises: what features of the bite mark and the suspect's dentition are consistent?

On arriving at the morgue, a forensic dentist dons sterile gloves and examines the pattern injury using a magnifying glass. If he or she considers the pattern injury to be a bite mark, a **dedicated dimensional standard** (ruler) is labeled with the case number and becomes the dimensional standard of record for the entire case. The same dedicated dimensional standard should be present in all photographs. The bite mark is swabbed with a sterile swab that is analyzed for traces of saliva (amylase, ABO blood group). A second sterile swab is collected from the bite mark for DNA analysis since saliva often contains traces of DNA. Both swabs are placed in ordinary envelopes that are dated, signed by the forensic dentist, and sent to the appropriate laboratories, taking care to maintain the chain of custody. A third sterile swab is collected from the bite mark and placed in an anaerobic microbiologic tube and analyzed for aerobic, facultative anaerobic, anaerobic, and yeast microorganisms. It is helpful to have these procedures photographed, again with the labeled dimensional standard in each photograph so that each procedure may be explained to the court.

The next step is the formal photographing of the bite mark. The first step is to attempt to reposition the bite mark in the manner it was inflicted. With the labeled dimensional standard in every photograph, the bite mark is photographed under conditions of varying light and exposure. It is very helpful to use both black and white and color film.

Using manufacturer's recommendation and normal dental techniques, impressions are made of the bite mark. Impressions are made first of the bite mark and surrounding areas using alginate impression material. The area of this impression should be large enough to give good anatomical representation. While the impression material is still soft, opened paperclips are placed in the alginate to assist in the retention of the quick-setting plaster. When the alginate impression has set, a mix of quick-setting plaster is applied to the impression for strength. Before the impression is removed, anatomic markings are made on the plaster backing to be later transferred to the casts that are constructed from the impression. In this manner, when the final casts are made, they can be oriented properly anatomically.

Again using manufacturer's recommendations, detailed impressions of the bite marks are made using dental crown and bridge impression material. Opened paperclips are placed in this material while the impression material is setting. Once set, a mix of quick-setting plaster is applied to the impression for strength. As before, anatomic markings are placed on the plaster backing to be later transferred to the final casts. Both the anatomical and the detailed casts of the bite mark allow the jury to examine the same evidence that the forensic dentist used in the analysis.

If the assailant is available, a cast of his dentition can be applied directly to the bite mark and photographed (Figure 6.10). While this may often produce dramatic photographs, the complete bite mark analysis should still be performed. If no suspect is in custody, the bite mark is excised for microscopic analysis. The forensic dentist uses routine and special stains (chemical reactions) on the tissue from the bite mark to look for traces of saliva, microorganisms from the mouth, and calculus or tartar that collects on the teeth.

Collection of Suspect Evidence

The gathering of evidence on the suspect follows the same procedure as described for the bite mark victim with two major differences. The suspect should be accompanied by at least two armed law enforcement officers for the forensic dentist's protection. The first thing that must be obtained is a waiver of

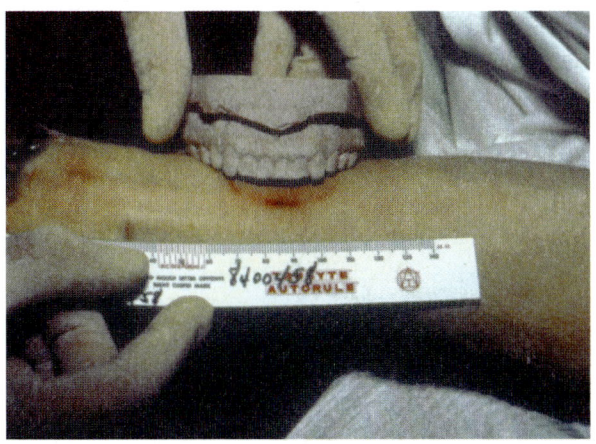

Figure 6.10 In this case, the suspect was apprehended within 3 hours of the homicide. Dental impressions were made of suspect's teeth and stone casts of his dentition were made from the impressions. The casts were applied directly to the bite marks and photographed. While this procedure has merit, it does not replace a complete bite mark analysis.

body search—a form that outlines each procedure to be performed by the forensic dentist—signed by the suspect. If the suspect refuses to sign the waiver, the procedure outline can be presented to the court and the court will issue an order commanding the suspect to submit to the examination, even if sedation is required.

With the dimensional standard in place, photographs are made of each procedure for the eventual presentation to the court. Salivary swabs of the suspect's mouth are obtained for amylase, ABO blood grouping, and DNA analysis. The suspect's mucosal and gingival (gum) tissues are swabbed for microbiology analysis. Photographs of the suspect's teeth and biting pattern are made. Impressions of the suspect's teeth are made so that casts can be made to compare with the casts of the bite marks. Finally, the biting pattern of the suspect is recorded in a semi-solid material, again to be compared with the bite mark.

Comparison of the Bite Mark and the Suspect's Dentition

The last step in bite mark analysis is the comparison of the bite mark with the suspect's dentition. The results of the salivary swabs are compared directly. The first conclusion that must be made is that the pattern injury is really a bite mark. Finding amylase in the pattern injury gives a positive correlation that saliva was present and confirms that the lesion is a bite mark. Finding a positive **association** between the ABO blood type in the bite mark and the saliva of the suspect means that the suspect cannot be excluded as the biter. Finding the suspect's DNA in the bite mark is even more compelling evidence that the suspect cannot be excluded. If the microorganisms in the bite mark are similar to the microorganisms in the suspect's mouth, again the suspect cannot be ruled out as the biter.

In a very interesting case, the microorganisms in the bite mark and the suspect's mouth were so unique that the suspect definitely could not be eliminated as the biter.[25] If a suspect is apprehended within 24 hours of a victim's death and casts are made of the suspect's teeth, the cast can be directly applied to the pattern injury (Figure 6.10).

The next part of the analysis is the most demanding. Using the case dimensional standard, 1:1 or 2:1 prints are made of the photographs of the bite marks. The casts of the assailant are marked along the biting surfaces and photographed from above, giving the outline of the biting pattern. Again using the dimensional standard, 1:1 or 2:1 photographic prints are made. Clear sheets of plastic are placed over the marked biting pattern photographs and the biting patterns transferred onto the plastic sheets. The biting patterns are then laid over the bite mark to see if there is a correlation. Also taken into consideration is the actual biting pattern of the suspect as shown by the bite registration and the curvatures of the anatomy. More recently, digital imaging has been suggested as an alternative to the more traditional use of photos.[25] Acceptance of this methodology in the forensic scientific community is not complete due to the potential for image manipulation. As with all other aspects of bite mark analysis, the objective is to rule out a suspect as the biter. If the suspect cannot be ruled out because of uniqueness of teeth and biting pattern, the suspect has to be considered the biter. Of course, if the DNA match is also positive, the evidence for confirming the suspect as the biter is even more compelling.

In closing this section, the author would like to stress the absolute necessity to have excellent postmortem data and complete data on the suspect. If either the postmortem data or the data from the suspect are of poor quality or inadequate, do not "stretch" to make a comparison. The comparison should include not only the suspect but also other cases taken from the files of the forensic dentist or from possible accomplices. Finally, remember that a positive bite mark does not convict a suspect of a crime. It only places the victim and the suspect in intimate proximity.

Personal Injury Analysis

The Role of the Forensic Dentist

Many dentists are involved in personal injury adjudications. As noted earlier, a dentist may be called to testify as a fact witness because he or she was the treating dentist and therefore, an advocate of the injured party. A dentist may be called to testify as an expert witness because he or she is recognized as an expert in a particular aspect of dentistry that pertains to the injury in question. It should be clearly understood, whether talking about dental personal injuries or personal injuries in general, that the treating doctor cannot be both the fact witness and the expert witness due to the bias inherent in treating a patient. It is also important to recognize that both the plaintiff (injured party) and the defendant (the party alleged to have caused the injury) may utilize expert witnesses. Expert witnesses, however, are supposed to be non-biased analysts of the facts in the case rather than proponents of one side or the other.

When a personal injury case goes to deposition or to trial, the treating dentist needs to prepare only by reviewing the plaintiff's dental records prior to the injury and the records of dental treatments required by the injuries. If the treating dentist was only involved in the case after the injury occurred, he or she can only testify concerning treatment, progress, and prognosis (predicted outcome). The

dentist who treated the patient prior to the injury will be called by the court to provide evidence of the plaintiff's oral conditions and health before the injury occurred.

In deference to the fact witness, the expert witness is asked by the court to consider both the facts in the case and the actual patient to determine:

1. Whether the injury was a result of the claimed incident or accident
2. Whether preexisting conditions were exacerbated by the claimed incident or accident
3. Whether the treatments were necessary and customary
4. Whether the patient returned to his or her preincident or preaccident state
5. What the prognosis (expected outcome) is for the patient

Independent Medical/Dental Examination

To answer the five questions above, an independent medical examination (IME) or dental examination may be conducted. While an IME may involve a records-only review to answer the questions, it is often helpful for a dental expert to interview and examine the injured party (the plaintiff). The author has found it important to have the IME-requesting party (plaintiff's attorney, defense attorney, or judge) pay the usual and customary fee for the IME in advance, thus assuring a more independent evaluation. For purposes of this discussion, a typical IME scenario will be presented.

The party requesting the IME contacts the dental expert directly. The dental expert makes an appointment with the injured party and asks that all pertinent dental and medical records, deposition transcripts, answers to interrogatories, and examination payment be sent in advance of the appointment. Again, even though the author receives all of these materials before the IME to eliminate bias, they are not reviewed until after the injured party's account of the incident is heard.

During the IME, the injured party is interviewed concerning the details of the incident, the nature of the injury, the treatments required, and the progress made. Questions about related issues such as previous dental treatments, previous accidents, smoking, alcohol use, habits, occupation, and hobbies are asked. The plaintiff's past medical and family medical histories are reviewed. After the interview, a physical examination is conducted that includes the muscles of the head and neck, the temporomandibular joints (TMJs), the teeth, and the entire oral cavity. Both positive and negative findings are recorded. As happens in many cases, if the injuries claimed deal with the TMJs, measurements of maximum opening, comfortable opening, and lateral jaw movements (excursions) are made. Depending on the injuries claimed and previous x-rays that may be available for comparison, comparable and appropriate x-rays are taken during the examination or are requested to be made at other medical facilities.

Final Report

When all data are gathered, the dental expert reviews the dental and medical records, the findings from the IME, and the x-rays. The questions asked earlier in this discussion are addressed and answered. A report is generated and sent to the requesting party. This report should contain all pertinent facts gathered from the IME and the record review so that the final conclusions are supported by the facts of the case. Even though certain elements of speculation are involved in attempting to predict the ultimate outcome with any patient, a statement should be included about further treatments that may be necessary and what the ultimate prognosis might be. Two important items are included in the final paragraphs of the report:

1. I reserve the right to change, modify, or correct any and all conclusions or opinions in the report if additional evidence is provided or if scientific research results in a need for such alteration.
2. I declare, under the penalties of perjury, that the information contained within the report was prepared by me and is my work product, and is true to the best of my knowledge and information.

Court Appearance

The final step for the dental expert witness is the court appearance. In preparation for court, the author has found it useful to organize all the important facts in the case into an outline. The court appearance may take several forms. First, the attorney representing either side of the legal action may request a discovery deposition, but this request is optional. At the discovery deposition, the requesting attorney asks a variety of questions to discover what the expert reviewed, how the expert performed the IME, what the findings were, and what conclusions the expert will render. The requesting attorney may explore the expert's credentials, experience, and continuing education that give him or her expert status. This deposition is conducted under oath. While the testimony rendered can be changed in the transcript or during court appearance, such indecision creates questions in the minds of the judge and jurors concerning the reliability of the expert.

Either party may decide to present the expert witness' testimony in a trial deposition, a video trial deposition, or in open court. Because a court appearance cannot be scheduled with any accuracy, many expert witnesses elect to have their testimony presented via deposition. So that the judge and jury can have a "feel" for the expert, most trial depositions are televised so that they may be played for the jury. Of course, most attorneys prefer to have their expert witnesses appear in court.

Dental Malpractice Analysis

Role of the Forensic Dentist

To say that we live in a litigious society is an understatement. This statement is certainly a reality in the areas of dental or medical malpractice. Unfortunately, as advanced as science has become, every patient and patient's family has an implicit expectation that the doctor will make the patient well or restore the patient's mouth to where it was before dental disease took its toll. Too often, either the patient or the family mistake a "bad outcome" for malpractice and legal action results. It is important to understand the issues involved in malpractice adjudications and what questions must be answered by a forensic dentist acting as an expert witness in this type of case.

First, a legitimate claim of malpractice must include (1) negligence by the treating dentist and his or her staff, (2) deviation from the community's **standard of care** by the treating dentist and his or her staff, and (3) injury of the patient. Negligence and deviation from the community's standard of care are often grouped together. In order to have cause for legal action, an action or lack of action on the part of the treating dentist and his or her staff and resultant injury to the patient are required. If negligence or deviation from the community's standard of care does not result in an injury, grounds for a legal action are voided. If no negligence or deviation from the community's standard of care occurred, but a patient suffered injury, even grievous injury, he or she has insufficient grounds for a legal action.

Once the legal action is filed with the court, the patient and the patient's family on behalf of the patient become the plaintiffs in the case. The dentist and often his or her staff become the defendants and also serve as fact witnesses because they treated the plaintiff. As with personal injury adjudications, both sides may utilize the services of expert witnesses.

Complaint Review and Report

After contact by the plaintiff's or defendant's attorney, the first thing a forensic dentist does is to review the complaint and the allegations. Most of the time this review process includes the dental records, the medical records, deposition transcripts, and the answers to the interrogatories posed by the opposing party. The decision whether to examine the plaintiff depends on the allegations and the facts of the case. For example, if the plaintiff claims failure of the dentist to diagnose a cancer and

refer him or her in timely manner, there is little point in examining the patient after the cancer has been removed other than to obtain a perspective of the alleged negligence.

Another example is a dentist who broke an instrument in a tooth extraction and pushed the instrument into a patient's maxillary sinus in attempting to remove it. The allegations state that the dentist did not tell the patient of the broken instrument, and she later developed a persistent maxillary sinus infection. Examining the patient clinically has merit in this case. Basically, the clinical examination follows the same procedure outlined for personal injury evaluations.

The next steps in the process of serving as an expert witness in a malpractice case can vary. Some attorneys request written reports on the case analysis. Such a report should contain the same two ending paragraphs used for personal injury reports: (1) the expert can change his or her mind and (2) under penalty of perjury, the information in the report is to the best of the expert's knowledge. An attorney may request an interview with the expert to discuss findings. The opposing attorney will often request a discovery deposition to determine what the dental expert will say in court. Attorneys in a discovery deposition represent both sides. A court reporter records the deposition and later provides all parties with transcriptions of the proceedings. The expert testifies under oath. While the expert may correct substantive issues in the transcribed deposition, such changes cast aspersions on the expert's competence and testimony.

Court Appearance

In malpractice suits more than personal injury cases, the requesting attorney will often want his or her expert witness to appear in court. Again, as mentioned earlier, the author finds it helpful to make an outline of the pertinent points in the case. Keep in mind that the expert witness' only responsibility is to answer the questions related to malpractice (negligence, standard of care, extent of injury). Qualifying an expert in a personal injury case is sometimes more difficult than either a criminal case or a personal injury. How does the court assess the ability of a witness to be an expert in negligence and issues of standard of care? Many times, having an academic appointment at a dental educational facility and holding offices in local, state, and national dental organizations provide that credibility.

Conclusion

Every dentist is, in part, a forensic dentist, because he or she is involved with treating patients' dental needs and recording dental findings. The discipline and essential aspects of forensic dentistry should be taught in every dental curriculum. Without understanding what is required of a dentist in medicolegal investigation, he or she cannot be prepared to actively participate in the adjudication process. Similarly, it is important for the general forensic science student to be aware of the various aspects of dentistry in order to completely utilize the available expertise. Additional readings are included in the reference section.

Questions

1. A forensic dentist might be called to:
   a. Perform a body identification by dental means
   b. Evaluate a case of alleged malpractice
   c. Determine whether a pattern injury on a child is a bite mark
   d. All of the above

2. The courts have accepted using dental means to identify a deceased individual only for the last ten years:
   a. True
   b. False

3. Any licensed dentist may act as a forensic dentist:
   a. True
   b. False

4. Because of the inherent bias in favor of a patient, a treating dentist in a personal injury case should be called as:
   a. A fact witness
   b. An expert witness
   c. A treating dentist may serve as both types of witnesses
   d. A treating dentist may not serve as either type of witness

5. In the permanent or adult dentition, tooth #19 is the:
   a. Upper right central incisor
   b. Upper left first bicuspid
   c. Lower left first molar
   d. Lower right cuspid

6. The antemortem records on a potential victim reveal that tooth #12 (upper left first bicuspid) is missing and the socket is well healed. The postmortem findings on the deceased victim in the morgue show that tooth #12 is present as are tooth #11 and tooth #13. Your report would:
   a. Accept the person identified in the antemortem records as the deceased victim (no inconsistencies)
   b. Accept the person identified in the antemortem records as the deceased victim because the treating dentist made a mistake in numbering the teeth (explainable inconsistencies)
   c. Reject the person identified in the antemortem records as the deceased victim on the basis of this inconsistency
   d. Be unable to make a statement based on this information

7. Dental identification in a mass disaster is complicated by:
   a. Dismemberment of the victims
   b. Decomposition of the victims
   c. Diversity of the victims and, therefore, their dental records
   d. All of the above

8. Only bite marks made in human tissue can be analyzed to determine the biter:
   a. True
   b. False

9. In a bite mark analysis, the objective is to exclude all possible suspects as the biter:
   a. True
   b. False

10. In a dental malpractice case, negligence by the dentist or injury to the patient has to be proven, but not both:
    a. True
    b. False

11. The first use of the dentition to identify a decedent is said to be in the first century A.D. when the jealous wife of an emperor identified his mistress by a discolored tooth:
    a. True
    b. False

12. The fully formed tooth is not a good source of DNA:
    a. True
    b. False

13. All that is required to become a forensic odontologist is an associate's degree in a related field:
    a. True
    b. False

14. All expert witness testimony should support the party that retained the expert witness—it is only fair:
    a. True
    b. False

15. When presented with a suspect in a bite mark case, it is acceptable practice to have the guards force the suspect to comply with the dental examination:
    a. True
    b. False

References

1. Glass, R. T., *Practical Forensic Dentistry,* American College of Forensic Examiners, Springfield, MO, 1997.
2. Luntz, L. L., History of forensic dentistry, *Dent. Clin. N. Am.*, 21, 7, 1977.
3. Cottone, J. A. and Standish, S. M., *Outline of Forensic Dentistry*, Yearbook Medical Publishers, Chicago, 1982.
4. Luntz, L. L. and Luntz, P., *Handbook for Dental Identification: Techniques in Forensic Dentistry*, J. B. Lippincott, Philadelphia, 1973.
5. Bernstein, M., Forensic odontology, in *Introduction to Forensic Sciences*, Eckert, W. G., Ed., 2nd ed., CRC Press, Boca Raton, FL, 1996.
6. Glass, R. T., Forensic dentistry: the dentistry you never expect to practice, *Cal. Dent. Inst. Cont. Educ. J.*, 63, 31, 1998.
7. Bowers, C. M., Jurisprudence issues in forensic odontology, *Dent. Clin. N. Am.*, 45, 399, 2001.

8. Alonso, A. et al., DNA typing from skeletal remains: Evaluation of multiplex and megaplex STR systems on DNA isolated from bone and teeth samples, *Croatian Med. J.*, 42, 260, 2001.

9. Smith, B. C. et al., A systematic approach to the sampling of dental DNA, *J. Forens. Sci.*, 38, 1194, 1993.

10. Nakayama, Y. et al., Forced oral opening for cadavers with rigor mortis: two approaches for the myotomy on the temporal muscles, *Forens. Sci. Int., 118, 37,* 2001.

11. Bell, G. L., Dentistry's role in the resolution of missing and unidentified persons cases, *Dent. Clin. N. Am.,* 45, 293, 2001.

12. Taroni, F., Mangin, P., and Perrior, M., Identification concept and the use of probabilities in forensic odontology—an approach by philosophical discussion, *J. Forens. Odonto-Stomatol., 18, 15,* 2000.

13. Sommer, H., Ranta, H., and Penttila, A., Identification of victims from the M/S Estonia, *Int. J. Legal Med.*, 114, 259, 2001.

14. Fixott, R. H. et al., Role of the dental team in mass fatality incidents, *Dent. Clin. N. Am.* 45, 293, 2001.

15. Glass, R. T., An industrial accident simulating a terrorist act: the Aerlex Firecracker factory disaster, *Am. Acad. Forens. Sci. Abstracts*, 1987.

16. Glass, R. T., Forensic dentistry in the Oklahoma City disaster, *Gen. Dent.*, 49, 554, 2001.

17. McGivney, J. and Fixott, R. H., Computer-assisted dental identifications, *Dent. Clin. N. Am.,* 45, 309, 2001.

18. Liston, P. N. et al., Bite injuries: pathophysiology, forensic analysis, and management, *New Zealand Dent. J.*, 97, 58, 2001.

19. Nambiar, P. et al., Identification from a bite mark in a wad of chewing gum, *J. Forens. Odonto-Stomatol., 19, 5,* 2001.

20. Pretty, I. A. and Sweet, D., The scientific basis for human bite mark analyses—a critical review, *Sci. Just.*, 41, 85, 2001.

21. Wright, F. D. and Dailey, J. C., Human bite marks in forensic dentistry, *Dent. Clin. N. Am.,* 45, 365, 2001.

22. Glass, R. T., Andrews, E. E., and Jones, K., Bite mark evidence: a case report utilizing accepted and new techniques, *J. Forens. Sci.*, 25, 638, 1980.

23. American Board of Forensic Odontology, Guidelines for bite mark analysis, *JADA*, 112, 383, 1986.

24. Glass, R. T. and Taylor, E. M., The use of microbiology techniques in investigation of bite marks: presentation of two cases demonstrating the range of use, *Am. Acad. Forens. Sci. Abstracts*, 1987.

25. Bowers, C. M. and Johansen, R. J., Digital analysis of bite marks and human identification, *Dent. Clin. N. Am.* 45, 327, 2001.

Suggested Readings

Bowers, C. M. and Bell, G. L., Eds., *Manual of Forensic Odontology*, 3rd ed., American Society of Forensic Odontology, Colorado Springs, CO, 1995.

Stimson, P. G. and Mertz, C. A., Eds., Forensic Dentistry, CRC Press, Boca Raton, FL, 1997.

Forensic Anthropology

Marcella H. Sorg and William D. Haglund

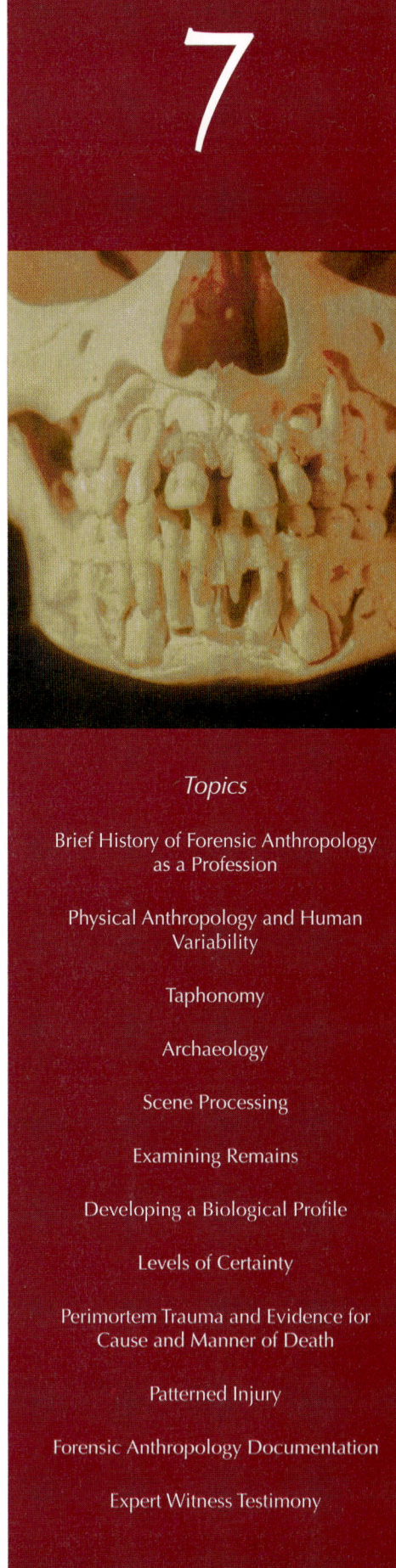

7

Introduction

Defining Forensic Anthropology

Forensic anthropology can be defined most broadly as the application of the theory and methods of anthropology to forensic problems. However, most forensic anthropologists have been specialists in physical anthropology, the study of human biological function and variation, particularly skeletal biology. Within the last two decades, forensic anthropologists with expertise in archaeological methods have played increasingly important roles in the recovery of human remains. Cultural anthropologists are not often involved in forensic casework.

The predominant forensic need for anthropological expertise emerged from the medicolegal context—that is, in the investigation of death and injury for criminal or civil legal purposes. The physical anthropologist's ability to understand the forms and variations of the human skeleton in individuals and populations complements the forensic pathologist's emphasis on soft tissue. Thus, the application of knowledge regarding human skeletal biology has been the foundation of forensic anthropology as a profession. However, this focus has been expanded by some specialists to include: (1) interpretation of primarily outdoor death scenes and postmortem processes (**forensic taphonomy**), (2) recovery of scattered or buried remains (**forensic archaeology**), (3) extrapolation of soft tissue form based on skeletal form (e.g., facial reproduction/approximation), and (4) biomechanical interpretation of sharp and blunt force injuries, primarily to bone.

Examination of human remains by the forensic anthropologist focuses on three tasks: (1) identifying the victim or at least providing a **biological profile** (age, sex, stature, ancestry, anomalies, pathology, individual features); (2) reconstructing the postmortem period based on the condition of the remains and the recovery context; and (3) providing data regarding the death event, including evidence of trauma occurring during the perimortem period. If the anthropologist has participated in the recovery, he or she will also be responsible for documenting the recovery processes and the forensic taphonomy of the site.

The forensic recovery and examination of human remains involves a multidisciplinary team. Competencies and levels of expertise often cross the disciplinary boundaries of those on the team. For example, both the criminalist and the forensic anthropologist may be trained to excavate buried bodies. Ideally, regular team members function in an interdisciplinary fashion, maximizing shared knowledge and skills, and working collaboratively.

In addition to serving as consultants to a medical examiner, coroner, or law enforcement body, forensic anthropologists frequently are asked to testify in court. Although a few forensic anthropologists work full-time in this capacity, most have academic employment at a college or university and do forensic work part-time. They also participate in forensic teams, often international in scope, mobilized to investigate mass fatalities, war crimes, crimes against humanity, and genocide. In the United States, the regional Disaster Mortuary Operational Response Teams (DMORTs) that process mass fatalities for the Federal Emergency Management Agency include forensic anthropologists. In addition, U.S. military death investigations routinely involve forensic anthropologists.

Brief History of Forensic Anthropology as a Profession

Forensic applications of physical anthropological expertise date back about a century in the United States, roughly as long as the field of **physical anthropology** has been recognized academically. A pivotal article on examining skeletal remains by W.M. Krogman was published in 1939 in the *FBI Bulletin*, marking the commencement of the modern period for the discipline.

During and after World War II and the Korean War, physical anthropologists became involved in the identification of war dead. The need for baseline information on skeletal development and variation in contemporaneous American populations stimulated substantial and systematic data gathering and analysis, mostly on young American males. These very important contributions in human

variability research increased forensic anthropology's visibility within physical anthropology and within the forensic community.

In 1972, the Physical Anthropology Section of the American Academy of Forensic Sciences was established. It has expanded from 14 to over 200 members (including many students), and is still growing rapidly. In 1977, section members formed the American Board of Forensic Anthropology (ABFA) to examine and certify forensic physical anthropologists at the postdoctoral level and to provide oversight in matters of professional conduct. A diplomate of the ABFA must have a Ph.D. in physical anthropology and a minimum of 3 years of postdoctoral practice in forensic work, in addition to passing a rigorous written and practical examination. There are now 59 active, board-certified forensic anthropologists, and the D-ABFA (diplomate of the American Board of Forensic Anthropology) credential is increasingly utilized nationally and internationally to qualify expert witnesses in court.

Physical Anthropology and Human Variability

Physical anthropologists study human physical variation and evolution in relationship to behavioral patterns, including culture. Explanations about patterns of human shape and size (skeletal, dental, and **soft tissue morphology**), include growth and development, pathology, and population differences. These explanations are derived from a theoretical understanding of how population **gene** pools change through time and space—that is, from evolutionary theory.

Because bones and teeth survive over time, many physical anthropologists specialize in human **osteology** and odontology, including detailed study of skeletal and dental functional anatomy, physiology, pathology, and their variations (Figure 7.1). This expertise has been of substantial value in forensic cases. It is important to emphasize that although knowledge of normal skeletal or

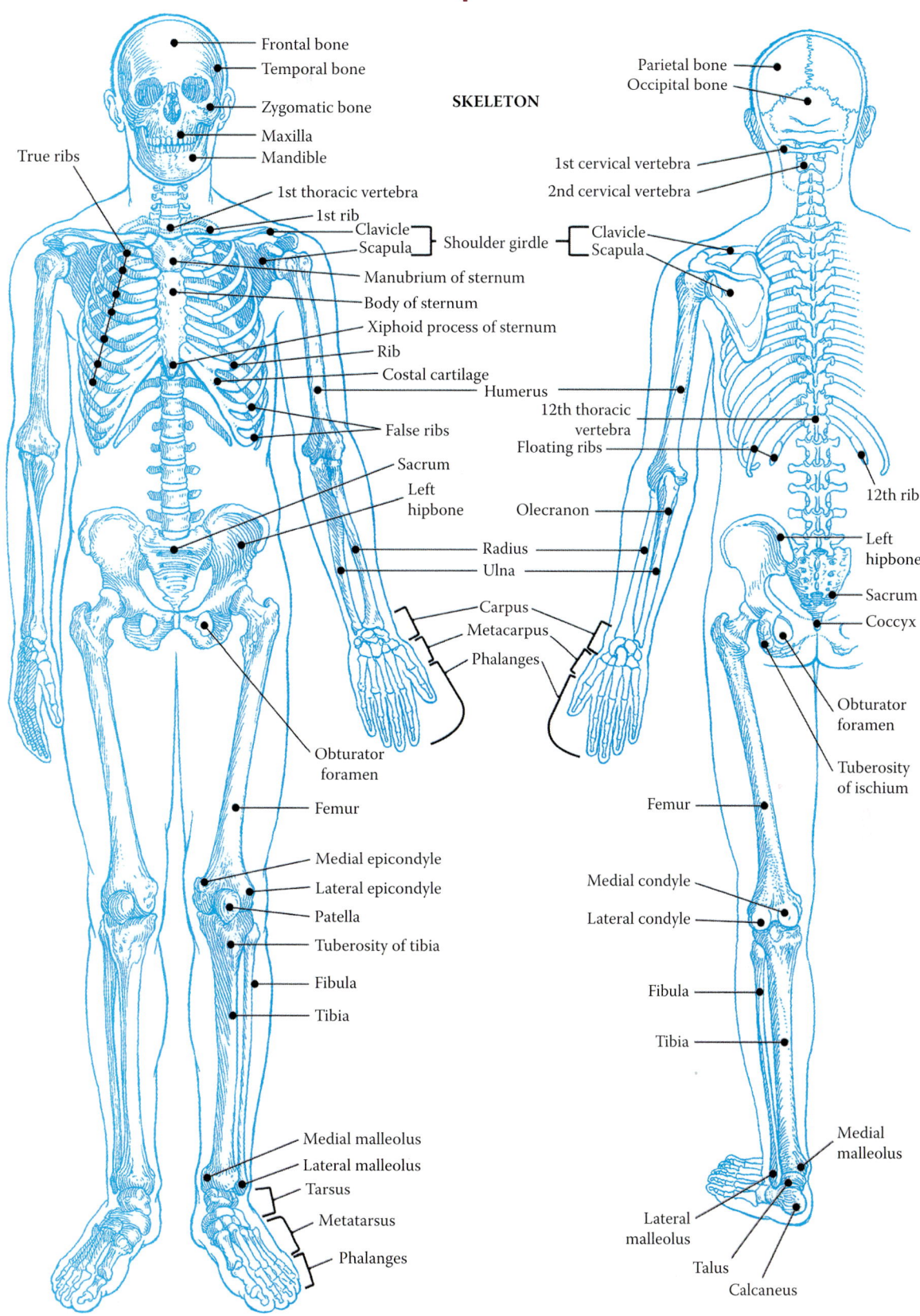

SKELETON

Frontal bone
Temporal bone
Zygomatic bone
Maxilla
Mandible
True ribs
1st thoracic vertebra
1st rib
Clavicle
Scapula
Manubrium of sternum
Body of sternum
Xiphoid process of sternum
Rib
Costal cartilage
False ribs
Sacrum
Left hipbone
Obturator foramen
Femur
Medial epicondyle
Lateral epicondyle
Patella
Tuberosity of tibia
Fibula
Tibia
Medial malleolus
Lateral malleolus
Tarsus
Metatarsus
Phalanges
Shoulder girdle
Humerus
Olecranon
Radius
Ulna
Carpus
Metacarpus
Phalanges

Parietal bone
Occipital bone
1st cervical vertebra
2nd cervical vertebra
Clavicle
Scapula
12th thoracic vertebra
Floating ribs
12th rib
Left hipbone
Sacrum
Coccyx
Obturator foramen
Tuberosity of ischium
Femur
Medial condyle
Lateral condyle
Fibula
Tibia
Medial malleolus
Lateral malleolus
Talus
Calcaneus

Figure 7.1 Adult human skeleton. (From Dox et al., *Melloni's Illustrated Medical Dictionary*, 4th ed., CRC Press, Boca Raton, FL, 2001. With permission.)

dental anatomy is critical, the practical and theoretical understanding of human variation is the expertise that allows forensic anthropologists to interpret individual cases.

Taphonomy

Knowledge about the human physical form and function must then be combined with scientific input regarding postmortem changes (taphonomy) in order to interpret the condition of human remains. In an outdoor scene, for example, such changes might include normal decomposition, **mummification, saponification,** alteration and scattering by **scavengers**, and movement and modification by flowing water, freezing, or mummification. Postmortem alterations must be differentiated from the antemortem condition of the body in order to estimate time since death, reconstruct the place of death, interpret data regarding cause of death, and sometimes to properly identify the remains.

Forensic anthropologists are frequently called upon to participate in or even direct body recoveries in outdoor settings. Such participation in body recovery is preferred whenever possible because it provides an opportunity to describe and interpret the taphonomic context firsthand. If the forensic anthropologist is not present for the recovery, the examination of the remains should nevertheless include a description and evaluation of the taphonomic condition of the remains, taking into account any contextual or environmental data gathered by the investigators. In some cases, it will be advisable for the forensic anthropologist to return to the scene for further exploration.

Archaeology

The methods of archaeology are important tools for the forensic anthropologist handling recoveries, particularly when remains have been buried or scattered, usually in outdoor scenes. Archaeologists have an extensive array of methods and techniques for recovering and interpreting material from previous events. Many forensic physical anthropologists are skilled in archaeological methods.

Remote sensing methods commonly used in archaeology can be used to locate human remains—techniques such as aerial and/or **infrared** photography, ground-penetrating radar, and metal detectors. Precise mapping, manual or computer-based, is an essential part of documenting the horizontal and vertical layout of a scene, including the position and location of the (buried or scattered) body and the spatial relationships of associated materials, as well as the layers of burial sediments.

Accurate methods of excavating buried remains can be critical in the location and interpretation of **trace evidence** associated with bodies. Differentiating primary from secondary disturbances within a grave, for example, is essential in order to correctly reconstruct the process of body deposition and eliminate extraneous data. Finally, since archaeological methods demand complete documentation of provenience for each artifact, maintaining a forensic chain of custody record for each piece of evidence recovered at a scene is a comfortable extension of normal excavation routine.

Scene Processing

Locating Remains

The forensic physical anthropologist frequently participates in searches by law enforcement or medical examiner officials, or in the recovery of remains in a mass fatality incident or human rights investigation. These searches may be focused on a particular location or a broad area; they may be terrestrial or over water. They may be done in conjunction with search and rescue teams, **cadaver dogs** (trained to find decomposition scent), or divers. In cases where partial remains have already been found, searches by forensic personnel and cadaver dog teams may be conducted to expand the recovery.

The role of the anthropologist ranges from search team member to team leader, depending on the type of case and the jurisdiction. The anthropologist generally does on-site identification of scattered remains as they are found to determine whether they are human and inventory which skeletal elements are present. On-site evaluation of remains by an anthropologist allows revision of the search strategy in response to emerging anatomical scatter patterns, as well as informing the search team when the remains are complete and the search can be discontinued.

Buried Remains

Processing a scene involving buried remains requires considerable effort and expertise, particularly if the remains are decomposed or skeletal (see case study in Figure 7.2). Access to and within the scene needs to be well marked and limited. If the grave has been disturbed by scavengers, it may be reasonable to set two scene perimeters: one for the immediate grave area and another to encompass the potential scatter area that must be searched. The area surrounding the grave or along the access to it should be examined for footprints or other evidence prior to investigatory disturbance.

Initially, a grid will be superimposed on the area to be intensively examined in order to preserve information about the spatial distribution of remains and artifacts within the scene boundaries. The grid may actually be marked off on the ground or provided with a computerized "total station," which digitally records three-dimensional locations of features and artifacts. The entire area must be photographed and documented before any work begins. The area should be examined for insects, larval or adult, that may be associated with the body; these are collected and preserved. Any living plants directly associated with the body and indicative of postmortem interval must be collected as well. It is frequently advisable to use a metal detector, particularly where vegetation or leaf cover is

present. If the metal detector signals a "hit," the location should be flagged.

A screening area is selected in a location that has already been thoroughly searched and is somewhat convenient to the grave. Here material from the grave and surrounding area will be systematically sifted through a screen to reveal human remains, artifacts, fibers, and associated insects (flies, beetles, larvae).

The next step is to clear a staging area for the excavation, usually consisting of at least several yards in every direction surrounding the grave. Leaf or other vegetation cover is removed and screened, section by section. Bushes and saplings can be cut and removed, unless they are associated with the burial. Sod covering over the grave, if present, can be carefully removed and examined. A second pass with the metal detectors is done to locate the sources of previous hits and new areas of interest. Metal artifacts are documented and removed. Sediment samples are taken from the perimeter area and the grave matrix.

An attempt is frequently made to learn the position of the body prior to excavation. Some graves may need to be excavated from the side to preserve vertical stratigraphic patterns; for example, in a mass grave. Excavation of the grave is done with small instruments such as trowels and brushes, taking care to preserve the original perimeter of the grave and any hairs, fibers, or artifacts associated with the body. If the body is fresh or decomposing, it will be necessary to prevent damage to the deteriorating soft tissue surfaces. As soon as the hands, feet, and head are exposed and photographed, it is a good idea to bag them to prevent loss of small bones, fingernails, teeth, or other evidence. In the case of burned or fragmented remains or remains of infants and children, it may be necessary to remove the body along with the surrounding sediment in order to prevent loss of tiny fragments of bone; the material can then be screened in the morgue or laboratory. In these cases, a finer mesh screen can be used.

All sediment removed from the grave is screened to search for additional remains and evidence. If the grave matrix is mud rather

Figure 7.2 Case study of a young woman who was killed in northern New England. The body was brought to a wooded area and buried in a shallow grave in July. Eighteen months later a witness directed investigators to the general search area. The nearly empty grave (A) was located by a cadaver dog. The area surrounding the grave, approximately one quarter mile in all directions, was searched by state police and warden service personnel using two cadaver dogs, followed by a shoulder-to-shoulder search (B). All materials located were flagged, mapped using a total station (C) and recovered; most were found within 50 feet of the grave. Prior to excavating, the leaf cover was removed and examined carefully (D). The immediate grave was excavated by the forensic anthropologist to expose the original grave contours, which were diagrammed and photographed. The excavation was then expanded to search for items that might have penetrated the grave floor. The soil within the grave and several inches of top-soil within 10 feet surrounding the grave were sifted through a one quarter-inch mesh screen. Three mandibular incisors were found within the grave, along with a ring. A large amount of long, light brown hair was found within and immediately adjacent to the grave. Successful DNA identification was done using one of the teeth. All bone elements found were heavily modified by carnivores, leaving only midshaft fragments of many long bones (E) and tiny fragments of cranial bones (F). The right ulna fragment (G) had evidence of a healed fracture, a swollen callus indicating poor alignment of the bone; this was consistent with the medical history of the victim. The pelvic fragment (H) revealed a wide sciatic notch (right side of this image). That, along with the small size and lack of muscularity, initially indicated the remains were female. Court testimony centered on opinion regarding whether all the remains were from one individual and whether the damage to the bone was perimortem trauma or postmortem scavenging.

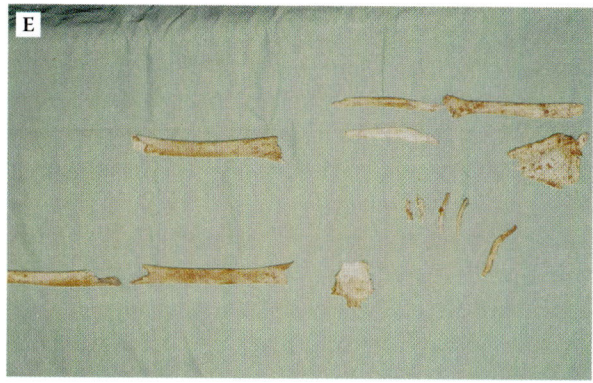

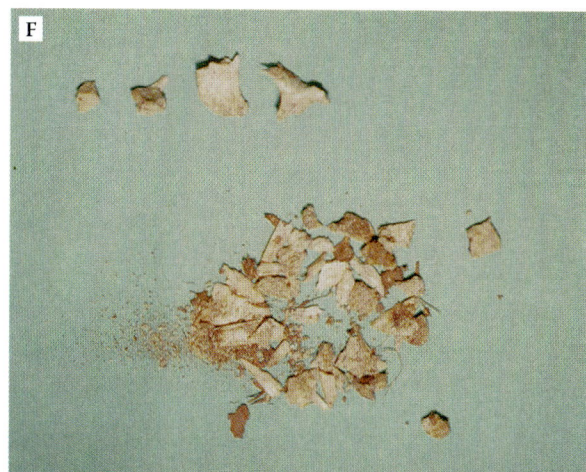

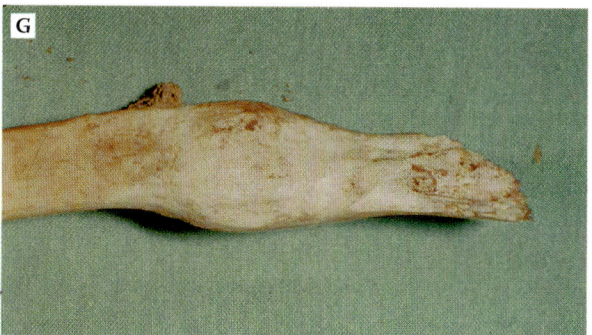

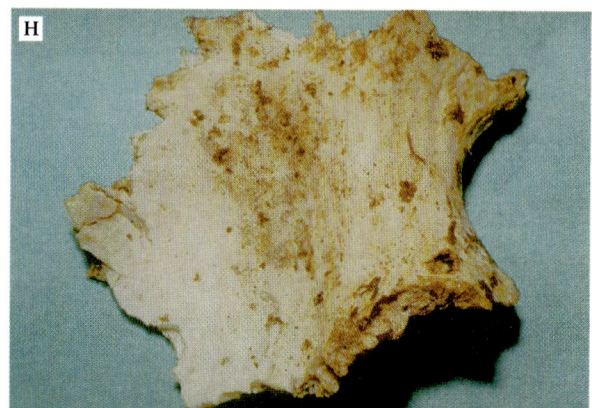

Figure 7.2 (Continued)

than dry soil, a water screening process may be necessary. In skeletal cases, small bones are easily missed visually; screeners should be experienced in identifying such materials. At regular intervals, photographs should be taken of the excavation process and the excavation may be videotaped.

If at all possible, the body or skeleton should be completely exposed prior to removal. In addition, the original grave perimeter, walls, and floor must be inspected for impressions of tools or footprints. If the remains are very fragile, they may be strengthened with preservatives or splints. The body is removed to a body bag placed beside the grave. Following another pass by the metal detector, the base of the grave is excavated and screened, going down several more inches in case small bones, teeth, artifacts, or projectiles have become embedded in the sediment.

Examining Remains

Are They Human?

Differentiating human from nonhuman animal bones and teeth is a critical function. Most laypeople can readily differentiate adult human from nonhuman skulls, but other elements are more difficult. The similar shapes of mammalian and avian long bones produce confusion among animals of similar adult size. Bear paws are notoriously similar to human hands and feet, particularly when they have been skinned and the claws are removed. Bones and teeth of immature humans and other animals can be particularly ambiguous. A newborn human skull, for example, consists of many bony elements that have not yet fused, and are not readily recognized by nonspecialists. Anthropologists can usually determine whether remains are human or nonhuman on first sight, and have access to reference collections for more difficult specimens.

Are They of Forensic Importance?

In most jurisdictions, skeletal materials found unexpectedly on private or public land are turned over to the medical examiner or coroner and then to a forensic anthropologist. Some of these cases turn out to be unmarked historic or prehistoric graves. Federal and state laws govern the final disposition of remains judged to be historic or prehistoric Native Americans. Some states have statutes regarding the handling of historic remains not known to be Native American. The forensic anthropologist can identify these nonmodern cases, usually by the appearance of the body and the grave and associated grave goods or coffin fragments. State-specific procedures are then followed, usually involving reburial.

Taphonomic Assessment

The first step in examining remains determined to be human is to perform a taphonomic assessment, including a full inventory, evaluation of the condition of the remains, and an estimation of time since death.

Inventory

The anthropologist will identify all skeletal elements found, whether right or left, and whether adult or immature. An assessment will be made of skeletal completeness, and any missing or unmatched pairs of bones are itemized. If any elements are duplicated, a determination will be made of the minimum number of individuals. In such cases, it may not be possible to fully sort all bone elements by individual.

If the remains are still fleshed or decomposed, it may be necessary to remove the flesh in order to complete the examination. In these cases, the remains should be x-rayed, photographed, and described prior to soft tissue removal.

Condition of Remains

Taphonomic assessment of the condition of remains, whether from terrestrial or aquatic settings, focuses primarily on natural postmortem processes: the decomposition stage; disarticulation sequence; evidence of scavenger modification including insects; botanic modification; weathering; and modification by water, sand or other sediment, and erosion. These taphonomic observations of the remains can only be interpreted in the context of the biological, physical, and atmospheric characteristics of the site. Regional published models of decomposition and nearby weather station data can be helpful in this process.

Taphonomic data, including input from other experts, can be used to estimate postmortem interval, reconstruct postmortem event sequences (such as movement of the body), and identify potential wounded areas of the body (which tend to attract insects disproportionately). The postmortem interval estimation should be given as a range, making it clear that such judgments are probabilistic and conditional.

Soft Tissue Examination and Processing

After the taphonomic documentation is complete, the remains must be processed for skeletal examination. In decomposed or partly skeletonized cases, some soft tissue examination may still be possible. Fingertips may be preserved enough for fingerprints. Sturdier and more internal soft tissues may yield information about age, sex, and medical conditions; collaboration with a forensic pathologist may be needed in these situations. X-rays and careful examination of soft tissue are necessary in many cases to locate embedded or hidden bullets or teeth. With fleshed or articulated infant or child remains, x-rays are needed to reveal the presence of bone formation centers, which can easily be displaced and are difficult to identify when out of anatomical position. When washing or boiling material, care must be taken to screen for small bones or artifacts, and not to overprocess the remains, which can destroy bone integrity.

Developing a Biological Profile

When the victim is unknown, the anthropologist plays a key role by developing a **biological profile**. Developing such a **profile** entails studying the remains, noting generic characteristics of shape and size, which may allow an estimation of age, sex, and population ancestry. Stature is estimated by measuring total body (or skeletal) length or by extrapolating from long bone lengths. Unique antemortem characteristics of the individual that would have been known to family or acquaintances, such as a healed bone fracture or an unusual dental configuration, are also included in the profile.

The goal of developing a biological profile is to describe the individual in such a way that law enforcement or acquaintances can narrow the range of possible identities. It is important to stay in a middle ground and not be too specific or too general. For example, it may be better to use a broader range of estimated stature and age to avoid erroneous exclusions at the outset of the investigation. The scope can be narrowed in subsequent stages.

The process of developing a biological profile requires the use of statistical descriptions of various populations that have been studied previously and reported in the professional literature. Difficulty can be encountered if the victim is unusual, has mixed ancestry, or belongs to a population that has not been well studied. Recent international investigations in human rights and crimes against humanity have pointed to problems using U.S. standards in non-U.S. populations and the need for population-specific studies, particularly for age and ancestry.

Human populations do not have fixed boundaries; they grade into one another. Hence, the ability to assign a set of remains to a specific age, sex, or ancestry can be significantly impaired if the individual is divergent or has unfamiliar characteristics. For example, size and muscularity are important indicators of sex, but it is well known that some males and some females fall outside of the normal range for their genders, and some population groups exhibit greater size differentiation between sexes than others.

Estimating Age

Growth and Development in Subadults

A forensic anthropologist must be familiar with dental and skeletal development at each stage of life, beginning with the fetus. Size is used as a common indicator of age in infants and children. The lengths of the long bone shafts, called diaphyses, can be compared to published tables of age-associated size or used in regression formulae to extrapolate stature. Obviously, such estimates are imprecise due to the range of variation of size in children, and should be reported as ranges.

Ossification Centers—The skeleton is formed by the development and growth of ossification centers, which gradually replace cartilage. In long bones, for example, bony tissue develops from a set of three main ossification centers: the shaft or **diaphysis**, and an **epiphysis** at either end. These three centers will ultimately grow together when the individual reaches full size. The timing of the formation, growth, and ultimate fusion of these and other ossification centers is patterned, depending on age, sex, bone element involved, nutritional and hormonal status, and individual variation. By the time a fetus is fully developed into a newborn, approximately 405 ossification centers are present (Figure 7.3). When an individual reaches adulthood (generally by his or her mid 20s), that number decreases to about 206 fully formed bones.

Patterns of bone development differ somewhat in males and females, with females developing a little earlier than males on average. Bone development sequences and timing also differ slightly from one population to another. Even within well-nourished, homogeneous populations, bone development may differ significantly from person to person. Thus, age estimates should always be expressed as ranges and should utilize as many indicators as possible for a single set of remains.

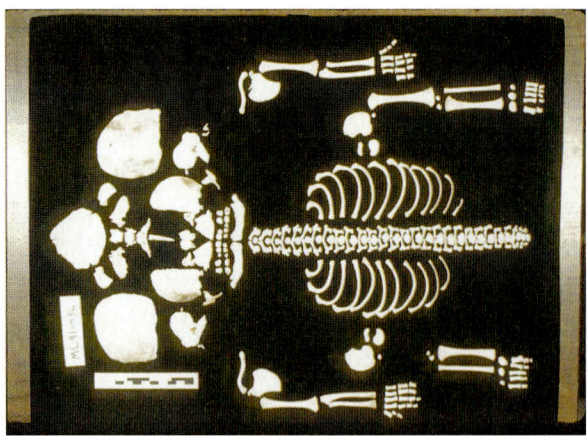

Figure 7.3 Remains of a small baby found in an abandoned car. A witness testified that the baby had been stillborn. The sizes of the long bones and the presence of ossification centers and developing teeth indicated the baby was probably a full-term newborn. The anthropologist found no evidence of pathology or perimortem trauma. The remains were too decomposed to ascertain whether the child had begun to breathe after birth.

Dental Development—Humans, like other mammals, have two sets of teeth: the deciduous dentition, so named because it is shed in childhood, and the permanent dentition (Figure 7.4). A human child (eventually) has 20 teeth. Each quadrant has two incisors, one canine, and two deciduous molars. Most human adults have 32 permanent teeth; each quadrant has two incisors, one canine, two premolars, and three molars: a dental formula of 2.1.2.3.

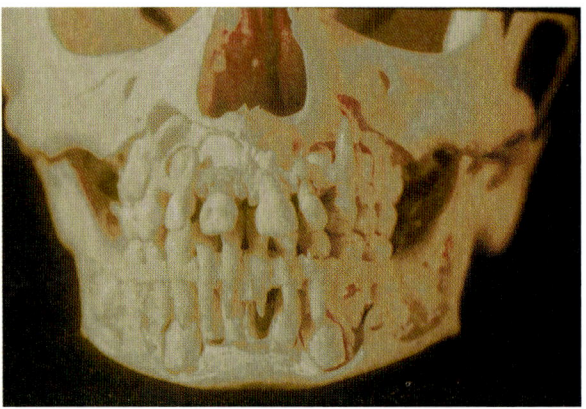

Figure 7.4 Dental development is the best indicator of age in a young child. In this 4- to 6-year-old child, the outer layer of bone has been dissected away to reveal developing and erupting teeth. Permanent teeth are still mostly unerupted and have not yet formed roots. (Courtesy of the Northeast Forensic Anthropology Association slide exchange.)

Tooth development begins in fetal life with the formation of the deciduous tooth crowns within sockets or crypts. Beginning usually around the sixth month after birth, the deciduous teeth begin to erupt. Meanwhile the permanent teeth begin to form underneath the deciduous dentition. By about the sixth year, the deciduous front teeth begin to be lost and the permanent teeth erupt to take their place. The last teeth to erupt are the third molars, often called "wisdom teeth," that begin to emerge in the 18th year in about 70 percent of the population. Many people have one or more third molars that never form (agenesis) or never erupt.

The patterns of tooth development, like those of bone, differ slightly by sex (females develop a bit earlier) and by population. The ranges of variation tend to overlap a great deal. Some dental traits are common in some populations and rare in others. Individuals of Asian or Native American ancestry, for example, commonly have a trait called shoveling (marginal ridges), whereby the anterior teeth are slightly thicker (ridged) around the margins of each tooth on the tongue (**lingual**) side (Figure 7.5). However, this trait is not uniform

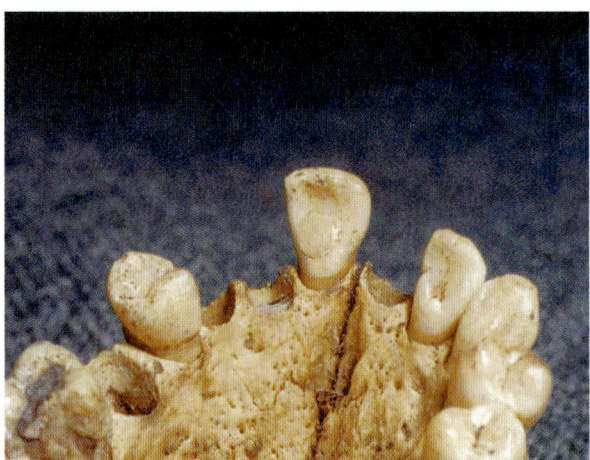

Figure 7.5 Some traits have higher frequencies in particular populations. Pronounced shoveling (marginal ridging) of incisors and canines is usually seen in individuals of Asian or Native American ancestry, although there are exceptions. This upper jaw is from a young woman of primarily European ancestry, whose remains were scattered by carnivores in a wooded area after her suicide, resulting in the loss of several teeth from their sockets. The right central incisor (center of photo), left lateral incisor, and left canine (right side of photo) show marginal ridging.

or universal in these populations, and members of other groups occasionally have it.

Reference Standards—The numbers of skeletal collections that include children with documented ages at death are few. Thus, our knowledge of population differences in bone and tooth development is somewhat limited, although the broad patterns are well known. There are comprehensive radiographic or x-ray studies, more extensive because they involve living subjects. It is important to note, however, that **macroscopic** (nonenhanced visual) and radiographic standards of bone development differ from each other. Stages of **epiphyseal union**, particularly when bones are partly or recently fused, have a different appearance in x-rays than they do to the naked eye.

Standards for dental development tend to be slightly more precise in estimating age in children, particularly young children, than are those for bone development. Hence, forensic anthropologists will generally emphasize dental standards in prepubescent remains. However, it is important to evaluate the entire set of indicators, dental as well as osteological, rather than depending on only one or two.

Growth and Development in Teens and Young Adults

During the teens and early twenties, the epiphyses of long bones undergo the process of fusion (Figure 7.6). The length of time from start to finish may be a matter of years for any particular site. Published anthropological and radiographic standards utilize various methods of evaluation. Some focus on the age at which 100 percent of the reference sample begins to fuse (at a certain site) or completes fusion. Others report the age at which the majority achieve these benchmarks. Some utilize only three states (unfused, partly fused, and fused); others use four (the partly fused category is divided into under and over 50 percent fused).

An individual's rate of epiphyseal union may be different, depending on the anatomical site observed. That is, some sites may mature earlier than average and others later than average, even within the same individual;

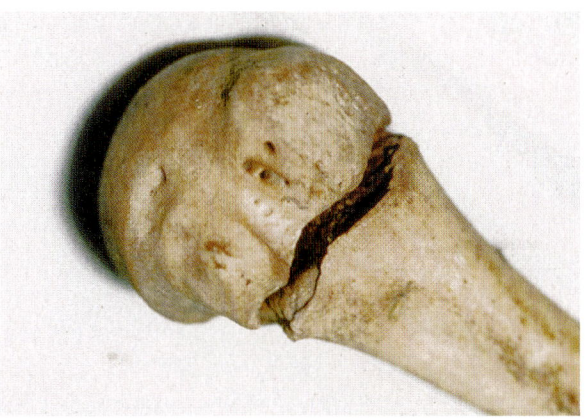

Figure 7.6 Bone growth and epiphyseal union occur in fairly regular patterns associated with age ranges and sex. This humerus is full-sized, but the epiphysis (head) at the shoulder end has not yet begun to fuse to the diaphysis or shaft, indicating an age of 15 to 20 years.

occasionally indicators may even conflict. It is best to analyze all of the sites available in the decedent and report the broadest range of ages for use in selecting prospective missing person identities.

Age-Related Patterns in Adulthood

Bone Density Changes—The bony skeleton is not fixed at adulthood; it changes continually until death, balancing the building and replacing of bony tissue at the cellular level. In general, bone density reaches a peak in the twenties and stays fairly high in the thirties, but begins to decline in the forties. Females experience a fairly precipitous bone density drop around menopause; this decline levels off in most women after the age of 55 or 60, but continues to decline. Adult males over the age of 40 experience a gradual decline into old age. A minority of individuals suffer serious bone density loss (osteoporosis), more common in females and more prevalent in some populations.

Bone density depends on factors other than age and hormonal status. Weight-bearing exercise and good nutrition (particularly calcium, magnesium, and vitamin D intake) can minimize bone loss in many individuals. Bone density can be observed macroscopically, radiographically, microscopically, or via photon absorptiometry (bone densitometry). Macroscopic assessment is very general, requires experience, and relies on evidence of thinning cortical layers, reduced concentration of **trabecular** or spongy bone, resorption and remodeling, and evidence of fractures. Radiographic methods depend on similar observations, but are done using x-ray images; CT scans and MRIs can be used in a similar fashion. Such observations are not standardized, however. Bone densitometer assessment is, on the other hand, standardized by site, sex, age, and population. Microscopic measures of bone density have also been standardized for some populations, requiring a fixed and decalcified thin section of one of the major long bones, and involving the counting of osteons and osteon fragments, cellular structures that increase in number with age.

Osteoarthritis—It is common also to see some deterioration in joint integrity connected with use-wear and exacerbated by inflammation. This is related to the reduction in bone density after the age of 40. The resulting condition, osteoarthritis, tends to affect the spine and joints that are over-used due to occupation or frequent patterned activity. It is infrequently seen before age 40, and varies a great deal among individuals. Osteoarthritis is significantly different from other forms of arthritis, e.g., rheumatoid arthritis, that tends to affect younger individuals.

Pelvic Joint Morphology—Two of the most reliable indicators of adult age are **pubic symphysis** and iliac **auricular surface** morphology. The area on the pelvis where the right and left pelvis halves join in the front of the body constitutes a flattened area at the end of each pubic bone, with a band of cartilage joining the two symphyseal faces (Figure 7.7). With age, these surfaces change from billowed to more flattened and rimmed. The changes have been divided into age- and sex-associated stages. Several published standards exist. The older the individual, the broader will be the estimated age range, however.

Standards also exist to estimate age from the auricular surface—that is, the joint surface that connects the **ilium** to the **sacrum**: the **sacroiliac joint**. Changes in density and texture are correlated with aging, and phase

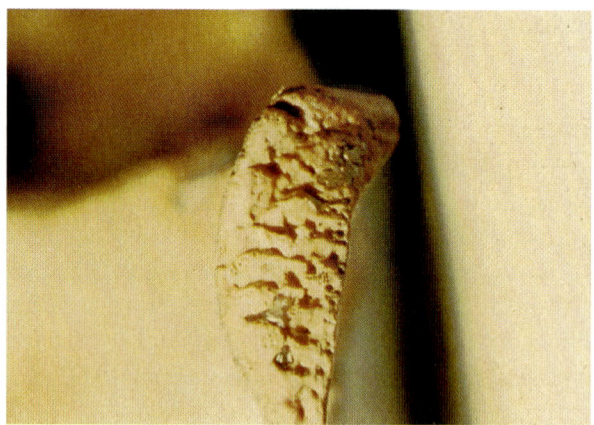

Figure 7.7 The pubic symphysis changes with age and can be analyzed to produce an estimated age range. The range of variation increases with age. This specimen from a 17-year-old shows a well-defined billowing surface, and no development of a rim around the edge.

models have been proposed. Age estimates from this joint tend to correlate highly with the pubic symphysis.

Sternal Rib Morphology—Sex-specific standards have also been published for age-related changes in morphology of the sternal (nearest the **sternum** or breastbone) end of the fourth rib. When ribs are fragmented, identification of the fourth rib is more difficult.

Cranial Sutures—With age, the bones of the braincase tend to fuse together along the suture connections between them; generally older persons have more fusion. However, the range of variation is great, moreso for older individuals and, hence, results are not precise. Nevertheless, cranial suture closure may be appropriately used as one of several measures, and occasionally is the only measure available in fragmentary remains.

Ossification of Hyaline Cartilage—The cartilage that connects the ribs to the sternum, as well as the laryngeal and thyroid cartilages in the front of the neck, tends to ossify (turn bony) with age. These changes constitute general indicators of age.

Dental Changes—As a person ages, wear and attrition affect the dentition. These processes are, however, dependent on population-related and **individual characteristics**

involving dietary practices, dental hygiene, and genetic background. Although population standards have been developed using **seriation** methods (e.g., placing each individual dentition in a progressive series of least to most worn), application to individual dentitions is problematic. Longitudinal sections of teeth may be examined to observe root **opacity** along with secondary **dentine** formation, receding gingival tissue, and wear of the **occlusal surface** of the crown; all gradually increase with age. Taphonomic changes in the postmortem period may, however, compromise this approach. Even in cases with well-preserved teeth, the estimated age range is about 14 years.

Sexual Dimorphism

Along with many other animal species, male and female *Homo sapiens* differ by size, with males larger on average. Dimorphism between the sexes is qualitative as well as quantitative. Size is an indicator of sex, but so are muscularity, robusticity, and the presence or absence of certain traits. DNA methods can produce very accurate determinations of biological sex.

Gender, the psychological and sociocultural attribution of sex, may be different from biological sex, which is based on genetic and hormonal differences. In fact, genotypic (genetic) and phenotypic (based on observed differences) sex attribution can be ambiguous in rare cases, particularly with abnormalities of sex **chromosome** number or endocrine disorders. In forensic anthropology, although the ultimate goal is to discover identity (including gender), the initial step is the assessment of biological sex, not gender.

The attribution of biological sex to skeletal remains using size, shape, and the presence or absence of skeletal markers is inexact. All methods are limited by morphological overlap between males and females. It is well known, for example, that males are, on average, taller than females when the entire population is described. However, the ranges of size overlap greatly, and individual males and females may deviate from any central tendencies. The same is true for observations of muscularity.

At puberty, when circulating sex hormones increase greatly, skeletal morphology between males and females differentiates significantly. Prior to that developmental stage, although differences exist, they are limited; prepubescent sex assignment is problematic and should generally not be done in individual forensic cases by morphology alone.

After puberty, differences become more salient, and lead to an accuracy rate well over 90 percent in sex attribution within well-studied populations. The skull and pelvis are the most sexually dimorphic skeletal areas, although it is necessary to examine the entire skeleton for indicators. The shape of the pelvis is critical for giving birth and enabling upright (**bipedal**) posture and locomotion. Both purposes are central to evolutionary success, hence, under strong influence of natural selection. The bowl-shaped human pelvis balances and sustains the weight of the upper body to permit upright posture and efficient movement. The female pelvis has additional breadth and increased diameter of the pelvic inlet and outlet, enabling the infants of our relatively large-brained human species to pass through the birth canal. Associated female traits include a broad, shallow **sciatic notch**, the U-shaped **subpubic angle** (Figure 7.8), the well-developed ventral arc, and the raised auricular surface. Generally broader pelvices tend also to display preauricular sulci; these are more frequently found in females but occasionally in males.

The typical male skull tends to be larger. It also tends to be more robust at areas of muscle attachment (**mastoid processes**) and biomechanical stress (brow ridges, chin), more right-angled at the lower jaw, and exhibits larger joint surfaces (**mandibular condyles**, **occipital condyles**). These differences are enhanced by later age at puberty. Hence, males undergo a longer period of **somatic** growth and hormonally mediated, increased muscle mass.

The postcranial (below the skull) skeleton is, on average, larger in males. The male tends to exhibit larger weight-bearing joint surfaces (e.g., the size of the hip ball and socket), more accentuated areas of muscle attachment, larger diameters of long bones, and greater

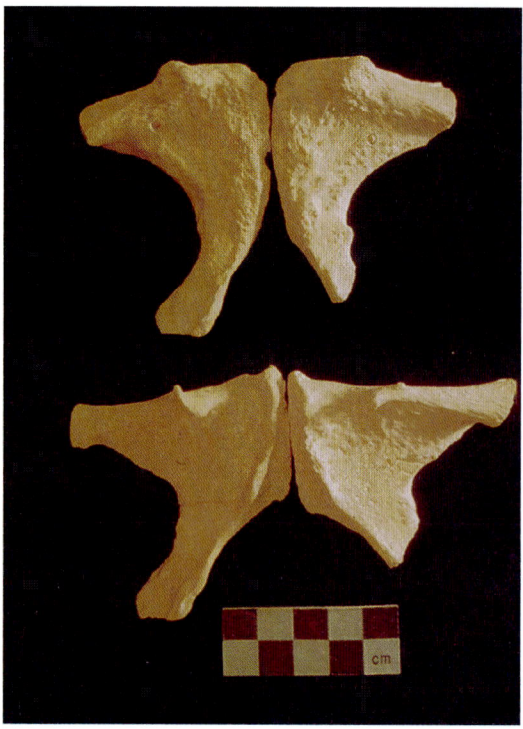

Figure 7.8 The pelvis has a number of traits associated with sex. These two pairs of partial pubic bones viewed from the front of the body illustrate the difference in the subpubic angle, the angle created where the bones come together at the midline. The upper pair has a narrower upside-down-V-shaped subpubic angle and is male. The lower pair has a wide, upside-down, U-shaped subpubic angle and is female. (Courtesy of the Northeast Forensic Anthropology Association slide exchange.)

stature. However, these traits can be strongly influenced by population affinity (some groups tend to have less sexual dimorphism), nutritional status (undernutrition can produce shorter statures, and obesity can increase robusticity), behavior (weight-bearing occupations, strength training, heavy chewing stresses), or simply genetic differences. No morphological indicator of sex is infallible. The best approach is to assess the entire skeletal pattern in the context of what may be known about the person or the population.

Population Ancestry

Concept of Race

Homo sapiens is a single species. The fabric of a single, shared gene pool extends throughout and varies continuously across all populations. There are no absolute physical or genetic reproductive barriers, even between the major human groups. Populations exhibit much more variation within themselves than exists between groups. Although regional populations, when viewed as a group, may appear to share more morphological traits with each other than with distant groups, individual group members do not necessarily exhibit all the expected traits. Decisions about population boundaries are arbitrary, and group membership is ultimately fluid.

Thus, the attribution of population membership based on skeletal characteristics is difficult, often impossible, due to the complexity of human mating and migration patterns. Our categories of human populations, sometimes termed races, are in fact socially constructed and arbitrary, without an empiric biological base. It may be possible to differentiate large populations from one another statistically based on sample morphology; but individual morphology frequently does not match the central tendency of any of our population reference samples, and population reference samples may or may not reflect social group membership, which can be a matter of personal choice or happenstance.

When the forensic anthropologist examines an unidentified body, the first task is to tease out a profile of biological and morphological characteristics from the details of the skeletal shape and size. The goal is to reduce the possibilities for group membership (age group, sex, or population) in order to increase the chances for identification. It is an exercise in statistics, sociocultural context, and judgment, and not at all exact. Since the goal is to place an individual within a modern sociocultural matrix, the methods should utilize comparable reference populations. The United States has only a handful of well-known skeletal collections (or reference databases) with reliable antemortem information, i.e., known age, sex, and ancestry. Each population has its limits. Data analysis on World War II and Korean War casualties, for example, tends to emphasize young males, mostly European-American and African-American, born in the first third of the 20th century. Some reference populations are American (such as the Terry Collection at the Smithsonian Institution) or European (such as the Hungarian fetal sample studied by Fazekas and Kosa). There are a few modern forensic collections and databases including, for example, the University of Tennessee Forensic Database.

It is essential to point out that skeletal morphology is produced by both genetic and nongenetic factors. This is true of both metric (continuous, measurable) traits and nonmetric (discontinuous—present or absent or graded) traits, making attributions of ancestry particularly problematic. Modern scientific methods of ancestry attribution have utilized various reference populations to develop statistical methods based on measurements or discrete trait frequencies. Each method focuses on a specific study population and develops a technique that works reliably within that study group. The application of the method to an unknown forensic case assumes the unknown individual fits the parameters of the study population, at least in general terms.

Many methods exist for forensic ancestry attribution. Metric statistical methods generally require that a prescribed set of measurements be made and plugged into a formula for **discriminant function** analysis. For example, the Gill method assesses midfacial flatness utilizing a specialized coordinate caliper called a simometer, includes several measurements of the upper face, and produces a numeric value that falls on either side of a sectioning point—thus, pointing to one of two populations being compared. Although ancestry assessment focuses on the skull, particularly the face, some methods have been developed for individual postcranial bones.

Observations of nonmetric traits can be used additionally (or alternatively). For example, Western European ancestry is frequently associated with facial morphology that includes a pinched nasal bridge and a narrow nasal opening, whereas in Asian populations and Native American populations, the nasal bridge is more apt to be flattened and the nasal **aperture** broader. Carabelli's cusp, a small additional cusp or pit on the maxillary molar teeth, is not unusual in Europeans, but

is rarely seen in Asians or Africans. However, no single trait should ever be used as the basis for a decision about ancestry. Rather, the entire complex of traits and metrics should be assessed. The level of experience of the forensic anthropologist with the regional population within which he or she practices is critical.

Estimated Stature

The estimation of living stature is not as straightforward as most people believe. First, stature is not a fixed trait. Adult stature changes (shrinks) from morning until night and over the course of the lifetime. The accurate measurement of living stature in an individual is actually a range. Secondly, stature measurement is frequently done incorrectly, resulting in fairly dramatic errors that can become incorporated into personal information. Missing person reports may include only a rough estimate of height. Interestingly, research has shown that American men tend to exaggerate their height on their driver's licenses by as much as 1 or 2 inches. Thus, the probability is fairly high that recorded statures will be inaccurate, particularly for males.

Stature can be calculated for decomposed or skeletonized remains using several reliable methods. First, if the body is still articulated, the length can be measured, keeping in mind that the loss of muscle support will loosen joints and lengthen the body somewhat. If a skeleton is **disarticulated** but the head, spine, pelvis, and at least one leg are present, the individual bone heights can be measured, and estimates made for vertebral disc thickness and total stature.

Allometry

Remains are often incomplete. In those cases, a forensic anthropologist will need to estimate stature by extrapolating it from the lengths of individual long bones, or combinations of long bones. This is done by using regression formulae developed for reference populations of known stature, mainly military casualties or modern forensic case databases. The ability to estimate stature from long bone lengths depends on the presence of patterned and proportional relationships between the sizes of body parts, a concept called **allometry**.

Allometric relationships between bone elements are systematic but not exact. They differ from population to population and from individual to individual. Individuals in any population may or may not conform to the central tendencies. Stature formulae usually specify a presumed ancestral population. They also require the application of particular measurements of skeletal elements. Forensic anthropologists are trained in **osteometry**, the measurement of the bones. Standards exist for how and where to measure each bone element. It is critical when applying stature formulae that the measurement is done exactly the same way as was done for the reference population. The resulting estimate is reported as a range.

Individuation and Identification

A description of one or more individual characteristics may help narrow the number of possible identities for an unknown set of remains. These may include evidence of medical conditions, congenital defects, handedness, and other markers of occupational stress. The body may have unique or unusual features that family and friends were aware of. An anthropologist may also observe features that may never have surfaced.

Congenital Anomalies

A **congenital anomaly** such as a cleft palate may be documented in a medical record or may have been apparent to others. Other types of anomalies such as an extra lumbar vertebra may not have been noticed.

Pathology

Antemortem medical conditions or pathology likewise may or may not have been known. For example, evidence of previous surgeries may include a medical apparatus, such as a pinned tibia, and the apparatus will probably have a unique identification number. On the other hand, evidence of bone inflammation or infection may or may not have produced clinical symptoms. Child or other domestic

abuse may be revealed by the combined presence of healed and partially healed patterned injuries.

Handedness and Markers of Occupational Stress

Despite suggestions in the popular media, the evaluation of markers of occupational stress is neither straightforward nor, in most cases, possible for most forensic cases. Individuals rarely maintain the same occupation throughout adulthood. Further, similar bone changes may be produced by more than one activity. Right- or left-hand dominance can often be ascertained, however, at least in individuals who exhibited that dominance behaviorally, and unusual or particular patterns of use-wear may suggest certain repetitious actions that can be helpful in narrowing identification possibilities.

Levels of Certainty

An anthropologist may contribute evidence toward making an identification, although the legal authority to make that identification rests with the medical examiner or coroner. Evidence often falls into one of three levels of certainty. The lowest level is "possible," where there is consistency with the circumstances, biological profile, and general information. The middle level is "presumptive," when the identification is more likely true than not: a "preponderance of the evidence" suggests identification. Presumptive or tentative identification requires consistency with witness testimony and one or more unique features, but not independent, scientific verification. A unique feature might be a specific disease or injury or congenital anomaly. This is the standard in the United States for civil matters. The third level is "positive" or confirmed identity, which is accomplished when objective characteristics of the remains can be independently verified by qualified experts. Positive identification requires matching information documented during life with information collected from the remains after death. Comparisons are usually made between fingerprints, dental records, x-ray or other medical imagery, or DNA. In the United States this is the standard for criminal matters.

Perimortem Trauma and Evidence for Cause and Manner of Death

The legal authority to determine cause and manner of death rests with the medical examiner or coroner. However, the anthropologist often contributes critical evidence for these determinations, particularly in the interpretation of skeletal trauma and judgments regarding the timing of the trauma.

Perimortem Injury

Human remains frequently exhibit signs of trauma, both antemortem (healed or healing prior to death) and perimortem (occurring at or around the time of death). By noting the presence of active or previous bone remodeling, e.g., the formation of a bony scar or **callus** at the site of a fracture, the anthropologist can assign a traumatic injury to the antemortem period.

Most remains examined by forensic anthropologists have an extended postmortem period, and are decomposed or skeletonized. Such remains may undergo postmortem modification by a wide range of agents, such as carnivores or transport by flowing water, which may damage bone. This taphonomic damage is not related to the cause of death and must be differentiated from perimortem trauma by noting the patterns of bone breakage in relation to moisture and fat loss, differential staining on fracture margins, and signature modifications of scavengers, plants, or geological processes.

Bone damage with no signs of healing, which apparently occurred when the bone was still fresh and for which a taphonomic cause can be ruled out, is described as perimortem trauma. However, it is not possible to be precise regarding whether bone damage occurred just before or just after death. Unlike soft tissue, bone does not exhibit a detectable vital reaction without several days of healing time.

Blunt and Sharp Force Trauma

Bone damage is conventionally divided into blunt or sharp force categories, depending on the presence or absence of cut surfaces. Blunt force damage produces impact marks or fractures, and can fragment bone. The force can be delivered at great speed (e.g., gunshot) or slower speed (bludgeoning by hand with a heavy object); these can often be differentiated by the amount of warping of the fragments, termed **plastic deformation**, which is more likely with slower loading. In both blunt and sharp force injuries, the pattern of impact scars, fractures, or cut marks can sometimes indicate object shape (e.g., hammer head), trauma type (e.g., frontal motor vehicle impact), or weapon class (e.g., single-bladed knife).

Gunshot Wounds—A gunshot wound is a special form of blunt force trauma. Firearm projectiles frequently create signature patterns in bone, particularly in the skull. An anthropologist will generally reconstruct a skull shattered by gunshot in order to evaluate the injuries. With many gunshot wounds, it is possible to differentiate the entrance from the exit by virtue of inward beveling on the former and outward beveling on the latter (Figure 7.9). Due to the relatively slower speed of the radiating fractures compared with the projectile speed, combined with the fact that fracture lines are halted upon encountering previous fracture lines, it is frequently possible to determine the sequence of multiple gunshot wounds.

Patterned Injury

In each case, an anthropologist looks at the pattern of antemortem and perimortem injury throughout the skeleton to render an

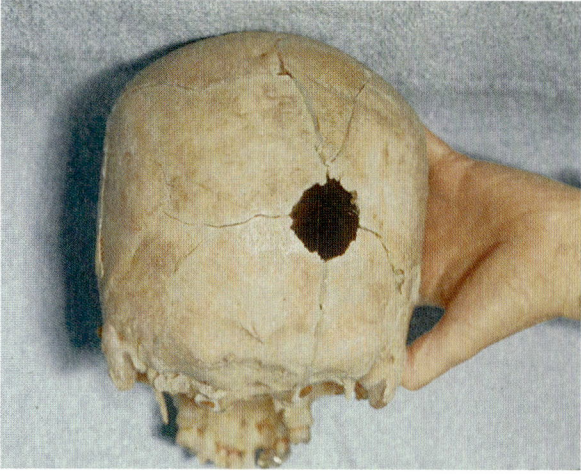

Figure 7.9 Example of a gunshot wound to the right posterior cranium viewed from the outside (left) and the inside (right). This entrance wound is beveled inward, with fractures that radiate away from the bullet hole under the stress of the projectile's high energy. The high speed of the applied force, or loading, is associated with an elastic response by the bone, generally producing fragmentation without permanent deformation or warping.

interpretation about cause, timing, trajectory, and weapon characteristics. Taphonomic damage must be excluded. With many suspected instances of trauma it is necessary to examine bone damage microscopically to rule out postmortem modification. It is possible for different agents of bone modification to produce similar types of damage, a concept known as **equifinality**. These mimics can generally be differentiated, however, by analyzing the pattern throughout the body rather than focusing on a single site.

Forensic Anthropology Documentation

Documentation and Testimony

A forensic anthropologist is held to high standards of procedure and documentation. The chain of custody must be clear. All examination procedures should be documented carefully. The forensic anthropology record should, in most cases, include a full range of anthropological measurements and determinations, although the formal report format requested in specific jurisdictions may be abbreviated. It is recommended that the basic forensic

anthropology report include four components: (1) taphonomy; (2) biological profile (age, sex, ancestry, stature, anomalies, pathology); (3) **individualization** characteristics and interpretation; and (4) evidence of perimortem trauma.

Expert Witness Testimony

The forensic anthropologist is often asked to testify as an expert witness in court, offering opinion testimony regarding his or her findings. It is essential that anthropologists uphold the principle of scientific neutrality and objectivity, regardless of whether they testify for the prosecution or the defense. Scientific certainty is generally restricted to levels of 90 percent or higher for determinations regarding biological profile, but the standard is much higher than 99 percent for individual identifications. Scientific probability is greater than 50 percent, generally between 67 and 90 percent. In many cases, the level of certainty is derived from the reference population with which a particular method is developed and tested. The appropriateness of the reference population, the method, and the conclusions may all be challenged in court.

Questions

1. What role does the forensic anthropologist play in a medicolegal death investigation? Differentiate the role of forensic anthropologist from those of the forensic pathologist and the medical examiner or coroner.

2. Recovery of human remains sometimes requires the skills of a forensic archaeologist. Name some of those skills and describe the situations in which those skills are likely to be applied.

3. The condition of the remains should be interpreted with reference to its taphonomic context. What does this mean and how does it affect the tasks of the forensic anthropologist?

4. When a forensic anthropologist generates a biological profile, what is that likely to include?

5. What factors may limit the ability of the anthropologist to attribute sex to a set of skeletal remains?

6. A child's long bones provide clues to age. What features of long bones are important in estimating age and why?

7. If bone growth and tooth eruption processes have ended, list several skeletal or dental features that can be used to estimate adult age. In very general terms, how precise are these methods?

8. The race concept is socially constructed and socially defined. How does this fact affect judgments the anthropologist might make regarding evidence of ancestry?

9. Stature can be estimated based on long bone length. What might limit the precision of these methods?

10. When a forensic anthropologist testifies as an expert witness, for which side might he or she testify? Explain your answer, including a discussion of the difference between science and advocacy.

References and Suggested Readings

American Board of Forensic Anthropology, http://www.csuchico.edu/anth/ABFA.

Bass, W. M., *Human Osteology*, 4th ed., Missouri Archaeological Society, Columbia, 1995.

Buikstra, J. E. and Ubelaker, D. H., Eds., *Standards for Data Collection from Human Skeletal Remains,* Arkansas Archaeological Survey, Fayetteville, 1994.

Cox, M. and Mays, S., Eds., *Human Osteology in Archaeology and Forensic Science*, Greenwich Medical Media, London, 2000.

Fazekas, I. G. and Kosa, F., *Forensic Fetal Osteology,* Akademiai Kiado, Budapest, 1978.

Galloway, A., Ed., *Broken Bones*, Charles C. Thomas, Springfield, IL, 1999.

Gill, G. W. and Rhine, S., Eds., *Skeletal Attributions of Race*, Maxwell Museum of Anthropology, Albuquerque, NM, 1990.

Haglund, W. D. and Sorg, M. H., Eds., *Forensic Taphonomy: The Postmortem Fate of Human Remains,* CRC Press, Boca Raton, FL, 1997.

Haglund, W. D. and Sorg, M. H., Eds., *Advances in Forensic Taphonomy: Method, Theory, and Archaeological Perspectives*, CRC Press, Boca Raton, FL, 2002.

Iscan, M. Y., Ed., *Age Markers in the Human Skeleton,* Charles C Thomas, Springfield, IL, 1989.

Iscan, M. Y. and Kennedy, K. A. R., Eds., *Reconstruction of Life from the Skeleton*, Alan R. Liss, New York, 1989.

Klepinger, L. L. *Fundamentals of Forensic Anthropology,* John Wiley & Sons, Hoboken, NJ, 2006.

Komar, D. A. and Buikstra, J. E. *Forensic Anthropology: Contemporary Theory and Practice.* Oxford University Press, New York, 2008.

Krogman, W. M. and Iscan, M. Y., *The Human Skeleton in Forensic Medicine*, 2nd ed., Charles C Thomas, Springfield, IL, 1986.

Reichs, K. J., Ed., *Forensic Osteology*, 2nd ed., Charles C Thomas, Springfield, IL, 1998.

Stewart, T. D., *Essentials of Forensic Anthropology*, Charles C Thomas, Springfield, IL, 1979.

Thompson, T. and Black, S., Eds., *Forensic Human Identification: An Introduction,* CRC Press, Boca Raton, FL, 2007.

Forensic Taphonomy

William D. Haglund and Marcella H. Sorg

8

Introduction

Taphonomy, as applied in forensics, deals with the history of a body after death. This is important in forensic death investigation because postmortem changes can affect estimates of time since death, the identification of the individual, and the ability to determine cause and manner of death. Taphonomy offers an organized body of information that deals with postmortem environments and the ability to better interpret factors that influence the preservation and degradation of human remains and other evidence.

The term taphonomy derives from the Greek taphos (burial or grave) and nomos (laws). This has been translated literally to mean the laws of burial. The history of taphonomy as a discipline finds its first application in paleontology where its goal was to explain "taphonomic histories," the environmental factors that affect organic remains between an organism's death and its final representation in the fossil record. Results of these efforts were then applied to the reconstruction of ancient environments. Archeologists used taphonomic data to better understand human behavior. A major line of their inquiry has been to determine human subsistence patterns; for example, were former populations scavengers or hunters?

In order to resolve such issues, it is critical to be able to sort out modifications of bones by humans such as butchering from changes made by animals or the physical environment. Other subjects of early taphonomic studies included weathering of bone, scavenging disarticulation sequences of mammalian carcasses, and transport of bodies by water. For an excellent overview of taphonomy, its history, research, and major topics of study, see Lyman's Vertebrate Taphonomy (1994).[1]

In the mid-1980s, the forensic literature had little to offer in addressing most taphonomic questions posed by death investigations. Very few references specifically covered "taphonomy." In both forensic articles and presentations, taphonomic issues had for the most part been treated anecdotally through case reports or based on limited numbers of examples. Surveys of larger numbers of cases involving human remains were limited, for example, to Warren,

Haglund et al. and Galloway et al.[2–4] The paucity of attention to taphonomy did not reflect a lack of interest on the part of death investigators, for they often are intrepid in their search for information that would lead to case resolution. The lack was a reflection of limited case experience.

The following overview of forensic taphonomy introduces the scope and utility of this rapidly growing field. A short introduction is given to contributions of method and theory. This is followed by a brief overview of major taphonomic factors and their influence with respect to human remains in a wide variety of environments.

Contributions of Methods and Theory

Forensic taphonomy requires interdisciplinary input of experts from many fields, potentially all the biological and geological sciences. Archeologists, anthropologists, entomologists, botanists, marine biologists, and geologists have made notable contributions. Forensic anthropologists have been prime movers of taphonomy's introduction to forensics, an advance that naturally paralleled the expansion of their roles in death investigation. Long recognized for their skills in identification of skeletal remains, forensic anthropologists now routinely aid in interpretation of trauma, locating and recovering human remains that are fleshed, decomposed, burned, and in other states of preservation.

Cross-fertilization between disciplines requires a meeting of minds—a process that often involves a learning curve on the part of all experts involved. Often it is necessary to realign discipline-significant priorities and determine the ground rules of working in a new context. Using archaeology as an example, a primary distinction of working in the medicolegal context is that "archaeological indecision, unlike forensic indecision, does not send people to prison."[5] The same can be said for other disciplines not usually applied in a forensic context. Medicolegal work requires familiarity with chain of custody issues and working in a legal context. Dealing with recent death and fleshed remains may present a considerable psychological hurdle to an expert accustomed to working with skeletons. Terminology applied in one disciplinary context may require redefining in another. For instance, terms such as *stab* or *cut*, when bone alone is under discussion, may have different implications to the pathologist who routinely deals with soft tissue.[6] Such concerns are the subjects of an excellent group of papers in a special edition of *Historical Archaeology* subtitled *Archaeologists as Forensic Investigators: Defining the Role*, edited by Connor and Scott.[7]

The thoroughness with which a scene is processed bears upon issues such as identification of the deceased, assessment of the time since death, and cause and manner of death. It may also aid investigators to develop suspects. Use of archaeological techniques in the recognition, search, recovery, and documentation of remains and related materials has proven invaluable to investigations. The utility of archaeological recovery has been demonstrated for various contexts in which human remains are found including remains that are on the surface, buried, and burnt.

A host of techniques have been borrowed from other disciplines and applied in forensic settings. These include three-dimensional mapping, basic principles of stratigraphy, botanical and entomological collection procedures, and techniques of conservation and exhumation techniques. Scott and Connor[8] applied computerized mapping of ballistic evidence as an aid in reconstructing human behavior in the archaeological site of the Battle of Little Bighorn and in forensic investigations of the Koreme execution site, an aftermath of the Iraqi government's 1988 offensive against the Kurds. Other specialized knowledge borrowed from archaeology has dealt with the degradation of organic materials, clothing, and other dress materials.

Field research and treating human remains as part of a complex environment has put forensic analysis of human remains on a theoretical footing. A focus on context underscores that a vast array of physical and biotic factors must be taken into consideration to fully appreciate the condition in which human remains are discovered, and also makes it apparent that the dynamics of postmortem change vary drastically in specific habitats with different biogeoclimatic regimes. This recognition has encouraged retrospective studies and research that serves to increase confidence and credibility of case interpretation.

Taphonomic Factors

Forensic taphonomic history includes the actual death event, bone exposure through loss of soft tissue, and bone modification, including the processes of recovery and curation. Of particular interest in medicolegal death investigation is the **perimortem interval**, the timeframe that includes death and often also includes soft tissue loss and bone exposure and before the loss of fats and moisture from bone. This occurs prior to the **postmortem interval**, during which bone is exposed to modifying agents. These intervals may potentially overlap and lack distinct boundaries. Another interval crucial to forensic death investigation is the elapsed time between death and recovery of remains.

Perimortem Interval

Estimating the timing of injury, specifically discriminating antemortem from postmortem injuries to bone, is particularly problematic for the forensic anthropologist. To distinguish these periods, we must be able to determine the conditions under which certain taphonomic processes come into play. Because these processes are not precise, the boundary between life and death becomes blurred when viewed retrospectively. The circummortem or perimortem interval is an ambiguous period into which are lumped our inabilities to distinguish antemortem from postmortem occurrences. Technically, time since death may include portions of the perimortem interval.[5]

To illustrate, the reaction of a bone to stress is dependent upon its moisture and fat content. Fractures to bone that are considered definitely perimortem are characterized by evidence of this moisture and grease content, such as the presence of greenstick or spiral fractures. Definite postmortem fractures are characterized by clean, brittle breaks, frequently parallel or at cross section to the long axis of the bone. Moisture loss occurs over time, and there is no distinct point after which it can be said to have occurred. Undoubted **antemortem** (prior to death) fractures, on the other hand, are characterized by healing.

Postmortem Interval

Estimating the postmortem interval can be similarly imprecise. In order to more accurately determine elapsed time since death, the observations (traces) used to mark the passage of time from death to recovery need to be specified. Certain processes lend themselves to partitioning as stages of postmortem change, including decomposition, carnivore-assisted disarticulation, and weathering. Such stages are conventions that artificially "freeze frame" a process by defining an array of observable attributes that mark its presence. It is ideal to know the triggering event for a particular process and the variables that regulate or influence the rate of movement through that process, as well as to be able to measure the rate of change.

Unfortunately, these processes are complex, potentially overlapping, context-specific, and are not well related to absolute time intervals. Although one may be able to judge where a set of remains falls within a sequence of stages, it is almost never possible to assign a narrow time interval. For example, in order to estimate postmortem interval for skeletal remains exposed outdoors, it is necessary to make judgments about the time needed for decomposition. Depending on the process chosen to measure time passage (e.g., temperature), it is ideal to know the triggering events of that process and be able to measure the rate and the variables that regulate the rate.

Converting "traditional taphonomic time" to "forensic taphonomic time" involves determining the triggering event for the taphonomic phenomena being observed, that is, the amount of elapsed time after death required to reach that triggering event, and the time necessary to reach the observed stage of the taphonomic phenomenon observed at the time of bone recovery. To illustrate, consider bone weathering. Exposure of bone to the atmosphere is the triggering event that initiates weathering of bone. In order to determine time since death, an estimation of the time it took the bone to be defleshed and exposed must be added to the time it took for the bone to reach the weathering stage observed at the time of its recovery.

Many processes can alter the condition of human remains. A basic perception is that remains simply decompose, skeletonize, and disarticulate, but the reality is more complex. During the course of these processes, body parts (and associated hair, fiber, or clothing evidence) may be modified, moved, preserved, or destroyed. How these processes occur is not always simple. They are dependent on an array of taphonomic factors. Taphonomic factors can be categorized as individual, cultural, or environmental, with the disclaimer that they are not mutually exclusive.[9] Each deceased person brings individual factors to the taphonomic context. They include the individual's age, body size, cause of death, and the physical properties of bone. Physical properties, for example microscopic structure and density, in turn influence how bone weathers, decomposes, is scavenged, or reacts to trauma.

Cultural factors are those attributed to human activity and include a range of intriguing topics. Among these are autopsies as well as funeral practices, including embalming, casketed burial, cremation, and burial at sea. Alteration of bones resulting from unusual circumstances, for example, United States war casualties in Vietnam and influences of cultivation on preservation and location of buried remains, have been studied. Aberrations of human activity resulting from dismemberment, human rights abuses, war, and acts of genocide also affect the fate of human remains.

Environmental factors are controlled by temperature, moisture content, pH, and properties of soil. Temperature is a primary regulator of the rate of chemical changes such as decomposition. Extremes of temperature may range from fire to freezing and thawing. As a rule, a chemical reaction doubles in rate for every 10-degree rise in temperature. Conversely, it halves for every 10-degree drop in temperature. Above certain temperatures, proteins become denatured, while at freezing temperatures normal chemical reactions slow to become practically negligible.

Carcass temperature equilibrates with that of its environment. Cooling is hastened if a body is exposed to a draft or a breeze, or is nude. Loss of blood or removal of body parts also accelerates temperature loss. A body holds its temperature longer if it is protected or clothed.

Atmospheric temperature may represent only a crude indication of temperature conditions to which various parts of a body are subjected. Different parts of a body may reflect a mosaic of temperatures, dependent on exposure to sun, shade, and the microenvironmental factors to which those parts are exposed. For example, air temperatures fluctuate more widely, and change more rapidly than temperatures beneath objects such as a body. Coe[10] compared temperatures of open ground surfaces, rocky surfaces, and forest areas for the same time period and found that temperature near the surface of a carcass has more influence on decomposition than air temperature. Microenvironments, such as the temperature of an accumulating maggot mass, affect rates of decomposition and insect consumption.

Obviously whether the body is deposited on the surface or in water or is buried will determine the changes it undergoes and the rates at which these changes take place. To illustrate, total skeletonization of human remains has been reported to occur in Florida as soon as 2 weeks. The same process may take 2 to 3 years in more northern climates. Buried human remains may retain flesh for decades, as noted in the investigation of the Nazi concentration camp at Belzec, Poland,[11] or even centuries.[12] Such disparity underscores the complexity of decomposition and the significance of the specific environmental contexts in which it occurs.

Animals and Human Remains

The most commonly encountered scavengers of human remains are insects, carnivores, rodents, and microbes. Insects are the champion scavengers, not only because of their ubiquitous presence and voracious destruction of tissue, but paradoxically for their utility in demonstrating that bodies have been moved or stored, the presence of drugs, the season of death, and locations of trauma. Their precise contributions as "biological time clocks" in determining the postmortem interval has endeared them to experts and captured the imagination of the lay public. The relationship between insect activity and estimates of the time since death is based on the maturation stages of flies and population dynamics of successive insect species associated with a carcass. The presence of two major orders of insects, **Diptera** (flies) and **Coleoptera** (beetles), have dominated studies in forensic entomology. Some of the best examples of taphonomic research have been in the field of forensic entomology.

Life cycles of flies are species-dependent and the rate of development is highly sensitive to temperature conditions, hence geographic location. In a generalized life cycle, adult flies lay eggs (ovadeposit) near a food source. For human remains, the most common locations are the moist mucosa of the body orifices, such as the eyes, mouth, and external nares (nostrils). Exposed genitals or axillary regions are often locations of egg deposits, as are open wounds. (Some species of flies develop their eggs internally.) Dependent upon the species of fly and ambient temperature, the eggs hatch into the first of three larval stages or instars within hours. Fly larvae are referred to as maggots. As the maggots feed and grow, the instars are separated by a molting or shedding of the old exoskeleton to accommodate their increase in size. Following feeding, they migrate and their outer exoskeletons harden into casings or puparia of the **pupal stage**. During this stage of several days, a metamorphosis takes place and an adult fly emerges.

Later in the postmortem interval, other insect species arrive at the carcass. The species succession within the community profile is used to estimate time since death. Insect succession is strongly impacted by geographical region, habitat, and season. Beetles become major players and are mostly involved with feeding on eggs and larvae of flies. However, sometimes they feed directly on the decomposing body.

Estimations of the rate of insect development also depend upon accurate documentation of the context and environment from which the samples are taken. The input of experts is often needed in the recognition, collection, and preservation of insect evidence, including eggs, larvae, pupae, and adults.

Eggs are laid in clumps or masses. They are usually found in wounds or natural orifices, but may be found on clothing. Eggs are only of value when no maggots or later insect stages are present. They can be collected with an artist's paintbrush dipped in water or with forceps. One half of the sample should be preserved in 75 percent ethyl alcohol or 50 percent isopropyl alcohol and the other half placed in a vial with a little damp tissue paper to prevent dehydration. Because eggs may hatch and newly emerged maggots can escape through small holes, a paper towel held over the top of the vial with a rubber band is useful, as long as the vial stays upright. No lid other than the paper towel is needed. This form of cover will also allow air to enter the vial. If more than a few hours will pass before the eggs reach an entomologist, a small piece of beef liver should be placed in the vial in the event the eggs hatch. If liver is added, tissue paper or sawdust should also be placed in the vial for insects to use as a refuge to prevent drowning.

Maggots will be found crawling on or near the remains or may be found in masses. A range of sizes should be collected. One half should be killed and preserved immediately to demonstrate in what stage the larvae were when collected. If many maggots are present on the body, preserve approximately half of all sizes collected. One half of the specimens can be killed by immersing them in hot water for a few minutes. If no hot water is available, they can be placed directly into a preservative of 75 percent ethyl alcohol or 50 percent isopropyl alcohol. The other half of the sample should be kept alive. This sample should contain about 100 maggots (of each size if possible). Living specimens should be placed in a vial with air and food (preferably beef liver). Only enough maggots to cover the bottom of the vial should be present. A small piece of paper towel in the vial will help prevent drowning.

Pupae and empty pupal cases are often found in clothing, hair, or soil near the body. Pupae like dry, secure areas away from the wet food source in which they pupate. Pockets, seams, and cuffs are likely hiding places. If the remains are found indoors, the pupae may have traveled some distance and be under clothing, rugs, or boxes. Both hatched and unhatched pupal casings should be collected. They should be placed in a vial, not in a preservative. They need air, so secure a paper towel over the vial as adult flies may emerge during transit.

Adult flies can be collected by net or by using an inverted vial. They can be left in the vial without air. Insect specimens are fairly fragile, and are best picked up with gloved fingers. Tiny or delicate specimens can be picked up using an artist's paintbrush dipped in water or alcohol depending on whether they are to be kept alive or killed. Documentation of insect specimens should include the site of collection, date and time of collection, name of the collector, and stage of development, for example, egg, larvae, or adult. If maggots are collected from a maggot mass, the temperature and estimate of the size of the mass should be included (Figure 8.1). Make sure all the vials are well sealed. The insects should be taken to an entomologist as soon as possible. They should be sent by courier or hand delivered to maintain continuity of custody.

Of mammalian scavengers, scavenging by **canids** (members of the dog family) and rodents receives the majority of attention. There have also been reports of scavenging by hogs and bears. The type and magnitude

Figure 8.1 Metabolism of maggots in a "maggot mass" produces heat that can be significantly higher than the ambient temperature. Adequate documentation requires an estimate of the size and temperature of the mass.

Figure 8.2 Coyote-scavenged remains. V-shaped tooth marks along soft tissue margins are hallmarks of the canine teeth of members of the dog and cat families.

of their effects depend on species, size, and behavior. Behavior is influenced by season, food supply, and competition with other animals. Mammalian scavengers impact forensic investigations by virtue of their of chewing, disarticulation patterns, and dispersal of human remains.

Various animal groups produce characteristic tooth marks and scavenging patterns. The canine teeth of the dog and cat families leave v-shaped defects in soft tissue (Figure 8.2) and bone. The soft tissue margins of eaten areas are irregular. Often, defects are undermined because of tugging and pulling that leave loose flaps of skin rimming the areas of damage. Collateral damage to unconsumed

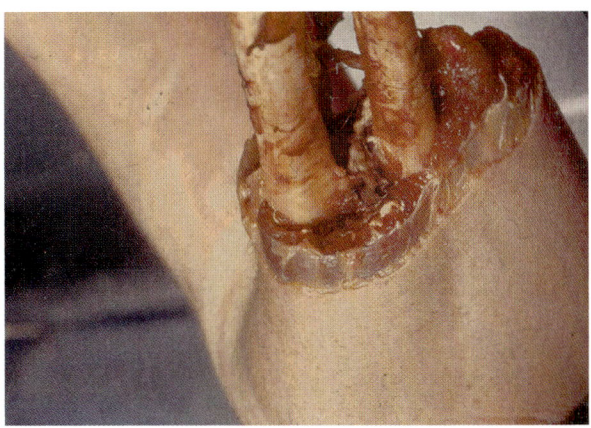

Figure 8.3 Rodent damage to soft tissue showing the "dainty" pattern progressing neatly through various levels of tissue.

Figure 8.4 Rat incisor marks at the superior margin of the left eye orbit demonstrate characteristic parallel incisor grooves.

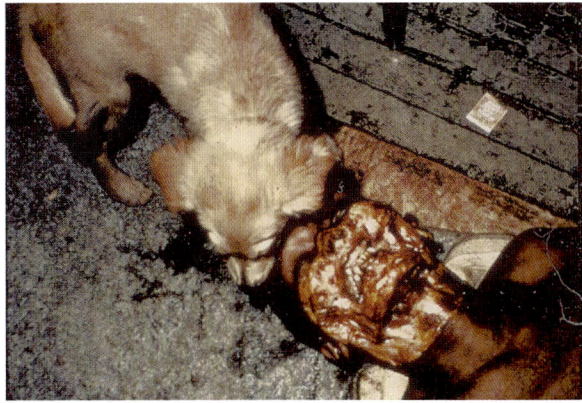

Figure 8.5 Scavenging by medium-sized canids (members of the dog family) is usually patterned and is initiated in the area of the face.

areas may involve scratch marks from claws. In contrast, the defects caused by rodent incisors in flesh tend to be tightly circumscribed with relatively even margins (Figure 8.3). The damage pattern is "dainty," progressing neatly through various levels of tissue. Rodent incisor damage to bone usually leaves characteristic parallel incisor grooves (Figure 8.4). Other evidence left behind by scavengers includes scat (feces), tracks, and hair.

Under normal conditions, medium and larger canids approach consumption in a patterned sequence, usually beginning at the head, removing soft tissue and scalp (Figure 8.5). The upper limbs, including the pectoral girdle are removed when the clavicles are chewed through. Next is evisceration of the thoracic, abdominal, and pelvic cavities with accompanying damage to the anterior bony portions of the thorax (Figure 8.6). Finally lower limbs are disarticulated. This general pattern is influenced by several factors: size and number of the animals scavenging, season of the year, and whether scavengers are prevented from having full access to the remains. As body parts are removed, they may be taken away and dispersed, sometimes over wide areas. Understanding the sequence of disarticulation and dispersal of remains allows a more informed search strategy to be implemented. Predation and scavenging patterns of various animal groups such as dogs, cats, and bears differ.

Plants and Human Remains

Plants, including algae and fungi, can contribute potentially complex environmental nuances to buried remains. As a result of seasonal shedding of leaves and needles, plant debris can cover remains. Roots can cause mechanical damage to the remains, particularly to bones. They can invade burial enclosures such as caskets, coffins, or vaults. "Root etchings" indicate that a bone has been in a plant-supported sedimentary environment for at least part of its taphonomic history (Figure 8.7). These dendritic impressions on bones have been attributed to the excretion of humic acid on bone surfaces. Others have

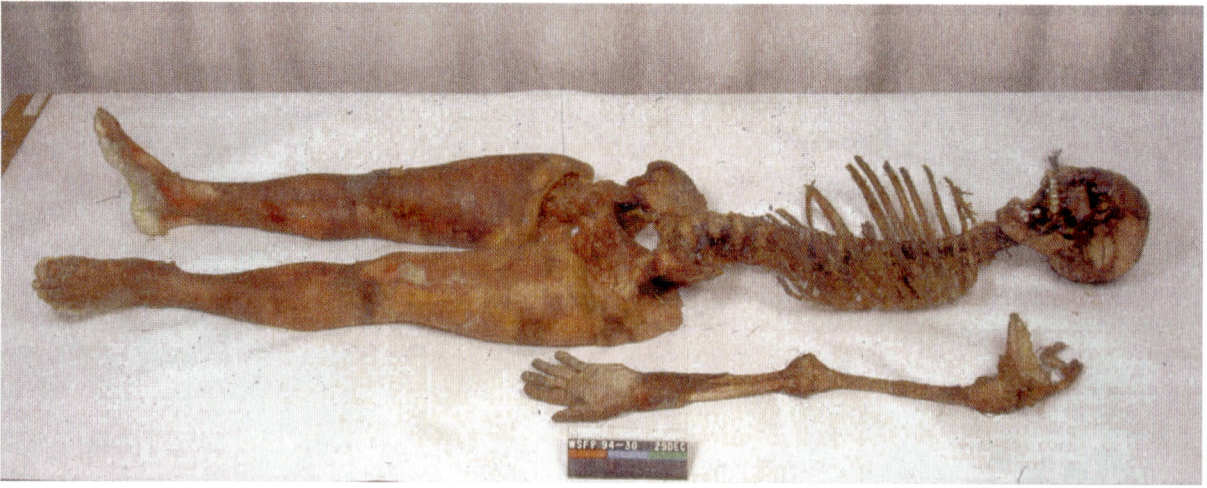

Figure 8.6 Farther along in the sequence of canid scavenging, the upper limbs, including the pectoral (shoulder) girdles, are removed and the thoracic and abdominal cavities eviscerated.

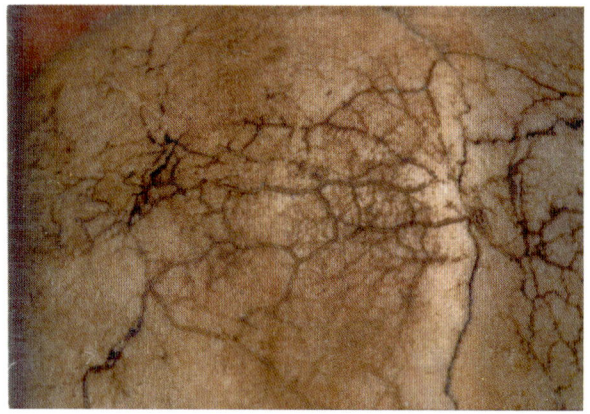

Figure 8.7 "Root etchings" caused by contact of plant rootlets on bone.

attributed root etchings to decomposing plants and to acids secreted by root-associated fungi.

Rootlet proximity to bones also introduces microenvironmental changes by affecting fluctuations in moisture and through the introduction of chemicals, fungi, and microorganisms.[13] The rhizosheath, a thick soil cylinder that surrounds plant roots, appears to be an adaptation to moisture conservation, but also provides an environment for extensive root–microbe interaction. Its size (extent) depends on the root structure. Plant roots and microbial populations contribute to rhizosphere environment. In turn, microbes produce plant growth factors and affect microbially mediated availability of mineral nutrients. There is a higher proportion of Gram-negative,

rod-shaped bacteria, and a lower proportion of Gram-positive rods, cocci, and pleomorphic forms in the rhizosphere than in root-free soil. The roots affect interactive modification of the soil of the rhizosphere by water uptake and release of organic chemicals to the soil. Plants release a succession of materials during their maturation. Initially, carbohydrate exudates and mucilaginous materials are released.

As with entomological evidence, plants require special considerations for collection and handling.[14] An easy way to preserve plant materials such as leaves and flowers is to press them between sheets of newspaper or a telephone book, catalog, etc. If placed directly into paper bags, they will easily fragment when dried. Small plant fragments can be carefully wrapped in folded papers. Fluid-soaked plants can be placed in plastic containers and refrigerated if the period between collection and analysis is short. Watery plants, such as succulents or juicy fruits, can be sliced or placed whole in plastic as long as they are wrapped in waxed paper.

Botanical evidence has been used to determine time and season of death and prior locations of the remains in cases where the body has been relocated. Various plant structures such as roots, stems, branches, leaves, fruits, or flowers may be of evidentiary value. It is essential that a botanist knowledgeable about plant taxonomy is involved in botanical analysis.

Human Remains in Water Environments

The fate of human remains in water is less well understood than remains that are buried or found on the ground surface. In part, this is due to the variety and complexity of water environments, such as rivers, lakes, sounds, oceans, and even bogs. The variability presented by local habitats of the same body of water often defies generalization. To date, major topics of study have been the decomposition, disarticulation, and movement of remains and strategies for search and recovery. Freedom of movement of remains in water presents frustrating questions to forensic investigators. As remains disarticulate, they may leave parts behind. If remains are found beached or floating on the surface, the question is where did the remains come from? If the victim was known to have disappeared in water, where did the body go?

Both oceanographic models and field data from forensic death investigations may set limits for searches, determine potential origins for floating or beached remains, and correlate information regarding points of water entry and sightings of remains. One novel approach is the use of floating debris from historical shipwrecks combined with known data from contemporary forensic cases to establish a predictive framework to help resolve origins and trajectories of human remains. This technique has been applied to remains found in the Pacific Northwest's Puget Sound. Of particular interest was the influence of storms on the reversal of normal surface currents.[15]

Nawrocki et al.[16] drew from case experience and seminal taphonomic literature from the disciplines of archaeology, geology, and paleontology to examine transport of human remains in rivers and the influences of decomposition, disarticulation, and sorting. They commented on unique patterns of damage to crania caused by fluvial transport. Brooks and Brooks[17] examined the influences of river floodwaters on the transport of and damage to a cranium. They remarked on differentiating taphonomic modifications of skeletal material versus modifications produced by the trauma that caused death. O'Brien[18] pointed out that movement of bodies in Lake Ontario is affected by season, wind, temperature, depth, and current. In a retrospective survey of remains recovered from California's Monterey Bay area, Boyle et al.[19] assessed the impacts of environmental factors and how they compared to typical sequences of postmortem changes in **aqueous** environments. The above studies highlight alterations to what might be assumed to be typical patterns when examined in the light of different locations.

Buried Remains

Buried remains comprise a relatively small number of domestic forensic cases in contrast to the recent experiences of investigators with human rights abuses, war crimes, and genocide where burial is the predominant environment of recovery. A major problem presented by burial is the difficulty of locating a grave. An array of techniques can be applied to grave location, including witness statements, visual clues, probing, trenching, area photography, cadaver dogs, and remote sensing. Excavation and recovery of buried remains are best accomplished by persons with expertise in archaeological techniques.

As a rule, buried bodies decompose at a slower rate than bodies found on the surface. The condition of buried remains is influenced by many factors. Temporal factors include those of the preburial and postburial periods. For example, the longer the preburial interval, the more vulnerable to exposure are the remains to scavengers and insects. Insects, if allowed to lay their eggs, can continue development and degradation of the remains after burial.

In deeper graves, temperature is relatively stable; shallow graves are susceptible to fluctuations that reflect surface temperatures. The vulnerability of shallow subsurface layers to warming is dependent on the amount and duration of exposure of the grave to the sun, the type of surface exposed, and its capacity to absorb heat. In graves deeper than 2 or 3 feet, temperatures remain relatively stable.

Compaction and depth of burial allow or prevent encroachment of the grave by scavengers. Except for shallow graves or those in poorly compacted soil, burial offers protection from scavengers. Compaction may also account for distortion of bone that can both mimic pathologies and distort statistical measurement. Subtle warping is often not detected unless all fragments of a skull are present and incorporated into a reconstruction. The presence of imperfections or gaps between articulating fragments would indicate distortion has occurred.

Preservation is poorer in acidic than alkaline soils. As bodies decay, soil surrounding the remains becomes more alkaline. The differences in **pH** of different types of leaves and needles and other ground covering contribute to the pH of the underlying substrate.

Moisture can alter the character of decomposition. Moisture may or may not be trapped in the substrate, depending upon porosity and percolation characteristics of the soil. Percolating ground water may transport acids and other chemicals into bone or affect leaching of fluids from a decomposing body and minerals from bones. At the molecular level, water hydrolyzes collagen proteins into smaller polypeptide units, disrupting the protein–mineral bond and leading to more rapid degradation of bone tissue.[20]

Moisture in and around the grave environment may originate from one or more of several sources and can be moderated by differences in relative humidity, mean annual precipitation, and drainage. A body will provide a major source of moisture via body fluids and the products of decomposition. Within the core of a mass grave, these fluids may become trapped in sequestered spaces between individual remains and sealed by bodies tightly pressed together. At the periphery of the grave, such fluids drain away into the surrounding matrix, thus enhancing the decomposition process. A major dynamic of mass graves is that remains at the periphery of the body mass are less well preserved than those toward the center (Figure 8.8). Excess water also confounds the quality of exhumation.

Figure 8.8 Skeletalized remains from periphery of mass grave, Ovcara, Croatia, 1996.

Case Studies

Case 1

A 22-year-old female homicide victim's remains were discovered in April in a rural area of woods and thick underbrush 2 months after her disappearance. Her killer had severely beaten her about the face and head. Blood and a maxillary left central and lateral incisor were found in the trunk of the suspect's vehicle. By the time the remains were discovered, (Figure 8.9) much soft tissue decomposition

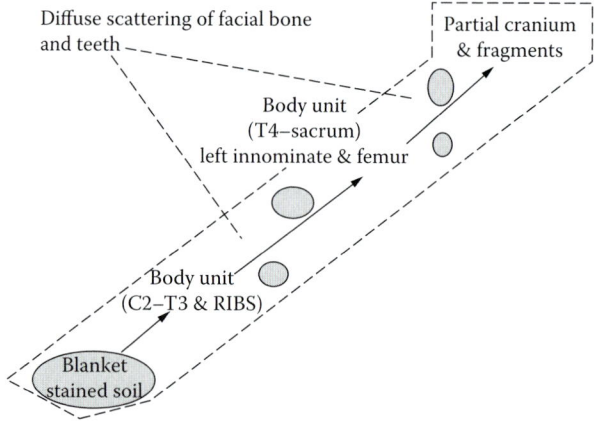

Figure 8.9 Location of human remains in a case involving extreme antemortem blunt trauma to the skull. Cranial fragments and disassociated teeth (darkly shaded areas) were dispersed along the trajectory (arrows). (From Haglund, W.D. and Sorg, M.H., *Forensic Taphonomy: The Postmortem Fate of Human Remains,* CRC Press, Boca Raton, FL, 1996.)

had occurred and the remains were scavenged and scattered by coyotes. The original body deposition site was marked by a blanket, clothing, and darkly stained, odor-laden soil. The extremes of scatter were marked by the decomposition site at one end and at the other end by a partial cranial vault that was missing the face from below the superior orbital margins. Two major articulated body units were present. The first body unit included the partially skeletonized and scavenged upper trunk inclusive of the second cervical vertebrae through the third thoracic vertebrae through the sacrum, left innominate bone, and femur. All remains were located along a 240-foot (61.5-meter) linear trajectory. Numerous cranial fragments were littered along this path. Where were the missing teeth?

A special circumstance complicates this case. Extensive antemortem fragmentation of facial bones and teeth was caused by blunt force trauma to the face. Because of this and the fact that scatter took place while decomposition was in progress, cranial bone fragments, including teeth and tooth-bearing bones, were potentially scattered between the extremes from where the body was deposited and where the mandible and farthest portion of the cranium were found. Indeed, in the weeks following the initial scene processing, further searches of the area were made separately by medical examiner and law enforcement personnel and family members. All teeth were eventually recovered along the path where they had fallen away from the mandible and cranium as they were moved. Careful screening of the path between the original site where the body rested to the section of cranial vault at the time of initial scene processing would have saved considerable time and embarrassment for law enforcement and unnecessary distress to surviving family members.

Case 2

A 27-year-old male was discovered approximately 3 days after his death in a dilapidated wooden shack. He was fully clothed. His head and thorax were inside a plastic garbage sack tucked into his belt. Both upper extremities were within the plastic bag. Nestled in the

deceased's left arm was a small propane tank with the valve in the "on" position. At the time of discovery, the plastic bag had been torn open by rodents, allowing them access to the upper portion of the body. Rat droppings were found on the chest and inside the bag. Rodent hairs adhered to exposed muscle tissues.

All soft tissue of the face and neck was absent. Also absent were both eyes and the soft tissues of the left temporal, frontal, and parietal areas. Both forearms were completely skeletonized. The metacarpals of the right hand were completely exposed and most digits were absent (Figure 8.10A). The bones of the left hand were completely absent. Damage to the thin cortical layers of bone of the fingers demonstrated no telltale parallel stria. Margins between rodent-damaged areas and nonaffected areas of soft tissue of the forearm exhibited scalloped edges, giving a finely-serrated appearance. Damage to soft tissue took place in a layered fashion with distinct differential destruction to the skin and underlying adipose tissue and muscle (Figure 8.10B).

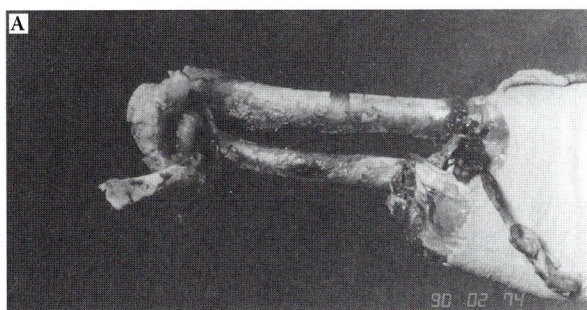

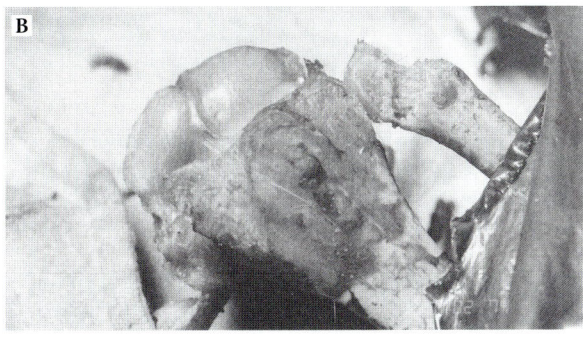

Figure 8.10 Rodent modification of soft tissue. (A) Layered rodent damage to forearm. (B) Removal of tissue from face and mandible. (From Haglund, W.D. and Sorg, M.H., *Forensic Taphonomy: The Postmortem Fate of Human Remains*, CRC Press, Boca Raton, FL, 1996.)

The Future

Forensic taphonomy is an exciting and rapidly evolving field. The complexity of factors that influence the fate of human remains during the postmortem period demands an interdisciplinary approach. Existing literature holds a wealth of information and numerous ways of looking at the postmortem issues presented by forensic cases and underscores the need for an ecological framework of questioning and for a detailed understanding of the idiosyncrasies of local environments.

Marshall[21] itemized five barriers to the progress of research in his overview of bone modification and taphonomy. They include:

1. Lack of standard nomenclature
2. Dearth of comparative case studies
3. Unsynthesized and scattered data sets
4. Limited data sets
5. Researchers who lack broad knowledge bases appropriate for this multidisciplinary field

These barriers affect forensic taphonomy. However, forensic anthropologists can make a real contribution in offering well-documented comparative case studies and by working to alleviate the other problems in their own research. In particular, forensic taphonomists should be closely examining existing taphonomic models with an eye to refining them to our foreshortened temporal needs and to control for ideographic variation. We need to look more closely at our use of terminology, for example, in descriptions of fractures, in what is meant by adipocere and saponification, and how we depict the condition of remains. We need to encourage the concept of collaborative case assemblages for research purposes. This might overcome some of the hobbling effects of low case volume and limited examples of particular types of cases in individual case loads.

The development of shared research paradigms is imperative. If a research plan is in place, data can be collected routinely in the normal course of forensic investigation without interfering with the medicolegal process. If data collection strategies are shared among practitioners, the data sets will be more comparable and broader based.

An optimistic forecast for taphonomy was expressed by Clyde Snow in his discussion of decomposition and the dispersal of evidence in Cattone and Standish's *Outline of Forensic Dentistry*[22]: "Eventually, as data accumulates, it should be possible to devise more efficient search procedures for the recovery of evidence and also improve our estimates of the time of death by a fuller understanding of the taphonomic factors involved." We share their hope that these and other taphonomic research goals can be achieved through careful work and cooperation as this new field develops

Conclusion

Taphonomy offers a body of knowledge particularly applicable to forensic questions. In turn, forensic cases offer human-specific models for which many variables affecting death assemblages may be assessed. Relative to archaeological and paleontological assemblages, forensic skeletal cases potentially provide refined resolutions of the variables immediately subsequent to death and allow actualistic settings for cross-sectional taphonomic observations. Such actualistic observations can provide needed understanding of the taphonomic pathways to which archaeological human remains have been subjected.

Both taphonomy and the forensic sciences are multidisciplinary. Because of this, there is no tidy way to limit and focus their union. Nevertheless, we think that forensic anthropologists, because of their links with both human biology and archaeology, are in the best position to lead the way. We advocate an expansion in the practices of forensic anthropology and forensic pathology to include more involvement in outdoor scene processing. The idea that methods from archaeology are useful in forensic investigations is augmented to suggest the models and approaches from taphonomy be integrated as well. Archaeology has been a leader among the historical sciences in the development of methods of searching for and collecting data in outdoor settings in both two- and three-dimensional contexts. Highly systematic methods are indeed finding increased acceptance in forensic scene processing. When a forensic anthropologist receives skeletal remains in the laboratory, she or he must infer the context of recovery. This inference is one step removed from the original context, and information will be lost. Indeed, any recovery process will bias the data collected at the scene, data that may be critical to the interpretation of the postmortem interval or other issues. Thus, participation in recovery is critical.

In conclusion, we propose that the theoretical perspectives from taphonomy are necessary to improve the interpretation of individual biology and history within the context of depositional environment, but we must go farther. In order to assure that such involvement will actually result in improved investigations, we must take steps to improve the rigor and reproducibility of our techniques. This involves increased focus on both diachronic and synchronic data collection within forensic casework, as well as attention to experimental research.

Acknowledgment

For much of the section on insects and human remains, I extend my gratitude to forensic entomologist Gail Anderson, Ph.D., of Simon Fraser University, British Columbia, for her excellent handout, *Forensic Entomology, The Use of Insects in Death Investigation.*

Questions

1. Name two major disciplines contributing to forensic taphonomy?
2. What adjustments in thinking are required of experts not accustomed to working in the medicolegal context?
3. List five cultural factors that affect decomposition.
4. How does scavenging influence scene investigations?
5. What is the typical pattern of canid scavenging and how might it affect search strategy?
6. Describe the general development of flies and their significance to forensic issues.
7. What important information can be obtained from botanical evidence?
8. What are the major topics of study relating to bodies in water?
9. What techniques and/or evidence are used to locate graves?
10. Discuss the application of archaeological techniques in the recovery of human remains from a forensic context.

References

1. Lyman, R. L., *Vertebrate Taphonomy,* Cambridge University Press, Cambridge, 376, 1994.
2. Warren, C. P., Verifying identification of military remains: a case report, *J. Forens. Sci.*, 24, 182, 1979.
3. Haglund, W. D., Reay, D. T., and Swindler, D. R., Canid scavenging/disarticulation sequence of humans in the Pacific Northwest, *J. Forens. Sci.*, 34, 587, 1989.

4. Galloway, A. et al., Decay rates of human remains: in arid environments, *J. Forens. Sci.,* 34, 607, 1989.
5. Haynes, G., Foreword, in *Forensic Taphonomy: The Postmortem Fate of Human Remains*, Haglund, W. H. and Sorg, M. H., Eds., CRC Press, Boca Raton, FL, xvii, 1997.
6. Symes, S. A. et al., Taphonomic context of sharp-force trauma in suspected cases of human mutilation and dismemberment, in *Advances in Forensic Taphonomy: Methods, Theory, and Archaeological Perspectives*, Haglund, W. D. and Sorg, M. H., Eds., CRC Press, Boca Raton, FL, 2001, p. 403.
7. Connor, M. A. and Scott, D. D., Archeologists as forensic investigators: defining the role, *J. Historical Archaeol.*, 33, 57, 2001.
8. Scott, D. D. and Connor, M. A., Context *delecti*: archaeological context in forensic work, in *Forensic Taphonomy: The Postmortem Fate of Human Remains,* Haglund, W. H. and Sorg, M. H., Eds., CRC Press, Boca Raton, FL, 1997, p. 27.
9. Komar, D. A. and Buikstra, J. E., *Forensic Anthropology: Contemporary Theory and Practice,* CRC Press, Boca Raton, FL, 2008.
10. Nawrocki, S. P., Taphonomic processes in historic cemeteries, in *Bodies of Evidence: Reconstructing History through Human Skeletal Remains,* Grauer, A., Ed., Wiley-Liss, New York, 1994.
11. Coe, M., The decomposition of elephant carcasses in the Tsavo (East) National Park, *J. Arid Environ.*, 87, 96, 1978.
12. Kola, A., *Belzec: The Nazi Camp for Jews in the Light of Archaeological Sources, Excavations 1997–1999*, Agencja Wydawnicza MakPrint, Warsaw, 2000.
13. Reeve, J. and Cox, M., Research and our recent ancestors: post-medieval burial grounds, in *The Loved Body's Corruption*, Downes, J. and Pollard, T., Eds., Cruithne Press, Glasgow, 1999, p. 159.
14. Atlas, R. M. and Bartha, R., Interactions between microorganisms and plants, in *Microbial Ecology,* Benjamin/Cummings, Menlo Park, CA, 1987, p. 133.
15. Hall, W. D., in *Forensic Taphonomy: The Postmortem Fate of Human Remains,* Haglund, W. H. and Sorg, M. H., Eds., CRC Press, Boca Raton, FL, 1997, pp. 353–366.
16. Ebbysmeyer, C. and W. D. Haglund, Floating remains on Pacific northwest waters, in *Advances in Forensic Taphonomy: Methods, Theory, and Archaeological Perspectives,* Haglund, W. D. and Sorg, M. H., Eds., CRC Press, Boca Raton, FL, 2001, p. 219.
17. Nawrocki, S. et al., Fluvial transport of human crania, in *Forensic Taphonomy: The Postmortem Fate of Human Remains,* Haglund, W. H. and Sorg, M. H., Eds., CRC Press, Boca Raton, FL, 1997, p. 529.
18. Brooks, S. and Brooks, R. H., The taphonomic effects of flood waters on bone, in *Forensic Taphonomy: The Postmortem Fate of Human Remains*, Haglund, W. H. and Sorg, M. H., Eds., CRC Press, Boca Raton, FL, 1997, p. 553.
19. O'Brien, T. G., Movement of bodies in Lake Ontario, in *Forensic Taphonomy: The Postmortem Fate of Human Remains*, Haglund, W. H. and Sorg, M. H., Eds., CRC Press, Boca Raton, FL, 1997, p. 559.
20. Boyle, S., Galloway, A., and Mason, R. T., Human aquatic taphonomy in the Monterey Bay area, in *Forensic Taphonomy: The Postmortem Fate of Human Remains*, Haglund, W. H. and Sorg, M. H., Eds., CRC Press, Boca Raton, FL, 1997, pp. 605–614.
21. Von Endt, D. W. and Ortner, D. J., Experimental effects of bone size and temperature on bone diagenesis, *J. Archeol. Sci.,* 11, 247, 1984.
22. Marshall, L. G., Bone modification and "the laws of burial," in *Bone Modification*, Bonnischen, R. and Sorg, M. H., Eds., Center for the Study of the First Americans, University of Maine, Orono, 1989, p. 7.
23. Cattone, J. A. and Standish, S. M., *Outline of Forensic Dentistry*, Standish Yearbook Medical Publishers, Chicago, 1982.

Suggested Readings

Advances in Forensic Taphonomy: Methods, Theory, and Archaeological Perspectives, Haglund, W. D. and
　　Sorg, M. H., Eds., CRC Press, Boca Raton, FL, 2001, p. 435.

Forensic Entomology: The Utility of Arthropods in Legal Investigations, Byrd, J. H. and Castner, L., Eds.,
　　CRC Press, Boca Raton, FL, 2001.

Forensic Taphonomy: The Postmortem Fate of Human Remains, Haglund, W. H. and Sorg, M. H., Eds.,
　　CRC Press, Boca Raton, FL, 1997.

Lyman, R. L., *Vertebrate Taphonomy*, Cambridge University Press, London, New York, 1994.

White, T. D., *Prehistoric Cannibalism at Mancos,* 5MTUMR-2346, Princeton University Press,
　　Princeton, NJ, 1992.

Forensic Entomology

Gail S. Anderson

Introduction

Entomology is the study of insects. **Forensic** (or medicolegal) **entomology** is the study of the insects associated with a dead body. Insects colonize a dead body almost immediately after death, assuming that the season and environment are appropriate. Their rate of development and the species dynamics over time can be used to accurately determine time since death, from a matter of hours up to a year or more postmortem.

In the very early postmortem interval, the forensic pathologist will use medical parameters such as rigor mortis, algor mortis, and livor mortis to estimate the elapsed time since death. However, medical parameters are impacted by numerous variables, many of which are unknown, and are of little or no use once 72 hour or more have elapsed (Henssge et al., 1995). After 72 h, entomological evidence is the most accurate and frequently the only method available to determine the elapsed time since death (Kashyap and Pillai, 1989). Although medical parameters are usually used to estimate time since death when only a few hours have elapsed, insects are normally attracted to a body immediately after death, so they, too, can be used in the very early postmortem interval (Anderson, 2004; Anderson and Cervenka, 2002).

Insects can be used to determine many other factors about a death scene, such as whether the body has been moved after death, whether it has been disturbed, the presence or position of wound sites, whether the victim used drugs or was poisoned, and the length of time of neglect or abuse in living victims. Insects can also be involved in the investigation of wildlife crimes.

Forensic entomology is a broad term meaning the application of entomology to legal cases, which can also include civil investigations such as urban and stored-products entomology (Haskell et al., 1997). However, in general, the term refers to the study of the insects associated with a dead body.

Importance of Determining Time since Death

Determining elapsed time since death is extremely important, whether or not the death is criminal. In all deaths, it is of great importance to the family to know when their loved one died. People almost invariably refer to the death of a loved one in terms of time, such as "It was a week ago … a month ago … a year ago." If certain factors about a death are unknown, particularly the timing, it is very difficult for family and friends to grieve and, therefore, eventually heal. This is particularly true when time has elapsed between when the person was seen alive and when their decomposed remains are found. Understanding how, when, and why a person has died can help to give closure to family and friends and allow them to move on with their lives. Therefore, for purely humanitarian reasons, it is vital to be able to determine elapsed time since death in all cases.

The timing of death may also have legal implications. A person might have a life insurance policy that expired after the person was seen alive but before their remains were discovered. Whether that death occurred before or after the insurance coverage ceased can be of paramount importance. In a suicide case, a man drove deeply into the woods and killed himself (Haskell, personal communication). His remains were found over a year later. Foul play was not suspected, but knowing the timing of his death was vital, as he was a soldier when he disappeared, but by the time his remains were found, his tour of duty had ended. If his death had occurred while he was still a soldier, then his family would receive a substantial pension. However, if his death had occurred after his tour of duty had ended, his family received nothing. It was extremely important to the family to know when he died. In this case, a forensic entomologist was able to determine the minimum elapsed time since death, proving that the suicide had occurred when he was still a soldier, so the family received their pension.

Timing of death may also indicate the length of time a fraud has been perpetrated. An elderly gentleman in a rural area was purported to be the only caregiver to his elderly sister and even more elderly aunt (Lord, 1990). He claimed senior pensions for himself and his two elderly family members. When he eventually died of natural causes, an examination of his house uncovered the mummified remains of both his aunt and sister. Although no foul play was suspected, authorities wished to determine the time of death of the women in order to determine the length of time the pension fund had been defrauded. Again, an entomologist was able to determine the time of the deaths (Lord, 1990).

When the death is a result of homicide, understanding when the victim died has a major impact on the success of the subsequent police investigation. It focuses police efforts, can support or refute a suspect's alibi, provides documentation of time since death, and helps in the identification of an unknown victim. Most importantly, time of death is vital in determining the victim's activities and associates in the period prior to the crime.

History

Forensic entomology is not a new forensic technique, despite its rise in popularity in the last few years. In fact, it is probably one of the oldest forensic sciences to be used in death investigations. Its first recorded use in a homicide investigation was in 13th-century China. Sung Tz'u, in his treatise "The Washing Away of Wrongs" (translated by McKnight, 1981), referred to a case in which a man was found murdered in a rural village. The murder weapon was deemed to be a sickle, so the men of the village who owned such a knife were asked to line up and lay their sickles on the ground in front of them. Each sickle was then examined. The killer had cleaned his weapon so that there was no incriminating evidence visible to the human eye, but enough blood and tissues remained to be still attractive to flies. Large numbers of flies massed to the murder weapon. The killer took this to be a message from God, and he dropped to his knees and confessed (McKnight, 1981).

Modern use of entomology in criminal investigations began in France in the mid-1800s with the case of a mummified baby found in the wall of a house near Paris (Bergeret, 1855; Smith, 1986). This was followed by the landmark French works of Megnin in the late 1800s, *La Fauna des Cadavres* (Megnin, 1894). This served as the first serious work in the field and initiated further research (Motter, 1898; Yovanovitch, 1888). The first reported investigation into forensic entomology in North America was in Quebec, when two pathologists attempted to repeat the works of Megnin (Johnston and Villeneuve, 1897).

Although forensic entomology became quite commonly used in Europe throughout the 20th century, it did not really begin to take hold in North America until the 1970s. A single case in Canada was reported in the early 1960s (Howden, 1964), but in the 1970s and 1980s several entomologists in both the United States and Canada were asked to assist in homicide investigations (Anderson, 2001a; Catts and Haskell, 1990). As police became more aware of the value of entomology at a crime scene, and as entomologists began to testify as expert witnesses, the science of forensic entomology grew.

In 1996, led primarily by the late Dr. Paul Catts and Dr. Lee Goff, the American Board of Forensic Entomology (http://www.missouri.edu/~agwww/entomology/index.html) was born. In 2001 the European Association of Forensic Entomology (EAFE) was formally convened (http://www.eafe.org). Today, forensic entomologists are frequently called into homicide investigations in which time of death is unknown, and entomology is commonly presented in court in expert testimony.

Training

Forensic simply means "as applied to law" (Saferstein, 2004). Therefore, all forensic experts must be scientists before they can apply their knowledge in a forensic or legal setting. In other words, it is not possible to be a forensic chemist without first having extensive training in chemistry. In the same manner, a forensic entomologist must have extensive training in entomology before he or she can begin applying that knowledge to a criminal investigation. Future forensic entomologists will begin their training with a B.Sc. in biology, zoology, or entomology, followed usually by an M.Sc. and Ph.D. in entomology, preferably in forensic entomology, insect ecology, and taxonomy. Board certification requires, among other things, five years of experience in case work after obtaining the Ph.D. As most active forensic entomologists today are university professors, their work includes ongoing research and training in the field. Graduate students frequently work under the supervision of a board-certified forensic entomologist.

Forensic entomologists do not, with a few European exceptions, work full time for a crime lab. Their primary employment is not in case work, but in research and teaching. This means that the majority of forensic entomologists are research scientists first and foremost. Therefore, although highly trained in entomology, they may have little or no experience with crime scenes, legal report writing, or court testimony. Hence, it is vital that they receive support and instruction from the police and the courts. University scientists are adept at writing scientific articles and books, and at presenting their work at academic conferences, but they have little experience of writing a forensic report that can easily be understood by the layperson or of explaining that report and their conclusions to a jury. Hence, an entomologist should be encouraged and invited by police to attend and submit reports on noncriminal death cases in order to become used to the application of their expertise in this specialized area. This way, when the entomologist's expertise is required in a homicide, that person will have scene, morgue, and forensic report writing skills that will be invaluable. Past experience has shown that, frequently, an entomologist is only considered after a high-profile homicide occurs, and the scientist is suddenly plunged into an intense police investigation. As this may be the entomologist's first view of a dead body, and is almost certainly their first view of a decomposed person who has died of violence, this can be very disturbing. Such can easily be avoided by simply involving an entomologist in nonthreatening cases before the high-profile case occurs.

Determination of Elapsed Time since Death

There are two methods of determining elapsed time since death using entomological evidence. The first method is based on the predictable development of larval Diptera, primarily **blow flies**, over time. This form of analysis takes advantage of the known passage of time from when the first egg is laid on the remains until the first adult flies emerge from the **pupal** cases and leave the body, making it most valuable in calculating time since death from a few hours to several weeks. The second method is based on the predictable, successional colonization of the body by a sequence of carrion insects. This can be used from a few weeks after death until nothing but dry bones remain. The method used will depend on the age of the remains and the types of insects collected.

Dipteran Larval Development

Flies are attracted to a body immediately after death (Anderson and VanLaerhoven, 1996; Erzinclioglu, 1983; Nuorteva, 1977). The first flies to be attracted to a body are usually the blow flies, the large, metallic flies seen near food or garbage cans in summer, although in some areas of the world, the flesh flies are also early colonizers (Early and Goff, 1986). Blow flies belong to the family Calliphoridae, in the order Diptera or "true flies."

Both male and female flies require a protein meal before the ovaries and testes develop and oogenesis and spermatogenesis can occur, so some adult feeding may occur at the body

(Erzinclioglu, 1996). However, the majority of flies attracted to the remains are mature females searching for suitable oviposition or egg-laying sites.

The blow flies develop from egg through first-, second-, and third-instar stages, and the pupal stage before becoming adults, following a predictable pattern that is primarily influenced by species and temperature (Anderson and Cervenka, 2002). Insects are cold-blooded, so their development is temperature dependent. As temperature increases, they develop more rapidly, and as it decreases they develop more slowly. This relationship, at temperature optimums, is relatively linear, making it predictable. As the development rates are predictable, an analysis of the oldest insect stage on the body, together with a knowledge of the meteorological conditions and the microclimatic conditions at the scene, can be used to determine how long insects have been feeding on the body, and hence, how long the victim has been dead (Anderson and Cervenka, 2002).

Once laid, the blow fly eggs will hatch, after a predictable period of time, into first-**instar** or first-stage larvae. These first-instar larvae are very delicate and prone to desiccation. They are unable to break skin, so they rely on liquid protein for their meals. Therefore, the female must be careful in choosing a suitable oviposition site, where the newly enclosed larvae can access food. Wounds are an excellent source of liquid protein, in the form of blood, so these are usually the first site of oviposition. In the absence of wounds, the female will lay her eggs near the natural orifices, on or near the mucosal layer, as this tissue is moist and much easier to penetrate than normal epidermis. The face is most usually colonized before the genitalia, simply because the genitalia are frequently covered with clothing. However, if sexual assault has taken place, the presence of blood and semen will rapidly attract flies and, subsequently, eggs.

After a brief period of time, primarily dependent on temperature and species, the first-instar larvae will molt to the **second instar**, shedding the first-instar larval cuticle and mouthparts. The second-instar larva is a slightly larger, less delicate stage that can

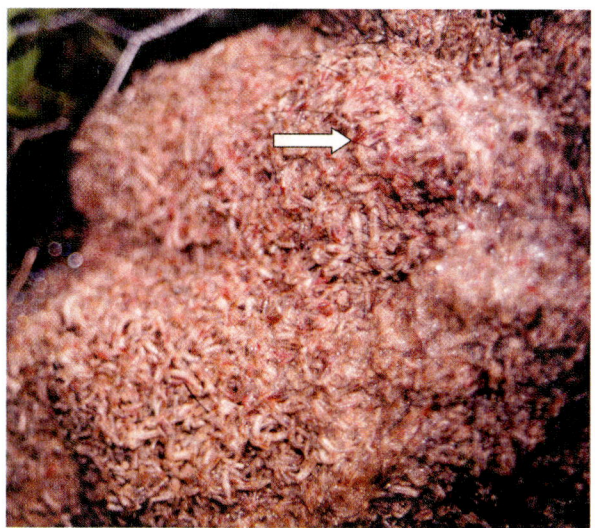

Figure 9.1 Large maggot mass on pig carcass. Insects are blow fly larvae. (Diptera: Calliphoridae). Note foam created by metabolism and motion of insects and red crops clearly visible in maggots (arrow).

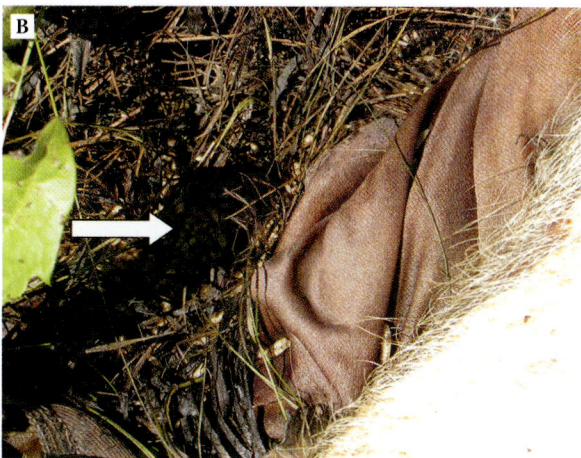

Figure 9.2 Nonfeeding third-instar larvae leaving food source in order to find suitable puparation site. (A) Crop no longer visible. (B) Large number of nonfeeding third-instar larvae burrowing into soil.

penetrate skin by using proteolytic enzymes and the rasping actions of its mouthparts. It feeds for a period of time, then molts to the third instar, again shedding the cuticle and mouthparts of the previous stage.

The **third instar** is a voracious feeder and frequently aggregates in large masses (Figure 9.1) that can generate a tremendous amount of heat. These masses can remove a large amount of tissue in a very short period of time. During this time, the crop, a food-storage organ in the foregut, can clearly be seen as a dark oval through the relatively translucent tissue of the **maggot** (Figure 9.1).

After a period of intense feeding, the third instar enters a nonfeeding or wandering stage. No physical change takes place at this time, although extensive internal physiological changes occur. However, a forensically significant behavioral change occurs. Once the third-instar larva enters the nonfeeding stage, it wanders away from the food source looking for a suitable pupation site, where it will not be vulnerable to predation (Figure 9.2). This can include the surrounding soil, or carpet, or even the hair or clothing of the corpse. They may burrow down several centimeters into the soil and may crawl several meters away from the remains until they find a suitable pupation site. During this stage, they utilize the contents of the crop, and toward the

end of this stage, they become shortened and translucent.

Once the prepupal maggot has contracted, it begins to pupate. It does not shed its third-instar cuticle as was done in its previous molts, but instead the cuticle loosens and the insect secretes a number of substances into this cuticle, which then hardens and darkens to form the puparium. The part referred to as a pupa is a living insect, with the outer hardened pupal case or puparium serving as a nonliving structure that encloses the pupa (Erzinclioglu, 1996; Fraenkel and Bhaskaran, 1973). However, it is common to refer to the outer puparium and the living pupa inside together as a pupa, and the empty case left behind after the fly emerges is known as a pupal case.

Inside the outer hardened pupal case, the insect metamorphoses or changes into an adult fly. During this time, the immature tissues are completely broken down and rebuilt to form the adult tissues. Once complete, the new adult breaks open the top of the pupal case by expanding and contracting a hemolymph (blood)-filled sac, or ptilinum, on its head (Anderson and Cervenka, 2002; Erzinclioglu, 1996). The top of the pupal case, or operculum, breaks off and splits into two halves. The upper cap still has the respiratory horns attached, and the lower cap bears the mouthparts (Erzinclioglu, 1996). The new adult fly emerges from the pupal case, then leaves behind the pupal case and the two halves of the operculum as evidence that this cycle has been completed. As the prepupal larva previously burrowed into dirt or some protected area in order to pupate, the newly emerged adult fly must now orient itself to emerge from this hiding place. Adult flies are positively phototropic, which allows them to orient correctly when they are buried and emerge into the air (Erzinclioglu, 1996). This positive phototropism also results in adult flies being mistakenly attracted to lightbulbs later in life.

The newly emerged adult fly does not resemble the beautiful green or blue metallic animal that it will become. Its wings are crumpled; the legs are thin, weak, and spindly; the body is a dull gray; and the head still appears misshapen by the ptilinum, which has not yet retracted. At this stage it is extremely vulnerable to predation, and although it cannot fly, it can run very fast and will immediately attempt to hide until it dries and is able to fly. This is the same process that occurs when a butterfly emerges from a chrysalis and expands and dries its wings and body. The newly emerged fly will similarly expand its wings and body by pumping hemolymph throughout the tissues. The wings expand, and the body begins to take on its usual metallic sheen. The fly may appear to be fully expanded and pigmented quite shortly after emergence, but it is usually approximately 24 hours before the process is complete and the insect can fly (Erzinclioglu, 1996).

Newly emerged flies are of great forensic significance, as they indicate that an entire blow fly life cycle has been completed on the body. Once the fly is dry and can fly, an adult fly cannot be linked directly back to the remains. Even if collected on the body, it may just have arrived minutes before. However, the newly emerged insect cannot fly and so is inextricably linked with the remains. In the same way, empty pupal cases indicate that an insect has completed its entire life cycle on the body.

The fly's entire life cycle is predictable. It is heavily influenced by temperature, as well as species, nutrition, humidity, etc. However, the major variable is temperature. When determining time since death using blow fly development, there are several factors that must be known:

1. *The oldest stage of blow fly associated with the body.* It is imperative to know how far the insects have progressed through the life cycle. As development is highly temperature dependent, the insects that are farthest through the life cycle are the oldest, as they arrived first. There is no point in determining the age of second-instar larvae when empty pupal cases are present. The empty pupal cases indicate that at least some of the specimens have completed the entire life cycle, so they were clearly laid on the body before the specimens who have only reached the second instar. This means that, when collecting insects from the remains, whenever a stage is collected, the investigator should immediately start searching for the subsequent stage. In other words, once feeding third instars are noted and collected, the investigator should then search the clothing, hair, and surroundings to determine whether any have entered the nonfeeding stage. If these are recovered and collected, then the investigator should search for pupae. If a search is made and no pupae are found, then the investigator can be confident that the oldest stage on the remains is the previously collected nonfeeding or prepupal third instar.

2. *The species of insect.* The entomologist must be able to identify the species of blow fly. Each species develops at a different rate, so the species must be determined. All living things have diagnostic morphological characteristics that are used to identify them down to the species level. Adult flies have more diagnostic features than larvae, but larvae can still be identified based on mouthparts and other morphological features. Recently, DNA has begun to be used to identify insects, and it is of particular value in old and damaged specimens and in first-instar larvae, which are hardest to identify (Sperling et al., 1994; Wells et al., 2001; Wells and Sperling, 2001).

3. *Temperature data.* As insects are temperature dependent, it is vital to be able to determine the temperature of the crime scene. This is usually determined from government weather station data. A drawback to this approach is that bodies are rarely dumped close to a weather station, so the weather data may come from some distance from the crime scene and thus may not apply to the insects. Therefore, it is good practice to place a portable datalogger at the scene and record scene temperatures for two to three weeks. These data can then be statistically compared with those from the weather station to determine whether temperatures at the scene are comparable with those from the weather station. A regression analysis can then be used to predict temperature at the crime scene based on temperature at the weather station.

4. *Developmental data.* In order to determine the age of the oldest insects, the entomologist must know the rate of development of the species in question, related to temperature. This information is obtained from published literature on the developmental rates of insects at different temperatures. Most entomologists develop such data for their own local species, and consequently there is a large body of literature on insect development rates (e.g., Anderson, 2000; Ash and Greenberg, 1975; Byrd and Butler, 1997; Byrd and Butler, 1998; Dallwitz, 1984; Davies and Ratcliffe, 1994; Goodbrod and Goff, 1989; Grassberger and Reiter, 2002; Greenberg, 1991; Greenberg and Tantawi, 1993; Melvin, 1934; Nishida, 1984; Nuorteva, 1977; Wall et al., 1992).

When all four of the above pieces of information are available, the simple question is, how long does it take this species to reach this stage, under these conditions? The answer might be, for example, at least seven days, in which case, the victim would have been deceased for at least seven days.

In most cases, forensic entomology will only determine a minimum time since death. The science of forensic entomology is based on determining the length of tenure of insects on a body, rather than the actual time of death. Death precedes insect colonization (except in rare instances), so the insects will indicate a time elapsed since death that is less than the actual time of death. It may be very precise, depending on how soon the insects colonized, but it will be less than the actual elapsed time since death. Although an entomologist can make an educated guess as to how long it took the first insects to colonize the body, this is an unknown.

Eggs may have been laid within minutes, or they may not have been laid until the next day, or if the body was buried or wrapped, or killed in the depths of winter, insect colonization might be delayed. Even when the condition of the remains and the scene suggests that a body would be colonized very shortly after death, unknown parameters may impact the rate of colonization. For instance, a person killed violently, in the heat of summer, during the day and left fully exposed with large quantities of blood present, would be expected to be colonized within minutes. However, it is still not certain. In one such case, the flies did colonize immediately, but the majority of the first eggs were consumed by predator wasps (Hymenoptera: Vespula, Vespa species). The large numbers of wasps successfully removed all the eggs laid on the first day, so that only younger specimens were present to be collected when the remains were discovered a few days later. Had an attempt been made to determine a maximum time since death based on the insects collected, the maximum time since death would have been underestimated by one day. This could have been dangerously misleading. The decedent had died three days prior to discovery, so the minimum elapsed time since death of *at least* two days, estimated by the entomologist, was still correct, although an underestimate. Therefore, it is always safer to give a minimum elapsed time since death, of which the entomologist can be sure.

As well, entomologists usually rely on the minimum or the average development rates of insects. Some insects in a cohort may take longer to develop; therefore using a minimum ensures accuracy. However, it may also underestimate the elapsed time since death. As such, entomological evidence will usually indicate that a person has been dead for at least a certain period of time. It could have been longer, but could not have been less. It is this date that the entomologist will stand on in court.

Time of death can be very valuable in a homicide investigation in supporting or refuting an alibi. When remains are found in an advanced state of decomposition and a person comes forward to say that they witnessed the killing, their testimony may not be believed, due to the length of time that has elapsed before the witness came forward. In many cases there are two sets of witnesses, one claiming that death occurred at a particular time and others claiming the victim was alive and well at a later date. The jury is then asked to give a subjective opinion on the veracity of the witnesses. Entomological evidence is not subjective; it is based on scientific principles, and although it cannot tell the court who is telling the truth, it can frequently indicate who is lying.

In a case in which two young women were found murdered, two sets of witnesses were presented in court (Anderson, 1999a). Two of the witnesses were young men who claimed they had seen the murders take place on May 3, and two were average citizens who claimed they had seen one of the victims alive on May 9. The insect evidence indicated that death had occurred on or before May 6 and had probably occurred several days prior to this time. This evidence supported that of the two eye witnesses, who might otherwise have not been believed, and it refuted that of the citizens, who must have been mistaken about the sighting. The entomological evidence resulted in the eye witness testimony being believed, and the defendant was convicted of two counts of first degree murder and is presently serving life without parole.

In many cases, the entomological evidence may be of great value in the actual police investigation but may never be required in court. In the case of a murdered young woman, her boyfriend rapidly became the chief suspect. Police believed she had been killed some three weeks prior to discovery, at a time when the boyfriend had no alibi. However, the suspect's friends claimed that they had seen the victim much more recently, when the suspect was out of town, providing him with an excellent alibi. Despite lengthy and repeated questioning, the friends stuck to their story. However, the entomological report clearly indicated that death had occurred three weeks prior to discovery. When faced with this scientific evidence, the friends recanted their statements, and the suspect confessed.

Although insect development is predictable, there are many variables that must be taken into account. Temperature of the crime scene is rarely known, but is extrapolated based on the nearest weather station. Although the use of data loggers can aid in determining whether these data are valid for the scene, and can be used to predict the temperature at the scene, the actual temperature is not precisely known. Even when temperature can be predicted on a macro level, it is not known on a micro level. Maggot masses can greatly elevate temperature on the body and will impact insect development. This must be taken into account (Catts, 1992; Wells, 2001).

Although insect development is relatively linear at optimum temperatures, it tails off at both lower and higher temperatures. This is particularly important at lower temperatures, as development can take much longer than predicted (Anderson, 2000). Also, at extremes of temperature, insect development ceases. These temperatures are referred to as developmental thresholds—the temperatures below or above which no development takes place. These thresholds are important but vary between species and also between stages. They also have been shown to vary within species from different geographic areas. Some researchers have reported very slow development at extremely low temperatures (Davies and Ratcliffe, 1994). Other factors can also impact insect development, such as drug use in the victim (Goff and Lord, 2001).

Successional Colonization of Body

A dead body, whether animal or human, is a rich but ephemeral nutritional resource that supports a rapidly changing ecosystem. Within minutes or even seconds of death, assuming conditions are suitable, insects, primarily the blow flies, arrive at the body to colonize (Anderson and VanLaerhoven, 1996; Erzinclioglu, 1996; Nuorteva, 1977). As decomposition progresses, the body becomes less attractive to these early colonizers and more attractive to other insects. As the body decomposes, it goes through a sequence of rapid biological, chemical, and physical changes. These changes in decomposition attract a dynamically changing sequence of colonizing insects that continue to feed until there is no nutritional value left in the remains.

The sequence of insects that colonize depends on the nutritional changes in the body and is greatly impacted by geographic region, habitat, season, meteorological conditions, and microclimate, but the sequence is predictable within those parameters (Anderson, 2001b). This predictable and sequential colonization of a body allows an entomologist to determine the tenure of the insects on the body, and therefore the minimum time since death.

When an entomologist studies the insects found on the remains, the first step is to determine the species that are present at that time. As long as experimental data are available for the same geographic area and scenario,

then the entomologist can say that the insects indicate that the victim has been dead for a certain period of time. For instance, the entomologist might say that a victim has been dead three to five months. Then the entomologist will look at the evidence of insects that are no longer on the body, having left when the body passed the stage of decomposition that made it a suitable source of nutrition for that particular species of insect. This indicates that it is past the time frame in which these species normally are found on a body in this scenario, and the entomologist can state that the victim has been dead for at least that period of time. For instance, it might indicate that the victim has been dead for more than four months. Finally, an entomologist who is very familiar with the geographic area and general scenario might observe that the next group of insects to be expected is not yet present, further refining the estimate.

In a case in the far north of North America, human remains were found in an enclosed area in the basement of a home (Anderson and Cervenka, 2002). Insect evidence included live pupae and empty puparia of many later colonizers such as flies in the families Fanniidae and Sphaeroceridae. It also included the cast larval skins, frass (feces), and peritrophic membrane from dermestid beetles (Coleoptera: Dermestidae). Peritrophic membrane lines the gut of insects and is passed out when the insect defecates. In most situations, it is simply lost in the environment, but in protected situations, it may survive. It can be found years after death, although it has been found in as little as four months (Haskell et al., 1997).

The presence of large quantities of peritrophic membrane as well as cast dermestid larval skins indicated that the remains had passed through the time of most dermestid activity. This indicated that death had occurred months before discovery, but of most interest in this case was the lack of one particular group of insects, the blow flies. By the time of discovery, the remains were mummified and would not have been attractive to early colonizers such as blow flies. However, when death first occurred, it would be expected that blowflies would have been attracted. The complete lack of empty pupal cases and any other evidence of blow fly activity indicated that death had occurred in winter the previous year, allowing normal forced-air heating to mummify the remains before the insect season began (Anderson and Cervenka, 2002).

Although insects can be used to determine time since death in this manner for months or even years after death, often when no other methods are available, great care must be taken in the interpretation of this evidence. Insect development is primarily dependent on temperature and species, making it relatively predictable. However, colonization of a body over time by insects is dependent on many variables, so this method should not be attempted without considerable local geographic knowledge (Anderson, 2001b). Factors such as geographic region, altitude, season, sun exposure, and habitat, as well as whether the remains were in a rural or urban environment, inside a house, buried or aboveground, in a vehicle, hanged, burned, or wrapped, can all impact insect colonization rates and the specific species involved (Anderson, 2001b). For instance, burying a body may eliminate some species that do not burrow while encouraging others that are less likely to be outcompeted by species that would normally colonize exposed remains (VanLaerhoven and Anderson, 1996; VanLaerhoven and Anderson, 1999). Therefore, local geographic information must be available for this technique to be of value.

Determining whether the Body Has Been Moved

Under certain circumstances, insects can be used to determine many other factors related to a death scene besides elapsed time since death. For instance, they can be used to determine whether a body has been moved after death. A killing may occur on the spur of the moment, leaving the killer with a body that must be secreted in some way. The adult murder victim is heavy, difficult to lift, frequently bloody, and may also be covered in excretory material, as sphincter muscles relax during violent death. Therefore, it is not uncommon for the killer to leave the remains at the death site for a period of time while the perpetrator goes away and considers how to remove the body. The killer may return later, possibly with assistance and a vehicle, and dump the body, usually in some remote area. If the body has remained at the original site for any length of time, and the conditions are appropriate, insects will colonize. At first, the only sign of colonization will be blow fly eggs in the wound, the corners of the eye, and in the nostrils. When the killer returns, the insects will probably not be noticed. If they are observed, it is unlikely that the killer will consider them as evidence. If the killer is aware of the value of entomological evidence, he or she might try to remove the eggs, but it is unlikely that all the eggs would be removed. The killer will then move the body and dump it in a remote location, thus moving the insects from their usual habitat to a new area.

Many carrion insects are ubiquitous, so moving them from one area to another may not provide any information. However, some species are associated with only rural areas, or are restricted to intertidal zones, etc. Therefore, the presence of a species on the remains that does not fit the scenario in which the remains have been found may indicate that the body has been moved, while also suggesting the likely scenario of the original death scene. Caution must be taken, as many species are more commonly found in one region or habitat, but may still be recovered in small numbers in other areas. Also, those species that are considered urban are associated with human refuse, whereas rural species are associated with carrion. However, human refuse may be found in rural areas, allowing insect transfer. Nevertheless, if the insect evidence suggests that the body has been moved, this may be of great value to the police investigators.

Body Disturbance

It is not uncommon for a killer to return to the scene of the crime and disturb the remains.

Killers return for a variety of reasons. They may go back because they were incapacitated by drugs or alcohol when the crime was committed and cannot remember if they committed the crime. They may also go back to fantasize about the killing, and finally they may go back to recover evidence they feel may incriminate them. If they disturb the body in some way, it might also disturb the insects. An entomologist may be able to determine not only time since death, but also when the body was disturbed. It is probable that the killer will have developed an alibi for the time the murder was committed, but it is unlikely that the killer will have developed an alibi for the time when he or she went back to the scene. As well, in many killings, the police have a strong suspicion of the person who is responsible, but not enough evidence for a warrant. If both the time of death and the time of disturbance are known and fit the activity pattern of the suspect, that may help to corroborate other evidence.

In the case of a burial, the remains of a man were found in a shallow grave (Anderson, 1999a). The careful police exhumation took several days. The remains were skeletonized, yet little evidence of blow fly activity in the form of empty pupal cases was recovered from the upper portion of the body. On the second day, however, very large numbers of pupal cases were found associated with the lower half of the body. This suggested that the upper half of the body had decomposed at the same site as the lower half, but had later been moved to a secondary position after skeletonization, perhaps because part had been exposed. The evidence suggested that the body had been reinterred several weeks after the original burial (Anderson, 1999a).

Position of Wounds

Blow flies are the first insects to be attracted to remains and usually lay their eggs close to a wound, so that the first-instar larvae have access to liquid protein for nutrition. Once a body is in a state of advanced decomposition, it is often difficult to determine whether wounds are present. If a wound does not impact the hard tissue, such as bone and cartilage, it may easily be missed, despite the fact that a wound in soft tissue alone can be fatal. Insects, however, can locate a wound, however small, as laying eggs close to a wound is a survival strategy that increases the chance of success of the female flies' offspring. Therefore, female flies are genetically programmed to be extremely efficient at locating a wound. They can even locate a venapuncture from a hypodermic needle when it is no longer visible to the pathologist.

Once decomposition is advanced, insect colonization patterns can be used to indicate the possible site of wounds. It is not up to an entomologist to state that an area is a wound site, but rather the forensic pathologist, who is qualified to identify a wound and the weapon that caused it. However, it is up to the entomologist to point out an irregular or atypical insect colonization pattern that might indicate a wound.

In a case that was originally considered to be a suicide or undetermined death, the body of a young woman was found several days after she had left home to go for a short walk (Lord, 1990; Rodriguez and Bass, 1987). The remains showed large maggot masses in the chest region, as well as the palms of both hands. No entomological evidence was recovered, and the victim was interred and listed as an undetermined death. Some time later, investigators showed photographs to Dr. Bill Rodriquez, a forensic anthropologist, who noted that the primary maggot activity was in the chest region and the palms of the hands, rather than the face (Rodriguez and Bass, 1987). This strongly suggested the presence of wounds in these areas. In particular, the presence of maggots in the palms of the hands, an area rarely colonized by insects, suggested the presence of defense wounds. Based on this evidence, a court-ordered exhumation was performed and the remains examined again. When the body was reexamined, many stab wounds were identified in the chest region, as well as severe slashing to the hands that had almost severed one thumb. The case is now reopened as a homicide, but it would never have been closed had an entomologist been called into the case at the beginning.

Caution must be taken when inferring wound sites from insect activity. For instance, it is often thought that the presence of maggot activity in the genital area of a victim indicates a rape. If the oldest insects on the remains are in this region, and the only other activity is clearly of a later date, this may indicate that a wound or semen was present at this site. However, experience has shown that if the genital region is colonized at the same time, or later than other areas, it may just be normal insect colonization. The orifices are also attractive colonization sites due to the presence of a mucosal layer, and they are quite normally colonized. Even the presence of the added attraction of semen is not indicative of rape, as it may be the result of consensual sexual activity that is unrelated to the death.

It should be pointed out that if rape is suspected, then a sexual-assault kit should be used to collect DNA swabs as soon as possible, as insects will remove this evidence rapidly. If the remains are placed in a morgue cooler with the intention of collecting the DNA evidence later, the evidence will frequently be destroyed by maggot activity. Although refrigeration will slow down insect activity greatly, if a body has large maggot masses, it will take a considerable time before the temperature of the maggots in a refrigerator will cool enough to immobilize them. In the meantime, vital evidence may be consumed.

Linking Suspect to Scene

Much of forensic science is based on Locard's exchange principle (Saferstein, 2004), named after Locard, of the French police or Sûreté. Locard observed that, in all cases, a criminal leaves something behind at a crime scene, and also takes something away. This exchange usually refers to trace evidence such as a footprint or fingerprint, hair, blood, semen, clothing fibers, paint, etc. However, it can equally well be applied to insect evidence. There have

been several cases in which a criminal has taken away entomological evidence from a crime scene that links him or her back to the scene or victim.

In a case from suburban Chicago, a woman was raped by a man wearing a ski mask (Lord, 1990). The rape occurred in early summer to midsummer. A suspect was identified, and a search of his home revealed the presence of a ski mask. The suspect admitted to owning the mask but swore he had not worn it since the previous winter, months before the rape. However, investigators noticed that there were several plant parts on the mask, including cockleburs. A forensic entomologist opened the cockleburs and discovered that they contained live caterpillars. An analysis of their life cycle indicated that, in order to have collected cockleburs containing larvae of this stage, the mask must have been outside in the early part of the summer of the present year. When shown the evidence, the suspect confessed (Lord, 1990).

One of the first published reports of forensic entomology in Canada describes how insects were used to connect a suspect to the crime (Howden, 1964). In 1963, a person broke into the house of an elderly man in order to rob him. The homeowner confronted the suspect and was brutally beaten, resulting in his death some days later. A suspect was arrested and searched. A "shinplaster" or Canadian bank note worth 25¢ was found in the pocket of the defendant. However, it was simply a bank note and could have come from anywhere as the defendant claimed. An examination of the note revealed that there were several unusual "fibers" stuck to the bank note. An entomologist was asked to look at the fibers. Instead of fibers, they were found to be the plumose hairs of a bumble bee.

The investigation revealed that the elderly victim had kept his money in a drawer in his house. When this drawer was examined, a dead bumble bee was discovered. It was considered that finding bumble bee hairs on a bank note was very rare, so this was part of the evidence used in the defendant's conviction and subsequent execution (Howden, 1964). Nowadays, one hopes that data would be collected to determine how common it is to find bumble bee hairs on bank notes to determine the probability of an innocent person having such notes. As well, depending on the specimen, it might actually be possible to perform a DNA analysis and determine whether the hairs came specifically from that bee.

In Southern California, the body of a young woman was found in a rural area in late summer (Lord, 1990; Prichard et al., 1986; Webb et al., 1983). When death investigators processed the crime scene, they found that they were all severely bitten by something. The bites were identified by a doctor to be those of the immature stage of a trombiculid mite, *Eutrombicula belkini* Gould, also known as a chigger. Chiggers are known to have a discrete geographic distribution and to leave very characteristic bites (Prichard et al., 1986). When a suspect was developed and examined, it was noted that he also bore similar characteristic dermal lesions resulting from chigger bites. An extensive search of the region found that large numbers of chiggers were found in only a very narrow area between a cultivated field and the edge of natural vegetation—the very site of the crime scene (Lord, 1990). The defendant was found guilty of first-degree murder, based primarily on this entomological evidence (Lord, 1990; Prichard et al., 1986; Webb et al., 1983).

Drugs

Insects feed on the tissues of the body. Therefore, they also feed on any unnatural substances that may be in the victim's body. This can include alcohol, poisons, and drugs. Alcohol is a normal product of decomposition (Henssge et al., 1995), so carrion insects are unlikely to be impacted by its presence. However, when the victim has been poisoned or used drugs, whether illicit or therapeutic, these may impact the development of the insects. When the body has decomposed to a point where normal toxicological procedures are no longer valid, an analysis of the insects themselves can reveal the identity of the toxin. In one rare case, an unknown victim was identified by inferring the level of pollutants, primarily mercury, in the body from an analysis of the insects feeding on her body. The levels and type of pollutants present indicated the probable area in which she had lived. Sketches of her face by a forensic artist were exhibited in that region, and she was identified (Nuorteva, 1977).

Drugs have been identified from the analysis of dipteran larvae feeding on the body both experimentally and in actual cases (Beyer et al., 1980; Duke, 2003; Goff and Lord, 2001; Hedouin et al., 1999; Introna et al., 2001; Kintz et al., 1990; Kintz et al., 1994; Sadler et al., 1995) as well as from dipteran puparia (Miller et al., 1994), larval and adult Coleoptera (Bourel et al., 2001), and Coleoptera exuviae (Miller et al., 1994). Therefore, insects can be important drug specimens, especially in cases in which decomposition is advanced.

Research has shown that the blow fly retains the toxin throughout the larval and puparial stages and eliminates it early in the adult stage (Nuorteva and Hasanen, 1972; Nuorteva and Nuorteva, 1982). Not only was the toxin bioaccumulated by the primary insects feeding on the toxin-laden tissue, but it was secondarily bioaccumulated by adult beetles (*Creophilus maxillosus* [L.], Coleoptera: Staphylinidae) that fed on these larvae (Nuorteva and Nuorteva, 1982). The successful recovery of drugs and toxins from fly puparia and beetle exuviae (Miller et al., 1994), which may survive for years after death (Gilbert and Bass, 1967), means that a toxicological analysis could potentially be performed even decades postmortem.

Drugs may have a major impact on insect development, either speeding up or slowing down development. Pioneer work by Goff in Hawaii indicated that many drugs can impact insect development, and the effect varies with drug and insect stage (reviewed in Goff and Lord, 2001). This work has been followed up by many other scientists in the world (Bourel et al., 1999; Carvalho et al., 2001; Duke, 2003; Hedouin et al., 1999; Introna et al., 2001; Sadler et al., 1995; Wilson et al., 1993).

As drugs can have a significant impact on insect development, it is important to ensure that the entomologist be aware of the information revealed in the toxicologist's report before submitting any analysis.

Abuse

The insects used in forensic entomology analyses are those that feed on dead organic matter. Living people and animals can have dead organic matter on their bodies. This can be in the form of an unhealed wound, a bed sore, gangrenous and dying tissue, or very poor personal hygiene. This material is just as attractive to carrion insects as a freshly dead body. Flies, primarily blow flies, oviposit or lay eggs on the person. The subsequent maggots will feed on the necrotic material. The infestation of living human or other vertebrate animals with dipteran larvae that, for at least a period of time, feed on the host's dead or living tissue, liquid body fluids, or ingested food is termed *myiasis* (James, 1947). In most cases, the insects do no damage as they are feeding on necrotic material; indeed, they are, in fact, cleaning the wound. As such, they may actually help the victim. However, their presence is unwanted and incurs feelings of revulsion. As well, insects can mechanically transmit diseases such as polio and tetanus (Baer, 1931; Nuorteva, 1959; Nuorteva and Skaren, 1960; Sherman and Pechter, 1988).

The unwanted presence of insect larvae on a living victim may indicate the length of time of neglect or abuse. In these cases, the victim is alive, so time of death is not the issue, but rather, length of time of neglect. The temperature to which the insects have been exposed is known, as the person is alive, and therefore the age of the maggots can be determined. The maggot age will indicate the minimum length of time of abuse or neglect. Such evidence was used in a case in Hawaii in which a baby was dumped in a remote area (Goff et al., 1991). The age of the maggots in the diaper feeding at the site of severe diaper rash was used to determine the timing of her abandonment and was used in the conviction of her mother.

Entomology provides powerful and convincing evidence of neglect in court. It is often difficult to prove neglect. However, when an entomologist can state in court that young children have not had a diaper change in almost five days because of the four- or five-day-old maggots feeding on their tissue, it clearly and scientifically proves neglect and abuse, and it is very compelling evidence (Lord, 1990). Entomology is increasingly being used as an evidentiary tool in child and senior abuse cases (Benecke and Lessig, 2001). It is equally valuable in cases of animal abuse.

A problem with dealing with a living victim is that the person or animal may be mobile. Therefore, nonfeeding third-instar larvae, which leave the food source, could be dropped anywhere and may not be recovered. This could lead to an underestimation of length of time of neglect. In such cases, careful searching of clothing and bedding may reveal further entomological evidence (Anderson and Huitson, 2004). When a victim dies and the body decomposes, it only remains attractive to blow flies for a short period of time. Therefore, by the time the first flies have completed their development on the body, the rest of the remains are no longer attractive to further blow fly colonization. However, in a living victim, successive generations of blow flies could colonize the tissue, again indicating that only a minimum time of death could be determined by the insect activity.

If the victim is colonized during life but subsequently dies prior to discovery, the oldest maggots on the remains will indicate a time of death several days earlier than the actual death. Therefore, great care must be taken to determine the site of colonization, as this could indicate whether it was likely that the victim was colonized prior to death (Goff et al., 1991). In the case of the young baby discussed above, the fact that the diaper region, covered in clothing, was colonized prior to the exposed facial orifices would have suggested to the entomologist that colonization had occurred in life (Goff et al., 1991). However, in many cases, colonization patterns may appear to be the same as that seen in normal colonization occurring after death. In a case in British Columbia, an elderly man was found unconscious with severe head injuries. His face and mouth were heavily colonized by third-instar

Calliphora vomitoria (L.) and *Phaenicia sericata* (Meigen) (Diptera: Calliphoridae). He later died in a hospital. The sites of larval colonization were the head wounds and the mouth (Anderson and Cervenka, 2002). As insects usually lay eggs at wound sites and natural orifices (Anderson and VanLaerhoven, 1996), this would not have appeared unusual had he been discovered after death. Careful examination of the body and the scene by the entomologist and pathologist may indicate whether the victim had been colonized in life.

Although the presence of fly larvae on a living person is repugnant to most people, the use of maggot debridement therapy has a long and successful history. Most species of blow fly feed exclusively on necrotic tissue, and medical practitioners discovered centuries ago that they could be used to cleanse wounds. Anecdotal reports of such use date back centuries, but this technique was first put into more traditional medical practice after the First World War, when it was used in the very successful treatment of osteomyelitis in children (Baer, 1931). The practice was discontinued due to side effects such as tetanus, due to the use of unsterile maggots, and due to the exciting discovery of antibiotics. However, with worldwide resistance to most antibiotics growing, maggot debridement therapy, as it is now known, was reinvestigated (Sherman and Pechter, 1988).

Maggots not only remove the necrotic tissue, cell for cell, without damaging the living cells, but they also remove and destroy the bacteria and pus. They secrete antibiotics and allantoin, and their presence in the wounds promotes natural granulation (Sherman, 1998). As methods are now available to surface-sterilize eggs (Sherman and Wyle, 1996), the treatment has become more popular, and it has frequently been used as an alternative to amputation (King and Flynn, 1991; Mumcuoglu et al., 1998; Sherman, 1995; Sherman, 1997; Sherman, 1998; Sherman and Pechter, 1988; Sherman et al., 1990; Sherman et al., 1993; Sherman et al., 1995; Sherman et al., 1996a; Sherman et al., 1996b; Sherman et al., 2000).

The illegal killing of wildlife for the wildlife trade is a very major crime worldwide. In fact, it is considered to be one of the biggest crimes in the world, second only to drug dealing in terms of profit margin. The illegal trade in animals and animal parts is a worldwide concern, and forensic entomology can be used to determine time of death, as well as many other factors in a wildlife crime, just as it can in a human homicide.

At a rural garbage dump, three adult female black bears (*Ursus horribilis* [L.]) were killed, disemboweled, and their gallbladders removed (Anderson, 1999b). The bear gallbladder has a very high value in traditional Chinese medicine (Espinoza et al., 1993). The killed bears had twin and triplet cubs, which were left orphaned at the garbage dump. Two of those tiny cubs were later found killed, disemboweled so that their minute gallbladders could be removed. At this young age, their gall bladders had no commercial value.

Fly eggs collected from the cubs indicated that death had occurred a matter of hours prior to discovery, and this information was used in the conviction of two men. They were convicted of two counts of poaching under the Provincial Wildlife Act of Manitoba and were sentenced to six months in prison (three months per cub). The maximum they could have received was six months per cub. The six months was appealed down to three months, but they did serve jail time, which was precedent-setting for Manitoba, and possibly for Canada, for future cases. In his summing up, the judge considered the entomological evidence to be the "most compelling" (Anderson, 1999b).

Collection of Entomological Evidence

Entomological evidence should be collected by an entomologist whenever possible. An entomologist is trained to identify and collect insects and will immediately understand what is important and what is not. However, this is not always possible due to time constraints and the distances frequently involved. In such cases, a trained police death investigator should collect the evidence. Entomological evidence is not difficult to collect and handle, nor is its collection particularly time consuming. However, it is essential that it be done correctly if it is to be used in the investigation and subsequent trial. Incorrectly collected, such evidence can never be retrieved. There are many descriptions of the correct procedure for collecting entomological evidence in a variety of texts (Anderson, 1999b; Anderson and Cervenka, 2002; Haskell et al., 2001), and there is at least one training video (RCMP, 1997). Therefore, only a brief overview will be provided here. Collection techniques are summarized in Table 9.1 and Table 9.2.

Collection at the Crime Scene

Entomological evidence should be collected at the crime scene. When a major crime scene is discovered, the police investigation of that scene is laborious, intensive, and meticulous. Every square centimeter of ground should be examined and sifted, and potential evidence will be continuously photographed, sketched, and collected. Therefore, searching for and collecting entomological evidence does not require extra time or extra searching. The area is already being searched for other evidence, so the crime scene investigator just needs to know how to recognize entomological evidence and how to collect it.

Before any collection is attempted, the remains and the environment should be observed and photographed. If maggot masses are present on the remains, then the temperature of each mass should be taken. A thermometer should be rested gently on top of the mass, with a small amount of pressure on the upper surface. This will allow the movements of the maggots to roll the thermometer into the center of the mass, thus reducing the chance of the thermometer causing any damage to the body.

Collecting Blow Fly Evidence

Blow fly development is the most important entomological method for determining time since death in the first few days and weeks after death. Each stage is important.

Eggs

Blow fly eggs are usually laid immediately after death. If the remains are found within a few hours, then the forensic pathologist will determine time of death using medical parameters, so the entomological evidence will not be required. In many cases, however, the pathologist would like the corroborative evidence of an entomologist. Also, in some cases, such as those of dismemberment or decapitation, normal medical parameters are unavailable. Therefore, eggs may be very important (Anderson, 1999a; Anderson, 2004). Eggs are only of value in the very early postmortem interval. Once maggots are present on the remains, these will be of much greater value; therefore, eggs do not usually need to be collected. However, if only eggs are present, then they will be very important.

Eggs are laid in clumps close to wounds and orifices (Figure 9.3 and Figure 9.4). The eggs are creamy colored. A small portion of eggs should be broken off from the clump, preferably close to the center, as those on the edge are more likely to be parasitized or desiccated. The eggs are partially stuck to each other, so breaking a small piece off is similar to breaking off a piece of a cookie. Some eggs will be broken, but more are being collected than needed. The clump should be approximately a quarter the size of a dime. This should be broken into two (or simply collect a second sample from the same site).

The first half of this sample should be preserved immediately in alcohol (75 to 90 percent ethanol, or 50 percent isopropyl alcohol). Evidence preserved at the scene will be used to determine the exact stage of development at that time. Preservation stops the clock, so the sooner preservation occurs, the more specific the entomologist can be. The vial should be sealed and fully labeled.

The second half should be kept alive and placed on beef liver in a plastic vial. First, a small piece of damp paper towel should be placed in the vial to prevent desiccation, then a small piece of beef liver, followed by the collected eggs. The vial should be covered with a double layer of dry paper towel, held in place by an elastic band, rather than replacing the lid.

Table 9.1
Summary of Blow Fly Collection Techniques for Forensic Entomology Analysis

Insect	Location	Live	Collection Preserved	Comments
Eggs	Usually laid near wounds or a natural orifice in clumps Break off a piece a quarter of the size of a dime	Keep half of each sample alive for later identification; place in vial on top of small piece of beef liver, and cover with a double layer of paper towel held in place with an elastic band. Label vial with collection site and time of collection.	Preserve the other half of the sample in a separate vial with 75–90% ethanol or 50% isopropyl alcohol immediately after collection at the scene. Label vial with collection site and time of collection	Collect separate samples from several areas observe and note time of hatch in living specimens. Eggs are not important when maggots are present
Feeding	Found on body or near wounds and orifices; may be in maggot masses; may be all over body. Collect 100–200 larvae per sample; collect from several different areas and keep separately. Collect with blunt forceps, small paint brush, or spatula	Treat exactly as for eggs	Preserve the other half of the sample in a separate vial with 75–90% ethanol or 50% isopropyl alcohol immediately after collection at the scene. If possible, larvae should be placed in hot water briefly before being placed in alcohol; if this is not possible, preserve directly in alcohol. Label vial with collection site and time of collection	Collect several exhibits larvae from different areas. Do not overfill vial with larvae. They should be no more than one maggot deep in the vial.
Prepupal	Found in soil, hair, clothes, carpet larvae	Treat exactly as for feeding larvae and for eggs	Treat exactly as for feeding larvae	These do not require nonfeeding food May be several centimeters deep and several meters from body
Pupae	Found at same sites as prepupal or nonfeeding larvae; very fragile	Place in vial, with small piece of damp paper towel to prevent damage; use dry paper towel held in place by an elastic band as a lid; no food required	Do not preserve; keep alive for identification and timing of emergence	Pupae are dark brown and are often found a distance from the remains; they are often mistaken for plant parts. May be very tiny, a few mm long, or up to 1.5 cm long
Empty	Found in same area as pupae pupal cases and nonfeeding larvae; or puparia; very fragile	Empty pupal case is not alive	Just keep dry in a vial; use paper towel to cushion puparia in vial, and use a normal vial lid	Adult flies have already emerged, so empty pupal case is indicative of entire life cycle completed on body
Adult blowflies	Found anywhere on body. Collect with net or with small, dampened paint brush	Can be kept alive in vial; no air required; place screw lid on vial	Do not preserve if wings still crumpled; place in dry vial and allow to dry; label as newly emerged fly	Only valuable if just emerged

Table 9.2
Summary of Collection Techniques for Later-Colonizing Insects

		Collection		
Insect	**Location**	**Adults**	**Immatures**	**Comments**
Other fly	Anywhere on body; may be found in clothing and in joints. Collect with a net or with a small, dampened paint brush	Can be kept alive in vial; no air required; place screw lid on vial	Keep alive in vial with damp paper towel and paper towel lid. Preserve some in alcohol. All pupae should be kept alive	Both adults and immature insects are valuable
Beetles	Anywhere on or under body or in close surroundings and clothing. Collect with a net or with a small, dampened paint brush	Can be kept alive in separate vials or all placed in alcohol	Keep alive in vial with damp paper towel and paper towel lid. Should be kept individually, as they will cannibalize other specimens. Preserve some in alcohol. All pupae should be kept alive	Both adults and immature insects are valuable, and both move fast; easiest to preserve all (except pupae) in alcohol
Soil sample	Collect ≈4 cups of soil from close to torso	Place in can, approximately twice the size of sample; rigid structure of can means that lots of air is also trapped, so insects can survive in soil for several days		Entomologist will examine soil in lab

Figure 9.3 Blow flies (Diptera: Calliphoridae) laying eggs inside the mouth of a pig carcass. Eggs can be seen at the edges of the mouth (arrow). These eggs were laid within less than 5 minutes of death (Anderson and VanLaerhoven, 1996).

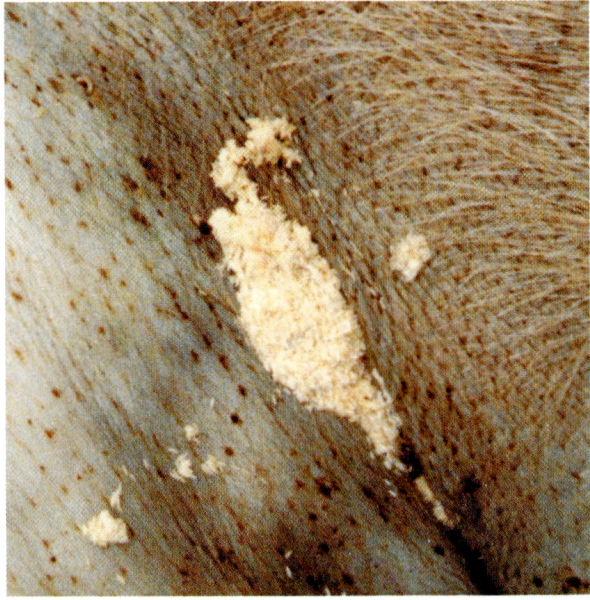

Figure 9.4 Large clump of blow fly eggs (Diptera: Calliphoridae) on a pig carcass. Mass is approximately 4 cm long. Eggs should be collected from central region and observed until hatched.

Figure 9.5 Appropriate vials for collecting insect evidence. The vial on the left contains alcohol and will be used to preserve half of an exhibit of larval Diptera or maggots. The vial on the right contains a piece of dampened paper towel to prevent desiccation and to protect larvae from damage in motion. The right vial is covered with a double layer of paper towel, held in place by an elastic band, to allow for air flow. A small piece of beef liver should be added to the vial on the right, as a food source. Both vials should be carefully labeled and transported upright.

A paper towel allows much more air flow than the lid of the vial punched with small holes, and it will prevent escape of even the smallest larvae, as long as it remains dry (Figure 9.5).

These eggs must then be observed on a regular basis to determine the time of hatch. If shipped to the entomologist as eggs, they will almost certainly have hatched by the time they are received, and the exact time of hatch will be unknown. As using eggs to determine elapsed time since death can be accurate to an hour, the actual timing of hatch is important. Therefore, the vial containing the live eggs should be placed in the shade and checked every hour or so. The investigator should note the first hatch (when movement is observed) and then note approximate percentage hatch over the following few hours—for example, 1200 hours, no hatch; 1300 hours, 10 percent hatch, 1400 hours, 70 percent hatch, etc. The time of hatch will allow the entomologist to count back to the time that the eggs were laid on the remains.

This is only necessary when eggs alone are present. Once maggots are present, eggs are of lesser value and do not need to be observed.

Larvae or Maggots

First-, second-, and third-instar larvae or maggots will be found in or near wounds, orifices, and generally throughout the body. They may be under clothing. They are often aggregated in masses (Figure 9.1). After taking the temperature of the mass (or at least estimating its size), the investigator should collect a sample of maggots from each major area in which they are observed. Each sample should be kept separate from the others. For instance, if a large mass is seen in the abdomen area and another in the face, these should be collected separately and maintained separately. If the entomologist determines that the maggots in the abdominal region are older than those in the face, it may suggest there was a wound in that region.

Each collection should consist of approximately 100 to 200 larvae. The largest are probably the oldest and, therefore, the most valuable, although smaller fly species will have smaller larvae. Hence, collect the largest maggots primarily, but also collect a sample of the smaller maggots, as it is possible that they are older specimens from a smaller species. Each sample should be divided into two: one preserved and one alive. The sample to be preserved should be placed in very hot, but not boiling, water very briefly to destroy the internal enzymes, which will allow better preservation. The samples should then be placed in alcohol for permanent preservation. They must not be left in the hot water, or they will decompose rapidly. If hot water is not available, then it is best to preserve them directly in alcohol rather than delay preservation.

If preservation does not occur until later, then the entomologist will have to attempt to take into account the temperature of the vehicle in which the insects were transported, the temperature of the office in which they sat, the plane in which they were couriered, etc. Each of these parameters provide a variable that is difficult or impossible to account for fully. Preservation at the scene means that only the scene temperatures need to be taken into account. Even when remains are moved immediately to the morgue and refrigerated until an autopsy, the temperature of the casket in the hearse is unknown. Although the temperature of the refrigeration unit at the morgue is known, the body will take a considerable time to drop to that temperature, particularly when large masses of maggots are present. The time it took for the body and, consequently, the maggots to drop to the temperature of the refrigerator is also an unknown variable. Preservation at the scene stops the clock.

The other half of each sample should be kept alive. The live sample should be placed in a vial with a piece of damp paper towel. This prevents desiccation and allows extra surface area for movement. Do not put too many specimens into a vial. A 100 milliliter vial (such as a urinalysis vial used in a hospital) is ideal. The insects should only be one maggot thick in the vial. Overcrowding will lead to death. Maggots are air breathers and will suffocate if the vial is too full. The paper towel allows more surface area for crawling. The maggots will require food, such as beef liver, and the small piece of paper towel will prevent drowning when blood leaches from the meat. The vial should be covered with a double layer of dry paper towel, held in place by an elastic band, as for the eggs (Figure 9.5). All vials should carry a detailed label including the time of collection (or the time of preservation). These live specimens will be raised to adulthood by the entomologist, allowing the insects to be identified and also providing further developmental data.

Once representative samples of larvae have been collected from the body, the surrounding area must be searched. Nonfeeding or prepupal third-instar larvae will be found in clothing, hair, soil, carpet, etc. This area is being searched for other evidence, and it is just a matter of the investigating officer also searching for insect evidence. The larvae may have burrowed several centimeters into the soil. If the body is indoors or on a hard surface, the larvae may have migrated several meters to find a source of refuge. Therefore, loose carpets, flower pots, bags, etc., that are on the ground may well have been sought out. These larvae should be collected as before, keeping half alive and half preserved. No food is required. Their site of collection should be noted.

Figure 9.6 A range of tanning stage in blow fly (Diptera: Calliphoridae) pupae. The pupae on the left have only just begun to pupate, so are at the beginning of the pupal stage. Those to the right have completed the tanning phase and are indistinguishable, outwardly from older pupae. Those in the middle began pupating several hours prior to the photograph being taken (photo by N.R. Huitson).

Pupae

Pupae will be found in the same areas as nonfeeding third-instar larvae. These should be collected and placed in a vial along with a paper towel to cushion them. No food is required, although air should be provided by using a double layer of paper towel, secured by an elastic band, as a lid. Pupae should not be preserved.

When the insect first pupates, it is a very pale color, similar to that of the maggot. The outer cuticle of the pupa hardens and tans over a period of hours (Figure 9.6). If the pupa is pale colored when collected, it should be photographed, collected, and kept separately from others. Although it will darken before the entomologist receives the evidence, its color will be important in determining the age of the insect. Pupae may be found in the dirt or in the general area surrounding the remains. They may be buried several centimeters deep or scattered up to several meters away from the body (Figure 9.7).

Empty Pupal Cases

Empty pupal cases will again be found in the same areas as pupae and nonfeeding third-instar larvae. They should be collected, but as they are not living tissue, they do not require air. They are, however, delicate, so they should also be given a piece of paper towel to cushion them during transit.

Figure 9.7 Blow fly (Diptera: Calliphoridae) pupae in leaf litter close to remains. They can also be recovered several meters from the body. Note variety of colors of pupae. The light pupae have only just pupated.

Adult Blow Flies

Adult blow flies are of little forensic value, as it is unclear whether they have just arrived at the scene or have developed on the body. Once a blow fly has dried its wings, it is very mobile and can travel great distances; therefore, it cannot be linked to the scene. However, if the blow flies have just emerged and cannot yet fly, they are of great value, as they must have emerged from the remains. They should be collected and kept alive in a dry vial. If preserved at this stage, they will crumple and be very difficult to identify; however, if allowed to dry, they will be very easily identified. They should be collected and labeled as an undeveloped fly.

Other Insects

Blow flies will be the main insects used to determine time since death in the first weeks after death. However, other insects will colonize the remains in a sequence over time, and these will be very valuable in determining time of death in the later postmortem interval. Larval insects, whether Diptera (flies) or Coleoptera (beetles), should be collected and placed in a vial with damp paper towel as previously described. Some can also be preserved in alcohol. Adult flies, such as skipper flies (Piophilidae), can be collected with a camel's

hair damp paint brush and placed immediately into alcohol. Adult beetles can run fast and are usually found under the remains. They can be collected by forceps or with a damp paint brush and again placed in alcohol. If kept alive, many beetles will eat smaller beetles or flies, so these must be stored individually.

Other Information Required

A small sample of soil collected from close to the torso of the remains (but not under) can be placed in a metal can (Figure 9.8). If only half full of soil, the can will also contain enough air for any insects in the soil to survive until examined by the entomologist. This will allow the entomologist to collect any very small specimens that may not have been observed.

Weather records from the nearest or most appropriate weather station should be requested, and a data logger placed at the scene. The logger should not be in direct sunlight, even if the remains were in the sun, as plastic and metal heat up differently from flesh.

A description of the scene and body should be provided to the entomologist, including photographs and video, as well as details of any other carrion that might have been present. The toxicologist's report should also be submitted to the entomologist.

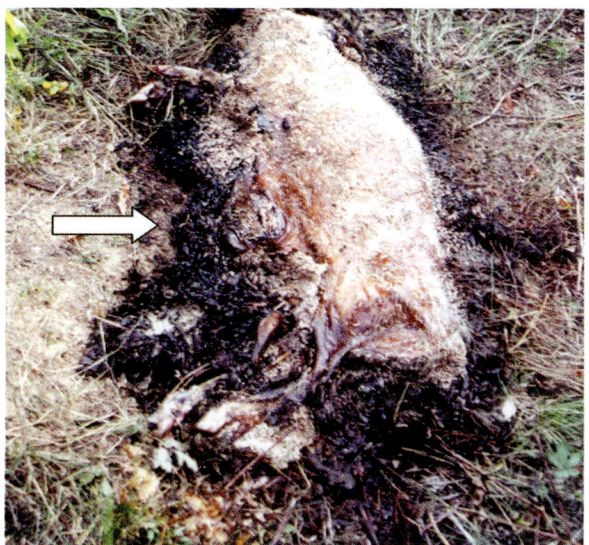

Figure 9.8 A sample of soil should be collected from just outside the area demarked by the body fluids (arrow).

Collection at the Morgue

Insects should be collected as soon as possible; therefore, it is preferable to perform the collection at the scene. However, sometimes this is not possible, or only a partial collection can be performed at the scene, and the rest of the collection is performed at the morgue. Insects should be collected and preserved as described for insects at the scene. Times of refrigeration should be provided, as well as temperature of refrigeration.

Challenges to Forensic Entomology

As with any science, there are many things that may limit the application of forensic entomology. Careful collection of entomological and environmental evidence may ameliorate the affect of most limitations.

Temperature

As has been mentioned several times, the actual temperature of the scene and that to which the insects have been exposed is unknown. The temperature can be estimated and extrapolated from a weather station in concert with data loggers placed at the scene. However, in order to determine elapsed time since death using maggot development, temperature data are required. Therefore, both macro- and microclimatic factors must be taken into account. The use of a data logger at the crime scene will indicate whether the data collected at the weather station is valid for the crime scene and whether the scene is, in general, warmer or cooler than that recorded at the weather station. A statistical analysis will allow the entomologist to predict the scene temperatures based on a comparison of the temperature data at the scene and weather station.

Season

Insects are seasonal. Therefore, forensic entomology is only of value in spring, summer, and fall in temperate climates, although it is equally valuable all year round in tropical regions. When recent remains are found in winter in temperate regions, there is little an entomologist can do. However, in some situations, the presence of a cold season may be of value in setting boundaries. For instance, if remains are found in winter and clearly indicate the presence of insect activity, then death must have occurred prior to the onset of the cold season.

In one case, the highly scavenged partial remains of a victim were discovered with almost no insect evidence associated, except for some empty puparia of a common blow fly species (Anderson, 1999a). The remains were found in early spring, when it was not yet warm enough for insect activity to begin, so the insects could not have proceeded from egg to adult during the spring, as indicated by the empty puparia. Therefore, the meager insect evidence indicated that the victim had died the previous fall. In fact, the victim had died early enough in the fall for development of the blow flies to be completed, indicating the time since death.

Exclusion of Insects

Insects may be excluded from the body by several means. The body may have been frozen after death. If this occurred naturally, during a winter season, then insects will colonize the remains when they thaw in spring. However, if the remains are artificially frozen in a freezer, insects will be excluded until the remains are removed from the freezer and allowed to defrost. Freezing can impact the rate of decomposition (Micozzi, 1986; Micozzi, 1997), which could influence the insect colonization, but insects would still colonize the remains, and time since thawing could be determined.

Burial can affect insect colonization, but the effect is strongly impacted by many factors such as depth and soil type (Rodriguez, 1997; Rodriguez and Bass, 1985; VanLaerhoven and Anderson, 1996; VanLaerhoven and Anderson, 1999). In most cases, colonization patterns are different, but insects are not excluded. Even deep graves can be colonized (Motter, 1898).

Wrapping a body can limit or delay insect activity, but it rarely completely impedes insects. Several cases of bodies being wrapped in plastic or other materials or locked inside containers have been documented in which the insects were able to penetrate the wrappings and colonize the body (Anderson, 1999a; Anderson, 2001b; Goff, 1992). In most cases, the wrappings themselves were not breached, but the insects were able to move between the wrappings.

The insects may have been delayed in reaching the remains, but their tenure on the body could still be determined. However, the delay is difficult to predict without completely recreating the scenario (Goff, 1992), and even then the result is uncertain. This underscores the reason why an entomologist can usually only provide the minimum elapsed time since death.

Report Submission

The entomological report will be of value to the police in their investigation, and it may also be admitted as evidence in court. This report will be the basis for the entomologist's testimony. It should, therefore, be understandable to such laypersons as the jury. It should not contain scientific jargon. Rather, the report should be written in a straightforward manner. The report should stand alone, allowing the reader to understand the scientific basis for the conclusions without recourse to further literature.

The report should begin with a brief description of the scene, site of discovery, victim, and collection as it pertains to the entomological conclusions. It should clearly explain how, when, and by whom the entomologist was contacted and how the entomological evidence was received by the entomologist. It should explain the procedure, the rearing data used (for timing the development of specific species under known conditions), and the identifications of the insects. It should also include sufficient background information about forensic entomology, outlining the species involved and their developmental or successional rates based on published literature. This information should be comprehensive enough for the reader to understand how conclusions were reached without including information that is not relevant to the case. Although it is always tempting for a scientist to write a report as if it were to be published in a scientific journal, it must be remembered that the report will be read and used by people with no entomological training and, frequently, little to no scientific training. This does not mean that information should be excluded, but it should be clearly explained, and primary references should be cited. This allows counsel to delve further into the appropriate literature, if desired.

The analyses should be clearly explained and the conclusions understandable. The report must be unbiased and based only on the scientific evidence, such as the entomological data, the meteorological conditions, and the published literature. With certain exceptions, the less the entomologist knows of the actual police investigation, the better. The entomologist should not be informed of the time the police believe that the victim died. Although this should not impact the analysis at all, it makes it clear to the reader that the report and conclusions are based on the scientific information alone and could not be influenced by other information, thus making the conclusions stronger in court.

There should be enough information in the report, and in the preserved specimens maintained from the case, that opposing counsel could present the report and specimens to an entomologist of their choice to request a second opinion. The preservation of the evidence and the entomologist's report must be comprehensive enough to allow another entomologist to analyze the data, if required. A review of entomological expert testimony is available in the literature (Greenberg and Kunich, 2002; Hall, 2001).

Testifying in Court

The expert witness is presented to the court as someone who has knowledge relevant to the case that will assist the jury or judge in understanding certain types of evidence. This knowledge is not expected of the layperson; therefore, by definition, an expert witness has more than the average person's knowledge of a specific subject. Expert witnesses are presented by either side, but they are impartial witnesses. An expert witness is permitted to give opinion evidence, as opposed to a layperson or material witness, who can only speak to something that he or she has actually seen or heard. Although a person may have a great deal of knowledge about a subject, that person is not legally an expert until pronounced so by the court. Once declared an "expert" in a case, the person is allowed to give opinion evidence and will be cross-examined. When the testimony has been completed and the "expert" has been dismissed by the judge, that person is no longer an expert in the legal sense until qualified as such in another trial.

Expert witnesses have a great responsibility to the courts and to justice, and they must be careful that they testify only on their own expertise and present entirely unbiased testimony. Although an expert is called by one side or the other, the expert does not testify for or against either side. The expert must be an unbiased professional who explains the science. Experts explain the methodology they used and explain the meaning of the results in terms a layperson would understand. They are not there to serve either side. Expert witnesses give opinion evidence, and they are the only witnesses allowed to do this. This is a great responsibility, and it must be treated as such by the scientist.

Forensic entomology is not a difficult science to explain to the courts. The basic principles are straightforward and easily understood. Jargon should be avoided, and visual aids that illustrate the life cycle can be helpful. Any caveats in the opinion evidence, such as the fact that temperature data may originate from a weather station several kilometers from the scene, should be clearly expressed and not left to be dragged out under cross-examination.

In most cases, the actual entomological evidence is examined by an entomologist who was called into the case by law enforcement agencies. They will, therefore, usually testify for the prosecution. This is not due to any sort of bias. It simply reflects the fact that if the police do not call an entomologist at the beginning of an investigation, any entomological evidence will invariably be lost or destroyed. Forensic entomologists also frequently testify for the defense, but usually as a rebuttal witness. By the time a person has been charged and a defense team is in place, the entomological evidence will have long since been destroyed, if it was not originally collected for the police case. Therefore, it is of the utmost importance that the primary entomologist maintain impeccable records and ensure that the preserved and reared entomological evidence be carefully maintained. This will allow opposing counsel to consult another forensic entomologist who can then examine all the evidence, despite having been brought into the investigation later.

Whether an entomologist or any other expert witness testifies for the prosecution or the defense, it must never be forgotten that the expert is an unbiased witness whose role is not to convict or exonerate, but to explain the science and its implications to the true triers of fact, the judge and jury.

Conclusion

Forensic entomology is an excellent tool in a death investigation. It is the primary method for determining time since death in the later postmortem interval and can be valuable for a year or more after death. It can also be used to determine other factors about the death. However, recognizing the value of insect evidence at the crime scene and accurate collection of this evidence are vital.

Questions

1. What is it about insect development that allows insects to be used to determine time since death?
2. Why is it important to preserve some specimens?
3. Why is it important to preserve specimens at the crime scene rather than waiting until the remains are at the morgue?
4. When are blow fly eggs of most value?
5. What information is required to determine elapsed time since death using maggot evidence?
6. What information is required to determine elapsed time since death using insect succession?
7. Why is it important to keep some specimens alive?

References

Anderson, G. S., Forensic entomology in death investigations, in *Forensic Anthropology Case Studies from Canada,* Fairgreave, S., Ed., Charles C. Thomas, 1999a, pp. 303–326.

Anderson, G. S., Wildlife forensic entomology: determining time of death in two illegally killed black bear cubs, a case report, *J. Forens. Sci.*, 44(4), 856–859, 1999b.

Anderson, G. S., Minimum and maximum developmental rates of some forensically significant Calliphoridae (Diptera), *J. Forens. Sci.*, 45(4), 824–832, 2000.

Anderson, G. S., The history of forensic entomology in British Columbia, *J. Entomol. Soc. B.C.*, 98, 129–138, 2001a.

Anderson, G. S., Insect succession on carrion and its relationship to determining time since death, in *Forensic Entomology: the Utility of Arthropods in Legal Investigations,* Castner, E. and Byrd, J., Eds., CRC Press, Boca Raton, FL, 2001b, pp. 143–176.

Anderson, G. S., Determining time of death using blow fly eggs in the early post mortem interval, *J. Legal Med.*, 118(4), 240–241, 2004.

Anderson, G. S. and Cervenka, V. J., Insects associated with the body: their use and analyses, in *Advances in Forensic Taphonomy: Method, Theory and Archeological Perspectives,* Haglund, W. D. and Sorg, M., Eds., CRC Press, Boca Raton, FL, 2002, pp. 174–200.

Anderson, G. S. and Huitson, N. R., Occurrence of myiasis in pet animals in British Columbia and some recommendations for treatment and determining length of time of abuse or neglect, *Can. Soc. Vet. Med.*, 45(12), 2004.

Anderson, G. S. and VanLaerhoven, S. L., Initial studies on insect succession on carrion in southwestern British Columbia, *J. Forens. Sci.*, 41(4), 617–625, 1996.

Ash, N. and Greenberg, B., Developmental temperature responses of the sibling species *Phaenicia sericata* and *Phaenicia pallescens*, *Ann. Entomol. Soc. Am.*, 68, 197–200, 1975.

Baer, W. S., The treatment of chronic osteomyelitis with the maggot (larva of the blow fly), *J. Bone Joint Surgery*, 13, 438–475, 1931.

Benecke, M. and Lessig, R., Child neglect and forensic entomology, *Forens. Sci. Int.*, 120, 155–159, 2001.

Bergeret, M., Infanticide, momification du cadavre; découverte du cadavre d'un enfant nouveau—né dans une cheminée ou il setait momifié; détermination de l'époque de la naissance par la présence de nymphes et des larves d'insectes dans le cadavre et par l'étude de leurs métamorphoses, *Ann. Hyg. Med. Leg.*, 4, 442–452, 1855.

Beyer, J. C., Enos, Y. F., and Stajic, M., Drug identification through analysis of maggots, *J. Forens. Sci.*, 25, 411–412, 1980.

Bourel, B. et al., Effects of morphine in decomposing bodies on the development of *Lucilia sericata* (Diptera: Calliphoridae), *J. Forens. Sci.*, 44(2), 354–358, 1999.

Bourel, B. et al., Determination of drug levels in two species of necrophagous Coleoptera reared on substrates containing morphine, *J. Forens. Sci.*, 46(2), 600–603, 2001.

Byrd, J. H. and Butler, J. F., Effects of temperature on *Chrysomya rufifacies* (Diptera: Calliphoridae) development, *J. Med. Entomol.*, 34(3), 353–358, 1997.

Byrd, J. H. and Butler, J. F., Effects of temperature on *Sarcophaga haemorrhoidalis* (Diptera: Sarcophagidae) development, *J. Med. Entomol.*, 35(5), 694–698, 1998.

Carvalho, L., Linhares, A. X., and Trigo, J., Determination of drug levels and the effect of diazepam on the growth of necrophagous flies of forensic importance in southeastern Brazil, *Forens. Sci. Int.*, 120, 140–144, 2001.

Catts, E. P., Problems in estimating the postmortem interval in death investigations, *J. Agric. Entomol.*, 9(4), 245–255, 1992.

Catts, E. P. and Haskell, N. H. E., *Entomology and Death: A Procedural Guide*, Joyce's Print Shop, Clemson, SC, 1990.

Dallwitz, R., The influence of constant and fluctuating temperatures on development and survival rate of pupae of the Australian sheep blowfly, *Lucilia cuprina*, *Entomol. Exp. Appl.*, 36, 89–95, 1984.

Davies, L. and Ratcliffe, G. G., Development rates of some pre-adult stages in blowflies with reference to low temperatures, *Med. Vet. Entomol.*, 8(3), 245–254, 1994.

Duke, L., Entomotoxicology: the Effects of Illicit Drugs on Insect Development and Implications for Time of Death Determinations, School of Criminology, Simon Fraser University, Burnaby, BC, 2003, p. 129.

Early, M. and Goff, M. L., Arthropod succession patterns in exposed carrion on the island of Oahu, Hawaii, *J. Med. Entomol.*, 23, 520–531, 1986.

Erzinclioglu, Y. Z., The application of entomology to forensic medicine, *Med. Sci. Law*, 23, 57–63, 1983.

Erzinclioglu, Z., *Blowflies*, Richmond Publishing, 1996.

Espinoza, E. O., Shafer, J. A., and Hagey, L. R., International trade in bear gallbladders: forensic source inference, *JFSCA*, 38(6), 1363–1371, 1993.

Fraenkel, G. and Bhaskaran, G., Pupariation and pupation in cyclorrhapous flies (Diptera): terminology and interpretation, *Ann. Entomol. Soc. Am.*, 66, 418–422, 1973.

Gilbert, B. M. and Bass, W. M., Seasonal dating of burials from the presence of fly pupae, *Am. Antiquity*, 32, 534–535, 1967.

Goff, M. L., Problems in estimation of postmortem interval resulting from wrapping of the corpse: a case study from Hawaii, *J. Agric. Entomol.*, 9(4), 237–243, 1992.

Goff, M. L., Charbonneau, S., and Sullivan, W., Presence of fecal matter in diapers as potential source of error in estimations of postmortem intervals using arthropod development patterns, *J. Forens. Sci.*, 36(5), 1603–1606, 1991.

Goff, M. L. and Lord, W. D., Entomotoxicology: insects as toxicological indicators and the impact of drugs and toxins on insects, in *Forensic Entomology: the Utility of Arthropods in Legal Investigations*, Castner, E. and Byrd, J., Eds., CRC Press, Boca Raton, FL, 2001, pp. 331–340.

Goodbrod, J. R. and Goff, M. L., Effects of larval population density on rates of development and inter-actions between two species of *Chrysomya* (Diptera: Calliphoridae) in laboratory culture, *J. Med. Entomol.*, 27, 338–343, 1989.

Grassberger, M. and Reiter, C., Effect of temperature on development of the forensically important hol-arctic blow fly *Protophormia terraenovae* (Robineau-Desvoidy) (Diptera: Calliphoridae), *Forens. Sci. Int.*, 128, 177–182, 2002.

Greenberg, B., Flies as forensic indicators, *J. Med. Entomol.*, 28, 565–577, 1991.

Greenberg, B. and Kunich, J. C., *Entomology and the Law: Flies as Forensic Indicators*, Cambridge University Press, London, 2002.

Greenberg, B. and Tantawi, T. I., Different developmental strategies in two boreal blow flies (Diptera: Calliphoridae), *J. Med. Entomol.*, 30(2), 481–484, 1993.

Hall, R. D., The forensic entomologist as expert witness, *Forensic Entomology: the Utility of Arthropods in Legal Investigations*, Castner, E. and Byrd, J., Eds., CRC Press, Boca Raton, FL, 2001, pp. 379–400.

Haskell, N. H., Personal communication, St. Joseph's College, Rensselaer, IN.

Haskell, N. H., Hall, R. D., Cervenka, V. J., and Clark, M. A., On the body: insects' life stage presence and their postmortem artifacts, in *Forensic Taphonomy: the Postmortem Fate of Human Remains*, Haglund, W. D. and Sorg, M. H., Eds., CRC Press, Boca Raton, FL, 1997, pp. 415–448.

Haskell, N. H., Lord, W. D., and Byrd, J. H., Collection of entomological evidence during death investi-gations, in *Forensic Entomology: the Utility of Arthropods in Legal Investigations*, Castner, E. and Byrd, J., Eds., CRC Press, Boca Raton, FL, 2001, pp. 81–120.

Hedouin, V. et al., Determination of drug levels in larvae of *Lucilia sericata* (Diptera: Calliphoridae) reared on rabbit carcasses containing morphine, *J. Forens. Sci.*, 44(2), 351–353, 1999.

Henssge, C. et al., *The Estimation of the Time since Death in the Early Postmortem Interval*, Arnold, London, 1995.

Howden, H. F., The mercure trial: a sideline of entomology, *Canadian Entomologist*, 96, 121, 1964.

Introna, F., Campobasso, C. P., and Goff, M. L., Entomotoxicology, *Forens. Sci. Int.*, 120, 42–47, 2001.

James, M. T., *The Flies that Cause Myiasis in Man*, U.S. Dept. Agriculture, Washington, D.C., 1947.

Johnston, W. and Villeneuve, G., On the medico-legal application of entomology, *Montreal Med. J.*, 26, 81–90, 1897.

Kashyap, V. K. and Pillai, V. V., Efficacy of entomological method in estimation of postmortem interval: a comparative analysis, *Forens. Sci. Int.*, 40(3), 245–250, 1989.

King, A. B. and Flynn, K. J., Maggot therapy revisited: a case study, *Dermatol. Nursing*, 3, 100–102, 1991.

Kintz, P. et al., Fly larvae: a new toxicological method of investigation in forensic science, *J. Forens. Sci. (JFSCA)*, 35, 204–207, 1990.

Kintz, P., Tracqui, A., and Mangin, P., Analysis of opiates in fly larvae sampled on a putrefied cadaver, *J. Forens. Sci. Soc.*, 34, 95–97, 1994.

Lord, W. D., Case histories of the use of insects in investigations, in *Entomology and Death: a Procedural Guide,* Catts, E. P. and Haskell, N. H., Eds., Joyce's Print Shop, Clemson, SC, 1990, pp. 9–37.

McKnight, B. E., *The Washing Away of Wrongs: Forensic Medicine in Thirteenth Century China by Sung T'zu*, 1981.

Megnin, J. P., *La Faune des cadavres: application et l'entomologie à la médecine legale*, Villars, Masson et Gauthiers, Paris, 1894.

Melvin, R., Incubation period of eggs of certain muscoid flies at different constant temperatures, *Ann. Entomol. Soc. Am.*, 27, 406–410, 1934.

Micozzi, M. S., Experimental study of postmortem changes under field conditions: effects of freezing, thawing and mechanical injury, *J. Forens. Sci.*, 31, 953–961, 1986.

Micozzi, M. S., Frozen environments and soft tissue preservation, Chap. 11 in *Forensic Taphonomy: the Postmortem Fate of Human Remains,* Haglund, W. D. and Sorg, M. H., Eds., CRC Press, Boca Raton, FL, 1997, pp. 171–180.

Miller, M. L. et al., Isolation of amitriptyline and nortriptyline from fly puparia (Phoridae) and beetle exuviae (Dermestidae) associated with mummified remains, *J. Forens. Sci.*, 39(5), 1305–1313, 1994.

Motter, M. G., A contribution to the study of the fauna of the grave: a study of 150 disinterments, with some additional observations, *J. N.Y. Entomol. Soc.*, 6, 201–231, 1898.

Mumcuoglu, K. Y. et al., Maggot therapy for the treatment of diabetic foot ulcers, *Diabetes Care*, 21(11), 2030–2031, 1998.

Nishida, K., Experimental studies on the estimation of postmortem intervals by means of fly larvae infesting human cadavers, *Jpn. J. Legal Med.*, 38, 24–41, 1984.

Nuorteva, P., Studies on the significance of flies in the transmission of poliomyelitis IV: the composition of the blowfly fauna in different parts of Finland during the year 1958, *Annales Entomol. Fennici*, 25, 137–162, 1959.

Nuorteva, P., Sarcosaprophagous insects as forensic indicators, in *Forensic Medicine: a Study in Trauma and Environmental Hazards,* Tedeschi, C. G., Eckert, W. G., and Tedeschi, L. G., Eds., W. B. Saunders Co., Philadelphia, 1977, pp. 1072–1095.

Nuorteva, P. and Hasanen, E., Transfer of mercury from fishes to sarcosaprophagous flies, *Annales Zoologici Fennici*, 9, 23–27, 1972.

Nuorteva, P. and Nuorteva, S. L., The fate of mercury in sarcosaprophagous flies and in insects eating them, *Ambio*, 11, 34–37, 1982.

Nuorteva, P. and Skaren, U., Studies on the significance of flies in the transmission of poliomyelitis V: observations on the attraction of blowflies to the carcases of micromammals in the commune of Kuhmo, East Finland, *Annales Zoologici Fennici*, 5, 188–193, 1960.

Prichard, J. G. et al., Implications of Trombiculid mite bites: report of a case and submission of evidence in a murder trial, *J. Forens. Sci.*, 31, 301–306, 1986.

RCMP, *"E" Division Training*, Mellis, J., dir., and Torp, N., prod., Anderson, G. S. and Andrews, B., Royal Canadian Mounted Police, Vancouver, BC, 1997, forensic entomology training video.

Rodriguez, W. C. and Bass, W. M., Decomposition of buried bodies and methods that may aid in their location, *J. Forens. Sci.*, 30, 836–852, 1985.

Rodriguez, W. C. and Bass, W. M., Examination of Badly Decomposed or Skeletonized Remains: Overlooked Evidence, paper presented at 39th annual meeting of American Academy of Forensic Sciences, San Diego, CA, 1987.

Rodriguez, W. C. I., Decomposition of buried and submerged bodies, in *Forensic Taphonomy: the Postmortem Fate of Human Remains,* Haglund, W. D. and Sorg, M. H., Eds., CRC Press, Boca Raton, FL, 1997, pp. 459–467.

Sadler, D. W., Fuke, C., Court, F., and Pounder, D. J., Drug accumulation and elimination in *Calliphora vicina* larvae, *Forens. Sci. Int.*, 71(3), 191–197, 1995.

Saferstein, R., *Criminalistics: an Introduction to Forensic Science*, Prentice Hall, Upper Saddle River, NJ, 2004.

Sherman, R. A., Maggot therapy, *Infection Control in Long-Term Care Facilities Newsletter (APIC)*, 6(3), 5, 1995.

Sherman, R. A., A new dressing design for use with maggot therapy, *Plastic Reconstr. Surg.*, 100, 451–456, 1997.

Sherman, R. A., Maggot debridement in modern medicine, *Infections Med.*, 15(9), 651–656, 1998.

Sherman, R. A., Hall, M. J. R., and Thomas, S., Medicinal maggots: an ancient remedy for some contemporary afflictions, *Ann. Rev. Entomol.*, 45, 55–81, 2000.

Sherman, R. A. and Pechter, E. A., Maggot therapy: a review of the therapeutic applications of fly larvae in human medicine, especially for treating osteomyelitis, *Med. Vet. Entomol. Oxford: Blackwell Scientific Publications*, 2(3), 225–230, 1988.

Sherman, R. A., Tran, J., and Sullivan, R., Maggot therapy for treating venous stasis ulcers, *Arch. Dermatol.*, 132, 254–256, 1996.

Sherman, R. A., Wyle, F., and Vulpe, M., Maggot debridement therapy for treating pressure ulcers in spinal cord injury patients, *J. Spinal Cord Med.*, 18(2), 71–74, 1995.

Sherman, R. A., Wyle, F., and Vulpe, M., Maggot therapy for treating pressure ulcers in spinal cord injury patients, *J. Spinal Cord Med.*, 18, 71–74, 1996.

Sherman, R. A. et al., The utility of maggot therapy for treating pressure sores, *J. Am. Paraplegic Soc.*, 16, 269, 1993.

Sherman, R. A. et al., Maggot debridement therapy for treating pressure sores, *J. Am. Paraplegic Soc.*, 14, 200, 1990.

Sherman, R. A. and Wyle, F. A., Low cost, low maintenance rearing of maggots in hospitals, clinics and schools, *J. Am. Soc. Trop. Med. Hyg.*, 54(1), 38–41, 1996.

Smith, K. G. V., *A Manual of Forensic Entomology*, Trustees of British Museum (Nat. Hist.) and Cornell University Press, London, 1986.

Sperling, F. A., Anderson, G. S., and Hickey, D. A., A DNA-based approach to the identification of insect species used for postmortem interval estimation, *J. Forens. Sci.*, 39(2), 418–427, 1994.

VanLaerhoven, S. L. and Anderson, G. S., *Forensic Entomology: Determining Time of Death in Buried Homicide Victims Using Insect Succession*, Canadian Police Research Centre, 1996.

VanLaerhoven, S. L. and Anderson, G. S., Insect succession on buried carrion in two biogeoclimatic zones of British Columbia, *J. Forens. Sci.*, 44(1), 32–43, 1999.

Wall, R., French, N., and Morgan, K. L., Effects of temperature on the development and abundance of the sheep blowfly *Lucilia sericata* (Diptera: Calliphoridae), *Bull. Entomol. Res.*, 82, 125–131, 1992.

Webb, J. P. J. et al., The chigger species *Eutrombicula belkini* Gould (Acari: Trombiculidae) as a forensic tool in a homicide investigation in Ventura County, California, *Bull. Soc. Vector Ecologists*, 81, 41–146, 1983.

Wells, J. D., Estimating the postmortem interval, in *Forensic Entomology: the Utility of Arthropods in Legal Investigations*, Castner, E. and Byrd, J., Eds., CRC Press, Boca Raton, FL, 2001, pp. 263–286.

Wells, J. D. et al., Human and insect mitochondrial DNA analysis from maggots, *J. Forens. Sci.*, 46(3), 685–687, 2001.

Wells, J. D. and Sperling, F. A., DNA-based identification of forensically important Chrysomyinae (Diptera: Calliphoridae), *Forens. Sci. Int.*, 120, 110–115, 2001.

Wilson, Z., Hubbard, S., and Pounder, D. J., Drug analysis in fly larvae, *Am. J. Forens. Med. Pathol.*, 14(2), 118–120, 1993.

Yovanovitch, P., *Entomologie Appliquée à la Médicine Légale*, Ollier-Henrey, Paris, 1888.

SECTION II

Evaluation of the Crime Scene

Chapters

Crime Scene Investigation

Forensic Digital
Photo Imaging

Recognition of
Bloodstain Patterns

Crime Scene Investigation

Marilyn T. Miller, Ed.D.

10

Introduction

From nanogram quantities of DNA to artificial intelligence databases capable of identifying latent fingerprints and biological fluids, forensic science and the analysis of very minute quantities of physical evidence have advanced and improved. Yet, this utilization of the scientific method has been underutilized at the crime scene and in its investigation. Crime scene investigation is the starting point for the successful use of physical evidence by the forensic laboratory and the criminal investigator. Now more than ever, the scene of a crime must always be properly managed and investigated in the best manner possible.

Successful, high-quality crime scene investigation is a simple, methodical process. It is not rigid, but follows a set of principles and procedures that adhere to guidelines ensuring that all the physical evidence is discovered and investigated; the result will be that justice is served. The basic crime scene procedures are physical evidence recognition, documentation, proper evidence collection, packaging, and preservation, and finally, scene reconstruction. Every crime scene is unique, and with experience, the crime scene investigator will be able to use this logical and systematic approach to investigate even the most challenging to a successful conclusion.

Defining the Crime Scene

The only thing consistent about crime scenes is their variety. Because of the diversity of possible scenes there are many ways to define or classify crime scenes. First, crime scenes can be classified according to the location of the original criminal activity. This classification of the crime scene labels the site of the original or first criminal activity as the **primary crime scene** and any subsequent crime scenes as **secondary**. This classification does not infer any priority or importance to the scene but is simply a designation of sequence of locations.

A second classification of crime scenes is based on the size of the crime scene. Using this classification, a single **macroscopic**

crime scene is possibly composed of many crime scenes. For example, a gunshot victim's body dumped in a field represents the following crime scenes within the overall crime scene of the field: the body, the body's wounds, and the ground around the body. The *microscopic* classification of the scene is more focused on the specific types of physical evidence found in the macroscopic crime scenes. Using the previous example, the **microscopic crime scenes** are the trace evidence on the body, the gunshot residue around the wound, and the tire tread marks in the ground next to the body.

Other classifications of the crime scene are those based on the *type of crime* committed (homicide, robbery, sexual assault, etc.); the crime scene *condition* (organized or disorganized); the *physical location* of the crime scene (indoors, outdoors, vehicle, etc.); and the type of criminal *behavior* associated with the scene (passive or active).

Even with these various crime scene classifications, no single definition will adequately work for every scene. Ultimately, it is a combination or adaptation of the classifications that is used by the investigator. The definition of the crime scene should never establish immoveable boundaries to the crime scene. The crime scene investigator must be constantly evaluating and often times changing the defined area called the crime scene.

Uses and Information from Physical Evidence in Criminal Investigations

The objectives of any crime scene investigation are to recognize, to preserve, to collect, to interpret, and to reconstruct all the relevant physical evidence at a crime scene. A forensic laboratory examines the physical evidence to provide the investigator with information to help solve cases. The integration of the crime scene investigation with the forensic testing of the scene's physical evidence forms the basis of scientific crime scene investigation.

The following are examples of the information that are obtained from the forensic testing and examination of physical evidence in a criminal investigation:

- *Linkage of persons, scenes, or objects.* This forms the principle of all crime scene investigations. The Locard exchange principle states that whenever two objects come into contact, there will be a mutual exchange of matter between them. Linking suspects to victims is the most important and common type of linkage by physical evidence in criminal investigations. Linking victims and suspects to objects and scenes can also be accomplished by use of physical evidence. A surviving victim may not always know the location of the crime scene. The physical evidence on the victim will help identify the scene.

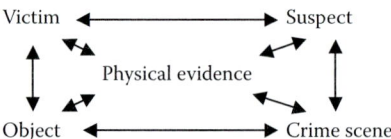

- *Provide investigative leads.* Physical evidence can provide direct information to an investigator. However, not all physical evidence at the crime scene will be directly linked to a suspect. Frequently, the physical evidence will provide indirect information or investigative leads to the investigator. This is an important and significant use of physical evidence in any criminal investigation. Not every crime scene has individualizing physical evidence, such as fingerprints, but every crime scene will have physical evidence that assists the investigator with information, such as a footwear impression's manufacturer, size or type of shoe worn by the suspect.

- *Information on the corpus delicti.* This is the determination of the essential facts of an investigation—the physical evidence itself, the patterns of the evidence, and the laboratory examinations of the evidence. The red-brown stains in a kitchen may be significant to an investigation, but may be more relevant if those stains are bloodstains with DNA matching a victim.

- *Information on the modus operandi.* Criminals repeat behavior and this particular behavior represents their "signature" or preferred method of operation. Burglars will frequently gain entry into scenes using the same technique or bombers will repeatedly use the same type of ignition device. The physical evidence they leave behind, once found at the scene, can be used to identify them.
- *Proving or disproving witness statements.* Credibility is an important issue with witnesses, victims, and suspects. The presence or absence of certain types of physical evidence will be useful in the determination of the accuracy of their statements. Crime scene patterns or patterned physical evidence (bloodstain patterns, fingerprints, gunshot residue, etc.) are especially well suited for determination of credibility.
- *Identification of the suspect(s).* Forensic examination is a process of recognition, identification, individualization, and reconstruction. Identification of a suspect is accomplished by using the first three steps that result in an individualization or determination of the source of an item of physical evidence. This individualization is facilitated by comparison testing. The best example of a comparison-type of individualization is fingerprint evidence. Recent advances in the use of the Automated Fingerprint Identification System (**AFIS**) or DNA databases (**CODIS**) will allow for a single fingerprint or small bloodstain found at a crime scene to identify or more properly individualize a suspect.
- *Identification of unknown substances.* As above, the identification of unknown substances is a common use of the physical evidence. Identification of controlled substances or poisons such as anthrax are good examples.
- *Reconstruction of a crime.* This is the final step in the forensic examination process. The crime scene investigator is frequently more interested in how a crime occurred than to identify or individualize the evidence at the scene. The "how" of a crime scene is more important than the "who" of the crime.

Science and Crime Scene Investigations

Crime scene investigation is not a mechanical process relegated to "technicians" to go through a series of steps to "process the crime scene." It is a dynamic, thoughtful process that requires an active approach to the scene investigation like being aware of the linkage principle of the evidence, using scene analysis and definition, and being able to offer opinions on the reconstruction of the scene. (See Figure 10.1.) Scientific crime scene investigation is grounded in the scientific method. It is methodical and systematic. It is based on the Locard exchange principle, logic, and the use of the scientific knowledge of forensic techniques of physical evidence examinations to develop investigative leads that will ultimately solve the crime.

General Crime Scene Procedures

Scene Management

In today's criminal investigations, crimes are solved by the teamwork of investigators and the crime scene investigators. It is the combined use of techniques and procedures recognizing the power of crime scenes, physical evidence,

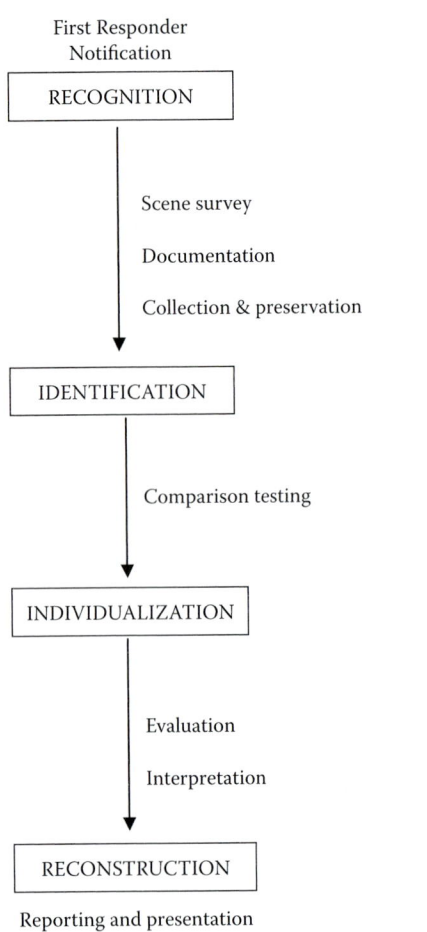

First Responder
Notification

RECOGNITION

Scene survey

Documentation

Collection & preservation

IDENTIFICATION

Comparison testing

INDIVIDUALIZATION

Evaluation

Interpretation

RECONSTRUCTION

Reporting and presentation

Figure 10.1 Steps to the scientific examination of a crime scene.

records, and witnesses. Unfortunately, numerous cases, routine and complex, have shown that despite available crime scene technologies and specially trained personnel, the productivity of crime scene investigations is only as good as the supportive management team.

The four distinctive but interrelated components of **crime scene management** are: (1) information management, (2) manpower management, (3) technology management, and (4) logistics management. Deficiencies, negligence, or overemphasis of any one of these components will imperil the overall crime scene investigation. These components are all based on the fundamental need for good and on-going communication between all personnel throughout the entire investigation process.

The components of crime scene management and the need for continual communication have resulted in some choices for appropriate crime scene investigation models. Each model has its advantages and disadvantages based on the allocation of personnel and resources, training and expertise, crime rates, types of crimes, jurisdictional issues, and the support services available. (See Table 10.1.)

Table 10.1
Crime Scene Investigation Models

Model Type	Description	Advantages	Disadvantages
Traditional	Uses patrol officers and detectives as crime scene technicians	Useful if resources and demand are relatively low	Minimal experience and time commitment conflicts with regular duties
Crime scene technicians	Specially trained, full-time civilian personnel	Continuity, specialization, scientific/technical training	Minimal investigative experience; lack global view of investigation
Major crime squad	Full-time, sworn officers	Primary assignment; increased experience	Transfers out of unit; may deplete investigation resources; only major cases
Lab crime scene scientist	Laboratory scientists	Superior technical and scientific skills; knowledge of current methods	No investigative experience; deplete lab resources
Collaborative team	Uses police, technicians, lab personnel, medical examiners, and prosecuting authorities	Advanced scientific, technical and investigative resources; shared responsibilities	Requires extensive resources and comprehensive procedures with continual communication

First Responding Officer

The first responding officers to a crime scene are usually police officers, fire department personnel, or emergency medical personnel. These first responders are the only people to view the crime scene in its most original or pristine condition. Their actions at the crime scene will form the basis for the successful or unsuccessful resolution of the investigation. They must perform their duties, but they should always keep in mind that they are part of the beginning effort to link victims to suspects to crime scenes. They must never destroy that link. It is imperative that they gain experience and receive continual training or education.

The first responder must always maintain an open and objective mind when approaching the crime scene. Upon arrival at the scene, safety is a primary concern for themselves and the victim. Once the scene or the victim is safe then the first responders must begin to thoroughly document their observations and actions at the scene. As soon as possible, the first responder should initiate crime scene security measures.

Duties of the First Responder

1. Assist the victim and prevent any changes to the victim.
2. Search for and arrest the suspect if that person is still on scene.
3. Detain any witnesses. As discussed above the witnesses possess valuable information about the crime scene. Keep the witnesses separated to preserve their objectivity. Do not take them back to the scene if at all possible.
4. Protect the crime scene. Begin the crime scene security measures by use of barrier tape, official vehicles, or other means, as required. Establish a crime scene security log to record any persons who enter/exit the crime scene. Do not smoke, drink, or eat within the secured crime scene and prevent unnecessary persons or officials from entering or contaminating the scene.

5. All movements, alterations, or changes made to the crime scene should be noted and communicated to the crime scene investigators.

Securing the Crime Scene

As shown above, the Locard exchange principle is the basis for linking physical evidence from or to the victim, suspect, and crime scene. Anyone entering the crime could potentially alter or change a crime scene and its evidence. For this reason alone, access to the crime scene must be restricted and, if possible, prevented except for crime scene personnel. The first responder must establish tapes or use any physical barriers like vehicles that will assist with the protection of the crime scene as soon as possible.

Once the scene barriers have been established an officer shall be designated as the scene security officer. That officer has the security responsibility for preventing any entrances into the crime scene by curious onlookers. A contamination log or security log is kept as a written record of all entry/exit into the secure areas of the crime scene. Use of a multilevel security approach can successfully prevent unwanted entries. (See Figure 10.2.)

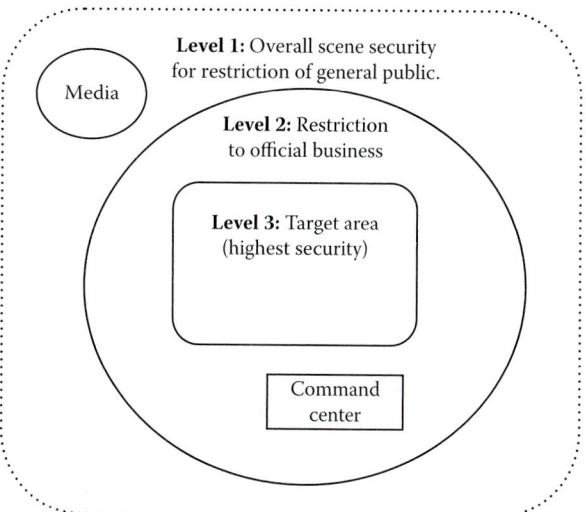

Figure 10.2 Multilevel crime scene security.

Only in rare situations will the crime scene investigator be the first responder to the scene but once on-scene the established secure areas should be evaluated by the crime scene investigator and changed if necessary.

Crime Scene Survey

Once the crime scene investigator has arrived at the crime scene and scene security has been evaluated, the preliminary scene survey or "walk-through" should be done. The crime scene investigator and the first responder will usually do the scene survey together. The lead investigator or detective, if available, can also benefit from the scene survey. At this time the use of digital or some form of instant photography for preliminary documentation can be helpful. The survey is the first examination or orientation of the crime scene by the crime scene investigator, and the following guidelines should be followed:

- Use the walk-through as a mental beginning for a reconstruction theory that can and should be changed as the scene investigation progresses.
- Note any transient (temporary) or conditional (the result of an action) evidence that might be present and requires immediate protection or processing.
- Be aware of the weather conditions, and take precautions if adverse weather is anticipated.
- Note any points of entry or exit and paths of travel within the crime scene that may require additional protection. Be aware of any alterations or contamination of these areas by first responder personnel.
- Record briefly initial observations of the answers to who, what, where, when, and how questions. This is not an appropriate time for detailed description of the scene.
- Access the scene for personnel, precautions, or equipment that will be needed.
- Notify superior officers or other agencies as required.

Crime Scene Documentation

Once the crime scene has been evaluated by the preliminary scene survey, the crime scene's condition must be recorded or documented (Figure 10.3). Documentation of the crime scene is the most important step in the processing of the crime scene. The purpose of crime scene documentation is to permanently record the condition of the crime scene and its physical evidence. It is the most time-consuming activity at the scene and requires the investigator to stay organized and systematic. Problem solving skills, innovation, and originality will also be needed. The four major tasks of documentation are: note taking, videography, photography, and sketching. All four are necessary and none is an adequate substitute for another. For example, notes are not substitutes for photography; video is not a substitute for sketching, etc.

Documentation, in all its various forms, begins with the initial involvement of the scene investigator. The documentation never stops; it may be slowed down, but documentation remains constant. **Crime scene documentation** will be presented below in the sequence by which it should be done at the crime scene. The systematic process presented is suggested in order to maintain the organized nature of scientific crime scene investigation.

Taking Notes at the Crime Scene

Effective notes as part of a crime scene investigation provide a written record of all of the crime scene activities. The notes are taken as the activities are done and not subjected to memory loss at a later time. A general guideline for note taking is to consider the "W"s (who, what, when, why, and how) but specifically include the following:

- *Notification information.* Date/time, method of notification, and information received.
- *Arrival information.* Means of transportation, date/time, personnel present

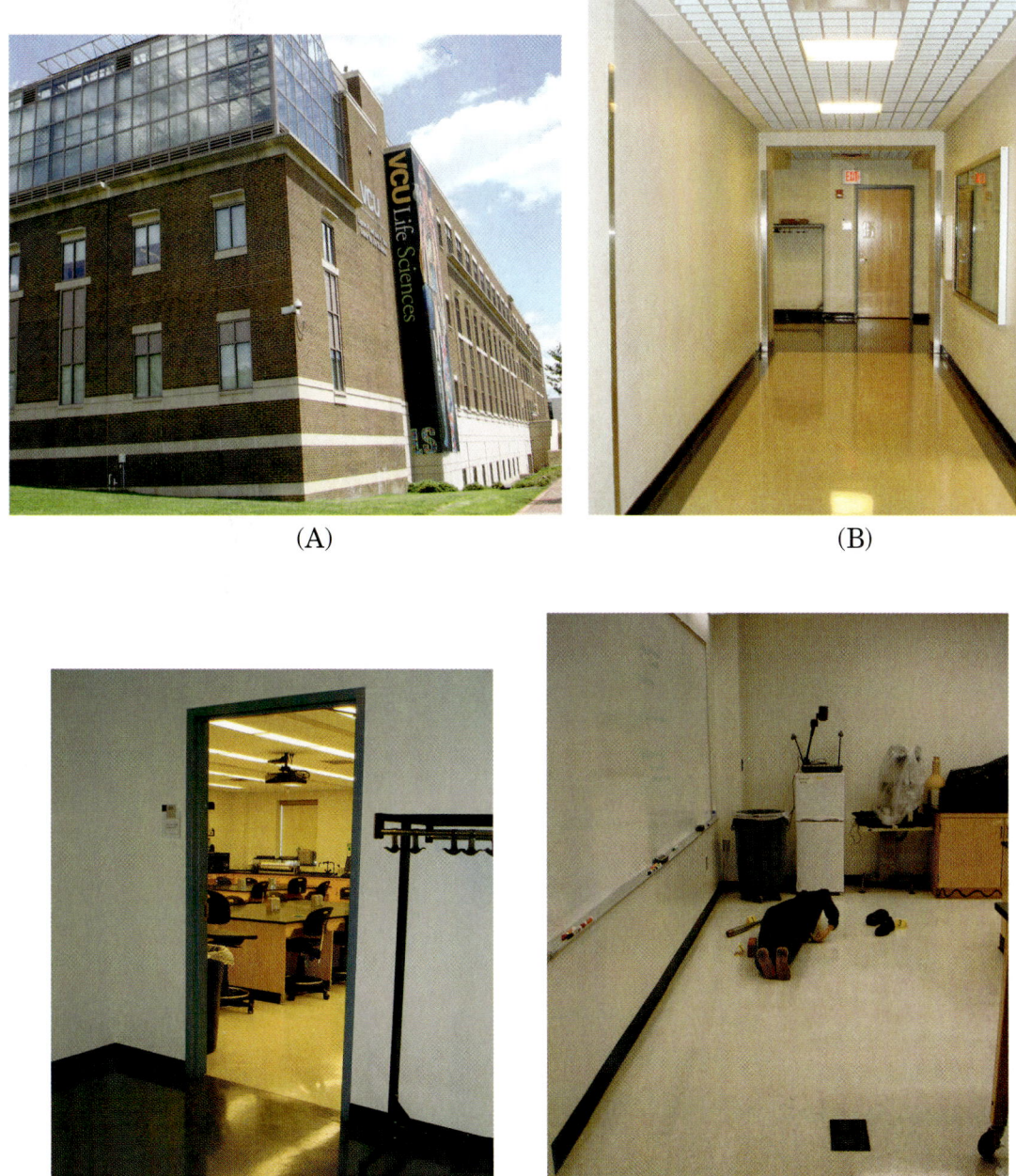

(A)

(B)

(C)

(D)

Figure 10.3 Photographic documentation at the scene. (A) Overall of scene building; (B) hallway to scene; (C) point of entry; (D) overall from point of entry. *(continued on next page)*

at the scene, and any notifications to be made.

- *Scene description.* Weather, location type and condition, major structures, identification of **transient** and **condi-** **tional evidence**, especially POEs, containers holding evidence of recent activities (ashtrays, trash cans, etc.), clothing, furniture, and any weapons present.

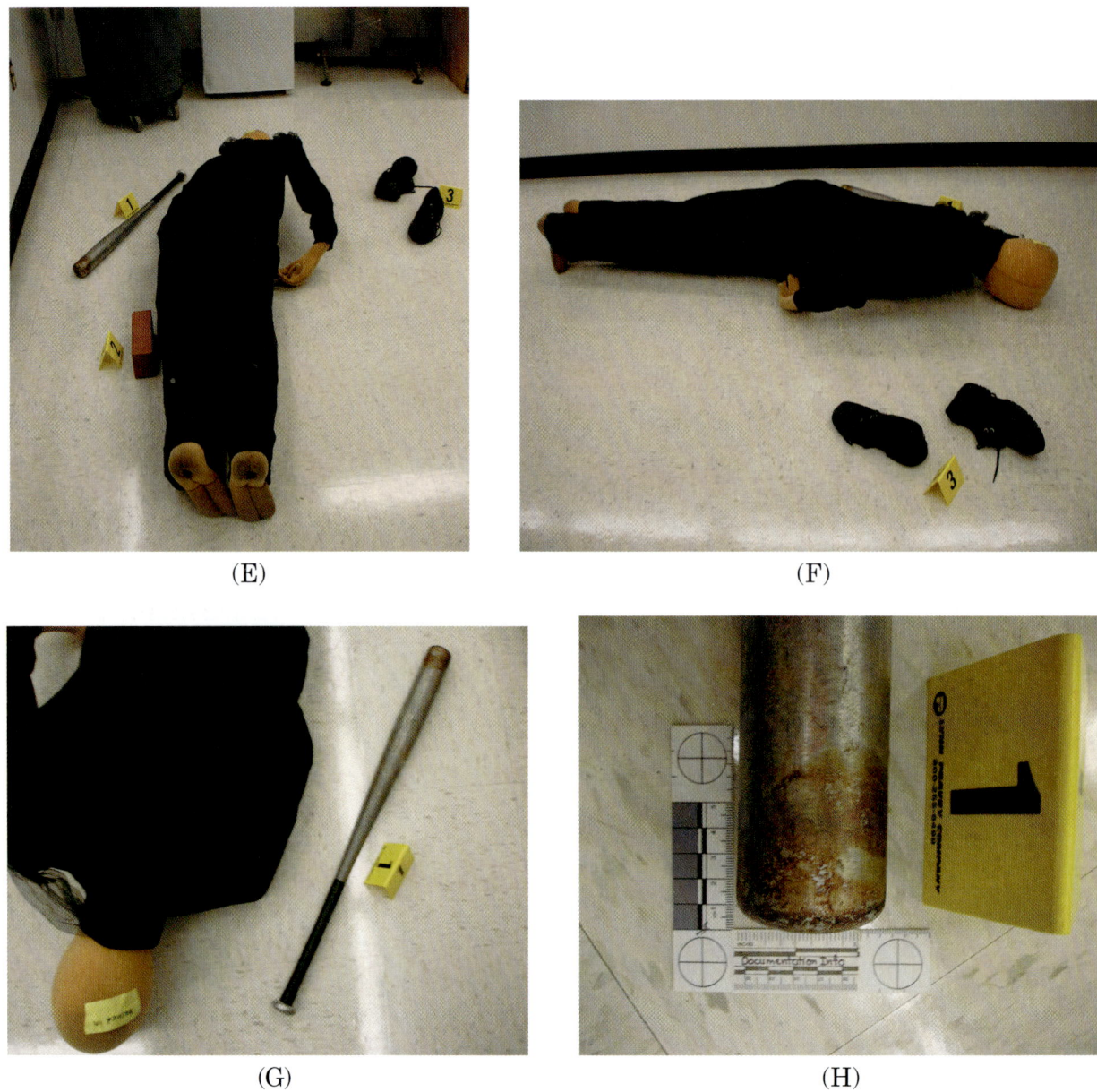

(E)   (F)

(G)   (H)

Figure 10.3 *(continued)*   (E) overall of body 1; (F) overall of body 2; (G) midrange of body and bat; (H) exam photo of blood on bat.

- *Victim description.* In most jurisdictions the body should never be moved or disturbed until the medical examiner has given approval. Once given permission then notes of position, lividity, wounds, clothing, jewelry or identification (its presence or absence).
- *Crime scene team.* Assignments to team members, walk-thru information, beginning and ending times, and evidence handling results.

Accurate crime scene note taking is crucial at the initial crime scene investigation, but it is essential for any subsequent investigations that may follow.

Video Recording of the Crime Scene

Video recording of the crime scene is a routine procedure for crime scene documentation in recent years. Its acceptance is widespread

(I)   (J)

Figure 10.3 *(continued)* (I) closeup of shoes; (J) exam photo of shoe.

and is due to a virtual appearance of the scene and increased availability of affordable equipment with user-friendly features like DVD recording, built-in stability, digital zoom lenses, and compact size. Jury acceptability and expectation has also added to the recognized use of video recording of the crime scene investigations.

Videography of the crime scene should follow the scene survey in scientific crime scene investigation. The video recording of crime scenes is an orientation format and should remain objective in its recording of the crime scene. It should not include any members of the crime scene team or their equipment. It should not be narrated and not contain any audio recording of subjective information at the scene. The following summarizes the process that should be followed for effective video taping of crime scenes:

- *Document the recording by use of a placard.* This is the documentation of the documentation technique; include case number, date/time, location, and videographer's name.
- *Begin with the scene surroundings.* Include roads to and from the scene before taping the general views of the scene itself; use the four compass points as a guide.

- *General orientation of the scene.* Tape the orientation of the items of evidence in relation to the overall scene; wide-angle views are especially useful; do not jump from one location to another, use a smooth transition that includes the overall locations of evidence.
- *Victim's viewpoint.* Move to a safe location near the victim and tape the four compass points viewed away from the victim.
- *Camera techniques.* Make smooth movements; use a tripod or monopod if possible. Use additional lighting for all interior scenes (most camcorders have low-light automatic aperture corrections but additional lighting is suggested); once a tape is completed, review it on scene and reshoot the scene as needed.
- *Original videos.* It is evidence and should not be edited or changed; make copies when needed.

Video recording of crime scenes is a valuable tool for easy perception or virtual reality of the crime scene that is often not accomplished by the other documentation tasks. However, it is never an adequate substitute for any of the other tasks.

Photographing the Crime Scene

The purpose of still photography documentation of the crime scene is to provide a true and accurate pictorial record of the crime scene and physical evidence present. As a result of this documentation, it records the initial condition of the scene, it provides investigators and others with a record that can be analyzed or examined subsequent to the scene investigation, and it serves as a permanent record for any legal concerns. Photography of a crime scene is normally done immediately following the videography of the scene or after the preliminary scene survey.

The systematic, organized method for recording the crime scene and pertinent physical evidence is best achieved by following a progressive general to specific guideline:

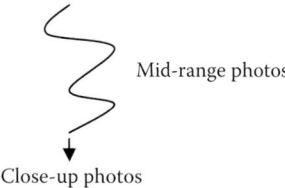

General or overall photos (with locator images)

Mid-range photos

Close-up photos

Adherence to this guideline will allow for orientation of the crime scene a whole, the orientation of the evidence within the scene and to provide for laboratory or **examination quality photographs** of specific items of evidence that may be used for testing purposes away from the scene. Any number of photographs should be taken at the crime scene and cannot be predetermined or limited. A crime scene investigator should never doubt whether a photo should be taken—it should always be taken (Figure 10.4).

See Table 10.2 for some of the guidelines that should be followed when photographing a crime scene.

As shown in Table 10.3, every photograph that is taken at the crime scene must be recorded in a photo log. This proof of a documentation photograph taken includes the time taken, the roll number, the exposure number, the camera settings used (f/stop, shutter speed), an indication of distance to subject, the type of photo taken, and a brief description of the photograph.

(A)

(B)

Figure 10.4 (A) Closeup photograph of tire tread mark in snow. (B) Closeup photograph of footwear impression in the soil..

The following is a list of basic equipment needed for photographic documentation of crime scenes:

- Camera—SLR 35mm is most common
- Normal lens—50 to 60 mm
- Wide angle lens—28 to 35 mm
- Close-up lens with accessories
- Electronic flash with cord
- Tripod
- Image card or film—color and black and white
- Label materials—cards, pens, markers
- Scales or rulers
- Flashlight
- Extra batteries
- Photo **log** sheets

Table 10.2
Guidelines for Photographing Crime Scene

Type of Photo	Guidelines for Photographing
Overall photos:	**Exteriors**—Surroundings; buildings and major structures; roads or paths of travel into or away from scene; street signs or survey markers; mail boxes or address numbers; take aerial photographs when possible; photograph before 10 am or after 2 pm if possible.
	Interiors—Use the four compass points or room corners; overlapping of views; doors leading into/from structure; use tripod in low light situations for increased depth of focus concerns.
Mid-range photos:	Follow a step-wise progression of views; use various lenses or change the focal length of the lenses to achieve a "focused" view of the individual items of evidence within the original view of the crime scene; add flash lighting to enhance details or patterned evidence.
Close-up photos:	Use documentation placards; use flash photography; flash must be detached from the camera; use proper side lighting effects; use filling-in with flash when harsh shadows present; take photos with and without scales.
All photos:	Record in log; use camera settings that achieve good depth of focus; no extraneous objects like team members, equipment, feet or hands; change point of view; be aware of reflective surfaces; when in doubt—*photograph it!*

Table 10.3
Photo Log

DATE: _____

NAME: _____   CASE # _____

EQUIPMENT USED:

    CAMERA _____   BODY SN _____

    LENS(ES) _____   LENS SN _____

    FLASH USED _____   FLASH SN _____

DIGITAL IMAGES DOWNLOADED: _____

Roll #: _____

Time	Exp. No.	Type*	Description*	Settings & Misc.*

* Type: overall, mid-range, specific
Description: object(s) in photo
Settings and Miscellaneous: camera f/stop, shutter speed, distance from camera to objects, and any other useful information.

Sketching the Crime Scene

The final task to be performed in the documentation of the crime scene is the sketching. All of the previous tasks for documentation record the crime scene without regard to the proportionality or actual size with measurement of the scene and its physical evidence. Sketching the crime scene is the assignment of units of measurement or correct perspective to the overall scene and the relevant physical evidence identified within the scene.

Sketching the crime scene is not difficult but does require some organization and planning by the investigator in order to assure that an accurate sketch results. There are two basic types of sketches as part of crime scene investigations: a rough sketch (see Figure 10.5) and a final or finished sketch (See Figure 10.6). There are many types of perspectives used for sketching crime scenes but the two most common types are: the overhead or bird's eye view sketch and the elevation or side-view sketch. Occasionally, a combination perspective sketch called a cross-projection sketch is used to integrate an overhead sketch with an elevation sketch. Three-dimensional sketches or scaled models are not common but can also be used as a form of crime scene documentation.

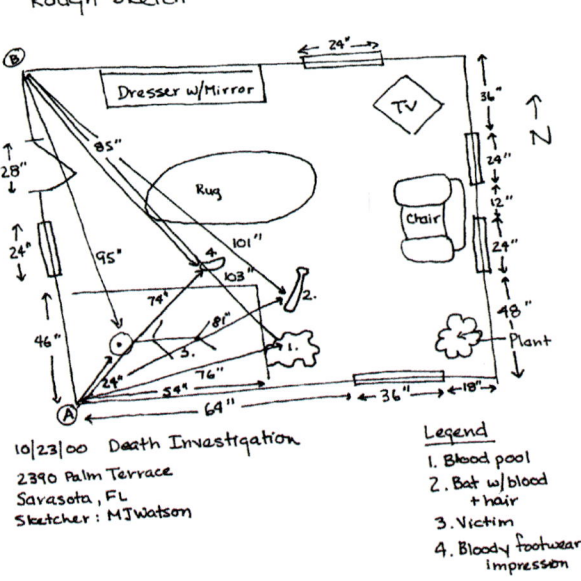

October 23, 2000
Death investigation
2390 Palm Terrace
Sarasota, FL
Sketcher: M J Watson
Drawn: November 10, 2000

Legend
1. Pool of blood
2. Baseball bat with blood & hair
3. Victim
4. Bloody impression

Figure 10.6 Finished or final sketch.

Finished Sketch

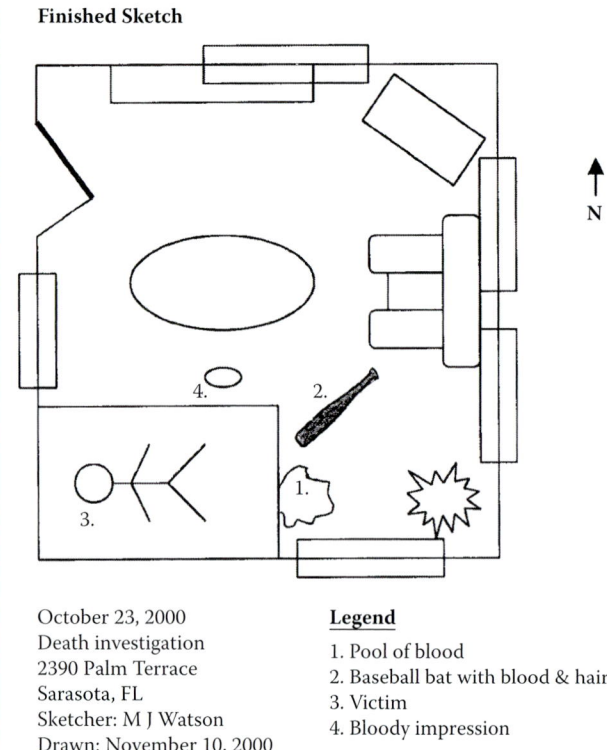

Figure 10.5 Rough sketch.

There are three techniques for obtaining measurements for the crime scene sketch: **triangulation**, **base-line** (fixed line), and **polar coordinates**. See Figure 10.7. All three techniques are based on identifying two starting, fixed points, and all subsequent measurements of the crime scene are in relation to those points. The fixed points are "fixed" because their location is known or can be precisely determined. This fixed nature is used for subsequent reconstruction. Good fixed points can be building corners, in-ground survey markers, large trees, or recorded utility poles.

All crime scene sketches require their own documentation. This documentation within the documentation includes a title or caption; a legend of abbreviations, symbols, numbers or letters used; a compass designation; if drawn to scale, then the scale used; and the documentation block with the case number, offense type, victim name(s), location, scene descriptor, date/time of sketch beginning, and sketcher's name(s).

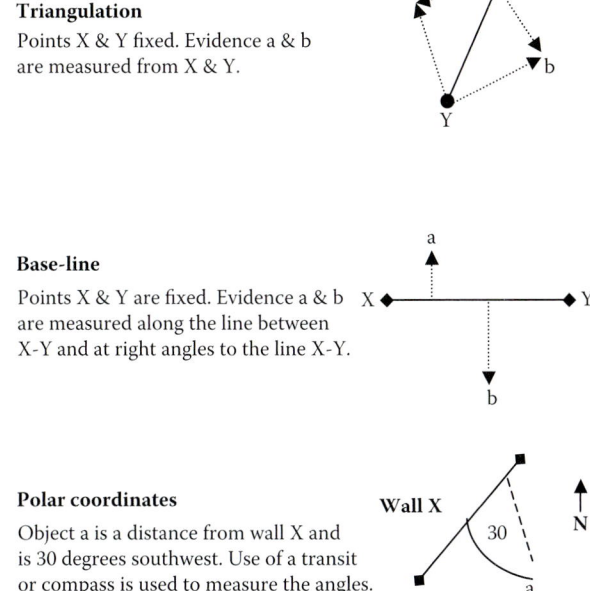

Triangulation

Points X & Y fixed. Evidence a & b are measured from X & Y.

Base-line

Points X & Y are fixed. Evidence a & b are measured along the line between X-Y and at right angles to the line X-Y.

Polar coordinates

Object a is a distance from wall X and is 30 degrees southwest. Use of a transit or compass is used to measure the angles.

Figure 10.7 Crime scene measurement techniques.

Digital Imaging at the Crime Scene

Digital image technology provides the crime scene investigator with powerful new tools for capturing, analyzing, and storing the record of the crime scene and its physical evidence. These digital image tools complement the traditional video and still photography used in crime scene documentation. The advantages of digital images include instant access to the images, easy integration into existing electronic technologies, and no need for the often-expensive film processing equipment and darkrooms. Some disadvantages for the use of digital image technology are centered on issues of court admissibility because of easy image manipulation. However, it is important to remember that it is the investigator that is testifying, not the image, and with written and implemented policies and procedures for using digital images, these disadvantages can be averted. It is agreed upon within the law enforcement community that the use of digital images in crime scene documentation can best be used as a supplemental technique and not completely replace the traditional techniques currently used.

Crime Scene Searches

The preliminary crime scene search was an initial quasi-search for physical evidence present at the crime scene. That search is for the obvious items of evidence, and it is done for orientation purposes before the documentation begins. Once the scene documentation as described above is completed then a more efficient and effective search for less obvious or overlooked items of evidence must be done. This intensive search is done after documentation but before the evidence is collected and packaged. If any new items of evidence are found, then they must be subjected to the same documentation tasks that were done earlier.

Crime scene search patterns are varied and different in style, but they share a common goal of providing organization and systematic structure to ensure that no items of physical evidence are missed or lost. There is no single method for specific types of scenes. The experienced crime scene investigator will be able to recognize and adapt the search method that best suits the situation or scene. It is important for the crime scene investigator to use that method. Simple reliance on their experience alone and omitting the search step in the investigation will produce mistakes and significant evidence can be missed.

Most commonly employed search methods are geometric patterns. The six patterns are: link, line or strip, grid, zone, wheel or ray, and spiral methods. Each has its advantages/disadvantages and some are better suited for outside versus indoor crime scenes. Table 10.4 summarizes the various patterns. Before any intensive crime scene search is done, care must be taken to instruct the members of the search party. It is very tempting for search party members to touch, handle, or move the items of evidence found during the search. Instruct the members to mark or designate the items found without altering the item. In the old West, firing a shot into the air was the common technique for identifying items when found. Today, that may not be the appropriate

Table 10.4
Crime Scene Search Methods

Search Type	Geometric Pattern	Information on Use
Link method		Based upon the linkage theory; most common & productive; one type of evidence leads to Another item; experiential, logical & systematical; works with large and small, indoor or outdoor scenes.
Line or strip method		Works best on large, outdoor scenes; requires a search coordinator; searchers are usually volunteers requiring preliminary instructions.
Grid method		Modified, double line search as above; effective method but is time consuming.
Zone method	1 2 / 3 4	Best used on scenes with defined zones or areas; effective in houses or buildings with rooms; teams are assigned small zones for searching; combined with other methods; good for search warrants.
Wheel or ray method		Used for special situations; limited applications; best used on small, circular crime scenes
Spiral method		Inward or outward spirals; best used on crime scenes without physical barriers (open water, etc.); requires the ability to trace a regular pattern with fixed diameters; limited applications.

method, but with proper training, diligence and care, no evidence will be mistreated during the search of the crime scene. Documentation of the found items must be done before any evidence can be moved or collected.

The practical application of the search methods to the crime scene may be a combination of methods. Searching will frequently require the use of field testing, and visualization and enhancement reagents for biological fluids or impression evidence. Also, keep in mind that the searching of the crime scene should never diminish or interfere with the other functions of the scene investigation like the proper documentation, collection, and preservation of the physical evidence. Do not avert established crime scene procedures. Chain-of-

custody issues with regard to the evidence is paramount and can be addressed by restricting the number of searchers and subsequent collectors of the evidence.

Collection and Preservation of Physical Evidence

After the completion of the crime scene documentation and the intensive search of the crime scene for the physical evidence, the collection and preservation of the evidence can begin. One individual shall be designated as the evidence collector. This appointment will insure that the evidence is collected, packaged, marked, sealed, and preserved in a consistent manner. No item of evidence will be missed, lost, or contaminated if only one person has the obligation for this important stage in the scientific crime scene investigation.

There is no rigid order for collection of the evidence but some types of evidence, by their nature, should be given priority of order. Transient evidence, fragile or easily lost evidence, should be collected first. Some items of evidence by their location within the crime scene may need to be moved or repositioned. If items are moved and new evidence is discovered then the previously discussed methods of documentation must be done immediately.

It is difficult to generalize about the collection of physical evidence. Different types of physical evidence will require specific or special collection and packaging techniques. The various chapters in this book discuss the individual types of evidence with the specific collection techniques appropriately place there. General collection guidelines are presented here.

Most items of evidence at the crime scene will be packaged into a primary container that is then placed inside a secondary container. **Druggist's folds** are especially well suited as primary containers for trace evidence collection and packaging. Larger pieces of paper can be folded around larger items to "hold" trace evidence in place. (See Figure 10.8) These folds are then placed inside outer containers like envelopes, packets, canisters, paper bags, and plastic bags. (See Figure 10.9) The outer containers are then completely sealed with tamper-resistant tape. The outer container should be marked with information about the item, identification about the collector, and date/time/location of collection of the item. The sealing tape or evidence tape should completely cover the opening of the outer container. It is marked with the initials of the collector, the date, and time of collection. (See Figure 10.10) It is a good idea to have a wide variety of packaging containers, sealing materials, and markers available at the crime scene.

Most items of evidence are solid and can be easily collected, stored, and preserved in the

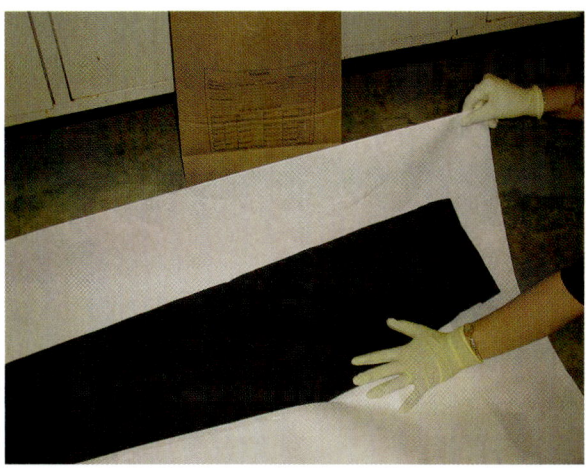

Figure 10.8 Proper use of primary container.

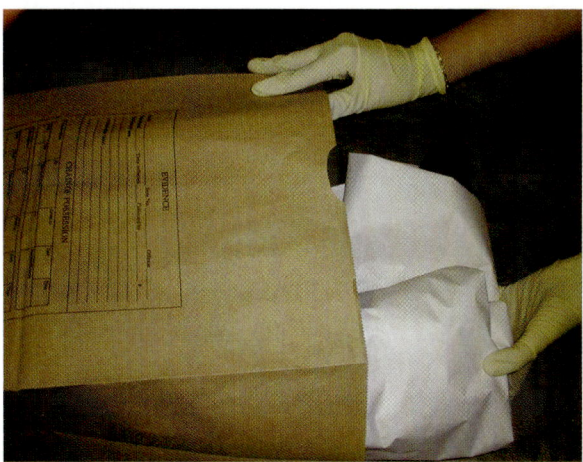

Figure 10.9 Packaging of primary container inside outer container.

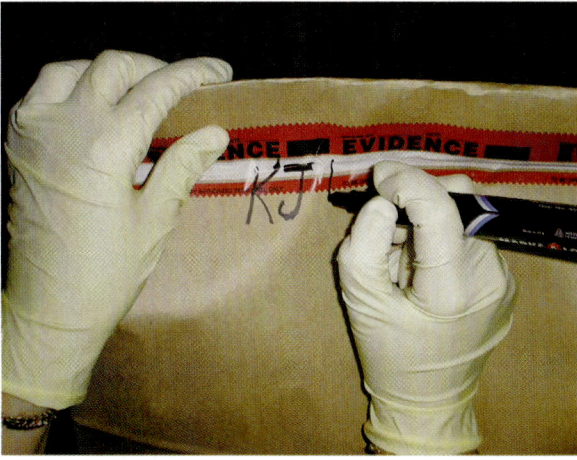

Figure 10.10 Proper marking of sealed evidence.

above manner. Liquid or volatile items of evidence should be placed in airtight, unbreakable containers. Wet, moist, or living biological evidence can be temporarily packaged in air-tight containers. It should then be allowed to air-dry in a controlled environment and then repackaged with the original containers in new nonair tight containers.

Each item of evidence should be packaged separately to prevent cross-contamination between items of evidence. The containers should be sealed and marked at the time of collection to prevent intermingling of evidence while being moved or transported to other locations. Control standards, alibi standards, and other control samples can be important to an investigation. The crime scene investigator should always be aware of the types of evidence being collected and determine the appropriateness and need for these controls to be collected. These controls are especially important in fire investigations, and with trace evidence, blood and body fluid stains, and **questioned documents**.

Forensic analytical techniques are improving, the amount of sample required for testing has been reduced, and the information about probable sources of the evidence has significantly improved. It is because of these improving techniques and sensitivities, the proper collection and packaging of physical evidence is extremely important. Consensus groups, scientific working groups or technical working groups (SWGs and TWGs), are publishing collecting, packaging, and preservation techniques specific for a variety of physical evidence types. Advanced lab techniques cannot be used if the evidence is lost or contaminated because of improper or poor collection and packaging at the scene.

Crime Scene Reconstruction

Introduction

Crime scene reconstruction is the process of determining or eliminating the events that could have occurred at the crime scene by the analysis of the crime scene appearance, the location and position of the physical evidence, and the forensic laboratory examination of the physical evidence. It involves scientific crime scene investigation, interpretation of the scene's patterned evidence, laboratory testing of the physical evidence, systematic study of related case information, and the logical formulation of a theory.

Nature of Reconstructions

Crime scene reconstruction is based on scientific experimentation and the past experiences of the investigator. Its steps and stages, like those in forensic science, follow basic scientific principles, theory formulation, and logical methodology. It incorporates all investigative information with physical evidence analysis and interpretation molded into a reasonable explanation of the criminal activity and its related events. Logic, careful observation, and considerable experience, both in the crime scene investigation and the forensic testing of the physical evidence, are necessary for proper interpretation, analysis, and the crime scene investigation.

Stages in Reconstruction

Crime scene reconstruction is a scientific fact-gathering process (see Figure 10.11). It involves a set of actions or stages. The stages are:

1. *Data collection.* All information or documentation obtained at the crime scene, from the victim or witnesses will be necessary. Data including the condition of the physical evidence, patterns and impressions, condition of the victim, etc., are reviewed, organized, and studied.
2. *Conjecture.* Before any detailed analysis of the evidence is accomplished, a possible explanation or conjecture of the actions involved in the crime scene may be done. It is not fixed or the only possible explanation at this point. There may be several possible explanations as well.
3. *Hypothesis formulation.* Additional accumulation of data is based on the examination of the physical evidence and the continuing investigation.

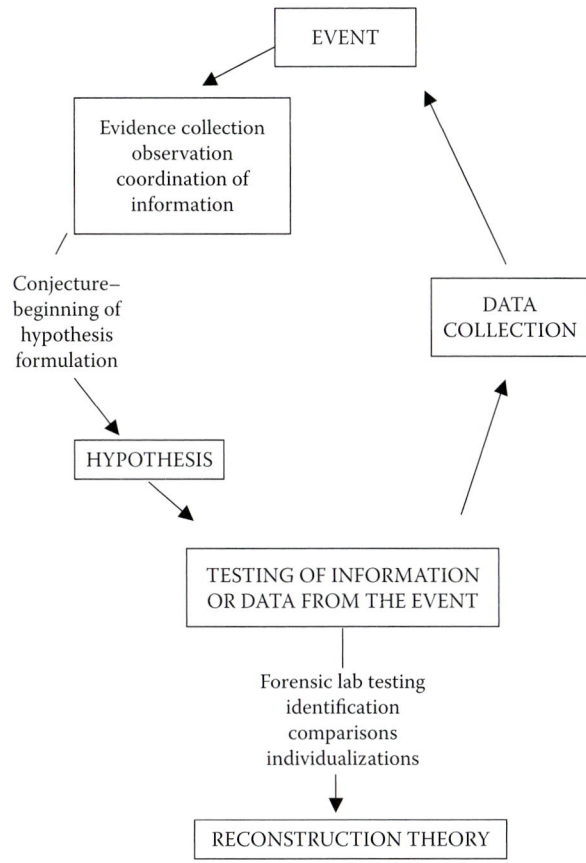

EVENT

Evidence collection
observation
coordination of
information

Conjecture–
beginning of
hypothesis
formulation

DATA
COLLECTION

HYPOTHESIS

TESTING OF INFORMATION
OR DATA FROM THE EVENT

Forensic lab testing
identification
comparisons
individualizations

RECONSTRUCTION THEORY

Figure 10.11 Diagrammatic representation of scientific fact gathering process.

Scene examination and inspection of the physical evidence must be done. Interpretation of bloodstain and impression patterns, gunshot residue patterns, fingerprint evidence, and analysis of trace evidence will lead to the formulation of a reconstruction hypothesis.

4. *Testing.* Once a hypothesis has been developed, then additional testing or experimentation must be done to confirm or disprove the overall interpretation or specific aspects of the hypothesis. This stage includes comparison of samples collected at the scene with known standards, chemical, microscopical, and other analyses and testing. Controlled testing or experimentation of possible scenarios of physical activities must be done to collaborate the hypothesis.

5. *Theory formulation.* Additional information may be acquired during the investigation about the condition of the victim or suspect, the activities of the individuals involved, accuracy of witness accounts, and other information about the circumstances surrounding the event. When the hypothesis has been thoroughly tested and verified by analysis, the reconstruction theory can be formulated.

Any reconstruction can only be as good as the information provided. Information may come from the crime scene, the physical evidence, records, statements, witness accounts, and known data. The information gathering process as shown above and its use in the crime scene reconstruction show the scientific nature of scene reconstruction and as a result, will allow for its successful use by the investigators.

Patterned evidence shown below is especially well suited for crime scene reconstruction:

- Impression location/position—fingerprints, footwear, or tire tread marks
- Glass fracture patterns—direction of force or order of fire
- Fire burn patterns—points of origin determination
- Wound location or dynamics
- Clothing location or damage
- Bloodstain patterns
- Shooting investigations-—range of fire or gunshot residue analysis

Case Study

If a crime scene is properly investigated with systematic methodology, then it will provide the means for completing criminal investigations and resolving the case. Some of the primary uses of physical evidence found at crime scenes are corroborating the statements of witnesses, assisting investigators in determining the credibility of eyewitnesses, and assisting in the reconstruction of the events leading to the crime including the way in which the crime was committed. The case presented here will illustrate that a properly investigated crime scene and evidence can disprove an eyewitness' account of a criminal act.

Late one Friday afternoon in April of 1995, a decomposing body was discovered in a bedroom. The dead man was dressed only in a bath towel and was shot twice at close range with a shotgun. Two television sets in the home were turned on and no signs of forced entry or ransacking of the premises were found. The state police crime scene unit responded and began the investigation. The investigators made sure the scene was secure, spoke with the first responders, proceeded with the preliminary scene survey, documented the scene, and collected and packaged the physical evidence found (Figure 10.12).

Within weeks of the discovery of the body, the investigation focused on two teenaged girls who were friends of the 50-year-old victim.

One of the girls periodically visited and drank alcohol with the victim during the 3 years before his death. Both girls knew the suspect. The second teenaged girl had been romantically involved with the suspect. At first, both girls denied knowledge of the death. However, when the investigation focused on them, they claimed that the suspect said he had shot the victim. The girl who drank alcohol with the victim gave an even more detailed statement. She said she was an eyewitness to the shooting and that robbery was the motivation.

Her detailed eyewitness account of the shooting specifically stated that the suspect hid behind the door to the bedroom and emerged to face the victim who was entering the bedroom from the hallway, and shot the victim twice. The first shot was in the hallway near the door to the bedroom. The second shot was fired as the victim stumbled forward into the bedroom. According to the eyewitness, the

Figure 10.12 (A to G) House and body in bedroom.

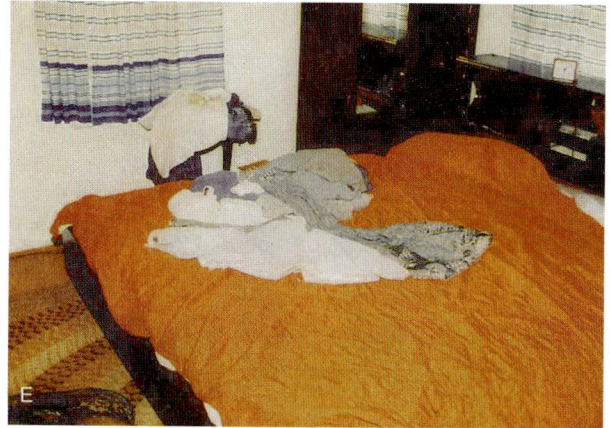

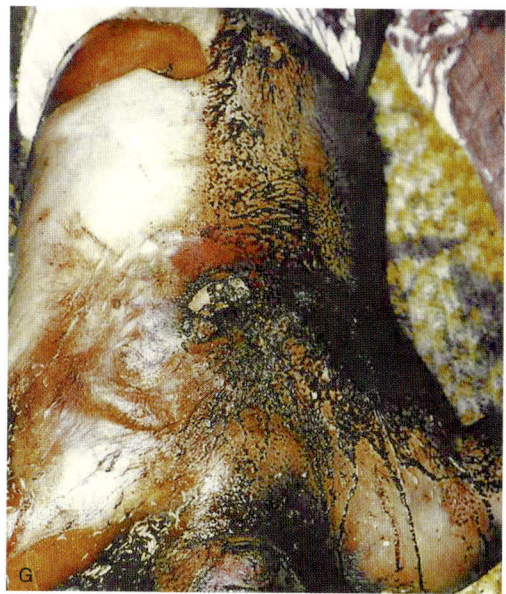

Figure 10.12 (continued)

victim then fell in the position in which investigators found his body. The eyewitness said the victim was facing toward the bedroom when he was shot and that the suspect faced toward the hallway entrance to the bedroom. She also told investigators that the suspect held a pillow in front of the shotgun when he shot the victim and that both blasts were fired while the suspect was partially hidden behind the bedroom door.

Use of Physical Evidence and Crime Scene Investigation

Several items of physical evidence documented and collected from the crime scene, including the shotgun pellets, bloodstain patterns, and tears in a pillowcase, were determined by crime scene investigators to be inconsistent with the eyewitness' claims. In fact, the physical evidence showed that the crime could not have occurred in the way described by the eyewitness.

Shotgun Pellets and Trajectory Marks at Crime Scene

A not-to-scale crime scene sketch prepared by an investigator showed the locations of various items of physical evidence (Figure 10.13). This sketch reveals shotgun pellets were found in the bedroom at locations marked 7 through 10. The photographs and crime scene videotape revealed additional shotgun pellets and trajectory marks (Figure 10.14). The crime scene report by the note taker at the scene shows that in the scene investigation, no shotgun pellets or wadding were found in the hallway and that all the pellets were found in the bedroom as documented in the photographs, videotape, and sketch.

The location of these shotgun pellets and the trajectory marks found in the bedroom led the investigators to believe that the shotgun was pointed not towards the hallway, as described by the eyewitness. It was fired into the bedroom from the hallway (Figure 10.15). The positions of the pellets, the trajectory marks, the position of the wadding, and the absence of pellets in the hallway, as documented by the crime scene investigators,

CRIME SCENE SKETCH

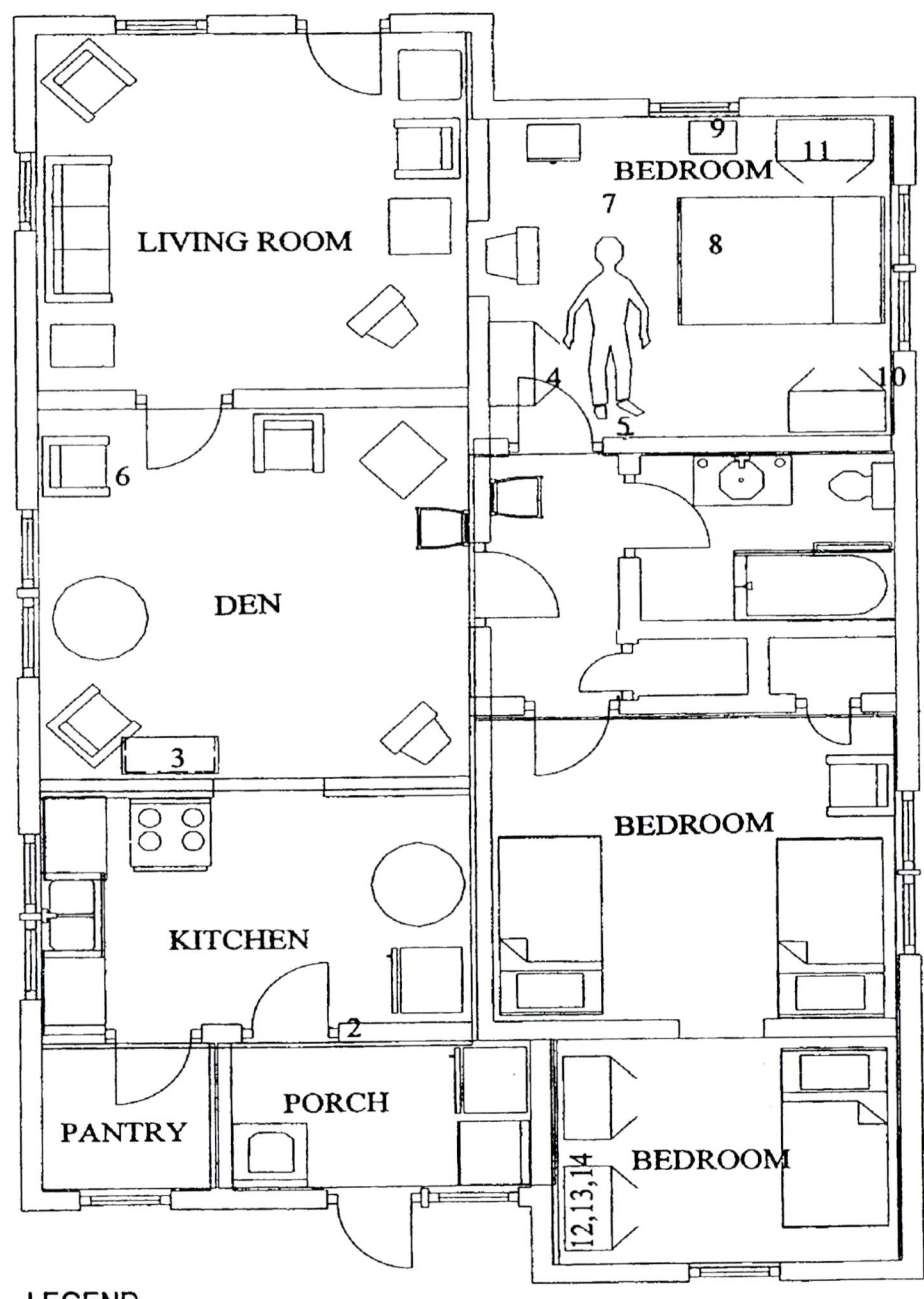

LEGEND

2-Bloodstain
3-Wirebound Notebooks
4-Pillow
5-Shotgun Wadding
6-Fingernail
7-Pellet

8-Pellet
9-Pellet
10-Pellet
11-Two Prescription Vials
12-Denim Jacket
13-Black Pants
14-Remington Shotgun

Figure 10.13   Crime scene sketch with legend.

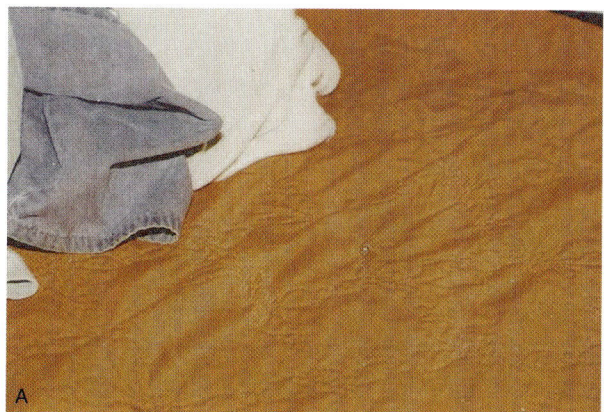
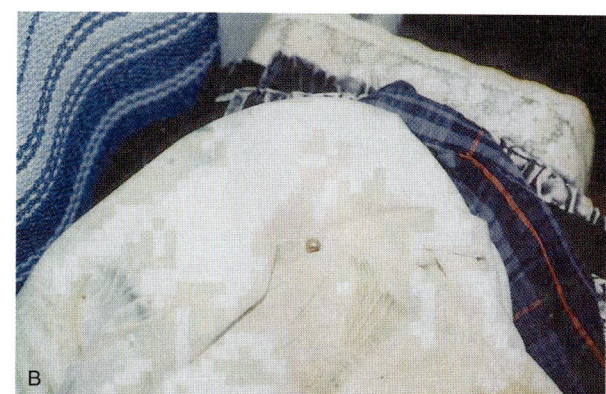
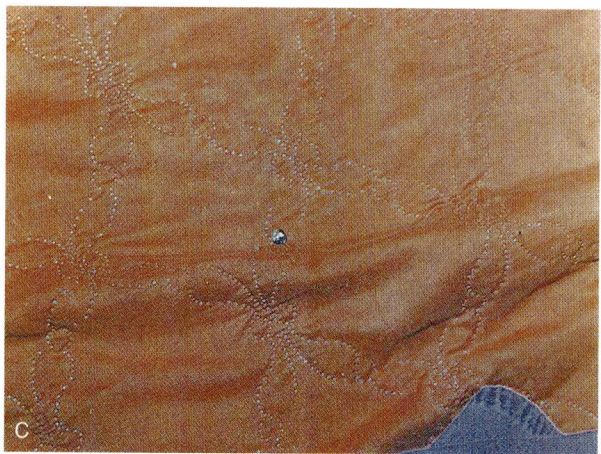

Figure 10.14 (A to G) Pellets and their trajectory marks.

contradicted the eyewitness' account of the criminal act.

Bloodstain Patterns at Crime Scene

The crime scene videotape and photographs showed various bloodstain patterns in the bedroom and in the area of the hallway nearest the bedroom (Figure 10.16). The bloodstain patterns are consistent with arterial gushes from wounds similar to those suffered by the victim. Their documented locations indicate that the victim was standing in the bedroom

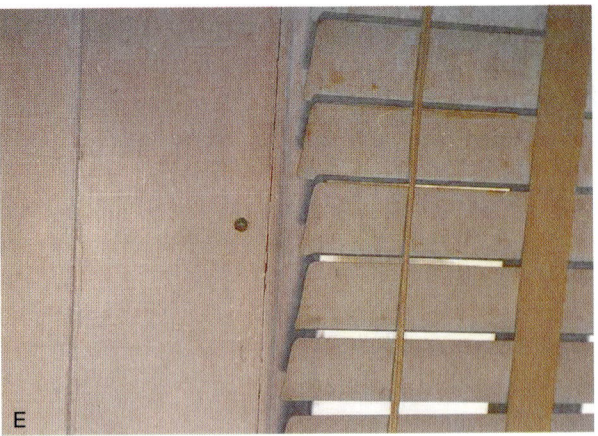

Figure 10.14   *(continued)*

facing the door into the hallway when shot. The bloodstains contradict the eyewitness' account. Also, the documentation of the bedroom door shows no bloodstains. If, according to the eyewitness, the victim received his chest wound as he entered the bedroom and the shotgun was fired from inside the bedroom toward the hallway, then arterial gush bloodstains should have been present on the door, the floor around the door, and the door frame opposite the area where they were found.

Other crime scene photographs and videotape segments of the crime scene showing bloodstains in the hallway leading into the bedroom were used by investigators to contradict the eyewitness' account of the criminal act (Figure 10.17). The bloodstains documented by the crime scene investigators were from the victim's chest wound and they confirm that the victim was inside the bedroom facing the hallway when he was shot.

Tears in Pillowcase from Crime Scene

A final piece of physical evidence documented at the crime scene was the pillowcase removed from the pillow. The pillowcase was found to have several small tears in an upper portion (Figure 10.18). The location, size, and shape of the tears were determined by the investigators to be consistent with a glancing blast of shotgun pellets and not with a full blast or discharge from a shotgun muzzle. Once again, the physical evidence and its documentation at the crime scene were used to disprove the eyewitness' statement.

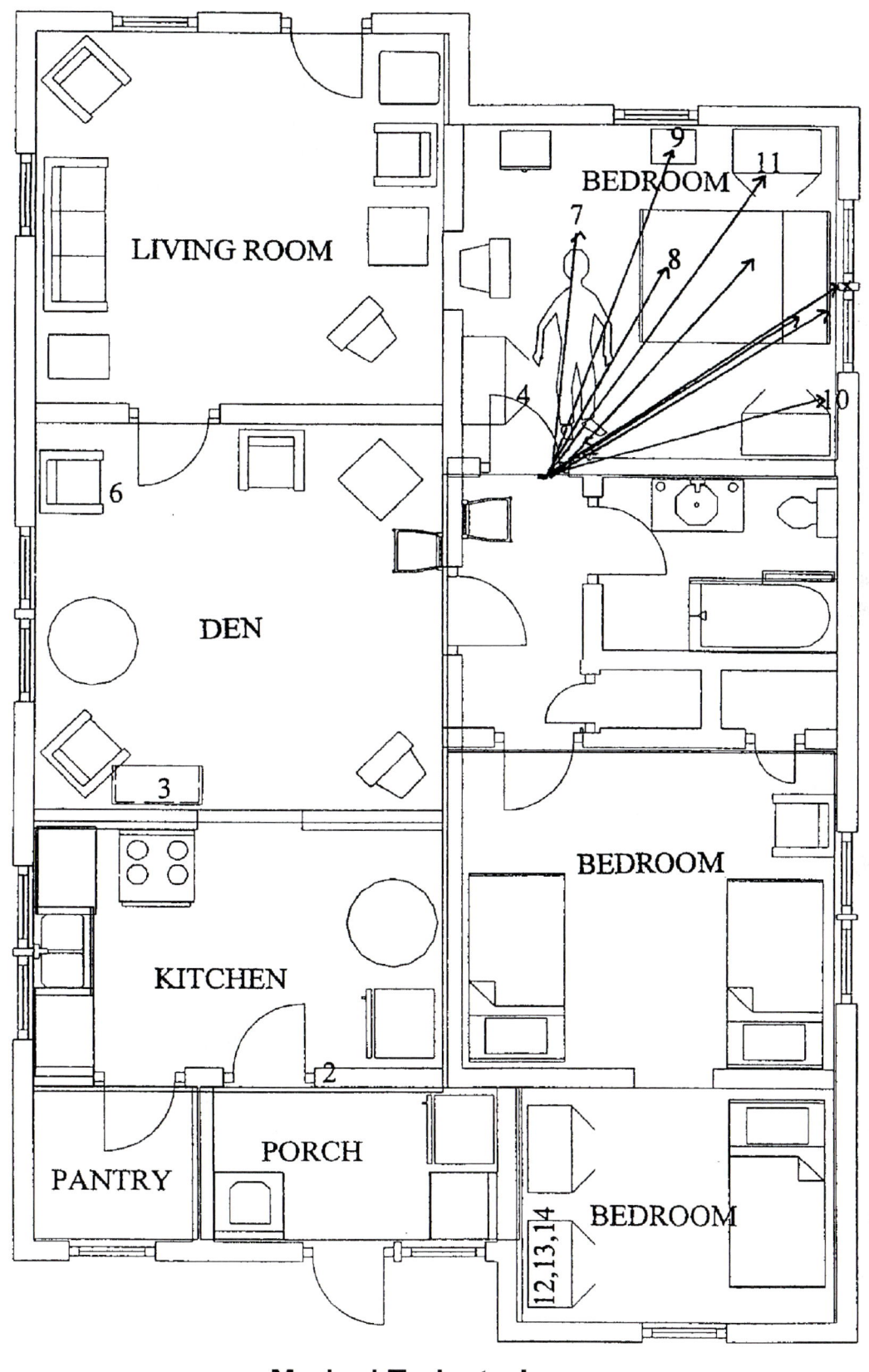

Marked Trajectories

Figure 10.15   Sketch of approximate trajectory pathways.

Figure 10.16 Arterial gush found on wall near door inside bedroom.

Figure 10.18 Pillowcase showing tears caused by a glancing blast of shotgun pellets.

Figure 10.17 Bloodstains on hallway rugs.

The proper documentation of the crime scene allowed the investigators to disprove the lone eyewitness' account of a homicidal act.

Her account of the shooting and the implication of the suspect were not substantiated by the physical evidence and, in fact, led investigators to conclude that her account was not even possible.

Conclusion

Scientific crime scene investigation is the best methodology to insure that an investigation is properly conducted and justice is served. Use of this methodology will prevent the abrupt end of an investigation and allow for the best and reliable use of the physical evidence found at crime scenes.

Review Questions

1. What are the basic steps of scientific crime scene investigation?

2. List and describe the definitions or classifications of crime scenes.

3. What are the eight types of information that can be obtained from the examination of the physical evidence found at crime scenes? Explain each type and give an example.

4. What are the four components of crime scene management?

5. What are the five crime scene investigation models? Describe them and give the advantages and disadvantages of each.

6. Discuss the duties of the first responder at the crime scene.

7. What is the multilevel approach to crime scene security?

8. What are the components or tasks of crime scene documentation? What is the purpose of each?

9. What is the basic process used for photographing crime scenes? Discuss each step.

10. What are the two basic types of crime scene sketches? What are the two types of perspectives used in sketches?

11. Describe and discuss the six types of search patterns used in crime scene investigations.

12. What are the general guidelines for the collection, packaging, and preservation of physical evidence?

13. List and discuss the stages of crime scene reconstruction.

14. The above method is packaged in a _____-_____ container.

15. A "walk-through" is the _____ scene survey.

16. Trace evidence like hairs and fibers are packaged in _____ _____ as primary containers and then placed in secondary outer containers.

17. _____ _____, although convenient and easily processed, are not replacements for conventional photographs of crime scenes.

18. A _____ _____ is used to maintain a list of all personnel entering/exiting a secured crime scene.

19. Documentation of crime scenes is _____ and never stops.

Suggested Readings

Books

ATF, *Arson Investigation Guide*, The Department of the Treasury, Washington, D.C., May 1997.

Bevel, T. and Gardner, R., *Bloodstain Pattern Analysis with an Introduction to Crime Scene Reconstruction,* CRC Press, Boca Raton, FL, 1997.

Bodziak, W., *Footwear Impression Evidence*, Elsevier Science, New York, 1990.

DeForest, P., Gaensslen, R., and Lee, H., *Forensic Science: An Introduction to Criminalistics*, McGraw-Hill, New York, 1983.

DiMaio, V., *Gunshot Wounds: Practical Aspects of Firearms, Ballistics, and Forensic Techniques*, Elsevier, New York, 1985.

Dix, J., *Handbook for Death Scene Investigators*, CRC Press, Boca Raton, FL, 1999.

James, S., Kish, P., and Sutton, T.P., *Principles of Bloodstain Pattern Analysis—Theory and Practice,* Taylor and Francis (CRC Press), Boca Raton, FL, 2005.

Fisher, B., *Techniques of Crime Scene Investigation,* 5th ed., Elsevier, New York, 1992.

Hawthorne, M., *First Unit Responder: A Guide to Physical Evidence Collection for Patrol Officers*, CRC Press, Boca Raton, FL, 1999.

Lee, H. et al., *Crime Scene Investigation*, Central Police University Press, Taoyuan, Taiwan ROC, 1994.

Lee, H. and Harris, H. *Physical Evidence in Forensic Science*, Lawyers & Judges Publishing Co., Inc., Tucson, AZ, 2000.

Lee, H. and Gaensslen, R., *Advances in Fingerprint Technology*, Elsevier, New York, 1991.

Lee, H., Palmbach, T., and Miller, M., *Henry Lee's Crime Scene Handbook*, Academic Press, London, 2001.

McDonald, J., *The Police Photographer's Guide*, PhotoText Books, Arlington Heights, IL, 1992.

McDonald, P., *Tire Imprint Evidence*, New York: Elsevier, 1989.

National Medicolegal Review Panel, *National Guidelines for Death Investigation*, National Institute of Justice and U.S. Department of Justice, Washington, D.C., December 1997.

Ogle, R. *Crime Scene Investigation and Physical Evidence Manual*, Robert Ogle, Jr., Vallejo, CA, 1995.

Redsicker, D., *The Practical Methodology of Forensic Photography*, CRC Press, Boca Raton, FL, 1994.

Technical Working Group for Bombing Scene Investigation, *A Guide for Explosion and Bombing Scene Investigation.* Washington, D.C.: National Institute of Justice and US Department of Justice (NCJ #181869), June 2000.

Technical Working Group on Crime Scene Investigation, *Crime Scene Investigation: A Guide for Law Enforcement*, Washington, D.C., January 2000.

National Institute of Justice and U.S. Department of Justice (NCJ #178280), Technical Working Group on Fire/Arson Scene Investigation, *Fire and Arson Scene Evidence: A Guide for Public Safety Personnel*, Washington, D.C.: National Institute of Justice and U.S. Department of Justice (NCJ#181584), June 2000.

Forensic Digital Photo Imaging

Patrick Jones

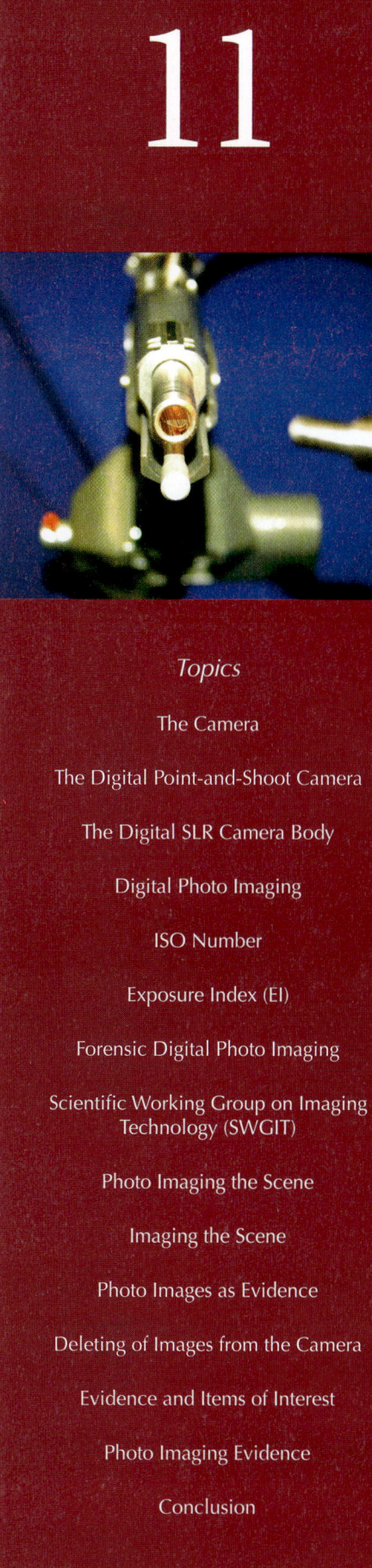

11

Introduction

Crime scenes, evidence, victims, and suspects must be documented. This is accomplished by measurements, notes, diagramming, and photo imaging. The old adage "A picture is worth a thousand words" is very true; in fact, in some cases a picture is worth many thousands of words. Photo imaging, then, is a very important part of the documentation process. It provides the judge and jury with a realistic visualization of the conditions, and relative positions of evidence and items of interest, at a crime scene. In this chapter, we will be addressing digital photo imaging. More and more agencies are choosing this better technology over the film camera. However, the 35 mm film camera still remains a very viable tool in the CSI's arsenal.

The Camera

The 35 mm film single lens reflex camera has for many years been the camera of choice for crime scene Investigators. It is versatile, relatively easy to use, and takes excellent images. Various types of film including black and white, color negative, and color positive, along with a wide variety of special films and film speeds allow the 35 mm film camera to be adapted to just about every situation.

With technology ever moving forward, we are now living in the era of digital photo imaging. There are two basic types of digital cameras. The "point and shoot" and the SLR (single lens reflex). The digital camera uses a sensor, a CCD or charged coupled device, to "see" the image. The image is then recorded on a memory card instead of film. An advantage to the digital camera is the ability to review the picture that was taken immediately. If there was a mistake made, if the image is out of focus, or improperly exposed, another image can be taken that is correct. Remember, you may not erase or delete an image taken at a crime scene, just as you cannot destroy the negatives of out-of-focus or bad pictures.

The Digital Point-and-Shoot Camera

The digital point-and-shoot camera is a good, basic, inexpensive camera. It is able to record digital images on internal memory, with most having the ability to record to a memory card of varying sizes. The camera has either a rangefinder-type aiming device or the camera may have a video display screen from which the image can be composed. Most cameras in this category have built-in flashes and are, for the most part, fully automatic. They produce an adequate image but lack the versatility of the digital single lens reflex (SLR).

These cameras do not have the ability to use interchangeable lenses. The lens that comes with the camera is what you have. The ability to zoom in and out to your subject is limited. In some cases, the zoom feature is not optically driven but rather software driven. Some have what is alleged to be macro capabilities or settings; however, these are software not optical functions.

The Digital SLR Camera Body

The digital SLR is a single lens reflex camera. This means that the viewfinder that is used to aim and compose the image uses the same lens through which the image is recorded. There is a viewing screen on the camera; however, it is not used to compose the image, but rather to view the image after it is taken and to view the menu setting for the camera. (See Figure 11.1.)

The digital SLR camera body is similar to that of the 35 mm film camera. Most of the controls are located on the body. There is usually a screen on which the menu can be displayed and various options are available. There are many options that can be set from the menu. Some will be specific to the particular brand of camera, and some are generic to most or all digital SLRs (see Figure 11.2). Some of these options include:

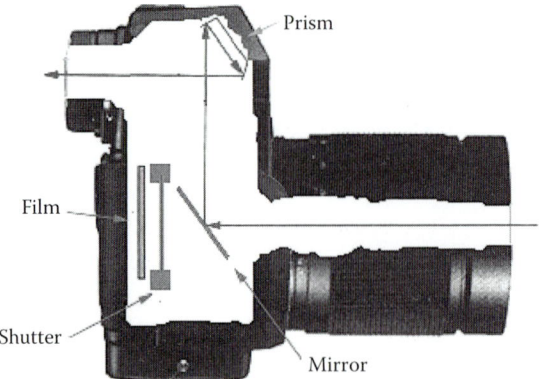

Figure 11.1 Single lens reflex (SLR) camera. When composing an image, the photographer looks through the viewer and the image is reflected off of a mirror, allowing him or her to see exactly what the camera sees.

Figure 11.2 SLR camera

- Image quality—You can set the quality for high, medium, or low.
- White balance—This allows the image taker to adjust to the type of light being used, i.e., fluorescent, daylight, incandescent, cloudy, or shade.
- ISO sensitivity—This allows the setting of a "virtual film speed." Since there is no film used, this setting allows the setting in many digital SLRs from ISO 100 to ISO 1600.
- Multiple exposure—With some digital SLRs with this setting, it is

possible to make double, triple, and more exposures.

- Monochrome—This setting allows the camera to record images in black and white.
- File number sequence—When set to continuous, this will identify the image with a unique number. If not set to "continuous," each time you remove the memory card, the identifying number starts at one.

Aperture

An aperture is a hole or an opening through which light is admitted through the lens and onto the sensor. The size of the aperture can be made larger or smaller to accommodate a proper exposure. The smaller the number, the larger the opening. The larger the number, the smaller the opening. The numbers that identify these openings are called *F stops* (Figure 11.3).

Shutter

A shutter is a device that allows light to pass for a determined period of time, for the purpose of exposing a light-sensitive electronic sensor to light to capture a permanent image of an object or of a scene.

Image Numbering

Image numbering is a setting that can be set to Reset or Continuous. This setting should be

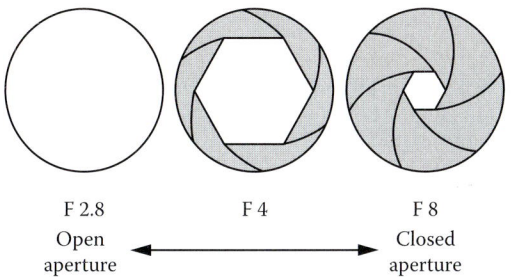

F 2.8
Open
aperture

F 4

F 8
Closed
aperture

Figure 11.3 This drawing depicts the aperture of a lens. The settings are in increments of "stops" or F-stops. The larger the number the smaller the opening. The smaller the number the larger the opening. F-32 would be a small opening, while F-2.8 would be a large opening.

set to Continuous. This is for several reasons. The photo image identifier should be unique. If set to continuous, it will assign a new number in successive order, not repeating any one number previously used. If the reset setting is used, the camera will reset the photo image identifier each time the memory card is removed and reinserted or after each time the pictures are uploaded to the computer and deleted from the camera. If additional images were taken and uploaded to the same folder, the computer would advise that there are already images with those numbers. It will then ask if you want to overwrite (and destroy) the original images. The image file name/number could be changed, but this would not adhere to the rule that we do not change a photo image. For both of these reasons, set this setting to Continuous.

Digital Photo Imaging

The terms *digital imaging* and *photography* are basically synonymous. They have the same meaning. They both can be defined as the act of recording light or reflected light, from an object or group of objects and recording that image on media. The media has traditionally been film, but in digital imaging it is a memory card. Very rarely is light directly recorded, for this would require the camera to be pointed directly at a light source such as the sun, a light bulb, or other visible spectrum light emitting source. While there may be a need for it in some specific instances, such as recording sun spots, the great preponderance of what we do in forensic digital imaging is the recording of reflected light.

Light from a source, a flash, light beam, or ambient light strikes an object or group of objects and is reflected off the objects and through the lens. The amount of light allowed through the lens is controlled by the aperture. The duration that the light is allowed through the lens is controlled by the shutter. The camera then "sees" the light through its lens and allows it to continue to the sensor, the charged coupled device. It is then recorded or saved on media such as a memory card.

These memory cards are manufactured in many different configurations, depending on what type of memory card the manufacturer has decided to build into the camera design. The memory cards are not interchangeable, however, so in order to share the images taken by the camera, you must upload them to a computer.

Several types of memory cards include SD (secure digital), memory sticks, compact flash, xD-Picture Card, the RS-MMC (reduced size multimedia card), mini-SD card, and many more.

ISO Number

ISO numbers indicate the level of sensitivity of the digital SLR camera to light. On a digital SLR, this sensitivity level is set from the menu. Each film for 35 mm film cameras has its own ISO number. It is necessary to set the film speed on the camera so the internal meter will compute the correct exposure. If you wanted to change ISO or film speeds, you would have to change film. In the digital SLR, one can change the ISO (virtual speed) at any time by accessing the menu. ISO stands for the International Organization for Standardization.

Exposure Index (EI)

Exposure index, or EI, refers to speed rating that you assigned to a particular photo imaging situation. Exposure index may or may not be the same as manufacturer's recommendation. A light meter can understate or overstate lighting conditions. In such cases one could adjust EI rating accordingly in order to compensate for these effects and consistently produce correctly exposed negatives. The digital SLR has a setting to compensate for extreme conditions by overriding the meter so that a proper exposure is obtained.

Forensic Digital Photo Imaging

Forensic is defined "as pertains to a court of law." So there is forensic digital imaging, forensic entomology, forensic physical anthropology, forensic engineering, etc.; all address the functioning of these fields in a legal context.

The addition of "forensic" to digital imaging adds a number of layers to the correct practice of this discipline. Good digital imaging requires knowledge of the technical aspects of recording objects and locations. Forensic digital imaging requires knowledge of the technical aspects of recording objects and locations but also requires this to be done with a proper method, a set policy, and procedures all approved and accepted by law, the courts, and judicial decision.

Scientific Working Group on Imaging Technology (SWGIT)

The Scientific Working Group on Imaging Technology (SWGIT) was created to provide leadership to the law enforcement community by developing guidelines for good practices for the use of imaging technologies within the criminal justice system. It should be considered *the* guide for any questions involving how images should be taken, treated, and stored so that they may be used in legal proceedings. The complete SWGIT document is available in PDF format at http://www.theiai.org/guidelines/swgit and is periodically updated.

Technical photo imaging or *technical photography* is the recording of an object or group of objects in the best manner possible to accurately represent a scene or that object, on the day and time in question. *Photo imaging* or *photography* is the recording of an object or group of objects to make an image that is pleasing to the eye.

Evidence or lack of evidence must be documented at a crime scene. In order to understand what happened at a crime scene, it is necessary to reconstruct the scene and the

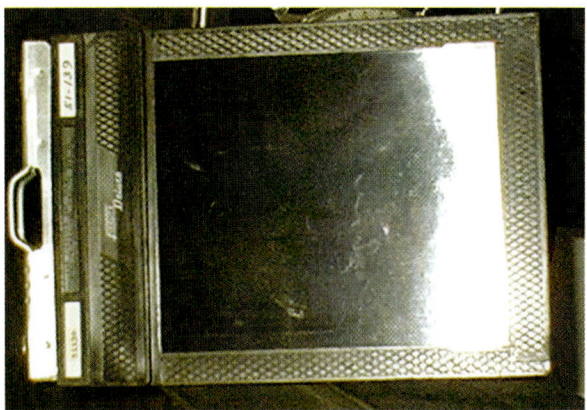

Figure 11.5 4 X 5 sheet film holder. Each holder had two sheets of film—one on each side.

Figure 11.4 Used in the 40's, 50, and 60s the old 4X5" sheet film camera was used by police detectives and newsmen alike. Each exposure required the photographer to change the film holder, focus the camera, cock the shutter, and press the shutter. The operator would then have to remove the film carrier, flip it over, reinsert it into the camera, then focus the camera, cock the shutter, and press the shutter.

scenario in order to investigate the crime and, if necessary, prosecute the perpetrator successfully.

We cannot just take several snap shots of a scene and expect those few images to be acceptable. We must tell the story. For example, the St. Valentine's Day Massacre of 1929 is a very poorly documented scene. There were only approximately five images taken. The camera used was a Speed Graphic 4 x 5 in. sheet film camera (see Figure 11.4 to Figure 11.7). The images were taken by a news photographer. In fact, as the story goes, the photographer thought the image would look better with a hat on one of the victim's chest, so he (the photographer) placed it there. So much for the integrity of that crime scene.

Do the five or so images taken tell the story? Do they show the wounds? Do they show the correct positions of objects? Do they show the position of the victims in relation to the shell casings? Are the images distorted because of the angle of the camera? What about livor mortis? Were scales used to size objects in the scene? What did the other sides of the room look like? Where was the door or doors? Were there windows and how many were there? What were the victims looking at before they were killed? What direction is north? These may have been sensational newspaper photographs, but they certainly did not tell the story; they were not good technical images.

Photo Imaging the Scene

You must identify the case and location. You must record the overall area of the scene. You must image evidence and items of interest, and their perspective to each other. You must capture each piece of evidence, record it correctly, scale it (show relative size), and you must do so without appearing to hide *exculpatory evidence*—evidence which could be used by the opposition in the case.

The use of a narrative to explain the condition, the exact location of items, and the existence of artifacts, of course, is also very important and necessary components of the overall documentation of a scene. They become a part of the case report.

Digital imaging allows us to visually recreate the scene for investigators, prosecutors, and jurors deciding the fate of individuals charged with a crime.

Figure 11.6 St. Valentine's Day Massacre, Chicago, Illinois. Notice the hat on the chest of one of the victims. The hat was allegedly placed on the man's chest by a newspaper photographer to make it a more interesting picture.

Figure 11.7 St. Valentine's Day Massacre, Chicago, Illinois. Only overall photos were taken. There were no close ups or photos of items of interest.

A picture is worth many thousands of words. If the narrative of a report states "The handgun was located on the table next to the phone," we can summon a visual picture in our mind. But as we look at the picture (see Figure 11.8), we observe that the image we conjured may be quite different from what was actually described.

The image shows the location of the handgun, which is a semiautomatic pistol with the slide open and locked to the rear position, indicating that the handgun does not have a cartridge in the chamber. The phone is a cell

Figure 11.8 The description of the table in the living room may conjure an image in our mind that is remarkably different from what existed at the scene. This is why the photo image is so important. The gun could be a rifle or shot gun, the phone could be a landline or cordless phone, and the table may not be thought of as in the middle of the floor.

phone which may or may not belong at this scene. Since it is a cell phone, it may belong to anyone—the victim, a witness, an EMT, or even the perpetrator.

Documentation of a scene or of evidence requires us to document the documentation. Sounds redundant, but it is something that should be done. The images should be documented, and I strongly recommend that they be listed in the case report (see Figure 11.9).

The first thing that must be done is to identify the images. We want to identify them as unique or individual. As an example, when an object is analyzed in a laboratory we can sometimes identify it as belonging to a class. We can identify a handgun as a .38 caliber. But, to make it unique, the handgun is further examined and found to have a serial number. Now the handgun is unique, one of a kind. It is the same with digital images. Images belong to the class of "images of a scene." We must make them unique by specifically identifying them. We do this by making our first image on the memory card, a color balance/ ID card, sometimes called a "gray card" (see Figure 11.10). On this card we place the case number, the image taker's name, the date, and the agency. This is done so that if the CD, magnetic media or memory card is misfiled, it can be identified as belonging to a specific

Crime Scene Investigator's Report

[Print Form]

Report Classification [   ]   Case Number [   ]

Date/Time [   ]   Type of Location [   ]   Agency [   ]   Investigator [   ]

Victim's Name [   ]   Victim's Address [   ]

Injuries [   ]   Taken for Treatment [   ]   Victim Rape Kit [   ]   Suspect In Custody [   ]

Suspect Rape Kit [   ]   Weapon [   ]   Gunshot Residue [   ]   Gun Sheet [   ]

Weather [   ]   Inside Temperature [   ]   Outside Temperature [   ]   Crime Scene Drawing [   ]

Alcohol [   ]   Drugs [   ]   Lighting Conditions [   ]

Vehicle [   ]   Make [   ]   Model [   ]   Year [   ]   Color [   ]

Vehicle Sheet [   ]   License Plate [   ]   License Plate State [   ]

Other Distinguishing Characteristics [   ]

Inventory Control Number [   ]   Images Taken [   ]

Evidence Collected				Photo Images			
1		23		1		23	
2		24		2		24	
3		25		3		25	
4		26		4		26	
5		27		5		27	
6		28		6		28	
7		29		7		29	
8		30		8		30	
9		31		9		31	
10		32		10		32	
11		33		11		33	
12		34		12		34	
13		35		13		35	
14		36		14		36	
15		37		15		37	
16		38		16		38	
17		39		17		39	
18		40		18		40	
19		41		19		41	
20		42		20		42	
21		43		21		43	
22		44		22		44	

Note: Start listing photo images on this report - if additional space is needed use Photo Listing Sheet

Signature [   ]

Page [   ]

Figure 11.9 A sample case report with fields for listing photo images.

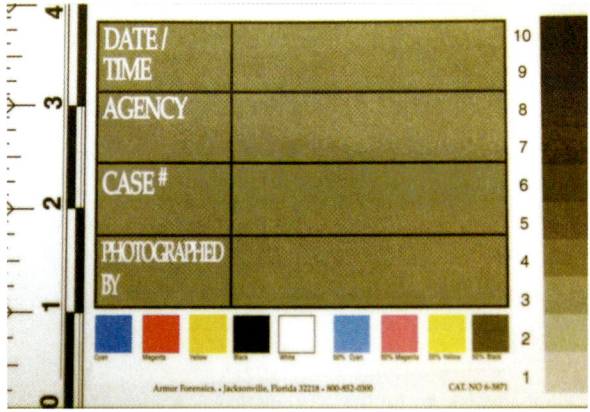

Figure 11.10 A color balance/ID card, sometimes called a "gray card." On this card we place the case number, the image taker's name, the date, and the agency. This is done so that if the CD, magnetic media or memory card is misfiled, it can be identified as belonging to a specific case. It also identifies the case and images at trial.

Figure 11.11 The "N" or north card. This is photo imaged in an overall photo image of the north wall or a view to the north (if exterior). This allows the image taker to acclimate him or herself to this and other images of a scene after a large amount of time has past, such as trial.

case. It also identifies the case and images at trial.

The ID/color balance card can be laminated. A dry marker can be used to print the appropriate information. It can then be reused by the CSI over and over again.

The second image on the roll of film or as the second digital image should be an image of the letter "N" for "north" (see Figure 11.11). This "N" can be held by an assistant or propped against the wall or an object. The important thing is that the north card is on the north side of the room or location. The image should include a good portion of the north wall or north side of the crime scene. This allows the CSI to document the north direction so that there is no confusion at trial. When cases go to trial, too often there is a large span of time between the time the images were taken and the actual trial. Sometimes years pass. This is compounded by the fact that since the CSI works many cases, this makes remembering which direction is north at a particular scene very difficult.

Images taken at a crime scene should be listed on the report form used by the CSI. If the amount of images exceeds the number of spaces on the report, an image continuation sheet can be used. The use of the log is a matter for each individual agency to set as its own policy and procedure. The FBI, for example,

advocates the use of the image log. There are advantages and disadvantages to the use of an image log. It is helpful for the prosecutor to have an inventory of the images available to him in order to prepare a solid case. It also eliminates any question as to whether all images taken were delivered to the defense during the discovery phase of the trial. It also identifies any images that may have been lost or misplaced.

The type of description placed on the log should be short but descriptive, such as: "gray card/image ID," "north shot," "handgun on end table," "overall living room s/n." Note: "s/n" is an abbreviation meaning from the south to the north.

A disclaimer should be added to the report form, indicating that directions, such as s/n (south to north) are approximate and are stated as such for the purpose of identifying each image. The disclaimer eliminates the potential for a defense attorney to nitpick the description, such as "Isn't it true, detective, that the image log states that this image was taken south to north and the image taken was actually southeast to northwest?" If the defense can call into question one image, he or she can then attempt to call all the images into question and move to have all images thrown out.

Imaging the Scene

At a crime scene, photo images are taken of the scene. Many are overall images or *perspective* shots. Perspective shots are images that show one object in relation to another object. These shots are very important in telling the entire story.

Is there a minimum number of photo images that can be shot at a scene? I would say yes with qualification. We try to shoot the "four points of the compass." From the center of the scene (or the most important object in the scene: North to South, South to North, East to West and West to East. Then from the outer perimeter of the scene shoot inward north to south, south to north, east to west, and west to east. The "north" card and the ID card should also be included. This is 10 shots. This is the absolute minimum.

There are exceptions to this rule. You have to use some common sense. If there is evidence, you will be photo imaging each piece with and without scale. You may wish to take additional perspective shots showing one piece of evidence in relation to another.

There may be a situation where a victim is on the floor in the center of an 8 x 8 ft bathroom. If you tried to photo image using the four points of the compass, you would end up with eight images of the four walls. In this type of situation, take several overall images showing as much of the room as possible.

Remember, 10 shots is the absolute minimum. Very rarely will you ever end up with just 10 photo images for a case.

Photo Images as Evidence

Evidence is normally thought of as something that is placed in a bag, signed, sealed, and locked away. Digital photo images can also be evidence. There are procedures that must be followed. After the images are taken at the crime scene, they should be uploaded to a computer file. The images may not be changed, modified, or manipulated. Once they are on the computer, burn them to a CD. This will become your "master." Seal the CD in an envelope, and mark it as you would any other type of evidence. These images are like the negatives from a film camera. They will never be manipulated or modified. This is your original and should not be opened unless you are directed to do so by the courts. If there is an issue concerning the images introduced as evidence, the master can be opened and a comparison can be made. Since the master is a digital copy of the images from the camera, they are identical. Any copies or printed photo images should be done from the file on the computer.

What actually makes the photo images evidence? It is your testimony. Attorney: "Detective, do these digital photo images truly and accurately represent the crime scene on the date and time in question?" Detective: "Yes, they do."

There is another issue of what are the "best images" or the "official photo images" of the crime scene. If 100 photo images are taken of the scene by the crime scene investigator 45 minutes after a news reporter took several photo images with a Polaroid camera, what are the best images? Unfortunately, the several Polaroid images that were taken by the reporter are the best images. This is because the photo images taken that are closer in time to the actual incident would more truly and accurately represent the crime scene on the date and time in question.

Deleting of Images from the Camera

Digital cameras have many benefits. You can see the images immediately after they were taken. A single memory card (4 gigabyte) can hold over a thousand photo images. It also has some features that we may not use. Specifically, we may not delete an image either because we didn't like it, it was improperly exposed, or it was out of focus. This is because it can be stated by the defense that the image erased may have shown exculpatory evidence.

Defense attorney: "Detective, the digital photo images that have been introduced into evidence jumps from 1234.JPG to 1236.JPG. Where is 1235.JPG? This also occurs several more times throughout the images submitted as evidence in the case." Detective: I deleted them because they were out of focus and the lighting was not good in some." Defense attorney: "How do I know that the photo images deleted didn't show that my client is innocent? Your, Honor, I move that all the photo images submitted by the prosecution be thrown out due to the fact that the missing images may contain exculpatory evidence."

Figure 11.12 The image depicts what appears to be an area without dust on this shelf, which may indicate that something has been removed from this area.

by its position, location, or condition, may help in furthering an understanding of the story of the crime scene. An image of the toilet seat being in the up position in a home with only women in residence, a kitchen knife in the bedroom, and a tire iron in a living room are all items of interest. Remember, tell the story with your images. You can never take too many images. *You can never go back to the original scene.* Document, document, document.

Evidence and Items of Interest

Documenting the scene also includes recording items of evidence out of place. Items out of place are called *artifacts*. An artifact, in forensic science, can be defined as "an object that does not belong." Artifacts can be a bullet hole in a mirror or glass window. It could be a letter clutched in the hand of a victim. It could even be a VCR in the middle of the living room floor. An item out of place could be a fireplace poker in a bathroom.

We must also document items which are not there. An example of this would be a table with one bottle of wine, one wine glass, and *two wine glass rings* on a glass table. Where is the second glass? Was it taken by the perpetrator? The two rings on the table should be imaged as well as the single glass, and an overall showing both. An area of dust on top of a shelf, with area that is dust-free could mean that there was a CD player there that was removed by the perpetrator (see Figure 11.12). What is else missing? Have the napkins been removed. Has an ashtray been emptied? It is very important to record negatives as well as positives. Items of interest must be documented. An item of interest is something that,

Photo Imaging Evidence

At a crime scene, nothing should be picked up, moved, or collected until it has been properly documented (measurements taken) and photo-imaged. You *cannot* pick up an item, examine it, then replace it and photo-image it. In court, if asked if the photo image truly and accurately represents the scene at the date and time in question, you would have to say no, due to the fact the object was lifted and replaced. This one small movement may cause all of your images to come under question. Photo-image, then collect and examine.

Angle

When photo-imaging objects of interest and evidence at a crime scene, images must be taken at 90 degrees. This is extremely important. As an example, if you photograph

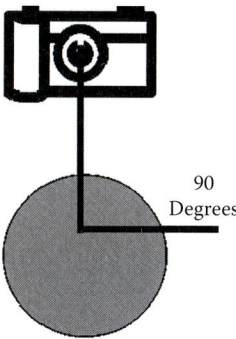

Figure 11.13 This is a circle, photo imaged at 90° to the subject. The photo examined shows a circle.

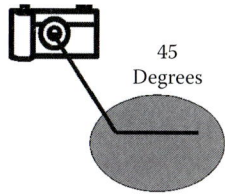

Figure 11.14 This is the same circle photo image at 45° to the subject. The photo examined shows an oval. This is distortion. The distortion is caused by the angle from which the camera was placed in relation to the subject.

a circle at 90 degrees, when you look at the photo image, you see a circle. If you photo image a circle at 45 degrees, when you look at the photo image, you see an oval. This is *distortion*. The circle is distorted by the lens because of the angle the camera and lens were placed relative to the circle (see Figure 11.13 and Figure 11.14).

Scale

A scale is nothing more than a small ruler, usually 6 in. in length. The scale is used to "size" items in the photo image. For example, if we photo-image a shoe print, and we fail to place a scale in the image, we have no idea if the shoe is a size 13 or a size 8. If a bite mark is photo imaged on a victim, if there is no scale, then a match cannot be accomplished because the measured distance between teeth cannot be measured on the photo image.

Just as we photo image items with scales, we also photo image those same items without scales. This is done so that there can be no accusation that the photo imager was hiding evidence, either intentionally or accidentally,

Figure 11.15 This is an example of a document, with scale. Is this information, a threat, or a suicide note?

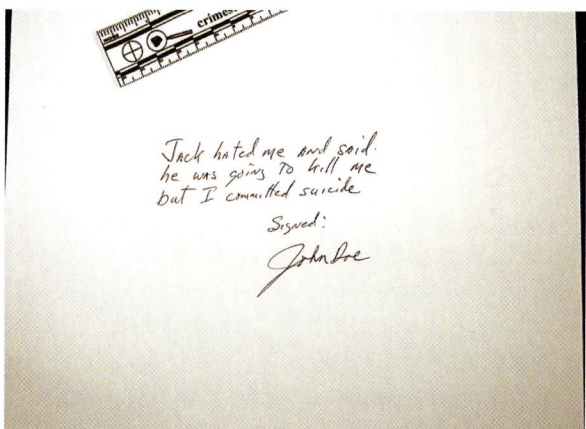

Figure 11.16 This is an example of the same note, without scale. You can see that part of the note was hidden, causing a different interpretation than when the first photo image was examined. This could be considered exculpatory evidence, which is why we photo-image evidence both with and without a scale.

that would prove that the suspect is innocent. This type of evidence that could work for the other side is called *exculpatory* evidence (see Figure 11.15 and Figure 11.16).

Depth of Field

Depth of field (DOF) is the portion of a scene that appears sharp in the image or in focus. In Figure 11.17, the camera and lens are focused on boy A. Boy A is in focus, and boy B is in focus also. That is because of depth of field or the depth of the field of view is in focus. The depth of field can be increased by closing down the aperture, making the aperture smaller. Remember the smaller the F-stop, the larger

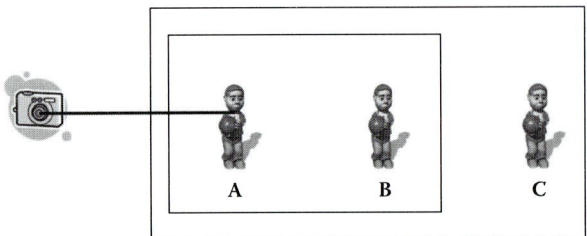

Figure 11.17 This diagram shows three boys. The camera and lens are focused on boy A. Boy A and Boy B are "in focus." Boy C is in "acceptable focus." Everything inside the blue square is "in focus." Everything inside the red box and outside the blue box is in "acceptable focus."

the opening of the aperture, and the larger the F-stop, the smaller the opening of the aperture. A rule of thumb can be used: when focused on an object, an area 1/3 in front of the object and 2/3 in back of the object will be in focus. "In focus" is sharp focus.

What about boy C? He is in an area that is considered "acceptable focus." "Acceptable focus" is not as sharp as "in focus" but is, well, acceptable. Boy C is in "acceptable focus" because of the hyperfocal distance of the lens and its specific lens setting (aperture). The hyperfocal distance is the closest distance at which a lens can be focused while keeping objects at infinity acceptably sharp. When the lens is focused at this distance, all objects at distances from half of the hyperfocal distance out to infinity will be acceptably sharp.

Light

With the exception of the camera, light is probably the most important factor in imaging. Without light, there would not be an image. The camera, with only a few exceptions, records the reflection of light rather than the light itself. Only when the object that is being imaged produces the light, such as a candle, a light bulb, or fire, does the camera record the light itself.

Ambient Light

Ambient light is another way of saying the light that exists or existing light. This is the level of light that will primarily affect your image. Many photographers shoot with ambient light

only. They use it to convey a mood or an idea. As forensic photographers, we wish to record, document, and convey facts. If the existing or ambient light can do this, then we use it. If the ambient light will not do this, then we must use other light sources to augment the ambient light.

Ambient light can fall into many categories: incandescent (common light bulbs), florescent, halogen, sodium vapor, daylight, dawn, and dusk to name a few. Some of the more exotic light categories cause different hues of color to change in your images. It is a good idea to indicate the type or types of ambient light present in your report form (see Figure 11.9).

Combinations making up ambient light can also exist. An example is a scene in which daylight is streaming through a window but there are fluorescent lights on in the room. In this case there is a combination of light sources which make up ambient light.

Color Temperature

A measure of the distribution of power in the spectrum of white, or colorless, light is stated in terms of the Kelvin temperature scale. The human eye is incredibly adept at quickly correcting for changes in the color temperature of light. Many different kinds of light all seem white to us. Photographic film is not so forgiving; daylight film is made to be exposed at 5500 K light, while indoor film requires light with a color temperature of 3400 K. Photographs taken indoors with incandescent light on ordinary daylight film will come out orange; photographs on indoor film taken in sunlight will be blue, as will photographs taken outdoors in shade illuminated by blue sky.

While a digital camera is not as adept as the human eye at quickly correcting for changes in the color temperature of light, it does a much better job than that of film cameras. In digital photography, calibration of the white point is accomplished with either an automatic or manual setting. The assumed white point can vary depending on the light conditions; the concept of "white" is not an absolute thing. Most digital cameras let you specify whatever white point you want, usually by pointing the camera at a white object illuminated by the current light used by the imager. Some cameras also can detect the ambient light and determine the white point from that: automatic white balance.

Flash/Strobe

The strength of the flash is measured and documented as the Guide Number. As you get further away from a light source, the intensity of the flash drops off.

As you change the f stop of a lens, the intensity of the light at the film plane or sensor also changes. They change at the same rate, so this statement is always true: distance times f-stop equals a fixed number (called the "guide number"). Every combination of flash and film speed has one guide number. Faster films and more powerful flashguns have higher guide numbers. Often manufacturers use the guide number as part of the model name of a flashgun. You can figure out what f stop to use by dividing the guide number by the distance from the flash to the subject. If a flashgun has a guide number of 80; at 10 feet use f8 (80 divided by 10=8), at 5 feet use f16 (80 divided by 5=16), and at 20 feet use f4 (80 divided by 20=4).

When the ISO setting on a digital camera is increased by a factor of 4, the guide number is doubled. If the guide number for ISO (speed of the film) 100 was 80, the GN for ISO 400 is 160. The increase as film speeds double is just like the 1.4 times increase in the f-stop numbers on your lens: f4, f5.6, f8, f11, f16.

Direct Flash

Direct flash is a flash that originates on or near the camera. This would be a camera's internal flash or an external one attached to the hot shoe on the camera body. This light lights the subject very well but it does create shadows (see Figure 11.18).

In forensic photo imaging, our main job is to document, record, and identify. A directly lit shot will do this very well. However, if the image would be more clear in "telling the story" of the scene by a softer or less shadowed light, then there are other options available.

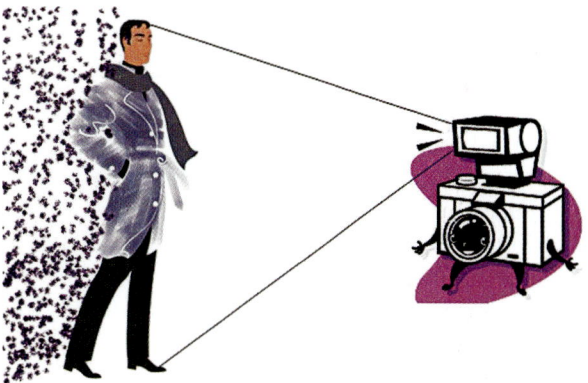

Figure 11.18   An example of direct flash.

Indirect Flash

We can use indirect lighting. Indirect lighting is sometimes called "bounced lighting." We use this if we wish to eliminate or reduce shadows.

In order to use indirect or "bounce lighting" we do not aim the light source at the subject. We aim the light source or flash at the ceiling or wall and bounce the light off the ceiling or wall, then onto the subject. This lights the foreground and the background, reducing and often eliminating shadows (see Figure 11.19).

Fill Flash

Fill flash is an additional light source used to highlight shadowed areas. This additional light source can be a second flash, a white reflecting board, or an incandescent light. In Figure 11.20 we see a cardboard box from our crime scene. As overall photo images are taken

Figure 11.19   An example of indirect flash.

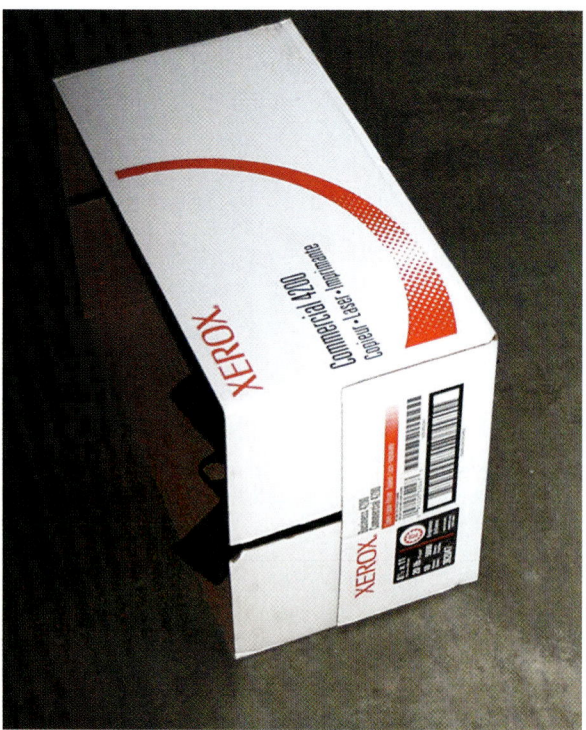

Figure 11.20 This image shows a cardboard box. The box is not necessarily our true object of interest, but because the primary light source is a flash located on the camera, what we see is the box with something dark in it.

at approximately 10 feet, the box is lit with the flash on the camera (see Figure 11.18). Because this is direct light coming from the top of the camera the light does not bend around the corner to light the inside of the box. If we use a secondary light source to light the inside, and our camera mounted flash to light the top, we obtain a clear image of what we want to show: the gun in the box (see Figure 11.19).

Macro Imaging

Macro imaging or macro photography is close-up photography that does not involve the use of microscopes. True *macro* photography is generally considered to have a ratio of 1:1 or greater, and the lens used will be able to focus on an object less than 8 in. distant. Many digital cameras (specifically, point-and-shoot type cameras) identify the macro setting on the camera with an icon of a flower. This is a

macro setting, and the camera does not have a macro lens.

Bench Imaging or Rephoto Imaging

Bench imaging is defined as close-up imaging, which may include macro imaging (1:1 ratio). This is usually done in the laboratory or at the CSI's office. Even though the evidence was photographed at the scene, additional imaging or rephoto imaging is sometimes desired.

The lab setting allows the imager much more control over the lighting, temperature, conditions, and posing of the object of evidence or interest. One example would be that of a handgun at the scene of a shooting. The scene is dark with relatively no ambient light. The gun was found at the base of a bush. Only three images were taken of the gun at the scene (one with and one without scale) and an additional image for perspective. Due to the brush, additional images were impossible to obtain. Only one side of the gun was recorded at the scene.

Back at the lab, the gun can be "posed" under controlled conditions. Light can be manipulated to provide a perfect recording of the gun. Both sides of the gun can be imaged. Use of a macro lens can be employed to take 1:1 ratio images of scratches, marks, blood, the serial number and other artifacts. Remember to photo image negative aspects (missing parts), the absence of blood, or the area from which a serial number was removed.

There will be occasions in which you will wish to photo-image the inside of the barrel. This could be important if the wound was a contact wound or made at a very close range. In these instances there is "blowback." This a phenomenon occurring in contact and very close-range wounds made with a gun. Blood and tissue are "blown back" into the barrel of the gun. This can be documented by photo imaging (Figure 11.21).

The gun should be secured in a vice or with a clamp. If the gun is a semiautomatic, the slide should be opened and locked to the rear. If the gun is a revolver, the gun should

Figure 11.21 This image shows our true object of interest. We have used the flash on top of the camera as our primary light source, but we have added a secondary light source to light the inside of the box. Now we can clearly see the gun. This is "fill flash."

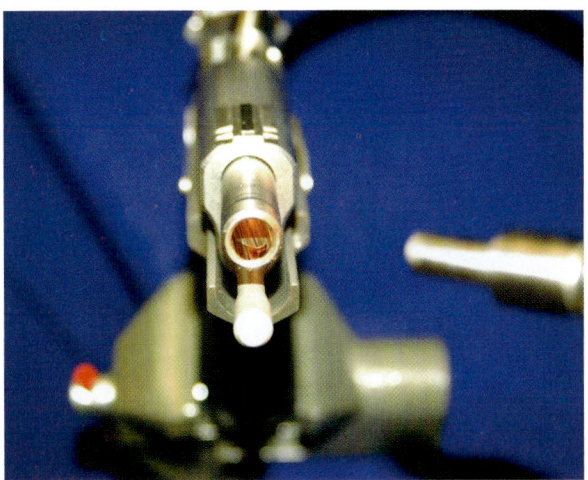

Figure 11.22 This is the barrel of a handgun, illuminated with a fiber optic light source. Notice the "blowback" of blood in the barrel. This may be produced when the barrel of the gun is discharged in contact or close range to the victim. The image was shot with a macro lens at F-32.

be photo imaged with the cylinder opened. Using an additional clamp, secure a light source and aim it into the chamber of the gun so that the light shines through and out the end of the barrel. A battery powdered flashlight with a "goose neck" type adjustable shaft works very well. In the picture, we are using a fiber optic light (dual source) with adjustable light intensity (see Figure 11.22).

Disable the flash on the camera when working with moveable light sources. If you cannot disable the flash on the camera, temporally cover it with black electrician's tape. If the flash is not disabled, the flash will overlight the area of interest, the barrel, and the blood and/or tissue will not be photo imaged clearly; you will observe that your image is a bright white blob.

Tool Marks

Tool marks are artifacts left when one object presses or is drawn against another. Often, on a door when it is "jimmied," an impression of the tool is left on the door jam or striker plate. The best evidence is, of course, to take the striker plate or the door jam; however, some victims are not so inclined to permit. Photo imaging is next best choice. A macro lens is helpful due to many tool marks being very small (see Figure 11.23). Oblique lighting

Figure 11.23 This image shows the lighting setup that was used to photo image the blood in the barrel of the handgun in Figure 11.21.

Figure 11.24 This is a toolmark that has been photo imaged using an oblique light source. This light source is 45° to the subject while the camera is 90° to the subject.

Figure 11.25 This is an example of a tool mark on a shell casing. The hammer of the gun makes a unique mark on the primer of the round when it strikes it. These marks can be identified as belonging to a specific gun. The image was shot with a macro lens at F-32.

(lighting from the side at approximately 45°) works well. This is because the light from the side creates tiny shadows in the "valleys" of the tool marks.

Markings on shell casings and on bullets are also tool marks. Examiners look at these items under comparison microscopes. Many of these very expensive microscopes have image capture capabilities. Some do not have these scopes and must rely on macro photo imaging. This is where the ability to use a good macro lens and adjustable lighting is very helpful.

Using a copy stand and with a macro lens set at F-32 or smaller, place the shell casings side by side using rope calk to secure them in place. Rope calk is a malleable clay-like substance available at most hardware stores. It allows the posing of items without damage to the item. Using a high intensity light source (a halogen desk lamp works well), set the focus to manual and use the copy stand rails to focus the camera by moving the camera up and down. Take several images, moving the light source each time. Use the best image for comparison (see Figure 11.25).

Blood

The photographic documentation of blood-stains and patterns has been discussed in Chapter 10, Crime Scene Investigation, and Chapter 12, Recognition of Bloodstain Patterns.

Extreme Conditions

Extreme conditions are those that will cause a bad exposure when using the internal light meter of the camera. The camera's meter averages the amount of light it sees reflected from the object or objects to be photo-imaged. The three most important extreme conditions are snow, sand, and the aftermath of a fire (see Figure 11.26).

Snow is all white. White reflects almost 100%, since the camera is attempting to average the light. This "fools" the meter in the camera. In order to compensate, if there is bright sunlight on snow, close down the aperture on the lens by three stops.

Sand reflects more than an average scene, but less than snow. Again, the camera is attempting to average the light, which "fools" the meter. If there is bright sunlight on sand, close down the aperture on your lens by one and a half stops.

The aftermath of a fire creates a scene where all the walls, ceilings, and floors are black with soot. Black absorbs light. Because the camera is trying to average the light, it is "fooled" again. In the scene of a fire, where there is 90% soot (black), open your aperture three stops. If you are using a camera-mounted flash, set the flash to manual or full power. If

Figure 11.26 The three main "extreme conditions" that require overriding the camera's internal meter.

you are using a flash or strobe, you may still have to make an adjustment. Remember, if you shoot a test shot, you may not delete or erase it from the camera. List it as a "test shot for lighting."

Conclusion

The goal of this chapter is to acquaint the reader with the basic contents of forensic digital photo imaging. While photography has been around for a while, digital imaging is relatively new. Although you can learn a lot from a book, photo imaging is a hands-on discipline. Read about a technique then try it with your own equipment. Experiment with tried-and-true methods, then modify the technique and try again. The good news is that with digital photo imaging, you do not have the expense of film, film development, and printing. You can take hundreds of images, upload them to your computer, delete them from your memory card, then do it all over again. Practice makes perfect, the more you practice the better you will get.

Questions

1. What are the three extreme conditions at a scene that requires you to override the automatic settings on your camera?

2. What is a SLR camera?

3. What is the best setting for your camera to assign numbers to the images that are taken?

4. What is aperture?

5. What is ISO?

6. How is the word forensic defined?

7. What is SWGIT?

8. What makes photo images evidence?

9. Can a forensic photo imager delete out of focus or improperly lit images?

10. Before evidence can be moved, examined, or collected, what must be done first?

11. What is a scale?

12. What is exculpatory evidence?

13. What is ambient light?

14. Explain the difference between direct and indirect flash and what is accomplished by each?

15. What is fill flash?

16. What is macro photo imaging?

17. When photo-imaging tool marks, what type of lighting is preferred and why?

18. Explain EI (exposure index)?

19. Explain the differences among art photo imaging, technical photo imaging, and forensic photo imaging.

20. How does one identify direction at a scene using photo imaging?

21. Why is evidence photo imaged both with and without scale?

22. Should we photo image items that are not there? Explain.

23. Explain depth of field.

24. Explain the differences between in focus and acceptable focus.

25. Explain the term long exposure.

References and Suggested Reading

Jones, P., *Practical Forensic Digital Imaging, Applications and Techniques*, Taylor & Francis/CRC Press, Boca Raton, FL, 2009.

SWGIT, Scientific Working Group on Imaging Technology, International Association for Identification, http://www.theiai.org/guidelines/swgit.

Recognition of Bloodstain Patterns

Stuart H. James, Paul E. Kish, and T. Paulette Sutton

12

Introduction

The geometric interpretation of human bloodstain patterns at crime scenes is not a new idea, but it has acquired much greater recognition over the past several decades. Bloodstain pattern interpretation should be viewed as a forensic tool that assists the investigator or the forensic scientist to better understand what took place and what could not have taken place during a bloodshed event. The information obtained from the interpretation of blood-stain patterns may assist in apprehending a suspect, corroborate a witness's statement, assist in interrogating suspects, allow for the reconstruction of past events, and most importantly, exonerate an accused. As with any tool, bloodstain pattern interpretation has its strengths and weaknesses. The interpretation will only be as valid as the information available and the ability of the examiner performing the analysis.

History of Bloodstain Pattern Interpretation

The interpretation of bloodstains and patterns has been documented in books, journals, and articles for centuries. Herbert Leon MacDonell, a prominent bloodstain analyst from Corning, New York, conducted a worldwide literature search over the past several years and compiled an extensive bibliography of over 550 references dating back as far as a trial held in London in 1514. One of MacDonell's more significant findings was an extensive study of bloodstain patterns by Dr. Eduard Piotrowski, an assistant at the Institute for Forensic Medicine in Krakow, Poland. His work in German, entitled *Uber Entstehung, Form, Richtung und Ausbreitung der Blutspuren* (Concerning origin, shape, direction and distribution of the bloodstains following head wounds caused by blows), was published in Vienna in 1895 (Piotrowski, 1895). According to MacDonell, "No one preceded Piotrowski in designing meaningful scientific experiments to show blood dynamics with such imagination, methodology, and thoroughness. He had an excellent knowledge of the scientific method and a good understanding of its practical application to bloodstain pattern interpretation."

Original research and experimentation with bloodstains and patterns was done by the French scientist Dr. Victor Balthazard and his associates, who presented the material as a paper at the 22nd Congress of Forensic Medicine in 1939 (Balthazard et al., 1939). The work was entitled *Etude des Gouttes de Sang Projecte (Research on Blood Spatter)*.

The use of bloodstain pattern interpretation as a recognized forensic discipline in the modern era dates back to 1955, when Dr. Paul Kirk of the University of California at Berkeley submitted an affidavit of his examination of bloodstain evidence and findings in the case of *State of Ohio v. Samuel Sheppard*. This was a significant milestone in the recognition of bloodstain evidence by the American legal system.

Herbert Leon MacDonell received a grant from the Law Enforcement Assistance Administration and conducted research to recreate bloodstains observed at crime scenes. His initial publication, entitled Flight Characteristics and Stain Patterns of Human Blood, appeared in 1971 (MacDonell and Bialousz, 1971). MacDonell followed that publication with his Laboratory Manual on the Geometric Interpretation of Human Bloodstain Evidence in 1973 (MacDonell and Bialousz, 1973). Both publications have been periodically updated.

MacDonell established a training program for basic bloodstain pattern interpretation and conducted his first bloodstain institute in Jackson, MS, in 1973. Since then, he and others have conducted numerous basic and advanced bloodstain interpretation courses throughout the United States and abroad.

Articles have appeared in scientific journals, and additional texts have been published over the past 20 to 25 years. The study of bloodstain pattern interpretation has been featured in numerous crime novels, forensic television programs, and the news media. Many high-profile cases shown on court TV have included expert testimony on bloodstain pattern evidence. This exposure has increased the public awareness of this important forensic discipline.

Properties of Human Blood

Biological Properties

Blood is the fluid that circulates throughout the body by way of the heart, arteries, veins, and capillaries. It transports oxygen, electrolytes, nourishment, hormones, vitamins, and antibodies to tissues, and transports waste products from tissues to the excretory organs. Blood consists of a fluid portion referred to as **plasma** that contains cellular components consisting of red blood cells, white blood cells, and platelets. When blood has had opportunity to **clot**, the fluid or liquid portion of clotted blood is referred to as **serum**. Red blood cells (erythrocytes) transport oxygen from the lungs via the arterial system and return carbon dioxide to the lungs for expiration via the venous system. White blood cells (leukocytes) assist with defense against foreign substances and infection. The nuclei of the white blood cells are the sources of DNA in the blood. Platelets are major components of the clotting mechanism of blood. In normal individuals, cellular components comprise approximately 45% of the total blood volume, which ranges in healthy adults from 4.5 to 6.0 liters.

Physical Properties

Exposed human blood is not unlike other common fluids. It will act in a predictable manner when subjected to external forces. Blood, whether a single drop or large volume, is held together by strong cohesive molecular forces that produce a surface tension within each drop and on the external surface. Surface tension is defined as the force that pulls the surface molecules of a liquid toward its interior, decreasing the surface area and causing the liquid to resist penetration. The surface tension of blood is slightly less than that of water.

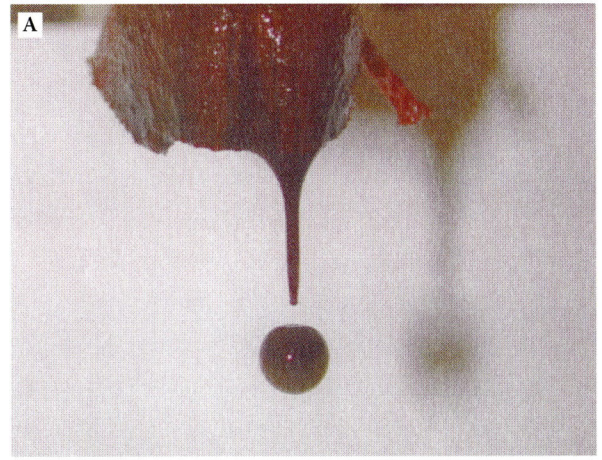

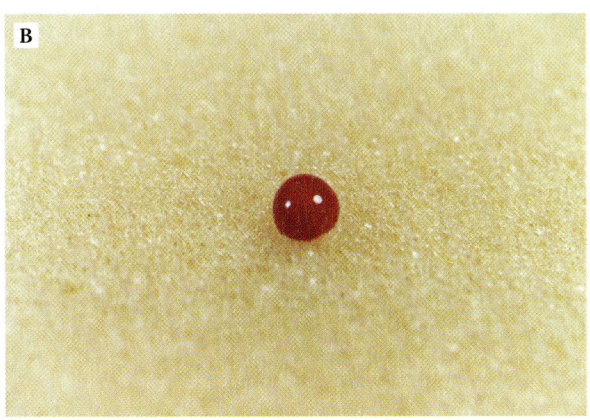

Figure 12.1 (A) Spheroid shape of a blood droplet due to the effect of surface tension as it falls through the air after separating from a blood-soaked cloth. (B) Spheroid shape of a blood droplet on cloth due to the effect of surface tension.

By comparison, one can appreciate the high surface tension of liquid mercury, which is almost 10 times greater than that of blood. To create spatters of blood, an external force must overcome the surface tension of the blood. The shape of a blood drop in air is directly related to the molecular cohesive forces acting upon the surface of the drop. These forces cause the drop to assume the configuration of a spheroid. Blood, like all fluids, does not fall in a teardrop configuration, even though many artists portray raindrops and other fluids in that manner on television and in newspapers (Figure 12.1A and Figure 12.1B).

A passive drop of blood in air is created when the volume and mass of the drop increase to a point where the gravitational attraction acting on the drop overcomes the molecular cohesive forces of the blood source.

The volume required to produce these free-falling drops of blood is a function of the type of surface and the surface area from which the blood drop has originated. For example, research and experimentation have shown that the volume of a passive drop of blood falling through air from a fingertip will be larger than a drop that originates from a hypodermic needle and smaller than a drop originating from a surface such as a baseball bat. The volume of a typical or average drop of blood has been reported to be 0.05 ml, with an average diameter of 4.56 mm (while in air). These reported measurements can vary as a function of the surface from which the blood has fallen.

The mutual attraction of the molecules of blood is due to cohesive forces. Viscosity is defined as resistance to change of form or flow. The more viscous a fluid, the more slowly it flows. Blood is approximately six times more viscous than water and has a specific gravity or density slightly higher than water. Specific gravity is defined as the weight of a substance relative to the weight of an equal volume of water. These physical properties of blood tend to maintain the stability of exposed blood or blood drops and cause them resist alteration or disruption.

A blood drop falling through air will increase its velocity until the force of air resistance that opposes the drop is equal to the force of the downward gravitational pull. At this point, the drop achieves its **terminal velocity**. In his early research, MacDonell established that the maximum terminal velocity for an average-sized (0.05 ml) free-falling drop of blood was approximately 25.1 ft/sec and was achieved in a maximum falling distance of 20 to 25 ft. The resulting diameter of the bloodstain produced by a free-falling drop of blood is a function of the volume of the drop, the surface texture it impacts, and, up to a point, the distance fallen.

One can easily demonstrate that free-falling drops of blood with a typical volume of 0.05 ml will produce bloodstains of increasing diameters when allowed to drop from increasing increments of height onto smooth, hard cardboard. The measured diameters range from 13.0 to 21.5 mm over a dropping range of 6 in. to 7 ft. Blood drops that fall distances greater than 7 ft will not produce stains with any appreciable increases in diameter. It is not possible to establish with a high degree of accuracy the distance that a passive drop of blood has fallen at a crime scene, since the volume of the original drop is not known.

Target Surface Considerations

Exposed blood has an invisible outer skin referred to as its **surface tension**. In order to create smaller blood droplets or spatter from a volume of blood, this surface tension must be disturbed in some manner. Although a single drop of blood falling through air is affected by the forces of gravity and air resistance, these forces do not overcome the surface tension of the blood. No matter how far a drop of blood falls, it will not break into smaller droplets or spatters unless something disrupts the surface tension. One factor in breaking the surface tension of a blood drop is the physical nature of the target surface the drop strikes. Generally, a hard, smooth, nonporous surface, such as clean glass or smooth tile, will create little if any spatter (Figure 12.2A), in contrast to a surface with a rough texture such as wood or concrete that can create a significant amount of spatter (Figure 12.2 B). Rough surfaces have protuberances that rupture the surface tension of the blood drop and produce spatter and irregularly shaped parent stains with spiny or serrated edges.

Size, Shape, and Directionality

The geometry of individual bloodstains will generally allow the analyst to determine their direction of flight prior to impacting an object. This is done by examining the edge characteristics of individual stains (Figure 12.3). The narrow end of an elongated bloodstain usually

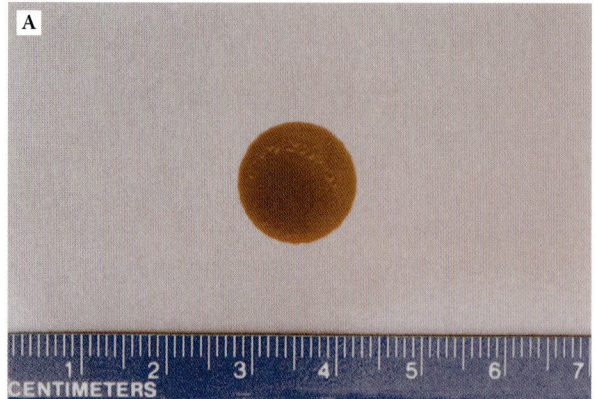

Figure 12.2 Effect of target surface texture on bloodstain characteristics and degree of spatter produced from single drops of blood falling vertically 30 in. onto (A) smooth polished tile and (B) wood paneling.

points in the **direction of travel**. After the **directionality** of several bloodstains has been determined, an area or point of convergence may be established by simply drawing straight lines through the long axes of the bloodstains (Figure 12.4). The point where these lines converge represents the relative location of the blood source in a two-dimensional perspective on the *x*- and *y*-axes. This area of convergence will be an area, not an exact point.

The **area of origin** or the location of the blood source in a three-dimensional perspective can also be determined. By establishing the impact angles of representative bloodstains and projecting their trajectories back to a common axis extended at 90 degrees up from the two-dimensional area of convergence along the z-axis, an approximate location of where the blood source was when it was impacted may be established. Diagrammatic

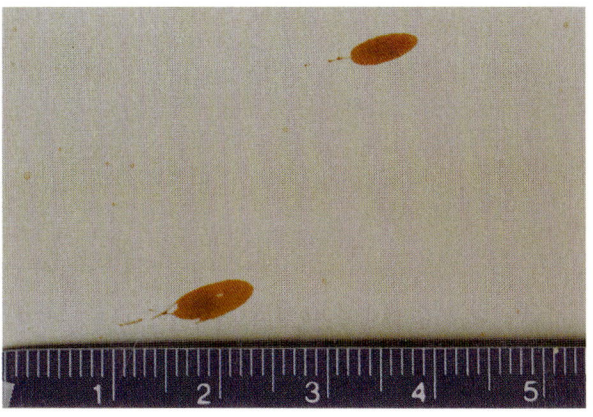

Figure 12.3 The direction of travel of these blood-stains is from right to left and downward, as determined by the characteristics of the leading edge of the stains.

Figure 12.4 Representation of the point or area of convergence of bloodstains on a wall by drawing straight lines through the long axes of the stains.

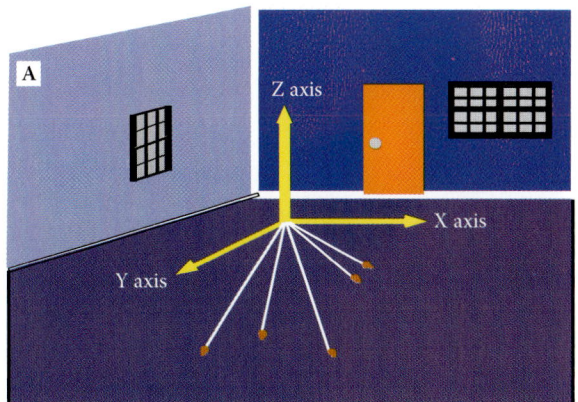

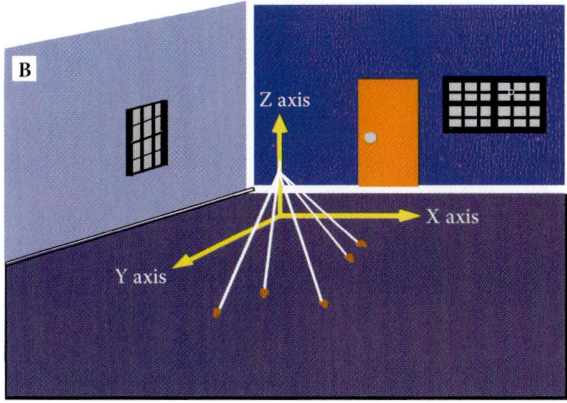

Figure 12.5 Graphical representation of (A) the point or area of convergence with the x- and y-axes and (B) the point or area of origin in space along the z-axis of the bloodstains with the use of the angle of impact of the stains.

representations of convergence and origin utilizing the x-, y-, and z-axes are shown in Figure 12.5A and Figure 12.5B.

If the **angle of impact** is 90 degrees, the resulting bloodstain generally will be circular in shape (Figure 12.6A). Blood drops that strike a target at an angle less than 90 degrees will create elliptical bloodstains (Figure 12.6B). A mathematical relationship exists between the width and length of an elliptical bloodstain that allows for the calculation of the angle of impact for the original spherical drop of blood. This calculation is accomplished by measuring the width and the length of the blood-stain (Figure 12.7). The width measurement is divided by the length measurement to produce a ratio number less than 1. This ratio is

the sine of the impact angle. The impact angle of the bloodstain can now be determined by either referring to the sine function in a trigonometric table or by using a scientific calculator with a sine function. With a calculator, after dividing the width by the length, utilize the function key, arc sin, sin-l, or inverse sin function, and the corresponding angle of impact will be displayed. For a circular blood-stain, the width and length are equal and thus the ratio is 1.0, which corresponds to an impact angle of 90 degrees. For an elliptical bloodstain whose width is one half its length, the width-to-length ratio is 0.5, which corresponds to an impact angle of 30 degrees.

After establishing the angle of impact for each of the bloodstains, the three-dimensional origin of the bloodstain pattern can be determined. One method is to place elastic strings at the base of each bloodstain and project

Figure 12.6 (A) The shape of a bloodstain resulting from a single drop of blood falling 30 in. onto smooth cardboard at 90 degrees. (B) The shape of a bloodstain resulting from a single drop of blood falling 30 in. onto smooth cardboard at 10 degrees.

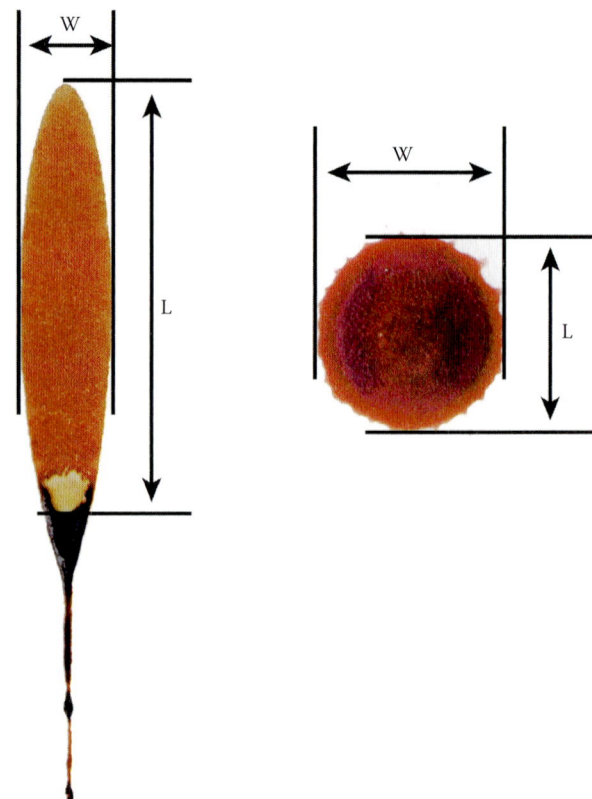

Figure 12.7 Measurement of the width and length of bloodstains.

these strings back to the axis that has been extended 90 degrees up or away from the two-dimensional area of convergence. This is accomplished by placing a protractor on each string and then lifting the string until it corresponds with the previously determined impact angle. The string is then secured to

the axis placed at the two-dimensional area of convergence. This is repeated for each of the selected bloodstains (Figure 12.8A and Figure 12.8B). Remember that this calculated area of origin is always higher than the actual origin of the bloodstains because of the gravitational attraction affecting the spatters while in flight. This gives the analyst the maximum possible height of the blood source. In practical terms, the analyst is attempting to determine whether a victim was standing, lying down, or sitting in a chair when the blood was spattered.

This method for determining the location of a blood source is not always necessary. For instance, if no blood spatter appears on a table top or chair seat, but spatter associated with a gunshot is found on the underside of the table and chair, the obvious conclusion is that the victim was on or near the floor when shot. Common sense and quality observations will often resolve the question of where someone was when injuries were inflicted.

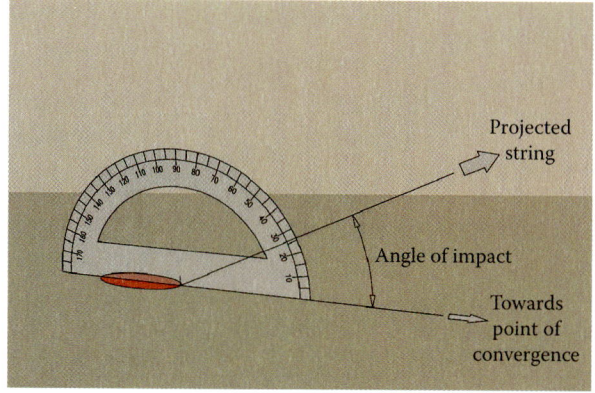

(a)

(b)

Figure 12.8 (A) Method of use of a protractor with the calculated angle of impact to determine the area of origin of a bloodstain. (Courtesy of Alexei Pace, Marsaxlokk, Malta.) (B) Elastic strings placed at the base of selected bloodstains and projected along the z-axis to represent the three-dimensional point or area of origin of bloodstains.

Spattered Blood

Spattered blood is defined as a random distribution of bloodstains that vary in size that may be produced by a variety of mechanisms. The quantity and size of spatters produced by a single mechanism can vary significantly. The amount of available blood and the amount of force applied to the blood affect the size range of spatters. Spatter is created when sufficient force is available to overcome the surface tension of the blood. The amount of force applied to a source of blood and the size of the resulting spatter vary considerably with gunshot,

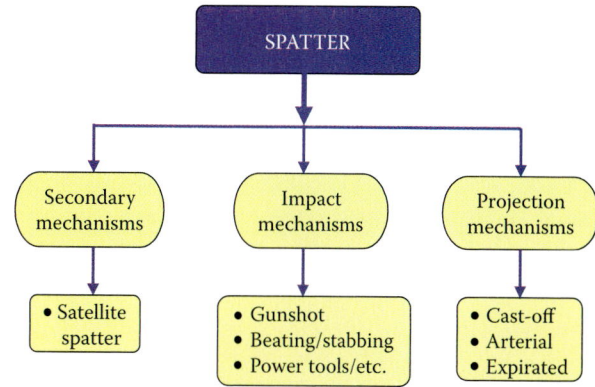

Figure 12.9 Categories of blood spatters based on the mechanisms by which they were created.

beating, and stabbing events. The size range of spatter produced by any one mechanism may also vary considerably. Frequently, a single mechanism will create spatters whose size will fit all the categories as outlined in Figure 12.9. Upon examination, the analyst must identify a pattern as a spatter before attempting to ascertain the specific mechanism that created it. A single small stain does not constitute a spatter pattern. Determining the mechanism that created a spatter pattern normally requires more information than merely a look at the pattern. Therefore, it is advisable to refer to bloodstain patterns as simply "spatter" until all available information has been reviewed. The identification and interpretation of spattered blood are significant for the following reasons:

- Spattered blood may allow the determination of an area or location of the origin of the blood source.
- If found on a suspect's clothing, spattered blood may place that person at the scene of a violent altercation.
- Spattered blood may allow the determination of the mechanism by which the pattern was created.

After identifying small bloodstains as a spatter pattern and gathering pertinent scene, medical, and case-related facts, the analyst may then be able to establish the specific mechanisms by which the pattern was

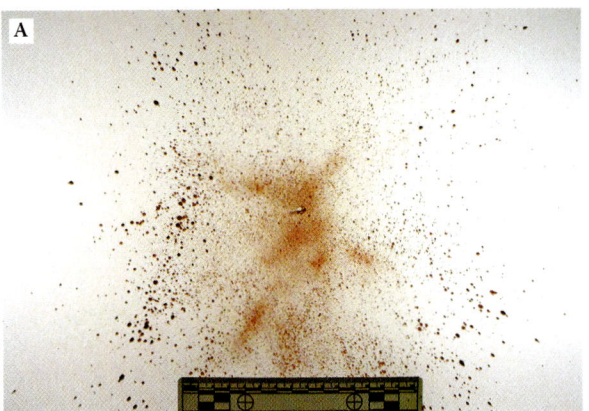

Figure 12.10 (A) Forward spatter associated with gunshot. Note the mistlike dispersion of the minute stains sometimes referred to as the "rouging effect." This misting is not often seen in casework. (B) Back spatter associated with gunshot. It is generally seen in less quantity than forward spatter.

created. The size, quantity, and distribution of these spatters vary depending upon:

- The quantity of blood subjected to impact
- The force of the impact
- The texture of the surface impacted by the blood

In a laboratory environment, the amount of force applied to a blood source and the quantity of blood impacted are easily controlled. However, in actual casework, these factors are not known. In the category of spatter created by an impact mechanism, all three of the mechanisms may produce the similar size ranges of spatter, depending on these factors.

Impact Spatter Associated with Gunshot

Impact spatter associated with gunshot may produce minute spatters of blood less than 0.1 mm in diameter that are often referred to by analysts as mist-like dispersions. This mist-like spatter is not frequently seen, but when observed, it is indicative of gunshot. This **misting** effect is not observed in spatter patterns associated with beatings, stabbings, or the production of **satellite spatter** created by blood dripping into blood. Impact spatters associated with gunshot often exhibit a wide size range from less than 0.1 mm up to several millimeters or more. The size range is dependent on the quantity of available blood,

the caliber of the weapon, the location and number of shots, and impeding factors, such as hair, clothing, etc.

Impact spatter of this type is most commonly associated with gunshot, but may also be produced in cases involving explosions, power tool and high-speed machinery injuries, and occasionally high-speed automobile collisions (Figure 12.10). In gunshot cases, two sources can account for impact spatter. When associated with an entrance wound, it is referred to as **back spatter** or **blowback**. This spatter may be found on the weapon and the shooter, especially on the hand and arm areas. Conversely, when the impact spatter is associated with an exit wound, it is referred to as *forward spatter* (Figure 12.10A, Figure 12.10B, and Figure 12.11). Generally,

Figure 12.11 Horizontal perspective of back spatter and forward spatter produced by gunshot mechanism. The arrow indicates the direction of the projectile. Note the larger quantity and distribution of forward spatter.

the mechanisms that create impact blood spatter will create a variety of sizes of bloodstains that would fit into the categories of impact spatter associated with beating and stabbing events, as well as spatter produced by expiration of blood.

Impact Spatter Associated with Beating and Stabbing

Impact spatter associated with beating and stabbing events generally exhibits a size range from 1 to 3 mm in diameter. The spatters may be smaller or larger than this general range, depending on the force of the impact and the quantity of available exposed blood (Figure 12.12A and Figure 12.12B). Exposed blood is necessary before spatter can occur. The blood itself does not have to receive the impact. The victim can be bloody and receive an impact on another site on the body. Spatter will be produced whether the blood itself was

hit or not The weapon used in the assault, whether a sharp object (knife, glass, etc.) or a blunt object (fist, bat, concrete block, etc.) and the number of blows inflicted have effects on the resulting pattern. Mechanisms other than a beating or a stabbing, such as gunshots, expiration of blood, and satellite spattering, may also produce spatter in the size range of less than 1 mm to 3 mm.

In certain situations, multiple mechanisms may exist, such as any combination of gunshot, beating, stabbing, expired blood, or secondary or satellite spatter resulting from blood dripping onto surfaces. The size ranges of spatters produced by these mechanisms can be similar, so it may not be possible to determine which mechanism produced a specific pattern. Figure 12.13A through Figure 12.13C show how the sizes of these spatters overlap with respect to gunshot, expired, and beating mechanisms. For this reason, the analyst should consider all possible mechanisms for the production of spatter and utilize all available information. Sometimes, two or three plausible explanations can apply to a spatter pattern, and that must be acknowledged by the analyst. Spatter created by a projection mechanism is produced by the disruption of the surface tension of the blood without an impact (e.g., castoff, arterial, and expired), as shown in the diagram in Figure 12.9. These patterns will be discussed in greater detail later in the chapter.

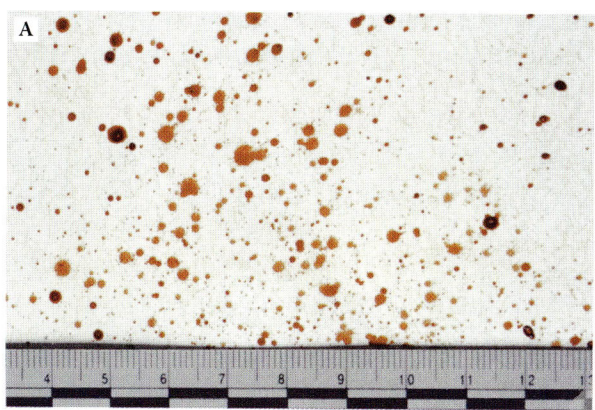

Figure12.12 Impact spatter produced by beating mechanism (A) on smooth cardboard vertical surface and (B) on smooth cardboard horizontal surface.

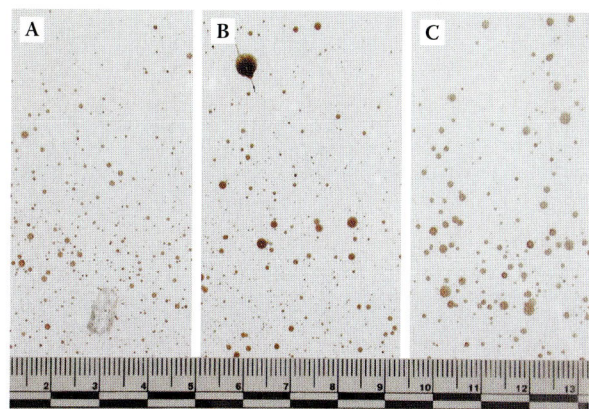

Figure 12.13 Comparison of size ranges of spatters produced by (A) gunshot, (B) expired blood, and (C) beating.

Significance of Secondary or Satellite Spatters Resulting from Dripped Blood

Single drops of blood will produce small spatters around the parent stain as a result of striking a rough target surface. Spatter produced in this manner is referred to as secondary or satellite spatter. When multiple free-falling drops of blood are produced from a stationary source onto a horizontal surface, **drip patterns** will result from blood drops falling into previously deposited wet bloodstains or small pools of blood. These drip patterns will be large and irregular in shape, with small satellite spatters around the periphery of the central stain on the horizontal and nearby vertical surfaces (Figure 12.14). Satellite spatters are the results of smaller droplets of blood that have detached from the main blood volume at the moment of impact. These satellite spatters of blood, circular to oval in shape, usually have diameters ranging from 0.1 to 1.0 mm in size or slightly larger.

Paul Kish conducted research on the topic of satellite spatter resulting from single drops of blood and the factors affecting their interpretation. He presented his preliminary findings at the annual meeting of the American Academy of Forensic Sciences in February 1996. He demonstrated the maximum height to which satellite spatters of human blood would impact a vertical surface. Factors affecting this are blood drop volume, freshness of blood, surface texture, and the distance of the vertical target from the **impact site**. He concluded that rough surfaces, such as concrete, produce substantial satellite blood spatter from a single drop impact as well as from blood dripping into blood.

The vertical height achievable by the satellite spatter created with a single drop impacting concrete can be as high as 12 in. He observed a greater concentration of spatter on the horizontal and vertical surfaces with blood dripping into blood on concrete. Investigators often interpret small spatters of blood on suspects' trouser legs, socks, and shoes as impact blood spatter associated with a beating or shooting due to their small diameters (Figure 12.15). The mechanism of satellite blood spatter causing these stains should be thoroughly explored before reaching a final interpretation and conclusion.

From a practical view, it is important that the investigator be able to recognize the types of bloodstains and patterns resulting from free-falling drops based on their size, shape, and distribution and then document their locations. The bloodstains should be categorized relative to the events that produced them, and they should be related to the possible sources and movement of these sources through the recognition of trails and drip patterns.

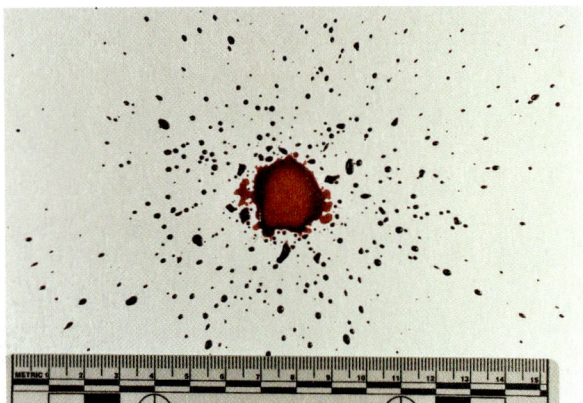

Figure 12.14 Drip pattern (blood dripping into blood) produced by single drops of blood that fell 36 in. onto smooth cardboard. Note the extent of the satellite spatter around the parent or central stain.

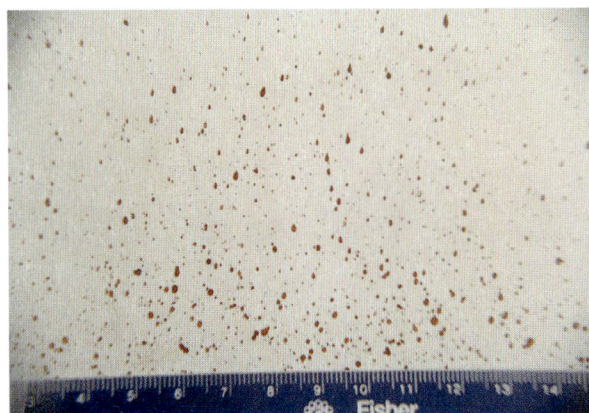

Figure 12.15 Satellite blood spatters produced on a vertical cardboard surface within 3 in. of blood dripping into blood on a concrete surface.

Castoff Bloodstain Patterns

During a beating with a blunt object, blood does not immediately accumulate at the impact site with the first blow. As a result, no blood is available to be spattered or cast from the first blow. Spatter and **castoff patterns** are created with subsequent blows to the same general area where a wound has occurred and blood has accumulated. Blood will adhere in varying quantities to the object that produces the injuries. A centrifugal force is generated as an assailant swings the bloodied object. If the centrifugal force generated by swinging the weapon is great enough to overcome the adhesive force that holds the blood to the object, blood will be flung from the object and form a castoff bloodstain pattern.

The blood that is flung (castoff) will strike objects and surfaces, such as adjacent walls and ceilings in the vicinity, at the same angle from which it is flung or cast. The size, distribution, and quantity of these castoff bloodstains vary. Castoff bloodstain patterns may appear linear in distribution, and the individual stains are frequently larger in size than impact blood spatters (Figure 12.16). Castoff patterns are often seen in conjunction with impact spatters, and a study of each may help determine the relative position of the victim and the assailant at the time the injuries were inflicted. Castoff bloodstains are not always present at scenes where blunt or sharp force injuries have occurred. The arc of the back or side swing may be minimal, especially in

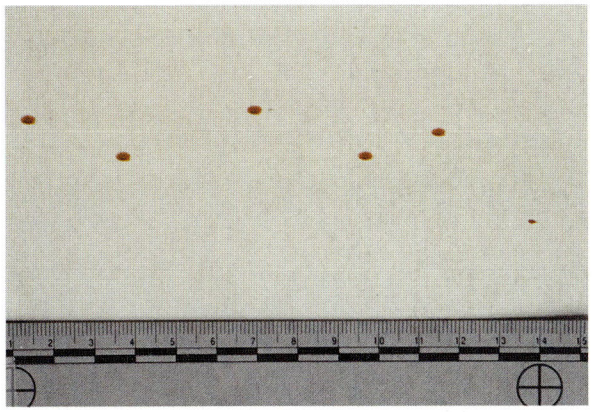

Figure 12.16 Linear distribution of castoff bloodstains on a vertical surface.

the case of a heavy blunt object. Occasionally, analysts will attempt to determine whether the person swinging the object was right- or left-handed. This is dangerous, since many individuals may swing objects effectively with either hand. The analyst must also consider the possibility of back-handed swings that may appear similar.

Bloodstain Patterns Resulting from Large Volumes: Splashed and Projected Blood

When a quantity of blood in excess of 1.0 ml is subjected to minor or low-velocity forces or is allowed to freely fall to a surface, a **splashed** bloodstain pattern will be produced. Splashed bloodstain patterns usually have large central areas with peripheral spatters appearing as elongated bloodstains. Secondary blood splashing or **ricochet** may occur as a result of the deflection from one surface to another of large volumes of blood after impact. When sufficient bleeding has occurred, splash patterns may be produced by the movement of the victim or assailant. These patterns are often created by large volumes of blood falling from a source such as a wound. Larger quantities of splashed blood will create more spatters.

A projected bloodstain pattern is produced when blood is projected or released as the result of force exceeding that of gravity. When blood of sufficient volume is projected horizontally or downward with force exceeding the force of gravity, the resultant bloodstains exhibit numerous spinelike projections with narrow streaking of the secondary spatters compared with splashed bloodstains (Figure 12.17A and Figure 12.17B). Vomiting blood is an example of projected blood in a large volume. Blood may also be projected from a source or pool by rapid movement or by running through the pool.

Expirated Bloodstain Patterns

As a result of trauma, blood will often accumulate in the lungs, sinuses, and airway passages of the victim. In a living victim,

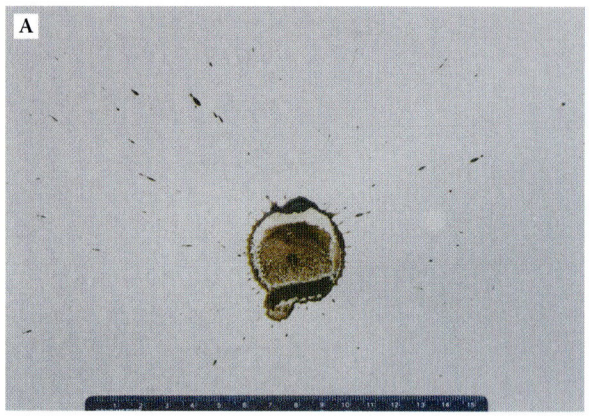

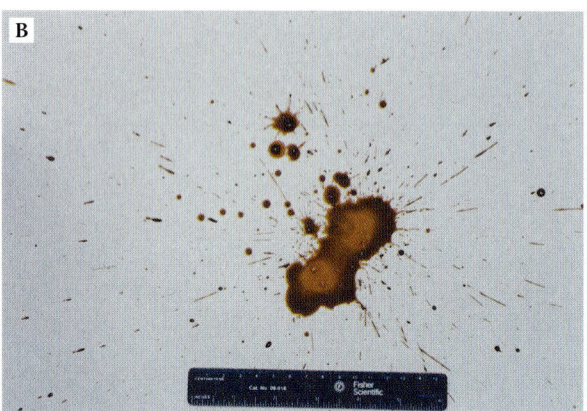

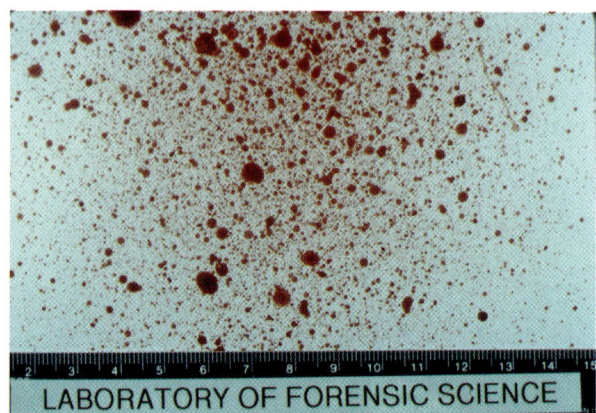

Figure 12.17 (A) Bloodstain pattern produced by 1 ml of blood falling downward 36 in. onto smooth cardboard. (B) Bloodstain pattern produced by 1 ml of blood projected downward 36 in. onto smooth cardboard.

Figure 12.18 (A) Bloodstain pattern produced by exhalation of blood from the mouth onto cotton cloth with air bubble in stain pattern. (B) Pattern produced in a similar manner with no evidence of air bubbles or vacuoles. In their absence, the pattern is similar to those produced by beating or shooting mechanisms.

this accumulation of blood will be forcefully expelled from the nose or mouth in order to free the airways. This type of bloodstain is referred to as an expired bloodstain pattern. The size, shape, and distribution of an expired bloodstain pattern are often similar to the patterns that are observed with impact spatter associated with beatings and gunshots (Figure 12.18A and Figure 12.18B). Since impact spatter due to a gunshot or beating mechanism can closely resemble expired bloodstain patterns, the deciding factor may be the case history. An expiratory bloodstain pattern cannot possibly be produced unless the victim has blood on their face, in their mouth or nose, or some type of injury to their chest or neck that involves the airways.

Expired bloodstains may appear diluted if mixed with sufficient saliva or nasal secretions. If the blood has been recently expelled there may be visible air bubbles within the

stains due to the blood being mixed with air from the airway passages or lungs. When the bubbles rupture and the bloodstains dry, the areas of previous air bubbles will appear as *vacuoles*. In the absence of these vacuoles within the stains, this type of bloodstain pattern may be misinterpreted. Air bubbles or vacuoles and dilution are not always present in an expired bloodstain pattern.

Arterial Bloodstain Patterns

When an artery is breached, blood is projected from it in varying amounts. The size of arterial bloodstains varies from very large gushing or spurting patterns to very small spray types of patterns (Figure 12.19). The type of arterial pattern observed is a function of the severity of the injury to the artery, the size and location of the artery, whether the

Figure 12.19 Arterial spurt pattern produced by victim who sustained a severed right carotid artery.

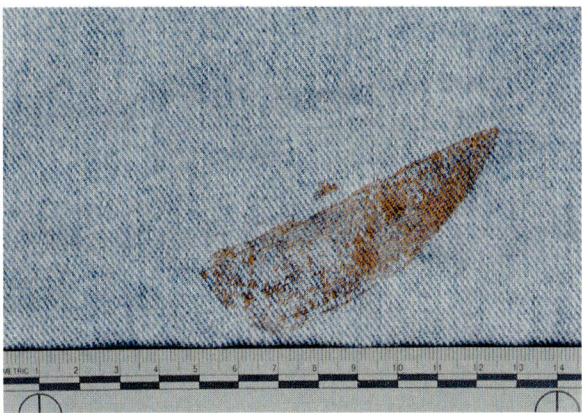

Figure 12.20 Transfer pattern on denim produced by contact with a knife blade containing wet blood.

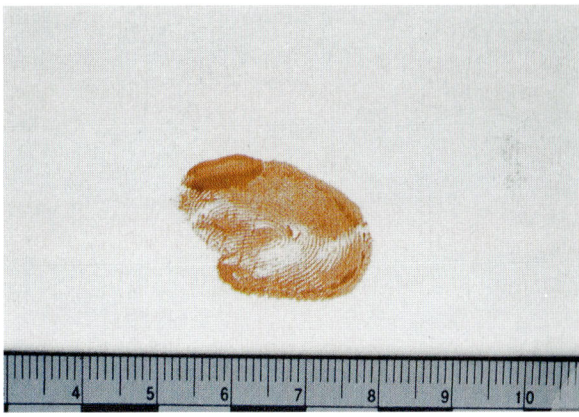

Figure 12.21 Partial fingerprint produced by contact of a finger containing wet blood with smooth cardboard.

injury is covered by clothing, and the position of the victim when the injury was inflicted. Obviously, arterial bloodstaining is accompanied by demonstrable arterial damage. The bloodstain pattern analyst should verify his or her hypothesis about an arterial bloodstain pattern by reviewing the autopsy report or speaking directly with the forensic pathologist who conducted the autopsy. These patterns are usually very distinctive due to the overall quantity of bloodstains observed.

Other Bloodstain Patterns

Transfer Bloodstain Patterns

When an object wet with blood comes into contact or wiping with an unstained object or secondary surface, a blood **transfer pattern** occurs. These patterns may assist an examiner in determining the object that made the pattern, i.e., hair, knife, shoe, etc., since a recognizable mirror image of the original surface or a portion of that surface may be produced (Figure 12.20). When attempting to determine whether an object could have produced a particular transfer pattern, it is usually necessary to conduct a series of experiments using items similar to those in question, since it is never good practice to add blood to an evidentiary object. Class or individual characteristics may be determined from distinct blood transfer patterns, such as finger and palm prints or foot and footwear impressions (Figure 12.21). Partial bloody impressions are often chemically enhanced to resolve additional detail.

The differentiation between a minute transfer pattern and an impact spatter pattern on a suspect's clothing may determine whether he or she could have been the perpetrator or merely someone who came into contact with the blood source. If spatter is identified on garments, that generally means that the wearer of the garment was in the immediate vicinity of the bloodshed event, i.e., beating, shooting, etc. The determination of whether the bloodstains on garments are the results of spatter or transfer is not always easy and often requires experimentation and microscopic examination of the garments.

Altered Bloodstains

Bloodstains deposited on surfaces at a scene are subject to various forms of change from their original appearance at the time the bloodshed occurred. Recognition of these alterations and an understanding of their significance are important for the reconstruction of the event. When blood exits the body, the processes of drying and clotting are initiated. The drying time of blood is a function of its volume, the nature of the target surface texture, and the environmental conditions. Small spatters and light transfers of blood will dry within a few minutes under normal conditions of temperature, humidity, and air currents. Larger volumes of blood may take considerable time to completely dry. Drying is accelerated by increased temperature, low humidity, and increased airflow. Initially, the outer rim or perimeter of the bloodstain will show evidence of drying, which then proceeds toward the central portion of the stained area.

When the center of a dried bloodstain flakes away and leaves a visible outer rim, the result is referred to as a skeletonized stain (Figure 12.22A). Another type of **skeletonized bloodstain** occurs when the central area of a partially dried bloodstain is altered by contact or a wiping motion that leaves the periphery intact (Figure 12.22B). This can be interpreted as movement or activity by the victim or assailant when or after injuries were inflicted.

As dried bloodstains age, they tend to progress through a series of color changes

Figure 12.22 (A) Skeletonized bloodstain created by drying and flaking away of the central area of the stain. (B) Skeletonized bloodstain produced by a wiping alteration of a partially dried bloodstain, indicating activity shortly after the blood was deposited. Note the remaining peripheral ring of the original bloodstain caused by drying around the edges.

from red to reddish brown and eventually to black. The estimation of the age of bloodstains based on color is difficult because environmental conditions and the presence of bacteria and other microorganisms affect the sequence and duration of color changes. Any estimate of time lapse should involve experiments utilizing freshly drawn human blood of similar volume placed onto a similar surface with environmental conditions duplicated as closely as possible.

The clotting process is also initiated when blood exits the body and is exposed to a foreign surface. The appearance and extent of clotted blood at a scene may provide an indication of the amount of time elapsed since the injury occurred. Normal clotting time of blood that has exited the body ranges from

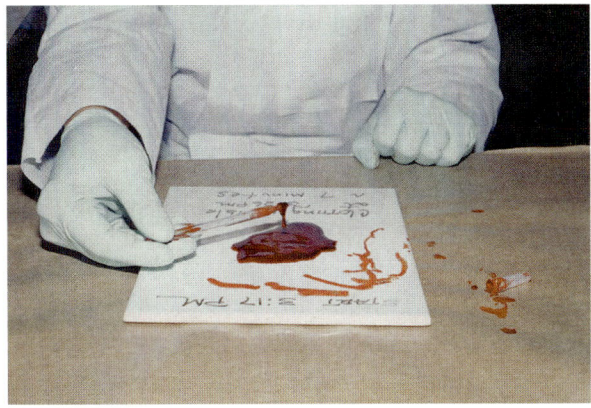

Figure 12.23 Demonstration of clot formation in freshly drawn human blood after 7 min.

Figure 12.24 An example of clotted spatter on fabric exhibiting dark central area with lighter peripheral area of the stain.

3 to 15 minutes in healthy individuals (Figure 12.23). As a clot progressively forms a jellylike mass, it retracts and forces the serum out of and away from the progressively stabilizing clot. Occasionally a bloodstain analyst will observe clotted impact spatters on clothing or other surfaces (Figure 12.24). Clots of blood may show drag patterns that indicate that additional activity, such as movement or further injury, occurred after a significant interval had elapsed from the initial bloodshed. Evidence of coughing or exhalation of clotted blood by a victim may be associated with post-injury survival time.

Existing wet bloodstains at a scene are also subject to alteration in appearance due to smudging, smearing, and wiping activities of the victim or assailant. Changes in the appearance of bloodstains and patterns and additional bloodstains may also be created by paramedical treatment of the victim or removal of the victim from the scene.

Another source of bloodstain alteration is the effect of moisture, such as rain or snow at a scene exposed to the outside environment that will dilute existing bloodstains. Investigators may also encounter indoor scenes and vehicles that have been cleaned with water and detergents or that have been painted after a bloodshed event. Diluted bloodstains may be difficult or impossible to evaluate without the use of a chemical enhancement process such as luminol treatment. The alteration of bloodstains by heat, fire, or smoke may also cause problems of interpretation. Bloodstains covered with soot may be entirely missed at the scene of a homicide that preceded a fire. Heat and fire may also cause existing bloodstains to fade, darken, or be completely destroyed.

Void Areas or Patterns

Void areas or patterns are absences of bloodstains in otherwise continuous patterns of staining. These patterns are commonly seen where items have been removed from an area previously spattered with blood. This permits the analyst to establish sequencing and identify alterations within a crime scene. At a scene containing a considerable amount of spattered blood, the void areas may be utilized to recognize the general location where the spatter event occurred.

Interpretation of Bloodstains on Clothing and Footwear

The clothing of a suspect is often a critical piece of evidence that can help link the suspect to the incident through the bloodstain patterns present on his or her garments. Generally, two questions arise with bloodstained garments: (1) Whose blood is on the garment? (2) How was the blood deposited onto the garment?

With the advances in DNA technology, the determination of whose blood is on clothing is normally not a problem. The bloodstain pattern analyst determines how the blood was deposited onto the garments. Generally, the deposition of blood onto garments falls into one or both of the following categories:

1. Passive bloodstaining, including transfer, **flow patterns**, saturation stains, and stains resulting from dripping blood
2. Active bloodstaining, including impact spatter, arterial spurts, expired bloodstains, castoff, etc.

It is necessary to identify and document the specific patterns prior to attaching any case-related significance to the patterns. Before drawing conclusions from bloodstain patterns on clothing or any other medium, an analyst should request and review the serological findings.

The interpretation of bloodstain patterns on clothing often centers on substantiating or refuting the suspect's version of how his or her clothing became stained with blood. A common example is where a suspect claims to have come into contact with the victim only after the injuries had been inflicted, i.e., the suspect was not present when the assault occurred. In such instances, the determination of the mechanism of stain creation may have more evidentiary value than identifying the source of the blood. If the bloodstains have transfer patterns in which the blood is deposited on top of the weave of the fabric, then the analyst may substantiate the suspect's claims. If dozens of small spatters of blood are embedded within the fibers of the garment, then the bloodstain evidence refutes the suspect's version of events.

The interpretation of bloodstains on clothing can be difficult and often requires experimentation and extensive experience. It has been the authors' experience that bloodstain patterns on textiles are unpredictable due to their varying compositions and textures. For this reason, one should be cautious when interpreting bloodstain patterns on garments. To facilitate the examination of clothing by a bloodstain analyst, the following steps should be taken:

1. Establish the manner in which the garments were collected, documented, and preserved prior to their examination.
2. Document the garments while the victim or suspect is still wearing them, when possible.
3. Allow the bloodstain analyst an opportunity to examine the stains before their removal for DNA analysis. The amount of bloodstaining is usually limited, and the geometry of the stains should be examined before they are consumed in serological analysis. The geometric interpretation of bloodstain patterns is a nondestructive examination.
4. Take photomacrographs and, if needed, photomicrographs before sample cuttings of stains are removed.
5. Obtain a history of where the garment has been and how it has been handled. An example would be a shirt collected from the emergency room floor after a suspect's injuries were treated. The significance of the bloodstain patterns on this shirt could have been compromised, since additional bloodstains may have been deposited on the shirt or existing bloodstains may have been altered.

Documentation of Bloodstain Evidence

On many occasions, the degree of significance that may be attached to a given bloodstain pattern is compromised due to insufficient documentation. When documenting bloodstain patterns, attention should be given to the following points:

1. Accurately document the size, shape, and distribution of the individual stains and the overall patterns.
2. Include measuring devices within the photographs.

3. Use more than one mechanism for documentation, i.e., photographs, video, diagrams, and notes. This overlap should prevent anything of significance from being overlooked.
4. If possible, collect articles of evidence that may contain significant or questionable bloodstain patterns.
5. Utilize overall, midrange, and close-up macrophotography when documenting bloodstain patterns. Photographs should overlap so that close-up photographs can be associated with their location within the pattern. Microphotography is also a useful technique to study small spatters. Bloodstain pattern interpretation is very visual, and high-quality photographs make it easy to illustrate the significance of bloodstain patterns to a jury.
6. Complete the documentation in such a manner as to allow a third party to utilize the photographs, notes, diagrams, and video to place the bloodstain patterns and articles of evidence back in their original locations.

A basic premise of crime scene and evidence documentation is that one can never take enough photographs or make enough drawings, videos, and notes. More is always better.

The Use of Luminol Photography for Bloodstain Pattern Analysis

Luminol is a chemiluminescent reagent that can be utilized both as a presumptive test for blood as well as a method of chemical enhancement of impressions in blood on various surfaces. It is an excellent search technique for latent bloodstains at crime scenes or those scenes where it is suspected that attempts were made to clean bloodstains from an area. Since luminol is applied by a spraying technique, wide areas can be efficiently searched

for blood. The spraying device should be capable of producing a fine mist. Overspraying with luminol should be avoided. The use of luminol as a means of searching for spatters of blood on clothing is not recommended. It must be remembered that luminol is not specific for blood, and further confirmation of blood species testing and ultimately DNA testing is essential. The serology department of the Forensic Science Laboratory of the Netherlands has developed and tested a luminol solution that has a clear and bright chemiluminescent reaction with minimum deterioration of the DNA with a detection limit of 1:50,000, which is adequately sensitive. Luminol is best used in a darkened environment and requires special photographic techniques.

Bluestar™ is a luminol preparation developed by Professor Loic Blum in France that is extremely sensitive and stable and produces a very bright, long-lasting chemiluminescence. It is reported to be DNA-typing compatible. It has been used successfully by Philippe Esperanca, MS of the French Gendarmerie Forensic Laboratory in France, as shown in the images in Figure 12.25 and Figure 12.26, which were taken with a Fugifilm Finepix S1® digital camera.

Photography of the luminol reaction with a 35-mm camera requires a wide open aperture (f-2.8 or f-3.5) at "B" (bulb) setting with an exposure time of at least 40 to 80 sec. Color film is preferred over black and white with ASA of at

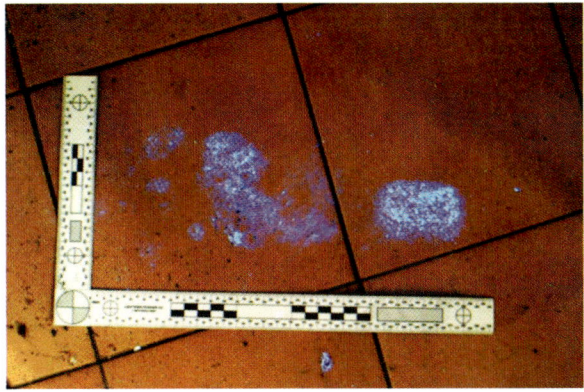

Figure 12.25 Enhancement of latent bloody footprint with Bluestar® luminol reagent and digital camera (30-sec exposure at f-2.8). (Courtesy of Philippe Esperanca, French Gendarmerie Forensic Laboratory.)

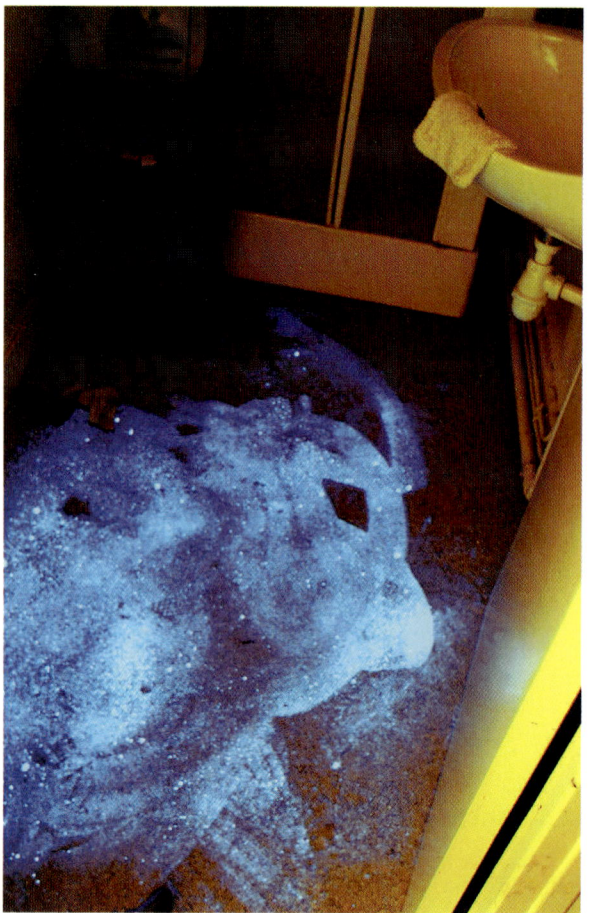

Figure 12.26 Enhancement of area of room subjected to prior cleanup of blood with Bluestar® luminol reagent and digital camera (30-sec exposure at *f*-2.8). (Courtesy of Philippe Esperanca, French Gendarmerie Forensic Laboratory.)

Figure 12.27 (A) Area of room prior to the application of luminol. (B) Luminol reaction of blood photographed with digital camera (28-sec exposure at *f*-2.8 and camera sensitivity at ISO 1600).

least 200 to 400. The camera must be stationary and fixed to a tripod with a cable release to avoid motion of the camera during filming. It should be set up perpendicular to the surface to be photographed if possible in order to minimize depth of field considerations and optical distortion of the image. Many examiners will photograph the area to be sprayed prior to the application of luminol. Measuring devices that glow in the dark are available for use as a scale in the photograph. Successful results have been demonstrated with the use of a flash during the process. Luminol photography is a two- to three-person procedure. One person sprays the luminol, and the second operates the camera. The third person is available to take notes. Dark clothing and a dark spraying apparatus are advisable to avoid reflection.

Martin Eversdijk of the Amsterdam-Anstelland Police Region, the Netherlands Institute for Criminal Investigation and Crime Science in the Netherlands and other forensic examiners share the opinion that the use of a 35-mm camera with color film is not suitable for luminol photography. A disadvantage of this type of camera is that it is impossible to check the quality of the photograph at the crime scene. A digital camera is a good option. The quality of the photograph can be checked immediately with or without a laptop computer, and new images can easily be taken immediately during the process. Examples of Martin Eversdijk's luminol photography taken with a Nikon COOLPIX® digital camera are shown in Figure 12.27A through Figure 12.27C.

Figure 12.27 (C) Exposure of the luminol reaction described in Figure 12.27B with flash after 27 sec. (Courtesy of Martin Eversdijk in the Netherlands.)

Absence of Evidence Is Not Evidence of Absence

In many cases, the presence of bloodstains originating from the victim and found on the clothing or person of a suspect is powerful evidence to link the suspect to the violent act. It must be pointed out that the absence of blood spatter on a suspect or his clothing does not preclude his or her active participation in a bloodshed event. It is possible to beat, stab, or shoot someone without being spattered with blood, and exceptions to this rule are few. Unfortunately, many defense attorneys attempt to offer the absence of blood spatter on their clients as proof of lack of participation. From a review of the scientific literature and from practical experience, it is not uncommon for an assailant to have little if any blood on his or her person after committing a violent crime. The absence of bloodstaining on an active participant in a bloodshed event has several explanations:

1. The directionality of the blows with a blunt object or thrusts with a knife may direct spatters of blood away from the assailant.

2. If the site of the injury is covered with clothing or other material during the assault, the amount of spatter may be greatly reduced or absent.
3. The assailant may have cleaned up or changed clothing prior to being apprehended.
4. The assailant may have worn protective outerwear.
5. The assailant may have removed his clothing prior to committing the assault.
6. The amount of blood present at a scene described as "covered in blood" or a "bloodbath" may be primarily due to active bleeding from a victim who is still alive or from the draining of blood from wounds of a deceased individual that occurred after the assailant left the scene.
7. Individuals have been known to confess to crimes that they, in fact, did not commit.

It is important to recognize that conclusions in bloodstain pattern analysis should not be based on bloodstains or spatter the analyst would expect to be present, but rather on bloodstains or spatter that are physically present. In most cases, the absence of bloodstains on the clothing of a suspect should neither exonerate nor implicate his or her involvement in a violent act.

Report Writing

The purpose of a bloodstain pattern report is to convey the findings of the analyst to the attorneys, the court, and ultimately the jury. In addition to being completely objective and unbiased, the report should be concise and worded in a manner that will allow it to be easily understood. A bloodstain pattern report is a systematic description of the involvement of the analyst and an explanation of the information and evidence examined in the case. It should include overall observations and experiments from which the final conclusions are rendered. A bloodstain pattern report should be divided into sections in a logical sequence.

Introduction

The report should include pertinent information about the case, including the name of the case (i.e., *State v. John Doe*), with case number, name of the agency or attorney requesting the examination, location of the crime scene, and a list of materials and evidence examined. The dates of receipt of documents and evidence examination should be included. Some bloodstain analysts include definitions of key bloodstain terms to be utilized in the report in this introductory section.

Body of Report

The body of a bloodstain pattern report should include detailed descriptions of items that were examined and experiments conducted:

- Crime scene
- Physical evidence collected at the crime scene
- Victim's body and autopsy information
- Victim's clothing and footwear
- Suspect's body
- Suspect's clothing and footwear
- Reconstruction with models
- Chemical enhancement techniques
- Experiments

Conclusions

The conclusions or findings should be listed in sequential order and substantiated by the information included in the body of the report. They should be limited to those that have scientific foundations and fall within the discipline of bloodstain pattern interpretation and the degree of the expertise of the analyst. Photographs and diagrams may be included with the report to assist with the understanding of the basis for the conclusions.

Case Report

Summary of the Case

Police responded to a single family residence. Upon arrival they were met at the front door by the victim's husband who indicated his wife had shot herself. Blood was observed on the husband's clothing. The victim was located lying on the bed in the master bedroom. She was clad in pajamas and was partially covered with blankets. The victim had sustained a single gunshot wound to the right side of her head. A Sig Sauer, P226, 9mm pistol was located on the nightstand adjacent to the bed, the clip had been removed and the slide was locked open. The husband acknowledged removing the clip and clearing the pistol and placing it on the night stand; he also indicated he washed blood off his hands and face in the bathroom, although he denied any involvement in the death of his wife.

Materials Evaluated in this Analysis

Scene examination on day of incident
Examination of bed linens at scene
Husband's clothing
Scene photographs
Autopsy photographs
Autopsy report
Crime scene reports
Scene diagrams
Laboratory reports

Injury Sustained by the Victim

The victim died as a result of a single loose contact gunshot wound to the right side of her face, just anterior to the right ear. The projectile did not exit her body.

Description of the Scene

Master Bedroom

Figure 12.28 depicts an overall location of the body of the victim in the master bedroom bed. A light blue blanket and a white quilted comforter are noted on the lower half of the bed.

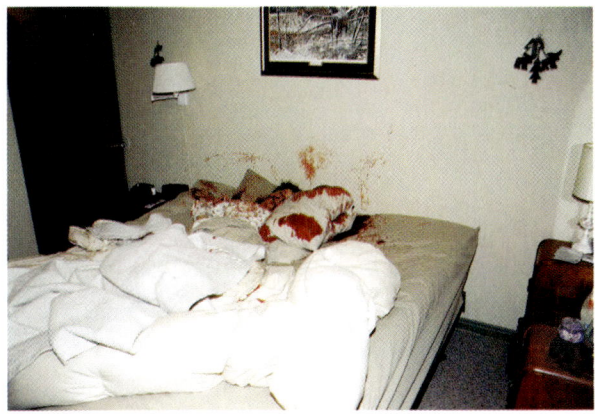

Figure 12.28 Overall location of the victim in the master bedroom.

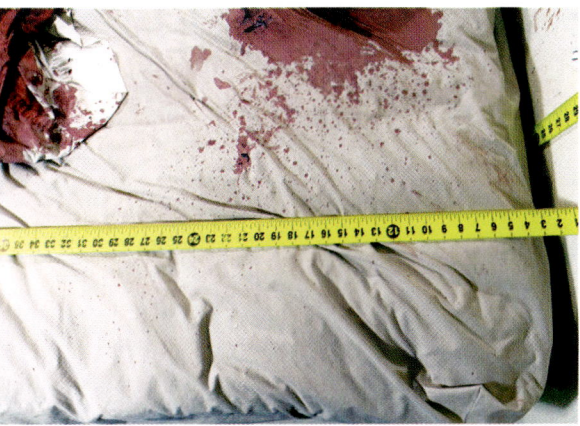

Figure 12.30 Closer view of the spatter distribution on the bed observed in Stain Area D of Figure 12.29.

Several pillows are located around the victim's head and torso. The east wall of the master bedroom and the entrance to the bedroom is located in the background. A nightstand is located along the east wall between the bed and doorway. A Sig Sauer, P 226, 9 mm pistol was located on top of this nightstand. The slide of the pistol was locked open and the clip had been removed from the pistol. The pistol, the clip, and a live round of ammunition were all on the nightstand. Transfer type bloodstains were located on the majority of the pistol. Bloodstains were present on the exterior of the clip. A dresser with a mirrored top is located on the south wall located to the right of the bed.

Figure 12.29 is a mid-range photograph of the victim's final position after the pillows and blankets were removed. The left arm of the victim was noted to be folded back beneath her torso. A large saturation pattern was located

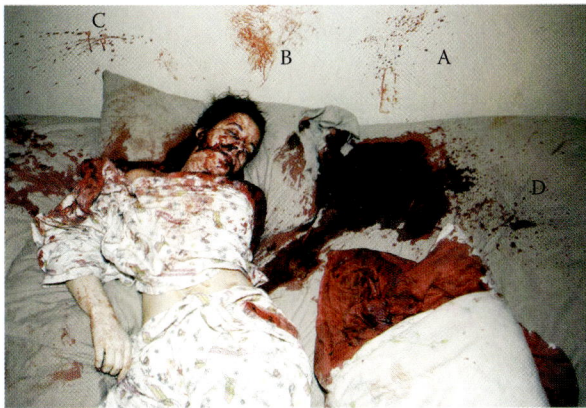

Figure 12.29 View of the victim's final position after the pillows and blankets were removed.

on the fitted sheet adjacent to the victim's left side. This saturation pattern along with the positioning of her right arm is consistent with the victim being on/over this area while bleeding heavily prior to her final position.

Three (3) distinct bloodstain patterns are visible on the east wall as shown in Figure 12.29. Reportedly, the DNA profile of the blood samples taken from these three areas matched the DNA profile of the victim. A large saturation pattern is located on the fitted sheet. A number of spatters are present on the fitted sheet to the right of this saturation pattern. Figure 12.30 depicts a closer view of the spatter distribution on the bed observed in Figure 12.29. Reportedly, human blood with the same PGM genetic marker subtype as the victim was identified in stains less than 1 mm in diameter from this area of the fitted bed sheet. The physical appearance, location, and distribution of these spatters are consistent with there being back spatter which emanated from the victim's entrance gunshot wound. Figure 12.31 depicts the overall appearance of stain patterns on the east bedroom wall.

East Wall of Master Bedroom

The stain pattern on the left in Figure 12.31 consists of a linear distribution of larger spatters, whose direction of travel is right to left and downward. The physical appearance and distribution of these spatters is consistent with their being produced by a projection mechanism secondary to the shooting.

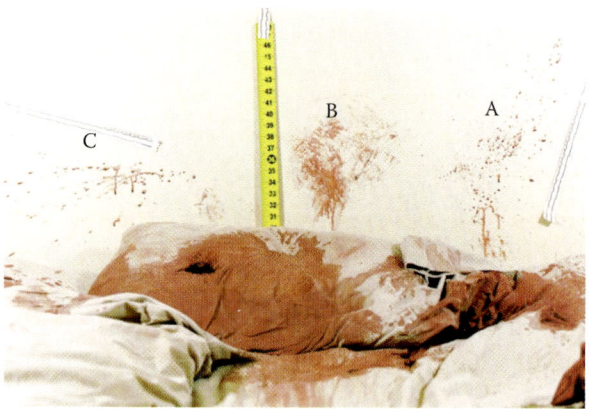

Figure 12.31 View of the overall appearance of stain patterns A–C on the east bedroom wall.

The physical appearance of the bloodstain pattern depicted in the center in Figure 12.31 is consistent with its being a transfer bloodstain pattern. This transfer pattern was created, secondary to the shooting, when an object wet with blood came into direct contact with this portion of the wall.

The physical appearance and distribution of the individual stains, which compose the spatter pattern on the right in Figure 12.31 are consistent with there being back spatter associated with the victim's entrance gunshot wound. The direction of travel of the spatters is left to right and upward (Figure 12.32). When a bullet strikes a blood source, the spatter

Figure 12.32 View of stain area A on east bedroom wall.

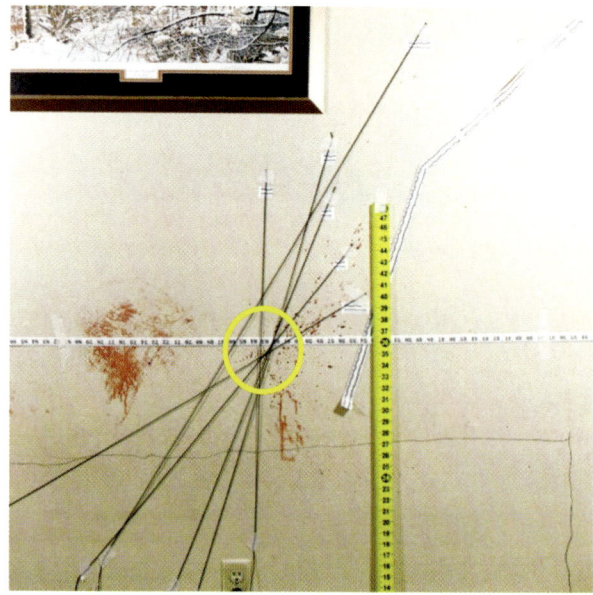

Figure 12.33 View of the area of convergence of the back spatter pattern on the east wall of the master bedroom.

which emanates from the entrance wound is referred to as *back spatter*.

The direction of travel of the spatters is left to right and upward. Representative spatters from this pattern were selected and utilized to establish an area of convergence. Figure 12.33 illustrates the area of convergence of the back spatter pattern on the east wall of the master bedroom. The area of convergence was located at approximately 35 inches from the floor and 64¼ inches from the southeast corner of the bedroom. The black line located below the area of convergence in Figure 12.33 depicts the location of the bed along the east wall prior to its removal. The uncompressed mattress height was determined to be between 26 and 28 inches This indicates the victim's entrance gunshot wound was located approximately 7 to 9 inches above the mattress when her wound was inflicted. This would place the left side of her head close to if not in contact with the mattress and/or pillow when the shot was fired.

Figure 12.34 depicts a view of the bedroom from the doorway of the master bedroom across the bed towards the mirror-topped dresser, which was located along the south bedroom wall. In addition, the overall location of the

Figure 12.34 View of the bedroom from the doorway of the master bedroom across the bed.

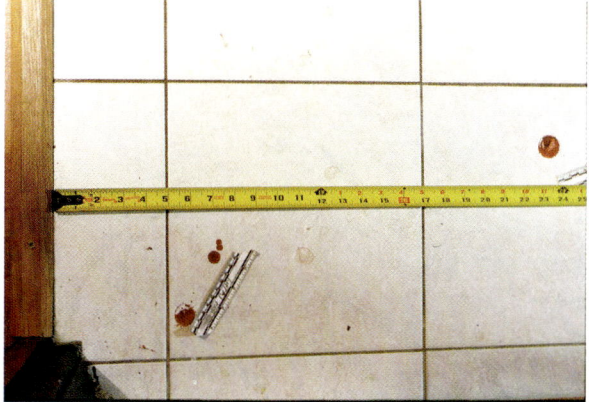

Figure 12.35 View of passive bloodstains on the bathroom floor.

back spatter patterns located on the fitted sheet and on the east wall can be observed.

Spatters were observed on both the right and left side front drawer areas of the dresser. The center most area of the dresser was void of any blood spatters. The large yellow arrows on the dresser were used to depict the distribution of the spatters on the front of the dresser. Samples were collected from both sides of the front of the dresser. Reportedly, both of these blood samples matched the DNA profile of victim. The spatters on the protruding drawer front areas of the dresser are consistent with back spatter from the victim's entrance gunshot wound.

Bathroom

Figure 12.35 depicts a series of bloodstains on the bathroom floor. These passive bloodstains are consistent with a blood source moving across the bathroom floor while dripping blood. Some of these bloodstains exhibited diluted appearances.

Figure 12.36 depicts diluted bloodstains which were located on the bathroom sink counter as well as within the sink basin. A bloodstained tissue/paper product was located on top of the counter. Reportedly, human blood with the same PGM subtype as the victim was identified in a sample from the sink counter as well as from the bathroom floor.

Figure 12.36 View of diluted bloodstains which were located on the bathroom sink counter as well as within the sink basin.

Kitchen

Figure 12.37 depicts the diluted bloodstains that were located within the kitchen sink as well as on the kitchen counter adjacent to the kitchen sink. A light blue plastic cup was also located on the counter adjacent to the kitchen sink. This cup also had diluted bloodstains present on its exterior surface. Reportedly, human blood with the same PGM genetic marker subtype as the victim was identified in the stains removed from the light blue plastic cup and a sample from the kitchen counter to the left of the kitchen sink.

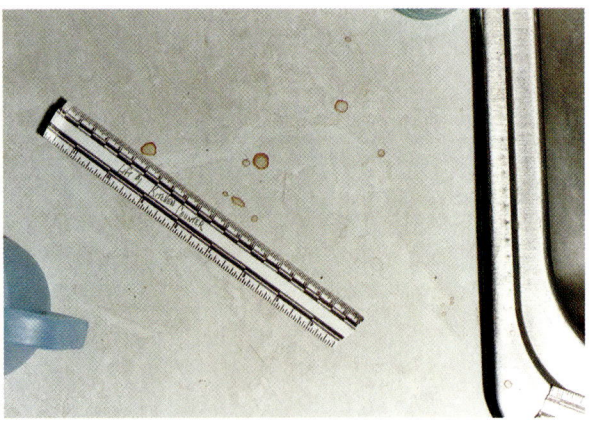

Figure 12.37 View of the diluted bloodstains that were located within the kitchen sink as well as on the kitchen counter adjacent to the kitchen sink.

Husband

Photographs taken of the victim's husband on the day of the incident depicted spatter on the right side of his face, cheek, and around his right eye. The husband was clad in a grey sleeveless T-shirt.

Figure 12.38 depicts the front of the husband's grey T-shirt. A distribution of small spatters is located over the front of both shoulders and over the top of the right shoulder area. A number of these spatters were encircled on the garment with a black marker to illustrate the distribution of the spatters. Reportedly, samples of these spatters from both the right and left shoulder regions matched the DNA profile of the victim. Two (2) transfer patterns are centrally located on the T-shirt with the higher and more dense pattern located 19 to 25 inches from the bottom hem. The lower less dense pattern was located 12.5 to 16 inches from the bottom hem. Additional spatters of blood were observed over the mid-section of the t-shirt.

Figure 12.39 is a view of the T-shirt from the top down that illustrates the distribution of spatters on the upper front of the T-shirt as well as on the back/top of the right shoulder region. Reportedly, a sample of the spatters from the back right shoulder region was identified as human blood with a PGM genetic marker subtype that matched the victim.

Figure 12.40 illustrates the location and physical appearance of the spatters located on the upper right shoulder region of the grey T-shirt worn by the victim's husband. The physical appearance and distribution of spatters on this T-shirt are consistent with their being the result of back spatter from the victim's entrance gunshot wound.

Figure 12.41 is a closer view of the transfer pattern located on the upper front chest region of the T-shirt in Figure 12.39. Clot-like material is located within this transfer pattern.

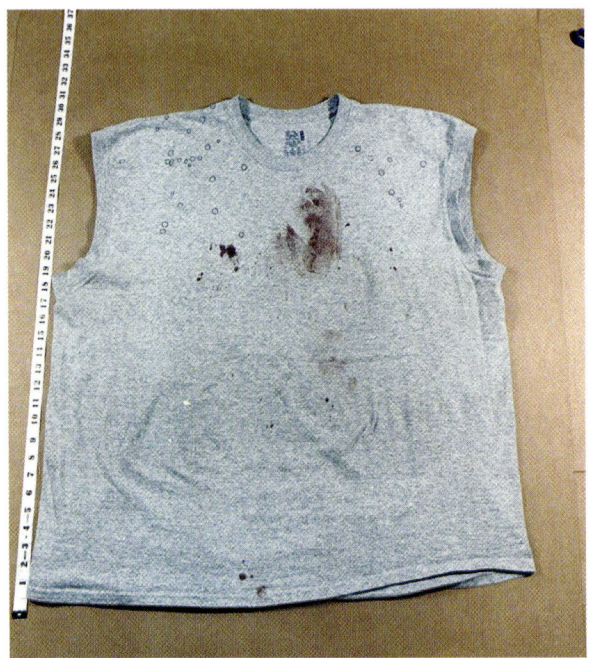

Figure 12.38 View of the front of the husband's grey T-shirt.

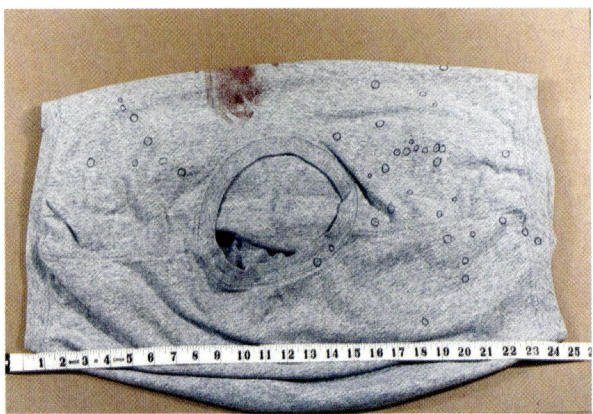

Figure 12.39 View of the gray T-shirt from the top down.

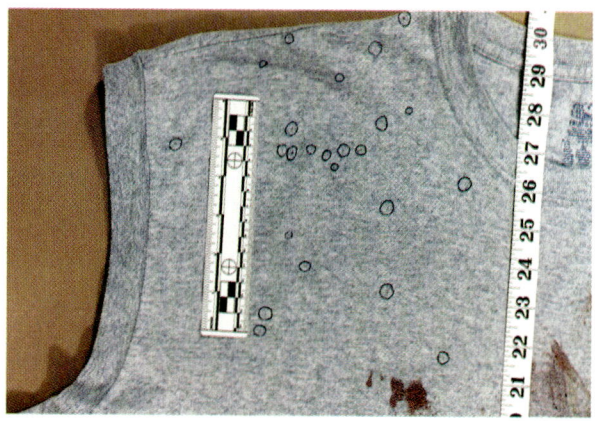

Figure 12.40 View of the spatters located on the upper right shoulder region of the gray T-shirt.

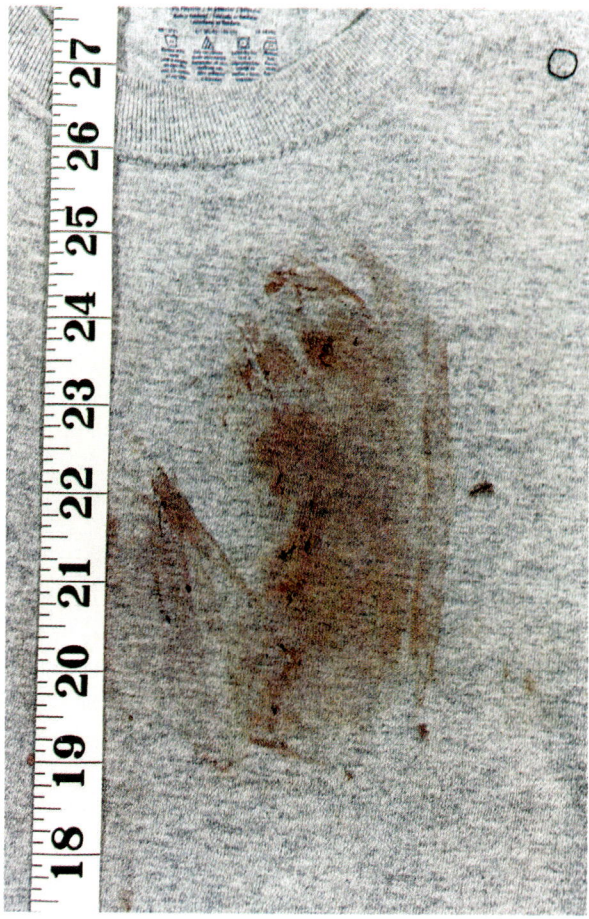

Figure 12.41 Closer view of the transfer pattern located on the upper front chest region of the gray T-shirt.

Conclusions

1. The distribution of spatter associated with gunshot on the east bedroom wall, the positioning of the victim's left arm under her body, the larger saturation pattern on the fitted sheet adjacent to her left shoulder, as well as, the distribution of spatter on the fitted sheet adjacent to the large saturation pattern indicate the victim's body was rolled and/or moved over to her final rest position after her gunshot wound had been inflicted.

2. The bloodstain patterns along with the forensic evidence of this case are consistent with the victim lying in bed on her stomach with the left side of her face against the bed.

3. The bloodstain evidence in the bathroom as well as in the kitchen is consistent with an individual, other than the victim, attempting to clean up a source of the victim's blood from their person and/or object(s) in their possession.

4. The clot-like material within the transfer pattern on the front of the husband's t-shirt indicates that the victim's blood had begun clotting prior to the T-shirt making contact with her blood.

5. The physical appearance and distribution of the spatters on the upper front shoulder region and back right shoulder region of the husband's T-shirt are consistent with back spatter associated with gunshot. Based upon the DNA and pathological results, this back spatter associated with gunshot is consistent with emanating from the entrance gunshot wound to the head of the victim.

6. The physical appearance and distribution of back spatter on the husband's T-shirt indicates he was in close proximity to victim's head when her gunshot wound was inflicted. This is further supported by the location and

distribution on stains of the right side of the husband's face.

7. The physical evidence is consistent with the husband being on the south side of the bed between the bed and dresser when the victim was shot in the right side of the head.

8. The bloodstain pattern evidence at the scene as well as on the husband's person placed him within the void area of the dresser on the south side of the bed when his wife's injury was inflicted.

The husband gave several versions as to where he was located when the shot occurred, none of which placed him on the south side of the bed in front of the dresser. The husband ultimately went to trial and was found guilty of murder.

Training and Education

Bloodstain analysts represent a range of forensic scientists and crime scene investigators with diverse levels of education. The courts have accepted testimony from individuals with strong backgrounds in chemistry, biology, and physics, many of whom possess degrees in science or forensic medicine. Many of these individuals are employed in crime laboratories or medical examiner offices that have crime scene responsibilities. Crime scene investigators, evidence technicians, and detectives who do not necessarily possess scientific backgrounds have also offered expert testimony.

It is highly recommended that individuals enroll in a basic 40-hour course in bloodstain pattern interpretation. These courses are usually taught over a five-day period and, occasionally, as a semester course at a college or university by qualified instructors at numerous locations in the United States and abroad. They provide instruction in the theory and practical aspects of bloodstain pattern interpretation, case presentations, and opportunities to perform laboratory experiments.

Participation in laboratory experiments is crucial for understanding the dynamics of bloodstain pattern production and the mechanisms involved. Students create stain patterns similar to those produced at crime scenes utilizing various types of apparatus. Patterns are created on cardboard and other surfaces that can be preserved and retained for future reference.

Advanced courses are also available following the successful completion of a basic course. The advanced courses are designed to review basic concepts and provide additional training in areas not explored in the basic courses. For example, an advanced course may concentrate on computer analysis, digital imaging, examination of bloodstained clothing and footwear, mock crime scenes, and in some cases, mock trials. Students are encouraged to present cases to the group for peer review sessions. Successful completion of basic and advanced courses in bloodstain pattern analysis does not imply that an individual is a qualified bloodstain analyst. The formal education must be coupled with years of experience with crime scenes and evidence examination along with regular attendance at scientific seminars. It is also important to stay abreast of the information in scientific journals and periodicals.

Membership in professional organizations is encouraged. Students who have successfully completed a basic course in bloodstain pattern interpretation are qualified to apply for membership in the International Association of Bloodstain Pattern Analysts (IABPA). This organization, currently with more than 900 members from the United States and abroad, was established in 1983. The group sponsors an annual meeting at which new techniques, research, and case studies are presented. IABPA also publishes a newsletter that contains current information for bloodstain analysts. The American Academy of Forensic Sciences holds an annual meeting that offers a diverse program covering all areas of forensic science, including bloodstain pattern interpretation. The *Journal of Forensic Sciences*, the official publication of the Academy, provides articles on current forensic issues.

Scientific Working Group on Bloodstain Pattern Analysis (SWGSTAIN)

The idea of a contemplative group assembled to discuss and intelligently address substantive matters concerning the forensic discipline of bloodstain pattern analysis (BPA) is not unique. The FBI Laboratory through its case-working units and Forensic Science Training Unit (FSTU) has developed a Scientific Working Group (SWG) program for assembling such contemplative groups. The Scientific Working Group on Bloodstain Pattern Analysis (hereinafter referred to as SWGSTAIN) serves as a professional forum in which local, state, and federal government bloodstain pattern analysis (BPA) practitioners and practitioners from related fields can share, discuss, and evaluate methods, techniques, protocols, quality assurance, education, and research relating to BPA. This forum shall address substantive and operational issues within the field of BPA and shall work to build consensus-based, or so-called "best practice," guidelines for the enhancement of BPA.

Conclusion

The goal of this chapter is to acquaint the reader with the basic concepts of bloodstain pattern interpretation. The interpretation of bloodstain patterns can become very complex, depending on individual case factors. Experimentation with specific case variables can be very critical and should not be overlooked. A major pitfall is oversimplification of this complex discipline. It is unfortunate that people fail to adhere to sound scientific practices and rely more on speculation than on fact. To become proficient in this discipline, one must have solid grasps of mathematics, physics, scientific method, and a great deal of practical experience. Anyone working in a field where bloodstains may later be used to incriminate or exonerate a suspect will find a solid working knowledge of what bloodstain pattern interpretation can and cannot do.

Questions

1. What significant physical properties of blood determine the shape of a blood drop in flight as contrasted to the oppositional forces acting upon a blood drop as it is being formed?

2. What is the most important factor governing the degree of distortion and spatter that results when a blood drop strikes a surface?

3. What factors influence the stain diameter produced by a free-falling drop?

4. How are the physical characteristics of spatter utilized to determine their angle of impact?

5. Compare the size ranges of the conventional low-, medium-, and high-velocity impact spatters.

6. What other mechanisms can create spatters in the same size range as impact spatter encountered in beating, stabbing, and gunshot events?

7. What variables can affect the size, quantity, and distribution of spatters created by an impact mechanism such as beatings and shootings?

8. Discuss the techniques employed for determining the area of convergence and origin of bloodstains within a pattern.

9. Name two types of bloodstain patterns that require confirmation by autopsy findings.

10. Explain why an assailant might not have any bloodstains on his or her person or clothing after participating in a beating death.

11. Explain the mechanism of castoff bloodstain patterns.

12. What are the features of progressive drying and clotting of blood?

13. How can bloodstains be physically altered at crime scenes?

14. What important information can be derived from the examination of bloodstain patterns?

15. What methods are commonly used to document bloodstain evidence?

References and Suggested Readings

Balthazard, V. et al., Etude des Gouttes de Sang Projecté, paper presented at the 22nd Congress of Forensic Medicine, Paris, 1939.

Bevel, T. and Gardner, R. M., *Bloodstain Pattern Analysis with an Introduction to Crime Scene Reconstruction*, 3rd ed., CRC Press, Boca Raton, FL, 2008.

Bluestar Latent Blood Reagent Manual, ROC Import, Monaco, France.

DeForest, P. R., Gaensslen, R. E., and Lee, H. C., *Forensic Science: an Introduction to Criminalistics*, McGraw-Hill, New York, 1983, p. 295.

James, S. H., Ed., *Scientific and Legal Applications of Bloodstain Pattern Interpretation*, CRC Press, Boca Raton, FL, 1999.

James, S. H. and Eckert, W. G., *Interpretation of Bloodstain Evidence at Crime Scenes*, 2nd ed., CRC Press, Boca Raton, FL, 1999.

James, S. H. and Edel, C. F., in *Bloodstain Pattern Interpretation: Introduction to Forensic Sciences*, Eckert, W. E., Ed., CRC Press, Boca Raton, FL, 1997.

James, S. H., Kish, P. E., and Sutton, T. P., *Principles of Bloodstain Pattern Analysis—Theory and Practice*, Taylor and Francis, Boca Raton, FL, 2005.

Kirk, P. L., *Crime Investigation*, 2nd ed., John Wiley and Sons, New York, 1974, p. 167.

Kish, P. E. and MacDonell, H. L., Absence of evidence is not evidence of absence, *J. Forens. Identification*, 46, 160, 1996.

Laber, T. L., Diameter of a bloodstain as a function of origin, distance fallen and volume of drop, *IABPA News*, 2, 12, 1985.

Laber, T. L. and Epstein, B. P., *Bloodstain Pattern Analysis*, Callen Publishing, Minneapolis, MN, 1983.

Lee, H. C. et al., *Henry Lee's Crime Scene Handbook*, Academic Press, San Diego, CA, 2001.

MacDonell, H. L., *Interpretation of Bloodstains: Physical Considerations*, Wecht, C., Ed., Appleton, New York, 1971, p. 91.

MacDonell, H. L., *Criminalistics: Bloodstain Examination*, Vol. 3 of *Forensic Sciences*, Wecht, C., Ed., Matthew Bender, New York, 1981, p. 371.

MacDonell, H. L., *Bloodstain Patterns*, revised edition, Laboratory of Forensic Science, Corning, New York, 1997.

MacDonell, H. L. and Bialousz, L., *Flight Characteristics and Stain Patterns of Human Blood*, U.S. Department of Justice, Washington, D.C., 1971.

MacDonell, H. L. and Bialousz, L., *Laboratory Manual on the Geometric Interpretation of Human Bloodstain Evidence*, Laboratory of Forensic Science, Corning, New York, 1973.

MacDonell, H. L. and Brooks, B., *Detection and Significance of Blood in Firearms*, Wecht, C., Ed., Appleton, New York, 1977, p. 185.

MacDonell, H. L. and Panchou, C., Bloodstain patterns on human skin, *J. Can. Soc. Forens. Sci.*, 12, 134, 1979.

Pex, J. O. and Vaughn, C. H., Observations of high velocity blood spatter on adjacent objects, *J. Forens. Sci.*, 32, 1587, 1987.

Piotrowski, E., Uber Entstehung, Form, Richtung und Ausbreitung der Blutspuren nach Heibwunden des Kopfes, K. K. Universitat, Wein, Austria, 1895.

Pizzola, P. A., Roth, S., and DeForest, P. R., Blood droplet dynamics I, *J. Forens. Sci.*, 31, 36, 1986.

Pizzola, P. A., Roth, S., and DeForest, P. R., Blood droplet dynamics II, *J. Forens. Sci.*, 31, 50, 1986.

Stephens, B. G. and Allen, T. B., Back spatter of blood from gunshot wounds: observations and experimental simulation, *J. Forens. Sci.*, 28, 437, 1983.

Sutton, T. P., Bloodstain Pattern Analysis in Violent Crimes, University of Tennessee, Memphis, 1993.

White, R. B., Bloodstain patterns of fabrics: the effect of drop volume, dropping height and impact angle, *J. Can. Soc. Forens. Sci.*, 19, 3, 1986.

Wonder, A. Y., *Blood Dynamics*, Academic Press, San Diego, CA, 2001.

Wonder, A. Y. *Bloodstain Pattern Evidence—Objective Approaches and Case Applications*, Elsevier, New York, 2006.

Yen, K. et al., Blood Spatter Patterns—hands hold clues for the forensic reconstruction of the sequence of events: *Am. J. Forens. Med. Pathol.*, 24(2), 132–140, June 2003.

SECTION III

Forensic Science in the Laboratory

The Forensic Laboratory

Linda R. Netzel

13

Introduction

Paul Kirk, in discussing physical evidence left by a criminal, best describes the purpose of the forensic laboratory in this passage from the introduction to his text titled *Crime Investigation:*

> Not only his fingerprints and his shoeprints, but also his hair, the fibers from his clothes, the glass he breaks, the tool marks he leaves, the paint he scratches, the blood or semen that he deposits or collects—all these and more bear mute witness against him. This is evidence that does not forget. It is not confused by the excitement of the moment. It is not absent because human witnesses are. *It is factual evidence.* Physical evidence cannot be wrong; it cannot perjure itself; it cannot be wholly absent. Only in its interpretation can there be error. Only human failure to find, study, and understand it can diminish its value. The laboratory must be devoted to this study and understanding if the all-important traces that can speak so eloquently of guilt or innocence are to be heard.[1]

Purpose

Kirk's words, originally written in 1953, are even more relevant today because laboratory capabilities are greatly enhanced by technology not available in the 1950s. For example, the structure of **deoxyribonucleic acid** (DNA) was first elucidated in 1953, and three decades later the use of this information revolutionized forensic science.

Advancing technology provides the forensic laboratory more tools with which to study and understand physical evidence. Because of the power of this nonbiased physical evidence, police, attorneys, judges, and juries are demanding more from forensic laboratories. Since so much more analysis can and should be done on each individual case, the number of cases analyzed today may be similar to or even fewer than the number analyzed a decade or more ago. Additionally, the introduction of new technologies such as the analysis of DNA requires laboratories to revisit old cases whether previously **adjudicated** or not. Unfortunately, these responsibilities of forensic laboratories are increasing disproportionately to their resources.

The forensic laboratory has not only the responsibility of analyzing physical evidence, but must also be involved in all aspects of its recognition, collection, and preservation. This includes training police and other crime scene investigators in the art and science of crime scene investigation, which is a critical stage in any criminal proceeding. Attorneys, judges, and juries must also be educated, not in the intricacies of the evidence, but in placing proper emphasis on the scientific results, which can be seriously compromised if a crime scene is mismanaged. Therefore, the forensic laboratory is charged with creating a team of highly skilled individuals to accomplish these tasks in a manner that will ensure the quality of the laboratory results.

Quality Assurance

To ensure the quality of all results reported, it is essential that a laboratory have an established **quality assurance** program. Quality assurance programs, when strictly maintained, are designed to ensure that the reported results are scientifically valid and that opinions are based only upon results that are deemed reliable. Quality assurance programs contain a number of requirements that must be adhered to by laboratory management and scientists. Such requirements may include required education of staff, peer review of scientific results or reports, specific case file documentation, distribution of reports, auditing of testimony, evidence handling, and security of the laboratory.

Quality control is a key component of quality assurance programs and is fundamental laboratory practice. Quality control includes specific procedural steps designed to alert the scientist to potential process failures. Specifically measures such as the use of positive and negative controls and reagent blanks used for quality control. Positive controls are used to demonstrate that a particular test is working properly. A known human blood standard run simultaneously with tests on unknowns for the presence of human blood is one example of a positive control. Negative controls and reagent blanks serve to detect false positive results and contamination respectively. A negative control is designed to determine if the test being used is not reacting specifically as designed. For example, if a saliva standard gives a positive reaction for a semen test, this should raise concerns about the test. Furthermore, if a chemical reagent

becomes contaminated with a foreign substance, such as a reagent blank in a DNA test (DNA reagent blanks contain no human DNA only the chemicals used in the process) this would potentially render any test results as suspect and possibly require that the test be repeated.

Monitoring equipment is also crucial to quality control. This involves the use of positive and negative controls as mentioned above but also regular maintenance and calibration. For example, a scientific scale used to weigh minute quantities of chemical reagents for reagent preparation or a scale used to weight illegal drugs must be accurate. To insure accuracy, standards must be used to determine if a particular piece of equipment is working properly, the equipment is regularly serviced by qualified technicians, repairs are made appropriately and the equipment is calibrated regularly. Calibration involves running certain standards and determining if the results are what is normally expected and is within an acceptable range. For some equipment, calibration may occur at each use and for other types, once a month, semiannually, or annually.

Quality assessment of any given analysis would involve a detailed review process. In crime laboratories this is typically referred to as a technical review. Technical review involves an independent examiner looking at the underlying test data and determining if the reported results and conclusions are supported by that data. The review process may also include verification, which involves a reexamination of the evidence by a second examiner. Verification is distinct from a technical review because it involves performing the actual laboratory examination and not just a review of the data. Additionally, an administrative review may also be performed as part of quality assurance. Administrative review is a process where a case is examined for completeness of all documentation. This would include verifying that all documentation is properly identified as being part of the case, all work is initialed by examiners, and that all pages are present for any given analysis.

As an important part of quality assurance, laboratory staff must routinely participate in proficiency testing. **Proficiency tests** are simulated forensic cases that can be produced internally or provided by an outside testing agency. The assigned criminalist must evaluate the proficiency test case as if it were genuine. When outside proficiency tests are used, the results are unknown to laboratory staff and management until all participating laboratories have completed testing and submitted their results. These results are then compiled and published so that interested parties may review them.

Staffing Issues

O'Hara and Osterburg, coauthors of *An Introduction to Criminalistics,* found the term *forensic* to be inadequate to describe the purpose of the police laboratory and the work of its scientists. They chose the term **criminalistics**, which was commonly used in Europe to describe scientific crime detection.[2] Although *forensics* and *criminalistics* are used interchangeably, they are distinct. *Forensics* is a more general term that can apply to any number of scientific disciplines, such as anthropology and odontology. A "criminalist," however, is a scientist who applies the principles of primarily biology, physics, and chemistry to evidence analysis and is also trained in crime scene investigation and reconstruction. Thus, a criminalist obtains scientific results and places them in context for a particular crime.

Maintaining the quality of results in a forensic laboratory begins with the quality of its staff. National guidelines[3–5] exist to ensure that forensic laboratories hire criminalists with adequate educational backgrounds. Furthermore, many laboratories perform extensive background investigations prior to hiring. Background investigation may include a polygraph examination, history of illegal drug usage, criminal record, driving record, and employment and residential histories. It is imperative that the laboratory hire qualified individuals who possess high degrees of integrity and honesty.

Depending on the size of a laboratory, a few or several levels of scientists may be involved in laboratory operations. Although a sworn law enforcement officer may manage the laboratory, it is essential that the laboratory director be a criminalist. He or she must have vast forensic experience and keep abreast of the latest advances in forensic science. This knowledge allows the director to anticipate staffing, equipment, and training needs. The director may not actively analyze cases, but must be aware of all quality assurance requirements and determine how they are to be met.

Supporting the laboratory director will generally be a quality assurance manager or section supervisors. These individuals should have extensive experience in one or more forensic disciplines. Their responsibilities may include hiring, training, and supervision of section scientists as well as case examination.

The staff of each laboratory section will have a range of experience and may be trained in several disciplines. In addition to examining cases, other staff duties may include training scientists, crime scene investigation, training police officers, attorneys, and other law enforcement personnel, and educating the public about the role of the crime laboratory.

Accreditation and Certification

The American Society of Crime Laboratory Directors (ASCLD) handles **accreditation** of crime laboratories in the United States. Accreditation is an endorsement of a laboratory's policies and procedures. ASCLD's Laboratory Accreditation Board (ASCLD/LAB) developed a comprehensive set of requirements to which laboratories seeking accreditation must conform.[6] Each laboratory must complete an extensive application process that includes a comprehensive inspection, interviews with staff, and review of all written procedures and quality assurance programs. Laboratory security and safety procedures are included in the inspection. The laboratory is also required to pay fees associated with this inspection. After inspection, the laboratory will be required to address any areas of weakness and document corrective actions. Laboratories that successfully complete the process will receive ASCLD/LAB certificates of accreditation.

ASCLD/LAB accreditation is awarded for 5 years. Each laboratory must submit an annual accreditation review report based on self-evaluation. Proficiency test results must also be provided to the ASCLD/LAB Proficiency Review Committee directly from approved proficiency testing agencies. At the end of the 5-year period, the laboratory will again be inspected for accreditation renewal.

Certification is awarded to criminalists, and it provides official recognition of professional development. The American Board of Criminalistics (**ABC**) and the International Association of Identification (**IAI**) are two organizations that provide certifications for criminalists. To obtain ABC certification, a criminalist must successfully complete one or more written tests. Advanced tests are also offered in specific disciplines. To maintain certification, a criminalist must provide an annual accounting of professional activities, which may include training he or she provided or received, participation in writing scientific publications, professional meetings attended, and proficiency tests completed.

Types of Laboratories

Government Laboratories

Government-sponsored forensic laboratories exist at city, county, regional, state, and federal levels. Many large metropolitan communities have forensic laboratories associated with local law enforcement agencies, such as police departments, county medical examiner offices, or sheriff's departments. A number of forensic laboratories are also associated with the federal government. The Federal Bureau of Investigation, Bureau of Alcohol, Tobacco, and Firearms, the Drug Enforcement Agency, and the U.S. Postal Service all have forensic laboratories.

Association with a law enforcement agency is critical to most government laboratories, but should never influence the outcome of a scientific investigation. Vital to a laboratory's role in an investigation is access to information and physical evidence at the crime scene. Without legal access to a crime scene, a laboratory will be severely hindered in performing its function during an investigation. Also, a free exchange of information between law enforcement agencies and the laboratory is necessary. Without an official association, this exchange of information could be difficult. Individuals who believe that a forensic scientist should work independently of investigative information are misled. No scientist in any discipline should choose to work without having as much *relevant* information as possible about his assignment. Knowledge of certain facts will assist the criminalist in determining what questions should be addressed by his examination.

Private Laboratories

Private laboratories may also perform forensic analysis in criminal cases. A number of such laboratories are available to police agencies and defendants for testing or retesting of **forensic evidence**. Many of these private laboratories are not strictly dedicated to forensic testing but utilize similar techniques of analysis for different applications. For example, DNA testing is routinely used in parentage testing. Although the technology is the same as in forensic testing, the application is quite different.

The adversarial judicial system of the United States makes the private forensic laboratory a necessity. Every defendant has the right to confront witnesses against him, and witnesses include forensic scientists testifying on behalf of the people. Although some police agencies may rely on private laboratories to analyze cases, the greatest contribution of the private lab is retesting of evidence already examined by a government or public lab. Retesting is generally performed on behalf of a defendant.

Qualifications of a Forensic Examiner

The formal education of a criminalist begins with a baccalaureate degree in a natural science. The emphasis of the degree is commonly chemistry, biology, physics, or forensic science.

Coursework in mathematics (including statistics), a public speaking course, criminal justice courses, writing, and logic courses are also valuable because they prepare the criminalist for different aspects of a career that are not purely scientific, such as testifying in court. Many forensic laboratories employ entry-level criminalists after completion of baccalaureate degrees.

Although an undergraduate degree provides a sufficient foundation for entry-level criminalists, on-the-job training will provide most of the useful skills required for a career. Accredited laboratories are required to have written training manuals for each section of the laboratory and for each specific examination handled within that section. For example, a training manual for trace evidence criminalists will include training modules in glass, paint, hair, and fiber examinations. The training will generally be supervised by an experienced criminalist and takes months to complete. An important part of the training includes reanalysis of previously adjudicated cases, past proficiency tests, and other simulated materials. Analysis of such cases provides "real world" experience. Most likely, a criminalist trainee will graduate to independent casework analysis after demonstrating competency and successful completion of proficiency tests.

Advanced college degrees are not necessary for an entry-level criminalist, but national guidelines require master's degrees for certain positions. The **DNA Advisory Board** (**DAB**) now requires DNA technical managers to have master's degrees.[5] In some situations, experience can substitute for a graduate degree. Master's degrees are desirable for many laboratories, but gaining experience and becoming productive in casework is much more desirable. A master's can be pursued as a part of continuing education.

Both accreditation and certification require that individuals be provided opportunities to continue their education by attending forensic meetings, training courses, and seminars, in addition to in-house training. The most beneficial aspect of attending meetings and seminars is meeting other criminalists and sharing experiences. This allows exchange of ideas and learning about other approaches to casework. An interesting aspect of criminalistics is that each case is different and, therefore, presents its own challenges. Learning about new techniques and meeting other criminalists can be rewarding training experiences.

Reading current forensic science journals and general scientific journals are other important aspects of a well-rounded continuing education program. One valuable resource is the *Journal of Forensic Sciences,* published six times a year. International forensic journals are also published regularly. The training of new criminalists should include the review of important historical forensic materials. The work of Hans Gross, Charles O'Hara, James Osterburg, and Paul Kirk, to name a few, demonstrates that the approaches used remain the same even though technology has advanced. Their contributions are both scientific and philosophical in nature.

Understanding the Role of the Criminalist

The role of the criminalist is to provide investigative leads through scientific evaluation of physical evidence and crime scene reconstruction, report results and conclusions of the scientific evaluations, and provide expert testimony concerning those results and conclusions. The criminalist cannot simply work at a lab bench and blindly generate reports of results. His peers, including experts for the opposing side, will potentially scrutinize every analysis performed and every conclusion drawn. This requires that the criminalist have sufficient training and experience to perform the task and also understand the significance of the results. He or she must not overstate the value of findings for the benefit of either side of the case.

Criminal Investigation: A Practical Textbook for Magistrates, Police Officers and Lawyers, based on the work of Dr. Hans Gross, was originally published in 1934. As with the work of Paul Kirk, much of this text contains valuable information for those involved in modern crime investigation. The following passage discusses the role of the expert:

> Experts are the most important auxiliaries of an Investigating Officer; in some way or other they nearly always are the main factor in deciding a case. True it is that the Investigating Officer has not skilled experts always at his immediate disposition, but, in important cases, he is able to refer to experts at headquarters; on the other hand, persons who are not strictly speaking "experts" at all but who have special knowledge, can give excellent results; everything depends upon knowing how to make use of them. Indeed, it is often less important to know who is to be questioned than to know how, upon what and when questions must be put.[7]

This text delves into the roles of medical experts, microscopists, chemists, physicists, and biologists, and the disciplines of firearms, fingerprinting, photography, and handwriting. The role of a modern criminalist is not very different from the historical one in that the objective is to answer questions. Because hundreds of "leads" may be generated in a case, the criminalist can help focus the investigation by eliminating incorrect theories and supporting viable ones. The technology with which the criminalist can explore those questions has changed substantially, but the entire process still requires close collaboration between the investigator and the criminalist.

Investigator

As a part of the criminalist's role in an investigation, he must utilize his own ability to ask questions and investigate the information already available. A solid approach to beginning a case examination, particularly a major case, would include review of crime scene reports, diagrams and photographs, evidence collected, statements of suspects or witnesses, and, of course, information in police reports. The criminalist must also be knowledgeable of the capabilities of other experts within his laboratory and experts outside the laboratory.

Most of the information that aids the criminalist comes from asking questions of the law enforcement investigator. Gaining as much information as possible about a situation will not alter scientific results. Kirk states the importance of this exchange in *Crime Investigation:*

> The efficiency of the laboratory usually bears a direct relationship to the willingness of the police officer to keep the laboratory workers informed of all pertinent facts. Complete frankness and confidence between the two types of investigators is most desirable and profitable, and unfortunately not so common as it might be. Only good understanding, by both, of their reciprocal functions can completely eliminate this barrier to the realization of the full benefits of a well managed crime laboratory.[1]

The police detective is not always informed of a criminalist's capabilities, so it is the responsibility of the criminalist to investigate and find the information he deems necessary.

Educator

A criminalist will have many opportunities to educate a variety of individuals involved in criminal investigations. Law enforcement officers, crime scene investigators, attorneys, judges, medical professionals, and others can all receive training from an experienced criminalist. The format of the training may be a seminar or workshop, a pretrial conference or hearing, one-on-one meetings, or published scientific articles. Well-written reports can also aid the criminalist in educating readers by clearly stating the results of the analysis that in turn provides the basis of the conclusions.

The conclusions of the criminalist may be all that is necessary to educate concerned parties on a particular case. Whittaker discusses the "varying degrees of specificity of identification" in his 1973 article titled "On the Role of the Criminalist."[8] His descriptions of the possible conclusions regarding forensic comparisons are (1) the totally positive and absolute identification, (2) a less absolute but positive identification, (3) a probable identification, (4) a possible identification, and (5) a negative or totally impossible relationship, usually referred to as an exclusion. In other words, the criminalist must educate as to the significance or probative value of the results. The results and conclusions will be limited by the quality, quantity, or type of physical evidence.

Student

Advancing technology and trends in crime require a criminalist to continually review current literature, perform experiments, and attend seminars, workshops, and professional meetings. Historically, many criminalists were trained to do a variety of analyses and were called "generalists." This designation is facing extinction because of advances in technology and increasing caseloads. Although it is necessary to be familiar with other forensic disciplines, many criminalists today specialize in a particular area. The quality assurance program, and accreditation and certification processes discussed earlier further foster specialization.

Sometimes a criminalist will be required to explore something unusual, such as locating the manufacturer of a rare type of cordage used as a binding. During this process, he may learn a great deal about cordage manufacturing, fibers, knot tying, intended use of certain types of cordage, and/or how common the cordage evidence is. Although this process can be time-consuming, individuals with insatiable curiosity will truly enjoy this aspect of a criminalistics career. Experienced criminalists will see common aspects of different crimes, but each case has its own unique set of circumstances that offers continual learning opportunities.

One of the most important training activities of a criminalist is attending crime scenes. This activity will enable the criminalist to determine what to look for, how and where to look for it, and ultimately the importance of the evidence to the case. Another way for criminalists to expand their knowledge bases is at professional forensic meetings. In addition to the routine presentations, meetings sometimes provide opportunities to tour manufacturing plants, such as fiber, glass, paint, paper making, and other types of production facilities.

Linking a Crime Scene to a Suspect

Reconstruction Evidence

Reconstruction evidence provides information about the events preceding, occurring during, and after commission of a crime. Reconstruction of a crime scene is particularly valuable in instances where a suspect admits to having been at the scene but did not play a role in the crime. Although scientific principles are applied, reconstruction of a crime involves observation, logic, experience, and evaluation of statements by key witnesses. This may provide the greatest opportunity for the criminalist and detective to work as a team in solving a crime.

Reconstruction of crime scenes was discussed in detail in Chapter 10. Three commonly used tools should be mentioned here. Luminol is a chemical that reacts with minute quantities of blood. It is sprayed in a fine mist over areas where blood may have been cleaned or otherwise diluted. In many instances, a positive reaction to luminol will produce a revealing pattern that may lead to the discovery of more bloodstain patterns (Figure 13.1). An alternate light source is another tool used to detect body fluid stains, fibers, or any of the thousands of compounds that fluoresce under various wavelengths of light. Alternate light sources are generally portable for use at crime scenes, and they are used extensively inside forensic laboratories. Finally, bloodstain pattern analysis, the topic of Chapter 11, can provide strong reconstruction evidence. The interpretation of bloodstains is generally very useful in cases where an individual claims to have aided or assisted an injured or dead person as an explanation for blood found on his or her person or clothing. Bloodstain evidence may also be of great assistance in determining whether a person died as a result of suicide, homicide, or accident.

Associative Evidence

The tools used for reconstructing a crime scene will generally locate evidence that can then be used to associate or disassociate a suspect to a crime. Hairs, fiber, blood, and other body fluids, paint, glass, firearms, bullets, fingerprints, and other imprint evidence are all examples of **associative evidence**. These items are considered of unknown or questioned origin until a comparison is made to a known standard or exemplar. A standard may

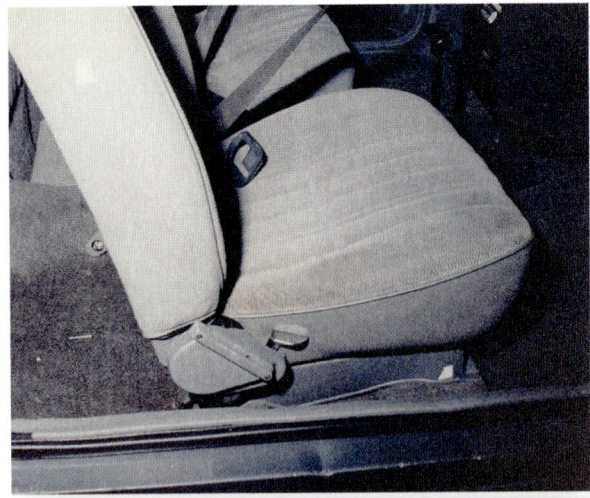

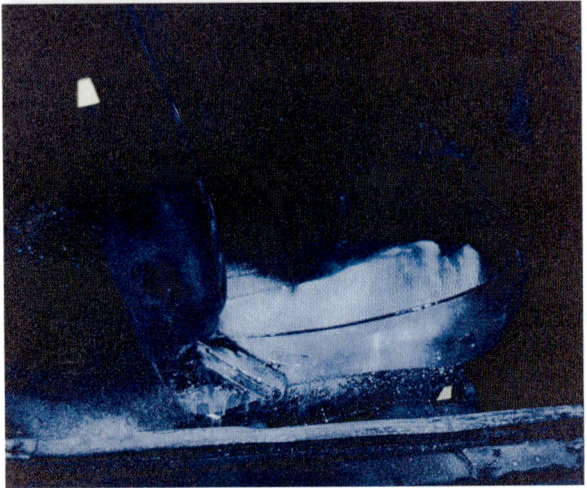

Figure 13.1 Reconstruction evidence. (Top) Murder suspect's car prior to the application of luminol. (Bottom) Positive luminol reaction for apparent blood in the suspect's car. (Photos courtesy of Police Crime Laboratory, Kansas City, Missouri.)

be collected from a victim, suspect, witness, or investigator. Although the examination of associative evidence is the primary focus of a forensic laboratory, the ability to reconstruct a crime is necessary to determine the significance of associative evidence.

Associative evidence can be further subdivided into class and **identification evidence**. **Class characteristic** evidence is not considered unique, and it is part of a limited class along with other potential members (Figure 13.2). Identification evidence, on the other hand, positively provides for identification of the source of questioned evidence (Figure 13.3). When comparing evidence of

both types, the examiner must convey the meaning or significance of the results in a written report.

When examining class characteristic evidence, the examiner must make all reasonable attempts to distinguish questioned samples from known standards. The result may be that the questioned sample is indistinguishable from the known standard, does not match the known standard, or the comparison is inconclusive. Conclusions regarding comparisons of class characteristic evidence are limited. The questioned sample, even when indistinguishable from the known, cannot be said to be from that particular standard to the exclusion of all others. For example, if green carpet fibers are indistinguishable from the carpet of a suspect's car, this evidence does not exclude all similar carpeting as the source of the fibers.

Individual characteristic evidence includes fingerprints, DNA profiles, some impressions, and evidence of fracture matches. Courts and juries have long accepted fingerprints as evidence of identification, although statistics regarding the significance of the match are generally not provided. Conversely, the use of DNA to identify individuals has been highly scrutinized and the forensic laboratory must provide statistics demonstrating the significance of the genetic profiling results. By examining a number of genetic locations, the frequency of occurrence of a genetic profile will become so small that the examiner may conclude that the evidence came from a particular individual. Fracture matches are made when an unknown fractured piece that may have come from an automobile grill is matched to a known by comparison of the fractured edges of both samples.

Whether the evidence is of the class or identification type, it is important that a scientist understand and acknowledge the significance of the final result. Determining the significance may include consideration of the location of the evidence, the type and quantity of evidence, the condition of the evidence, and crime scene reconstruction. If an item of evidence is identified as coming from a suspect, the suspect is associated with the evidence but that does not prove he or she committed

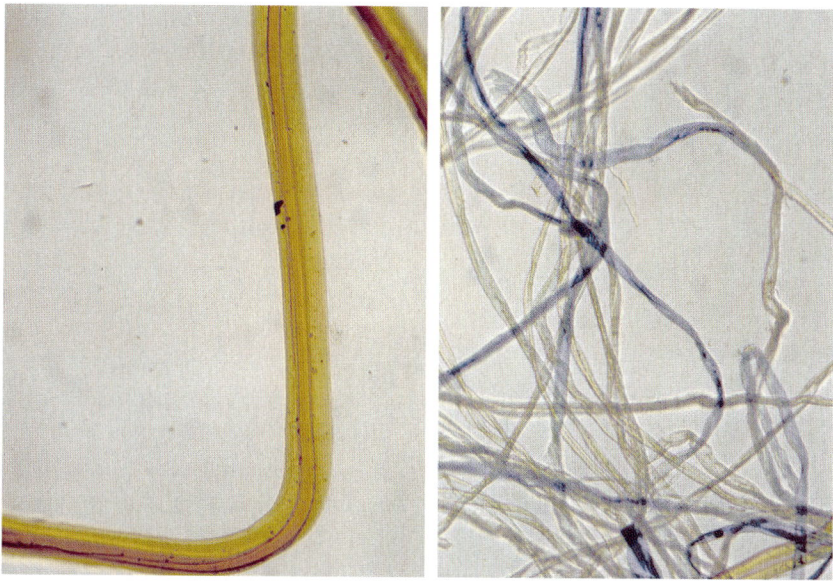

Figure 13.2 Examples of evidence with class characteristics. (Left) Trilobal nylon fiber typically used in carpet manufacturing. (Right) Dyed cotton fibers. (Photos courtesy of Police Crime Laboratory, Kansas City, Missouri.)

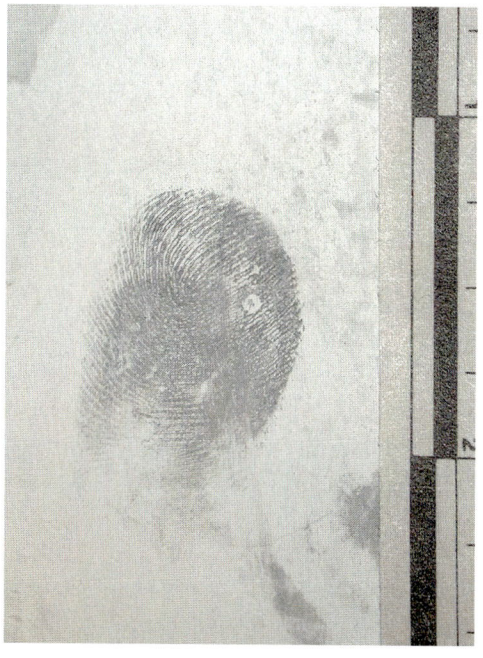

Figure 13.3 Fingerprints can provide positive identification of an individual. (Photo courtesy of Police Crime Laboratory, Kansas City, Missouri.)

the crime. The criminalist must responsibly report his conclusions in a timely and consistent manner and let the trier of fact, judge, or jury, determine guilt.

Laboratory Sections, Analytical Instruments, and Specialized Equipment

Forensic laboratories can differ greatly in the numbers and types of sections they contain. Funding, equipment, and staffing define the capabilities of a forensic laboratory. Some laboratories may employ a single criminalist who has many different responsibilities. Other laboratories may have larger staffs that specialize in particular disciplines or subdisciplines. For example, trace evidence analysis may be divided into hair and fiber sections, and paint and polymer sections. Therefore, it may be more proper to describe areas of emphasis within the laboratory rather than define specific sections.

Biological Evidence

Identification and individualization of human tissues constitute the focus of forensic biology. Conventional serology, such as ABO blood

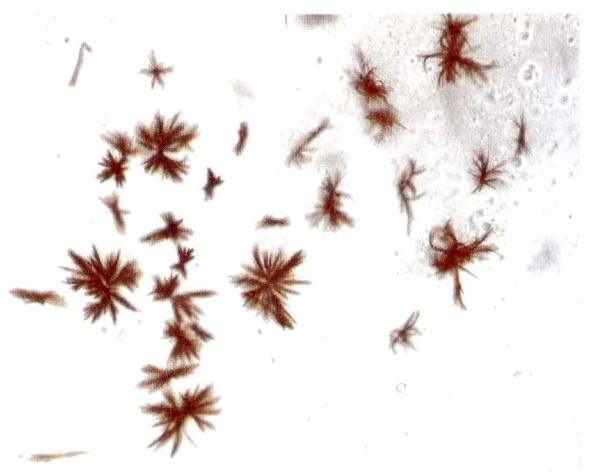

Figure 13.4 Positive Takayama microcrystal test for the confirmation of blood. (Micrograph courtesy of Police Crime Laboratory, Kansas City, Missouri.)

grouping has largely been replaced by DNA testing. Current DNA technology is primate-specific, utilizes minimal samples, and allows for identification of individuals. Conventional serology is now generally limited to the identification of tissues present in a sample. This limitation does not diminish the importance of serology because knowing which body tissue DNA came from can be critical to drawing correct conclusions (Figure 13.4). The more commonly encountered substances are blood, semen, saliva, and hair, but many other tissue samples are amenable to DNA testing.

Restriction fragment length **polymorphism** (**RFLP**) was the first method used to analyze forensic DNA samples. The method involved the extraction of DNA from the sample, restriction enzyme cutting of the DNA into smaller fragments, gel **electrophoresis** to separate the fragments by size, transfer of the DNA from a **gel** to a solid membrane, attachment of labeled DNA fragments called *probes*, and finally exposure of the membrane to a piece of film. The image captured on the film, known as a lumigraph or *autoradiograph*, represented the final results of the RFLP test. The disadvantages of the RFLP method are numerous. The DNA must be of sufficient quantity and quality to produce a result, and the process can take several months to complete.

As many forensic laboratories started implementing the RFLP method, the **polymerase chain reaction** (**PCR**) was beginning

to show promise as a method of analyzing genetic material in a variety of applications. PCR allows a DNA sample to be amplified in vitro by mimicking the process of in vivo DNA replication. PCR has substantial advantages over RFLP, particularly in analyzing samples with extremely low quantities of DNA or severely degraded DNA. Additionally, the method is much faster, primarily due to automation, allowing for more genetic locations within the DNA to be analyzed in less time.

PCR of short tandem repeats (STRs) is the DNA method now utilized by most forensic laboratories. After DNA has been extracted from a sample, it is amplified using PCR and a thermal cycler, an instrument designed to control temperature changes during repeating cycles. Each cycle has three steps: denaturing of the double stranded DNA, attachment of a fluorescently tagged primer DNA (for detection by a **laser**) to each parent DNA strand, and finally extension of the two new strands of DNA. Following the PCR, separation of the DNA fragments of interest is performed in one of two ways. The DNA may undergo gel electrophoresis or capillary electrophoresis. In either case, an instrument is used to carry out the electrophoresis. Attached to the instrument will be a computer with the necessary software used to analyze the results. The PCR and electrophoresis steps are now fully automated, which is one reason the process is faster. Obtaining a full genetic profile from a questioned DNA sample and matching that profile to a known DNA standard from a particular individual is now widely accepted as an identification of the donor.

Trace Evidence

In the early 1900s, Edmund Locard of Lyons, France, became convinced that traces of dust found on a person could provide evidence of where that person had been. Locard used a microscope to study dust particles and eventually was able to solve a number of crimes by identifying certain dust particles.[9] **Locard's Exchange Principle** basically states that whenever two objects come in contact with one another, a transfer of material will occur. The transfer may be tenuous, but it will certainly occur.[10] In other words, a suspect will leave something at and take something away from a crime scene.

Trace evidence, simply defined, is microscopic physical evidence. Obviously, this can encompass a large number of substances including body fluids, paint, glass, fibers, hair, soil, plant debris, and cosmetics. An abundance of microscopic evidence is likely to be present at every crime scene and is probably the most perishable type of physical evidence.

As part of a trace evidence examination, a criminalist must consider how the trace evidence was collected and preserved and what evidence should be the focus of the examination. Crime scene preservation is most critical to trace evidence since each person entering a crime scene has the potential to contaminate it. This includes the person who discovered the scene and others with legitimate first responder functions, such as first aid personnel and those securing the crime scene. The trace examiner must consider that fact when determining what items will be of value to the case. The ideal situation is for the trace evidence or item containing possible trace evidence to be collected and preserved as soon as possible.

Several methods are utilized in the collection of trace evidence. Vacuuming, hand picking, and tape lifts have all been used to collect fine particles. Tape lifts are probably most commonly used because they tend to collect surface traces or the most recently deposited trace evidence. Hand picking trace evidence from an item is more convenient within the laboratory and is generally avoided at crime scenes. Vacuuming may collect substantial quantities of trace evidence along with significant quantities of other debris and may actually serve to dilute the value of traces later deemed pertinent.

The workhorse of the trace evidence section is the microscope. Several microscopes will be utilized during the course of trace examinations. The use of the low magnification stereomicroscope will follow a visual, or "naked eye," examination. The stereomicroscope is invaluable in locating trace evidence on a variety of substances as well as investigating the nature

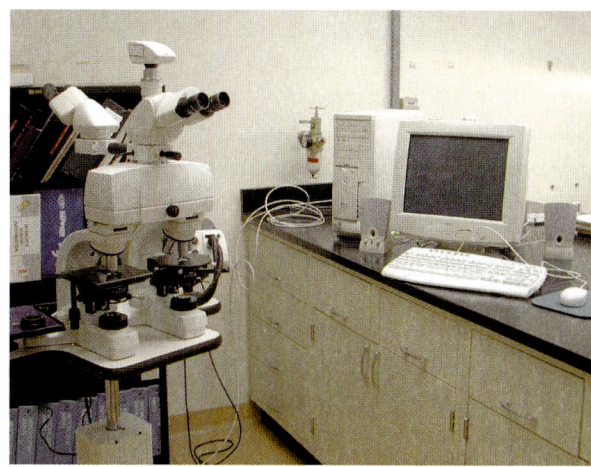

Figure 13.5 A comparison microscope is used to compare items of evidence simultaneously. A digital camera and computer are attached to this microscope. (Photo courtesy of Police Crime Laboratory, Kansas City, Missouri.)

of its deposition. Photographic documentation of this evidence is readily accomplished and is extremely valuable as a permanent record of the evidence. Although the stereomicroscope is not sufficient to identify the trace evidence, a variety of high magnification microscopes can further characterize the evidence. A compound light microscope, a polarizing light microscope, a phase contrast microscope, and a **comparison microscope** (Figure 13.5) are all potential tools of the trace criminalist.

The trace evidence criminalist may also utilize a variety of spectrometers. For example, Fourier transform infrared spectrometry (FTIR) may be used in **trace analysis**. This technique provides chemical fingerprints of many inorganic and organic substances. A microscope attached to an FTIR provides an even more powerful combination. Components of substances, such as adhesives, lubricants, paints, and cosmetics, may all be characterized by FTIR. Other ultrahigh power microscopes, such as the **scanning electron microscope** (**SEM**), may also be utilized.

Fingerprint Evidence

Two aspects of fingerprinting are generally associated with the forensic laboratory: **latent print development** and fingerprint identification. The development of latent fingerprints can be accomplished via chemical methods, physical methods (such as powders), and lighting and photographic methods. The method used depends on the latent print and the surface on which the print is located. For example, cyanoacrylate or Superglue™ fuming is commonly used on difficult surfaces such as plastics and firearms. Fluorescent powders may be used to enhance a latent print on a particularly colored surface or for use with an alternate light source. Alternate light sources and ultraviolet photography are used as nondestructive methods of latent print development.

Digital imaging is increasingly utilized in a variety of ways in forensic laboratories and is especially valuable to the latent print examiner. This technology allows for a developed print, partially obscured by background, to be further enhanced using a scanner and computer software. The print image is scanned into the computer and the software is used to remove background and, thus, clarify details within the print.

After a latent print is developed, it must be compared to known prints. A person trained in fingerprint identification performs this comparison. The latent print, often very limited in detail, is compared to known prints from inked impressions or from a computer database known as the Automated Fingerprint Identification System (AFIS). A number of "hits" may be obtained from AFIS, especially on poor quality latent prints. The examiner must then retrieve each set of known prints that were "hit" and perform a side-by-side comparison to the unknown latent print. Alternatively, if a suspect is known, his prints can be compared directly to the questioned print; in addition, an AFIS search may be performed.

Impression Evidence

Footwear and **tire impressions** are examples of impression evidence examined in the forensic laboratory. Impression evidence can provide substantial information, including an identification, depending upon the quality of the impression and the number of unique features present. For example, a tire impression may lead only to a list of possible tire brands or a specific model of tire. An identification of the source of the tire impression results when sufficient unique wear patterns in a questioned impression are also seen in an impression made by a known tire. The same can be said of footwear impressions. A newer shoe may lead only to identification of the **sole** pattern or brand of shoe, but an impression made by a well-worn shoe can potentially lead to identification of the shoe that made the impression.

Firearm and Tool Mark Evidence

This discipline in the laboratory is largely responsible for examining fired bullets, cartridge casings, and shot shells. A variety of markings or impressions are left on these items when a firearm is discharged, and the markings provide points of comparison to other ammunition fired from the same weapon.

A firearms examiner will utilize a stereomicroscope and comparison microscope during examination. When examining a bullet or casing from a crime scene, the stereomicroscope will provide general information about the item and possibly reveal trace evidence such as blood or fibers. The comparison microscope designed for firearms examinations is equipped with special stages to hold the evidence in a variety of positions and also provides an external light source.

The firing tank, which is a large, sturdy container of water, is used to collect fired bullets. The fired bullets can then be compared to bullets from the crime scene (Figure 13.6). Conclusions regarding this comparison depend on the amount and quality of the striae and other markings present on the unknown bullet. Another device used by the firearms examiner is a firing target that allows study of gunpowder or pellet patterns produced at various distances from the target. Distance determinations are commonly necessary when a person claims that a shooting was in self-defense and occurred during a struggle. Identification of residue pattern and gunshot residue by chemical tests can also assist the firearms examiner in a shooting reconstruction.

Impressions made by tools can also include **striation**-type markings. Tool marks are commonly examined in the firearms section.

Figure 13.6 Photomicrograph of a match between a known bullet fired in the laboratory using the suspect's gun and a bullet recovered from the crime scene. (Photo courtesy of Police Crime Laboratory, Kansas City, Missouri.)

A burglar who uses a screwdriver, bolt cutter, or other common tool may leave impressions of the tool at the crime scene. The item containing an impression or a cast of the impression is collected and when the suspected tool is located, an impression is made by the tool and compared to the crime scene evidence. The burglar's screwdriver is a common example of tool mark evidence, but any tool can create impressions on a variety of surfaces. Knife wounds to flesh and bones, and cutting tools used to cut wires, cords, and bindings are only a few examples of other tool marks that may be of value.

Comparison of physical fractures found in particular items of evidence and restoration of stamped serial numbers are two other functions performed by firearms examiners. A piece of headlight glass left at the scene of a hit and run may fit, much like a puzzle piece, back into the suspect vehicle's broken headlamp. Fracture matches have been made on a variety of materials such as paint chips and glass. When grinding obscures a serial number stamped into metal, restoration is necessary. Common surfaces with stamped serial numbers needing restoration are guns and automobile engines.

Questioned Documents

A variety of examinations can be performed in the analysis of questioned documents. The more common examinations include handwriting comparisons, alterations, **obliterations**, and erasures. Characterization of inks may also provide valuable information. The documents examiner is responsible for determining the authenticity of writing by comparing writing on a questioned document to writing standards or **exemplars** taken from possible authors. When a document such as a check contains an alteration, the document examiner attempts to identify the alteration and the original contents of the document. Determination of alterations, obliterations, and erasures may involve a number of tests, including microscopy, spectral analysis, and chromatography.

Document examination is not limited to handwritten documents. It includes typewritten and photocopied documents. Specific marks, sometimes created by defects in the device creating a document, can be examined and compared. Also, the paper may provide key information in a document's examination through its composition and/or **watermarks**.

Chemical Evidence

The chemistry section of a forensic laboratory focuses on but is not limited to identification of illegal drugs. This section may also be responsible for arson and explosives analysis, blood alcohol determinations, poison testing, and any number of examinations requiring the identification of chemical compounds. Some laboratory chemistry sections may perform limited numbers of toxicology tests, such as the analysis of urine and blood for drugs of abuse or date rape drugs.

Many chemical analyses, especially for the identification of illegal drugs, begin with a **presumptive test** commonly referred to as a "spot" test—generally a simple color test which, if positive, indicates the possible presence of a particular substance. Because spot tests are presumptive, additional tests must be performed to confirm the identity of the substance. When necessary, the purity or concentration of the illicit drug can also be determined.

An examiner in a chemistry section will have a great deal of experience in instrumental analysis. The bulk of chemistry evidence will be identified using the combination of gas chromatography and mass spectrometry (GCMS). GCMS analysis will first separate a compound into its individual components after which the mass spectrum of each component is determined. Attached to the GCMS is a computer with libraries containing spectra of thousands of compounds. A comparison of the evidence to these libraries can aid in the identification of unknown compounds.

Other instruments and methods utilized by the chemistry section include FTIR, ultraviolet spectrophotometry, scanning electron microscopy, x-rays, and high performance liquid chromatography. Additionally, the stereomicroscope and polarizing microscope can also be used extensively in examining chemistry evidence.

Photographic Evidence

The value of photographic evidence cannot be overstated. Every crime scene must be clearly documented by photographs in addition to notes and diagrams. The photographs are necessary for documentation, but are also extremely important for reconstruction of the crime and prosecution. Photographs are usually routinely taken during evidence examination by the criminalist. These photographs can be used to document patterns or other important aspects of the evidence and are very helpful to the criminalist in preparing for and presenting trial testimony.

The photography section of a forensic laboratory may be required to perform a number of duties. Processing of film, enlarging photographs, taking ultraviolet or infrared photographs, photography of Luminol reactions, and utilizing a variety of camera formats to obtain the best possible image are some of these duties. Although the crime scene investigator is responsible for taking most crime scene photographs, many laboratories also utilize professional forensic photographers in criminal investigations.

Digital imaging has expanded the capabilities and responsibilities of the photography section. Digital enhancement can be valuable to the criminalist for documenting

and also clarifying images. Digital imaging has not entirely replaced film photography, but provides many advantages. The process is faster and user friendly, and images are easily stored, transmitted, and preserved.

Case Studies

Case 1

In April of 1999, a 67-year-old female was found dead at the foot of a bed in an upstairs bedroom of her home (Figure 13.7). Her body was covered from head to ankles with a comforter taken from the bed. The cause of death was later determined to be strangulation. Autopsy also revealed that the victim had been raped and sodomized.

Several important items of evidence were noted on a path leading from the front door to the victim's bedroom. The exterior screen and

Figure 13.7 Body of 67-year-old homicide victim. (Photo courtesy of Police Crime Laboratory, Kansas City, Missouri.)

Figure 13.8 Point of forced entry into victim's home. (Photo courtesy of Police Crime Laboratory, Kansas City, Missouri.)

Figure 13.9 Contact bloodstains on stairs leading to victim's bedroom. (Photo courtesy of Police Crime Laboratory, Kansas City, Missouri.)

interior front doors of the house were damaged when the assailant entered (Figure 13.8). An impact blood spatter pattern was observed at the base of the stairway leading to the upstairs hall and bedroom. Transfer bloodstains were observed on the stairs, walls, and landing at the top of the stairs (Figure 13.9). A large clump of strongly pulled, typically Caucasian head hair was found on one stair. More contact bloodstains and vomit were found in the upstairs hall outside the bedroom.

After collecting the comforter that covered the victim's body, it was discovered that the victim was wearing only a nightgown and most of her head was tightly wrapped in duct tape. Her nightgown was covered in contact bloodstains. A roll of unused duct tape found in the room also had contact bloodstains on it. Fluid was observed in the victim's vaginal area and on her thighs. This fluid was later identified as semen, which was also identified on the carpet directly beneath the victim's buttocks. A number of possible foreign hairs

were also observed in the victim's pubic hair, on her nightgown, and on the carpet beneath her body. Apparently foreign hairs were also collected from a bathroom adjacent to the victim's bedroom. No probative fingerprints were located.

An examination was performed on the hairs collected from the body and crime scene. The results of the examination identified the hairs as foreign to the victim and having Negroid characteristics. Seven individuals from the surrounding neighborhood voluntarily provided head and pubic hair standards for comparison. Within days of the homicide, each of the seven individuals was excluded by microscopic comparison of his standard to the foreign hairs.

Within several weeks of the homicide, a DNA profile of the semen donor was entered into the Combined DNA Index System (CODIS). CODIS is a national database of DNA profiles from forensic unknowns and convicted offenders. The profile obtained from the semen did not match others at the local and state levels after the initial CODIS comparison. The semen evidence, because of its location on the carpeting beneath the victim and in conjunction with the hair evidence, was determined to be generated contemporaneously with the rape and homicide. The unknown DNA profile continued to be routinely searched in CODIS.

Several weeks after the homicide, a man was arrested in a nearby city after a routine traffic violation. He was detained when it was determined that a warrant existed for his arrest. He had been convicted of a sodomy offense 2 years earlier and was released on bond pending the outcome of his appeal. The appeals court upheld the trial court's conviction, but the suspect was not apprehended after the appellate ruling. The April 1999 homicide occurred while he was free.

After the man was sent to prison for the sodomy conviction, his blood was collected for routine DNA analysis and submission to the convicted offender database. After testing and input into the database, a CODIS match was made to the profile of the semen found on the victim's body. The convict became a suspect in the rape and homicide of the 67-year-old woman. A microscopic comparison of his head and pubic hairs to the crime scene evidence hairs revealed they had similar characteristics. The suspect had connections to the victim's neighborhood, but was a stranger to the victim.

The large clump of pulled head hair similar to the victim's hair played an interesting role in the case. The hairs were microscopically matched to the head hair standard of the victim. The presence of dozens of hairs, all covered in blood and strongly pulled, led to an additional charge of kidnapping.

Reconstruction of the crime scene including evidence of forced entry, blood and pulled hair on the stairway, blood and vomit in the hall outside the bedroom, and severe visible injuries to the victim indicated that a struggle occurred. The defense chose to argue the value of the identification provided by the DNA evidence, rather than argue that the defendant had consensual relations with the victim, but was not involved in the homicide. The jury convicted on all counts, including first degree murder, armed criminal action, burglary, kidnapping, sodomy, and rape. The defendant was sentenced to several life sentences without the possibility of parole.

Case 2

A woman visiting a large metropolitan city for a professional conference was attacked at the

Figure 13.10 Hotel room where victim was raped. (Photo courtesy of Police Crime Laboratory, Kansas City, Missouri.)

door of her 17th floor hotel room (Figure 13.10), struck on the head, and forced into her room. Once inside the room, the assailants bound her hands and began searching the room for property. The victim overheard an assailant say they only had one condom. She was then raped by both assailants and sodomized by one. After the sexual assaults, the suspects washed the victim with several towels and washcloths and wiped the surfaces of the room. They bound her legs with cut-up bed sheets and covered her with the mattress of the bed. Before leaving, they stole her cellular phone and a couple hundred dollars in cash from her purse.

The victim was able to free herself and immediately ran to an adjacent room to dial 911. The police responded and took her to a hospital for medical attention and a rape examination. Her clothing and evidence of the sexual assault were collected from her body.

Numerous items of value were collected from the crime scene. The sheets used as bindings, the towels used to wash the victim, a condom wrapper, the victim's purse, and several fingerprint lifts were collected. The used condom was found floating on the surface of the toilet bowl water (Figure 13.11). Transfer bloodstains were observed on the bindings, particularly in the areas where knots had been tied. When the knots were untied, foreign hair and a piece of chewed gum were located. Blood was also found on the victim's

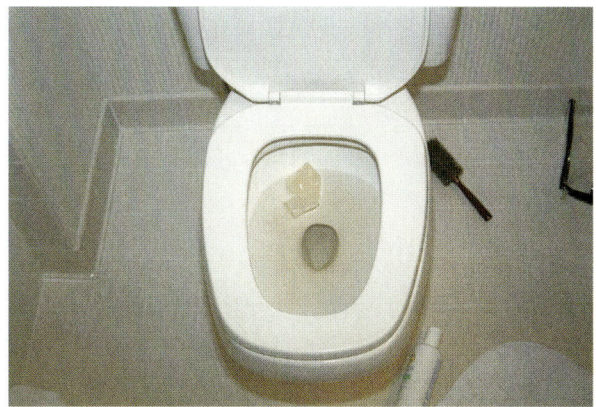

Figure 13.11 Condom found in hotel room toilet. (Photo courtesy of Police Crime Laboratory, Kansas City, Missouri.)

purse. No semen was located on the towels or sheets. Additional foreign hairs were found. A partial fingerprint was developed on the condom wrapper. Examination of the rape kit revealed the presence of semen on the rectal samples only. Also, condom trace evidence was found on the vaginal samples but not on the rectal samples.

The chewed gum, rectal samples, and blood lifted from several items resulted in a DNA profile of one male. Several attempts to detect a DNA profile on the condom produced the victim's profile and a partial profile of the second assailant. The fingerprint evidence matched the suspect whose DNA was found on the condom. Questioned hair was also consistent to the second assailant. All this evidence was crucial to the prosecution because the victim was unable to identify her attackers.

One suspect was quickly apprehended after he used the victim's cellular phone to make several calls before he sold the phone. The person who bought the phone contacted police after he realized that the phone might have been stolen from the rape victim in this highly publicized case because of the out-of-state area code on the phone. The fingerprint on the condom wrapper did not produce an AFIS match and neither DNA profile resulted in a CODIS match. The tracing of the cellular phone calls and the subsequent alerting of the police by the phone purchaser resulted in the apprehension of the suspects.

The physical evidence identifying the suspects was overwhelming but both suspects chose jury trials. The law precluded trying the defendants together so all of the evidence, including the victim's testimony, had to be presented twice. Both trials resulted in convictions on burglary, kidnapping, forcible rape, forcible sodomy, and sexual abuse charges. The defendants were given multiple life sentences.

Acknowledgment

Appreciation is extended to John Wilson, Quality Assurance Manager (retired) of the Kansas City (Missouri) Crime Laboratory, for his assistance with this chapter.

Questions

1. Describe the role of the forensic laboratory with respect to the recognition, collection, and preservation of physical evidence.

2. Describe the key components of a quality assurance program.

3. Describe the purpose of positive and negative controls and calibration in the crime laboratory.

4. Define proficiency tests as they relate to the forensic laboratory.

5. Define *criminalist*.

6. Describe the role of the criminalist.

7. What is the difference between laboratory accreditation and certification?

8. Define reconstruction evidence and describe what is involved in reconstructing a crime.

9. Define and provide one example each of class characteristic evidence and identification evidence.

10. List three areas of emphasis within a forensic laboratory.

11. Describe why attending crime scenes is valuable experience for a criminalist.

12. What are the criteria for the significance of analytical results utilized by the experienced criminalist?

13. What two items should be included in a criminalist's report of analysis?

14. When considering the role of the modern criminalist, how has his role changed from the historical role and where are the true differences in his approach to case analysis?

References and Suggested Readings

1. Kirk, P. L., Introduction, in *Crime Investigation*, 2nd ed., John Wiley and Sons, New York, 1974.
2. O'Hara, C. E. and Osterburg, J. W., Preface, in *An Introduction to Criminalistics*, McMillan, New York, 1949.
3. Technical Working Group on DNA Analysis Management, Guidelines for a quality assurance program for DNA analysis, *Crime Lab. Digest*, 22, 21, 1995.
4. National Research Council, *The Evaluation of Forensic DNA Evidence*, National Academy Press, Washington, D.C., 1996.
5. DNA Advisory Board, *Quality Assurance Standards for Forensic DNA Testing Laboratories*, U.S. Department of Justice, July 1998.
6. American Society of Crime Laboratory Directors, Laboratory Accreditation Board, *ASCLD/LAB Accreditation Manual*, 1997.
7. Adam, J. and Adam, J. C., *Criminal Investigation: A Practical Textbook for Magistrates, Police Officers, and Lawyers*, Sweet and Maxwell, London, 1934, p. 102.
8. Wittaker, E., The adversary system: role of the criminalist, *J. Forens. Sci.*, 18, 184, 1973.
9. Thorwald, J., *Crime and Science*, first American edition, Harcourt, Brace and World, New York, 1966, p. 280.
10. Palenik, S., Microscopy and microchemistry of physical evidence, in *Forensic Science Handbook*, Saferstein, R., Ed., Prentice Hall, Englewood Cliffs, NJ, 1988, chap. 4.

The Identification and Characterization of Blood and Bloodstains

Robert P. Spalding

Introduction

In recent years, a profusion of television programs devoted to the various fields of forensic science and the widespread reporting of a number of criminal trials have increased both public awareness of and interest in this area of criminal investigation. Technologies such as DNA analysis have captured our attention for both the uniqueness involved and the certainty with which an individual may be identified. A newcomer to the scene might even ask what we did prior to the arrival of DNA. The answer? Forensic serology.

A study of serology involves the examination and analysis of body fluids and, among those fluids, blood. In the medical setting, blood is analyzed to assess one's state of health. In the forensic realm, blood might be analyzed to determine its source at a crime scene or on an item of evidence. The difference doesn't end here. A clinical serologist typically deals with samples that are fresh, normally liquid, and usually recently acquired from a source individual. Forensic serology deals not only with a variety of body fluids (blood, saliva, semen, and urine), but, and more frequently, with samples that are in stain form and often degraded or deteriorated, often making successful analysis more difficult.

The quantity and condition (degree of degradation or putrefaction) of evidentiary stains may depend on a number of factors, many of which are not within the power of the forensic analyst to control. Sample quantity and condition often dictate the manner or strategy adopted by an examiner to deal with a specific piece of evidence or, indeed, whether analysis can be done at all. Requirements of the legal system require protection of the evidence to enable its introduction in court. This integrity of the evidence itself may easily be compromised by failure to maintain continuous control over it or by allowing degradation to occur under inadequate storage conditions. Uncontrolled exposure of evidence to heat and humidity can, in a biochemical sense, destroy much of the information contained in a stain by enhancing the breakdown of the chemical substances of importance to the analyst. This can occur before or after receipt of the evidence by a laboratory.

The popularity of DNA analysis in criminal investigations today would reasonably lead one to wonder why a chapter such as this is of any value. Forensic serology was the predominant field dealing with blood and body fluids in forensic laboratories worldwide from the 1950s until the late 1980s when the analysis of evidentiary material for deoxyribonucleic acid (DNA) became a reality. A rapid growth in the number of laboratories employing DNA methods has resulted in laboratories, both law enforcement and private, such that it is now available to most investigations. With this increase in DNA analysis, there has been a decline in the number of laboratories using the standard serological procedures. Indeed, many laboratories without DNA capabilities perform the basics and forward samples to DNA laboratories to have additional analysis completed. Even with this shift, however, it can be seen that the most basic serological procedures are used as screening techniques to locate/identify material for analysis, even in those laboratories employing DNA analysis.

That a review of basic serology is of importance actually comes from the use of DNA analysis. The reliability and discriminatory power contained in DNA has led to the reanalysis of a number of cases in which the innocence of a convicted defendant has been established and the individual released. The superior power of DNA in individualization has been able to greatly extend the biochemical information provided by serology. This is not to say that the original serological work was wrong or performed with any fraudulent purpose, but simply says that employing of DNA can provide more specific information, far exceeding that of serology. In situations such as these, the evaluation of the original serological data is hopefully made easier with a reference such as this text. From the foregoing, it should be evident that a knowledge of serology or at least an availability of information on the subject is important to thorough forensic study. It is the intent of this text to review and discuss the more commonly used techniques of forensic serology regarding blood, provide sources of information, and present newer information.

Analysis of Blood in Forensic Serology

Proper scientific approach and legal requirements dictate that the identity of a stain as blood be more than a simple visual observation and that such an identification be established to a scientific certainty before it can be presented in court. Further, a disciplined scientific approach dictates the use of a routine protocol incorporating a logical series of testing procedures to provide the basis for such an identification. Use of a protocol is intended both to prevent inadvertent elimination of a step in the process and ensure similar treatment of all items of evidence under consideration. An example of such a protocol is presented below. An important point to note, however, is that variation in evidence with regard to type, condition, and quantity requires flexibility in its application. Therefore, it is possible that one or more steps may be eliminated based on careful, scientifically based examination strategy as a foundation. Such a move would be carefully documented as to what and why any changes were made. The use of such a protocol, or a variation of it, can be applied to any of the body fluids most frequently encountered in forensic serology. It would incorporate these procedures:

1. Careful visual examination of the item of evidence to locate any stains or material visibly characteristic of blood
2. Application of a suitable presumptive screening test
3. Application of a specific and sensitive test to confirm blood presence
4. Determination of biological or species (animal or human) origin
5. Characterization of the blood using one or more genetic markers or DNA

Identification of Blood

Identification of blood using the above approach has often employed a presumptive chemical-screening test frequently followed by a confirmatory test to clearly establish the identification. A visual observation of the untested stain, coupled with positive chemical presumptive and confirmatory tests, then provides sound data to support the identification of blood. Some of the more common tests used for this are discussed below.

Presumptive Tests for Blood

A presumptive test is one that, when positive, would lead the forensic examiner to strongly suspect blood is present in the tested sample. When negative, the test often helps to eliminate stains that need no further consideration. In the event of a positive test, when sufficient sample remains, further action to confirm the presence of blood is usually taken, since no single test is absolutely specific for blood. Presumptive tests may be recognized as those that produce a visible color reaction or those that result in a release of light. Both types rely on the catalytic properties of blood to drive the reaction.

Catalytic Color Tests

Catalytic tests employ the chemical oxidation of a chromogenic substance by an oxidizing agent catalyzed by the presence of blood, or more specifically, the hemoglobin or red pigment in the blood. Those tests that produce color reactions are usually carried out by first applying a solution of the chromogen to a sample of the suspected material/stain followed by addition of the oxidizing agent (often hydrogen peroxide in a 3 percent solution). The catalyst is actually the peroxidase-like activity of the heme group of hemoglobin present in the red blood cells (erythrocytes). A rapidly developing color, characteristic of the chromogen used, constitutes a positive test. Some methods employ the chromogen and the oxidant in a single solution. There is a potential disadvantage in this, however, as the order of addition of the specific reagents can be important. A nonblood sample capable of producing a color reaction will normally do so without the addition of the oxidizing agent so that when the color reagent is added first, there is a reaction (without the addition of the oxidant). With a single solution, a reaction due to a nonblood material might be incorrectly interpreted. Twenty or more substances have been investigated over the years as chromogens, and those discussed below seem to have been used more than the rest.

It is important to recognize that these tests are presumptive, that is, they indicate, but are not absolutely specific for, the identification of blood. Further, any discussion of presumptive tests would be incomplete without mention of false positive reactions. Sutton (1999) has correctly pointed out that a false positive is an apparent positive test result obtained with a substance other than blood. This eliminates many results that consist of an uncharacteristic color when nonblood materials are tested. Accordingly, it is possible that some of the tests discussed below may display some reaction when testing nonblood samples, but experience, careful observation, and routine application of confirmatory testing will usually prevent errant identification of a nonblood sample as blood.

Misleading results usually can be attributed to chemical oxidants (often producing a reaction before the application of the peroxide), plant materials (vegetable peroxidases are thermolabile and can be destroyed with careful heating) or materials of animal origin (to include human) which are not blood but may contain contaminating traces of blood. Microscopic examination may give insight as to the true nature of the material. Sutton (1999) presented the results of testing a wide variety of nonblood materials with several different presumptive tests.

Despite all the work done on false positive reactions, there is surprisingly little information available concerning false negative reactions. Recognizing this, Ponce (1999) studied the effect of reducing agents on presumptive tests, using a representative reducing agent (ascorbic acid) to contaminate bloodstains tested with *o*-tolidine, phenolphthalein, leucomalachite green, and tetramethylbenzidine. They demonstrated that the ascorbic acid did, in fact, result in false negative results, although there was variation in the degree to which the individual tests were affected.

A common method of applying the presumptive tests involves sampling a questioned stain with a clean, moistened cotton swab, and adding a drop of the color reagent solution followed by a similar amount of hydrogen peroxide. With this procedure, the immediate development of the color typical of the particular reagent used indicates the presence of blood in the test sample. Alternatively, the evidence could be sampled by removal of a thread or fragment of dried material and testing it with the above reagents in a spot plate. Color development would then be observed in the spot plate as well. Immediate (within a few seconds) reading and recording of results is an important aspect of test result interpretation. A clearly negative result may appear positive several minutes after the test is completed due to a slow oxidation that often occurs in air. These tests are not usually affected by stain age, and it has been the author's experience that benzidine, phenolphthalein, and tetramethylbenzidine have given positive results with blood crusts or stains up to 56 years old.

Benzidine (Adler Test)

Benzidine has been used probably more extensively than any other single test for the presumptive identification of blood. Much of the earlier forensic work dealing with benzidine was done by the Adlers (1904). The reaction (Figure 14.1), normally carried out in an ethanol/acetic acid solution, results in

a characteristic blue to dark blue color. The blue, in turn, may eventually turn to a brown. Benzidine, however, is recognized as a carcinogen as documented by the Occupational Safety and Health Administration (*Federal Register*, 1974) and is seldom used in forensic laboratories today. Its inclusion here is mainly of historical interest. The test has largely been abandoned in favor of safer color reagents.

Phenolphthalein (Kastle–Meyer Test)

A test procedure commonly used in many forensic laboratories today involves the simple acid-base indicator phenolphthalein. Investigated by Kastle (1901, 1906), phenolphthalein produces a bright pink color when used as above in testing suspected blood. The reagent consists of reduced phenolphthalein (phenolphthalin) in alkaline solution, which is oxidized by peroxide in the presence of hemoglobin in blood. The reaction shows phenolphthalin (colorless in alkaline solution) being oxidized to phenolphthalein (pink in an alkaline environment) (Figure 14.2). Normally the reagent preparation involves reducing the phenolphthalein over zinc in potassium hydroxide solution to phenolphthalin. As with any of the catalytic tests the result is read immediately and a positive result a minute or more after the test is performed is usually not considered reliable. False positives with phenolphthalein usually are not really positives, in that the reaction is often not the characteristic pink but usually some other color change.

Figure 14.3 shows both positive and negative results with phenolphthalein.

o-Tolidine

o-Tolidine, or *ortho*-tolidine, is the 3,3–dimethyl derivative of benzidine (Figure 14.4A). The reaction, similar to that of benzidine, is

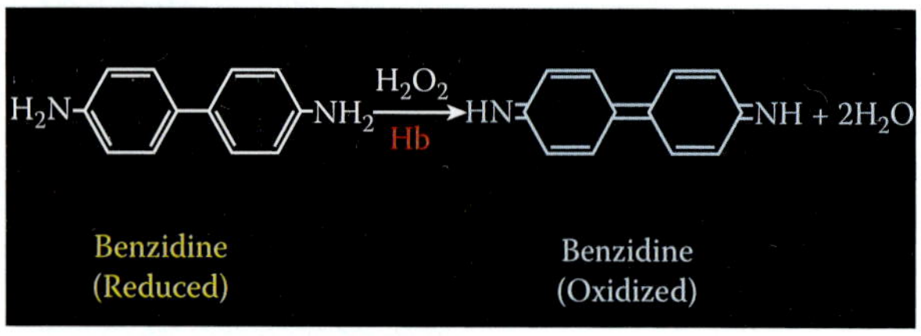

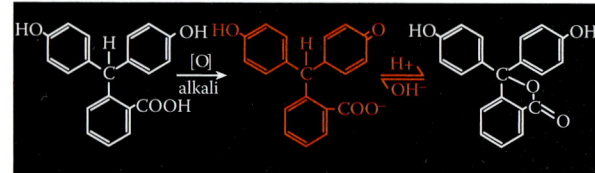

Figure 14.1 Benzidine oxidation.

Figure 14.2 Phenolphthalin oxidation. Phenolphthalin (left, colorless) is catalyzed to phenolphthalein (pink in alkaline solution, center), which is colorless in acid (right).

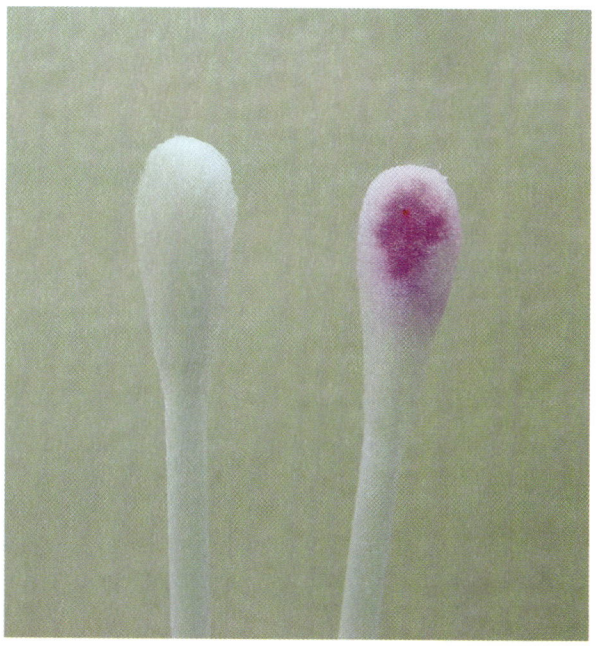

Figure 14.3 Negative and positive phenolphthalein reactions.

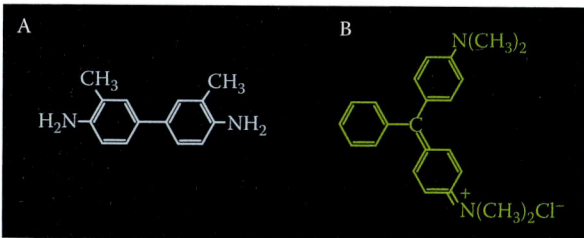

Figure 14.4 (A) *o*-Tolidine. (B) Malachite green.

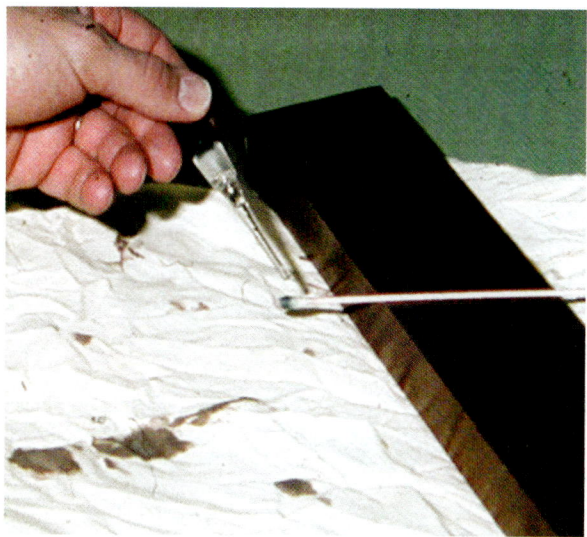

Figure 14.5 *o*-Tolidine positive reaction.

conducted under acidic conditions and produces a blue color reaction resembling that of benzidine when testing blood. Earlier workers proposed it as a replacement for benzidine (Hunt, 1960). However, the compound was reported to be carcinogenic in rats by Holland (1974), and evidence exists (Bell et al., 1980) that *o*-tolidine-based dyes may be metabolized to the parent compound in humans. Figure 14.5 shows a positive reaction with *o*-tolidine obtained when testing blood.

In 1981, *o*-tolidine was the active ingredient in the Hemastix® Test (Miles Laboratories), a diagnostic test used in a medical context for the detection of blood in urine. The test with was used with positive results on the fabric of a wounded Canadian Militia Officer's uniform coat from the War of 1812 (Macey, 1979). During the 1980s, interest continued to develop in the test for forensic work; however, the carcinogenicity of *o*-tolidine eventually resulted in its being replaced by tetramethylbenzidine (TMB) by 1992.

Leucomalachite Green (LMG)

Another compound investigated by the Adlers (1904) was the reduced or colorless form (leuco) of the dye malachite green (Figure 14.4B). Often referred to as McPhail's Reagent (Hemident®), LMG oxidation is catalyzed by heme to produce a green color. Like benzidine and *o*-tolidine, the reaction is usually carried out in an acid (acetic acid) medium with hydrogen peroxide as the oxidizer, although there are procedures that employ sodium perborate with the LMG in a single solution (RCMP, 2001). Figure 14.6 displays positive and negative LMG reactions. An interesting point is that the Adlers (1904) preferred the benzidine and LMG tests over the 30-plus reagents they tested.

Tetramethylbenzidine (TMB)

With the recognition of benzidine as a carcinogen (*Federal Register*, 1974), the search for replacement substances was under way. A number of laboratories changed to phenolphthalein as a primary presumptive reagent, but interest also centered on tetramethylbenzidine, synthesized by Holland (1974). TMB is the 3,3–,5,5–tetramethyl derivative of benzidine (Figure 14.7). It had been observed that

Figure 14.6 Negative and positive leucomalachite green reactions.

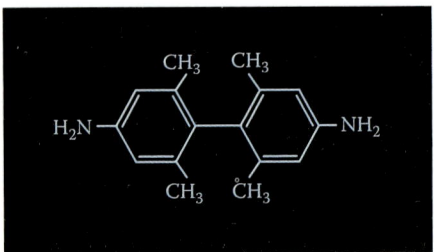

Figure 14.7 Tetramethylbenzidine.

the 3,3-dihydroxybenzidine metabolite of benzidine was more carcinogenic than benzidine itself. It was thought that *ortho*-methylation of benzidine would prevent the in vivo *ortho*-hydroxylation. Holland did, in fact, observe that the methylated benzidine gave a negligible yield of tumors in rats as compared with benzidine and *o*-tolidine. As with benzidine, *o*-tolidine, and malachite green, TMB is used in an acid medium (acetic acid) when employed as a solution, and the resultant color is green to blue-green.

The Hemastix® Test (Miles Laboratories, 1992) mentioned above has been adopted for field use by a number of laboratories, particularly at crime scenes when containers of

Figure 14.8 Hemastix® negative and positive reactions.

solutions can be hazardous or, at best, inconvenient. The test itself consists of a plastic strip with a reagent treated filter paper tab at one end. The tab contains TMB, diisopropylbenzene dihydroperoxide, buffering materials, and nonreactants. Testing a bloodstain may be accomplished by moistening a cotton swab with distilled water, sampling the stain, and touching the swab sample to the reagent tab on the strip. The reagent tab is originally yellow, and a normally immediate color change to green or blue-green indicates the presence of blood (Figure 14.8). The test kit (Figure 14.9) is often available in drug stores and from some forensic suppliers, and only swabs and distilled water are required. Experience has shown, however, that swabs pretreated with cosmetic substances may contribute false positive results.

Comparing the Tests

A reasonable search of the literature will show that more than a single method of preparation exists in the literature for each of the presumptive test reagents discussed above. Various protocols for applying these tests can also be found (Sutton, 1999; Lee, 1982). Accordingly, varying sensitivities and specificities are reported based on various authors' different experimental designs. As a means of providing some sense of how the various tests

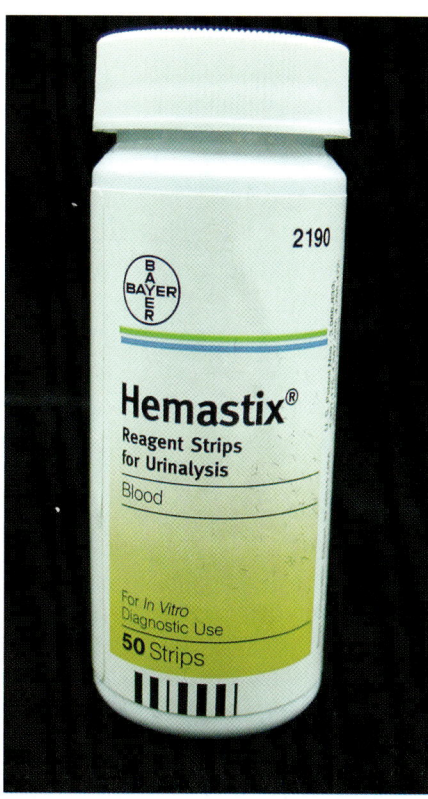

Figure 14.9 Hemastix® test kit.

Tests Using Chemiluminescence and Fluorescence

Often the presence of blood is suspected, based on witness information or expected in a particular location, but under normal lighting and viewing little is to be seen. A drag pattern across a floor that has been cleaned up or a washed spatter pattern on a wall might be typical examples. At this point the luminol and fluorescein tests may come into play. These tests are presumptive and do not identify blood. They involve spraying a chemical mixture on a suspected bloodstained area, usually *in situ*, and observing (sketching, photographing, etc.) the result, either in darkness or in reduced light with the aid of an alternate light source (ALS). The observed result is a production of light which often enables the observer to determine the limits, shape, and some degree of detail in the original bloodstained area, often including an enhancement of blood patterns already present.

The presence of blood and patterns displayed by blood can provide information of value, and one should be alert to both. Further, the nature of these tests makes them potential sources of contamination of the blood. Thus, if the stain can be seen and collected, these tests probably should not be used. Since these tests do add material to the tested area, their use should be carefully considered, often as a last resort. These tests are of more value in locating and defining blood than in specifically identifying it.

compare on a qualitative level, the results of several authors have been reviewed and are presented in Table 14.1.

Differences in specificity and sensitivity become of less consequence when it is realized that subsequent confirmatory testing and further characterization of the stains will usually disclose any irregularities in the earlier stages of testing.

Table 14.1
Comparative Sensitivities and Specificities of Presumptive Test Reagents

Author	Tests Examined/Discussed	Most Sensitive	Most Specific	Overall Most Suitable
Hunt et. al. (1960)	Benzidine, o-tolidine, LMG, phenolphthalein, Luminol	Not stated	Not stated	Benzidine or *o*-tolidine
Kirk (1963)	Benzidine, phenolphthalein, LMG, Luminol	Luminol	Phenolphthalein	Not stated
Higaki and Philip (1976)	Phenolphthalein, benzidine	Benzidine	Phenolphthalein	Phenolphthalein
Garner et al. (1976)	Benzidine, TMB	Equal	Equal	TMB (due to carcinogenic nature of benzidine)
Cox (1991)	Phenolphthalein, TMB, LMG, o-tolidine	*O*-tolidine and TMB	Phenolphthalein and LMG	Phenolphthalein

A word should be said about these substances regarding safety. Both luminol and fluorescein are classified as irritants. Material Safety Data Sheets (**MSDS**) (Mallinckrodt Baker, 1999; Fisher Scientific, 2007; Fluka Chemical Corporation, 2001) toxicological data do not indicate the substances to be carcinogenic at this time. However, while neither compound is classified as a carcinogen, safety recommendations are described for each and they should not be treated carelessly.

Luminol and fluorescein each produce light but in different ways. It should be understood that the terms *chemiluminescence* and *fluorescence* are two entirely different phenomena that result from the chemical mechanisms for each of these tests. Chemiluminescence is the process by which light is emitted as a product of a chemical reaction. No additional light is required for the reaction to take place. Luminol relies on this process. Fluorescence occurs when a chemical substance is exposed to a particular wavelength of light (usually short wavelengths, such as ultraviolet) and light energy is emitted at longer wavelengths. Light produced by fluorescein is a result of this irradiation, usually by ultraviolet light in the range of 425 to 485 nanometers.

Luminol

Since the early 1900s (Curtius and Semper, 1913; Huntress et al., 1934) it has been known that luminol, or 3-aminophthalhydrazide (Figure 14.10A), would luminesce after oxidation in acid or alkaline solution. It was found that various oxidizing agents could cause the luminescence in alkaline solution and positive results were observed with old as well as fresh bloodstains. It became apparent that the compound, which came to be called "luminol," could be useful in forensic investigations involving blood, and the years have confirmed its suitability in this regard. Luminol reacts in a fashion similar to the color tests discussed above wherein luminol and an oxidizer are applied to a bloodstain. The catalytic activity of the heme group then accelerates the oxidation of the luminol, producing a blue-white to yellowish green light (depending on the reagent preparation) where blood is present.

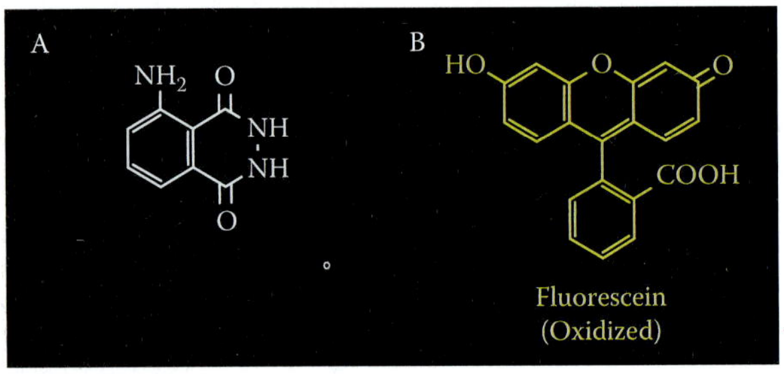

The forensic application of luminol at crime scenes involves spraying a mixture of luminol and a suitable oxidant in aqueous solution over the area thought to have traces of blood present. A resultant blue-white to yellow-green glow will indicate the presence of blood. Outlines and details are often visible for up to 30 seconds before additional spraying is required. Excessive sprayings will usually result in stain pattern diffusion.

The effects of luminol on stains destined for future analysis has been a concern. The more contemporary studies of Laux (1991) and Grispino (1990) found that pretreatment of bloodstains with luminol failed to significantly affect the presumptive, confirmatory, species origin, and ABO tests but interfered with several of the enzyme and protein genetic marker systems, notably erythrocyte acid phosphatase (EAP), esterase D (EsD), peptidase A (PEP A), and adenylate kinase (AK). Even more recently, Gross et al. (1999) determined that the effects of pretreating bloodstains with luminol did not adversely affect polymerase chain reaction (PCR) analysis of DNA, and Budowle et. al. (2000), as well as Jakovich (2007), reported that both luminol and fluorescein do not interfere with STR (short tandem repeat) analysis. Hochmeister et al. (1991) found that in bloodstains pretreated with ethanolic benzidine, phenolphthalein, or luminol, the recovery of high molecular weight DNA for restriction fragment length polymorphism (RFLP) analysis was unaffected with the exception of the phenolphthalein where the yield was only slightly lower.

Perhaps the greatest advantage of luminol is its sensitivity. Proescher and Moody (1939) found that luminol would detect hematin from

Figure 14.10 (A) Luminol (3-aminophthalhydrazide). (B) Fluorescein.

hemoglobin in dilutions up to 1 in 10,000,000, and Kirk (1963) reported that blood in dilutions of 1 in 5,000,000 could be detected. Luminol is more than capable of detecting blood not present in sufficient amounts to be seen with the naked eye. Indeed, Bily and Maldonado (Bily and Maldonada, 2006) have reported visualization of bloodstains under eight layers of paint with luminol. Additionally, it is important to record observed luminol reactions with photography (Zweidinger, 1973; Lytle, 1978; Gimeno, 1989a, 1989b; Young, 2006).

A new variation of luminol has recently become available in the form of Bluestar®. This reformulated reagent is reported to be easy to use, to present minimum health risks, and not to exhibit any harmful effects on STR DNA analysis (Blum et. al., 2006; Jakovich, 2007). Other workers (Young, 2006 and Dilbeck, 2006) have examined Bluestar® and consider it to perform as well as or better than traditional recipes for luminol.

Fluorescein

The use of fluorescein (Figure 14.10B) to detect blood was recognized as early as 1910 by Fleig. It appears that Cheeseman (1995, 1999, 2001) was the first to have considered the fluorescin/fluorescein system for larger area application at crime scenes. As with luminol, careful consideration of the ultimate aim of the investigation should guide fluorescein application at a scene. If there is blood that can be seen and collected without the aid of fluorescein, then its use may be unnecessary. On the other hand, if the purpose is to define or enhance patterns not visible but thought to be present, any visible blood (desired for subsequent analysis) should be covered to shield it from chemical contamination or secured before the area is tested.

Fluorescein is prepared for use much like phenolphthalein. Fluorescein is reduced in alkaline solution over zinc to fluorescin which is then applied to the suspected bloodstained area. The catalytic activity of the heme then accelerates the oxidation by hydrogen peroxide of the fluorescin to fluorescein which will fluoresce when treated with ultraviolet light. Fluorescein and luminol are similar in that they produce light to indicate the presence of

blood; however, practical differences exist in the use of the two. Where an aqueous solution of the luminol reactants is simply sprayed on a surface bearing suspected bloodstain residues, a common fluorescein system includes a commercial thickener, which effectively causes the mixture to adhere to the surface and is thus more effective on vertical surfaces. Once the mixture is sprayed, visualization of fluorescence requires the use of an alternate light source (ALS), typically set at 450 nanometers. It also may be noted that fluorescein does not fluoresce with household bleach where luminol will react (Cheeseman, personal communication). Studies regarding the interference of fluorescein with subsequent short tandem repeat (STR) testing of blood for DNA) were conducted by Budowle (Budowle et al., 2000) and Jakovich (Jakovich, 2007). They observed that the results from STR typing showed no evidence of DNA degradation.

Other Tests

While the bulk of forensic testing to identify blood has been limited to the methods above or similar ones, more novel innovative approaches have been used. Microscopic examination of reconstituted residues identified erythrocytes on archeological materials determined with radiocarbon dating between 7400 to 6900 B.C. (Loy, 1989). Procedures involving crystallography, chromatography, spectrophotometery, immunology, and electrophoresis have all been used. The scope of this writing precludes a detailed discussion of each of them, and the reader is directed to works by Lee (1982) and Gaensslen (1983).

Confirmatory Tests for Blood

Crystal tests are regarded by many as confirmatory to the presumptive tests of Hunt (1960), Dilling (1926), and Kerr (1926), to name a few. These tests involve the nonprotein heme group of hemoglobin, the oxygen carrying protein of erythrocytes that belongs to a class of compounds called porphyrins. The heme structure contains a hexavalent iron atom. Nitrogen atoms within the ring structure bind four of the iron coordination positions and one is bound to histidine nitrogen in the globin protein. In hemoglobin, the remaining coordination position is normally bound by water or, in oxygenated hemoglobin, oxygen. In dried bloodstains, these last two positions are used in the formation of crystals that are the basis for confirming the presence of blood.

Teichmann Test

First described by Teichmann (1853), the test consists of heating dried blood in the presence of glacial acetic acid and a halide (usually chloride) to form the hematin derivative (Figure 14.11A). The crystals (Figure 14.11B) formed are observed microscopically, usually rhombic in shape and brownish in color. Stain age is not an obstacle with experienced examiners, and the test has even appeared in basic biochemistry student laboratory manuals (Harrow et al., 1960). The crystals are formed by placing a sample of suspected blood on a microscope slide and adding a small amount of chloride-containing glacial acetic acid followed by heating. In the experience of this author, the greatest difficulty in performing the test is controlling the heating of the sample, as it is easy to over- or under-heat the preparation, resulting in no crystal formation, even when blood is known to be present.

Takayama Test

If heme is gently heated with pyridine under alkaline conditions in the presence of a reducing sugar such as glucose, crystals of pyridine ferroprotoporphyrin or hemochromogen (Figure 14.12A) are formed. The reaction was examined by Takayama (1912), who examined several mixtures and found best results with a reagent containing water, saturated glucose solution, sodium hydroxide (10 percent), and pyridine in a ratio of 2:1:1:1 by volume.

The normal procedure is to place a small stain sample under a cover slip and allow the reagent to flow under and saturate the sample. After a brief heating period, the crystals

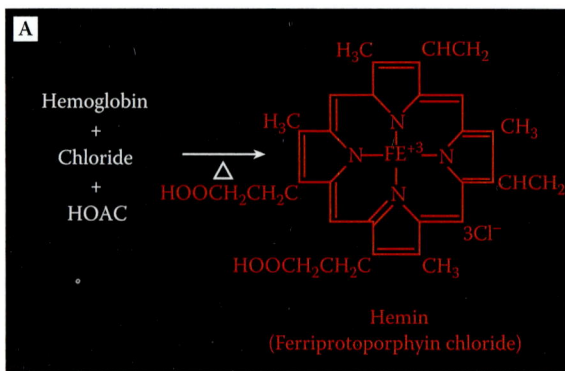

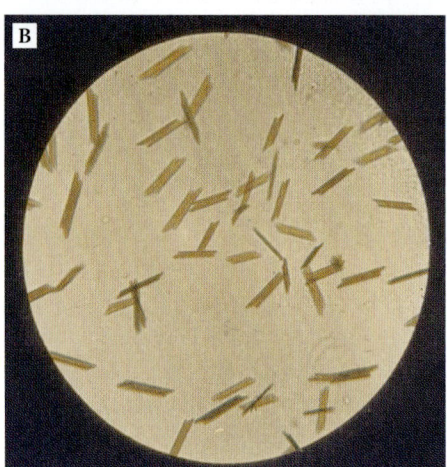

Figure 14.11 (A) Teichmann reaction. (B) Teichmann crystals.

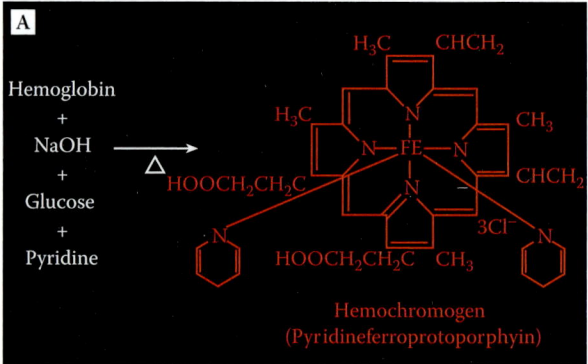

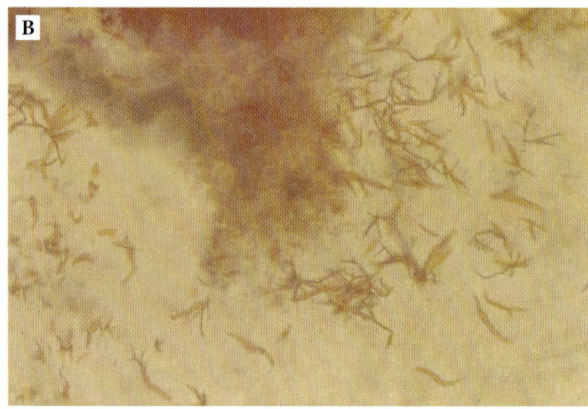

Figure 14.12 (A) Takayama reaction. (B) Takayama crystals.

are viewed microscopically (Figure 14.12B). A very small cover slip (2 mm square or smaller) allows the test to be carried out on a small stain quantity. The test has been effective on aged stains and crusts, and has produced crystals from blood on the uniform coat of a wounded Canadian Militia officer from the War of 1812 (Macey, 1979). Further, the test has given positive results on bloodstains that failed to test positively with the Teichmann test.

Probably the most recent improvement in the test procedure is that offered by Hatch (1993). It is known that oxygen and pyridine compete for the same binding site on the heme molecule. Hatch used Cleland's reagent (dithiothreitol) in the reagent to reduce this competition and shift the reaction in favor of pyridine, increasing the rate at which the hemochromogen crystals are formed.

Species Origin Determination in Bloodstains

Serum Protein Analysis

The serum proteins are a large collection of proteins in serum and consist of five classes: albumin, α-1 globulins, α-2 globulins, β globulins, and γ globulins. Albumin is present by far in the highest concentration, and the γ globulin fraction contains antibodies that are important in much of the following discussion.

General Methods

The following methods discussion provides a foundation for considering both origin determination and the analysis of genetic markers to be covered below. A wide variety of tests are available for the determination of species origin of an identified bloodstain, and most employ immunoprecipitation to effect a result. If a host animal (i.e., a rabbit) is inoculated with a serum protein (i.e., human), the immune system of the animal will normally recognize the protein as foreign and produce antibodies (gamma globulins) against it. Harvesting

the antibodies provides an antiserum to the protein (antigen), and when a sample of the antiserum and the antigen are brought in contact, a precipitin reaction normally occurs. The precipitate may form in a gel or in a solution, but its formation signals a reaction has occurred. The antibody and antigen are then said to be homologous.

Antibodies are also proteins but differ in structure and characteristics from the antigen proteins (mainly albumin). Key to the use of this in a forensic context is the fact that the antigen proteins bear species-specific characteristics that enable the serologist to identify the animal source of the stain extract protein. Several common test systems employing this reaction are discussed below and additional test systems are discussed elsewhere (Lee, 1982; Gaensslen, 1983).

Electrophoresis is a technique in which charged molecules (i.e., proteins) are caused to migrate in an electric field in a suitable support medium under controlled conditions that include temperature, pH, voltage, and time. Support media often used include starch gels, starch-agarose gels, cellulose acetate membranes, and polyacrylamide gels. Positively charged molecules will migrate toward the cathode and negatively charged ones toward the anode. The individual charges of the molecules will play a role in the rate of migration. If two molecules of differing charge are placed in a field at the same point, they will migrate at different rates and be separated. Once the separation is accomplished, the bands are visualized with staining or through a series of chemical reactions. Forensic electrophoresis is largely done in a gel support medium on a glass plate and samples as blood-soaked threads or stain extracts are placed in the gel. The characteristics of the sample, the gel itself, and those conditions cited above all play a role in how well separated and resolved the final result will be.

Isoelectric focusing (IEF) is an electrophoretic method that takes advantage of the fact that at a certain pH a protein in aqueous medium will exhibit a point of no net charge (the isoelectric point, IEP). In electrophoresis, the buffer in the gel controls the pH of the system uniformly throughout the gel. However, with IEF there is a pH gradient set up in the gel with the low end at the anode and the high end at the cathode. When current is applied, proteins will migrate to the point where they encounter their IEP and stop. The result is that the migrating molecules focus at their IEPs and form sharp, well defined bands that can be much more easily observed than results sometimes seen with electrophoresis. As with electrophoresis, different band patterns are detected through immunological or chemical techniques to define the phenotypes present.

Ring Precipitin Test—The ring precipitin test employs simple diffusion between two liquids in contact with one another in a test tube. The two liquids are the antiserum and an extract of the bloodstain in question. If the antiserum (antihuman) is placed in a small tube, and a portion of the bloodstain extract (human) is carefully layered over the denser antiserum, dissolved antigens and antibodies from the respective layers will begin to diffuse into the other layer. The result will be a fine line of precipitate at the interface of the two solutions (Figure 14.13). In the case where the bloodstain extract is not human, there should be no reaction. Occasionally, there will be more than one precipitin band, indicating different reacting components with different diffusion coefficients. A standard operating procedure would be to include necessary positive and negative controls with a questioned sample.

Ouchterlony Double Diffusion Test—Named after the man who devised it (Ouchterlony, 1949, 1968), the test is carried out in a gel on a glass plate. A hot agar solution is plated on small glass plates or a petri dish and allowed to cool, producing a gel layer usually 2 to 3 millimeters thick. Holes (wells) are punched in the gel layer at specified locations and the gel is removed. Commonly used patterns are a square of four wells or a rosette of six wells surrounding a center well or two rows of parallel wells. When the antiserum and stain extracts (antigen) are prepared, the antiserum is placed in the central well and the different stain extracts in the surrounding wells. The system is then incubated at constant temperature (4, 18, or 37 degrees Celsius are commonly used). After a given period of time, the diffusion of the reactants through the gel results in the formation of immunoprecipitate lines in the gel between the wells (Figure 14.14). This precipitate is a stable antigen–antibody complex that has grown beyond the limits of its solubility.

Antihuman serum from the center well forming an immunoprecipitate with sample from one or more of the surrounding wells indicates the presence of human protein in the respective outer well(s). The precipitate may be stained with a suitable protein stain. While the point at which the immunoprecipitate forms is controlled by the diffusion coefficients of the reactants, its formation is

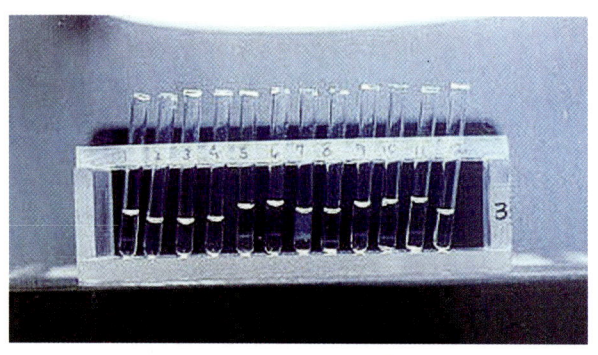

Figure 14.13   Ring precipitin test.

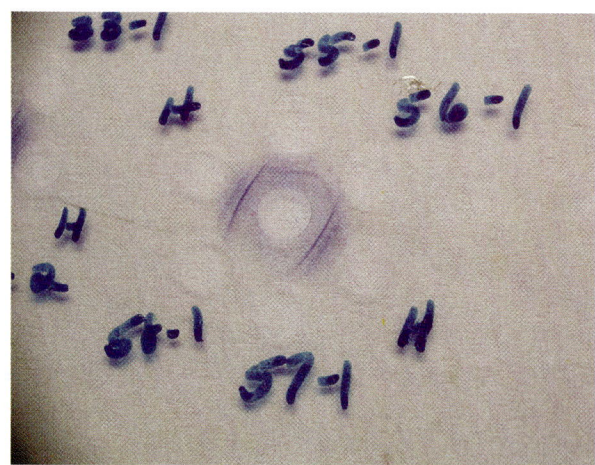

Figure 14.14   The Ouchterlony double diffusion test.

dependent on the concentration of reactants and the identities of the reactants. When antigen and antibody are present in relatively equivalent amounts, the line grows in an arc between the two wells. Incubation at cooler temperatures is a slower process but has the advantage of producing sharper lines and sometimes additional precipitin lines, due to reduced solubility at lower temperatures.

Crossed-Over Electrophoresis—Up to now, methods discussed have relied on simple diffusion to move antigens and antibodies toward one another. The crossed-over method uses electrophoresis to achieve this. Rows of opposing wells are cut in an agarose gel plate so that well pairs can be arranged with one well toward a cathode source and the other nearer the anode source. The stain extract (antigen) is placed in the cathodic well so that it can migrate toward the anode (because of the pH of the gel buffer used). The antiserum is placed in the anodic well to migrate toward the cathode. When the current is applied, extract proteins move toward the anode.

Antibodies are carried toward the cathode by a phenomenon known as electroendosmosis, which involves anions integral to the gel itself (such as sulfate and pyruvate), the associated cations (sodium), and their water of hydration. When electrophoresis is applied, the cations and water of hydration tend to move in a cathodic direction. This flow carries the antibodies along with it to meet the stain extract proteins between the two rows of wells and result in the formation of immunoprecipitate bands at a site between the wells. This technique has the advantage of forcing the reactants to move in directions optimal for maximum reaction, as opposed to diffusion in 360 degrees around a diffusion plate well.

Nonserum Protein Analysis

Antihuman Hemoglobin—Hemoglobin is a protein that has species-specific immunological characteristics and is unique to blood. If both characteristics could be addressed in a single test, the work of the serologist would be considerably simplified. Lee and DeForest

(1977) presented their work on this, offering a procedure capable of producing a high titer antihuman hemoglobin serum. Such an antiserum would potentially be very effective in any of the systems described above. An obstacle with such a test is the cross-reactivity exhibited by proteins from other higher primates which, of course, is balanced by the relative expectation of encountering nonhuman primate blood in an investigation.

The efforts of the commercial world are not without merit in this regard. Hochmeister et al. (1999) evaluated an immunochromatographic system, Hexagon OBT1®, and found the one-step test suitable for laboratory or field use, sensitive, and primate specific. They concluded the test was useful with aged and degraded material, produced easy-to-interpret results, and was unaffected by environmental insults except for detergents, household bleaches, and prolonged exposure to luminol. The test could be used in conjunction with current forensic DNA extraction techniques without loss of DNA. Johnston et al. (2003) examined the ABAcard® HemaTrace®, Abacus Diagnostics, Inc. (Figure 14.15), another

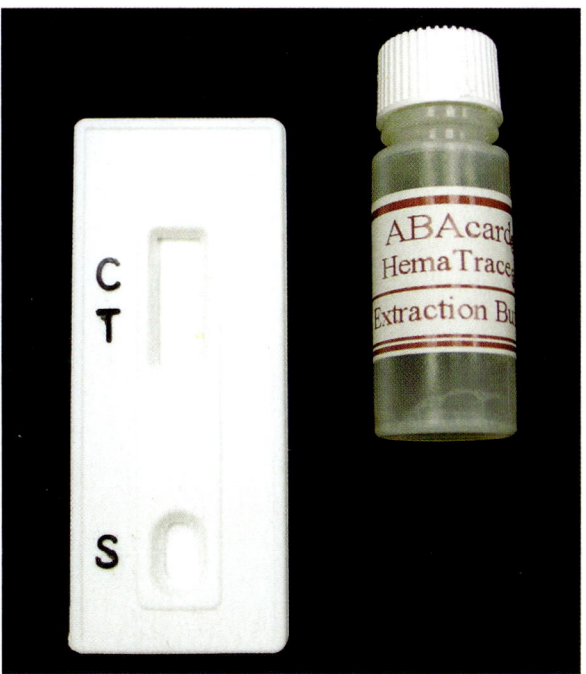

Figure 14.15 ABAcard Hematrace kit components.

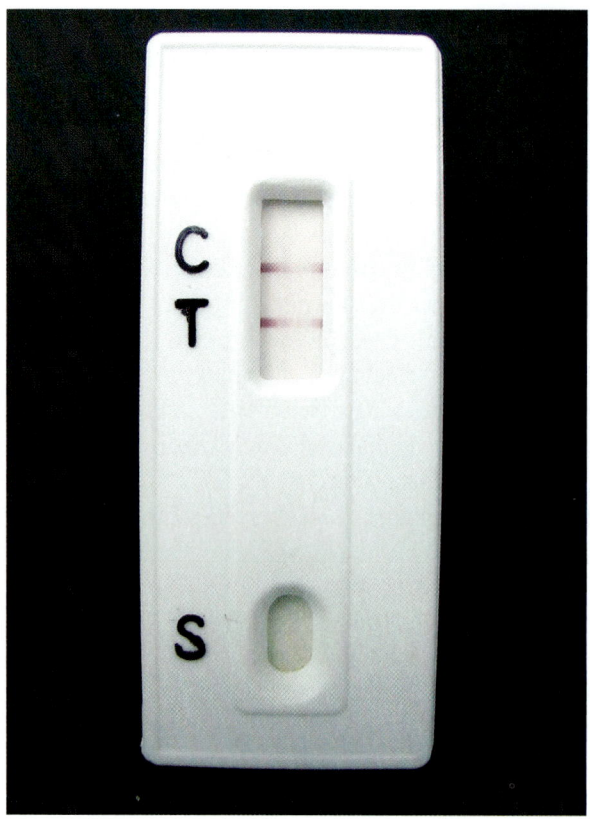

Figure 14.16 ABAcard Hematrace-positive reaction.

immunochromatographic test device for potential use in forensic investigations, and found it sensitive and specific for human, higher primate, and ferret bloods (Figure 14.16). Other investigators (Vandenberg and McMahon, 2004; Silenieks et al., 2004) have found the ABAcard® HemaTrace® to be rapid, sensitive, and accurate in the identification of human (higher primate) blood. Reynolds (2004) evaluated the ABAcard® HemaTrace® as well and concluded the test to be reliable for the identification of human blood.

Other Techniques—While additional origin determination methods are available, those presented above are frequently encountered in forensic serology. Techniques such as immunoelectrophoresis, rocket (Laurel) electrophoresis, variations on single and double diffusion, and tests involving agglutination of red blood cells (hemagglutination) have been investigated and provide interesting and valuable approaches to the problem of origin determination.

Genetic Markers in Blood

In discussing genetic markers, two things need to be established at the outset. The American Association of Blood Banks (AABB) (1981) defined a blood group as a group of antigens produced by allelic genes at a single locus and inherited independently of any other genes. Thus, a blood group is genetically controlled and is a lifelong characteristic we can identify in an individual. Secondly, it is perhaps beneficial to gain some understanding of blood itself. Blood consists of two major fractions: plasma and formed elements.

The plasma portion comprises approximately 55 percent of the total blood volume and contains a great many dissolved substances, among them a wide variety of carbohydrates, lipids, hormones, inorganic salts, and the serum proteins, including the blood group antibodies. Plasma also contains those substances involved in the blood-clotting mechanism, and it is an easily demonstrated fact that plasma can form a clot in the absence of the formed elements. The fluid portion remaining after the formation of a clot is called *serum*.

The formed elements or particulate portion comprise the remaining volume of 45 percent. This fraction is largely made up of cells and platelets. Red blood cells (RBC) or erythrocytes normally number from 4.5 to 6×10^6 cells per cubic millimeter, are anucleate, and are primarily responsible for transport of oxygen from the lungs to tissues of the body. RBCs are formed in the myeloid or red marrow tissue of bones. The white blood cells or leucocytes (WBC) normally occur in amounts ranging from approximately 10 to 15×10^3 per cubic millimeter, and are distinguished by the different nuclear shapes they exhibit as well as the different manner in which they accept histological stains. WBCs are primarily formed in the lymph nodes and play an important role in the body's immune response and antibody production. Platelets or thrombocytes appear as fragmentary cells and exist in blood in numbers ranging from 1.5 to 3×10^5 per cubic millimeter. Their primary purpose is that of initiating and participating in the blood clot-ting process.

Antigen Based Markers: Blood Groups

The ABO System

The first and best-known blood grouping system is the **ABO** system, discovered by Karl Landsteiner in 1900. The types A, B, O, and AB refer to the antigens (agglutinogens) on the surface of the red blood cells, which are glycoprotein substances and an integral part of the cell membrane. These substances have been well-characterized biochemically, and the specific antigenic nature resides with certain immunodominant sugar residues terminating the carbohydrate chains of the glycoprotein macromolecule. The corresponding antibodies (agglutinins), anti-A (α–A) and anti-B (α–B), of the system are present in the plasma. These characteristics are summarized in Table 14.2.

A person of blood group A will have anti-B (α-B) antibody in his or her plasma, and if that plasma is mixed with group B cells, the two are said to be homologous with agglutination being the result. The characteristics of the person who is group O present a little different picture. There are no routinely occurring antibodies in humans for the red cell H antigens possessed by the group O person. However, certain seed extracts called *lectins* are capable of achieving agglutination with blood cells. A lectin preparation from the seeds of *Ulex europaeus* (gorse, a spiny evergreen shrub) exhibits anti-H activity.

ABO antigens on red cells are produced as the ultimate expression of a series of genetic alleles specific for the ABO characteristics. The H antigen associated with group O is a result of the expression of the H allele in a homozygous condition. H antigen is produced when the H-allele-controlled fucosyl transferase enzyme places an l-fucose on the end of the basic carbohydrate chain protruding from the surface of the red cell membrane during the formation of the cell. The H antigen is a precursor to the A and B antigens. The A antigen is produced when an A allele controlled *N*-acetylgalactosamine transferase enzyme adds that sugar to the preexisting H chain. The point of this addition is on the subterminal *N*-acetylglucoasmine sugar. The same is true for the B antigen, where the B-allele-controlled transferase adds D-galactose at the same point. Thus, ABO characteristics are the result of the expression of both the ABO and Hh genes. This simplified view of ABO antigen production provides a basis for a brief discussion of the Lewis system below.

Forensic testing for the ABO system in dried bloodstains centers on identifying both the antigen and antibodies present. A variety of methods have been devised to accomplish both tasks. Those most commonly used are the absorption elution for antigenic characteristics and the Lattes Crust test for the presence of antibodies in stained material.

Absorption Elution

Briefly described, this technique involves the exposure of a portion (a cutting or thread) of the stain bearing the blood (and antigen) to absorb

Table 14.2
ABO Blood Group Characteristics

Blood Group	Antigen Present	Specific Sugar	Antibody Present	Population Frequency
A	A	*N*-acetylgalactose amine	Anti-B	40%
B	B	D-galactose	Anti-A	10%
AB	A, B	D-galactose and N-acetylgalactose amine	None	45%
O	H	L-fucose	Anti-A and Anti-B	5%

the homologous antibody. Unreacted antibody is then washed away and the absorbed antibody is eluted by warming and mixed with a cell suspension to be identified. As an example, a group A stain exposed to anti-A (α-A), anti-B (α-B) and anti-H lectin (α-H lectin) antibodies in separate containers (tubes, wells, etc.) would absorb the anti-A and not the anti-B or anti-H antibodies. After allowing sufficient time for absorption, the unreacted antibodies (and lectin) would be washed away and gentle heating would be applied to release (elute) the absorbed anti-A. This anti-A antibody would be detected by addition of group A cells, which would then agglutinate and be viewed microscopically. The other two containers would exhibit no reaction as no antibody or lectin was absorbed and then eluted to react with the B and O cells added.

Lattes Crust Test

Testing for the respective antibodies in the stain is a matter of exposing three separate stain samples to dilute suspensions of A, B, and O cells (usually on glass microscope slides), allowing a suitable period of time for elution of the antibodies from the stain and agglutination of the cells. A positive result or agglutination on the slide with the B cells would indicate anti-B antibody in the stain and thus confirm the stain as blood group A. The O cells should always be negative to serve as a control for any nonspecific agglutinins that might be encountered, as naturally occurring anti-H antibodies are exceptionally rare. This test is often referred to as the Lattes test, after Leon Lattes (1914). With the antigen (A) and the anti-B antibody in a stain so identified, the conclusion as to the blood group present becomes apparent: Group A.

Secretors

It is well known that some individuals secrete their ABO biochemical (antigenic) characteristics (A, B, and H blood group substances) into body fluids such as saliva, semen, and vaginal fluid. Such individuals are called secretors and occur in approximately 80 percent of the general population. The remaining 20 percent of the population are nonsecretors. The secretor phenomenon is intimately related to the Lewis blood group antigens.

The Lewis System

The Lewis (Le) antigens differ from the ABO antigens in that they are absorbed on the surface of the red cell from plasma, and are not a part of the membrane structure. Their primary importance to the serologist is that they provide a secretor status indicator in the blood. Testing a known blood sample for ABO and Lewis groups usually allows a conclusion as to ABO group and secretor status (whether or not an individual's ABH blood group substances [bgs] should be found in nonblood body fluids).

The presence or absence of the Lewis a or b antigens (Le^a and Le^b) is a result of intimate interaction between the ABO, Hh, Secretor (Sese), and Lewis genes, which are inherited independently of one another. A person may have either or both Lewis antigens on his or her red cells, and the association with secretor status may be summarized as follows.

Le (a–b+) individuals are ABH secretors and represent approximately 72 percent of the general population. Le(a+b–) individuals are ABH nonsecretors and represent approximately 22 percent of the general population. Le (a–b–) individuals cannot be stated on this basis alone to be ABH secretors or nonsecretors but are approximately 80 percent secretors and 20 percent nonsecretors. Le(a+b+) individuals have been identified but are rare.

Lewis testing involves subjecting the red cells of the blood sample in question, cells from a known secretor, and cells from a known nonsecretor to antisera specific for the Lewis antigens. The known secretor and nonsecretor cells serve as controls to ensure the test is performing correctly. A reagent control is usually run with each cell type.

The Rhesus (Rh) System

Discovered by Landsteiner and Wiener in 1940, this system consists of a number of antigens and is described by three different systems of nomenclature: Wiener (rh–hr), Fisher–Race (C, c, D, E, e) and Rosenfield (Rh:l, Rh:2, Rh:3, etc.). The two more prominent ones are Wiener and Fisher–Race. The Wiener system is based on a single gene locus being occupied (normally) by any of eight alleles. These genes are responsible for an agglutinogen on the red cell surface that has two or more antigenic specificities identifiable with specific antisera. The Fisher–Race system defines five antigens, the genes for which are located at three closely linked gene loci and are inherited as a three-gene complex. These genes then code for the individual protein antigens on the surface of the erythrocyte membrane. It should also be noted that in both these systems the genes are underlined or italicized when in print.

The Rh system has proven valuable in forensic work in spite of the larger quantity of sample required for dried stain analysis and the degree of sophistication of the available techniques. The primary method used in grouping dried stains is an absorption elution technique. However, the development of methods for the analysis of blood enzymes and proteins enabled a more effective use of limited samples and interest in Rh grouping dwindled. Rh antigenic characteristics are not present in nonblood body fluids, a fact that has limited biochemical studies into the nature of the antigenic structures. Also, natural Rh antibodies are not common in serum. These facts have hindered the study of this system from a biochemical point of view and detailed structural knowledge concerning these antigens is not as plentiful.

Other Antigen Markers

The following systems, while not primarily antigenic systems as discussed above, involve the use of immunospecific techniques to identify the different phenotypes present.

Gm and Km Systems—The Gm and Km antigen systems consist of antigenic specificities that exist on the polypeptide chains of immunoglobulin molecules (the gamma globulins) in the serum proteins of plasma. The five types of immunoglobulin molecules consist of both large and small protein chains. The small chains are known as kappa chains and the larger chains are specific to the type of antibody in question. Immunoglobulin G (IgG) would have gamma (γ) chains; immunoglobulin A (IgA), alpha (α) chains; immunoglobulin M (IgM), mu (μ) chains; immunoglobulin D (IgD), delta (δ) chains; and immunoglobulin E (IgE), epsilon (ε) chains.

Known as allotypes or allotypic markers, Gm specificities (more than 20) reside on only the large gamma chains of immunoglobulin G (IgG) molecules. The Km allotypes (three) are found on the kappa chains of all five classes of immunoglobulins (IgG, IgA, IgM, IgD, and IgE). Both Gm and Km allotypes are inherited independently of each other as autosomal, codominant alleles; however, the two differ in the way each system is inherited. Km allotypes are passed on through simple alleles. The alleles controlling the production of Gm sites on IgG heavy chains are linked on a single chromosome and are inherited as sets called *haplotypes*, which represent specific gene complexes.

The Gm and Km systems have represented distinct advantages to the forensic serologist because of the stability of the antigens and the variety of types possible (especially with Gm). The antigens are stable at moderate heat, and may be stored at room temperature for extended periods and frozen for years. They also exhibit a relatively high concentration in dried blood. A major disadvantage, however, is availability of antisera. Anti-Gm is naturally occurring but not common and samples of antisera from different sources have been observed to react differently. Informative reviews of Gm and Km are presented by Kipps (1979) and Dival (1986).

Group-Specific Component (Gc)—Group-specific component (Gc) is a vitamin-D-binding glycoprotein in the α-2 globulin fraction of serum proteins. Its specific biological function is not completely clear. Three common phenotype (Gc 1-1, Gc 2-2, Gc 2-1) band patterns are identifiable with electrophoresis. The use of IEF makes it possible to distinguish 9 Gc subtypes (1F, 1S, 2, 1F1S, 2-1F, 2-1S, 1F-1A1, 1S-1A1, and 2-1A1) (Budowle, 1987). Either method requires applying immunofixation (immobilizing the protein bands within the gel with specific antibody) to detect the separated protein bands. The phenotypes are first separated by migration through the gel, and the resulting band patterns are visualized applying an anti-Gc preparation. Usually the immunoprecipitate can be made more visible by staining with a protein stain such as Coomassie Brilliant Blue.

Protein Markers

Hemoglobin (Hb)

Hemoglobin is the major protein in erythrocytes comprising 95 percent of the dry cell weight, and is responsible for oxygen transport in the body. It exhibits 180 or more variants but only four (Hb A, Hb F, Hb S, and Hb C) are readily distinguished based on the positioning of the protein bands with electrophoresis or IEF (Budowle, 1986). The different alleles are expressed codominantly, and it is possible to observe the heterozygous Hb AS, Hb AC, or Hb AF types. Hb S is associated with the sickle cell trait and an individual homozygous for the S variant normally does not survive beyond childhood. Hb A and Hb C are adult forms, with Hb C being far less common, and Hb A being the most common of all. Hb F is fetal hemoglobin and is normally seen in fetal blood after approximately 3 months of gestation. It is not normally seen in blood from individuals older than 6 months.

Haptoglobin (Hp)

Haptoglobin is a glycoprotein of the α-2 globulin class that forms stable complexes with hemoglobin controlling hemoglobin excretion from the body. These complexes are stable, tightly bound, and irreversible. Electrophoresis is usually carried out in vertical polyacrylamide gel apparatus. Prior to electrophoretic separation of Hp types, hemoglobin is added to the samples to establish the Hp-Hb complex, which then migrates as a unit. When separated, the individual bands are stained to show the peroxidase-like activity of the bound hemoglobin. The reaction usually involves *ortho*-dianisidine, which stains the Hp-Hb complexes a brown color.

The most commonly observed phenotypes are Hp 1, Hp 2, and Hp 2-1 (Figure 14.17), although a Amodified@ Hp2-1M has been observed in casework in which some of the 2 bands are intensified and while some of the other normally stronger 2 bands are reduced.

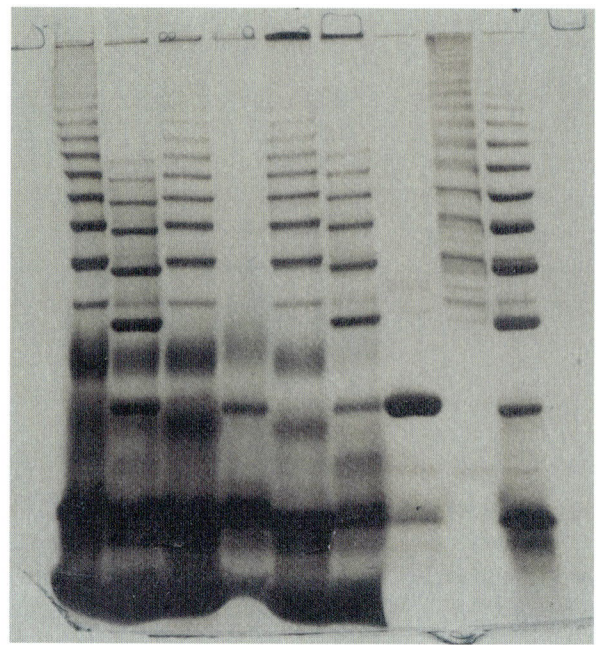

Figure 14.17 Haptoglobin phenotypes. From left to right: Hp 2, Hp 2-1, Hp 2, Hp 1, Hp 2, Hp 2-1, Hp 1, Hp 2, Hp 2-1 (anode at bottom). Note darkly stained hemoglobin at bottom of gel.

Hp is a relatively stable protein in dried bloodstains, which makes it a valuable marker for the serologist.

Enzyme Markers

Between 1971 and the coming of age of DNA in the mid to late 1980s, interest in polymorphic enzyme systems and their analysis was high. The result was a flourishing of technology designed to identify the different types of a host of enzymes, centering mainly on electrophoresis and IEF methods. This advancement in technology led to a greater ability to discriminate between individuals and evidentiary material, and paved the way for the more advanced technologies of DNA analysis. While methods have been developed for numerous enzymes with regard to their use in individualizing bloodstains (Table 14.3), space precludes a complete discussion of them. Phosphoglucomutase (PGM) is taken as a representative enzyme for discussion.

While the development of techniques enabled a greater discriminatory capability, there was still the basic nature of the macromolecules to contend with. The basic structure of proteins and enzymes is complex, both in

Table 14.3
Forensically Important Enzymes

Enzyme	Abbreviation	Common Phenotypes	Important Forensic Sources	Remarks
Adenosine deaminase	ADA	1, 2, 2-1	Blood; not in semen[a]	Rare variants exist (3-1, 4-1, 5-1,6-1, 7-1)
Adenylate kinase	AK	1, 2, 2-1	Blood; not in semen[a]	Rare variants exist (3-1, 4-1, 5-1)
Carbonic anhydrase	CA II	1, 2, 2-1	Blood; not in semen[a]	Takes 4 to 6 years to reach adult levels
Erythrocyte acid phosphatase	EAP, ACP	A, B, BA, C, CA, CB	Blood; not in semen[a]	Stability of bands C>B>A
Esterase D	ESD	1,2, 2-1, 1-5, 2-5, 5	Blood; not in semen	Five allelic products observable with IEF
Glucose-6-phosphate dehydrogenase	G6PD	A, B, AB	Blood; not in semen[a]	Mainly polymorphic in blacks; AB normally
Glyoxalase	GLO I	1, 2, 2-1	Blood and semen	In sperm and plasma
Peptidase A	PEP A	1, 2, 2-1	Blood and semen	In plasma more than sperm

physical structure and chemical nature. While many of the systems mentioned above are sensitive to unfavorable environmental conditions, proteins and enzymes are much more susceptible to such antagonism. A change in the molecule's physical structure or the ability to perform its chemical task will often result in an inability to detect any phenotype in evidentiary stains. Thus, the enemies heat and humidity become even more significant when dealing with these markers. A discussion of unfavorable environmental conditions often brings up the question of altered phenotypes being observed in the analysis of improperly stored evidentiary stains. In other words, would a PGM 1+1B become a PGM 2+2B? In fact, it may be shown that the result of subjecting evidence to unfavorable conditions is likely to be the loss of any detectable phenotypes rather than the introduction of new ones.

A term one will find mentioned in connection with forensic enzyme assays is *polymorphism*. Polymorphism may be described as the occurrence in a population of two or more genetically determined alternative phenotypes with frequencies greater than could be accounted for by mutation or drift. That is, two or more alleles for the production of the enzyme exist in the population. If enough samples are analyzed, they will be observed. A summary of information regarding forensically important polymorphic enzymes is presented in Table 14.3. Since space prohibits a detailed discussion of each enzyme, phosphoglucomutase, probably the best known, will be discussed here. For more information the reader is directed to Gaensslen (1983) and Sensabaugh (1982).

Phosphoglucomutase (PGM)—PGM is a phosphotransferase enzyme that catalyzes the reversible conversion of glucose-1-phosphate to glucose-6-phosphate, an essential reaction in carbohydrate metabolism in the body. It is found in many tissues of plants, animals, and microorganisms. In humans the enzyme exists in significant concentrations in blood and semen, and in small amounts in vaginal secretions and cervical mucus. It is known to be inhibited by heavy metals and fluoride, a significant point as blood samples taken by medical personnel may contain fluoride. When stored under cool dry conditions the enzyme survives well. It should always be the concern of the forensic investigator, however, to avoid exposing biological evidence to prolonged heat and humidity.

Figure 14.18 PGM phenotypes, (electrophoresis). From left to right: PGM 1, PGM 2, PGM 2-1, PGM 2, PGM 1, PGM 2-1, PGM 2-1, PGM 2, PGM 1 (anode at top).

There are actually three genetic loci that control PGM polymorphism. Locus 1 is on chromosome 1, locus 3 is on chromosome 6, and locus 2 is thought to be on chromosome 4. Only polymorphism originating at locus 1 is considered forensically important. Initially, electrophoresis of forensic samples for PGM activity routinely detected three phenotypes, PGM 1, PGM 2, and PGM 2-1 (Figure 14.18). With the application of newer IEF methods for PGM analysis this number was soon increased (Dival, 1984) to 10 phenotypes: PGM 1+, PGM 1−, PGM 1+1−, PGM 2+, PGM 2−, PGM 2+2−, PGM 2+1−, PGM 2+1+, PGM 2−1+, and PGM 2−1− (Figure 14.19). Each of these types is present in the population with a specific frequency with some being more common than

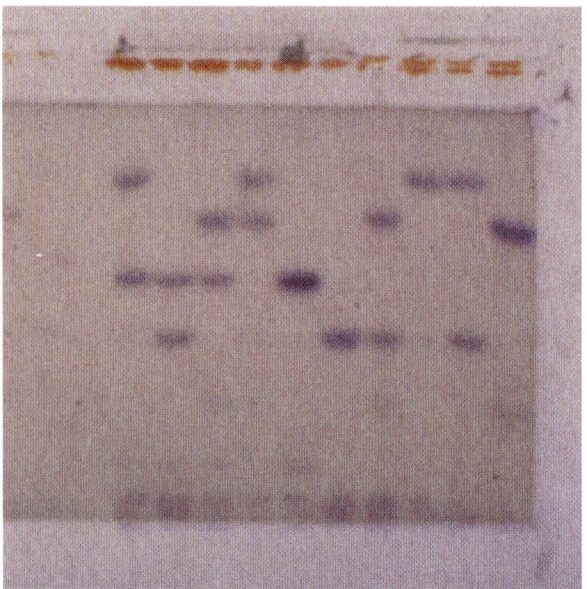

Figure 14.19 PGM phenotypes, (isoelectric focusing). From left to right: PGM 1+2+, PGM 1+1-, PGM 1+2-, PGM 2+2-, PGM 1+, PGM 1-, PGM 1-2-, PGM 2+, PGM 1-2+, PGM 2- (anode at top).

others. With this ability to place the subtypes of PGM in ten different population groups (usually expressed as percentages) instead of three, the enzyme presents the highest discrimination probability of any enzyme system commonly used in forensic serology.

This subtyping has also been useful in demonstrating additional types in other markers. It is the objective of forensic characterization of blood to as nearly as possible associate stains with the individual(s) from whom they could have come. An example of the use of population frequencies to show this is given with the following case example.

Case Example

During the early morning hours of a June day in a northeastern community, the teenaged daughter of the headmaster of a prestigious girl's school was awakened in her own bed, carried downstairs and into the back yard, and there, sexually assaulted and forced to commit sodomy. She had limited vision without her glasses and was later unable to identify her attacker except to provide a general description. When released, she fled to wake her parents who immediately reported the incident to the police. The girl's injuries included a vaginal laceration, which bled profusely.

Shortly after the police were notified, an individual who fit the general description given by the girl was located walking on the street about five blocks from the scene. The individual explained that the blood-like stains around the fly area on his blue jeans came from a nosebleed suffered by his girlfriend's daughter earlier that day.

The blue jeans, the girl's nightgown, and liquid blood samples from all three individuals were subjected to forensic examination. Human blood was, indeed, identified on the nightgown and the blue jeans. ABO analysis of the blood samples disclosed that the girl, the suspect, and the blood on the pants and the nightgown were all of blood group A, while the girlfriend's daughter was group O. Additional analysis of the known samples disclosed the girl to be PGM 2-1, EAP BA, Hp 1; the suspect, PGM 1, EAP A, Hp 1, and the girlfriend's daughter, PGM 1, EAP B, Hp 2-1. The blood on the pants and nightgown was PGM 2-1, EAP BA, and Hp 1, indicating the young girl as a possible source but eliminating the others. All additional testing was inconclusive. No semen was identified on the evidence. The blood on the jeans could have come from the victim but not the suspect or his girlfriend's daughter.

The significance of these results becomes clearer when we look at the relative frequency of the victim's set of types in the population. The population frequencies for the victim's characteristics are ABO A, 40 percent; PGM 1, 35.9 percent; EAP A, 41.4 percent; Hp 1, 16.5 percent. The combination of these frequencies ($0.40 \times 0.359 \times 0.414 \times 0.165$) gives us 0.0098 or 0.98 percent of the population can be expected to have all the characteristics of the victim. Stated differently, there is roughly 1 chance in 100 that we can randomly select a person from the population of that town with that combination of phenotypes. Had this case been examined when PGM subtyping was available and the victim had been the most common subtype of PGM 1, the figure would have been 0.006 or 0.6 percent, or 1 chance in 167. The result for the least common PGM subtype would have been 0.0004 or 0.04 percent, or 1 chance in 2500.

Conclusion

The field of forensic serology has experienced many changes in technology, ranging from simple improvements in technique to the addition of newer methods of analysis and the recognition of additional biochemical sources of information. Twenty years ago the possibility of establishing the degree of identity possible today with DNA and doing it with the minute sample quantity currently used seemed an impossibility. These kinds of changes are ongoing and do not show any indication of stopping. However, by knowing the past, we can better understand the future, and since the material presented here is a basic primer and not by any means all-inclusive, the reader is heartily encouraged to consult additional references for more in-depth information.

The author wishes to express grateful appreciation to T. Paulette Sutton, who assisted in no small regard with photographs of presumptive tests.

Questions

1. What two environmental factors are most important to consider in preserving blood evidence?

2. List three possible causes of "false" positive reactions with presumptive screening tests for blood.

3. Define the term *isoelectric point*.

4. Describe the principle behind a presumptive test for blood (what is done and what it means).

5. True or false: Benzidine, phenolphthalein, and *o*-tolidine are classed as carcinogens.

6. Forensic Phil conducted a phenolphthalein test on a stain he thought looked like blood. He observed a positive test immediately after adding the phenolphthalein reagent. He concluded blood was, in fact, present. Was he correct?

7. What is the role of pyridine in the hemochromogen test?

8. True or false: Luminol is so sensitive that a positive reaction can be taken to prove that blood is present.

9. Describe electroendosmosis and how it impacts on origin determination of bloodstains.

10. How many blood group systems are indicated by the following antigens: A, B, C, D? _____

11. A person having the A antigen on his or her red cells will have serum antibodies that will agglutinate cells from:

   _____ a group "A" person
   _____ a group "B" person
   _____ a group "O" person
   _____ a group "AB" person

12. When called to testify regarding evidence he had examined, Forensic Phil was questioned about the possibility of errant enzyme results being introduced into the evidence as a result of inadequate preservation. How should he have responded?

13. Define polymorphism.

14. What are the five classes of serum proteins?

15. The combined frequency of a person who is group B, PGM 1-2+, EAP B, Hp 2-1, Gc 2 is: _____. This represents 1 person in how many? _____. Use the frequency table below to calculate your answer(s).

References and Suggested Readings

General References

DeForest, P., Gaensslen, R. E., and Lee, H. C., *Forensic Science: An Introduction to Criminalistics*, McGraw-Hill, New York, 1983, chap. 9.

Gaensslen, R. E., *Sourcebook in Forensic Serology, Immunology and Biochemistry*, U.S. Government Printing Office, Washington, D.C., units 2, 4, 5, 6, 7, 9, 1983.

Saferstein, R., *Criminalistics: An Introduction to Forensic Science*, 6th ed., Prentice-Hall, Englewood Cliffs, NJ, 1997.

Spalding, R. P., *FBI Laboratory Serology Unit Protocol Manual*, U.S. Department of Justice, Federal Bureau of Investigation, May 1989, sections 2, 3, 4, 5, 11.

Sutton, T. P., Presumptive blood testing, in *Scientific and Legal Applications of Bloodstain Pattern Interpretation*, James, S. H., Ed., CRC Press, Boca Raton, FL, 1999, chap. 4.

Specific References

ABACard, HemaTrace, Abacus Diagnostics, product literature, 1998.

Adler, O. and Adler, R., The reaction of certain organic compounds with blood, with particular reference to blood identification, *Hoppe-Seyler's Z. Physiol. Chem.*, 41, 59, 1904, translation in Gaensslen, R. E., *Sourcebook in Forensic Serology, Immunology and Biochemistry*, U.S. Government Printing Office, Washington, D.C., 1983, unit 9.

Bell, G. H., Breslin, P., and Lemen, R., Eds., Health Hazard Alert: Benzidine-, o-tolidine, and o-dianisidine-based dyes, DHHS (NIOSH) Publication No. 81-106, U.S. Department of Labor and U.S. Department of Health and Human Services, 1980.

Bily, C. and Maldonado H., The application of luminol to bloodstains concealed by multiple layers of paint, *J. Forens. Ident.*, 56, 896, 2006.

Blum, L. J., Esperanza, Philippe, and Rocquefelte, Stephanie, A new high-performance reagent and procedure for latent bloodstain detection based on luminol chemiluminescence, *Can. Soc. Forens. Sci. J.*, 39, 81. 2006.

Budowle, B., Leggitt, J. L., Defanbaugh, D. A., Keys, K. M., and Malkiewicz, S. F., The presumptive reagent fluorescein for the detection of dilute bloodstains and subsequent STR typing of recovered DNA, *J. Forens. Sci.*, 45, 1090, 2000.

Budowle, B. A method for subtyping group specific component in bloodstains, *Forens. Sci. Int.*, 33, 187, 1987.

Budowle, B. and Eberhardt, P., Ultrathin-layer polyacrylamide gel isoelectric focusing for the identification of hemoglobin variants, *Hemoglobin,* 10, 1661, 1986.

Carcinogens, *Federal Register*, 39(20), 3779, 1974.

Cheeseman, R. and DiMeo, L. A., Fluorescein as a field-worthy latent bloodstain detection system, *J. Forens. Ident.*, 45, 631, 1995.

Cheeseman, R., Direct sensitivity comparison of the fluorescein and luminol bloodstain enhancement techniques, *J. Forens. Ident.*, 49, 261, 1999.

Cheeseman, R. and Tomboc, R., Fluorescein technique performance study on bloody foot trails, *J. Forens. Ident.*, 51, 16, 2001.

Cox, M., A study of the sensitivity and specificity of four presumptive tests for blood, *J. Forens. Sci.*, 36, 1503, 1991.

Curtius, T. and Semper, A., Verhalten des 1-äthylesters der 3-notro-benzol-1,2-dicarbonsäure gegen hydrazin, *Ber. Dtsch. Chem. Ges.*, 46, 1162, 1913.

Dilbeck, L., Use of Bluestar Forensic in Lieu of Luminol at Crime Scenes, *J. Forens. Ident.*, 56, 706, 2006.

Dilling, W. J., Haemochromogen crystal test for blood, *Brit. Med. J.*, 1, 219, 1926.

Dival, G., Red cell markers, in *Proc. Int. Symp. Forens. Electrophoresis*, Federal Bureau of Investigation, Quantico, Virginia, 1984.

Dival, G. B., The biochemical genetics and methodology for the analysis of the Gm and Km antigens, in *Proc. Int. Symp. Forens. Immunol.*, Federal Bureau of Investigation, Quantico, Virginia, 1986.

Fleig, C., Nouvelle reaction, a la fluorescine, pour la recherche du sang, en particulier dans l'urine, *C.R. Soc. Biol.* 69, 192, 1910.

Garner, D. D, Cano, K. M., Peimer, R. S., and Yeshion, T. E., An evaluation of tetramethylbenzidine as a presumptive test for blood, *J. Forens. Sci.*, 21, 816–821, 1976.

Gimeno, F. E. and Rini, G. A., Fill flash luminescence to photograph luminol blood stain patterns, *J. Forens. Ident.*, 39, 149, 1989a.

Gimeno, F. E., Fill flash color photography to photograph luminol blood stain patterns, *J. Forens. Ident.*, 39, 305, 1989b.

Grispino, R. R. J., The effect of luminol on the serological analysis of dried human bloodstains, *Crime Laboratory Digest*, 17, 13, 1990.

Gross, A. M., Harris, K. A., and Kaldun, G. L., The effect of luminol on presumptive tests and DNA analysis using the polymerase chain reaction, *J. Forens. Sci.*, 44, 8837, 1999.

Harrow, B., Borek, E., Mazur, A., Stone, G. C. H., and Wagreich, H., *Laboratory Manual of Biochemistry*, 5th ed., W. B. Saunders, Philadelphia, 1960, p. 75.

Hatch, A. L., A modified reagent for the confirmation of blood, *J. Forens. Sci.*, 38, 1502, 1993.

Higaki, R. S. and Philp, W. M. S., A study of the sensitivity, stability and specificity of phenolphthalein as an indicator test for blood, *Can. Soc. Forens. Sci. J.*, 9, 97, 1976.

Hochmeister, M. N., Budowle, B., and Baechtel, F. S., Effects of presumptive test reagents on the ability to obtain restriction fragment length polymorphism (RFLP) patterns from human blood and semen stains, *J. Forens. Sci.*, 35, 656, 1991.

Hochmeister, M. N., Budowle, B., Sparkes, R., Rudin, O., Gehrig, C., Thali, M., Schmidt, L., Cordier, A., and Dirnhofer, R., Validation studies of an immunochromatographic 1-step test for the forensic identification of human blood, *J. Forens. Sci.*, 44, 597, 1999.

Holland, V. R. et al., A safer substitute for benzidine in the detection of blood, *Tetrahedron*, 30, 3299, 1974.

Hunt, A. C., Corby, C., Dodd, B. E., and Camps, F. E., The identification of human blood stains C a critical survey, *J. Forens. Med.*, 7, 112, 1960.

Huntress, E. H., Stanley, L. N., and Parker, A. S., The preparation of 3-aminophthalhydrazide for use in the demonstration of chemiluminescence, *J. Am. Chem. Soc.*, 56, 241, 1934.

Jakovich, C. J., STR Analysis Following Latent Blood Detection by Luminol, Fluorescein, and Bluestar, *J. Forens. Ident.*, 57, 193, 2007.

Johnston, S., Newman, J., and Frappier, R., Validation study of the Abacus Diagnostics ABAcard® HemaTrace® membrane test for the forensic identification of human blood, *Can. Soc. Forens. Sci. J.*, 36, 2003.

Kastle, J. H. and Shedd, O. M., Phenolphthalein as a reagent for the oxidizing ferments, *Am. Chem. J.*, 26, 526, 1901.

Kastle, J. H. and Amoss, H. L., Variations in the peroxidase activity of the blood in health and disease, Bulletin No. 31, U.S. Hygienic Laboratory, U.S. Government Printing Office, Washington, D.C., 1906.

Kerr, D. J. A. and Mason, V. H., The haemochromogen crystal test for blood, *Brit. Med. J.*, 1, 134, 1926.

Kipps, A. E., Gm and Km typing in forensic science—a methods monograph, *J. Forens. Sci. Soc.*, 19, 27, 1979.

Kirk, P. L., *Crime Investigation*, Interscience Publishers, New York, 1963, p. 105.

Landsteiner, K. and Weiner, A. S., An agglutinable factor in human blood, recognized by immune sera for Rhesus blood, *Proc. Soc. Exp. Biol. Med.*, 43, 223, 1940.

Lattes, L., On the practical application of the test for agglutination for the specific and individual diagnosis of human blood, translation in Gaensslen, R. E., *Sourcebook in Forensic Serology, Immunology and Biochemistry*, U.S. Government Printing Office, Washington, D.C., 1983, unit 9.

Laux, D. L., Effects of luminol on subsequent analysis of bloodstains, *J. Forensic Sci.*, 36, 1512, 1991.

Lee, H. C. and DeForest, P. R., The use of anti-human Hb serum for bloodstain identification, presented at 29th Annual Meeting, American Academy of Forensic Sciences, San Diego, CA, 1977.

Lee, H. C., Identification and grouping of bloodstains, in *Forensic Science Handbook*, Saferstein, R., Ed., Prentice Hall, Englewood Cliffs, NJ, 1982, chap. 7.

Loy, T. H. and Wood, A. R., Blood analysis at Çayönü Tepesi, Turkey, *J. Field Archeol.*, 16, 451, 1989.

Lytle, L. T. and Hedgecock, D. G., Chemiluminescence in the visualization of forensic bloodstains, *J. Forens. Sci.*, 23, 550, 1978.

Macey, H. L., The identification of human blood in a 166 year old stain, *Can. Soc. Forens. Sci. J.*, 12, 191, 1979.

Material Safety Data Sheet, Fluorescein, Mallinckrodt Baker, Phillipsburg, NJ, 1999.

Material Safety Data Sheet, 3-Aminophthalhydrazide (Luminol), Fluka Chemical Corp., Milwaukee, WI, 2001.

Material Safety Data Sheet, 3-Aminophthalhydrazide (Luminol), Fisher Scientific, Fair Lawn, NJ, 2007.

Miles Laboratories, Product Literature, Hemastix, 1981 and 1992.

Ouchterlony, O., Antigen-antibody reactions in gels, *Acta Pathol. Microbiol. Scand.*, 26, 507, 1949.

Ouchterlony, O., *Handbook of Immunodiffusion and Immunoelectrophoresis*, Ann Arbor Science Publishers, Ann Arbor, MI, 1968, chap. 5.

Ponce, A. C. and Verdu Pascual, F. A., Critical revision of presumptive tests for bloodstains, *Forens. Sci. Comm.*, 1, 2, 1999.

Proescher, F. and Moody, A. M., Detection of blood by means of chemiluminescence, *J. Lab. Clin. Med.*, 24, 1183, 1939.

RCMP, Fingerprint Development Techniques, www.rcmp.ca/firs/recipes/leucomalachite_e.htm, 2001.

Reynolds, M., The ABAcard®–Hematrace®. Confirmatory identification of human blood located at crime scenes, *International Association of Bloodstain Pattern Analysts News*, Vol. 20, No. 2, June 2004.

Sensabaugh, G. F., Biochemical markers of individuality, in *Forensic Science Handbook*, Saferstein, R., Ed., Prentice Hall, Englewood Cliffs, NJ, 1982, chap. 8.

Silenieks, E., Atkinson, C., and Pearman, C., Use of the ABAcard® Hematrace® for the detection of higher primate blood in bloodstains, 17th International Symposium on the Forensic Sciences, Wellington, New Zealand, 2004.

Spear, T. F., and Brinkley, S. A., The HemeSelect™ test: a simple and sensitive forensic species test, *J. Forens. Sci. Soc.*, 34, 41, 1994.

Sutton, T. P., Presumptive blood testing, in *Scientific and Legal Applications of Bloodstain Pattern Interpretation*, James, S. H., Ed., CRC Press, Boca Raton, FL, 1999, chap. 4.

Takayama, M., A method for identifying blood by hemochromogen crystallization, *Kokka Igakkai Zasshi*, No. 306, 463, 1912, translation in Gaensslen, R. E., *Sourcebook in Forensic Serology, Immunology and Biochemistry*, U.S. Government Printing Office, Washington, D.C., 1983, unit 9.

Teichmann, L., Concerning the crystallization of organic components of blood, *Z. Ration. Med.*, 3, 375, 1853, translation in Gaensslen, R. E., *Sourcebook in Forensic Serology, Immunology and Biochemistry*, U.S. Government Printing Office, Washington, D.C., 1983, unit 9.

Young, T., A photographic comparison of luminol, fluorescein, and bluestar, *J. Forens. Ident.*, 56, 906, 2006.

Vandenberg, N. and McMahon, K., Evaluation of OneStep ABAcard®–p30 and Hematrace® tests for use in forensic casework, 17th International Symposium on the Forensic Sciences, Wellington, New Zealand, 2004.

Widmann, F. K., Ed., *Technical Manual of the American Association of Blood Banks*, American Association of Blood Banks, Washington, D.C., 1981, 105, 106.

Zweidinger, R. A., Lytle, L. T., and Pitt, C. G., Photography of bloodstains visualized by luminol, *J. Forens. Sci.*, 18, 296, 1973.

Identification of Biological Fluids and Stains

15

Andrew Greenfield and Monica M. Sloan

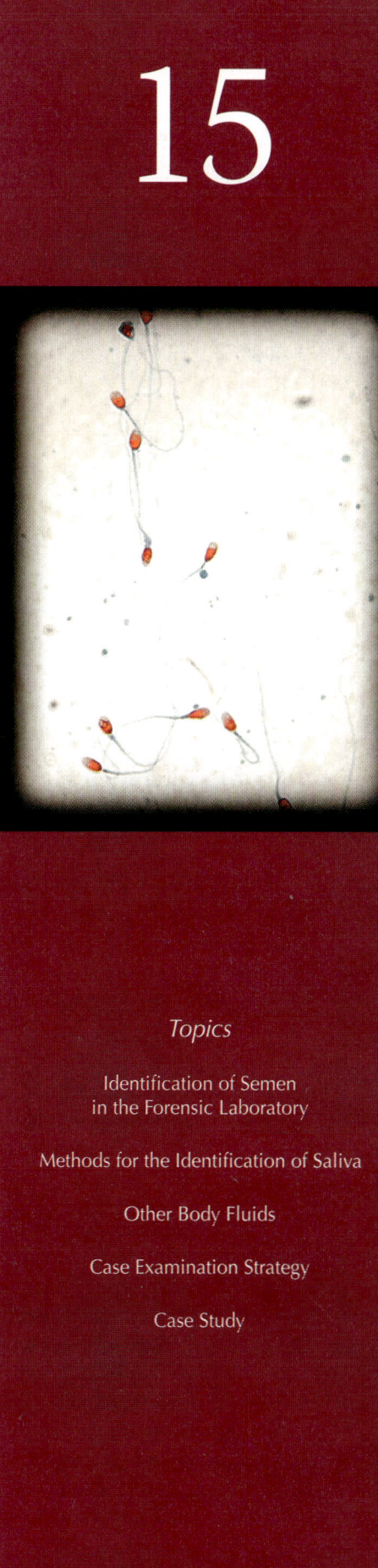

Introduction

Sexual assault crimes involve physical contact between perpetrator and victim, and consequently the transfer of materials, such as hairs, fibers, and body fluids, particularly seminal fluid and saliva, from one person to another. In many cases, the perpetrator is not known to the victim and the assault occurs in seclusion or at times when witnesses are absent. It is of paramount importance to locate and identify stains to assist with further investigation of the complaint.

Laboratory methods available today enable forensic scientists to detect and identify extremely small quantities of body fluids. Comparative DNA analysis may then be used to unequivocally eliminate an individual as a possible source of the fluids or attribute the origin to a particular individual with practical certainty.

The identification of saliva is also important to a wide variety of criminal acts. In sexual assault cases, detecting saliva may help corroborate an allegation of cunnilingus or fellatio. Saliva on a balaclava or cigarette butt discarded at a robbery scene may ultimately lead via DNA analysis to the perpetrator. Saliva can also be recovered from marginal sources such as stamps and envelope flaps during investigation of stalking or hate crimes.

Identification of Semen in the Forensic Laboratory

Semen Description

Semen is produced by postpubescent males and ejaculated following sexual stimulation. It is a semifluid mixture of cells, amino acids, sugars, salts, ions, and other organic and inorganic materials elaborated as a heterogeneous gelatinous mass contributed by the seminal vesicles, the prostate gland, and Cowper's glands. Ejaculate volumes of human males range from 2 to 6 milliliters and typically contain between 100 and 150 million sperm cells per milliliter. Certain disease states, genetic conditions, excessive abuse of alcohol or drugs, prolonged exposure to certain chemicals, and elective surgery procedures may result in a drastically reduced sperm count or complete absence of sperm cells from semen.

Sperm Cells

The principal cellular component of semen is the **spermatozoan** or sperm cell, a specialized, flagellated structure approximately 55 micrometers in length. The human sperm cell head is typically ovoid in shape with approximate dimensions of 4.5 micrometers in length, 2.5 micrometers in width, and 1.5 micrometers in thickness. The head contains the cell nucleus which is packed with deoxyribonucleic acid (DNA). The anterior portion of the head is capped with the acrosome. This structure is rich in enzymes to assist in penetrating the cell wall of the female egg during fertilization. A flagellated tail is attached to the head via a short midpiece and accounts for about 90 percent of the total length of a sperm cell.

A Brief History of Semen Identification

Surprisingly, medicolegal research into methods for semen identification was not widely conducted until the turn of the 19th century. The presence in semen of spermatozoa was communicated by Antonie Van Leeuwenhoek in 1679, more than 100 years earlier. Leeuwenhoek, in fact, credited this discovery to a Dutch medical student named Johan Hamm 2 years before that. Early tests involved studying suspected semen stains for odor, visual appearance, tactile properties, and chemical reactivity. Observations were compared with those from substances such as mucus and animal fat whose appearances were similar. By the mid to late 1800s, the microscope was the method of preference for semen identification, the observation of spermatozoa providing unambiguous information.

The morphologies of human and other animal spermatozoa were well documented, as was the ability to selectively stain the structural components of these and other cells.

The condition of azoospermia (semen lacking spermatozoa) and the consequent fallibility of tests that relied solely on the presence of sperm cells was acknowledged. Crystal tests in which chemical components of semen such as spermine and choline were detected and developed to assist with the identification of such samples. Characterization of enzymes in animal body fluids and tissues began in the 1920s and by the mid 1930s, a test was developed to test for seminal acid phosphatase (SAP). Modifications of the test remain in widespread use today. By the 1970s, analysis of stains for protein components of seminal fluid was commonplace.

Acid Phosphatase (AP)

Acid phosphatase are a class of enzymes that can catalyze the hydrolysis of certain organic phosphates. These enzymes are ubiquitous in nature and may be found in materials as diverse as mammalian liver and cauliflower stem juice. In 1935, Kutscher and Wolbergs discovered that human semen contains uniquely high levels of seminal acid phosphatase (SAP) compared with other body fluids and plant tissues. Ten years later, Lundquist suggested utilizing this fact as the scientific basis for the presumptive identification of semen. In males, puberty stimulates the large-scale synthesis of SAP by secretory epithelial cells that line the prostate gland. SAP levels remains high until the age of about 40, after which it gradually declines. No correlation exists between the level of SAP and the number of sperm cells present in an ejaculate, and no variation has been found between males with *normal* sperm counts and those who are clinically infertile or who underwent vasectomies.

Brentamine Fast Blue Test

Over the years, many methods for the presumptive identification of semen have been devised. Only one, the test for SAP activity, has withstood the test of time and is now used for this purpose in forensic laboratories worldwide, practically to the exclusion of all others. For this reason, it is the only method described in this text.

The concept of the test has been embraced in the form of numerous variations that all have a common thread—promoting a color reaction by using a substrate whose hydrolysis product reacts with a diazoniumsalt chromogen. In forensic laboratories, alpha-naphthyl phosphate is the preferred substrate and **Brentamine Fast Blue B**, the color developer. These two components are prepared separately in anhydrous sodium acetate and mixed to create a working solution or reagent that is sensitive enough to produce a positive result with semen diluted 500 times. This test is especially practical since it does not require that a suspected semen stain first be localized visually before applying it.

Application of the test is a very simple two-step procedure. Since SAP is readily soluble in aqueous media, a piece of absorbent paper (filter paper is ideal since it comes in a variety of sizes) or cotton swab is moistened with sterile water and applied to the questioned stain. The reagent is added to the paper or swab and development of an intense purple color noted (Figure 15.1). If an obvious purple coloration does not develop within 2 minutes, the test is recorded as negative. The intensity of the purple color of a positive reaction and the time taken for the color to develop should be recorded. The one-step process of adding reagent directly to a stain or swab in situ conveys no advantage if the method described above is correctly applied. The Brentamine

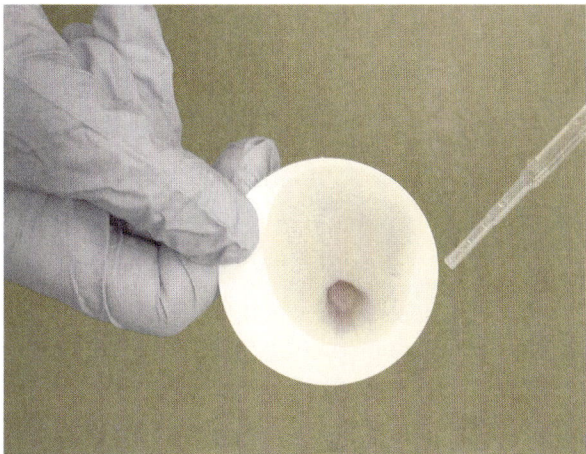

Figure 15.1 The Brentamine Fast Blue reaction. An intense purple color appearing within 30 seconds of application is a strong indication of a seminal stain.

Fast Blue B salt is potentially carcinogenic and liberal application directly onto items of clothing or areas of flooring is not recommended. The technique has several variations, depending on the size of the area or the type of item under investigation. The following list covers most situations commonly encountered in the laboratory.

Swabs

Swabs from orifices and skin constitute standard elements of most sexual assault kits. Swabs are also collected from substrates such as floors and walls and from the surfaces of condoms and other objects. A swab should be enveloped in a dampened 5-centimeter diameter filter paper and squeezed firmly with thumb and forefinger for 10 seconds or until the outline of the swab has clearly indented the paper. One drop of reagent is then applied to the indentation. Alternatively, if the swab has undergone a procedure for a microscopic sperm cell search (see section on confirmatory tests for semen), a drop or two of the extraction supernatant can be placed onto filter paper and a drop of reagent added on top. The practice of cutting away a piece from the swab for this purpose is acceptable but not recommended because it is unnecessarily destructive.

Underwear

An item of clothing such as underwear should be stretched across a clean Plexiglas™ sheet. Such sheets are available in many sizes. If the case history indicates that drainage of semen from the vagina into the underwear may have occurred, logic dictates that the most suitable area to test is the inside crotch panel. Dried semen appears as a yellowish-white crust, so, depending on the color of the garment, it may be readily visible. If so, a 5-centimeter diameter piece of filter paper can be folded twice (to make a pie shape), dampened with sterile water, and pressed firmly against the stain for 10 seconds or until the area under test is visibly moistened. The paper is then opened out and one drop of reagent applied. If no staining is visible, a single piece of filter paper of suitable dimensions to cover the area of search can be placed on the item, dampened with water, pressed firmly for 10 seconds, then lifted and exposed to reagent, which is applied in drop form. Markings should be made on the paper and item to ensure exact replacement to locate an area exhibiting a positive reaction.

Other Clothing

Sweaters, pants, jackets, and shirts are often submitted to the laboratory following a complaint or suggestion of rape where external (not into a body orifice) ejaculation is alleged or suspected, or where a victim states he or she wiped or spat the ejaculate. Generally, these items are fairly large and require a detailed search, especially when the case history does not specify where to look. Items should be visually scanned using either enhanced lighting or, if preferred, an alternative light source. Clothing items should be laid flat onto a clean surface and systematically searched. Unless the case history indicates otherwise, the search will be for stained areas that are clearly visible to the trained eye. For this reason, a logical approach should be taken when searching garments. A search should not take an inordinate amount of time to complete. Details such as the relative positions of victim and assailant should also be taken into account when performing a potentially lengthy item search. For example, if the victim faced the perpetrator throughout the incident, it makes no sense to search the back of the garment and the scientist is best advised to examine another item.

Bedding, Flooring, and Large Items

Sheets, duvets, mattresses, carpeting, and floored areas (often examined in situ) are ideal for prescreening using an alternative light source. Stains that become apparent following this nondestructive treatment may be tested as previously described. Since materials, such as washing detergents, contain agents that fluoresce or luminesce, a more efficient strategy is to use a slightly modified version of the press test. This applies also to items of clothing that are too large to search inch by inch in a time-efficient manner or items where stain definition may be problematic because of factors such as extreme soiling, pattern, and construction of the fabric.

The item should be laid flat on a clean surface. Individual strips of filter paper, available as precut rectangles of 46 centimeters by 57 centimeters, are placed over the area to be searched. Two to four well placed markings on the paper, each approximately 2 inches from the border, extending to the edge and onto the item, will enable the scientist to replace the paper, if necessary, back in the exact position where it was originally laid down. The outline of the item, parts thereof, seams, and other identifiers also should be marked. If more than one sheet is used, all should be numbered, corresponding to locations indicated by a schematic diagram in the notes. The paper is then saturated with sterile water using a spray bottle and firmly pressed down onto the item. The scientist should always be aware of certain characteristics of materials that may be tested. The degree of contact between the material and the paper depends on the nature of the material. Firmer, longer presses may be required to ensure close contact. The paper is removed and hung over an examination bar inside a fume hood. The reagent is then sprayed over the paper and areas of purple coloration outlined.

Alternatively, the reagent may be poured liberally over a large Plexiglas sheet and the filter strips slowly dragged through. The absorbency of the paper will ensure that the entire area is exposed to reagent. Positive areas on the papers provide a visual facsimile of where potential semen staining is located on the item. The paper may be placed back over the item and the now-located stains excised for further testing. Potential waste of reagent is an accepted disadvantage of this method, offset by the speed with which an examiner can complete the testing of large items.

Interpretation of Results

A strong reaction within 30 seconds is diagnostic for semen presence, notwithstanding that identification of spermatozoa or **prostate-specific antigen (PSA)** is the generally accepted standard (see the section below on confirmatory tests for semen). A positive reaction observed in less than 2 minutes in a stain located in an area with little likelihood of containing a contaminant (for example, the knee area of a pair of pants) should also merit confirmation. The forensic biologist must consider the presence of other body fluids, the time elapsed since alleged intercourse, and the fabric type when weak-to-moderate positive reactions are observed after a minute or so. Vaginal secretions, perspiration, feces, urine, or any combination of these fluids, particularly when repeatedly deposited, will typically exhibit such a reaction, especially in the crotch areas and seams of undergarments. Laboratory policy and individual discretion following thorough review of these issues will dictate whether a confirmatory test is undertaken in these situations.

Confirmatory Tests for Semen

Microscopic identification of sperm cells provides unambiguous proof that a stain under scrutiny contains semen. It is unusual for a forensic scientist to examine semen in which sperm cells are motile since motility is lost within 3 to 6 hours of ejaculation. However, established staining techniques greatly assist the trained eye to easily distinguish sperm cells from extraneous material such as epithelial cells. The most commonly encountered staining technique uses picroindi-gocarmine (PIC) and **Nuclear Fast Red** dyes and is colloquially referred to as the **Christmas tree stain**. It was developed specifically for sperm cell visualization. Different parts of the sperm structures are singularly colored and contrast well with the colors taken up by epithelial cells.

Preparation of Sample for Staining

Procedures for removing sperm cells from items may be modified in order to answer one of two questions: (1) Is this a semen-containing stain? (2) Is this a semen-containing stain and, if so, does it require DNA testing?

The identification of a single intact sperm cell or isolated sperm head may be sufficient to report the presence of semen or attempt DNA testing. If an answer to the second question is sought, it is not necessarily advisable that all available spermatozoa be removed

for microscopic examination. Only half of the genetic information of the donor is present in a single sperm cell, so approximately 80 cells are required to generate a full male DNA profile using modern techniques of DNA analysis. It is essential that the forensic scientist not only locates and identifies human semen on items, but retains an amount sufficient to allow DNA testing to be performed successfully. This is particularly important when dealing with small stains on clothing or internal body swabs taken at postcoitus intervals approaching the maximum period of spermatozoa detection. As a last resort, the material on a prepared slide can be collected for further analysis.

A suitable extraction procedure is to soak swabs and cuttings in a tube of sterile water for 5 to 10 minutes while applying periodic gentle agitation of the sample using a wooden applicator stick. The swab or cutting is then placed inside a filtering receptacle that can be piggy-backed onto the tube, covered, and centrifuged for 5 minutes or until a pellet is formed. The supernatant is removed and the pellet reconstituted in a few microliters of sterile water.

The suspension is spotted onto a labeled microscope slide situated on a slide heater. Repetitive spotting of the same area is acceptable if the pellet is small, although care should be taken not to overconcentrate the area to be searched. Once dry, the slide is removed from the warmer and waved over a Bunsen flame for a few seconds to heat fix the stain. After cooling, the spot is saturated with Nuclear Fast Red for 10 to 15 minutes, after which the excess is carefully washed away with sterile water. The spot is then saturated with PIC reagent and the excess carefully washed away with methanol after 10 to 15 seconds. The spot is air dried, protected with a cover slip, and prepared for examination under the microscope at 400× magnification for screening or 1000× magnification for detail.

More vigorous extraction procedures, such as vortexing or ultrasonication, are not necessarily any more efficient at extracting sperm cells than more passive agitation, particularly where it really matters—at the low end of sample availability. When a sample is saturated with spermatozoa, the extraction of 100 or 100,000 sperm for observation is irrelevant, particularly when only a single sperm cell head may be required to report the presence of semen. Retaining sufficient sample for retesting at the earliest practicable stage of the examination should also be considered. Aggressive extraction methods may unnecessarily compromise the integrity of the sample; in particular, cleaving tails from the head.

Identification of Sperm Cells

Sperm cell heads are ovoid and exhibit characteristic differential staining. The anterior portion is pink and the posterior is dark red or purple and often appears shiny (Figure 15.2). Sperm cell tails stain yellow-green and the midpiece stains blue. Epithelial cells also take up the stain and appear blue-green with red nuclei (Figure 15.3). Morphologies of sperm cells of other animals are documented and reference collections are available. It is unusual to encounter the spermatozoa of other animals during a search. Their possible presence should be apparent from the case history.

Proteinase K

To make searching for sperm easier, a working solution of **proteinase K** (ProK) enzyme, named for its keratin hydrolyzing activity, and sodium dodecyl sulfate (SDS) may be added to the extraction pellet if the sample is known

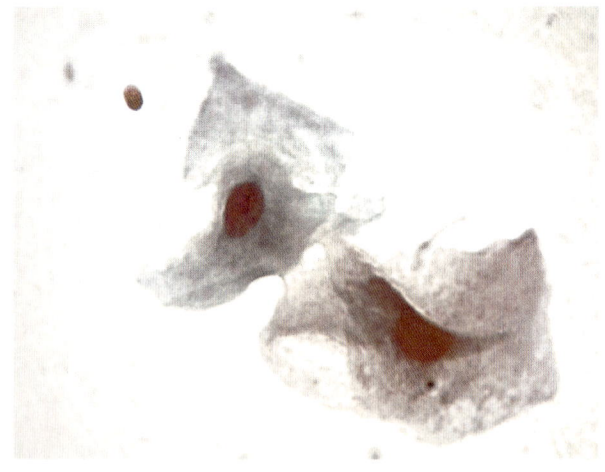

Figure 15.3 Human spermatozoon and nucleated epithelial cells stained with Christmas tree stain (Nuclear Fast Red and picroindigocarmine). (Original magnification 1000×.)

or expected to be abundant in epithelial cells. SDS denatures proteins, unfolding them into polypeptide chains which ProK breaks down into smaller molecules. Sperm head proteins are rich in sulfur-containing amino acids, making them largely unaffected by this treatment and resulting in selective lysis of epithelia. However, sperm cell tails will also be lysed.

Recording Observations

A common method of recording sperm cell density under magnification is to use a six-point scale first proposed by Kind in 1964. The exact definitions have evolved over the years to suit individual preference, but essentially the classifications are such that, starting from a negative result, a gradual increase in sperm cell density is defined. A typical scheme is shown below:

0   Negative; no sperm detected in the extract.

+FEW   Fewer than 20 sperm cells identified following search of an entire spot or **smear**.

1+   More than 20 sperm cells identified; sparsely distributed.

2+   One or more sperm cells identified in some fields; total exceeding 20 where it is not necessary to search the entire spot or smear.

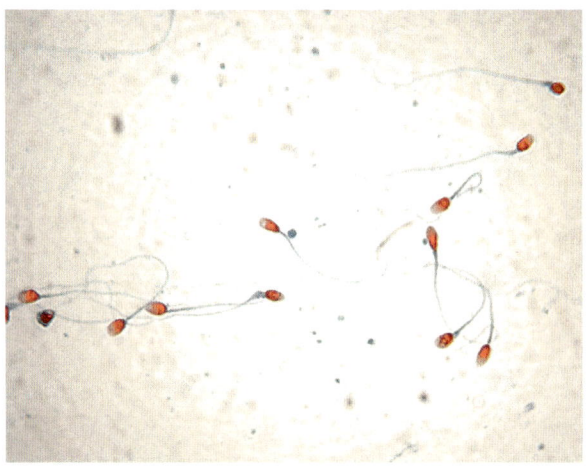

Figure 15.2 Human spermatozoa stained with Christmas tree stain (Nuclear Fast Red and picroindigocarmine). (Original magnification 1000×.)

3+   One or more sperm cells identified in many fields.

4+   One or more sperm cells identified in most, if not all, fields.

Prostate-Specific Antigen (PSA, p30)

In 1978, Sensabaugh identified PSA or p30 as a suitable marker for demonstrating the presence of semen. In the absence of spermatozoa in an acid phosphatase-positive specimen, the detection of PSA is proof positive for semen. PSA is secreted into seminal plasma by the prostate gland and varies in concentration from 300 to 4000 nanograms per milliliter.

Since the prostate gland lies distal to the customary point of interruption in a vasectomy, this procedure has no effect on the elaboration of PSA into the semen. The urine and serum of males, breast milk, and sweat glands contain levels of PSA that are usually below the limits of forensic detection. PSA is found in elevated levels in the serum of males with prostate cancer but since a positive AP reaction is usually required in tandem with PSA detection for semen to be unambiguously identified, this, too, is of little consequence forensically.

Classically, the detection of p30 was achieved using immunoelectrophoresis. This method is time-consuming, has limited sensitivity, and has been largely superceded by a screening test developed by Hochmeister et al. in 1999. This test is also based on an antibody–antigen reaction, but is quick, sensitive, and simple to use. A small portion of material to be tested is immersed for 10 minutes in sterile water and centrifuged. Two hundred microliters of this liquid are placed in a receiving well, and the liquid is drawn over the membrane by capillary action. The appearance of three bands within 10 minutes, two of which are the control and internal standard, is a positive result for increased levels of PSA (Figure 15.4).

Time Since Intercourse (TSI)

Can an estimate of postcoital interval be made based on laboratory findings? If so, how reliable is it? Expert opinion as to the length of

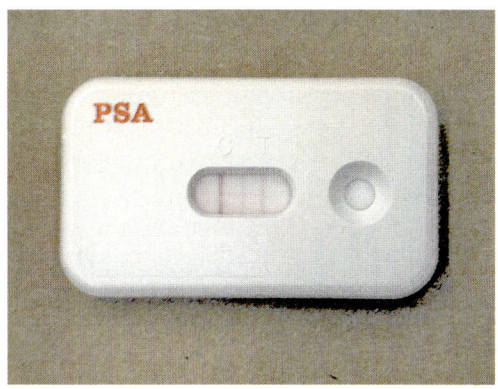

Figure 15.4 Seratec™ PSA test kit. The line closest to the sample well indicates a positive result. The middle line is an internal standard. The line on the left is the positive control.

time elapsed between a sexual act and collection of the evidence has been sought since the day methods for the detection of semen were devised. The question remains one of medicolegal importance, on which the outcome of a trial or a decision to proceed with criminal charges may hinge.

Since this text is largely devoted to the identification of semen, we make no attempt to regurgitate the studies, results, and conclusions in the forensic literature. However, comprehension of the salient points and limitations that underpin the opinions of a forensic biologist is essential, particularly since these issues have a direct bearing on the ability to identify semen. The following is a summary of the variables and issues for consideration in this important area of body fluid analysis. The References and Suggested Readings section at the end of this chapter should be consulted for a more thorough understanding of the subject.

A review of the literature reveals remarkable variation in quoted maximum survival times of spermatozoa in body orifices. This is not altogether surprising since the data are gathered from victims whose accounts of events and timelines may be understandably inaccurate, whose allegations may be false, and whose level of trauma may be such that examining physicians and nurses may not be able to take samples with consistency. Some experimental subjects are laboratory volunteers who are not always willing to fully disclose their sexual behaviors and whose postcoitus actions may not be typical of sexual assault victims.

The generally accepted maximum times for sperm heads to be detected in living persons are 7 days in the vaginal cavity, 2 to 3 days in the anus and rectum, and 24 hours in the mouth. Intact (tail attached) spermatozoa are commonly observed on vaginal swabs taken between 0 and 26 hours after intercourse and are rarely observed in oral, anal, and rectal samples after 5 hours. AP and PSA results are not used to approximate TSI due in part to the presence of AP in vaginal secretions and variations in detection methods of PSA.

High levels of AP may be detected in the vagina up to 72 hours postcoitus. PSA is rarely detected in the vagina 24 hours after intercourse. Any TSI estimate is made on the assumption that the levels of seminal constituents drop over time in a predictable manner. Variations in the seminal constituents of ejaculates within and among donors, the initial number of spermatozoa deposited, testing and preparation techniques, ambiguity in sperm cell scoring, the actions of the victim postcoitus, and whether the victim was menstruating, had defecated, or voided, may all render a TSI approximation exactly that. When asked to provide an opinion on TSI, hypotheses and scenarios should be proffered that illustrate the limitations of the results and account for all possible explanations for the findings on which the opinion is based.

Semen Stains on Clothing

Since SAP is readily water soluble, laundering will remove the enzyme to the point that it cannot be detected with the Brentamine Fast Blue test. SAP activity may be, however, detected after clothing is dry cleaned. Laundering does not necessarily remove all traces of spermatozoa, so it may be appropriate to prepare a sample for a microscopic search from a cutting of an area where semen was potentially deposited, such as the crotch of underwear, or from any area of interest as indicated by the victim or investigator.

When semen-free garments are laundered with those that are semen-stained, it is possible

for small numbers of sperm cells to transfer from the latter to the former. Identification of a few sperm cells in the absence of a positive AP result should be reported with caution. The persistence of SAP in dried semen stains varies, depending on the storage history of the item. Under favorable conditions of low temperature and humidity, little or no loss of activity may occur for several years.

Examination Strategy

In many cases, items in a sexual assault kit are submitted to a laboratory and routinely analyzed in sequence according to laboratory policy. A smear prepared from an orifice swab is usually examined first. Liquid samples (aspirates or washes) should be examined next to preserved swabs for retesting or DNA analysis. If these samples are negative for sperm cells, appear to have been collected correctly and well within the accepted maximum survival time for spermatozoa, no condom was used, and the sample was collected within a matter of hours of the incident, it is unnecessary to process a vaginal swab for a sperm search. A more appropriate examination would be to test the swab for the presence of SAP and, if positive, PSA. If the victim states that a condom was worn, examination of internal samples is not necessary. Testing samples collected after the accepted maximum survival times for seminal constituents also falls into this category, and it may be more appropriate to examine the underwear. Items of clothing should be examined for the possibility of external ejaculation in a logical sequence, based on the case history. Laundered garments generally should not be examined (see Semen Stains on Clothing).

After semen is identified, depending on circumstances, it may not be necessary to examine samples from more than one orifice even though penetration of more than one orifice is alleged, unless the case involves multiple assailants. Gravitational drainage of semen from the vagina to the anal area usually renders finding semen on swabs of both orifices meaningless. If sperm cell density is clearly greater on the anal swab compared with the vaginal sample, inference of anal penetration may be reported. However, it is essential that the samples are processed identically in the laboratory, assuming the examining nurse or physician took them consistently.

Methods for the Identification of Saliva

Saliva Described

Saliva is a slightly alkaline secretion comprised of water, mucus, proteins, salts, and enzymes found in the mouth. Humans produce 1 to 1.5 liters of saliva per day. Its primary purpose is to aid in the initial stages of digestion by lubricating food masses for easier swallowing and initiating the digestion of starches using the enzyme amylase.

Amylase

No test is specific for saliva. Traditionally, forensic tests for saliva relied primarily on the detection of alpha-amylase, an enzyme sometimes referred to as *ptyalin* in older references. **Amylases** are ubiquitous enzymes found in both animals and plants. They are responsible for the catalysis of the components of starch, amylose, and amylopectin into smaller, less complex sugars. Amylases can be subdivided into alpha and beta categories. Beta-amylases are found in plants and alpha-amylases in animals, including humans. Beta-amylase attacks only the bonds at the ends of polyglucan chains, unlike alpha-amylase, which catalyzes bonds within the chains. Primates, pigs, elephants, and certain rodents have high levels of amylase.

In humans, two DNA loci, AMY1 and AMY2 found on chromosome 1, code for amylase. The AMY1 **locus** codes for the amylase found in saliva, breast milk, and perspiration. The AMY2 locus codes for the amylase found in the pancreas, semen, and vaginal secretions. In the mouth, alpha-amylase is secreted primarily from the parotid glands as a component of saliva. While alpha-amylase is found in many body fluids and tissues, it still serves as a good marker since it is found in levels 50 times higher in saliva than in most other body fluids and is relatively stable. Some activity can be detected for up to 28 months.

Starch–Iodine Test

One of the earliest methods for saliva detection was the starch–iodine test devised by W. Roberts in 1881 (see Gaennslen, 1983, under General References). In the presence of iodine, starch appears blue. As amylase acts on starch to break it down, the color changes and subsides. This test has several drawbacks; for example, the presence of proteins, particularly albumin and gamma-globulin originating in other body fluids, such as blood and semen, compete with starch for iodine and produce false positive results. This test is also difficult to use as a locator test for stains on items. One version of this method is still in use as the radial diffusion test. An agar gel containing a known concentration of starch is prepared, and iodine is poured over the gel. An extract of the questioned sample is added to the wells in the gel. As the starch diffuses, it breaks down, leaving a circular void area proportional to the amount of amylase present.

Phadebas™ Reagent

More common is the use of commercial products in which starch is linked to a dye molecule to form an insoluble complex. When starch is cleaved from the dye by amylase, the dye molecule becomes soluble, producing a colored product that can be measured with a spectrophotometer. The degree of coloration is proportional to the amount of amylase in the sample. This procedure is commonly referred to as a *tube test*. Alternatively, the reagent can be dissolved and applied to large sheets of filter paper, which can then be placed on an item to map the location of amylase-containing stains for further analysis. This procedure is referred to as a *press test*. Procion red amylopectin (PRA) or **Phadebas reagents** are the most commonly used. The sensitivities of the two methods are similar in that they detect dilutions of approximately 1/128 in stain format. PRA has the disadvantage of requiring the use of ethylene glycol monomethyl ether, a noxious agent, to make the test paper.

The Phadebas reagent is manufactured by Pharmacia and is produced in tablet form. It is used clinically to detect alpha-amylase in urine, serum, and plasma, which may denote certain medical conditions, such as pancreatitis.

The forensic use of Phadebas was proposed by Willott in 1974. He standardized the clinical method that uses a known test volume instead of a given stain size. He also developed values for the ranges of amylase in saliva and other body fluids, such as semen and vaginal secretions. Thus, if a mixture of body fluids is encountered, it is possible to subtract the value of the amylase contribution from the other body fluid and determine whether the remainder is indicative of saliva.

In 1981 Willott went on to elucidate a press test method for saliva using the Phadebas reagent. It was found to have equal sensitivity to the PRA test without the use of noxious chemicals because the tablets simply are dissolved in distilled water then sprayed onto filter paper.

Press Test

Phadebas paper is made by dissolving six Phadebas tablets in 30 milliliters of distilled water for every sheet of paper to be made. Whatman #1 grade filter paper sheets measuring 46 by 57 centimeters (18 by 22 inches) or its equivalent are used. The solution is sprayed evenly onto one side of the paper using an atomizer and allowed to air dry (although the sheets may also be used immediately). Dried sheets can be stored up to one year in a dark plastic bag at room temperature.

The item to be tested must be fairly flat to ensure good contact between it and the paper. This can be accomplished by stretching the item over a frame of appropriate size. Polycarbonate (Plexiglas) or glass plates work well.

A piece of Phadebas is placed over the entire area to be tested. The paper is dampened slightly by spraying with distilled water, taking care to avoid overwetting the paper. A rough outline is drawn on the paper to aid

Figure 15.5 Phadebas™ press test. The position of the test paper is marked on the item for accurate relocation.

in accurate relocation of stains of interest (Figure 15.5). A piece of plastic wrap is placed on top to prevent the paper from drying during testing. Another Plexiglas plate is placed on top and a weight is applied to ensure good contact between the test paper and the item.

The test is observed for 40 minutes, with notes made of the time of the initial appearance of a reaction and its development over the 40-minute period. The reaction appears as a light blue smooth area in contrast to the grainy appearance of the test paper (Figure 15.6). The test should be observed every minute for the first 10 minutes and every 5 minutes thereafter through the layers of the "sandwich" to

Figure 15.6 Phadebas™ press test. A positive reaction appears as a smooth blue region. These areas are circled on the paper and on the item.

avoid dislocating the test paper. It may not be necessary to observe the test for a full 40 minutes if strong positive reactions are noted earlier. Continuing the test may serve to dilute the amylase and other cellular constituents, potentially hindering further analyses. At the end of the test period, the weight, Plexiglas, and plastic wrap are carefully removed. Any positive areas are outlined on the paper and on the item. A Brentamine Fast Blue test for acid phosphatase can also be done by spraying the paper after the testing and documentation for amylase are complete. The paper is allowed to dry and can be kept indefinitely.

Tube Test

A 15-milliliter, sterile, conical tube is prepared for each sample and control area to be tested, along with tubes for negative and positive controls. A 3 by 3 millimeter sample is excised from each stain (or its equivalent, depending on the material) and placed in a tube. If a Phadebas press identified the stain area, the corresponding 3 by 3 millimeter piece of Phadebas also should be included. Add 4 milliliters of distilled water to each sample and incubate in a water bath at 37 degrees Celsius for 5 minutes, ensuring that the tubes are capped while incubating. One Phadebas tablet is added to each tube and mixed to create a suspension. Samples are incubated for exactly 15 minutes at 37 degrees Celsius, then 1 milliliter of 0.5 molar sodium hydroxide solution is added to stop the reaction. The tubes are then centrifuged to remove the granular remnants of the tablet and the original stain sample from the supernatant, which is then transferred to a clean 13 by 100 millimeter glass tube. Using a spectrophotometer, such as the Spectronic 20, the optical density of the supernatant is read. The negative control can be used to zero the machine or alternatively its value can be subtracted from all test values.

The optical densities are converted to units per liter by consulting the standard curve tables specific to each batch of Phadebas tablets. This figure is divided by 5000 to convert it to a unitless measure. This must be done to standardize the value to the size of the test cutout rather than to a sample volume as is done with clinical samples. Typically, a corrected value of 0.03 is indicative of saliva in the absence of the other contributing body fluids.

Limitations

One of the drawbacks to using amylase as an indicator for saliva is its presence in other body fluids. Although typically found at much lower levels than in saliva, it is still possible for the range of values to overlap. One such body fluid is feces, where amylase, originating from the pancreas, is found in high levels. Despite the fact that the amylase results from two different gene products, they cannot be distinguished by the tests discussed above. Repeated depositions of fluids, such as vaginal secretions, semen, perspiration, or combinations, may also increase the levels of amylase to those considered indicative of saliva.

Another limitation is the sensitivity of the test. The Phadebas test may not detect low levels of amylase or may indicate levels that are more consistent with other body fluids. Low levels may result from very small stains, stains that have been diluted, or from individuals whose amylase levels are naturally low.

Sampling Strategy for DNA Typing

It is not always necessary to perform both the press test and the tube test. This decision will be based on the case history, the importance of body fluid identification, and the importance of DNA typing for source identification. It is important to remember no correlation exists between the level of amylase and the ability to obtain a DNA profile.

If no obvious contaminating material is present and an item appears clean, a press test may be the only one required to identify amylase and localize a stained area for DNA testing. However, if an item is heavily contaminated with other body fluids and the identification of saliva is important, it will be necessary to do a tube test to quantify the amylase and assess the levels as indicative of saliva or simply the product of other body fluids. In-house validation done at the Centre of Forensic Sciences (Toronto) revealed that only saliva stains reacted within 10 minutes, while other body fluid stains, including repeated deposition stains of perspiration and vaginal secretions, did not react for at least 15 minutes with a press test. Validation work of this type may reduce the need for the tube test, especially if DNA analysis is more important.

Body fluid identification is not always necessary, depending on the location of the stain or the type of item examined, and it may be detrimental to DNA analysis by consuming or diluting cellular constituents. Stamps and envelope flaps are good examples of situations where the body fluid present can be inferred, thus allowing retention of as much material as possible for DNA analysis to identify the source. A swab from a bite mark is similar since the bite mark provides corroborating information for the identity of the body fluid. The collection of a bite mark by swabbing may have already diluted a limited sample and made the amylase undetectable. Further testing for amylase will reduce the amount of material available for DNA analysis. If the identity of the saliva donor is important, it is prudent to preserve as much material as possible for DNA analysis.

Other Body Fluids

Urine

While the chemical detection of urine may play a role in sexual assault, harassment, and mischief cases, it is performed less frequently by forensic laboratories because of the insensitivity of the tests and the low success rate with DNA profiling. Like many other analyses, detecting urine relies on visual examination for stains with characteristic appearances. Stains may fluoresce or luminesce under alternate light sources, but diluted stains are harder to detect. Odor is another characteristic of urine, but it generally permeates an entire item and is not localized to the urine stain itself.

Historically, methods for the identification of urine have relied on identifying inorganic ions or **organic compounds** that concentrate in urine. It is advisable in the forensic context to attempt to identify more than one component for confirmation due to the ubiquity of the substances in question. Urine detection relies on identifying two organic compounds: **urea** and **creatinine**. Both components are found in other body fluids, such as perspiration, blood, saliva, and semen. Urea is present in high levels, approximately 1400 to 3500 milligrams per 100 milliliters. Creatinine is present at about one tenth of these values, with average concentrations of 105 to 210 milligrams per 100 milliliters. While urea and creatinine are found at relatively high levels in liquid urine, they may be difficult to detect in stains, the most commonly encountered forensic samples. As liquid urine is absorbed into fabric surfaces, it spreads across the surface, effectively diluting the test components.

Testing for urea relies on the use of urease, an enzyme that breaks down urea and releases ammonia and carbon dioxide. The ammonia is then detected using an indicator chemical, such as **Nessler's reagent** (mercuric iodide in potassium iodide) or DMAC (***p*-dimethylaminocinnamaldehyde**). **Azostix®**, a commercial test strip used for the clinical detection of urea in blood, relies on the same principle, except it measures a shift in pH caused by the formation of ammonia hydroxide.

Creatinine is detected by applying a saturated solution of picric acid in toluene or benzene to a stain extract. It combines to form creatinine picrate, an easily detectable colored product.

Attempts have been made to use immunological methods to identify urine components. One such method uses crossed-over electrophoresis to detect Tam–Horsfall glycoprotein (THG), a protein thought to originate in the kidney. It is specific to urine, but not to humans, having been detected in other animals. This would necessitate another immunological test for species determination. Based on the time-consuming nature of the testing and the infrequency with which urine detection is required, it is unlikely that further research will be conducted.

DNA Typing of Urine

Urine is comprised primarily of water and salts. Some cellular content comes from epithelial cells washed from the urinary tract, along with erythrocytes and leucocytes. Urine stains are very dilute and have low cell concentrations. This reduces the likelihood of successful DNA analysis. The bacterial content of urine may also hinder DNA analysis through degradation. Liquid urine samples give a better rate of success; larger volumes of urine will contain more cells that can be concentrated through centrifugation.

Feces

Feces are the end products of digestion after absorption of nutrients and reabsorption of water. Feces consist of undigested or undigestible food residues, mucosal cells, and bacteria. Bacteria comprise approximately one third of the dry weight of feces. The average fecal output for an adult is approximately 100 grams per day.

Detection of fecal stains is dependent on stain appearance characterized by a green-brown color caused by bile pigments and their breakdown products. Another distinguishing characteristic is the odor caused by the breakdown of amino acids into indole, skatole, and methyl mercaptan by bacteria. Feces can also be identified by the presence of microscopic components of foodstuffs. Undigested remnants of plant material, such as cell walls, starch grains, and striated muscles from meat can be identified.

The most common means of identification is the detection of the **urobilinogen** and urobilin products of bilirubin metabolism. These pigments are produced in carnivorous and omnivorous animals but not by herbivores. Infants under the age of 6 months produce little to no urobilinogen. When combined with alcoholic zinc acetate, urobilinogen is oxidized to urobilin, which is soluble in alcohol. A urobilin–zinc salt com-pound that fluoresces a bright apple green in long-wave ultraviolet light is formed. The method is not very sensitive. It requires stains concentrated enough to be visible and to have a characteristic odor.

DNA Typing of Feces

Although about 1010 cells are lost from the gastrointestinal tract per day, nuclear DNA testing of feces has proven relatively unsuccessful due to the inhibitory effects of the bile pigments even at low concentrations. The large amounts of bacteria and digestive enzymes also serve to degrade the DNA. Mitochondrial DNA testing has proven more successful with these kinds of samples.

Vomitus

No test detects the presence of vomitus. Various components that can be tested for include low pH resulting from stomach acids, amylase, and the microscopic and macroscopic identification of foodstuffs. Low pH is difficult to detect in stain form using pH strips as the addition of water to moisten the stain or strip for the transfer of material is often sufficient to neutralize the acid. Identification of microscopic components of foodstuffs is similar to the procedure outlined in the section on feces. It may be possible to identify undigested macroscopic components, depending on the interval between ingestion and vomiting. It may help to determine time since death if the details of a last meal are known.

Vaginal Secretions

The identification of vaginal secretions is especially important when a case involves an allegation of a foreign object inserted into the vagina as part of a sexual assault. Vaginal secretions are usually identified on the basis of detecting glycogenated epithelial cells. A **periodic acid–Schiff** (PAS) reagent serves to stain the **glycogen** in the cellular cytoplasm a bright magenta color. The cells can then be rated based on staining intensity on a scale of 1 to 4—4 being the most intense.

The test is not conclusive as the amount of glycogenation varies depending on the stage in the menstrual cycle; glycogenation levels are highest around the time of ovulation. Glycogenated cells are absent from pre-pubescent females and are uncommon in postmenopausal women, although estrogen replacement therapy will affect the measurement. Glycogenated epithelial cells are also not unique to the vaginal tract and can be found in smaller numbers in the mouth and in the urethral tracts of males. Thus, the finding of only a few cells is problematic for interpretation.

The test may also consume a large amount of the questioned material and reduce the amount available for DNA testing. It may be prudent to forego PAS testing in favor of retaining material for DNA testing. However, if the origin of the cells is particularly important (e.g., does a DNA profile from a bottle come from someone drinking from it or from vaginal insertion), this testing may need to be done.

Case Examination Strategy

The forensic biologist conducts relevant, scientifically accepted procedures to determine whether a body fluid is or is not present, then records and reports his or her findings in legally admissible form. Details provided in the case history, notes from examining nurses and physicians, and information from investigating officers should dictate how the case is processed in the laboratory. Preliminary results, such as the recording of a strong positive AP result after a few seconds, should also be used to determine the most appropriate action to take. In this example, only a portion of the swab (especially if it is the only one available) or a small cutting of material should be prepared for a sperm cell search. The items available for examination, retention of sample, and deciding the question to be answered—body fluid identification, association by DNA analysis, or both—will factor in the decision.

Some of the factors to consider are case type and issues (whether the victim is alive to act as a witness to the events; whether the suspect denies intercourse occurred or claims that intercourse was consensual); awareness of the victim (the age and mental capacity of the victim; whether drugs or alcohol were involved); the number of people involved (whether to look for body fluids from more than one person); relationship of the principals (whether they had a previous relationship that would explain certain evidence); time elapsed since the incident (in a sexual assault examination; some orifice samples may no longer be relevant with the passage of time); and actions that occurred after the event (whether victim bathed and whether clothes and bedding have been washed).

Review the items available for examination and select the items that will best answer the questions set out in the purpose statement. If you have multiple samples from an orifice (for example, a vaginal smear, rinse, and swabs), examine in a stepwise fashion for the presence of semen. An effective systematic strategy minimizes the number of examinations undertaken and allows the scientist to retain as much material as possible for DNA analysis and retesting.

Generally, only those tests that support the purpose of the examination should be conducted. Good laboratory practice should be followed at all times. Semen and saliva may contain biohazardous elements. Also, the importance placed on forensic scientists to ensure that they do not contaminate items cannot be overstressed. A clean mask, gloves, and laboratory coat are essential items of apparel.

Light Sources

The first step in a forensic examination is the visual examination of items using a good light source to detect any visible staining. Alternative light sources use different wavelengths of light that may cause a body fluid of interest to be more visible through fluorescence or luminescence. This is especially useful on dark colored or highly patterned articles or when searching for staining on large items (Figure 15.7).

Long-wave (320 to 400 nanometers) ultraviolet lights were once the most common

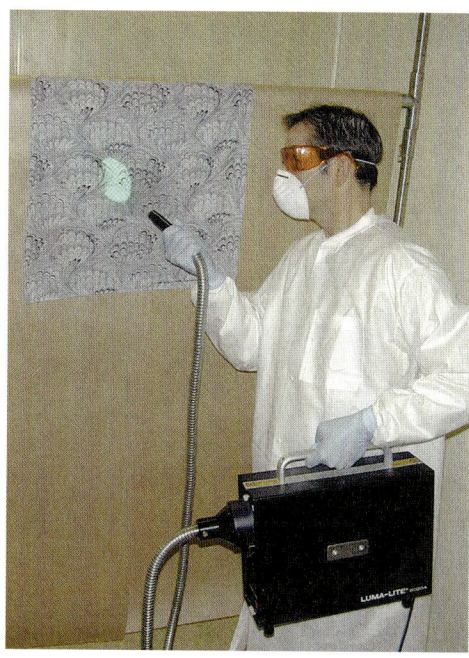

Figure 15.7 The detection of a semen stain using Luma-Lite™. Alternate light sources are helpful for searching large or highly patterned items where staining is difficult to see in normal light.

alternate light sources. Ultraviolet light is very good for detecting semen, saliva, and blood, even in dilute concentrations. Its disadvantages, however, are many. Short-wave ultraviolet light (180 to 280 nanometers) is very harmful to humans. It can cause burns to exposed skin and impair vision if viewed without protective eyewear. Short-wave ultraviolet light is also known to damage DNA through the formation of **thymine**–thymine dimers. In fact, ultraviolet light is often used as a sterilization/decontamination method for inactivating extraneous DNA within laboratories. With the advent of today's DNA technologies, it is clear that ultraviolet light is not necessarily the best choice for an alternate light source.

The other alternatives are laser light sources and high intensity lights used with filters that effectively render the emitted light as a single wavelength similar to long-wave ultraviolet light. While ultraviolet light seems best at detecting blood, the alternate light sources often outperform it in detecting semen and saliva, especially in diluted form. It is important to remember that a positive reaction using these methods is not confirmatory for the presence of any body fluid since other substances in conjunction with the substrate may also fluoresce.

There are numerous examples of alternate light sources such as the Spectra-Physics™ Model 171-19 argon ion laser, Plasma Kinetics™ Model 151D copper vapor laser, Omnichrome Omniprint™ 1000 tuneable wavelength light source, and Payton Scientific Luma-Lite™ high intensity quartz arc tube. These instruments are all comparable in their abilities to detect the body fluids of interest. The choice often comes down to factors, such as price, portability, ease of use, beam width (for example, the Luma-Lite has a larger beam width than the Omniprint, making it more efficient for searching large items), and maintenance costs. They also have the added advantage of not damaging DNA, and they are less harmful to human health.

Case Study

A city transit ticket collector was stabbed to death in his booth at a subway station. Surveillance camera footage showed a man wearing a hooded sweatshirt, which he subsequently threw in a garbage can. The garbage was removed to a disposal site. A search of the site revealed the sweatshirt. DNA testing detected blood that matched that of the ticket collector. Further testing of the item showed two semen stains that were a mixture of DNA from an unknown male and DNA from two females. Amylase detected on the drawstrings of the sweatshirt hood matched the DNA profile of the male semen donor. Interviews with the two female victims led to the arrest of a suspect. When his DNA was profiled, it matched the profile from the semen and the drawstrings of the sweatshirt. He subsequently proffered a guilty plea.

Questions

1. Where in the male reproductive tract is acid phosphatase produced?

2. Name two tests considered confirmatory for the presence of semen.

3. Give reasons for the absence of tails on spermatozoa on a microscope slide preparation.

4. How long might seminal fluid constituents be detectable after deposition (a) inside the vagina, (b) in dried form, and (c) on fabric after laundering/dry cleaning?

5. Beta-amylase is found in plants. Will plant extracts or stains react with the Phadebas test? Why or why not?

6. Assume you detected amylase using a Phadebas press test on a pair of panties worn after a sexual assault in which cunnilingus is alleged. Discuss the factors that would determine your choice of area for DNA analysis.

7. When would it be acceptable to omit the use of a test to identify amylase?

8. Why is it preferable to identify both urea and creatinine in suspected urine stains?

9. Describe two methods for identifying fecal material.

10. If only a few glycogenated epithelial cells were detected from a beer bottle, what other test(s) might help you determine the origin of the cells?

References and Suggested Readings

Semen

Allard, J. E., The collection of data from findings in cases of sexual assault and the significance of spermatozoa on vaginal, anal and oral swabs, *Sci. Justice*, 37, 99, 1997.

Chapman, R. L, Brown, N. M., and Keating, S. M., The isolation of spermatozoa from sexual assault swabs using proteinase K., *J. Forens. Sci. Soc.*, 29, 207, 1989.

Crowe, G. et al., The effect of laundering on the detection of acid phosphatase and spermatozoa on cotton T-shirts, *J. Can. Soc. Forens. Sci.*, 33, 1, 2000.

Davies, A., Evaluation of results from tests performed on vaginal, anal and oral swabs received in casework, *J. Forens. Sci. Soc.*, 17, 127, 1977.

Divall, G. B., Identification and persistence of seminal constituents in the postcoital vaginal tract, in *Proceedings of a Symposium on the Analysis of Sexual Assault Evidence*, Federal Bureau of Investigation, Quantico, VA, 1983.

Enos, W. F. and Beyer, J. C., Spermatozoa in the anal canal and rectum and in the oral cavity of female rape victims, *J. Forens. Sci.*, 23, 231, 1978.

Eungprabhanth, V., Finding of the spermatozoa in the vagina related to elapsed time of coitus, *Z. Rechtamedizin*, 74, 301, 1974.

Hochmeister, M. N. et al., Evaluation of prostate-specific antigen (PSA) test assays for the forensic identification of seminal fluid, *J. Forens. Sci.*, 44, 1957, 1999.

Iwasaki, M. et al., A demonstration of spermatozoa on vaginal swabs after complete destruction of the vaginal cell deposits, *J. Forens. Sci.*, 34, 659, 1989.

Kafarowski, E., Lyon, A. M., and Sloan, M. M., The retention and transfer of spermatozoa in clothing by machine washing, *J. Can. Soc. Forens. Sci.*, 29, 7, 1996.

Kind, S. S., The acid phosphatase test, in *Methods of Forensic Science*, Vol. 3, 267, 1964.

Nata, M. et al., Stability of acid phosphatase activity and spermatozoa in semen stains washed with water, *Act. Crim. Jpn.*, 59, 247, 1993.

Oettle, A. G., Morphologic changes in normal human semen after ejaculation, *Fertil. Steril.*, 5, 227, 1954.

Poyntz, F. M. and Martin, P. D., Comparison of p30 and acid phosphatase levels in postcoital vaginal swabs from donor and casework studies, *Forens. Sci. Int.*, 24, 17, 1984.

Schiff, A. F., Reliability of the acid phosphatase test for the identification of seminal fluid, *J. Forens. Sci.*, 23, 833, 1978.

Sensabaugh, G. F., The acid phosphatase test, in *Proceedings of a Symposium on the Analysis of Sexual Assault Evidence*, Federal Bureau of Investigation, Washington, D.C., 1985.

Sensabaugh, G. F., Isolation and characterization of a semen-specific protein from human seminal plasma: a potential new marker for semen identification, *J. Forens. Sci.*, 23, 196, 1978.

Shaler, R. C. and Ryan, P., High acid phosphatase levels as a possible false indicator of the presence of seminal fluid, *Am. J. Forens. Med. Pathol.*, 3, 161, 1982.

Stubbings, N. A. and Newall, P. J., An evaluation of GGT and p30 determinations for the identification of semen on postcoital vaginal swabs, *J. Forens. Sci.*, 30, 604, 1985.

Willott, G. M., The role of the forensic biologist in cases of sexual assault, *J. Forens. Sci. Soc.*, 15, 269, 1975.

Willott, G. M. and Allard, J. E., Spermatozoa: their persistence after sexual intercourse, *Forens. Sci. Int.*, 19, 135, 1982.

Willott, G. M. and Crosse, M. A., The detection of spermatozoa in the mouth, *J. Forens. Sci. Soc.*, 26, 125, 1986.

Saliva

Auvdel, M. J., Amylase levels in semen and saliva stains, *J. Forens. Sci.*, 31, 426, 1986.

Nelson, D. F. and Kirk, P. L., The identification of saliva, *J. Forens. Med.*, 10, 14, 1963.

Phadebas Amylase Test, Product Insert, Pharmacia, Uppsala, Sweden, 1997.

Searcy, R. L. et al., The interaction of human serum protein fraction with the starch–iodine complex, *Clin. Chim. Acta.*, 12, 631, 1965.

Tietz, N. W., Ed., *Fundamentals of Clinical Chemistry*, W.B. Saunders, Philadelphia, 1976.

Whitehead, P. H. and Kipps, A. E., The significance of amylase in forensic investigations of body fluids, *Forens. Sci.*, 6, 137, 1975.

Whitehead, P. H. and Kipps, A. E., A test paper for detecting saliva stains, *J. Forens. Sci. Soc.*, 15, 39, 1975.

Willott, G. M. and Griffiths, M., A new method for locating saliva stains: spotty paper for spotting spit, *Forens. Sci. Int.*, 15, 79, 1980.

Willott, G. M., An improved test for the detection of salivary amylase in stains, *J. Forens Sci. Soc.*, 14, 341, 1974.

Urine

Brinkmann, B., Rand, S., and Bajanowski, T., Forensic identification of urine samples, *Int. J. Legal Med.*, 105, 59, 1992.

Medintz, I. et al., DNA analysis of urine-stained material, *Anal. Lett.*, 28, 1937, 1995.

Poon, H. H. L., Identification of human urine by immunological techniques, *J. Can. Soc. Forens. Sci.*, 17, 81, 1984.

Feces

Frankel, S., Reitman S., and Sonnenwirth, A., Eds. *Gradwohl's Clinical Laboratory Methods and Diagnosis*, 7th ed., C. V. Mosby, St. Louis, 1970.

Giertsen, J., Faecal matter in stains: identification, *J. Forens. Med.*, 8, 3, 1961.

Hopwood, A. J., Mannucci, A., and Sullivan, K. M., DNA typing from human faeces, *Int. J. Legal Med.*, 108, 237, 1996.

Nickolls, L. C., *The Scientific Investigation of Crime*, Butterworth and Co., London, 1956.

Price, C., Tests for the identification of faeces, *MPFSL Memo 11*, 1984.

Vomitus

Yamada, S. et al., Vomit identification by a pepsin assay using a fibrin blue–agarose gel plate, *Forens. Sci. Int.*, 52, 215, 1992.

Vaginal Epithelial Cells

Hafez, E. S. E. and Evans, T. N., Eds., *The Human Vagina*, North Holland Publishing, Amsterdam, 1978.

Hausmann, R. and Schellmann, B., Forensic value of Lugol's staining method: further studies on glycogenated epithelium in the male urinary tract, *J. Legal Med.*, 107, 147, 1994.

Keating, S. M., Information from penile swabs in sexual assault cases, *Forens. Sci. Int.*, 43, 63, 1989.

Luna, L. G., Ed., *Manual of Histologic Staining Methods of the Armed Forces Institute of Pathology*, 3rd ed., McGraw-Hill, New York, 1968, pp. 158–160.

Randall, B. and Riis, R. E., Penile glycogenated epithelial cells as an indicator of recent vaginal intercourse, *Am. J. Clin. Pathol.*, 84, 524, 1985.

Randall, B., Glycogenated squamous epithelial cells as a marker of foreign body penetration in sexual assault, *J. Forens. Sci.*, 33, 511, 1988.

Light Sources

Auvdel, M. J., Comparison of laser and high-intensity quartz arc tubes in the detection of body secretions, *J. Forens. Sci.*, 33, 929, 1988.

Auvdel, M. J., Comparison of laser and ultraviolet techniques used in the detection of body secretions, *J. Forens. Sci.*, 32, 36, 1987.

Maloney, M. S., Housman, D. G., and Rowe, W. F., A comparative study of the effectiveness of forensic light sources in the detection of biological stain evidence, presented at 48th annual meeting of the American Academy of Forensic Sciences, Nashville, TN, 1996.

Muskopf, D. S. and Brown, M. L., Comparison of light sources for the detection of body secretions and blood, Newsletter, Midwestern Association of Forensic Scientists, July 1989.

General

Chayko, G. M. and Gulliver, E. D., *Forensic Evidence in Canada*, 2nd ed., Canada Law Books, Aurora, Ontario, 1999.

Gaennslen, R. E., *Sourcebook in Forensic Serology, Immunology, and Biochemistry*, U.S. Department of Justice, Washington, D.C., 1983.

Guyton, A. C., *Textbook of Medical Physiology*, 8th ed., W.B. Saunders, Philadelphia, 1991.

Saferstein, R., *Forensic Science Handbook*, 2nd ed., Vol. 1, Prentice-Hall, Englewood Cliffs, NJ, 2002.

Forensic DNA

Technology, Applications, and the Law

Susan Herrero

16

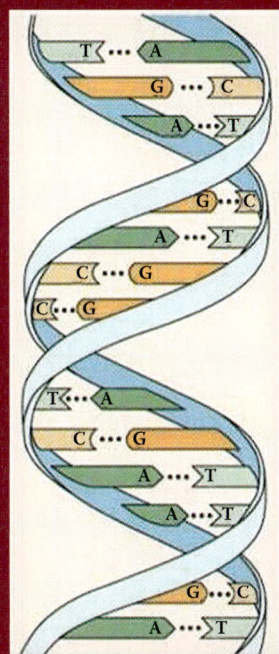

Genetics: The Beginnings

In 1865, Gregor Mendel presented two lectures to the Brno Natural Science Society that summarized the results of his experiments on heredity in the garden pea, *Pisum sativum*. Until the rediscovery of this work in 1900, the material basis of inheritance was thought to be fluid in nature, and terms such as "true-blood" and "half-blood" were thought to be scientifically correct. One result was a general perception that hereditary materials that were mixed could not be separated. Mixing was thought to alter the hereditary units or genes.

Mendel's major intellectual contribution was his demonstration that the material basis of inheritance was particulate and that mixing did not alter genes. His evidence in support of this conclusion was based on the second generation reappearance of characteristics present in one of the grandparental peas used for his studies. For example, he crossed round peas and wrinkled peas to produce a progeny generation that consisted entirely of round peas. The wrinkled characteristics were not seen in the children. The grandchild peas were, however, both round and wrinkled. The results (round children, and round and wrinkled grandchildren) were clearly at variance with the hypothesis of a fluid basis of inheritance that suggested that the children and grandchildren should have been sort of wrinkled and sort of round—fluid mixtures of both parents' characteristics.

Inheritance of DNA Characteristics

Today, we can provide a reasonably complete explanation of Mendel's results with peas, in particular, and the inheritance of characteristics controlled by single genes.

The reappearance of wrinkled peas in the second generation meant that the information (gene or deoxyribonucleic acid [DNA]) to produce wrinkles was present in the first generation plants but was not expressed. Since it was expressed in the second generation plants, the information or DNA was transmitted unchanged. The genetic information was shown to be particulate rather than fluid. It is the **particulate**, unchanging nature of the DNA molecule that allows **DNA fingerprinting**.

We inherit the information required to produce all our characteristics in the form of DNA. We do not inherit the characteristics—only the information to produce them. The rules of inheritance are simple. We inherit half our genetic material (DNA) from each parent. The production of sperm and eggs that occurs in parents' gonads results in the production of **gametes** that carry only half of the DNA that made the parents unique. When the egg and sperm unite at fertilization, a new, unique individual is created. Like her parents, this individual, if female, has two copies of all the genetic material and she can produce eggs that will contain only one copy of each gene. If the child is male, he, too, has two copies of each gene but can also pass only one of the two to each of his progeny. Most cells contain a nucleus made up of chromosomes. Normally we all have 23 pairs of chromosomes including the gender chromosome called the XY chromosome. Why pairs? For each chromosome, we inherit one from our mother and one from our father. If we are female, we will inherit an X from both parents. If we are male, we will inherit one X and one Y chromosome (Figure 16.1). Nuclear DNA is found in our chromosomes. Where are our chromosomes? Our chromosomes can be found in cells with nuclei throughout our bodies. Nuclear DNA

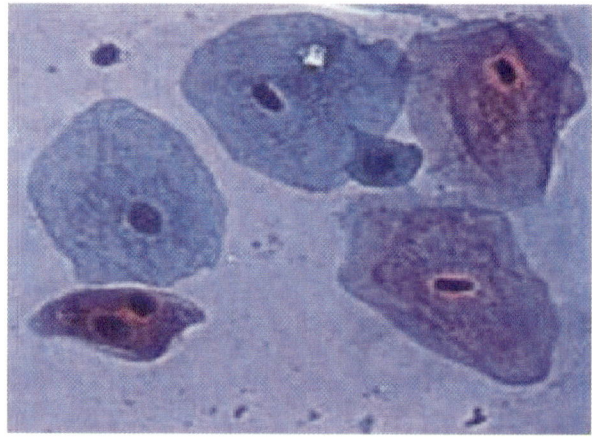

Figure 16.2 Epithelial (skin) cells are one example of nucleated cells.

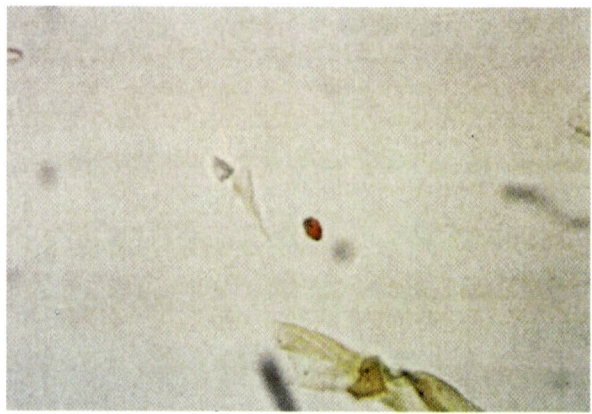

Figure 16.3 Spermatozoa are an example of nucleated cells. They function in the male reproductive system.

comes from nucleated cells: skin cells, spermatozoa, and white blood cells. Mature red blood cells, although important, do not have a nucleus (Figures 16.2–16.4).

One of our many 20 to 25 thousand genes specifies the ABO blood group. ABO blood grouping has been used for a long time in forensic science and paternity testing. There are four major ABO blood group classifications: A (AA and AO), B (BB and BO), O (OO), and AB. The technical name for genetic classifications based on appearance or behavior is **phenotype**; these four blood group classifications are the four possible ABO phenotypes (genotypes are in parenthesis). Each of the four ABO phenotypes is produced by DNA information carried in the sperm and egg, which act together to produce the phenotype in the new individual. The genetic constitution of the combined

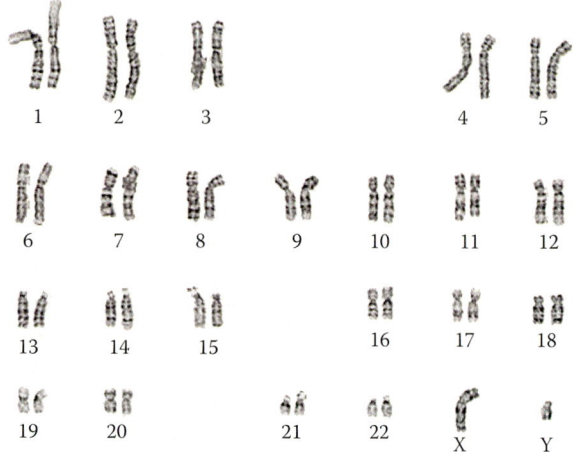

Figure 16.1 The human genetic code, determined by our DNA, is contained within 23 pairs of chromosomes.

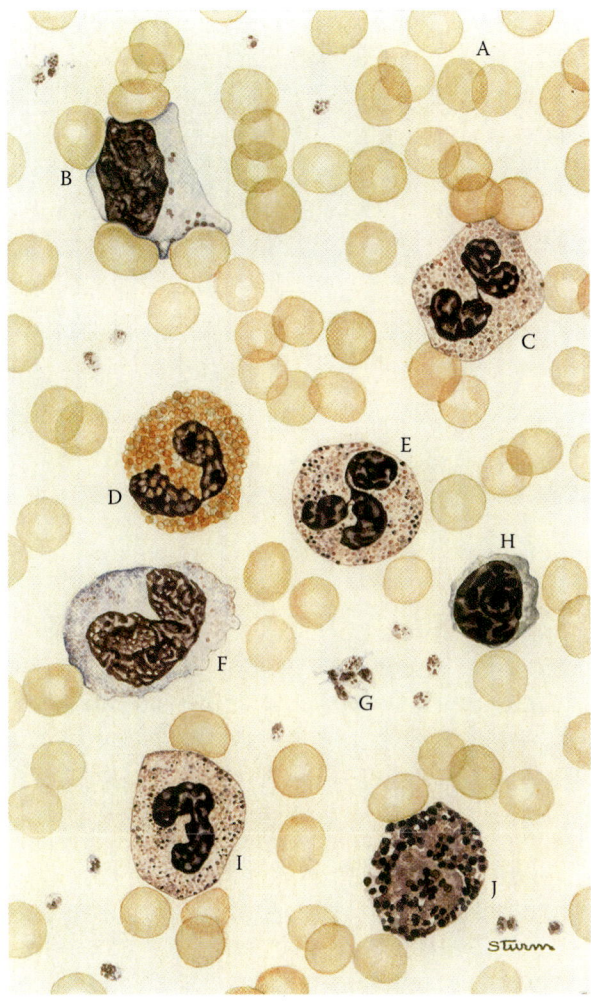

Figure 16.4 White blood cells are an example of nucleated cells. They function in the human immune response system. (Courtesy of Abbott Laboratories.)

	A	B	O
A	**AA**	**AB**	AO
B	**AB**	**BB**	BO
O	AO	BO	**OO**

A great deal of laboratory testing including separation and identification of DNA depends upon the mirror-image structure of DNA. Its double-stranded complementary structure encodes the information required to make new copies of itself. The intact DNA molecule is composed of adenine, thymine, cytosine, and guanine nucleotides. In essence, each nucleotide is made up of a base, a pentose sugar, and a phosphoric acid group that makes up the backbone of the helix. The complementarity of the two halves of the molecule or the two strands depends on the simple fact that A only pairs with T, and C only pairs with G, except during mutation.

About 50 years ago, an English graduate student named Francis Crick and an American postdoctorate researcher named James Watson first proposed that the structure of DNA was a double-helix. They thought that the two halves of the molecule were complementary with A-T and G-C nucleotide base pairs holding the two strands together. Crick, Watson, and Maurice Wilkins were awarded the 1962 Nobel Prize in Medicine for their "discoveries concerning the molecular structure of nucleic acids and its significance for information transfer in biological material."[1] The significance of Watson and Crick's DNA model has been enormous in the field of biology (Figure 16.5). In addition to opening the doors for nearly every breakthrough concerning our understanding of DNA replication, transcription, and translation, Watson and Crick have helped us to understand how we can manipulate DNA to perform the tests that have become central to any forensic investigation.

From Mendel's round or wrinkled peas through ABO blood groups and Watson and Crick's DNA model, the rules of inheritance are the same. Each individual receives half of his or her genetic heritage from each parent; in other words, each parent produces gametes that contain one or the other allele of one

parental and maternal DNAs responsible for producing the different ABO phenotypes are referred to as **genotypes**. In this system, the gene may take three different forms (A, B, or O) and any individual can only have two of these three forms. Different forms of a gene are called **alleles**. At the DNA level, alleles can be distinguished because the sequences of molecular subunits or **nucleotides** (A-adenine, C-cytosine, T-thymine, and G-guanine) are different. Thus, we can identify the A, B, and O alleles by studying blood cell characteristics, phenotypes, or DNA genotypes.

The Mendelian inheritance pattern for ABO blood groups is shown below. Letters in red indicates the dominant phenotype. O is recessive and is only expressed when two O genes are inherited (homozygous recessive).

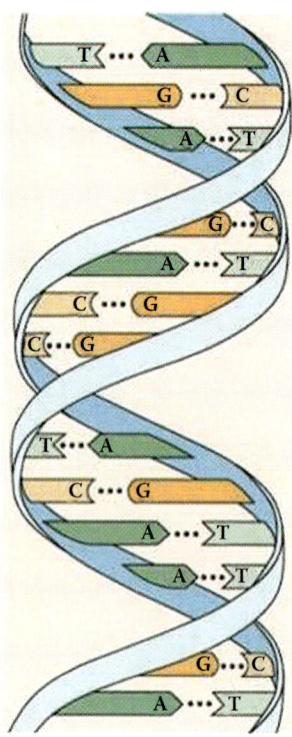

Figure 16.5 Watson and Crick's DNA model was an enormous breakthrough for the field of biology.

parent's genotype. The extreme specificity of one's entire genetic code allows for its use as another type of "fingerprint" in forensic analysis. It would be highly improbable, if not nearly impossible, to find two unrelated individuals with exactly the same inherited genetic code. This permits positive identification of individuals and enormously increases the probability of excluding others when DNA analysis is brought to bear on evidence in a court of law.

How DNA Made It Into the Courts

The Blooding is a true story written by author Joseph Wambaugh. A former Los Angeles detective, Joseph Wambaugh presents the true story of two murders that took place near Leicester, England. In November 1983, a 15-year-old schoolgirl Lynda Mann was raped and strangled on the Black Pad footpath in Narborough, Leicestershire. Despite a massive

Figure 16.6 Sir Alec Jeffreys was responsible for significant developments in the area of DNA testing.

manhunt, the murder went unsolved. A few years later, on July 31, 1986, another 15 year-old schoolgirl, Dawn Ashworth, was raped and strangled nearby on Ten Pound Lane near a psychiatric hospital. A cook at the psychiatric hospital eventually confessed to the second murder. Police collected over 4,500 voluntary samples from local men between the ages of 16 and 34 from three villages. Sir Alec Jeffreys, a geneticist at Leicester University had recently developed DNA testing along with Peter Gill and Dave Werrett of the Forensic Science Service (Figure 16.6).

Jeffreys performed DNA tests at Leicester University on the evidence samples and found that the same man had committed both murders and that man was not the cook.

Colin Pitchfork, a local baker and convicted flasher, married with children, had avoided taking the voluntary blood test by having a coworker take it for him. No matches were found between the sperm taken from both victims and the 4,500 men tested. Then, in September 1987, another coworker tipped off police. The police immediately arrested Colin Pitchfork. When they asked him why he had done it, he stated because they were there. In January 1988, Colin Pitchfork became the first criminal caught with DNA evidence and was sentenced to two life sentences. The judge said that without DNA, Colin Pitchfork would still be out there murdering young girls. The cook who had confessed to the second murder became the first suspect exonerated with DNA.

By the time Colin Pitchfork was being sentenced to life in prison, DNA was being introduced in courtrooms in the United States. By 1986, a few commercial companies in the United States were able to perform DNA profiling online. By 1988, the FBI also had that capability. The earliest DNA profiling tests, identical to the tests performed in the Leicestershire murder inquiry, required large samples and worked poorly with small or degraded samples. The FBI began training local state crime laboratory analysts in DNA profiling and local state crime laboratories soon ran their own DNA tests. They compiled convicted offender databases and tied into the FBI's national system known as CODIS (combined DNA information system). DNA had become the greatest crime-solving tool in history.

DNA in Forensic Analysis

RFLP Testing

The first DNA tests developed by Sir Alec Jeffreys in 1984 were known as **Restriction Fragment Length Polymorphism (RFLP)** tests. RFLP analysis allowed for the measurement of the size of DNA fragments. Analysts could compare the sizes of fragments from a known reference and a crime scene sample in order to "match" two DNA profiles (Figure 16.7). These tests were labor intensive, required large sample sizes (blood, spermatozoa, skin), and took several weeks to perform. Because of the difficulty and expense of RFLP testing, it would soon become obsolete. Many crime scene samples were too small for RFLP tests—a speck of blood on a suspect's shoe or skin cells deposited on the handle of a murder weapon would give no RFLP results. However, advances in biotechnology would soon allow for development of more sensitive tests and evolve into today's automated/computerized DNA profiling.

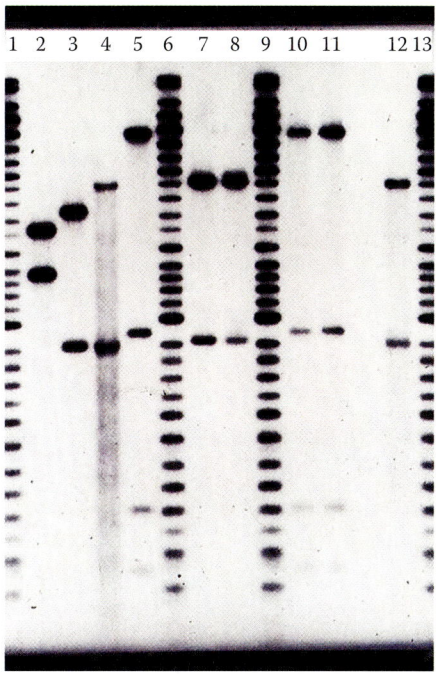

Figure 16.7 Restriction Fragment Length Polymorphism (RFLP) testing allows analysts to match two DNA profiles based on the size of DNA fragments.

PCR

In 1987, biochemist Kary Mullis and molecular biologist Henry Elrich invented the polymerase chain reaction (PCR), a method to make multiple copies from a small sample of DNA. PCR is one of the most powerful molecular biology techniques developed to date because numerous copies of specific target DNA can be made in a short time. The secret to developing the technique was finding an enzyme that could withstand the heat needed to unzip the two strands of the DNA double helix for replication without also breaking apart the enzyme itself. They were able to accomplish this using a thermo-stable enzyme from *Themus aquaticus*, a thermophilic bacteria discovered by Thomas Brock in a geyser in Yellowstone National Park. Kary Mullis won the Nobel Prize in 1993.

With any DNA test using PCR, contamination can become a problem. When amplifying tiny samples, the evidence must be handled carefully to avoid depositing extra cells onto the sample before amplification takes place.

Analysts must use sterile gloves and work under a special sterilized flow-through hood when performing PCR. Even laughing or sneezing near the evidence could contaminate it. With the ability to amplify small crime scene samples, newer tests were needed that could visualize and profile amplified DNA.

Short Tandem Repeat (STR)

By 1998, the FBI added a newer Multiplex PCR test called **STR (short tandem repeat),** which is currently in use today. Just as the previous PCR tests, STRs work well with small samples and degraded samples. The test is automated, color-coded, rapid, and computerized. Three different fluorescent dyes are used to distinguish STR alleles with overlapping size ranges. The technology uses capillary electrophoresis and includes 13 different genetic markers and a gender test called *amelogenin.*

Interpretation of Results: What is the Meaning of a DNA Match?

Prior to DNA analysis, experts testified that an item found at the crime scene was compatible with the suspect, was consistent with the suspect, was similar to the suspect, had the same general characteristics, could have originated from the suspect, or had identical characteristics as the suspect. Today with DNA profiling, we have a DNA match. But what is a match?

Prosecution DNA analysts typically rely on FBI statistics when they estimate how common or rare a DNA profile is. First we must assume the suspect does not have an identical twin (with identical DNA). Next, a calculation called a *random match* probability is performed. However, the probabilities offered of finding another person with this same DNA profile anywhere in the world typically exceed the total number of people who have ever lived on the earth. But are these estimates sometimes exaggerated? Can two unrelated people have the same DNA? In 2005, several partial matches were discovered in an Arizona CODIS (convicted offender database) database: Out of 65,493 samples, 122 people matched another person in the database at 9 loci (the specific locations of genes on a chromosome), 20 people matched at 10 loci, 2 people matched each other at 11 loci and another 2 people matched each other at 12 loci. Legal challenges can be brought when the meaning of a match is exaggerated.

Mixtures

Another area that can be subject to interpretation is when there is a mixed sample. Identification can be complicated if the scene is contaminated with DNA from several people. Many evidence samples are mixtures. For example, the vaginal swab in a rape kit will contain the victims' skin cells and the rapists' sperm cells. Or imagine a killers' clothing containing his or her shed skin cells stained with some of the victims' blood on it.

Y-STRs

In many sexual assault cases, it would be beneficial to be able to distinguish male DNA from female DNA. A set of STR markers associated with just the Y chromosome has recently been developed. Recall that within our 23 pairs of chromosomes, women have an XX and men have an XY pair. Unlike conventional STRs, where two alleles per locus is the norm, a man only has one Y in his XY gender chromosome. And since women do not have Y chromosomes as part of their genetic material, **Y-STR typing** is the only test that can unambiguously determine what a male has contributed to a mixed sample, like those collected as part of most rape investigations. Y-STRs are especially invaluable when very few sperm are detected in the sample, when the rapist has had a vasectomy or is sterile, when previous STRs show no Y signal at the amelogenin (gender) locus, when differential extraction is unsuccessful, when there are multiple semen donors, or when the ratio of female to male DNA is so large that the female DNA masks the male DNA. For the same reason, Y-STRs are advantageous in analyzing fingernail scrapings from a female assault victim, separating out the male skin cells from a ligature used in the strangulation of a female victim, or testing microscope slides from cold case rape kits. Y-STRs were used to confirm the identity of Saddam Hussein after his capture in a spider hole outside his hometown of Tikrit, Iraq, on December 13, 2003. Saddam's capture was verified with DNA profiling by the Armed Forces DNA Identification Laboratory (AFDIL) in Rockville, Maryland. His profile was compared to his sons, Uday and Qusay Hussein, who had been killed in a raid in Mosul in northern Iraq on July 22, 2003.

Nonhuman DNA in Criminal Cases

While the use of human DNA in criminal cases is widespread, nonhuman DNA found at a crime scene can be also be important in helping to solve a case. DNA evidence from plants and animals has been examined to solve cases and admitted as evidence. Animal DNA is routinely collected in horses and cows to prove pedigree. In a Washington case, two burglars entered a home, killing the family dog at the front door. Blood spatter evidence found on one of the suspects' jackets was determined to contain canine (dog) blood. The bloodstains were subjected to STR testing and the canine DNA match was introduced at trial.

On the Horizon

There are a few DNA tests that use nuclear DNA testing that have not been widely accepted as of yet. They include SNPs (single nucleotide polymorphisms), low copy number (LCN) DNA, and laser copy DNA.

SNPs (single nucleotide polymorphisms) are DNA sequence variations that occur when only one single nucleotide (an A, T, C, or G) in the genome sequence is altered. SNPs are helping medical researchers identify genes associated with complex diseases.

The Forensic Science Service (FSS) in the United Kingdom developed **low copy number (LCN)** DNA profiling in 1999. It is sensitive enough to generate a profile with just 15 to 20 cells, and involves a greater amount of copying from a smaller amount of starting material. That is what makes it time consuming and expensive. The LCN method can be used to analyze samples that have simply been touched. LCN is used in the United Kingdom, the Netherlands, and New Zealand. It has proven helpful in cold cases, but because of the small sample size, it is important to consider the risks to accuracy from allelic dropout, as well as contamination from a laboratory source. Because of the small amount of starting DNA, many more cycles of replication are necessary (see PCR above), and the contaminants will also be replicated, creating a greater risk of inaccurate or contaminated results.

Laser Copy DNA

In a new technique, medical researchers have been able to profile a very limited number of cells from micro-dissected tissues. Laser micro-dissection (LMD) techniques allow scientists to see several kinds of cells present in both pathological and forensic samples by cutting small tissue fragments, as well as single cells with an ultraviolet laser beam under direct microscopic view. The low number of target cells among a wide spectrum of cell types may complicate crime scene samples, but analysis of low copy number DNA from a few cells harvested by laser micro-dissection makes it easier to solve forensic problems.

Mitochondrial DNA

Some types of DNA testing do not come from the cell nucleus. **Mitochondrial DNA (mtDNA)**, is one of those. Mitochondrial DNA is the DNA located in the mitochondrion, a specialized subunit within a cell that functions as the powerhouse of the cell, and is separately enclosed within its own lipid membrane (Figure 16.8). Unlike nuclear DNA which is inherited from both parents, mtDNA is maternally inherited.

The likelihood of recovering mtDNA in small or degraded biological samples is greater than for nuclear DNA because mtDNA molecules are present in hundreds to thousands of copies per cell compared to the nuclear complement of two copies per cell. Therefore, muscle, bone, hair, skin, blood, and other body fluids, even if degraded by environmental insult or time, may provide enough material for typing the mtDNA locus.

In addition, mtDNA is inherited from the mother only, so that in situations where an individual is not available for a direct comparison with a biological sample, any maternally related individual may provide a reference sample. However, mtDNA profiles are not as informative or sensitive as STR profiles. More individuals will have profiles in common when compared with mtDNA profiling. MtDNA is extracted from the cell just like nuclear DNA extraction. MtDNA is amplified under sterile

conditions, just like nuclear DNA amplification. Computer software programs visualize the mtDNA for analysis.

Mitochondrial DNA, or mtDNA, is present in:

- Rootless shed hairs
- Hair fragments
- Skin
- Blood
- Muscle
- Body fluids
- Bone/skeletal remains
- Teeth and other samples too minimal and/or degraded for STR testing

---

Medical and Biological Applications of DNA

DNA sequencing and profiling has revolutionized many fields of medical research. It has contributed to profound advances in disease research, success in organ transplants, and the development of cutting edge gene therapies to change and correct inherited disease genes. DNA sequencing and profiling is applied in amniocentesis of the fetus to determine embryonic health. It is used in the functional analysis of genes, the diagnosis of hereditary diseases, and the detection and diagnosis of infectious diseases.

DNA profiling is used after organ transplants. In addition to their organ donors, transplant patients can thank advances in DNA technology for transplant success rates. Organ transplants save lives, but even after finding a compatible donor organ, there is great risk. When a donor organ is surgically placed into the living patient, one of the greatest risks to success is that the patient's immune system will recognize the new organ as foreign and the immune response will lead to rejection. During this transition, after transplantation, patients are carefully monitored and medications can be used to prevent

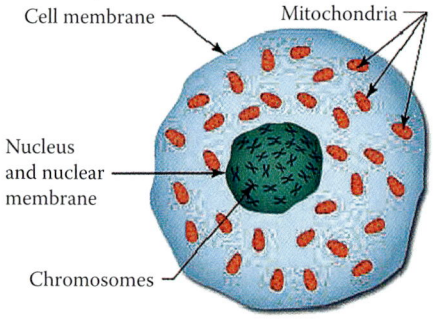

Figure 16.8 Mitochondrial DNA (mtDNA) is found in a separate organelle from nuclear DNA and has distinct characteristics.

rejection. As the patients' body accepts the new organ, the DNA in the donated organ will actually change from the DNA profile of the donor to that of the recipient. DNA tests on the donated organ indicate to medical experts when that critical threshold has been crossed and the patient is out of the woods.

Gene therapy is another medical area that is making great advances using DNA technology. According to the National Human Genome Research Institute, every person probably has six or more defective genes or genetic mutations that place them at risk for some disease. However, according to researchers, that does not mean that a disease will develop, only that the person is more likely to contract the disease than another person without the genetic mutation. We may go through life never knowing that we have genes susceptible to different diseases. About 1 in 10 people will develop an inherited genetic disease. Most of us do not suffer from these diseases because we have two copies of nearly all genes, one from our mother and one from our father. Normally, the healthy regular gene will protect us from disease. However, if we inherit a defective gene that is dominant, it can produce the disease. If we inherit a defective recessive gene from both parents, disease can develop. But what if we could replace the faulty gene with a normal gene? Because almost every disease has a genetic component, medical researchers have been working to identify the faulty genes and to replace them in an emerging medical field known as gene therapy. Genes in many neurologically inherited diseases have already been identified. With the mapping of the human genome in 2003, 3.1 billion bits of DNA have been decoded, and their sequence listed. People now have access to far more information about their hereditary disposition to such crippling afflictions as cystic fibrosis, Huntington's disease, or Lou Gehrig's disease worldwide. DNA profiles have also been used in the fields of animal conservation and zoology. Animal geneticists track migration and breeding of endangered species. In captive breeding programs, DNA is used to make sure the animals mating are not directly related.

Historical Applications of DNA

Historical investigations include genealogical investigation, adoption maternity, ancient DNA (the emerging field of molecular archaeology—the molecular biological analysis of ancient remains), and identifying war dead remains in the current war in Iraq, as well as remains from previous wars. DNA can be used for ancestry tracing in the field of genetic genealogy. Similarly, DNA has proved useful in genealogical investigation. For example, DNA tests indicated that Thomas Jefferson fathered at least one child with his slave concubine.

Some of the most noted historical DNA cases include identifying the remains of German SS officer Joseph Mengele and the Russian royal Romanov family. In the case of German SS officer and physician Josef Mengele, DNA was used to identify his remains found in Brazil. Known as the "Angel of Death" for performing human experiments on camp prisoners in Hitlers' Auschwitz-Birkenau concentration camp, Mengele escaped to South America after World War II. Wanted for war crimes, he was hunted for years. He reportedly lived under an assumed name to avoid prosecution. When he reportedly died, holocaust survivors wanted confirmation of his death. In 1985, his body was exhumed and in 1992, his identity was confirmed using DNA testing. Finally, holocaust survivors and families of holocaust victims had peace of mind knowing the Angel of Death was dead.

In the case of the Romanov family, who belonged to the last imperial dynasty of Russia, DNA was also used to confirm the identity of their suspected remains, found in 1979 hidden in a mine shaft in Yekaterinberg, Russia. Czar Nikolas II ruled Russia from 1894 until the pressures of World War I contributed to the Bolshevik Revolution of 1917. He and his family were arrested, held prisoner by the Bolsheviks, and were reportedly executed in 1918. Their bodies were burned so they could not be identified, and then were buried at an unidentified site. There was ongoing speculation that the family had survived, especially the Grand Duchess Anastasia who a number of people fraudulently claimed to be in subsequent years. (DNA testing in 1994 proved one of the most believable was unrelated to the family.) It was only in 1991, with approval from newly elected Russian President Boris Yeltsin, that the remains were exhumed. DNA tests performed on the bones eventually identified them as those of the Czar, his wife, three daughters, their doctor, valet, cook, and maid. However, the bodies of the fourth daughter and the only son were missing. Then, in September 2007, the remains of the two missing children were discovered by archaeologists in a burned field near Yekaterinberg. On April 30, 2008, Russian forensic scientists announced that DNA testing proved the remains belonged to the Tsarevich Alexei and to one of his sisters. Historians will finally be able to write the last chapter in the tragic story of this family.

Identification of War Dead

The U.S. military has unidentified remains dating back to the Civil War. The task of identifying remains of soldiers from the Civil War, World War I, World War II, the Korean War, and the Vitenam War is left to the Armed Forces DNA Identification Laboratory (AFDIL) located in Rockland, Maryland. From 1991 to 1998, the lab has worked hand-in-hand with the Army's Central Identification Laboratory (CILHI) in Honolulu, to identify the remains of Americans unaccounted for from the Vietnam War and earlier conflicts. As of April 1998, the lab had made 93 mitochondrial DNA matches: 72 cases from Southeast Asia, three from Korea, 15 from World War II, and three from the Civil War.

In 1998, the AFDIL and CILHI gained broad public recognition when the family of Air Force 1st Lt. William Blassie suggested that his remains were contained in the Tomb of the Unknown Soldier of the Vietnam War.

Defense Secretary William Cohen directed disinterment and DNA identification of the remains. It took the DNA laboratory about one month to report that DNA from the remains matched specimens provided by the Blassie family.

In 1990, with war in the Persian Gulf imminent and U.S. forces expecting many casualties, American military leaders launched a major DNA collection project and created the Department of Defense DNA Registry. Although the United States suffered relatively few casualties in the Gulf War, the Defense Department forged ahead with the creation of the AFDIL and the Armed Forces Repository of Specimen Samples for the Identification of Remains, both located in Rockville, Maryland. The goal of these and affiliated organizations is to attach a name to every fallen soldier, sailor, airman, and marine. Identification is extremely important to the families. The first step was to create a database containing the DNA of all members of the military, active and reserve. Vacuum-sealed with a drying agent, these 3.6 million blood samples are now stored at minus 20°C (−4°F), ready to be matched to DNA for a soldier killed in battle, accident, or terrorism attack.

The AFDIL has assisted with identifying the dead in plane crashes, the mass suicides of Branch Davidian members in Waco, Texas, and the U.S. embassy bombing in Nairobi, Kenya. As recently as January 2008, remains returned from Korea Korean War were identified.

DNA in Noncriminal Cases

DNA analysis in noncriminal cases includes paternity cases, probate issues, immigration cases, and victim identification in mass disasters. Paternity, probate, and immigration cases are relatively simple because they include a low number of known samples for comparison. Mass disasters include a larger number of unidentified dead and a long list of missing persons who may or may not be dead.

Figure 16.9 The use of DNA identification in the aftermath of Hurricane Katrina in 2005 allowed investigators to match the DNA profiles of victims with their families.

Two disasters in recent U.S. history have challenged the nation's ability to handle mass-fatality identification: the terrorist attack on September 11, 2001 and Hurricane Katrina.

On August 29, 2005, Hurricane Katrina struck the Gulf Coast causing over 1,599 lives to be lost. More than 1,000 persons died in the New Orleans area alone. Water levels exceeded 10 feet high in some neighborhoods, and rescue efforts were delayed. Some victims who perished while trapped in flooded homes or washed from roofs were not discovered for weeks or months later, when floodwaters receded and debris was removed. Indeed, some persons are still missing (Figure 16.9).

There were three major challenges to making these identifications. First, many medical and dental records typically used to match victims with names of the missing were destroyed in the flood. Secondly, many of the personal items of the victims, which are typically used to compare to victims' remains, were destroyed by the flood, preventing genetic

reconstruction. Third, the long-term evacuation made identification of human remains even more difficult, as evacuated family members, whose DNA is necessary to make a match, were displaced and difficult to locate. The solution would require a monumental effort of skilled detective work combining both genetic medicine and forensic medicine.

Several months after Hurricane Katrina, the Louisiana State Coroners' Office and the Louisiana State Police hired a multiinstitution team of experts supported by professionals associated with the National Institute of Health (NIH) and the National Human Genome Research Institute (NHGRI). Some of the experts, from various disciplines including epidemiologists and geneticists, had gained previous experience from the DNA analysis of human remains working on the September 11, 2001, Trade Center crash sites in New York City. After Katrina, genetic counselors and clinicians interviewed family members, collected data on family history, and constructed complex family trees. Investigators located evacuated biological family members to conduct DNA tests. The DNA profiles were then stored in a special Katrina database. Using the latest software, the genetic profiles from the victims' remains were then compared with close family members or with personal effects of the missing.

DNA in Criminal Cases: Criminal Guilt and Innocence

DNA has been an effective crime-solving tool since 1989 in the United States. Forensic uses of DNA analysis include victim identification, homicide investigations where a suspect may be linked to a crime scene, cold cases, missing persons investigations, and postconviction innocence cases.

Cold Cases

Criminals who thought they had gotten away with their crime now have reason to be worried. Victims and victims' families who thought their attacker would never be identified and convicted now have reason to anticipate their attacker will be brought to justice. Every law enforcement department has old, unsolved cases, and some of those cases contain biological samples that could be analyzed through new advances in DNA technology. Many local police departments now have a special unit called a cold case squad specializing in investigating those old, unsolved cases.

There are two reasons. First, they are the most serious cases and victims' family members have paid the ultimate price. Second, in cases of lesser serious crimes, there may be a statute of limitations that bars criminal prosecution after several years. A statute of limitations refers to the time period from the time of the offense in which a prosecution must begin. A prosecution after that time period is not allowed, even if the defendant is factually guilty. Every state has a statute of limitations for various crimes, but the laws differ from state to state. In addition, prosecutions of different crimes are limited by different periods of time. In general, the more serious the crime, the longer period of time the authorities have to begin prosecution. However, no state has a statute of limitations on first degree murder. It can be prosecuted at any time. Since many of the less serious cold cases will be barred by the statute of limitations, cold case units that concentrate on homicides are the most effective.

Missing Persons Investigations

Many of the people who go missing each year are victims of homicide. There are as many as 100,000 active missing persons cases in the United States at any given time. There are two reasons why so many missing persons cases go unsolved. First, every year, tens of thousands of people vanish under suspicious circumstances. Over a 20-year period, the number of missing persons can be estimated in the hundreds of thousands. Second, in addition to this volume, many cities and counties bury unidentified remains without first collecting a DNA sample. Even in cities and counties that do collect DNA samples, many labs are not able to process very old, degraded samples. This work can only be done with the aid of a national clearinghouse.

The National Institute for Justice (NIJ) has a Missing Persons Task Force. The Center for Human Identification located at the University of North Texas Health Science Center is funded by the NIJ. Nationwide, state and local law enforcement agencies can use this program.

Since 2000, the FBI has maintained the National Missing Persons DNA Database (NMPDD). It is also known as CODIS-MP (Combined DNA Index System for Missing Persons). It is a database specifically designed to assemble data on missing persons and unidentified human remains cases. The searchable database includes information on nuclear and mitochondrial DNA obtained from unidentified remains, relatives of missing persons, and personal reference samples. Having both types of DNA profiles maximizes the potential for a successful identification.

Admissibility of DNA Evidence

There are many advances taking place in DNA profiling. However, as with any emerging science, the technological advances may not be ready for use in the courtroom. DNA evidence is typically presented in serious criminal cases, such as murder and rape. In 1993, the U.S. Supreme Court considered the questions of scientific evidence and expert testimony in *Daubert v. Merrell Dow Pharmaceuticals*. It was agreed that the trial court has a gate-keeping function—to make sure that admitted scientific evidence and expert testimony are relevant and reliable. *Daubert* set forth a nonexclusive checklist for trial judges to use in assessing the reliability of scientific expert testimony. The U.S. Supreme Court suggested factors that might prove helpful in determining the reliability of a scientific theory or technique. Currently, most states rely on these factors, called the Daubert test, when deciding the admissibility of DNA evidence. Federal courts use Federal Evidence **Rule 702** and the **Daubert test**. The *Daubert* questions are:

1. Has the scientific theory or technique been tested?
2. Has the scientific theory or technique been subjected to peer review and publication?
3. What are the known or potential error rates of the theory or technique when applied?
4. Do standards and controls exist and are they maintained?
5. Has the theory or technique been generally accepted in the relevant scientific community?

In 2000, Evidence Rule 702 was amended in response to the Supreme Court's directive in *Daubert and Kuhmo Tire Co. v. Carmichael*, a Supreme Court case decided in 1999. The amendment affirms the trial court's role as gatekeeper and provides some general standards that the trial court must use to assess the reliability and helpfulness of proffered testimony. The revised Rule 702 states:

> *If scientific, technical, or other specified knowledge will assist the trier of fact to understand the evidence or determine a fact in issue, a witness qualified as an expert by knowledge, skill, experience, training, or education, may testify thereto in the form of an opinion or otherwise, if (1) the testimony is based upon sufficient facts or data, (2) the testimony is the product of reliable principles and methods, and (3) the witness has applied the principles an methods reliably to the facts of the case.*

A few states depart from Daubert. In Virginia, for example, the admissibility standard remains the Spencer standard, *Spencer v. Commonwealth,* 240 Va 78, 393 S.E. 2nd 609 (1990): The court must make a threshold finding of fact with respect to the reliability of the scientific method offered (i) unless it is of a kind so familiar and accepted as to require no foundation to establish the fundamental reliability of the system; or (ii) unless it is so unreliable that the considerations requiring its exclusion have ripened into rules of law; or (iii) unless its admission is regulated by statute. In past cases, the Virginia Supreme Court has referred to fingerprint evidence as the kind so familiar as to require no foundation; lie detector tests as the kind so unreliable that its exclusion has ripened into rules of law; and blood alcohol tests as the kind regulated by statute.

Although subject to some legal challenges, the STR tests and the earlier DNA tests explained above (i.e. RFLP, PCR, etc.) are now generally held admissible in most courts throughout the country. However the newer technologies (i.e., LCN DNA, SNPs, Laser copy DNA) have not yet been held admissible.

What Can Go Wrong and What Can Be Done to Prevent It?

Laboratory errors have been known to happen. Sometimes the error is a mistake involving the chain of custody and evidence samples are inadvertently switched. This can happen during collection, handling, processing, or even during the labeling of samples. For example, errors can be caused by clerical mistakes, such as reversing the order of samples loaded on a data sheet. Sometimes samples get lost. Sometimes samples get contaminated. This can happen during collection at the crime scene, or during processing at the crime laboratory. Occasionally, interpretation of test results can be biased. Statistics presented to give meaning to the match may be exaggerated or misleading.

How can we avoid these mistakes? One way is to have DNA guidelines and standards. In October 1998, the FBI implemented the DNA Advisory Board Guidelines and renewed guidelines in 2000. Since then, a technical working group of scientists has studied validation of newer technologies. The federal government has also audited the laboratories that receive federal funding to improve DNA testing. We know that lack of oversight can lead to negligence and misconduct. Negligence and misconduct can lead to wrongful convictions of innocent people. As a requirement under the Justice for All Act of 2004, the Office of Inspector General now reviews grant recipients for external investigation certification. Because negligence and misconduct in forensic laboratories can undermine public confidence in the criminal justice system, the external investigation certification is intended to provide a safeguard. In 2005, the Office of the Inspector General (OIG) found that the Office of Justice Program (OJP) had not enforced or exercised oversight to ensure the labs had the ability to conduct independent investigations of wrongdoing in forensic laboratories. Although changes were made, the OIG found that not all laboratories receiving funding were capable to independently investigate allegations of serious negligence or misconduct in 2006. Despite additional changes being made, the OIG found the same problem in January 2008: "The OIG contacted 231 of 233 labs and found that at least 78 (34%) did not meet the external investigation requirement because they lacked either the authority, capabilities and resources, or the appropriate process to conduct independent external investigations into allegations of serious negligence miscon-

duct by the forensic laboratories that received the FY2006 Coverdell Program funds."

An association of forensic laboratories from across the country has established an accreditation body named the American Association of Crime Laboratory Directors (ASCLD). Many laboratories have applied for and received accreditation through ASCLD. ASCLD's accreditation process requires that applying and renewing forensic laboratories comply with accreditation standards. These accreditation standards also serve as guidelines to minimize errors. However, as long as DNA profiling is performed by human beings, there will always be room for error.

Forensic Casework

The Disappearance of Jessica O'Grady—Missing Body

Is it possible to prosecute someone for murder when the victim's body has not been recovered? It is difficult, but not impossible. On May 10, 2006, a 19-year-old University of Nebraska student Jessica O'Grady was last seen leaving her apartment in Omaha, Nebraska, to visit a former boyfriend to inform him that she was pregnant with his child. The Omaha Police Department and the Douglas County Sheriff's Office conducted a series of searches around Omaha and Douglas County but found no trace of Jessica. Her body was never recovered. Was she murdered or did she simply vanish (Figure 16.10)?

Jessica had recently been dating Christopher Edwards. Edwards lived with his aunt and had another girlfriend at the time of Jessica O'Grady disappeared. The Douglas County Sheriffs' office received the case and began their investigation. They interviewed Christopher Edwards. On May 16, 2006, the Douglas County Sheriff's office CSI unit searched Edwards's aunt's home. In Christopher Edwards's bedroom, they noticed a red stain on the foot of the mattress. Lifting the mattress, they discovered a

MISSING
http://SomeonelsMissing.com
Jessica O'Grady

AGE - 19
EYES - Hazel
HAIR - Brown
HEIGHT - 5' 9"
WEIGHT - 135
LAST SEEN -
May 10th, 2006
OMAHA,
NEBRASKA

Jessica O'Grady made a call from her cell phone just before midnight Wednesday May 10th, 2006 saying she was headed to the 120th and Blondo streets area. She has not been seen since.

Douglas County Sheriff
402-333-1000
PLEASE HELP
http://SomeonelsMissing.com

Figure 16.10 University of Nebraska student Jessica O'Grady's body was never recovered after vanishing in May 2006 (Courtesy of David Kofoed, Douglas County Sheriff's Office, Omaha, Nebraska).

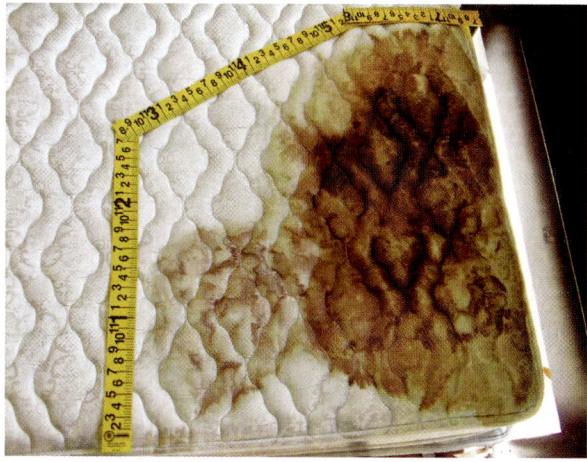

Figure 16.11 Bloodstains found on the underside of Christopher Edwards's mattress matched the DNA profile of Jessica O'Grady, linking Edwards to her death.

large, saturated bloodstain on the underside (Figure 16.11).

The Douglas County CSI unit also observed blood spatter on the headboard of the bed, the box spring, and the bookcase in the bedroom. They found cast-off bloodstains on the ceiling. A Bangkok battle sword with reddish staining was located in his closet. In the trash at the residence, the CSI unit found a bloodstained towel. In Christopher Edwards' car, the CSI

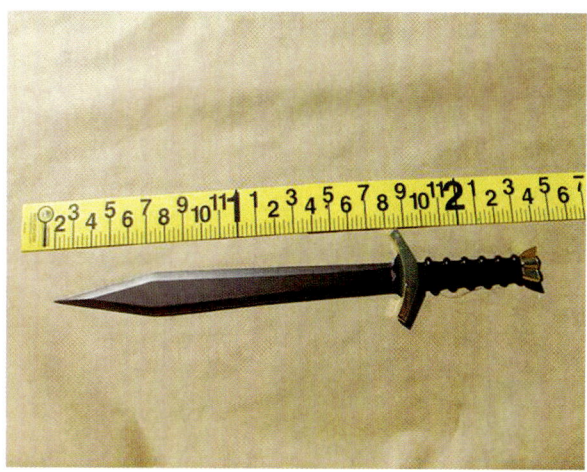

Figure 16.12 A mixture of both Christopher Edwards' and Jessica O'Grady's DNA profiles on the handle of the sword linked Edwards to O'Grady's death.

unit found a pair of bloodstained hedge shears and bloodstains on the metal frame of the trunk. The CSI unit collected all of this forensic evidence.

DNA testing was performed at the University of Nebraska Medical Center's Human DNA Identification Laboratory. When there is no body recovered, the forensic laboratory needs something belonging to the victim or a DNA sample from a close family member to inform what the missing victims' DNA profile is. In this case, O'Grady's personal items such as a hairbrush, underwear, and a razor were used to create a DNA profile. Her DNA profile was then compared to the DNA profiles obtained at the crime scene and the DNA profile of Christopher Edwards. Not all the samples collected yielded a result. However, the DNA profile results from the bloody mattress matched the DNA profile of Jessica O'Grady. Other blood evidence found on the ceiling, walls, mattress, hedge clippers, and bath towel provided profiles of DNA that matched Jessica O'Grady's. The DNA profile on the black handle of the Bangkok sword was consistent with a mixture of DNA from Jessica O'Grady and Christopher Edwards (Figure 16.12).

This was the first murder case prosecuted without a body in Douglas County, Nebraska. Although Jessica O'Grady's body has never been found, Christopher Edwards was convicted of second-degree murder in 2007 and was sentenced to 100 years in prison.

The Brown's Chicken Restaurant Massacre—Cold Case

On January 8, 1993, seven people were shot to death at Brown's Chicken Restaurant in Palatine, Illinois. The next day, the police found a back door open and found seven bodies—the two owners and five employees—in the restaurant's cooler and freezer. All seven victims had been shot in the head from behind or above. Although the motive appeared to be robbery, three victims were found with cash or credit cards. Despite approximately 100 full-time officers investigating the case initially, the case remained unsolved for 9 years. In 2000, partially eaten chicken recovered from the crime scene was tested for DNA in saliva, and the DNA profile did not match any of the seven victims, former employees, or former suspects. Police then took DNA samples from a former employee, Juan Luna. In 2002, the DNA profile of Juan Luna matched the DNA profile from the partially eaten chicken, and Juan Luna was arrested with one other man, James Degorski. Juan Luna was convicted and Degorski is awaiting trial.

State v. John Nicholas Athan— Cold Case

On November 12, 1982, in Seattle, Washington, police found the body of a 13-year old girl inside a cardboard box dumped behind a Seattle store. Except for a pair of socks, her body was nude from the waist down and the cause of death was strangulation. A ligature was found around her neck. Although no DNA was found under her fingernails, there was evidence of rape occurring near the time of the murder. The medical examiner found spermatozoa on the rape kit vaginal swabs. The girl's body was found near the residence of an acquaintance, 14-year-old John N. Athan. Although John N. Athan's brother reported that he had seen John pushing a large box on a shopping cart near the area where the victim's body was found, John N. Athan denied having sex with the victim. There was no confession, there were no eyewitnesses, his fingerprints did not match fingerprints found near the crime scene, nothing placed the two

together, and DNA profiling was not yet available. Therefore, the case remained unsolved for over 20 years until 2003. Twenty-one years later, with advances in DNA technology and a new Cold Case Homicide Unit, Seattle Police Detectives submitted the old crime scene evidence for DNA testing and a male profile was developed. Next, posing as a fictitious law firm, the Seattle police detectives sent John N. Athan a letter, along with a return envelope, stating he was eligible for money and inviting him to join a fictitious class action lawsuit over parking tickets. He returned the letter, licked the envelope, and detectives opened the letter without a warrant or court order. The Washington State Patrol Crime Laboratory obtained a DNA profile from saliva on the return envelope and compared the DNA profile to a DNA sample from the 21-year-old crime scene rape kit. The DNA profile obtained from the letter returned to the fictitious law firm matched the sperm fraction, a male DNA profile generated from the vaginal swabs, and John Athan was charged and arrested in New Jersey. A search warrant for his DNA obtained a DNA profile, which also matched the DNA profile on the envelope, and the male sperm fraction DNA profile on the vaginal swab. Back in Washington, John Athan was convicted of second-degree murder the following year, in 2004. In 2007, the Washington Supreme Court upheld John N. Athans' conviction and held that, under these circumstances, any privacy interest in his saliva was lost. The court decided that the envelope, and any saliva contained on it, became the property of the recipient.

Admissibility of Evidence: Technology Collides with the Fourth Amendment

Cold Hits and CODIS (Combined DNA Index System)

Since the 1980s, when DNA markers for identifying biological samples were first developed, the use of DNA evidence to convict defendants and to exonerate the wrongfully accused and wrongfully imprisoned has greatly increased. But the increase in databanks for storing DNA information on individuals convicted of certain crimes raises important legal and ethical issues on the use, collection, and storage of DNA evidence. These issues have been the subject of a recent national commission, which will, hopefully, broaden public discourse about the future uses of DNA forensic technology. CODIS began as a pilot project in 1990, serving 14 state and local DNA laboratories. The FBI launched the **CODIS (Combined DNA Index System)** database in 1998. Today, 44 states collect DNA from all felons, 28 from juvenile offenders, and 39 from those who commit certain categories of misdemeanors. Congress enacted the DNA Identification Act of 1994 authorizing the FBI to maintain a centralized, national DNA database and to develop a software system to allow for the sharing of information within and between the states. By 2004, the resulting system, CODIS, connected the databases of all 50 states, which at that time were limited to profiles from those convicted of serious, violent crimes including felony sex crimes. Signed into law by President George W. Bush on October 30, 2004, the Justice For All Act (P.L. 108-405) greatly expanded the CODIS system, allowing collection of DNA from all federal felons and enabling states to upload to CODIS profiles from *anyone* convicted of a crime. The FBI has over 3,000,000 entries in CODIS.

Searching CODIS for cold hits has been very successful. Many cold cases have been solved by comparing the DNA profile of a crime scene sample with the DNA profiles of convicted offenders. Although the evidence of a match has been admissible in court, the statistics used to state how common or rare the coincidence is are subject to legal challenges.

Familial Searching

Familial searching of databases expands the idea of searching CODIS for a cold hit. It is a new method of creating suspects in the absence of an immediate cold hit. Familial searching is based upon the fact that siblings and other closely related individuals share more common genetic material than nonrelated individuals, so they have similar DNA profiles. Familial searching involves searching DNA databases of convicted offenders, looking for near misses. A DNA profile of a convicted offender that is close, but not identical, to the crime scene profile might indicate a family relationship between a known convicted offender in the database and an unknown sample from a crime scene. Current methods of familial searching involve generating a list of possible relatives of the owner of DNA from an item of evidence collected at a crime scene. Close relatives of those matches are then tracked down and asked to "voluntarily" provide a DNA sample. Performing either a "low stringency" profile search to look for partial matches between crime scene evidence and offender family profiles or by conducting a rare-allele search, the suspects are narrowed down.

Although the practice of familial searching is accepted in England, the FBI had a policy that prohibited the release of any identifying information about an offender in one state's database to officials in another state unless the offender's DNA was an exact match with the DNA evidence found at the scene of crime. However, the FBI has recently relaxed its policy. In 2006, the FBI changed its policy in response to a request from Denver authorities that found a close match between evidence taken from the scene of a rape and a convicted felon in Oregon, indicating that he was a potential relative of the actual perpetrator. The interim policy, effective July 14, 2006, allows for states to share information related to partial matches, upon FBI approval.

Abandoned DNA

Assume you have just killed your unfaithful lover. You have bleached the murder weapon to destroy any DNA, removed all fingerprints, destroyed all incriminating evidence, and perfected your alibi. Then, the police find skin under your dead lover's fingernails. You are the number one suspect and you refuse to voluntarily give a DNA buccal swab (cotton swab rubbed against inside of your cheek). When a suspect refuses to give his or her DNA, police detectives can either get a search warrant signed by a judge or resort to trailing the suspect and collecting the suspect's DNA without his knowledge. Police officers may wait to collect abandoned DNA or offer the suspect a cigarette or beverage while in the interrogation room and secretly collect the discarded item later. This investigative tool can allow detectives to exclude anyone whose DNA profile does not match the crime scene evidence or confirm that they have the true perpetrator as a suspect. When is a search warrant needed to collect DNA? What about discarded DNA?

The Constitution of the United States was constructed to safeguard the rights of American citizens. One of the rights that people could expect, as stated by the Fourth Amendment, was the right "to be secure in their persons, houses, papers, and effects against unreasonable search and seizure." While the Fourth Amendment took a step towards guaranteeing privacy, its protection is limited by the interpretation of the phrase "unreasonable search." Unfortunately, this phrase was never well defined, and its definition has become increasingly blurred by the recent technological advances.

To conduct a legal search generally requires probable cause and a judicial warrant, or at least individualized suspicion. Recently, there have been Fourth Amendment challenges to the methods that law enforcement officers have used to obtain DNA samples from suspects without a warrant. The Fourth Amendment guarantees a citizens' right to privacy. However, in many cases around the United States, courts have ruled that if a suspect discarded or abandoned an item outside of the privacy of his or her home, then the suspect did not have a reasonable expectation of privacy. Therefore, discarded cigarette butts, gum, discarded eating utensils and coffee cups, discarded napkins and tissues are fair game.

DNA Dragnets

DNA testing has advanced since the first DNA dragnet took place in Leicestershire, England.

What has happened since *The Blooding*? DNA dragnets have been used in the United States with very limited success. They are used more widely in Europe. But are they legally permissible? In the United States, DNA dragnets ensnare thousands of innocent people, but most often fail to find their intended targets.

DNA collection through buccal swabs may violate the fourth amendment protections against unreasonable search and seizure. In particular, DNA dragnets may violate the Fourth Amendment because consent obtained in such dragnets is not always legally voluntary. First, sometimes police use coercive measures to obtain consent in DNA dragnets, and often the individuals asked to give samples in DNA dragnets are unaware of their right to refuse. Police will sometimes indicate that if the individual does refuse, they must have something to hide and will become a suspect. Second, sometimes subjects of DNA dragnets may reasonably, but falsely, believe their consent will produce no incriminating evidence. Third, consent to provide a DNA sample in a dragnet is not consent to use DNA for other purposes.

Two recent cases have highlighted problems with this controversial police practice. Over the last several years, police have used DNA dragnets to look for a serial murderer in Baton Rouge, Louisiana, and a serial rapist in Omaha, Nebraska. In Baton Rouge, Louisiana, police collected over 1,000 samples with no success. In Omaha, Nebraska, police trying to catch a rapist in 2004 used a DNA dragnet, and those who refused were forced to give sample by warrant. Yet the Omaha DNA dragnet produced no suspects. These DNA dragnets have resulted in law suits.

DNA dragnets have been challenged as a violation of the Fourth Amendment right to privacy and a violation of the Fifth Amendment right against self-incrimination. To ensure that DNA dragnets are constitutional, police should only use dragnets as a last resort. When they do use DNA dragnets, police should first limit the scope of a dragnet to those who match the description of the perpetrator or who have access to the victim. Then police should inform the potential donors of their right to refuse to volunteer a

DNA sample. Police should never be permitted to threaten potential donors with increased scrutiny, with embarrassing publicity, or with search warrants. Once voluntary samples have been tested, the police should destroy all samples gathered from donors exculpated in a DNA dragnet or disclose how the samples will be used. Police should protect the privacy of both innocent donors (excluded as a suspect) as well as those who exercise their right not to provide DNA voluntarily.

In the United States, DNA dragnets have rarely been successful in identifying suspects. According to a study conducted in 2004 by Samuel Walker, a criminal justice professor at University of Nebraska at Omaha, only one DNA dragnet of 18 reported in the U.S. resulted in arrest and conviction. That case involved targeting 25 employees of a nursing home in Lawrence, Massachusetts, where a resident had been raped.

DNA Dragnet Case Study—The Cape Cod Murder Case

On January 6, 2002, 46-year-old freelance fashion writer Christa Worthington was raped and stabbed to death in her bungalow home in Truro, a small Cape Cod beach community in Massachusetts. Her two year-old daughter, Ava, was found clinging to her body, unharmed. After 3 years and a $25,000 reward, the police had still not solved the case. They had focused their investigation on several suspects, including a former boyfriend who found the body, and a local fisherman and married father of six who had fathered a child with Worthington and then refused to pay child support. These suspects had been excluded. The case generated international publicity and a book, *Invisible Den*, was written containing explicit details of Ms. Worthington's love life and violent death. The F.B.I. advised local police that the killer had Truro ties and suggested trying to match the semen in a DNA sweep, a DNA dragnet. Authorities collected buccal swabs from local men by asking all male residents. Samples were collected from local men at the gas station, local eateries, the post office, and the town dump. Many residents cooperated,

while a few objected. The ACLU objected. Due to processing backlogs, the sample languished in a crime lab for eight more months. In April 2005, after more than 175 innocent men had been randomly tested, the suspect was identified as the killer by DNA testing. Richard McCowan, a local garbage collector familiar with the Worthington home, was convicted in 2006 and sentenced to life when the DNA sample he gave matched skin and semen samples collected from the victim's body.

Protecting the Innocent

Peter Neufeld and Barry Scheck established the first Innocence Project at the Benjamin Cardozo School of Law at Yeshiva University in New York. The Innocence Project represents clients seeking postconviction DNA testing to prove their innocence. The Innocence Project of Cardozo School of Law has become a national organization and serves as a model for local Innocence Projects, which have sprung up around the country.

Exonerations

As of May 2008, there have been 216 post-conviction DNA exonerations in the United States. The first DNA exoneration took place in 1989. DNA exonerations have been won in 32 states. The true perpetrators have been identified in 82 of the DNA exoneration cases. Of the 216 people exonerated, 16 were sentenced to death and served time on death row. One third of the Innocence Project cases remain unsolved due to lost or missing evidence. Therefore, finding old evidence is crucial to a DNA exoneration. However, only 23 states and Washington D.C. have statutes compelling the preservation of physical evidence. Therefore, 27 states do not have statutes requiring preservation of physical evidence. In addition to Washington, D.C., the following states have statutes that compel preservation of evidence: Arkansas, California, Colorado, Connecticut, Florida, Georgia, Hawaii, Illinois, Kentucky, Louisiana, Maryland, Michigan, Minnesota,

Missouri, Nebraska, New Hampshire, New Mexico, North Carolina, Oklahoma, Rhode Island, Texas, Virginia, and Wisconsin. Unfortunately, even when a state has legislation requiring the preservation of evidence, the evidence may still be destroyed.

Postconviction DNA Testing

Even with preserved evidence, postconviction DNA testing is not an automatic, absolute right. Not all states have postconviction DNA access statutes. Seven states have not passed legislation guaranteeing the right to DNA testing. Even in the 43 states which have postconviction DNA testing access statutes, this right is limited and inmates have been refused testing where the results might have affected the death sentence, even if not the determination of their guilt. As of May 2008, the seven states with no DNA access statutes are: Alabama, Alaska, Massachusetts, Mississippi, Oklahoma, South Carolina, and South Dakota.

Prosecutors have opposed testing even in cases where the convicted person had died in prison maintaining his innocence in various states. The Justice for All Act of 2004 (H.R. 5107) enhances protections for victims of federal crimes, provides resources and DNA training for law enforcement, reauthorizes and expands the DNA Analysis Backlog Elimination Act of 2000, and also includes the Innocence Protection Act of 2004. The Innocence Protection Act of 2004 provides funding for federal postconviction DNA testing, incentive funding to state prosecutors to conduct postconviction DNA testing to exonerate the wrongly convicted, and funding to improve defense representation in capital cases.

Genetic Privacy

There are many benefits of genetic testing. Increased genetic testing makes it more likely that researchers will come up with early, life-saving therapy for a wide range of diseases with hereditary links such as breast and prostate cancer, diabetes, heart disease, and Parkinson's disease. Genetic testing also will help doctors catch problems early, perhaps leading to preventive treatment and lower medical costs.

However, there have been risks associated with genetic testing. Many individuals and families have been experiencing discrimination by being denied health insurance or by losing their jobs because of the perceived risks attributed to their genetic status. In the 1970s, several insurers denied coverage to blacks that carried the gene for sickle cell anemia. The Lawrence Berkeley National Laboratory in California secretly tested workers for sickle cell trait and other genetic disorders from the 1960s through 1993; workers were told it was routine cholesterol screening.

The public's concern over privacy and fear of misuse of private genetic information was undermining the remarkable progress made in genomic research. Researchers recognized that Americans were reluctant to take advantage of new breakthroughs in genetic testing. This problem was confirmed in a 1996 Georgetown University study involving 332 members of genetic support groups with one or more of 101 different genetic disorders in the family. The researchers found that as a result of a genetic disorder, 25 percent of the respondents or affected family members believed they were refused life insurance, 22 percent believed they were refused health insurance, and 13 percent believed they were denied or let go from a job. According to the study, fear of genetic discrimination resulted in 9 percent of respondents or family members refusing to be tested for genetic conditions, 18 percent not revealing genetic information to insurers, and 17 percent not revealing information to employers.

This level of perceived discrimination pointed to the need for legal protection from genetic discrimination. Although federal law already bans discrimination by race and gender, until recently there has been no federal law banning genetic discrimination by employers outside of the federal government and by health insurance companies. "Your skin color, your gender, all of those are part of your DNA," said Francis Collins, head of the National Human Genome Research Institute. "Shouldn't the rest of your DNA also fall under that protective umbrella?"

In an effort to protect federal employees, in February 2000, President Clinton issued executive order 13145 prohibiting discrimination in federal employment based on genetic information. This executive order banned the federal government and its agencies from demanding that employees undergo any sort of genetic test or from considering a person's genetic information in hiring or promotion decisions. However, private employers were not prohibited from testing. A 2001 study by the American Management Association showed that nearly two-thirds of major U.S. companies require preemployment medical examinations. Fourteen percent conduct tests for susceptibility to workplace hazards, 3 percent for breast and colon cancer, and 1 percent for sickle cell anemia, while 20 percent collect information about family medical history.

With Americans refusing to take genetic tests or using false names and paying cash because they didn't want the information used against them by their employer or insurance company, researchers recognized the need for legal protection. In May 2008, Congress passed the first federal law to protect employees from mandatory genetic tests or to guarantee patients' privacy. The Genetic Information Nondiscrimination Act prohibits health insurance companies from using genetic information to set premiums or deny health insurance, and guarantees that employers cannot use genetic information in hiring, firing, or promotion decisions. "We will never unlock the great promise of the Human Genome Project if Americans are too afraid to get genetic testing," said Rep. Judy Biggert, R-Ill., who co-sponsored the bill.

But while Congress has been protecting our genetic privacy in civil areas, what has Congress done to protect our genetic privacy in criminal cases? On January 5, 2006, President Bush signed a little-noticed piece of legislation entitled the DNA Fingerprint Act of 2005 into law, greatly expanding the government's authority to collect and permanently retain DNA samples. The DNA Fingerprint Act of 2005 was signed into law as Title X of the Violence Against Women Act (VAWA), H.R. 3402, 109th Cong. (2006) (enacted). This brief addition to the popular, extensive Violence Against Women Act (VAWA) reauthorization bill slipped through Congress unnoticed, without public reaction, and without policy debate. Among other provisions, it grants the government authority to obtain and permanently store DNA from anyone who is arrested—but not convicted—of a crime as well as non-U.S. citizens detained under federal authorities. This represents a departure from the earlier CODIS (see above) that collected DNA samples from convicted offenders, typically felons convicted of violent crimes. This new law raises extraordinary questions for the future of genetic privacy and civil liberties. There will no doubt be constitutional challenges to this new law. Innocent people do not belong in a criminal DNA database.

DNA in the Future

Ethnic DNA Profiling Case Study—Derrick Todd Lee

In 2002–2003, seven women were raped and murdered in the Baton Rouge area of Louisiana. The attacks occurred during the day. Eyewitnesses described a Caucasian 20–25-year-old male driving a white pickup truck. However, using newer genetic markers to perform DNA ethnic profiling, DNAPrint Genomics, a Florida lab, concluded the DNA from five evidence samples indicated the suspect was African American.

After the police were advised that the attacker was African American, another woman in the area was attacked. She survived when her son came home unexpectedly, and she identified Derrick Todd Lee at trial. Accused of murdering seven women in 2002–2003, DTL was charged with four murders, convicted in 2004 in Baton Rouge, Louisiana, and sentenced to death.

But what is the future of ethnic profiling? In 2005, the same Florida lab released their

Retinome assay to improve their ability to construct a physical portrait of the person who left their DNA at the crime scene. Using human pigmentation gene SNPs and ancestry markers, Retinome can determine the persons' eye color based upon their DNA sample. The biotechnology company is working on methods to predict hair color, skin pigmentation, and other physical characteristics.

Looking at current advances in DNA in medical research allows us a glimpse into the future of forensic DNA. These advances present many global scientific and ethical issues. What will happen with genetically engineered food? What about genetically engineered children? Cloning? Gene patenting? Scientists around the world are grappling with these very issues today, anticipating the changes that we will be faced with tomorrow.

Questions

1. Where in the cell is DNA located?

2. Name three types of human cells that contain DNA.

3. If the father is blood group B and the mother is blood group O, what are the possible blood groups for their children?

4. What was the contribution of Watson and Crick to the understanding of DNA?

5. Name the scientist who first utilized RFLP for DNA analysis in a criminal case.

6. What is the purpose of the Combined DNA Information System (CODIS)?

7. Why is polymerase chain reaction (PCR) important in DNA analysis?

8. Where is mitochondrial DNA found in the body?

9. How is DNA analysis utilized in noncriminal cases?

10. What are the factors considered in a Daubert hearing for DNA evidence?

11. How are forensic laboratories accredited?

12. Who were the organizers of the Innocence Project?

References

1. Watson, J. D. and Crick, F. H. C., Genetic implications of the structure of desoxyribonucleic acid, *Nature* 171, 964, 1953.
2. Watson, J. D. and Crick, F. H. C., Molecular structure of nucleic acids, *Nature* 171, 737, 1953.
3. Jeffreys, A. et al., Individual-specific "fingerprints" of human DNA, *Nature* 316, 76, 1985.
4. Wambaugh, J., *The Blooding*, Bantam Book Company, New York, 1989.
5. Inman, K. and Rudin, N., *An Introduction to Forensic DNA Analysis*, CRC Press, Boca Raton, FL, 1997.
6. Mullis, K. B., The unusual origin of the polymerase chain reaction, *Scientific Am.*, 4, 56, 1990.
7. *State of Washington v. Kenneth John Leuluaialii and George J. Tuilefano*, 118 Wash. App. 780, 77 P.3d 1192, Wash. App. Div., 1,2003. October 13, 2003.
8. *Kuhmo Tire v. Carmichael*, 526 U.S. 137, 119 S. Ct. 1167, 1999.
9. *Daubert v. Merrell Dow Pharmaceuticals*, 509 U.S. 579, 113 S. Ct. 2786, 125 L. Ed. 2d 469, 1993.
10. *State of Washington v. Kenneth John Leuluaialii and George J. Tuilefano*, 118 Wash. App. 780, 77 P.3d 1192, Wash. App. Div., 1, 2003. October 13, 2003.
11. Jeffres, A. J., and Allen, M. J. et al., Identification of Skeletal Remains of Josef Mengele by DNA Analysis, *Forensic Sci. Int.,* 56(1), 65–76, 1992.
12. http://www.ojp.usdoj.gov/nij/journals/256/missing persons.html#sidebar_federal.

13. http://www.cstl.nist.gov/biotech/strbase/dabqas.htm (DNA Advisory Board Guidelines).

14. *State V. John Nicholas Athan*, 160 Wash.2d 354, 158 P.3d 27, Wash., May 10, 2007 (No. 75312-1).

15. Bieber, F. R., Science and technology of forensic DNA profiling: Current use and future directions, *DNA and the Criminal Justice System: The Technology of Justice*, David Lazer, Ed., 2004.

16. Grand, J., *The Blooding of America: Privacy and the DNA Dragnet,* 23 Cardozo L. Rev. 2277, 2002; Aaron B. Chapin, Note, *Arresting DNA: Privacy Expectations of Free Citizens Versus Post-Convicted Persons and the Unconstitutionality of DNA Dragnets,* 89 Minn. L. Rev. 1842, 2278, 2005.

17. *United States V. Prescott*, 581 F.2d 1343 (9th Cir. 1978).

18. Walker, S. *Police DNA "Sweeps" Extremely Unproductive: A National Survey of Police DNA "Sweeps"*, Police Professionalism Initiative, Dept. of Criminal Justice, University of Nebraska, 2004.

19. info@innocenceproject.org.

20. http://www.deathpenaltyinfo.org.

21. Lapham, E. V., Kozma, C., and Weiss, J. O., Genetic Discrimination: Perspectives of Consumers. *Science* October 25, 1996: Vol. 274. No. 5287, pp. 621–624.

Microanalysis and Examination of Trace Evidence

Thomas A. Kubic and Nicholas Petraco

17

Introduction

Microanalysis is the application of a microscope and microscopical techniques to the observation, collection, and analysis of microevidence that cannot be clearly observed or analyzed without such devices. Microanalysis today generally deals with samples in the milligram or microgram size ranges. Microscopes and the techniques to be discussed will be limited to those that employ light in the visible, ultraviolet (UV), and infrared (IR) frequency ranges or use electrons for illumination. Analysis with a microscope may be limited to observations of morphology or involve the collection of more sophisticated analytical data, such as optical properties, molecular spectra, or elemental analysis.

The definition of *trace* as in the phrase *"trace evidence"* is more problematical. Historically, criminalistics and forensic sciences used *trace evidence* to describe any evidence small in size, particularly evidence that would be analyzed with microscopical techniques. In the not-too-distant past, even the analyses of small blood samples and bulk soil or dust samples were considered trace analyses.

It is better to define *trace analysis* as the qualitative or **quantitative analysis** of the minor or ultraminor components of a sample. "*Sample*" means an entire submitted exhibit or a subsample of the exhibit. The section of a forensic laboratory where trace materials were submitted generally depended on the historical development of the laboratory. Large laboratories, for example, may have included separate sections for fiber, hair, mineralogy (soil), paint, serology, **controlled substances**, and firearms analyses. Small laboratories often grouped the sections differently, possibly as chemistry, biology, microscopy, drugs, and ballistics. The types of evidence examined by each area depended on the history of the laboratory. Hair examination cases were sometimes assigned to the biology group, while in other labs they may have been handled by the microscopy unit or chemistry section. Natural and synthetic fibers were assigned similarly to different units.

In this chapter, we will treat all types of classical *trace evidence,* if examined predominantly by microscopical techniques and methods, as examples of microanalysis. The purpose of these analyses is to determine whether an association of persons, places, and things can be established, and the strength of that association. The association is predicated on the comparison of materials found and the drawing of the conclusion that they are of common origin. The criminalist is drawn to the common origin conclusion, if after having examined the samples in sufficient detail, he is unable to establish a forensically significant difference in the materials. The weight of this association is a function of the level of individualization that resulted from his examinations.

Instruments of Microanalysis and Sample Types

A wide variety of microscopes are available for use in a forensic laboratory, and they can examine a wide variety of materials (evidence). Because this is an introductory text, we will limit our discussion to the application of the light microscope, particularly the **polarized light microscope (PLM)**, visible and infrared spectrophotometry via a microscope, and basic scanning electron microscopy (SEM) with energy dispersive x-ray spectroscopy (EDS). Concentration will be on the light and polarized light microscope methods. We will discuss evidence types, such as glass, fibers, hairs, and paint, with brief mention of pollen, soils, and gunshot residue (GSR).

The microscope most likely to be employed first in the examination of evidence is the **stereo binocular microscope**. It is often employed in the preliminary evaluation of submissions, and for the location and recovery of microscopic particles and materials from their substrates. Examples are the recovery of fragments of red wool fibers from a victim's sweater found on the denim jacket of a suspect in an assault case and the recovery of glass particles from the jeans of a burglary suspect. This microscope is a compound type. Total magnification is computed by the power of the objective (OBJ) or first lens multiplied by that of the eyepiece (EP) or finial lens.

A *lens* is an optical component that may be composed of one or multiple elements. The stereo microscope is constructed from two similar but separate optical microscopes for observation by each eye simultaneously. The views are separated by a small angle, usually about 15 degrees, so that each eye sees the subject from a slightly different perspective. This renders the appearance three-dimensional. It is by this mechanism that humans see nearby objects as three-dimensional. Most observations performed with stereo microscopes are carried out with reflected light analogous to how we normally see objects. Figure 17.1A is a photo and an optical diagram of a common stereo microscope.

Many significant preliminary and other analytically important observations are made with this microscope. The layer structure of a recovered paint chip including the color of each layer and an estimate of the curliness of a human hair are only two examples. The stereo microscope is also frequently employed for viewing an object while it is being prepared for further, more advanced forms of analyses, such as SEM observation or infrared **microspectrophotometry**.

The second most common type of microscope encountered in the laboratory is the **compound binocular microscope**. Figure 17.1B illustrates this type of microscope. Most people are familiar with such microscopes because they are routinely found in schools and medical laboratories. Although this microscope employs two eyepieces (EPs), both eyes see the same image because each eyepiece magnifies an image formed by a single common objective. Most often this microscope employs transmitted, bright field illumination for viewing. In transmitted light, the sample is transparent or mostly transparent. Most of the illumination passes through the subject and some passes around it. A number of alternative illumination methods can aid in detail observation by enhancing contrast and analytical sensitivity.

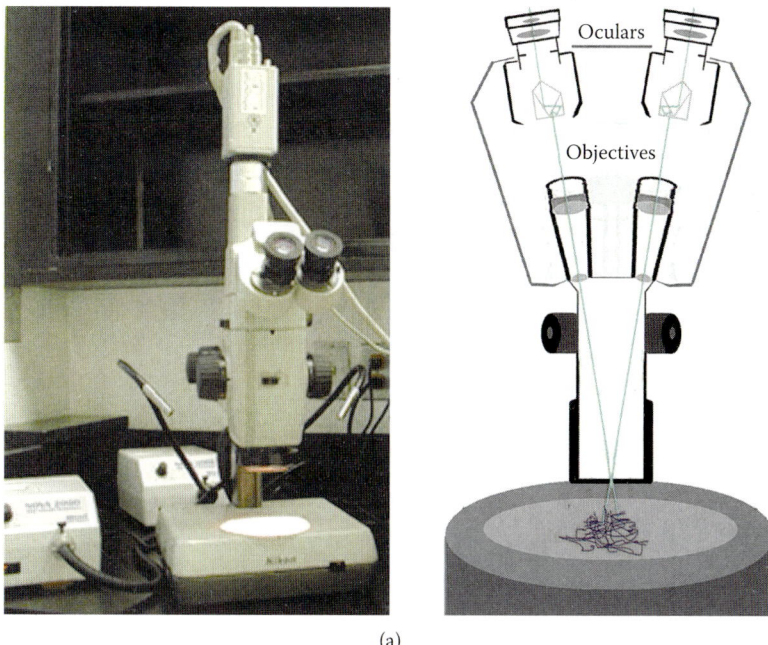

(a)

Figure 17.1A   Right: Optical diagram of a stereo microscope showing the two distinct optical paths that lead to a stereoscopic (three-dimensional) view. Left: A modern stereo microscope with fiberoptic illuminators and trinocular viewing head. The third position is equipped with a television camera.

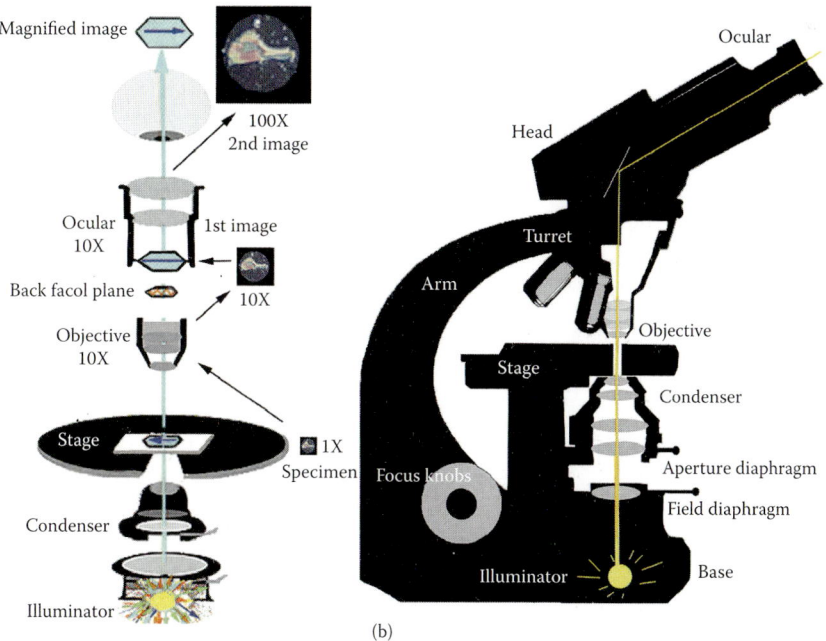

(b)

Figure 17.1B   Right: Optical diagram of a transmitted light compound microscope showing the most important parts and a light path through the microscope from illuminator to ocular. Left: A more stylized version showing important optical components and their positions and the position of the specimen and its images.

Methods such as dark field, reflected light or phase, modulation, and interference contrasts are commonly employed, but their discussion is best left to more advanced texts.

This microscope is capable of total magnifications in the range of 25 to 1200 times (×) greater than the object, with 40 to 400× magnification commonly encountered in forensic laboratories. The total magnification (TM) is the product of the OBJ magnification multiplied by EP magnification: OBJ× multiplied by EP× = TM×. For example 10× OBJ × 10× EP = 100× TM×. The most important information obtained with this instrument is morphological. It shows the visual appearance and details of construction of the subject. The use of the microscope allows the examiner to view the exhibit at higher magnification and therefore in more detail. The revelation of detail, in reality, is a function of the resolving power (RP) of the microscope, which is related to the numerical aperture (NA) of the microscope objective. The numerical aperture of the objective, the first lens of the microscope, generally increases with the magnification. The NA determines the operational characteristics of a particular objective, its practical use, and the information content of the image produced by the compound microscope utilizing it. The more important of these characteristics and practical considerations are magnification (×), working distance (WD), depth of field (DF), angle of acceptance (AA), NA, and RP. For definitions and a diagram, see Figure 17.2.

Of similar importance is analytical information about a sample that can be obtained without the use of sophisticated techniques or the addition of complex accessories to a microscope. Analytical information is obtained by measuring a particular characteristic that is observed. The color and layer structure (number of layers and their order) of a paint chip sample obtained by the simple viewing of the object is an example. Valuable additional information can be added to the basic data, if the information is further qualified by comparison to a standard; for example, detailing of the color of a soil by comparing it to a reference like the Munsell system.[1]

Quantification of characteristics such as its physical dimensions by actual measurement

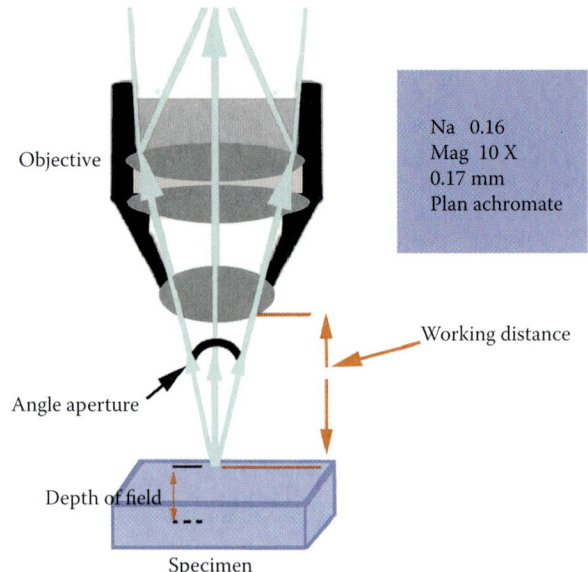

Figure 17.2 Stylized drawing of a microscope objective indicating important relationships when an object is focused.

can be even more useful. A scale calibrated with a **stage micrometer** can be placed in the EP of a microscope so that its image is superimposed on the view of the subject, thereby allowing evaluation of characteristics such as length, width, and thickness. This method is known as **micrometry**. The thicknesses of the layers of a paint chip, the average length of scales on a hair (scale count), and the modification ratio of a synthetic fiber determined by viewing a cross section are examples of quantitative information that can be obtained with a relatively unsophisticated bright field compound microscope.

Chemical tests that aid in the identification of a material can be performed on micro- and ultramicro-sized samples if the results are observed with a microscope. Chemical color and micro-crystal tests employed for the identification of controlled substances and the solubility or reaction of a paint to various solvents are routine applications of this technique.

When a compound microscope is fitted with certain accessories, it is converted to a polarized light microscope (PLM). Figure 17.3 shows an example of a PLM, also called a petrographic or chemical microscope. The power of the analytical measurements it can make and its usefulness in forensic science which is always perpendicular to the direction of propagation (travel). Wavelength and frequency are inversely related. The longer the wavelength, the lower the frequency of vibration. Shorter wavelength light, violet light, and ultraviolet light have higher energy than longer wavelength red light. Normal light is randomly polarized. That means the vibration direction of the light is in all directions, 360 degrees perpendicular to the propagation direction. If the vibration is restricted to only one direction, it is referred to as plane polarized light.

Light can become partially or totally polarized in a number of ways including reflection, adsorption, and propagation through an anisotropic material. The earliest devices employed to generate plane polarized light were obtained by the cutting and polishing of particular anisotropic materials along certain directions within a crystal and cementing them together in a certain orientation so that light transmitted along the optic axis of the microscope was plane polarized.

Today, plane polarized light is obtained by the use of polymer films in which the molecules are very highly oriented and have been treated with a dye so that they almost totally absorb light vibrating in all but one direction. This single direction is called the *privileged direction*. A portion of the light in the privileged direction is also absorbed, but this loss of intensity is small in comparison to losses of other directions. These polymer filters are known as Polaroid' filters or films. When two polarizers are placed in such a way that light passes through one and then the second and privileged directions of each are perpendicular, no light will emerge from the second. This condition is referred to as *crossed polars* and results in complete extinction of transmitted illumination.

When an object is placed in the illumination path of a PLM and between the two polarizers, it may affect the vibration direction of the plane polarized light reaching it from the first polarizing element. If this is the case, the material is called *anisotropic,* and it will resolve the original vibration's intensity into two perpendicular vibration directions. Each of these resulting rays, except in certain

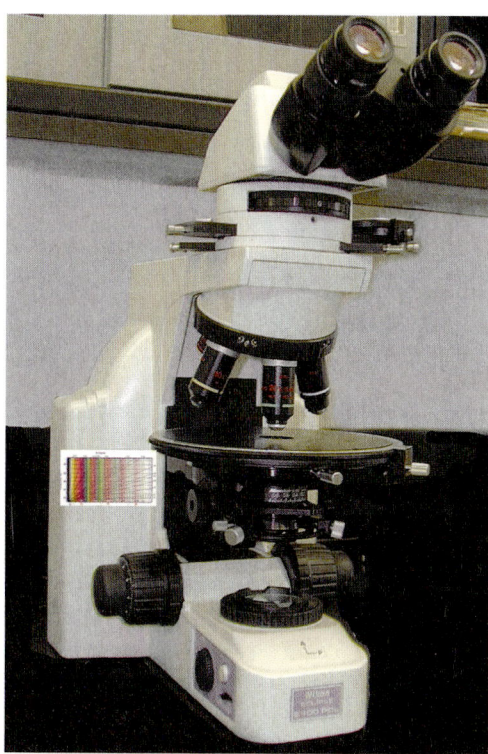

Figure 17.3 A polarized Nikon Eclipse E400 light microscope designed for transmitted light illumination, typically used by journeymen criminalists to examine microscopic evidence. It has a circular rotating stage, strain-free objectives, accessory slot and compensators, a Bertrand lens, and a binocular observation head. (Photo courtesy of Morrell Instruments, Melville, New York.)

are substantial in the hands of an experienced and trained criminalist. The basic requirements are that two polarizing elements are positioned in the optical path of the microscope. The first, called the *polarizer*, is placed prior to the sample, normally in the condenser mount just prior to the lenses. The second, called the *analyzer*, is positioned in the body of the microscope, usually in an intermediate accessory tube between the objectives and the viewing head that holds the EPs. This accessory tube also has a slot cut into its body prior to the analyzer, so that additional devices can be placed into the optical path at 45 degrees to the position of the viewer and the privileged directions of the polarizers. This is known as the accessory slot.

Light is a wave phenomenon. Its characteristics are velocity (c), wavelength (λ), and frequency (υ) related to color, amplitude (a) related to **brightness**, and vibration direction, special directions of propagation, will have a different refractive index (RI) and the difference of these indices is referred to as the *birefringence* (ΔRI). The maximum birefringence is an analytical and identifying characteristic of a material. Both rays travel through the material at different velocities, and this results in a phase shift of the rays when they emerge from the material. This phase shift is known as the *optical path difference* (OPD) and is calculated as the difference in RI (ΔRI) multiplied by the thickness. When the thickness is in micrometers and the resultant OPD is multiplied by 1000, the retardation (R) of the sample measured in nanometers is obtained.

$$R_{nm} = \Delta RI \times thickness_{\mu m} \times 1000$$

When the rays with the two separate vibration directions pass through the second polarizer (analyzer), they are both resolved into rays vibrating in the same direction and are then able to interfere with each other. The amount of interference and the resultant intensity depend on the phase difference between the rays. If the illumination is white light, the various wavelengths interfere in different amounts and certain colors are intensified and others decreased or even eliminated due to destructive interference. The result is an interference color associated with sample retardation. An analytical working tool referred to as the Michel–Lévy chart relates the birefringence, thickness, and retardation properties. If the microscopist directly measures any two components, the third can be easily determined.

The most common accessories used with a PLM and placed into the accessory slot are called *compensators*. Compensators are anisotropic materials of known birefringence constructed so that the thickness is controlled and the orientations of the vibration directions are known. The direction of vibration of the slow ray, high RI (Z), and the total retardation in nanometers are normally marked on the compensator. Compensators are fixed or variable, and are capable of measuring retardation from one to thousands of nanometers.

The most important analytical capability of the PLM is determining the characteristic

RIs of anisotropic materials. These are divided into uniaxial and biaxial classes. Uniaxial materials have one optic axis and two characteristic RIs called ε and ω. Biaxial materials have two optic axes and three characteristic RIs, identified as α, β, and γ, where α is the lowest value and γ the highest. An optic axis is a direction through an anisotropic material such that the resulting vibration directions have the same RIs and the OPD is zero. Most anisotropic materials are crystalline and although fibers are not really crystals, these materials can be considered to behave like uniaxial crystals.

A great number of materials are isotropic. An isotropic material exhibits only one RI no matter which direction light propagates through the item or what the vibration direction is. Isotropic materials do not affect the vibration direction of light. Vacuum, gases, most liquids, amorphous solids, and isomorphic crystals are all isotropic. Vacuums have none or very few atoms or molecules to react with light. In gases and liquids, the molecules are free to move about with no specific orientation. Therefore, light interacts similarly, no matter in what direction it travels.

The same explanation can be employed for amorphous solids. Although they are solid, the various atoms or molecules of which they are composed are arranged in a random-like pattern so that light traveling through the material encounters similar interactions in every direction. In isomorphic or cubic crystals, the atoms or molecules that compose the crystal lattice are arranged similarly along each of the three crystal axes so that light encounters the same atmosphere and interactions in all propagation directions. The reader is referred to any of a number of texts for details concerning polarized light microscopy and optical mineralogy.[2–7]

The RIs of the materials discussed above are characteristics rather than universal constants unless certain parameters of the measurement are controlled and stated. Except in a vacuum, the velocity and, therefore, the RI varies with the wavelength of light and is called *optical dispersion*. The RI also varies with temperature; if the temperature is lowered, the RI increases. This is referred to as the −dn/dt or change in RI with temperature. Liquids have markedly greater −dn/dt levels than solids and the values are on the order of 100 to 1000 times as great. By convention, and likely for practical reasons, when data were first organized into analytical data bases, the RI was given at the λ for the sodium D line at 589 nanometers and 25 degrees Celsius. The refractive indices at other λs for many materials have been compiled. They can be important analytical characteristics, but the most accepted reporting λ remains 589 nanometers (N_D). The reported reference temperatures for RIs are more varied because in many substances, particularly certain solids, the RIs may change substantially with temperature. When an RI is reported with a reference temperature and λ, the value is considered an optical constant.

The PLM is the instrument of choice to characterize many forms of microscopic materials, especially because analytical measurements can usually be made nondestructively. These measurements can lead to clear, unambiguous identifications and can aid significantly in the goals of association and individualization. A full length treatment of all the applications of PLM to evidence evaluation and methods employed is beyond the scope of this work and the reader is directed to quality texts on criminalistics.[7–10] The following is a brief description of a number of microscopes other than PLM that can be employed for the examination of microscopic evidence and the information that can be obtained with them. Readers should be impressed with their utility and value.

Comparison microscopes and macroscopes vary in their design and application to evidence analysis. At least one type of comparison microscope will be found in most broad service crime laboratories. They are all similar in one design principle. They are in reality two microscopes linked by an optical bridge so that the observer can simultaneously view two independent images in one field, each from a separate objective. The optical bridge often has a mechanical screen to provide a split field of view with a variable point of demarcation. These bridges also allow the superimposition of the two images.

The lowest total magnifications are found on what are referred to as macroscopes. These are like dual stereo microscopes but lack the three-dimensional imaging common with dual stereo devices. Large tool marks and fabrics are often examined with this instrument using reflected light. The public is most familiar with the firearms examiner's ballistics comparison microscope with which an examiner will attempt to establish that two projectiles were fired from the same weapon by examining the microscopic stria found on their surfaces. These stria are placed there by the contact of the softer projectile metal with the hardened surface of the weapon bore.

Other tool marks found on an ejected cartridge case and which originate from the firing pin, extractor, and ejector of the weapon can all be used to establish association with a weapon in question. Examinations are conducted with reflected light where control of the angle of incidence may be critical. The magnifications employed seldom rise above 60 diameters, with high contrast; low power quality optics favored over the need for great resolving power. This microscope is employed in the evaluation of tool marks left by all but the largest items.

The classical, transmission illumination, bright field microscopes, and even PLMs are often linked with a bridge so that very small samples, such as hairs and fibers, can be critically examined side by side in the field of view. Many experts opine that this is the only valid manner of comparison of two pieces of microscopic evidence. It is their position that data recording, sketches, photographs, and even videos are insufficient for the proper, ultimate comparison. Examples of reflected light, ballistics, and transmitted light PLM comparison microscopes are found in Figures 17.4A and B.

Microspectrophotometry is an area of microscopy that over the past 25 years has become very important. These instruments are commonly found in industrial and academic research laboratories. The technique is almost a required capability for forensic laboratories offering full service analysis of microscopic evidence. The principal types are visible and infrared microspectrophotometers, and

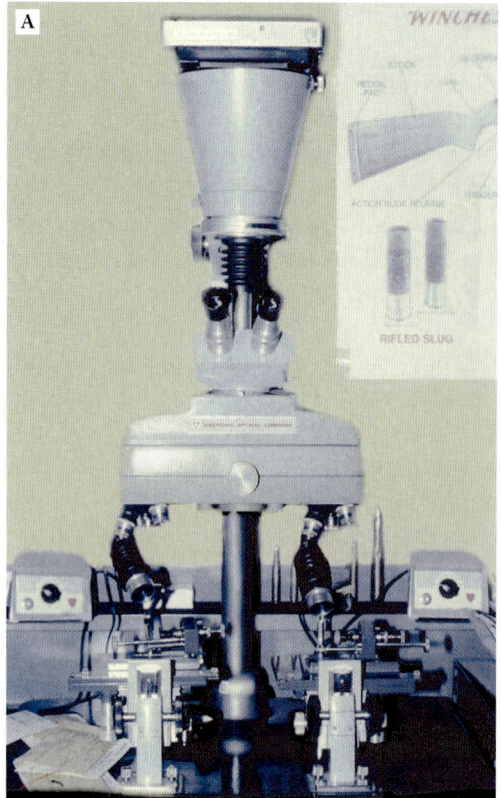

Figure 17.4A A firearms examiner's (ballistics) comparison microscope. Note the optical bridge, separate objectives, dual focusing stages, specimen holders, and two illuminators, the angles and intensities of which are independently adjustable.

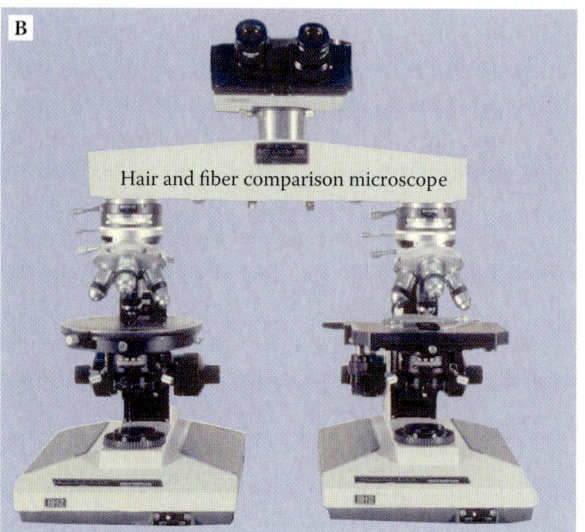

Figure 17.4B Olympus BH-2 polarized light microscopes linked by a bridge. The polarized light feature aids in hair and fiber examination, especially examination of synthetics. Note the circular rotating stage that aids in aligning fibers or hairs for more precise comparisons.

each requires a different instrument design because of the nature of the radiation employed to characterize the exhibit. Some microspectrometers employ UV light for imaging and measurement, but they are of less importance in forensic laboratories.

Visible microspectrophotometers lend themselves well to the accurate measurement of color by eliminating possible errors that can be made by analyst observation. These errors are due to lack of discrimination by the eye of similar hues or the problems that **metameric** samples present. These instruments generate transmission, reflection, or absorption spectra from various translucent and opaque samples. Examples of the most common applications are spectra obtained from colored fibers and paint surfaces. These spectrophotometers can be attached to fluorescence microscopes and employed to measure spectra from materials that fluoresce when illuminated with light of sufficient energy.

The infrared microspectrophotometer has become a highly utilitarian and valuable instrument in modern forensic laboratories. This device is capable of routinely collecting by transmission, reflection, or scattering measurements the vibrational spectra on samples as small as 20 micrometers.[11] Organic and inorganic materials or mixtures such as paint can be investigated. These spectra are referred to as *fingerprint spectra* and are valuable sources of structural information leading to chemical classification, generic grouping, and specific identification in many cases. Because of the inherent discrimination power of **infrared spectroscopy**, these instruments are invaluable aids in comparisons where association of materials is the goal. Figure 17.5 is a photograph of an infrared microspectrophotometer.

Infrared and Raman spectroscopy are classified as types of molecular vibrational spectroscopy. In molecules atoms are held together by bonds. These bonds do not act like static rigid rods that hold the atoms unmoving in space in relation to each other. Rather these bonds act more like springs which allow the atoms to move in various ways, called vibrations, in relation to each other. The atoms that make up multiatomic molecules continually alter the distances between themselves

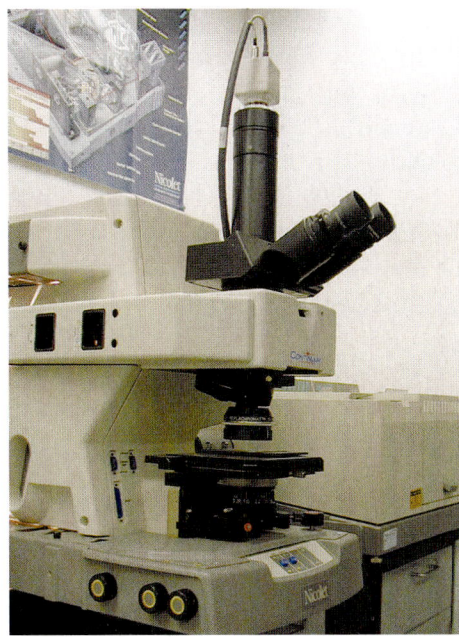

Figure 17.5 A Nicolet Continuum infrared microscope attached to a Magnum® 550 FT/IR spectrophotometer. Note the large reflected light objective just above the stage. It has an approximate 15× magnification factor and is very costly.

as they vibrate but are held from separating too far by the strength of the bonds between the atoms. These vibrations can be described as *stretches, bends, wags,* and *scissoring.* The frequency of these vibrations take place in the range of electromagnetic radiation (EMR) known as *infrared.*

The electron clouds that make up the bonds are not always uniform around their respective atoms. This nonuniformity results in a small separation of charge or dipole about the atoms and the bond between them. When a particular vibration causes a change in this dipole it is referred to as infrared active in that the bond is able to absorb radiation energy of that vibrational frequency.

The spectra obtained are displayed as a plot of the amount of light absorbed as a function of the wavelength, frequency, or wavenumber (cm^{-1}) of the radiation. These spectra can become quite complex when the molecule consists of three or more atoms. The spectra collected in the range of 4000–200 wavenumbers can be used to predict structural patterns or the functional groups contained in the molecules and the previously mentioned finger print region of 1800–200 wavenumbers can be

employed to make an unambiguous identification of a substance. Thus, vibrational spectra are a powerful tool for the forensic scientist to employ for his analysis of drugs, paints, inks, fibers, minerals, and other substances.

It has been over 80 years since the theoretical prediction of the inelastic scattering of light by molecules and since the first reported observation of this scattering by Dr. Chandrasekhara Venkata Raman, which now bears his name and for which he was awarded a Noble Prize.

Raman spectra are generated when light from a source is scattered by the electron cloud of a molecule. A molecule is considered Raman-active if this scattered light induces a dipole within the molecules bond. The photons that are scattered will either have less energy than the original source photons (Stokes shift) or more energy than the original photons (anti-Stokes-shift). Which one occurs depends on the vibrational energy state of the molecule at the time of scattering with Stokes being more probable because more molecules are in the vibrational ground state at room temperature at which most of these measurements are conducted. These shifts are directly related to the vibrational energy levels of the bonds of the molecules, and therefore the spectra obtained contain similar if not the same information as infrared spectra.

Within the last 10 to 15 years Raman spectrometers have become available to well equipped, full-service crime laboratories as costs have decreased to levels within reach. This is mainly the result of improvements in laser and filter technology, and exponential growth of computing power with the concomitant improvements in software, improved detector sensitivity, and more advanced digital electronics, all at a lower price tag.

Raman spectra are complementary to infrared spectra and supply more information concerning the backbone of the structure of the molecule while infrared excels at functional group identification. The recent availability of large commercially available quality data bases has also made Raman spectroscopy attractive to forensic scientists.

The principal advantages that make Raman spectroscopy attractive are: (1) it is essentially a nondestructive technique, (2) it requires very little if any sample preparation, (3) it is fundamentally a micro technique with the small laser spot illuminating only the sample of interest, and (4) it is able to measure vibrational frequencies without difficulty in the 700–200 wave-number region, which is not only an extended range for organic molecules but very useful for the analysis of inorganic moieties.

There are two fundamental instrumental instrument designs for Raman microspectrometers both of which use high quality filters to isolate the excitation wavelength and to block the elastically scattered source radiation from overwhelming the spectrum generated.

The first, called a *dispersive instrument,* employs a laser source in the range of visible light. The scattered Stokes and anti-Stokes resulting photons are at frequencies that remain in the visible area of the spectrum. Monochrometers are often employed to resolve the various shifted frequencies, while photo multiplier tubes (PMT) or high quality solid state diodes are employed as detectors. Additionally, two dimensional array detectors such as charged coupled devices (CCD) with two dimensional dispersing units are being employed to generate images of the samples based upon their spectral characteristics. These are sensitive instruments capable of determining the spectrum of the dye contained in a 0.5 center meter length of a commercial fiber. Raman spectrometry also excels is determining the polymorphic state of substances that exist in these multiple states. Rutile and anatase are two distinct solid state forms of titanium dioxide used in paints, and each can be identified and quantified in a mixture by Raman spectrometry.

A complicating factor in employing visible lasers as sources is sometimes the fluorescence they cause. Often, this can be eliminated by changing the laser wavelength or by using advanced computer software routines.

When this is not possible analysts often employ the second type of Raman instrument, the Fourier Transform Raman Spectrometer. With these instruments the source in a near-infrared laser, for example, at 1064 micrometers. These are not energetic enough to cause fluorescence, but the longer wavelengths are less efficiently scattered, and the scattered frequencies lack the energy to stimulate the high sensitivity visible light detectors. Therefore, the through-put and multiplex advantages of the Fourier transform spectrometer based on an interferometer and equipped sensitive infrared detectors, such as the germanium type, are used. Because FT Raman optical benches can be added to existing infrared FT/IR instruments resulting in units that determine the infrared spectrum and the Raman spectrum of a sample, these units are very popular (Figure 17.6).

As is often the case in analytical chemistry neither design solves all problems, and the well equipped laboratory will have one of each design.

The scanning electron microscope (SEM) is a powerful addition to a forensic laboratory that permits the viewing of samples at much greater magnification and resolution than is possible by light microscopes. Magnification is possible in the range of 10 to 100,000 times. In forensic labs, the lower magnifications are of more import with few samples requiring more than 5000× magnification. Very rarely is magnification above 25,000× needed. When the SEM is combined with an energy dispersive x-ray spectrometer (EDS), the usefulness of the technique becomes consummate. The SEM/EDS combination can readily resolve a particle or structure smaller than 1 micrometer in size, while generating spectra revealing the elemental composition of the object. An example of a modern SEM/EDS combination can be found in Figure 17.7.

The principle of operation is that an electron beam generated by a thermionic source is accelerated by a high potential difference, usually 10,000 to 30,000 electron volts. This beam is then focused by the use of electromagnetic lenses to a small beam spot and swept over the sample. The beam causes a number of interactions slightly below and at the surface of the sample. Back scattered electrons (BSEs) and secondary electrons (SEs) are emitted

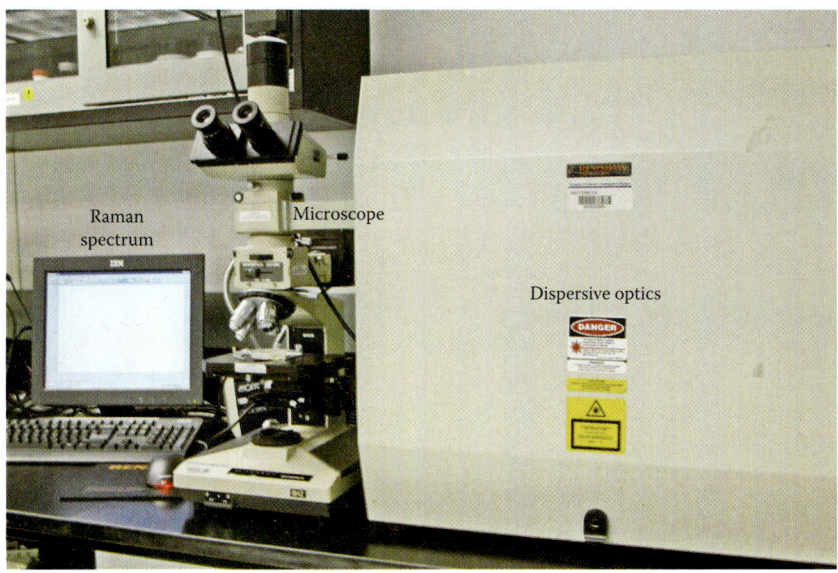

Figure 17.6 Photograph of a dispersive Raman spectrometer.

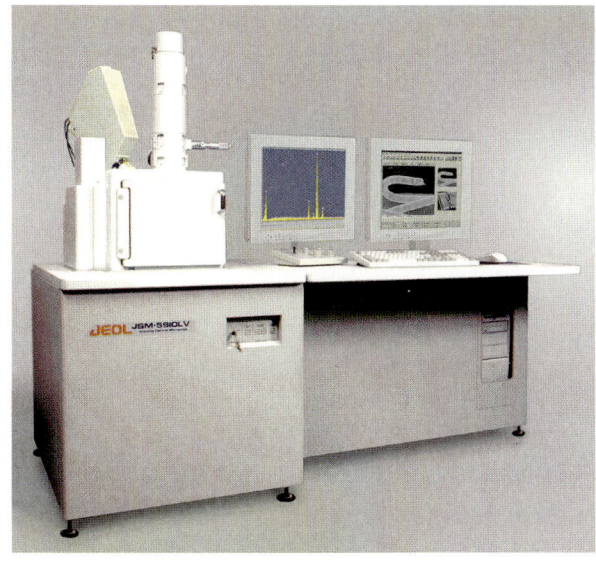

Figure 17.7 A JEOL JSM 5910 LV scanning electron microscope equipped with an x-ray spectrometer for elemental analysis. It is capable of operation in low vacuum mode, thus eliminating the need for coating most samples with a conductive material. This feature is very popular with criminalists. (Photo courtesy of JEOL USA, Inc., Peabody, Massachusetts.)

from the surface and converted to an electrical signal by an appropriate detector. The position of the sweeping beam is coordinated with the sweep of a cathode ray tube observation screen, and the intensity of the signal from the detector is converted to brightness on the tube. This results in a image similar to that from a television. The screen size is fixed, and the analyst uses the controls to vary the size of the portion of the sample scanned. The relationship of this scanned area to the viewing screen is the magnification of the microscope. A schematic of this process can be seen in Figure 17.8.

This electron beam causes many other interactions with the sample, two of which generate x-rays. The first interaction occurs when electrons penetrating the surface of the sample decelerate. This causes the release of energy as a continuum of x-rays and is referred to as the Bremsstrahlung or breaking radiation. This results in a background upon which an analytical signal is superimposed.

The analytical signal is formed when high energy electrons from the beam strike and cause an inner shell electron from an atom of the sample to be ejected. This results in an unstable electronic configuration for this atom, which is stabilized by electrons from higher level shells filling the voids. When these electrons fall to the inner shells, they need to release energy, which is done via the ejection of an x-ray photon. The energy released is quantized, identifiable to specific atoms, and, hence, useful for **qualitative analysis**. These specific energies are called characteristic x-rays.

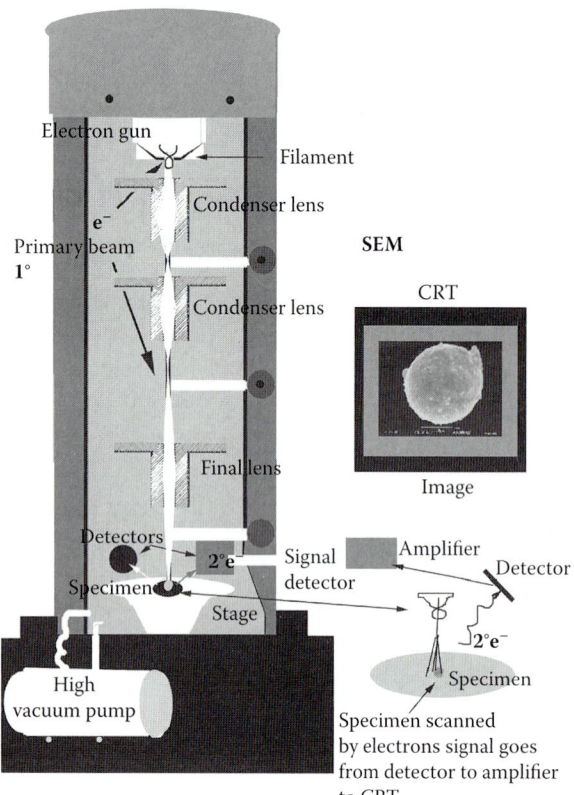

Figure 17.8 Stylized representation of a scanning electron microscope showing the source, condenser lenses, beam with its rastering of the specimen, sample position, and detectors. The monitor that displays an amplified signal for the image is linked to the scan generator. The image is a classic gunshot residue particle.

The strength or number of x-ray events (counts per second) under given conditions is proportional to the amount of the element present in the sample and, thereby, leads to methods of quantitative analysis. An x-ray detector is employed to sense these photons and convert them to electrical impulses. Electronic hardware and software sort and display the data so that qualitative and quantitative analyses can be performed on very small particles or limited portions of a sample. This can be of tremendous value to the microscopist dealing with microscopic evidence.

Historically, a SEM required a sample to be contained in a chamber at high vacuum. This caused problems in examining many samples of interest to the crime laboratory. In the last decade, a technique has been developed so that the sample to be studied need not be kept at such great reductions of pressure. This is often referred to as low vacuum, low

pressure, or environmental SEM. This development made the SEM/EDS system more valuable to the forensic analyst. The student who desires an in-depth discussion of SEM, EDS, and other topics is directed to any of a number of basic or advanced texts.[12-15] Kubic discusses detailed examples of a number of applications of SEM/EDS to forensic analysis in a recent text edited by Li.[16]

Other Instrumental Techniques of Value to a Microanalyst

Although not truly microscopical techniques in that they do not employ microscopes, a number of other methods must be briefly mentioned for the sake of completeness. All of the following can be conducted on very small samples when necessary, but are most often employed on milligram or larger samples.

X-ray diffraction (**XRD**) involves the targeting of a beam of monochromatic x-rays on a sample so that the radiation is scattered. The part of the scattering attributable to the interatomic spaces between lattices in crystals results in the development of constructive interference. This technique indicates how the atoms or molecules are arranged in a given crystal, which can vary even in materials of identical chemical composition. XRD can discriminate between the three polymorphic forms of titanium dioxide—an important issue in paint analysis. This is called *phase identification.*

In pyrolysis gas chromatography and pyrolysis gas chromatography/mass spectrometry, an organic, often polymer, material is broken into fragments by heating in an inert atmosphere that prevents combustion. The fragments are separated in a gas chromatograph and the recorded pattern of the eluting peaks contains both qualitative and comparison information for associative purposes. When a mass spectrometer is added to the instrument, a greater volume of information can be elucidated from the results.

Knowing the elemental composition, particularly the quantitative profiles of a sample, can be of immense value. Many of these profiles are most useful for associative purposes when trace elements are included, whether the samples are bulk or microscopic in size.

X-ray fluorescence, EDS from SEMs, **atomic absorption, atomic emission**, and **atomic mass spectrometry** are all very useful and routinely employed in full-service crime laboratories. The advanced reader is directed to instrumental and advanced criminalistics texts for treatment of these topics.[17–19]

The fact that testing may be destructive in nature should always be of concern. Even if one or a number of advanced techniques must be employed to strengthen the evidentiary value of a total analysis, the first approach should always be to complete as careful and detailed a nondestructive examination by microscopical techniques as possible, especially when a sample is microscopic in size.

In the following sections, we will discuss a number of physical evidence types that are often encountered as micro-sized samples, and some of the basic observations and measurements that can be performed with the aid of a microscope.

Microscopic Evidence and Its Analysis

Glass

Glass, a common type of microscopic evidence, is a reasonably hard, transparent or translucent material composed of fused inorganic materials. Upon cooling, it is amorphous in nature on all but ultramicro or atomic scales. Over the sizes normally analyzed, it lacks real symmetry, is not crystalline, fractures conchoidally, and is isotropic in its optical properties. Glass is found in many shapes, sizes, colors, and types. Its uses range from containers to optical devices. It has a wide variety of chemical compositions, by both design and happenstance. Variation of its elemental formulas

can alter significantly its characteristics and, therefore, often its ultimate uses.

For example, glass with high boron content is resistant to thermal shock and is employed in laboratory glassware and cookware. Inexpensive soda lime glass is usually high in sodium and calcium content and is found as containers, windows, and many other products. The addition of high atomic number elements increases the RI of glass, causing it to sparkle and serve decorative and aesthetic purposes. Because glass has so many uses, possesses different qualities, breaks easily, and ejects very small fragments in different directions that are retained by garments, it is frequently encountered as **transfer evidence**. Whether flat, container, decorative, optical, or other glass, varying its composition allows it to be discriminated by physical, optical, and elemental characteristics.

A first examination of glass should be directed to physical properties that can be evaluated macroscopically or with a stereo microscope. Examination of a broken window can reveal whether the impact that caused the fracture was a low-velocity blunt trauma or a high-velocity point trauma. Figure 17.9A shows breaks in sheet glass caused by multiple impacts. One can observe **radial cracks**— those originating from the impact point and propagating away, and those that seem to make a circle around the point (**concentric cracks**). By noting that some cracks terminate at their intersections with others, one can conclude that terminated cracks were caused by a later impact. Figure 17.9B displays a

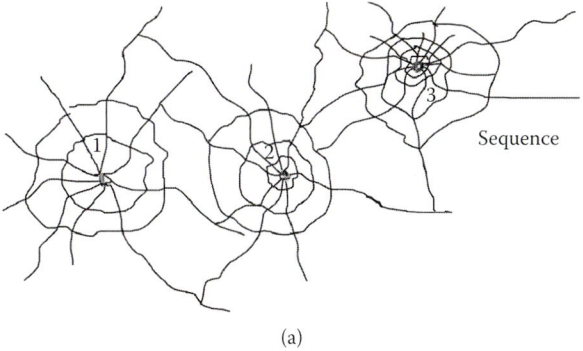

(a)

Figure 17.9A Radial and concentric fractures. A series of impacts on glass. The sequence is indicated by the terminations of newer cracks at existing fractures.

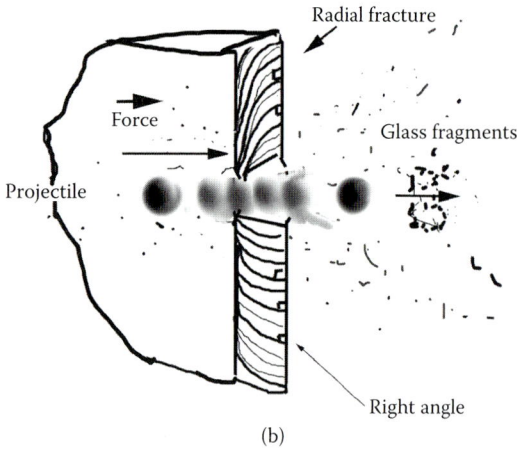

(b)

Figure 17.9B Coring effect fracture. The result of the impact of a high velocity projectile on glass. The fragmentation, coring, and fracture lines that confirm the direction from which the force originated can be seen.

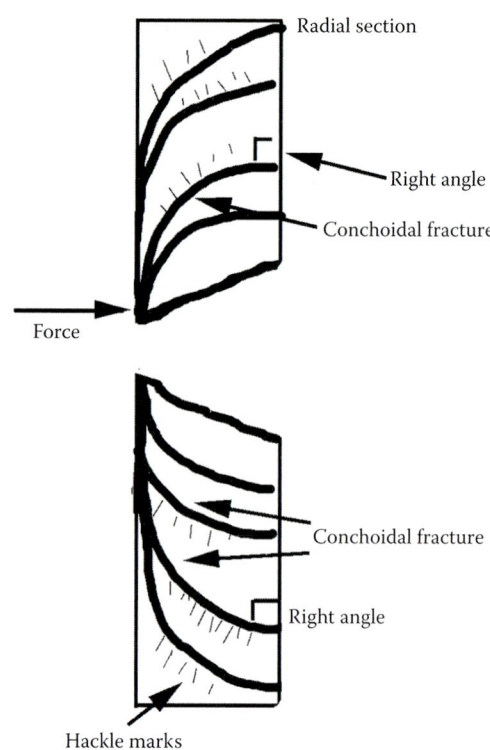

Figure 17.10 Expanded view of the conchoidal marks that appear on a fractured glass edge. The smaller hackle marks and the missing core area mentioned in Figure 14.8B are also shown.

cross section of a flat glass impacted by a high velocity projectile. If enough of the impact point is intact, one will note evidence of the core ejected from the far side of the glass upon impact.

Another fact that can be often ascertained from examination of such fractures is the side from which the force that caused the fracture was applied. When glass fractures, the edges often show characteristics referred to as **conchoidal lines**. The lines shown in Figure 17.10 reveal important information. When the fracture examined is located prior to any concentric crack and is not too far from the point of impact, then the surface opposite that part of the mark that appears to contact the surface at or near a right angle is the side from which the impact force originated. The acute marks point back toward the propagation point of the crack. Careful examination of these characteristics can allow a criminalist to determine whether a window was broken from the inside in an attempt to disguise an "inside job" as a burglary.

When microscopic glass chips are examined in an effort to associate two items, the physical, optical, and elemental properties are all very important. The most discriminating technique has been shown to be elemental profiles. This approach is very time-consuming and requires expensive instrumentation and highly trained analysts. Physical examinations and measurement of optical properties

may eliminate samples as having possible common origins. Because these measurements are more easily conducted, they should, in most cases, be performed first. If they are not capable of discriminating the samples and the values of the data obtained are common, then the evidentiary value of the match obtained is weak, and more sophisticated analyses are dictated.

Some physical observations that should be undertaken are thickness, color, uniformity, curvature, and surface conditions, such as tinting, soiling, and imperfections. Flat glass should be examined with an ultraviolet light so that float glass can be discriminated from double-ground and polished plate. Float glass is manufactured by "floating" molten glass onto the surface of a bath of melted tin. Some of the tin diffuses into the glass and results in a product that fluoresces when excited by ultraviolet light. The pale blue to yellowish glow does not appear on both surfaces.

If the recovered glass evidence consists of small fragments, many of the aforementioned tests may be precluded. However, the optical properties of small fragments can be successfully evaluated, and these characteristics can be reasonably discriminating. Small fragments of glass can be removed from larger pieces and the optical properties measured. The property that is most significant is the RI. As previously mentioned, this is customarily reported as N_D at 25 degrees Celsius. Other indices are also measured and the variation with λ reported as the relative dispersion (V), where $V = (N_D - 1)/(NF - N_C)$ or as a full dispersion curve normally with N on the abscissa and $1/\lambda^2$ as the ordinate.

The RIs of small glass chips cannot be measured directly, but may be measured indirectly by one of the immersion methods based on submersion of a sample in a liquid, usually referred to as an *oil* even if it is not organic in origin. Employing one of the immersion methods allows an analyst to determine when the RI of the sample matches that of the liquid medium at a given λ, and then he or she measures the liquid or reports the predetermined RI of that liquid. When the RI of the sample is near or matches the RI on the medium, the contrast of the sample will be low and it will be difficult to see in the liquid.

When the RI differs by an appreciable amount, the contrast will be significant and the sample will be easy to observe. See Figures 17.11A and 17.11B for examples of high and low contrast. At first impression, one might not want to attempt this measurement as it seems to require good luck or extended effort to find the matching liquid by trial and error. However, a number of methods can determine whether the liquid medium or the solid sample has the higher RI, thereby giving direction for the next choice of liquid. Oblique illumination, dispersion staining color, and the movement of the **Becké line** are the most common techniques. The most popular is the Becké line method. A microscope, preferably set for **axial illumination**, is critically focused on the sample and then the focus is raised. The distance between the sample and the microscope objective is increased. When this is done, a halo or brightness near the edge of the sample, called the Becké line, will move into the material of greater RI, whether it is the sample or the mounting medium. Figure 17.12 shows examples of a Becké line's appearance in and out of a sample.

When this measurement is made with a well maintained, quality microscope, the accuracy of the measurement can be in the area of 0.0005 RI units. Better accuracy is required for advanced criminalistics work. This can be accomplished by a number of methods, all of which require a phase contrast microscope

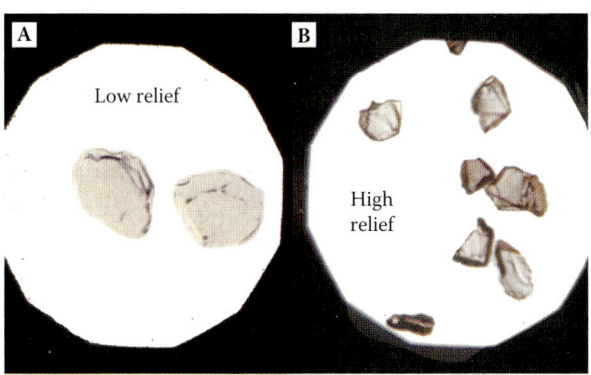

Figure 17.11 An example of low contrast (relief). (A) This is the appearance when the RI of a sample is similar to that of the immersion medium. (B) An example of high contrast (relief) that appears when the RI of a sample is significantly different from that of the immersion medium. Note that which RI is higher does not matter; the contrast principle remains the same.

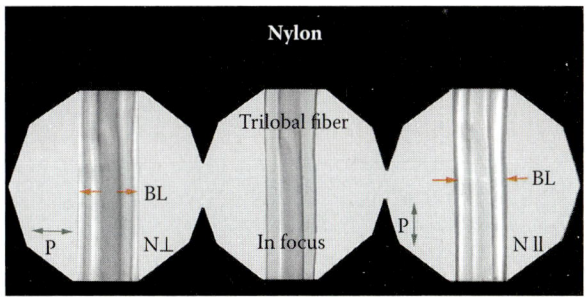

Figure 17.12 Fiber Becké line movement. Center: A nylon fiber in critical focus, mounted in a medium of RI 1.53. Left: When this fiber is viewed in plane polarized light and oriented so that the microscopist is viewing the perpendicular (cross-wise vibration) RI of the fiber, the Becké line or bright halo will move into the medium when the focus is raised. Right: Photomicrograph of the view with the parallel (lengthwise vibration) RI of the fiber, indicating the movement of the Becké line into the fiber when the focus is raised. Because the same immersion liquid is used in both cases, the sign of elongation of the fiber is readily determined to be positive, that is, the RI is greater along the length

that makes the detection of the minimum contrast or match point more accurate.

Improvements in accuracy and precision result from employing a phase contrast microscope, monochromatic light, and a well characterized oil, where the RI is varied with a microscope heating stage. Known as the *single variation method,* the RI is determined by recording the match temperature and employing the calibrated −dn/dt of the oil to calculate the RI of the sample. The double variation method employs a heating stage to maintain oil temperature, and a monochrometer is employed to determine the match wavelength. The oil temperature is then changed and the match λ determined again. This is repeated for four or more points. The data are plotted and the N_D determined.

The above methods have the ability to determine RI with an accuracy of 0.0001 and a reproducibility of approximately 0.00005. Today's modern manufacturing methods produce glass with significantly less RI variation than in the past. This has necessitated improvements in the measurement of the optical properties and the adoption of elemental profile analysis methods for testing glass samples.

An automated method for determining the match point of a glass chip employing commercial instrumentation has been available for over a decade.[20] It employs computer control of the heating stage and a video detector to determine the match point. Employing precalibrated −dn/dt data for the oil, the computer calculates the RI. Along with automating the calculation, the main advantage is the removal of the operator's subjectivity in determining the match point. With this instrument, measurements of RI can be made with an accuracy of 0.00005 and precision of 0.00002 or 0.00003. Figure 17.13 shows a GRIM automated RI measurement instrument. For a more comprehensive review of the forensic analysis of glass and its value as evidence see Koons et al.[21]

Hairs and Furs

Hairs and furs are additional examples of evidence types amenable to microscopical

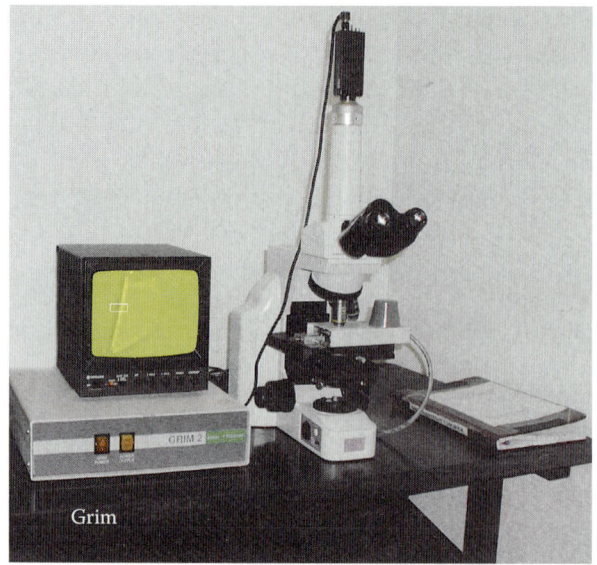

Figure 17.13 The GRIM automated RI measurement instrument by Foster and Freeman, Worcestershire, England employed for the precise and accurate measurement of the RIs of glass chips. The microscope is a phase-contrast instrument with a hot stage positioned on the microscope stage and a television camera mounted on the trinocular head. The monitor shows a magnified image of a glass specimen and the measurement point can be seen as a bright rectangular window superimposed over the bright edge of the glass.

analysis. They are natural fibers of animal origin. In this section, we will define hairs as animal fibers that originate from humans, while furs originate from other species. Hairs can generally be grouped by racial origin and often body location. Hair examiners can often conclusively eliminate a person as a source of a hair, but rarely can an examiner absolutely associate a hair sample to a given individual.

When this occurs, it is usually based on a factor or factors beyond those characteristics observable and measurable with a light microscope. Today's DNA technology is an example of a factor that may allow such an unequivocal association. Certain rare diseases, when detected and combined with microscopical observations, can similarly result in an unequivocal association. DNA testing requires a hair exhibit to have cellular material from which DNA can be extracted. This is not always possible, in which case the criminalist must revert to microscopical methods.

Furs can generally be classified by species with a microscope, but subclassification can be problematic unless an extensive reference collection is available. Even then, specific iden-tification is not always possible. Some steps toward individualization are possible with animal furs, but these are not well developed. However, basic characteristics such as color, length, and curliness can be valuable.

Hairs and furs are principally composed of keratins, which are sulfur-containing proteins that are interlinked to form stable fibrils and pigment composed of **melanin**. Trace metals are also present, having been deposited in the fiber during its growth stage or collected from contamination in the environment. One of the more important linkages is the disulfide bond that is present between sulfur atoms in adjacent keratin chains. The shaft of a grow-ing hair extends out of the skin, with the root imbedded in this tissue. The lower end expands to form the root bulb where growth takes place at the papilla. Except at this point, the fibers are composed of dead cornified cells. The root portion is referred to as the **proxi-mal end**. The tip away from the root is known as the **distal end**. After a period of growth known as the **anagen stage**, the hair or fur will enter the **telogen** *or dormant stage*, and eventually be sloughed from the body. The intermediate or transition phase is known as the **catagen stage**.

The structure of a hair can be consid-ered to have three parts, similar to a graph-ite pencil. The center portion is known as the **medulla**, and is usually amorphous and vacant of material in human hair. It appears dark when the exhibit is mounted in a liquid and viewed via a microscope. In animals, the medulla contains cells are arranged so that their appearance varies in a manner that assists in species determination. The next and predominant portion is the **cortex** that corresponds to the wood part of a pencil, and contains many important microscopic fea-tures, such as pigment, color, size, and distri-bution, tiny air pockets called **cortical fusi**, and **ovoid bodies**. These characteristics are especially important in human hair examina-tion. The third outermost portion of the hair, equivalent to the painted area of the pencil, is the **cuticle**. This consists of a layer of scales covering the shaft in such a way that they always point away from the proximal end and

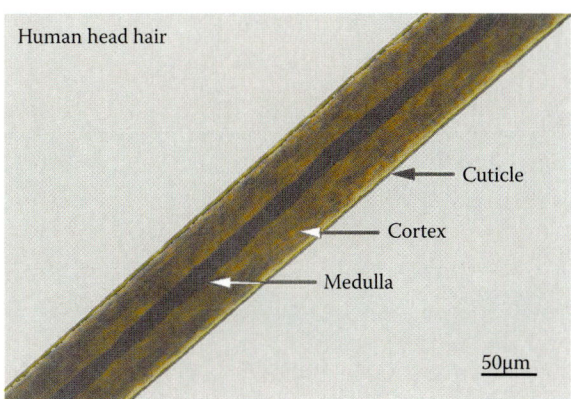

Figure 17.14 Photomicrograph of a dark brown human head hair. Note the three principal parts: the innermost medulla, the cortex, and the outer surface cuticle.

toward the distal end. The basic structure of a hair is shown in Figure 17.14.

The scale structures can be divided into three basic types. Coronal or crown-like scales resemble stacks of paper cups and are charac-teristic of very fine hairs. This type of struc-ture is normally found on small rodents and bats and rarely in humans. Spinous or petal-like scales are triangular in shape and usu-ally protrude from shafts. They are not found on humans; they are found on cats, seals, near the roots on minks, and others. Imbricate or flattened scales overlap, similar to shingles on a roof. They are found on humans and many other animals. The general appear-ances of different scale types can be found in Figure 17.15A.

The medulla of a hair can be continuous, discontinuous, or fragmentary, or may not be observable. In nonhumans, the structures are usually regular and well defined. The princi-pal types are (1) the uniserial and multiserial ladders found in rabbit hairs, (2) lattices found

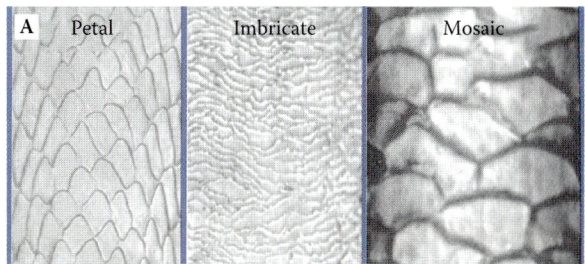

Figure 17.15A Cuticles. The petal, imbricate, and mosaic types of scales that comprise the cuticle are clearly differentiated.

Figure 17.15B Medullas. Three of the different kinds of medullas found in hairs and furs are displayed.

in the deer family, and (3) vacuolated or cellular types; the shapes vary and are commonly found in many animals. See Figure 17.15B for descriptions of three medulla types. Usually for fur examination, the determination of species and the development of basic information such as color are sufficient for the analysis.

The microscopic examination of hairs of human origin can produce an extensive amount of additional information. A number of these factors will be briefly mentioned but the reader seeking more detailed information is referred to the publications by Hicks, Bisbing, and Petraco.[22–25] The racial origin of hair can be determined and classified as Caucasian, Negroid, Mongoloid, or of mixed origin. The body locations from which the hair originated can be determined include the head, pubic area, limb, beard, chest, axillary, and other areas. After the factors cited above have been determined, careful examination of the characteristics of the tip, root, diameter, scales, pigment, medulla, and cortex; detection of artificial treatment, damage, and the presence of vermin or disease; and the determination of the method of cutting are all of value in the proper determination of association of a recovered hair with an individual.

Attempts have been made to associate hairs by trace metal composition, but this has met with limited success. Recently, the use of infrared microspectrophotometry to evaluate surface treatments on hair has met with some success. These topics are too complex to be treated here, but the reader should be aware of these approaches.

Fibers

Fibers constitute another common class of microscopic transfer evidence. Because they are multitudinous in major classifications and generic subtypes, are physically different, processed in many ways, transfer easily, and have significant persistence, they are treated as valuable forensic microscopic evidence. One manner of grouping fibers is as animal, vegetable, or mineral. Other categories are naturally occurring, manufactured, and synthetic.

Natural fibers are those found in nature that have not been greatly altered in physical composition or characteristics by processing. Coloring and treatments that improve merchandising or performance do not change the classification. Manufactured fibers are produced from fiber-forming substances that can be synthetic polymers, transformed or modified natural materials, and glass. Processing is necessary for the forming of the fiber. Synthetic fibers are manufactured from synthesized chemical compounds, such as nylon, that are then formed into fibers.

Although fiber evidence can be examined by powerful instrumental techniques, the use of microscopy for the initial examination and collection of the first analytical data is the accepted forensic procedure. A number of the microscopes mentioned in the first portion of this chapter are routinely applied to fiber analysis.

A stereo microscope is first employed, and the size, crimp, color and luster, possible cross section, damage, soil, and adhering debris are documented. Initial classification may be performed. If the exhibit is a yarn, thread, or fabric, additional information is recorded. This is followed by examination with a PLM from which a wealth of additional information can be obtained.

Examination of the fiber is carried out under crossed polars and also with plane polarized light. It should be obvious that the morphological characteristics of the fiber, no matter how they are observed, are important. The appearance of natural vegetable fibers is noteworthy. Certain characteristics allow the identification of the source plant. A number of characteristics can be determined with crossed polars. The first is whether the fiber is isotropic or anisotropic. Almost all isotropic fibers are made from glass. A few synthetic fibers appear to be isotropic because their birefringence (ΔRI) is so small and the fibers are not very thick.

The retardation of anisotropics is estimated and then determined with greater accuracy by compensators. The relative refractive indices can be estimated so that the **sign of elongation** is determined. If the thickness can be measured accurately, the ΔRI—a valuable analytical parameter—of the fiber can be calculated. When viewed with plane polarized light, the natural color is determined, and any dichroism or variation of color due to the selective absorption of light depending on the light's vibration direction can be recorded. With plane polarized light, the RIs of the fiber with light vibrating parallel and perpendicular to its length can be measured. These parameters, known as $N_{\parallel}$ and $N_{\perp}$, are important analytical parameters found in tabulated databases from which generic and sometimes subclass and brand can be determined.

Figure 17.16 displays the difference in appearance of a fiber, depending upon its

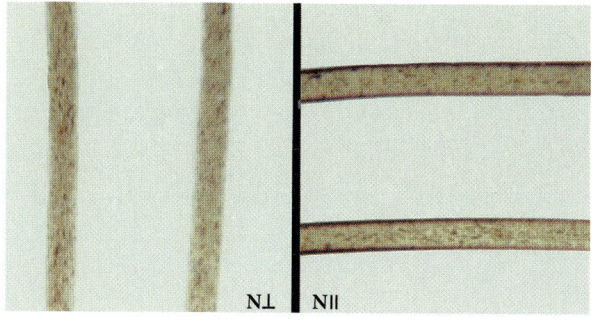

$N\perp$ | $N\parallel$

Figure 17.16 Photomicrograph of a synthetic polyester fiber mounted in a medium of approximately 1.66 RI. The fiber is displayed in two orientations in a PLM employing plane polarized light. The vibration direction of the light is left to right (east to west) in both representations. The RI of the fiber with the vibration direction crosswise (left) to the fiber is nearer the medium, rendering the internal characteristics more visible. When a fiber is parallel to the vibration direction of the light its index (N_D), is more distant from the medium and generates more contrast at the periphery of the specimen.

orientation when viewed in plane polarized light. This is due to the differences in RI. If a sample can be cross-sectioned, very important data can be collected, especially about synthetic fibers. Size and shape can be determined unequivocally, and the modification ratio (MR) is an aid in brand identification. The MR is the ratio of the smallest circle that contains all the lobes of a noncircular fiber compared to the largest circle that can be drawn in the core of the fiber. Cross-sectional shape alone can aid in determining the manufacturer and the end use of the fiber, for example, in clothing or carpets.

Decision trees or flow sheets for the forensic identification of fibers have been published.[26] Figure 17.17 is one such flow sheet. A wealth of quality references deal with fiber identification and the reader is directed to two.[27,28]

When an analyst is most interested in the fine surface structure and requires a higher resolution image of a fiber, he or she may decide that the advantages of SEM observation are warranted. Along with the advantages mentioned earlier, SEM observation also allows a sample to be viewed with a greatly increased depth of field. This can be very useful when examination of a piece of textile is the task or a more in-depth study of fiber morphology is desired. Examples include evaluation of a puncture in a garment to determine whether it is truly a bullet hole, and examination of individual fiber ends in an attempt to determine the mechanism of failure or the method of severing.[29]

The microprobe abilities of SEM/EDS that allow the determination of elemental composition can also assist the criminalist in identification and comparison. The presence of chlorine in a preliminarily identified acrylic fiber indicates that the fiber is a modified acrylic. The detection of appreciable amounts of titanium dioxide in a fiber is indicative of the mineral's addition to the fiber to act as a delusterant. When tin and bromine are found in a fiber in which they are not part of the expected chemical formulation, the presence of fire retardant is suspected. When fibers are colored by incorporation of inorganic pigments

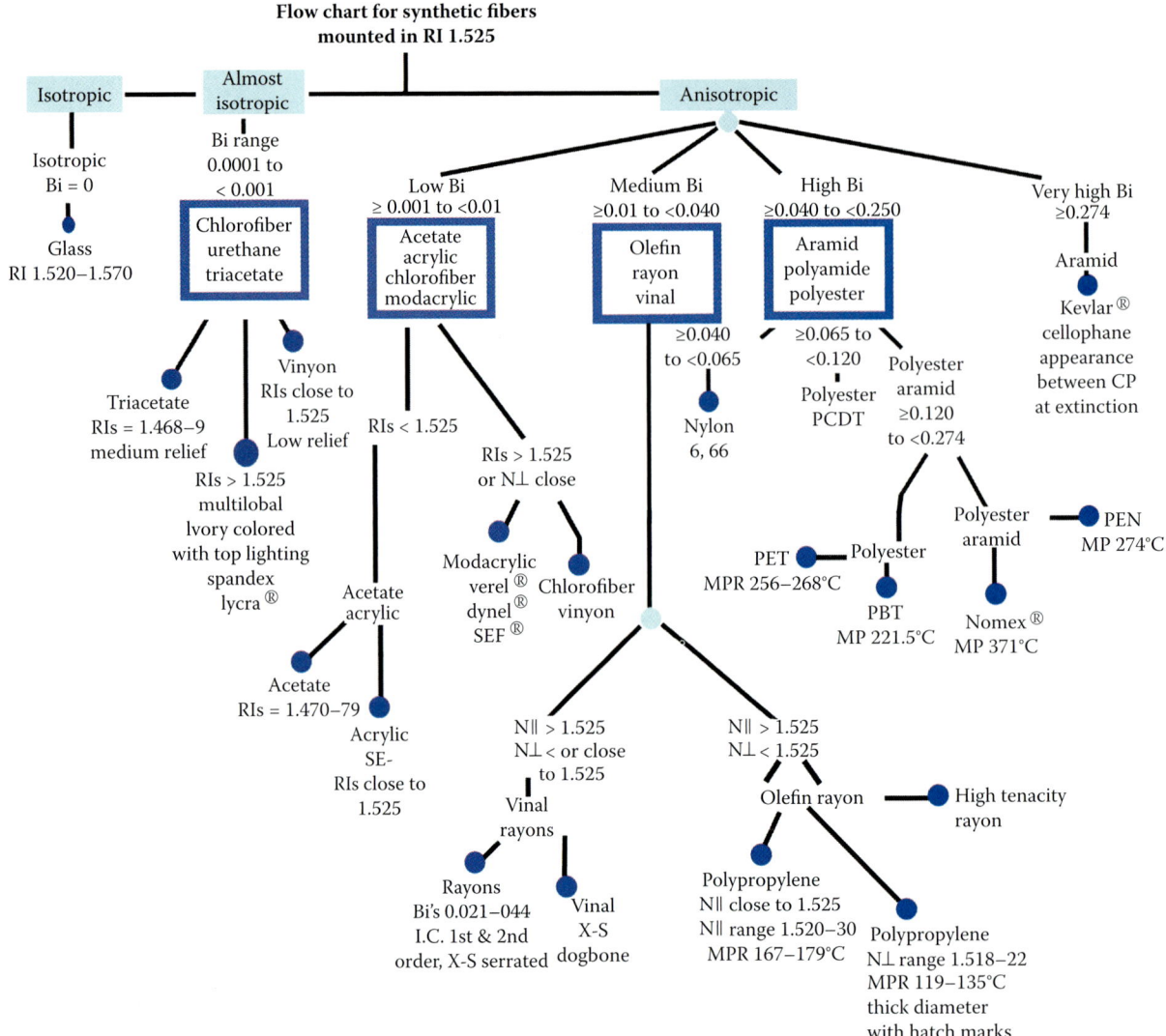

Figure 17.17 Flow chart or decision tree that can be employed for the systematic identification of synthetic fibers based on their optical properties, such as anisotropic character, birefringence, relative RI, and morphology.

in a polymer prior to extrusion, the elemental profile determined by x-ray spectra can aid identification.

The abilities of microspectrophotometers have been discussed. They are frequently employed in forensic fiber analysis and comparison. The visible spectrometer is employed to unambiguously and objectively determine the color of an exhibit. Infrared spectra determined from fibers as small as 1 millimeter long allow confirmation of the determination of generic class established by the PLM. Some fiber classes can be classified into subgroups by their infrared spectra. The specific type of nylon, for example, 6 or 6.6, can be more readily elucidated by this technique than by the

application of light microscopy techniques.[30] The collection of dichroic spectra of fibers is suggested as a method of detecting possible differences in submissions. The application of micro-Raman spectrometry to fiber analysis allows an analyst to identify the dye in a fiber that is usually present in too small a concentration to be determined by other techniques.[31]

All these methods are nondestructive. However, situations may arise that call for the application of other methodologies to strengthen evidence obtained by a comparison. There is little objection to employing destructive methods when an analyst has in his custody sufficient samples of the questioned and known samples. Pyrolysis gas chromatography, often employed with mass spectrometry, can add to the discriminatory ability of an analysis. Likewise, the application of elemental analysis to determine the trace elemental profiles of samples is carried out on occasion. Dye identification and comparison can be attempted by extraction of the dyes from the fibers and analysis by thin layer or liquid chromatography.

The information that can be obtained from the application of the methods mentioned and others for the forensic analysis of even micro-sized samples of fibers can be overwhelming. The reader is reminded that nondestructive techniques should be employed first, and before that, a judicious determination of what information is required or desired should be made.

Paint

Paint samples are a major portion of micro samples submitted to crime laboratories. This class of transfer evidence can play an important role in investigations and possible prosecutions. Forensic paint analysis and comparisons for common origin are distinguished from those performed by industrial laboratories by the size of the samples submitted for characterization. Forensic samples are not pristine; they are subjected to uncontrolled environmental and collection effects.

Paint submissions usually involve vehicular accidents where contact between two objects is sought to be established or investigative information such as make, model, and color of a vehicle involved in a hit-and-run is desired. Less often, paint from an architectural source is submitted. These exhibits are usually related to investigations of crimes against property. On some occasions, samples of an artistic nature are submitted. Any of these submissions may also involve crimes of a much more serious nature, such as assaults, rapes, and homicides.

Paints are applied for protective value, aesthetic purposes, or both. In this discussion, paint will include a range of materials from thin, translucent stains to heavy, opaque films. A complete discussion of the formulations, manufacturing steps, methods of application, properties, uses, and analyses of paint films is beyond the scope of this work. The reader is directed elsewhere for details on these topics.[32,33] A brief treatment of topics of interest to the investigator and analyst is appropriate.

Paint is composed of three principal parts. The vehicle is the binder that holds all the components together and is usually of polymeric nature, consisting of natural or synthetic resins. The binder can form a surface film in a number of ways. When the film forms by the simple evaporation of the solvent system of the liquid, the paint is normally classified as *lacquer*. Characteristic of lacquers is the fact that they resolubilize when subjected to many organic solvents.

When the film is formed by chemical cross-linkage of a number of its components, it is usually referred to as an *enamel*. The cross-linkages can be initiated by elevated temperature, oxidation by exposure to oxygen in air, chemical reactions of the components or special initiators, or a combination of these factors.

Latex paints form films by the coalescence of dispersed latex particles upon loss of water. These working definitions are not exclusive and combinations of the film-forming mechanisms are common.

Pigments supply paint with color, hue, and saturation. Pigments may be organic or inorganic. Blues and greens are predominantly organic, while whites, yellows, and reds are inorganic. This is not a strict rule and crossovers and mixtures are common in modern formulations. Pigments are expensive and manufacturers seek to minimize their use to lower costs.

Extenders are generally less expensive inorganic materials that are added to the paint to increase its solid content and, thereby, its opacity and hiding ability. Other advantages may ensue by their addition. A number of materials, such as titanium dioxide—not inexpensive but known for its hiding capacity—can fill the roles of pigment and extender.

Of equal importance in paint formulations, although not principal parts by concentration, are modifiers. They can affect the resultant film's durability, gloss, flexibility, hardness, resistance to ultraviolet radiation, and other characteristics. Other modifiers are added to aid in manufacturing, application and drying, or film formation.

Paint films can be investigated according to a number of their physical and chemical characteristics.[34,35] The size and shape of the exhibit, its surface condition, color, layer sequence, and thickness are physical attributes that can be readily assessed by macroscopic and microscopic examination.

The chemical compositions of the major components and modifiers can be evaluated individually or in combination by a number of chemical, microscopical, and instrumental methods. The size and condition of the sample and the information needed will guide the forensic analyst in choosing methods and techniques to be employed for the physical and chemical analysis of paint evidence.

The thrust of analysis is the attempt to find forensically significant differences in the questioned and known samples so that a hypothesis of common origin is rejected. However, differences in some physical and chemical characteristics have been found in samples of paint known to originate from the same source. It is the responsibility of the paint examiner to evaluate the meaningfulness of any differences so that a false exclusion does not result. It cannot be emphasized strongly enough that adequate documentation is necessary at every point of the examination, from sample collection and submission to the final test performed and the conclusion reached.

The first steps in the analysis are documentation, collection, preservation, and submission of samples to the laboratory. It is problematic that the analyst often has little control over these aspects of the process. The old computer axiom "garbage in results in garbage out" applies here. Samples may be found on a wide variety of substrates, clothing, tools, automobiles, and other fixed or movable objects.

When in doubt or when the retrieval of a sample may need specialized skills or equipment, it is best to submit the entire object to the laboratory. Smeared samples that may contain intermingled materials from a number of layers are problematic and the foregoing suggestion should be followed. When this is impractical, the paint must be removed for submission. A general rule is to collect a least one complete sample from an area very near, but not exactly adjacent to, the area of alleged contact. Additional samples from the known should also be collected, packaged separately, and submitted. The undermost layer of a paint chip, especially in an automotive paint sample, can be very useful and care should be taken to ensure that known samples contain this layer. When collecting samples, especially knowns, keep in mind the possibility of a physical match, which is the strongest association that can be established. Care should be taken not to alter or damage the sample's shape or surfaces.

Because a physical match is the most conclusive, the first part of a paint examination should be an attempt to establish it. A physical "jigsaw" fit of edges or a match of surface striae on the questioned and known samples is strong evidence. The quantity and quality of the characteristics that match should be sufficient to establish uniqueness. These examinations are generally conducted macroscopically, using an illuminated desk magnifier, a stereo microscope at its lower range of magnification, and reflected light illumination at various incident angles.

If a physical match is not attained, the layer structure order, color, thickness, and other details should be documented. Some manipulation of the sample may be necessary in order to collect the needed information in sufficient detail. Angle cuts and thin sectioning with a clean (new) scalpel blade can clearly reveal the layer structure. It may be necessary to embed the sample in a resin and employ microtomy techniques to obtain high quality thin sections. Embedded samples make it possible to grind and polish a sample so that fine physical details such as pigment size and distribution can be evaluated by higher resolution microscopes.

If a sample is sufficient, destructive tests based on chemical reactions can serve as sources of additional data. The dissolution, swelling, or generation of colors with various solvents or reagents is informative about the possible identity of resins, pigments, and extenders. These tests can be performed in a porcelain spot plate, small disposable test tube, or on a glass microscope slide. Observing the tests with a microscope allows them to be successfully performed on micro-sized samples. Because the pigments and extenders found in paint have been ground to such a small size, their unambiguous identification by use of a PLM is beyond the expertise of all but the most highly trained microscopists. Therefore, most analysts will resort to various instrumental techniques to further characterize samples.

Fluorescence microscopy is used to elucidate layer structures, especially on multi-layered "white" architectural paints. Visible microspectrophotometry is less popular with paint than for fiber analysis because sample preparation is more involved for transmission measurements and reflection spectra are more difficult to interpret.

Infrared microspectrophotometry is routinely employed for paint analysis.[36] Transmission measurements can be obtained on thinly sliced or rolled samples and by compressing the paint in a diamond cell. Attenuated total reflection objectives allow for the collection of spectra from the surface of a paint sample without the need to prepare a thin specimen. The DuraScope™ is a new accessory that has become available and allows for the collection of quality ATR spectra from samples as small as 100 micrometers. Reflection data varies slightly from transmission data, but successful comparisons, interpretations, and database searches can be carried out. The infrared microscope has a limited spectral range because of the detectors employed and the optics available, but is applicable to the analysis of the major organic components in paint. If data on the inorganic constituents is desired, additional tests with extended range spectrometer optics are required.

The recent introduction of commercially available Raman microspectrometers added to the information that can be obtained from a paint sample. This technique is based on scattering and can supply data complementary to those of absorption infrared. It can

be employed to assist in the analysis of the inorganic components of paint samples. These instruments, available for small bulk and **microanalysis**, remain expensive and only the larger laboratories have the resources to utilize this technique.

The analysis of paint by SEM/EDS is reasonably straightforward after sample preparation is complete. The availability of the new eco-SEMs eliminated most sample preparation problems and allows most samples to be placed directly into a chamber and analyzed. Not only can the layer structure be further elucidated by the higher resolution and the atomic number contrast available, but elemental analysis can be performed with the x-ray spectrometer attached to the instrument. Although the operation of this combination has been simplified on modern equipment, collection of proper data and correct interpretation, especially when quantitative analysis is involved, require more than rudimentary training, knowledge, and experience on the part of the operator.

Classic x-ray diffraction (XRD) can be performed on samples just slightly larger than micro-sized. If submissions are smaller, the analyst may need to employ a newly developed micro-XRD instrument to accomplish an unambiguous identification of a phase or resort to analytical electron microscopy (AEM), which uses electron diffraction on single particles of micrometer size or smaller to accomplish this goal.

X-ray fluorescence (XRF) instruments that generate and collect x-ray spectra in a number of different ways are available. The advantages of this technique over SEM/EDS are its higher sensitivity and lower detection limits for the higher atomic number elements. These features can be very useful in paint and glass analysis. The problem in the past was the inability to handle micro-samples. Instruments called capillary XRFs are available and have exciting x-ray beams as small as 20 micrometers. This technique is expected to be adopted by more forensic laboratories.

Pyrolysis combined with gas chromatography (PGC) or PGC linked to a mass spectrometer (PGC/MS) can be applied to the analysis of paints. The technique is applicable to the organic portion of the paint and can, when used with MS, supply information concerning chemical components. The **chromatograms** of the questioned and known samples, when compared by pattern recognition techniques, can be powerful aids in individualization and the establishment of **common origin**. PGC is more sensitive to formulation differences than infrared or Raman spectroscopy.

Other methodologies too numerous to mention may be applied in specific situations, and the techniques discussed above should not be considered exclusive. They are, however, those most commonly employed, but a prudent examiner should always keep an open mind to innovation. Additional applications of microscopy can be used for evidence collection and evaluation.

Soils

Soils are complex mixtures of materials of mineral, animal, and vegetable origin at various levels of change and decay. Many of the components are common. Some have been deposited by natural forces, while others have been delivered through the intervention of man. The great variation of these combinations leads some to believe that soil has a unique composition in any given area and changes detectably every few feet.

Light microscopes, particularly PLMs, lend themselves well to the investigation of forensic soil samples. Many other techniques, some instrumental, are also applicable but will not be covered in this section. Physical characteristics, such as color, pH, and particle size, can be relevant. As a very basic introduction to this topic, one could consider the pollen content and mineral assemblages present in an exhibit. Pollens can be readily identified by their morphology using light microscopy or SEM. Keys for identification are available.[37,38]

Although pollens are small and can be windblown, any reasonable concentration of a certain type can be a strong indicator of a location that becomes more specific when a number of pollens are identified. The reader should consider the implications of other uses of pollen analysis; for example, autos parked near certain plants or the clothing of a burglar who made contact with a number of flowering plants during entry through a window.

The identification and quantitative estimation of the mineral content of soils have long been accepted as indicators of location. In these analyses, the more common minerals referred to as the light fraction are considered much less important than those of greater density. Separations are first conducted, and then the minerals are identified by colors, shapes, and optical properties (RIs, ΔRI, fracture, pleochroism), and information obtained by conoscopic observation.[2,5–7] Quantitation is accomplished by one of the point-counting techniques.[39] When all the microscopically obtained data and those developed by other methods are considered, it is possible for a trained examiner to supply valuable investigative and probative information concerning soils.

Gunshot Residue (GSR)

GSR analysis should be mentioned whenever SEM is considered in the discussion of forensic analysis. We define GSR as a mixture of organic and inorganic materials originating from the projectile, cartridge case, propellant, and primer that emerge from the barrel and other openings of a firearm and are deposited on the hands, hair, face, or clothing of persons in close proximity to the weapon when it is discharged. To avoid confusion, we will not consider these materials when they are deposited on the victim, and their residue is used to determine weapon-to-target distance. Such materials should be called "firearm discharge residues" or "muzzle blast." Analysis of firearms discharge residue and distance determinations will not be considered here.

The goal of GSR residue analysis is to determine that a residue is indeed GSR and to ultimately place the discharging firearm into the hands of a shooter. This final goal remains to be accomplished. Reports may state, "… indicates that the individual recently discharged or was in close proximity to a discharging firearm." Classically, the analytical problem has been approached by collecting samples from the hands of a suspect and analyzing them by a bulk elemental analysis method.

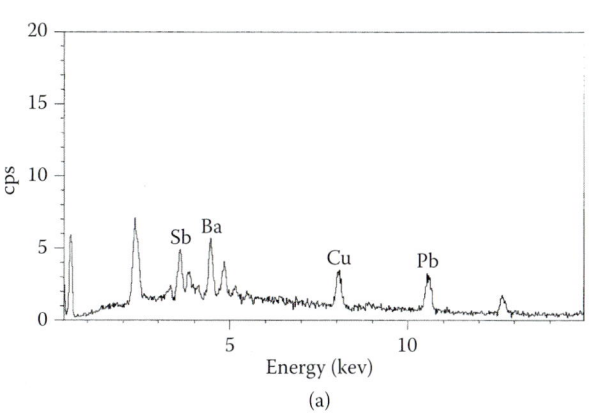
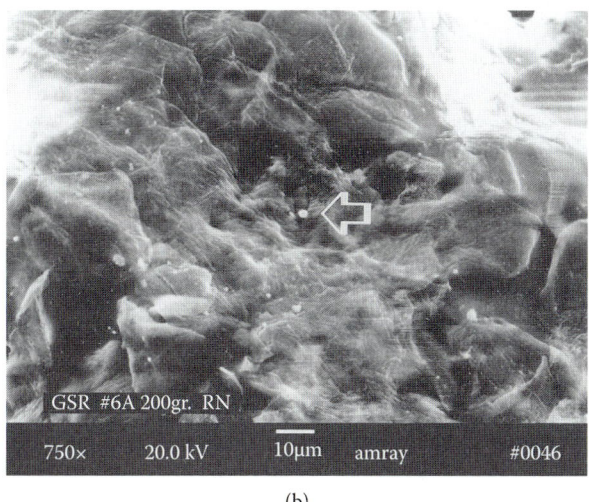

Figure 17.18 (Left) Adhesive lift with GSR particle. This photomicrographic indicates the bright GSR particle (arrow) found on human skin. (Right) EDS spectra. The spectra of a classic GSR particle with Pb, Sb, Ba, and Cu from a projectile jacket.

Instrumental neutron activation analysis, graphite furnace atomic absorption, and inductively coupled atomic emission spectroscopy have all been utilized for this analysis. A sample is considered positive when certain target elements, usually barium and antimony, are found together above a baseline level. Whether other environmental contaminations could result in false positive findings and whether samples with levels below the baseline cutoff values are false negatives are open questions.

A particle technique first reported by Wolton et al.[40] uses a SEM/EDS system to address these problems. The subject is sampled with a SEM stub coated with a sticky substance that collects the GSR, which is in the form of particles. These particles are predominantly in the 0.5 to 2 micrometer size range, with some as large as 10 micrometers. The particles are located by viewing in an SEM employing a backscatter electron image (BSE) formed by the higher energy electrons that are elastically scattered from the sample. BSE signals are sensitive to differences in atomic number with higher average atomic number materials appearing brighter on the instrument's screen. This allows the operator to either manually ignore duller particles or set a brightness threshold for further analysis by the computer automated instruments.

The brighter particles are analyzed for their elemental content by EDS, and those with particular compositions, especially if spherical, are classified as GSR or probable GSR. Particles containing barium, antimony, and lead or barium and antimony are considered characteristic of GSR. Others such as lead and antimony are indicative of GSR. Other compositions and methods of classifying the particles are acceptable, but they are too lengthy to be discussed here. Figure 17.18 is a photomicrograph of a GSR particle and its spectra.

Practitioners' opinions vary concerning the number of particles and composition required to determine that a sample is truly GSR and did not originate from environmental contamination. Consideration should be given to the location from which the sample was collected—for example, the web of the hand or the palm. The goal of unambiguously placing a particular weapon into the hands of a specific individual still remains to be attained. Further study and advances in techniques continue to be pursued.

It should now be evident that analysis of micro-trace evidence is wide ranging and complex. One should recognize the value of applying microscopy to this evidence class and appreciate the knowledge, skill, and experience required of the criminalist who applies these techniques.

Questions

1. What characteristic separates microscopic evidence from other evidence?

2. What instrument is employed for the collection and first evaluation of small evidence?

3. How is the total magnification of a microscope determined?

4. What is the most important factor in determining the resolving power of a microscope?

5. What are the important factors for the reporting of refractive indices?

6. What are the characteristics that firearms examiners evaluate for matching a projectile to a weapon?

7. Explain plane polarized light.

8. What information is gained by x-ray or electron diffraction techniques?

9. What determinations about a glass fracture can be made by macroscopic examination?

10. What are the three major portions of a hair or fur fiber?

11. What is the value of visible microspectroscopy for fiber comparisons?

12. What information about a paint sample can be obtained by use of infrared microspectroscopy?

13. What data are obtained from a paint sample by use of SEM/EDS?

14. What fraction or type of mineral is of most value for soil comparison?

15. Why does the stereo binocular microscope seem to give a 3-D image?

16. What is working distance and how does it vary with objective magnification?

17. How could one increase the magnification of a compound light microscope and not change the working distance?

18. What stage of hair growth usually results in the loss of hair?

19. What elements, when found in a spherical particle, are considered necessary to conclude that the particle is characteristic for a gunshot residue?

References

1. Munsell Soil Color Charts, MacBeth Division of Kollmorgan Instruments Corporation, Baltimore, MD, 1988.
2. McCrone, W. C., McCrone, L. B., and Delly, J. G., *Polarized Light Microscopy,* Ann Arbor Science, Ann Arbor, MI, 1978.
3. Hartshorne, N. H. and Stuart, A., *Crystals and the Polarising Microscope,* 4th ed., Edward Arnold Ltd., London, 1970.
4. Hallimond, A. F., *The Polarizing Microscope,* 3rd ed., Vickers Instruments, New York, PA, 1970.
5. Bloss, F. D., *An Introduction to the Methods of Optical Crystallography,* Saunders College Publishing, New York, 1961.
6. Stoiber, R. E. and Morse, S. A., *Crystal Identification with the Polarizing Microscope,* Chapman and Hall, New York, 1994.
7. Nesse, W. D., *Introduction to Optical Mineralogy,* 2nd ed., Oxford University Press, New York, 1991.
8. DeForest, P. R., Gaensslen, R. E., and Lee H. C., *Forensic Science: An Introduction to Criminalistics,* McGraw-Hill, New York, 1983.
9. Saferstein, J., Ed., *Forensic Science Handbook,* 2nd ed., Prentice-Hall, Englewood Cliffs, NJ, 2000.

10. Petraco, N. and Kubic, T. A., *Basic Microscopy for Criminalists, Chemists and Conservators,* CRC Press, Boca Raton, FL, 2002.

11. Reffner, J. A., Uniting microscopy and spectroscopy, In *Practical Guide to Infrared Microspectroscopy,* Humecki, H. J., Ed., Marcel Dekker, New York, 1995, chap. 2.

12. Bertin, E. P., *Principles and Practice of X-Ray Spectrometric Analysis*, 2nd ed., Plenum, New York, 1975.

13. Goldstein, J. I. et al., *Scanning Electron Microscopy and X-Ray Microanalysis*, 2nd ed., Plenum, New York, 1992.

14. Newbury, D. E. et al., *Advanced Scanning Electron Microscopy and X-Ray Microanalysis*, Plenum, New York, 1986.

15. Postek, M. T. et al., *Scanning Electron Microscopy: A Student's Handbook*, Ladd Industries, Burlington, VT, 1980.

16. Kubic, T. A., Forensic applications of scanning electron microscopy with x-ray analysis, in *Industrial Use of Electron Microscopy,* Li, Z. R., Ed., Marcel Dekker, New York, 2002, chap. 12.

17. Skoog, D. A., Holler, F. J., and Neiman, T. A., *Principles of Instrumental Analysis,* 5th ed., Saunders College Publishing, New York, 1998.

18. Willard, H. H. et al., *Instrumental Methods of Analysis,* 7th ed., Wadsworth Publishing, Belmont, CA, 1988.

19. Saferstein J., Ed., *Forensic Science Handbook.* Vols. I, II, III, Prentice-Hall, Englewood Cliffs, NJ, 1982, 1988, 1993.

20. GRIM-2 and GRIM, Foster and Freeman, Evesham, Worcestershire, England.

21. Koons, R. et al., Forensic glass comparisons, in *Forensic Science Handbook,* Vol. I, 2nd ed., Prentice Hall, Englewood Cliffs, NJ, 2002, chap. 4.

22. Hicks, J. W., *Microscopy of Hairs: A Practical Guide and Manual*, U.S. Department of Justice, Washington, D.C., 1977.

23. Bisbing, R. E., The forensic identification and association of human hair, in *Forensic Science Handbook,* Saferstein, R., Ed., Prentice-Hall, Englewood Cliffs, NJ, 1982, chap. 5.

24. Petraco, N., Protocol for human hair comparison, in *Forensic Science,* Wecht, C., Ed., Matthew Bender, New York 1984, chap. 37.

25. Petraco, N., A microscopical method to aid in the identification of animal hair, *Microscope,* 35, 83, 1987.

26. Petraco, N., A guide to the rapid screening, identification, and comparison of synthetic fibers in dust samples, *J. Forens. Sci.,* 32, 768, 1987.

27. Scientific Working Group for Materials Analysis, Forensic Fiber Examination Guidelines, U.S. Department of Justice, Washington D.C., 1998.

28. American Society for Testing and Materials, *Three Fiber and Textile Standards,* ASTM International, West Conshohocken, PA.

29. Choudry, M., The use of scanning electron microscopy for identification of cuts and tears in fabrics: observations based on criminal cases, *Scan. Micros.,* 1, 119, 1987.

30. Tungol, M. W., Bartick, E. G., and Montaser, A., *Forensic Examination of Synthetic Textile Fibers*, in *Practical Guide to Infrared Microspectroscopy,* Humecki, H. J., Ed., Marcel Dekker, New York, 1995, chap. 7.

31. Keen, I. P., White, G. W., and Fredericks, P. M., Characterization of fibers by Raman microprobe spectroscopy, *J. Forens. Sci.,* 43, 82, 1998.

32. Lambourne, R., Ed., *Paint and Surface Coatings*, Ellis Horwood, Chichester, U.K., 1987.

33. Flick, E. W., *Handbook of Paint Raw Materials,* Noyes, Park Ridge, IL, 1989.

34. Scientific Working Group for Materials Analysis, *Forensic Paint Analysis and Comparison Guidelines*, U.S. Department of Justice, Washington, D.C., 1998.

35. American Society for Testing and Materials, Standard E 1610–95, ASTM International, West Conshohocken, PA.

36. Ryland, S. G., Infrared microspectroscopy of forensic paint evidence, in *Practical Guide to Infrared Microspectroscopy,* Humecki, H. J., Ed., Marcel Dekker, New York, 1995, Chapter 6.

37. Kapp, P. O., *How to Know Pollen*, Wm. C. Brown, Dubuque, IA, 1969.

38. Moore, P. D. and Webb J. A., *An Illustrated Guide to Pollen Analysis,* John Wiley and Sons, New York, 1978.

39. Graves, W. J., A mineralogical soil classification technique for the forensic scientist, *J. Forens. Sci.,* 24, 323, 1979.

40. Wolton, G. M. et al., Final Report on Particle Analysis for Gunshot Residue Detection, U.S. Department of Justice, El Segundo, CA, 1977.

Fingerprints

R. E. Gaensslen

Fingerprints as Evidence

Fingerprints as an evidence category is one of the oldest and most important in all forensic science. The use of these curious, highly individual **friction ridge skin** patterns on the end joint of the fingers as a means of personal identification dates back many centuries. Further, we have reached a point where fingerprint individuality is an article of faith among the public, and is almost universally accepted among scientists and forensic scientists as well. Accordingly, a fingerprint match is widely accepted as certain evidence that identifies a particular person.

Physical evidence can be divided broadly into four major categories: drugs and chemicals (Chapters 5 and 23), trace (Chapter 17), biological (Chapters 14–16), and pattern evidence. The major members of the pattern group are fingerprints (this chapter), questioned documents (Chapter 22), tool mark and firearms evidence (Chapter 21), and other patterns such as footwear and tire impressions (Chapters 19 and 20). All of this kind of evidence consists of various patterns that might be called "individualization" patterns. Under favorable circumstances, individualization pattern evidence can be attributed to a unique source. It is thus potentially quite valuable in an investigation because it may uniquely associate an item or person with a scene or evidentiary item. There is another kind of pattern evidence that could be called "reconstruction" patterns. This consists mainly of patterns usually found at scenes, such as blood spatter (Chapter 12), which are often used to help reconstruct events. Reconstruction patterns could also include the investigation of structure failures, suspicious fires, and transportation accidents (Chapters 24–26).

Fingerprints as a Means of Identification

Because fingerprints are unique, they are used to identify people. In forensic science, we think of fingerprints as being used primarily to help locate, identify, and eliminate suspects in criminal cases. But fingerprints, along with dentition (Chapter 6), are also important in making unequivocal identifications of human remains when more conventional methods of postmortem identification cannot be used. Fingerprints may also be thought of as one member of a class of biometric identifiers that would also include retina or iris patterns, face thermography, and some others. As the technology for rapid scanning and storage of these biometric patterns develops, they are becoming more important as security features to help avoid problems associated with forged identification documents.

The two features of fingerprints most important for their use as a means of personal identification are: (1) Every fingerprint is unique (to an individual), and (2) fingerprints do not change during a lifetime (unless there is damage to the dermal skin layer).

What Fingerprints Are

It has long been recognized that the fingers, palms of the hands, and soles of the feet of humans (and some other primates) bear *friction ridge* skin. These areas are characterized by a complicated pattern of "hills and valleys." The "hills" are called *ridges*, and the "valleys" are called *furrows*. On the end joint of the fingers a number of basic patterns are formed by the friction ridge skin. Within each basic pattern, there are numerous possible variations. The patterns form on these skin surfaces early in embryonic development, and remain constant throughout embryonic life, birth, and the life of the individual. An individual's genetic make-up probably plays a part in determining the sizes and basic shapes of the patterns and ridges, but it is not the only factor. We know this because identical twins, who come from the same fertilized egg and thus have identical genetic make-up, have different and distinguishable fingerprints.

Fingerprint Patterns

There are three basic fingerprint patterns: arch, loop, and whorl (Figure 18.1). Within

| Arch | Loop | Whorl |

Figure 18.1 Basic fingerprint patterns.

these major classes, fingerprint examiners commonly recognize other categories. Arches, for example, can be plain or tented. Loops can be radial or ulnar, depending on whether the slope of the print pattern is in the direction of the inner arm bone (radius) or outer arm bone (ulna). Whorls are the most complex of fingerprint patterns, and there are several whorl categories, such as central pocket, double loop, and accidental. Loop and whorl patterns contain definable features called the core and the delta. These are important in ten-print fingerprint classification, and in comparisons. Figure 18.2 shows some of the additional fingerprint patterns. The core and delta of a fingerprint pattern are also indicated in one of the frames.

Within fingerprint patterns, there are a number of features called **minutiae**. Once evidentiary and reference (inked) fingerprints have been oriented and found to have the same general ridge flow or pattern type, these features are used to actually compare the fingerprints and decide whether they are or are not from the same source. The ridges of the fingerprint form the minutiae by doing one of three things: ending abruptly (ending ridge), splitting into two ridges (bifurcation), and being short in length like the punctuation mark at the end of a sentence (dot). Combinations of these minutiae also have names. For example, two bifurcations facing each other form what is called an *island*. Some minutiae are shown in Figure 18.3. The process used in fingerprint identification is discussed in more detail below.

History of Fingerprints

The use of fingerprint and handprint patterns as means of personal identification dates back

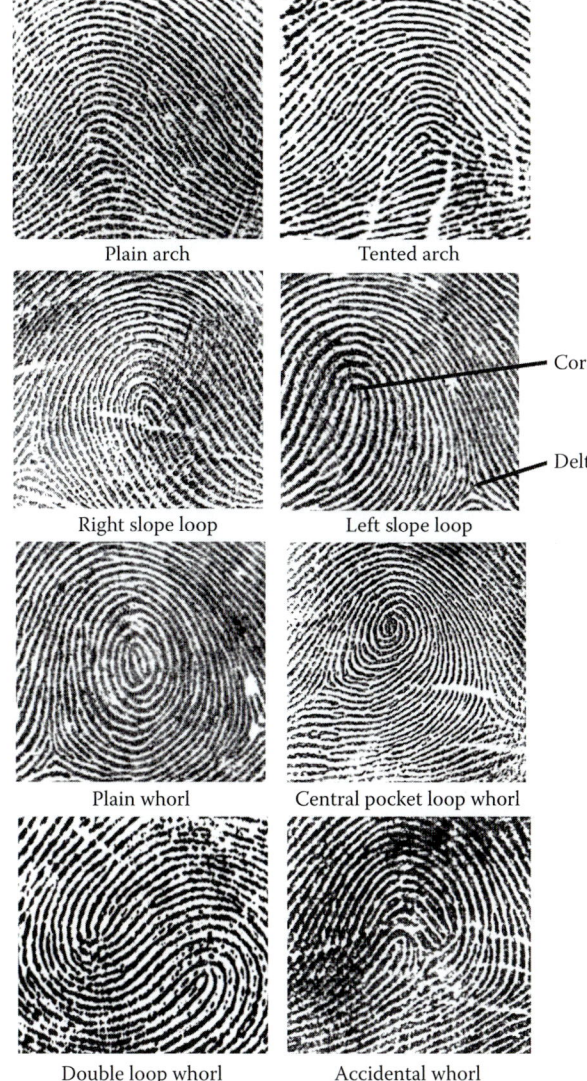

Figure 18.2 Variations of fingerprint patterns. One print shows the core and the delta.

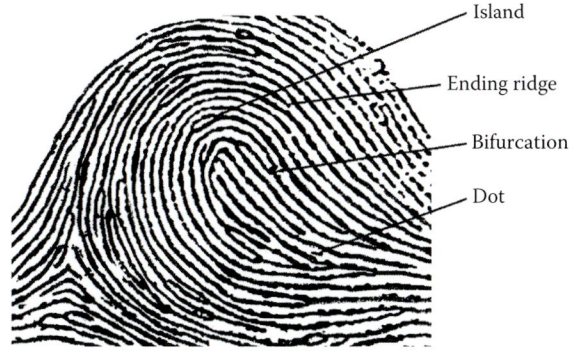

Figure 18.3 Fingerprint minutiae.

potters may have used them to "sign" their work. There are records of using fingerprints and handprints as marks of authenticity in China at least 2000 years ago.

Development of fingerprint science in Europe dates back to the 17th and 18th centuries. The first scientist to describe fingerprint ridges was an English plant morphologist called Nehemiah Grew (1641–1712). He also published very accurate drawings of the ridge patterns. Another early pioneer was Johannes Evangelista Purkinje (1787–1869), a Czech physiologist known for a number of important anatomical and physiological discoveries, wrote in detail about friction ridges and described a number of fingerprint patterns in his thesis in 1823. Thomas Bewick (1753–1828) was primarily known as a wood engraver whose engravings appeared in books, but he is often mentioned in connection with fingerprint history. First in a book published around 1800, and elsewhere, he "signed" his work with his own thumbprint. He apparently believed that his thumbprint was as unique as his signature.

Sir William Herschel (Figure 18.4) is often credited with being the first European to recognize the value of fingerprints as a means

Figure 18.4 Sir William J. Herschel (1833–1917). (From Berry, J. and Stoney, D.A., History and development of fingerprinting, in Lee, H.C. and Gaensslen, R.E., Eds., Advances in Fingerprint Technology, 2nd ed., CRC Press, Boca Raton, FL, 2001. With permission.)

thousands of years. There are indications of this in artifacts recovered from archaeological excavations of ancient civilizations. Early of personal identification. Herschel was a British administrator who went out to work in Bengal, India, in 1853. He later joined the Civil Service of India. In his civil service capacity, he entered into a contract with a local businessman, and asked that the man put his right hand print on the back of the contract. Herschel would later say that this action represented the first use of friction ridge impressions as a means of personal identification. It was apparently quite common, however, for parties to sign contracts in that area of India at that time by placing a hand or finger mark impression next to their signatures. And, as noted earlier, there is ample evidence that fingerprints were recognized hundreds of years earlier in China as means of identification. Herschel developed the use of fingerprints as a means of controlling fraud in contracts, false impersonations in government pension distributions, and other matters. He tried to convince others in the government to implement his practices but was unsuccessful. He nevertheless demonstrated the persistence of the ridge patterns in his own fingerprints taken periodically over a period exceeding 50 years.

Dr. Henry Faulds (Figure. 18.5), a Scottish physician by training and profession, went to India as a medical missionary in 1871. The next year, he traveled to Japan. By 1879, he is known to have been involved in the study of fingerprints. In 1880, Faulds wrote a letter to evolutionary theorist Charles Darwin, who forwarded the letter to Francis Galton. Around 1880, he noted that fingerprints could be classified, and that ridge detail is unique. He further mentioned apprehending criminals by locating fingerprints at scenes. He appears to have been familiar with the Chinese and Japanese (perhaps even Egyptian) use of fingerprints. Many of his observations were recorded in a letter to the British journal *Nature* in 1880. There was some dispute between Faulds and Herschel over the priority of fingerprint use as a means of personal identification. In point of fact, neither of them could make such a claim. The originators of fingerprint science are lost in antiquity. It is

Figure 18.5 Dr. Henry Faulds (1843–1930). (From Berry, J. and Stoney, D.A., History and development of fingerprinting, in Lee, H.C. and Gaensslen, R.E., Eds., *Advances in Fingerprint Technology*, 2nd ed., CRC Press, Boca Raton, FL, 2001. With permission.)

fair to say that Herschel and Faulds were very influential in introducing the idea of fingerprints to continental Europe of the later 19th century.

Thomas Taylor, a microscopist for the U.S. Department of Agriculture noted in a scientific journal article in 1877 that fingerprints and palm prints might be used as identification features, especially in criminal matters.

The first truly scientific method of criminal identification—called **anthropometry**—is due to Alphonse Bertillon. Today, we could look at anthropometry as an older kind of biometry. Bertillon is always mentioned in discussions of the history of fingerprint sciences, not because he contributed directly to fingerprints as an identification method, but because he laid down a foundation (with anthropometry) for the eventual acceptance of fingerprints as a scientific method for personal identification. Bertillon used his position in the Police Prefecture in Paris, where his duties included filling out and filing criminal information cards, as a means of developing his identification method. His father was an anthropologist and had spent considerable time thinking about individual differences in human physical features. It occurred to young Bertillon that these differences might have utility in uniquely identifying people in criminal records. Criminals at the time were skilled at avoiding identification by using disguises and aliases. Bertillon developed a biometric system based on head size, finger length, and so forth, ultimately choosing 11 measurements. The information was carefully recorded on file cards and a system devised for organizing the files. His superiors initially thought he was merely eccentric, but the system soon proved its worth in identifying and helping to convict persons with prior arrest or conviction records. He went on to become director of the identification bureau of the Paris police, and police agencies in many places began using his system. Bertillonage was ultimately undone by unequivocal proof that different individuals could have the same anthropometric measurements. One of the most well known examples of chance duplication came at the Leavenworth Prison. A new prisoner named Will West was having his measurements taken when a staff member began to suspect that he had encountered this profile before. Upon checking the files, he discovered that the same measurements fit a William West who was already incarcerated at the prison (and unrelated to Will West though they looked a lot alike). The two did have different fingerprints. This incident and probably others came to convince police agency identification personnel that fingerprints represented the basis for a superior method of personal identification.

One of the greatest scientists of the 19th century known for his contributions to many fields, Sir Francis Galton (Figure 18.6) became involved in the correspondence between Faulds and Darwin (Galton's cousin). A person of eclectic interests, he looked extensively at Bertillonage, and even visited Bertillon in Paris. He also looked at fingerprints in detail, especially at Herschel's data, which had been shared with him. His book, *Finger Prints*, published in 1892,

Figure 18.6 Sir Francis Galton (1822–1911). (From http://www.mugu.com/galton/start.html.)

is regarded as a classic work. Around 1893, the British Home Office, the parent agency of the Metropolitan Police, commonly referred to as Scotland Yard, decided to add fingerprints to the Bertillon cards for criminals in the identification system. Soon the success of fingerprints overshadowed that of Bertillonage in criminal identification, and anthropometry was abandoned in 1901. Fingerprint minutiae are sometimes called "Galton features" in recognition of his contributions.

Juan Vucetich (Figure 18.7) might be considered the western hemisphere's fingerprint

Figure 18.7 Juan Vucetich (1855–1925). (From http://www.argiropolis.com.ar/ameghino/biografias/vuc.)

pioneer. An employee of the police department in La Plata, Argentina, he became convinced of the value of fingerprints as a means of criminal identification and wrote a book on the subject in 1894. By 1896, the Argentine police had abandoned Bertillonage in favor of fingerprints in criminal records. The first recorded case in which fingerprints were used to solve a crime took place in Argentina in 1892. The illegitimate children of a woman named Rojas were murdered on June 18, 1892. She acted distraught and accused a man called Velasquez, who she said commited the act because he wanted to marry her and she had refused. Velasquez maintained his innocence and had an alibi. An investigator from La Plata, the provincial capital, assisting with the case found out that another of Rojas' boyfriends had made statements about being willing to marry Rojas except for the children. The investigator, a man named Alvarez, having been trained by Vucetich, found a bloody fingerprint at the scene and collected it. After comparing it to Rojas' fingerprints, the bloody print turned out to be of her right thumb. She confessed when confronted. Vucetich devised a classification system for fingerprints that was used in Argentina and throughout South America.

Sir Edward Henry (Figure 18.8) was in the British Indian Civil Service beginning around 1873. In 1891, he became Inspector General of Police for Bengal province, where they were using anthropometry. He became interested in fingerprints and read Galton's book. He corresponded with and later visited Galton in 1894. Galton shared everything he had about fingerprints with Henry, including materials he had obtained from Herschel and Faulds. Henry conceived of, and is widely known for, a fingerprint classification system that was adopted in British India. He presented it in the United Kingdom in 1899. He wrote a book titled *Classification and Uses of Fingerprints*. In 1901, he became assistant commissioner of police for criminal identification at New Scotland Yard in London, and commissioner in 1903.

In North America, fingerprints were in use by the New York City civil service (to prevent impersonations during examination) by

Figure 18.8 Sir Edward Henry (1850–1931). (From Berry, J. and Stoney, D.A., History and development of fingerprinting, in Lee, H.C. and Gaensslen, R.E., Eds., *Advances in Fingerprint Technology*, 2nd ed., CRC Press, Boca Raton, FL, 2001. With permission.)

1903, and fingerprints were introduced about the same time in the New York State prison system and at Leavenworth Penitentiary. A number of police departments began using fingerprints as identifiers in criminal records as well. The 1904 St. Louis World's Fair provided the venue for a chance meeting between Inspector Edward Foster of the Royal Canadian Mounted Police and Detective John Ferrier of Scotland Yard. As a result of what he learned in St. Louis, Foster convinced his superiors in the R.C.M.P. of the utility of fingerprints. In 1910, a man called Thomas Jennings was arrested in Chicago and brought to trial for murder. The primary evidence against him was fingerprints. The state wanted to try to insure that the fingerprint identification evidence would survive appeals to the Illinois Supreme Court, and they called Edward Foster as an expert witness. The defendant was convicted, the evidence did survive, and the Jennings case is considered a landmark fingerprint case in the courts.

Fingerprint Classification

It was apparent to the early fingerprint pioneers that a manageable, consistent classification system was necessary if large sets of fingerprint files were to be useful for criminal identification. In the United Kingdom and the United States, the classification systems are variants of the one developed by Sir Edward Henry. In Argentina and other South American countries, a different system based on the one developed by Vucetich has been used.

The modified **Henry system** as used in the United States is a scheme for the classification of ten-print sets, or a fingerprint card, for one individual. Figure 18.9 shows a typical fingerprint card of the kind long used by law enforcement and other agencies to maintain fingerprint records. Use of the classification system enabled efficient searching of large files. Keep in mind, though, that classification

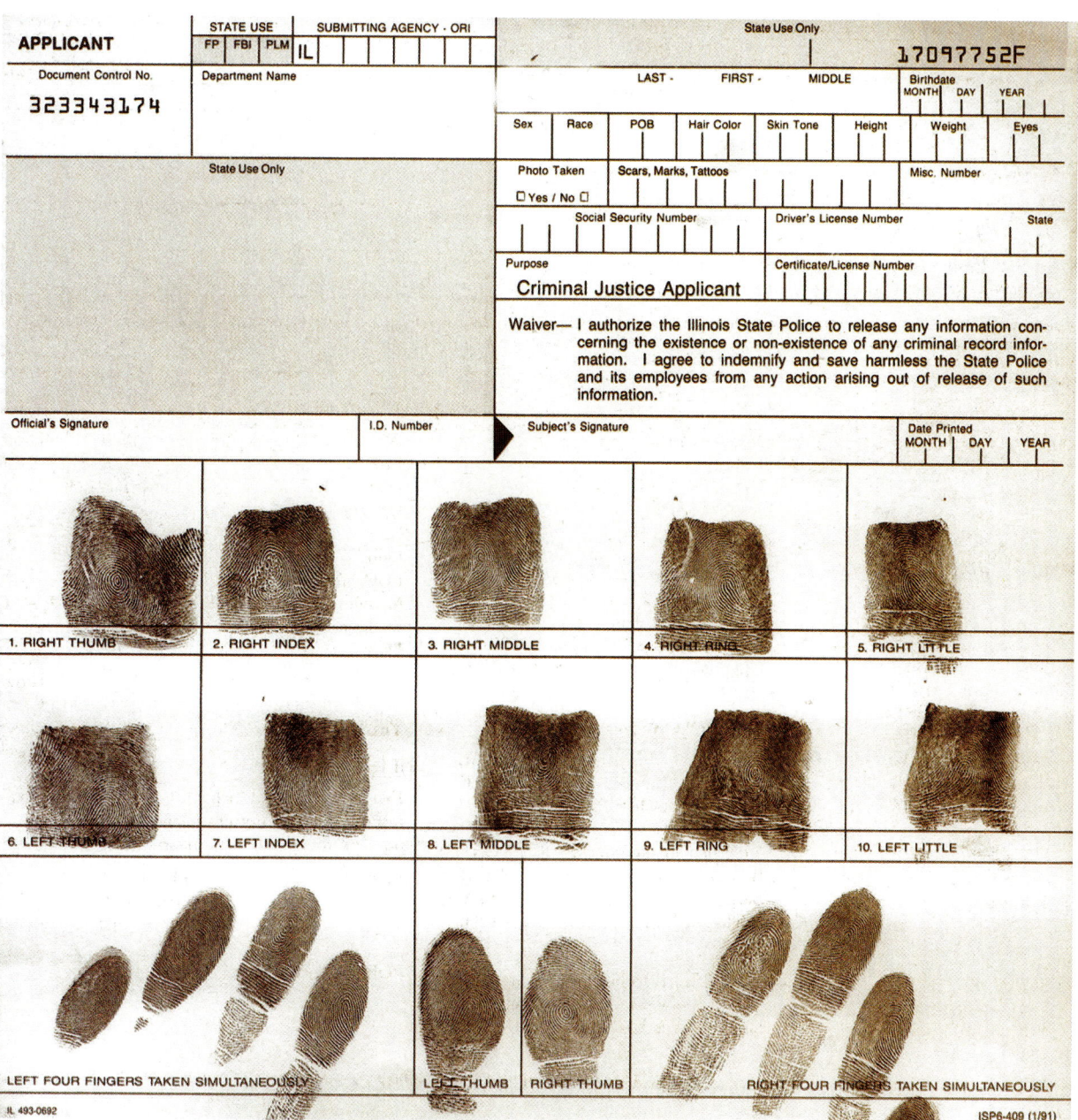

Figure 18.9 Typical 10-print fingerprint card.

is based on having all 10 prints. Thus, a 10-print card from a fingerprinted person can be classified and filed, but you would need all 10 prints from a person to use his/her classification to search in the large file. The system therefore allowed organized maintenance of the large files maintained by many law enforcement agencies, but the files could not easily be searched manually for a single print. It is typical to recover single prints, or sometimes even partial single prints, from crime scenes. Until the development of computerized fingerprint search systems, it was impractical to search large files for a single print because it took so much time. Partial prints could obviously be compared with prints from cards on file, but it was necessary to have a possible suspect or suspects first to know which cards to retrieve for manual comparison.

For years, the fingerprint classification of wanted suspects was shown on wanted posters distributed to police agencies and displayed in public places such as post offices. Figure 18.10 shows such a wanted poster, and the fingerprint classification can be seen in small print.

The development and fairly widespread deployment of computerized fingerprint storage and retrieval systems has made searching large files for single and partial prints routine. It has also rendered classification largely unnecessary. At one time not long ago, all fingerprint examiners and identification personnel were extensively trained in fingerprint classification using the modified Henry system, as were many police officers. Figure 18.11 summarizes the modified Henry classification of a 10-print set as used by the FBI.

Computer-Based Fingerprint Files

Computer storage and retrieval systems for fingerprints were originally developed for law enforcement applications. Efforts to develop these systems began in the early 1960s. In

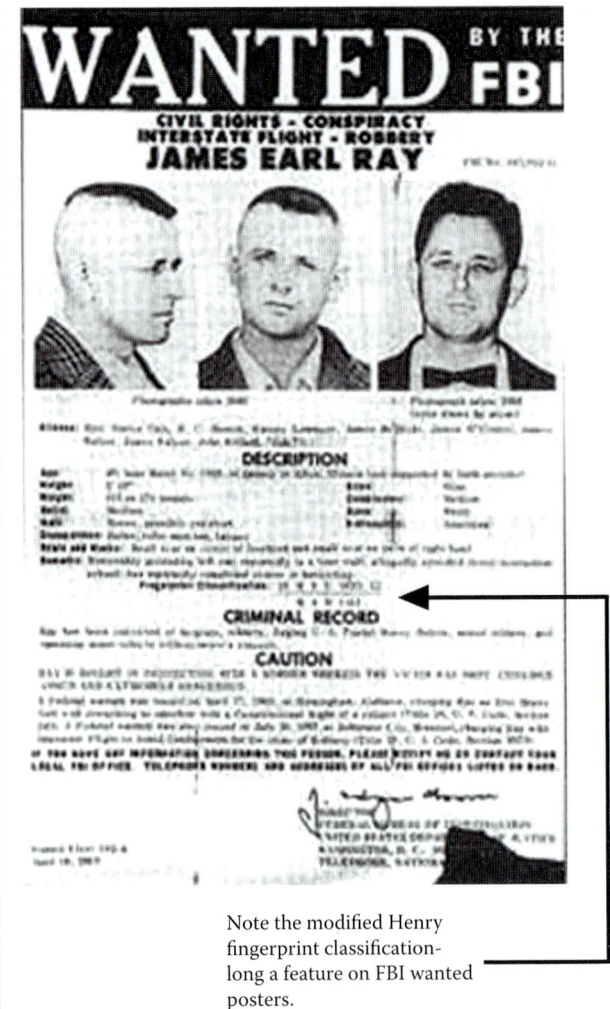

Note the modified Henry fingerprint classification—long a feature on FBI wanted posters.

Figure 18.10 Wanted poster for James Earl Ray, the man later captured and convicted of assassinating the distinguished civil rights leader Dr. Martin Luther King, Jr., in Memphis, Tennessee, on April 4, 1968.

the United States, these efforts were the result of collaboration between the FBI, which maintains the largest (and the only national) fingerprint database, and scientists at the National Bureau of Standards (later became the National Institute of Standards and Technology, or N.I.S.T.). The law enforcement–based automated systems are commonly called Automated Fingerprint Identification Systems or AFIS. There are two principal applications. The first is searching large files for the presence of a 10-print set of prints (taken from a person). The second is searching large files for single prints, usually developed latent fingerprints (see below) from crime scenes. Developing the appropriate scanning and

Key	Major	Primary	Secondary	Subsecondary	Final
10	M	19	U	IOM	1
	14	21	W	OII	

Key = The ridge count of the first loop pattern excluding the little fingers.

Major = Value of the ridge counts of the loop patterns or the tracings of the whorl patterns on the thumbs (fingers #1 and #6).

Primary = Summation of the value of the whorl patterns for fingers numbered 2, 4, 6, 8, and 10 for the numerator (top). Summation of the value of the whorl patterns for fingers numbers 1, 3, 5, 7, and 9 for the denominator (bottom). Add 1 to both the numerator and denominator.

Secondary = Pattern types located in the index fingers (#2 and #7).

Subsecondary = Value of the ridge counts of the loops or the tracings of the whorls for fingers #2, #3, and #4 in the numerator (top) and #7, #8, and #9 in the denominator (bottom).

Final = The ridge count of the loop in the right little finger (#5), if it is not a loop then use the left little finger (#10). If there is no loop in either of the little fingers, then there is no final.

Figure 18.11 Summary of modified Henry system fingerprint classification.

storage technologies and computer algorithms for these systems was far from trivial.

By the late 1980s, there were at least five operational AFIS, at least four of which had been commercialized. Most if not all, large jurisdictions had systems in place by the 1990s. Because different commercial vendors use different technologies, the systems are not intrinsically compatible with one another. Another important point about AFIS is that a given person's fingerprints may be in one system, but not in others. Depending on the criteria for including a set of prints in the files, a large city system could have someone's prints, but the corresponding state system might not have them, for example.

AFIS can be seen as one of the three significant classes of electronic databases for law enforcement purposes that are now a permanent part of the landscape. The other two are CODIS (Combined DNA Indexing System), which holds DNA profiles (Chapter 16), and **NIBIN** (National Integrated Ballistic Information Network), which holds searchable image information from fired bullets and cartridge cases (Chapter 21). Generally it can be said that each of these databases holds two types of files or profiles. One type is the knowns. In the AFIS case, this file contains the prints of known individuals. Any questioned specimen, image or profile can be searched for in the "known" database, and if it is found, its source is thereby identified. The other type is often called the "forensic" file or database. It

consists of images or profiles from unsolved cases, the sources of which are not known. In the AFIS case, for example, the forensic file contains images of developed latent single fingerprints from unsolved cases. The person whose fingerprints these are has not yet been identified. The file is valuable to investigators, however, in that it allows cases to be connected by the fingerprints that were not obviously related. Such connections can allow investigators to share information and leads, thus increasing the probability of apprehending a suspect.

AFIS has become a successful tool in the apprehension of unknown offenders. Traditionally any person arrested would have a 10-print card that is entered into the AFIS system. Many job applicants will also have a 10-print card in the system. The number of cases solved by AFIS in the United States is difficult to determine, but the national average of latent prints entered into AFIS that will be deemed an identification by a latent print examiner is around 15%. This average of latent prints identified by AFIS can vary among jurisdictions. It is dependent on the knowledge and skill of the individual entering the latent or unknown impressions (usually forensic scientists), and also by the individuals that submit and register the known fingerprint cards (usually technicians or police personnel).

The FBI has made their criminal database of known fingerprint cards available to other law enforcement agencies. This system is called the Integrated Automated Fingerprint Identification System, IAFIS. Using the IAFIS system, a latent print examiner can search unknown latent impressions in a neighboring state or several states. This is the national criminal database maintained by the FBI of all of the 10-print cards received from all over the country. To date this system has been very successful in developing leads and sometimes solving unresolved cases.

Today, AFIS must be seen as part of a much larger picture that includes an array of automated systems for human **biometrics**—the use of some type of body metric for identification. It may be fair to say that the Bertillon system of anthropometry described above was the first well-defined system of human biometry, leaving aside the pre-19th century uses of fingerprints as means of human identification.

While criminal identification and related law enforcement applications of automated fingerprint systems are obviously very important, fingerprints along with other biometrics (such as retinal or iris patterns, voice recognition, hand geometry, etc.) have increasingly been incorporated into automated identification systems for other purposes. These systems may control entry and/or access into computers or structures, identification of persons for security purposes, to prevent identity theft, and to help control welfare or social services fraud.

Types of Evidentiary Fingerprints

At crime scenes and/or on items of evidence, there are essentially three types of fingerprints that may be encountered: **patent**, **plastic**, and **latent**.

A patent (or visible) print is one that needs no "enhancement" or "development" to be clearly recognizable as a fingerprint. Such a print is often made from grease, dark oil, dirt, or even blood, rendering it visible and recognizable, and possibly even suitable for comparison without additional processing.

A plastic print (might be called "impression" or "indented" print) is a recognizable fingerprint indentation in a soft receiving surface, such as butter, silly putty, tar, etc. Such prints have distinct three-dimensional character, are immediately recognizable and often require no further processing.

A latent print is one that by definition requires additional processing to be rendered visible and suitable for comparison. Processing of latent prints to render them visible, and hopefully suitable for comparison, is called *development*, *enhancement* or *visualization*. An enormous amount of literature has grown in this subject in the past 30 years or so. Many great strides have been made in the area due to clever applications of chemistry and physics principles coupled with better understandings of the composition of latent residues. In many ways, these advances are as impressive and important as the progress made in applying DNA typing techniques to biological evidence, though they have not received even a fraction of the publicity that has been given to DNA.

Development (Enhancement) of Latent Fingerprints

Composition of Latent Print Residue

The starting point for understanding latent print development techniques is the fingerprint residue itself. Friction ridge skin (Figure 18.12), as found on the surfaces of the fingers, hands, and bottoms of feet, has pores through which small sweat glands can empty their contents onto the skin surface. These sweat glands, called *eccrine*, produce the watery-type sweat composition of which forms a basis for latent fingerprint residue. Another type of sweat gland called *apocrine* located in other parts of the body, produces sweat that contains organic molecules (lipids and proteins) and pheromones. This material can become part of the latent print residue from a person touching those areas of the body. In addition, an almost unlimited variety

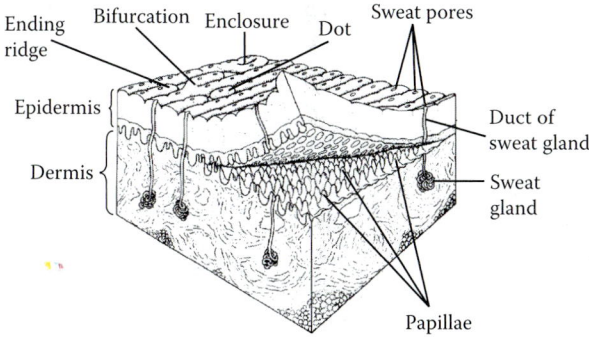

Figure 18.12   Friction ridge skin—diagram of longitudinal section. (From FBI, Science of Fingerprints, Rev. 12–84, U.S. Government Printing Office, Washington, D.C.)

Table 18.1
Constituents of Fingerprint Residue from Sweat

INORGANIC	
Major Components	**Minor Components**
Na^+	Mg^{++}
K^+	Zn^{++}
Ca^{++}	Cu^{++}
Fe^{++}	Co^{++}
Cl^-	Pb^{++}
F^-	Mn^{++}
Br^-	
I^-	
HCO_3^-	
PO_4^{3-}	
SO_4^{2-}	
NH_4OH	

ORGANIC	
Proteins	Pyruvate
Amino acids[a]	Creatine
Lipids[b]	Creatinine
Glucose	Glycogen
Lactate	Uric acid
Urea	Vitamins Sterols

[a] Major amino acids: serine, glycine, ornithine, alaine, asparticacid.
[b] Major lipids: fatty acids, triglycerides, wax esters, squalene.

of substances from the environment can get onto the friction ridge skin and can then be deposited into latent residue when the person touches a surface. Thus, although there are a number of common constituents of most latent print residues based on the composition of sweat, the proportions may vary, and there are many other compounds and materials from the environment that can be present as well. Table 18.1 shows some of the common constituents.

Most methods for the development of latent prints were developed based on knowledge of the latent print residue composition. Usually, a method known to be capable of detecting or visualizing one of the compounds or elements present in latent residue has been applied to target that compound or element. For the application to be successful, it has to be possible to apply the method to evidentiary fingerprints on the variety of surfaces where they are found, and without destroying the integrity of the impression pattern. The methods commonly employed can be broadly divided into three groups: physical, chemical, and special illumination or a combination of these methods, many of which involve laser or narrow band pass illumination.

It should be noted here (and will be reiterated below) that fingerprints developed in situ at a scene must be photographed prior to any lifting or other collection effort. Prints developed in the laboratory or identification unit must also be photographed to document the results as well as the location of the print on the evidence item. Investigators often face difficult decisions about whether to employ fingerprint enhancement procedures at a scene or seize the evidentiary item and submit it to the lab or identification unit. Many factors will be involved in resolving the matter, including the training and experience of the investigator, how well equipped he/she is at the scene, and whether the evidentiary item can be readily seized and transported.

Physical Methods

Classically, physical methods are those that do not involve any chemicals or chemical reactions. They work by applying fine particles to the fingerprint residue where it adheres preferentially, thus creating contrast between the ridges and the background.

The most well known one is powder dusting—a mainstay of latent fingerprint detection for a century or more. Most common powders

Figure 18.13 Technique of powder dusting a latent fingerprint and transparent tape lifting.

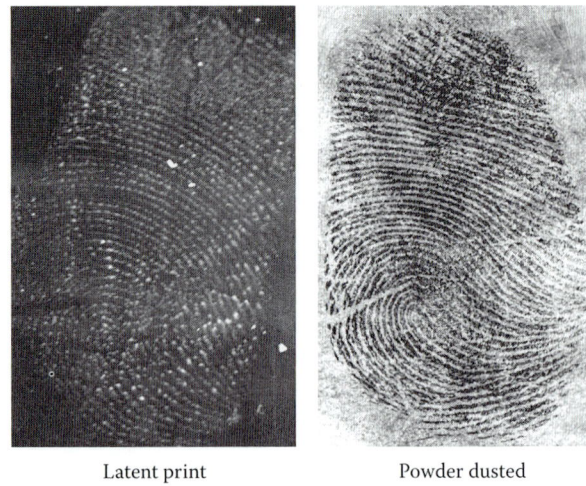

Latent print   Powder dusted

Figure 18.14 Latent print before and after powder dusting.

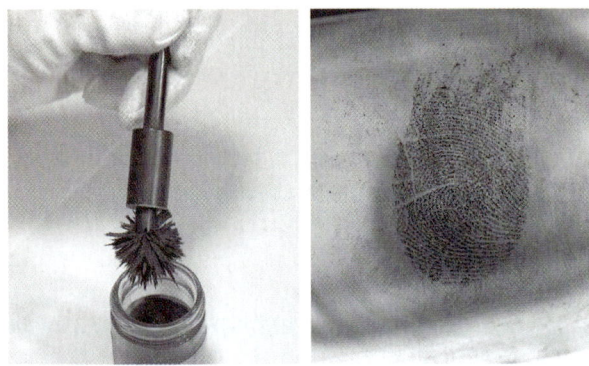

Figure 18.15 Magna Brush and magna powder (left) and developed latent print (right).

are inorganic and come in several colors. There is also a large variety of fingerprint brushes available. The principle of powder dusting is simply that the powder particles adhere to the latent residue. Careful use of the proper brush and powder often results in the development of excellent prints. Black powders are generally superior to the other colors. Black powders are produced in a way that yields more uniform particle size and generally produces better results. The technique is illustrated in Figure 18.13. Also illustrated is the lifting of a developed latent impression using transparent lifting tape. The tape lift is mounted on a backing card with a color maximally contrasting to that of the powder (e.g., white backing for black powder). One-piece lifters also known as "hinge lifters" are commercially available for this purpose. A latent fingerprint developed by powder dusting is shown in Figure 18.14.

A variant of the simple brush and powder combination is the magnetic brush (Figure 18.15), the original trademark version of which is called the Magna Brush. Actually a small retractable magnet, not a real brush at all, the magnetic brush uses special magnetic powders that also can be obtained in several colors. The principle of magnetic powder enhancement is the same as for conventional powder, namely adherence of the fine particles to fatty components of the residue. The magnetic brush technique is more useful on some surfaces than conventional powder dusting, mainly because the magnetic wand can be used to remove any excess powder from the substratum. It also has the potential to be a gentler technique, in the sense that there is no brush, and thus no bristles, so it is less likely to damage the latent print ridges in the brushing process.

Another physical latent print developing procedure involves small particle reagent (SPR). Typically applied by spraying or immersion, the most common formulation of SPR is a fine suspension of molybdenum disulfide in a detergent solution. The particles adhere to the lipid components of the residue. SPR is most commonly used on evidence that has been formerly wet.

Chemical Methods

The greatest progress in latent fingerprint visualization techniques in the past quarter century has involved chemical and instrumental/special illumination techniques.

Classically, the chemical techniques were silver nitrate, **iodine fuming**, and **ninhydrin**. Silver nitrate is not used any longer because there are much better techniques available. The principle of its use lay in the reaction of the silver with the chloride in the fingerprint. The silver chloride was then photoreducible to silver, which contrasted with the background. A somewhat related method that still involves silver is called *physical developer* (PD), and is discussed below.

Elemental iodine is one of the compounds in nature that sublimes, that is, it can pass from solid to vapor without becoming liquid. Iodine sublimes easily with only moderate warming. The vapor is directed toward the fingerprint residue with a so-called iodine fuming gun (essentially a plastic blowpipe containing some iodine), or an object to be fumed can be placed in a closed cabinet, which is then filled with iodine fumes. Even though it has long been placed under the "chemical methods" category of latent print development, the iodine probably does not react with any of the components in the residue. It probably interacts with the lipid components in such a way that it is trapped in the residue, giving the ridge features a dirty-brown colored appearance. The iodine-developed color is not stable in the latent print, however, and iodine prints have to be quickly photographed. There are chemical methods for rendering iodine prints a permanent color that will not fade. The traditional method involved using starch solution for that purpose, but 7,8-benzoflavone (α-naphthoflavone) treatment is the preferred method today. Iodine fuming is used primarily on inherently valuable items precisely because of its impermanence.

Since around 1910, ninhydrin has been known to react with amino acids, forming an adduct called Ruhemann's purple (named for one of the original observers). In the mid-1950s, Swedish scientists noted that ninhydrin reacted with the amino acid components of latent fingerprint residue, and took out a patent on the process. Ninhydrin is applied by spraying, painting, or dipping. The ninhydrin reaction is slow unless accelerated by heat in the presence of humidity. The classical ninhydrin preparation was made up in Freon 113 (1,1,2-trichlorotrifluoroethane). Concern over the effect of these compounds on the Earth's ozone layer culminated in the signing of the Montreal Protocol in 1987 in which many countries agreed to ban the use of chlorofluorocarbons (CFCs) of which Freon 113 is an example. The most recent ninhydrin formulations have thus avoided Freon as a solvent.

Ninhydrin develops bluish-purple fingerprints (Figure 18.16), and is extremely useful on porous surfaces (such as paper), but occasionally the bluish-purple color may not show sufficient contrast with the background substratum. Ninhydrin is also useful as a preliminary treatment in a sequential process followed by other chemicals and viewing under laser or alternative light source illumination. Another significant factor in ninhydrin use for latent prints has been the design, synthesis, and evaluation of a series of ninhydrin analogues, perhaps the most important of which is 1,8-diazafluoren-9-one (DFO). Treatment of ninhydrin-developed prints with $ZnCl_2$ (zinc chloride) provides the basis for the use of laser and special illumination methods. The metal salt treatment converts Ruhemann's purple to another compound which can be excited by blue-green light using a laser or alternate light source to yield strong fluorescence. Many modifications of postninhydrin latent print treatments have been designed to maximize fluorescence under illumination by commercially available lasers or broadband pass filtered alternate light sources.

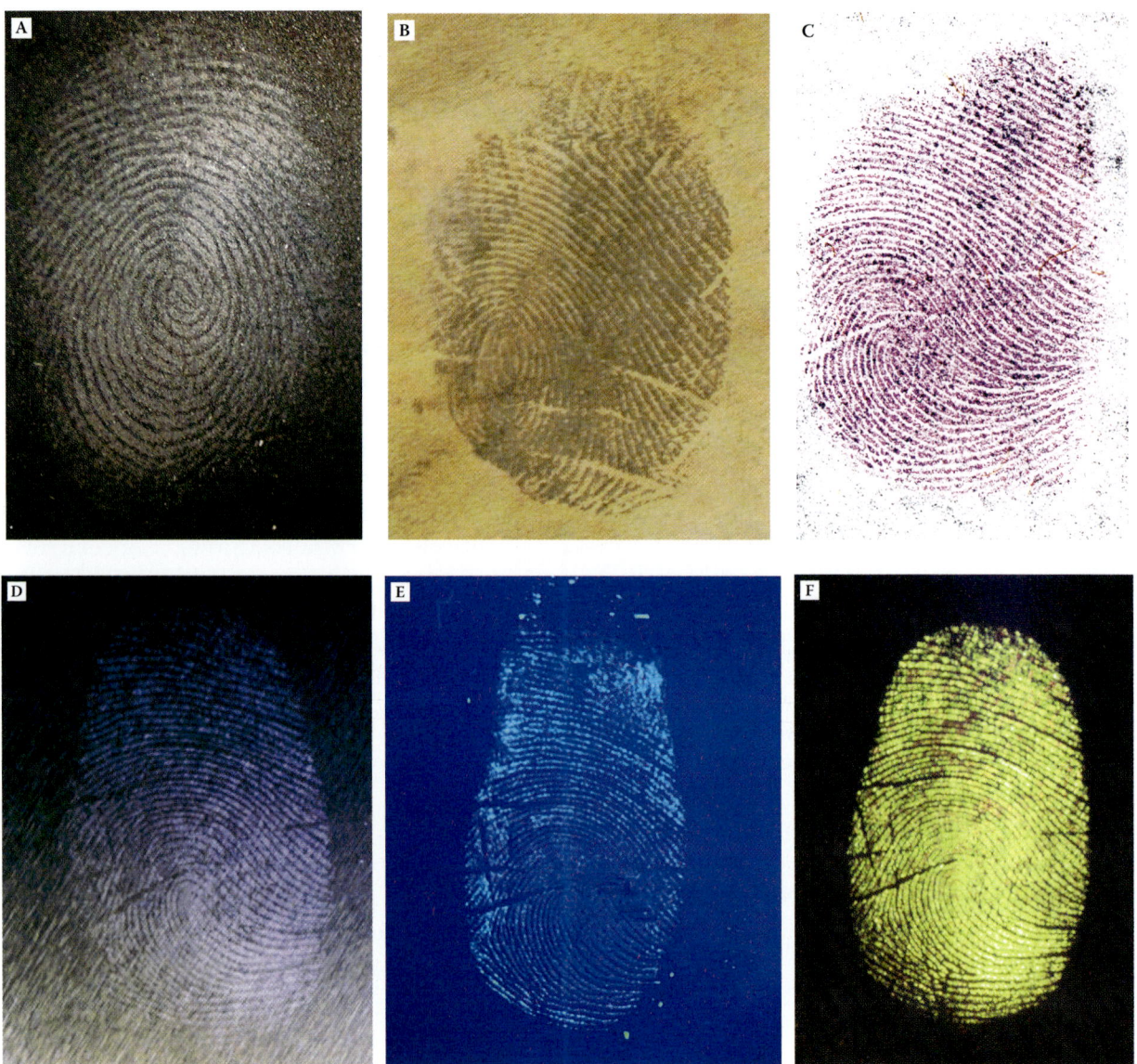

Figure 18.16 Visualization of latent fingerprints by various methods. (A) Ambient (oblique) white light illumination; (B) physical developer; (C) ninhydrin; (D) cyanoacrylate ester (super glue); (E) super glue–developed print, post-treated with ardrox dye, under UV illumination; and (F) super glue–developed print, posttreated with Rhodamine 6G, under UV illumination.

There are a host of other chemicals that react with amino acids yielding colored or fluorescent products. However, none have been as important in operational latent print work as ninhydrin (and posttreatments), and DFO.

The most important other chemical procedure is treatment with **cyanoacrylate** esters ("**super glue**"). Super glue "enhancement" was first observed by Japanese scientists in the National Police Agency. The method quickly caught the attention of latent print examiners all over the world. In the early 1980s, fingerprint examiners working in the U.S. Army Criminal Investigation Laboratory in Japan, and a little later in the U.S. Bureau of Alcohol, Tobacco and Firearms Laboratory, introduced alkyl-2-cyanoacrylate ester (super glue) fuming as a method for latent print development in the United States

Super glue can be induced to fume, and the fumes will interact with latent fingerprint residue by polymerizing in situ, yielding a stable, even robust, friction ridge impression off-white in color (Figure 18.16). Items to be processed by glue fuming are usually placed into well-sealed cabinets where the glue is induced to fill the cabinet with vapors. The process requires a certain level of relative humidity, and a moisture source is always placed in fuming cabinets. Glue fuming can be accomplished in a closed cabinet or container without any "acceleration," but it is very slow. Typically, strong alkali or heat is used to accelerate fuming. The progress of latent residue development in the fuming cabinet can be monitored by placing a test latent print onto a piece of aluminum foil, or similar surface, and placing the object into the cabinet where it can be viewed during fuming. Latents underdeveloped with super glue can simply be further fumed. But overfumed latents may be ruined.

Like ninhydrin, glue fuming is an excellent method of developing latent fingerprints. But it may be most useful operationally as an initial step in protocols where the cyanoacrylate-developed prints are further treated and examined. The simplest enhancement of a cyanoacrylate-developed print is dusting it with powder. Other posttreatments of cyanoacrylate-developed prints include "dye stains," which induce luminescence or fluorescence in the residue when it is illuminated with laser or alternate light source at the appropriate wavelength. Gentian violet, Coumarin 540 laser dye, Ardrox (Figure 18.16), Rhodamine-6G (Figure 18.16), and other treatments have been used for the enhancement of cyanoacrylate-developed print luminescence under alternate light or laser illumination.

The last chemical method we will discuss is physical developer. In spite of the name, physical developer (**PD**) is a chemical method. It has become common in many laboratories to use PD as a follow-up method to ninhydrin or DFO on porous surfaces, such as paper. While ninhydrin or DFO react with water-soluble components in the latent print residue, PD reacts with the lipid and other water-insoluble components (Figure 18.16).

PD is a photographic-type process based on the deposition of silver onto latent fingerprint residue from a ferrous/ferric redox couple and silver salt mixture in solution. There is a modified procedure developed at the U.S. Secret Service laboratory that employs colloidal gold in addition to silver salt.

Combination/Special Illumination

Sometimes, a "latent" fingerprint can be visualized by simple **oblique lighting** (Figure 18.16). This technique might be considered physical, since it involves no chemicals or special light sources. A latent amenable to visualization by oblique lighting also illustrates that the definition of a "latent" fingerprint is operational to large extent. Some kind of pattern was visible in order to have prompted the observer to obliquely illuminate the surface, thus "enhancing" it.

Latent prints may also be viewed under alternate light (525 nm) or laser illumination. Alternate light is generally superior to incident white light in revealing ridge detail. "Alternate light sources" are exceptionally bright (such as xenon arc) white light sources. White light is a mixture of wavelengths between about 300 and 800 nm on the electromagnetic spectrum. Alternate light sources are supplied with colored filters, which serve to filter the source light so that the developed latent print can be viewed with light of a narrow wavelength range. The filter is selected to induce illumination at the wavelength known to excite the chemical compounds used in the latent development procedure. Other filters may be used to view the luminescence resulting from the alternate light source excitation. The luminesced light will be of longer wavelength than the excitation, and the filter may provide sufficient contrast between the background and the luminescence of the latent print ridges to give a useful ridge detail image that can then be photographed and used for comparison. A laser is also a very high intensity light source, but it emits light of a single wavelength (monochromatic light). There are several commercially available lasers. Methods have been developed over the years to take advantage of the excitation wavelengths afforded by these lasers. The undisputed primary developer of laser methods for latent fingerprints has been Dr. E. Roland Menzel of

Texas Tech University. The most recent development in this context is development of luminescent nanoparticles for latent fingerprint enhancement. Refer to the appendices of this volume for more background on the electromagnetic spectrum, and the concepts of light absorption, luminescence and fluorescence.

Bloody Fingerprints and Other Special Situations

Latent fingerprints in blood or on certain unusual types of surfaces require special techniques, enhancement modalities or approaches.

Bloody fingerprints are almost by definition not "latent." Some type of pattern is visible in the blood, and investigators may recognize that the pattern is probably a fingerprint (or palm or foot print), but that the ridge characteristics are not sufficiently well defined to make the print suitable for comparison. In these circumstances, the use of a bloody fingerprint enhancement reagent might be considered. In this context, an investigator needs to think about whether it will be necessary to attempt to do DNA profiling of the blood that is forming the apparent ridge patterns. There is substantial published evidence that latent fingerprint chemical enhancement procedures, including those designed for bloody prints, do *not* interfere with subsequent DNA profiling by PCR for the now-universal (at least in the United States) CODIS core STR loci.

Bloody fingerprint enhancement reagents can be sprayed or squirted from a wash bottle. Some of the formulas result in a reagent that is not very stable in solution and, thus, has almost no shelf life. Therefore, they need to be prepared shortly before use. Further, there are serious chemical hazards associated with some of the ingredients, so some training is required in the proper preparation and use of these reagents. Many bloody fingerprint enhancement reagents are based on the "peroxidase reaction" chemicals (phenolphthalin, leucomalachite green, tetramethylbenzidine, etc.) commonly used as presumptive tests for blood (Chapter 14), tetramethylbenzidine along with hydrogen or sodium peroxide probably being the most common. There are also formulas and procedures based on general protein staining dyes such as Amido Black 10B and Coomassie Blue R250. They are generally less hazardous and easier to use than the "peroxidase reaction" chemicals.

Another special situation involves fingerprints deposited on tape, especially on the adhesive surface. Techniques involving staining with crystal violet were traditionally used for this kind of latent impression. Today, the most commonly used method for developing this kind of impression is a material called "sticky side powder." This so-called powder is actually composed of lycopodium (a plant) pollen mixed with a detergent and water. The sticky side powder slurry is painted onto the adhesive side of the tape with a brush, and the tape then rinsed off with water. The process can be repeated until the desired contrast has been achieved, and it often yields identifiable prints. Sticky side power can be used on almost any kind of tape, and works especially well on duct and electrical tape.

Yet another "special situation" deserving brief mention is the development of latent prints on human skin. A variety of different techniques has been tried, but with only modest success and even that in selected cases. A generally useful procedure has not emerged thus far.

Systematic Approaches

This concept is probably followed by most latent fingerprint examiners even if they do not identify it as such. The idea is to apply latent development techniques in a way that maximizes the number of identifiable prints. The least destructive technique is applied first, and techniques are generally applied in a sequence that allows the maximum number to be used if necessary.

Systematic approaches vary according to the surface or substratum on which the latent is located. Porous surfaces, such as paper, for example, call for a different set of techniques applied in a different sequence than nonporous surfaces.

Recognition, Collection, and Preservation of Fingerprint Evidence

Fingerprints are among the best and most probative of all types of physical evidence for associating people with locations or objects. Accordingly, fingerprint evidence should be sought at any type of scene, and particularly at scenes of crimes committed by unknown perpetrators.

Recognition of fingerprint evidence at a scene is not unlike recognition of evidence in general (Chapter 9). It requires training and experience. Once recognized, the location of fingerprint evidence should be appropriately documented by photography, notes, and sketches as appropriate.

As a general rule, objects believed to have latent fingerprints on their surfaces should be collected intact and submitted to the laboratory's latent print section for examination. If collection of the object or surface is impossible or impractical, investigators must consider applying latent development techniques at the scene, or requesting assistance from personnel trained to do so. Sometimes, developed prints may be collected using tape-lift or other methods. Other times, the developed print may have to be carefully photographed because the photograph will be the only permanent record of the evidentiary print. Like other evidence, collected fingerprint evidence such as lifts or photographs should be carefully documented to preserve the chain of custody from the scene to the courtroom.

Fingerprint Identification— The Heart of the Matter

Everything in this chapter so far—indeed, the reason there is a chapter on this subject at all—comes down to the use of fingerprints as a means of identification of persons. As noted earlier, the uniqueness of fingerprints is a matter of common knowledge. Advertisements and commercials commonly use phrases like "… as unique as a fingerprint," etc. Even the term "DNA fingerprints," as undesirable and sometimes misleading as it is, was coined to reflect the notion that a DNA profile might be as individual as a fingerprint. Infrared spectra of pure compounds are sometimes called "chemical fingerprints." But most people have never given much thought to the process by which fingerprint identification is actually done.

David Ashbaugh, among others, has noted that fingerprint individuality, and therefore fingerprint identification, rests on four premises:

1. Friction ridges develop during fetal growth before birth in their definitive form.
2. Friction ridges remain unchanged throughout life with the exception of permanent scars.
3. The friction ridge patterns and their details are unique and not repeated.
4. The ridge patterns vary within certain boundaries, which allows the patterns to be classified.

Fingerprint examiners are generally extensively trained and required to accumulate significant experience before being entrusted with the responsibility of making identifications. Thus, in addition to the general principles and approaches used to make identifications, the knowledge, training and experience of the examiner also comes into play.

In a law enforcement context, identifications are always made by trained, and often certified, examiners. Sometimes, inked prints from a person may be compared with a set of inked prints on file to determine if they came from the same individual. But more commonly, the examiner will be comparing a developed latent print with inked prints from a known person or persons. AFIS searches can quickly narrow down the number of possible matches to a manageable size, but an examiner, not a computer, makes the actual identification.

In the case of latent examinations, perhaps the first issue that comes into play is determining what is called "suitability" of the latent fingerprint for identification. Here, the examiner must decide if sufficient quality and quantity of the ridge detail is present in the latent to make it possible to compare with a known print. This determination also requires training and experience. Once a latent print is evaluated and determined to be suitable for comparison, it will then be compared to the known prints on fingerprint cards. Such knowns might be obtained through an AFIS search, or because certain persons are suspected and their prints are taken or are already on file.

The overall process an examiner uses has been described by Ashbaugh as "**ACE-V**," for the analysis, comparison, and evaluation that comprise the formal process, followed by verification. The examiner must first analyze the latent, determine its proper orientation, decide if there are any color reversals or other unusual circumstances, decide suitability, then proceed to the comparison. Comparison with the known takes place at several levels. The overall pattern and ridge flow (what Ashbaugh calls level I) must be examined. Next, the individual characteristics (minutiae) are compared as to type of features and their locations (what Ashbaugh calls level II). Finally another level of detail (Ashbaugh's level III), consisting of pore shape, locations, numbers, and relationships, and the shape and size of edge features, is compared. Any unexplained difference between the known and latent print during this process would result in the conclusion that the known is *excluded* as a source of the latent. This is one possible outcome of the evaluation decision. If every compared feature is consistent with the known, and there are enough features sufficiently unique when considered as a whole, the examiner concludes an individualization. Since peer review is a feature of most scientific endeavors, Ashbaugh notes that verification of conclusions by an independent examiner is a necessary practice.

There has been considerable discussion in the identification literature about defining the criteria for individualizing fingerprints. For quite a few decades, a "minimum number of points rule" was followed. This "rule" meant that if 12 points were required to make an identification, the examiner had to find at least that many points of comparison before making the identification. Over time, it became clear that such an absolute rule was not a proper basis for making decisions in every situation. The International Association for Identification (IAI), the organization that, by and large, sets peer standards for the fingerprint community, adopted a resolution in 1973 that no minimum number of features is required for making a fingerprint identification. This position was reiterated in a slightly modified form in 1995. Most fingerprint examiners in most countries subscribe to this principle. Discussions of the criteria for making a fingerprint identification lead naturally into discussions of the basis for fingerprint individuality. Although the subject has been discussed in the literature, it is well beyond the scope of this chapter.

Thus, the comparison of a latent with a known inked print could result in one of three possible conclusions: insufficient ridge detail to form a conclusion; exclusion; or individualization.

The process of comparing latent prints to knowns has occasionally resulted in an incorrect identification, even when experienced examiners conducted the comparison. In 2004, suspected terrorists set off bombs on several commuter trains in Madrid, Spain. Spanish police recovered some latent prints from evidence believed to be closely associated with the bombers. Images of the latents were distributed internationally to law enforcement agencies in an attempt to identify them. The FBI was one of the agencies that examined the latents and searched its IAFIS files for potential matches. The FBI eventually identified one of the latents as belonging to a man called Brandon Mayfield. Mayfield lived in Portland, Oregon, and was Muslim, although he was a U.S. citizen and had been in the Army. The fingerprint identification eventually led to Mayfield's arrest. Later, it became clear that the latent print was not Mayfield's but from an Algerian national named Daoud. The incident resulted in an investigation by the U.S. Department of Justice's Office of Inspector General, which issued a report in March,

2006. The FBI has taken steps to change and improve its internal procedures to prevent such an occurrence happening again.

The Fingerprint Identification Profession

In the past several decades, a number of trends have worked to professionalize fingerprint examiners more than they have ever been in the past. People entering the profession have more formal education today than was once true. The extraordinary progress in latent development and recovery methods noted above has demanded that fingerprint examiners understand much more chemistry and physics than ever before. The IAI has been a positive force, through the facilitation of discussions, encouragement of research and scientific approaches, and through its professional publication, the *Journal of Forensic Identification*. Journals are commonly and routinely the primary, peer-reviewed source of original research in scientific fields, and the IAI journal rose to fulfill that role for the fingerprint and identification sciences especially under the editorship of David Grieve. The IAI has also been the primary peer standard-setting group in the science of fingerprints. Today, there is a scientific working group on friction ridge analysis, study, and technology (**SWGFAST**) that started in 1995. Like all technical working groups, its purpose is to arrive at consensus standards for the professional area (or arena). In the 1990s, DNA typing and profiling became a major feature of forensic science. That, coupled with the Supreme Court's *Daubert v. Marion Merrell Dow* et al. decision, has forced many forensic disciplines besides DNA to take a careful look at their underlying assumptions and criteria for evaluating matches and identifications. These trends are likely to continue.

Fingerprint examiners today have a wide range of educational backgrounds. The SWGFAST recommends that trainees hired in 2005 and later should have a 4-year degree. There is also a considerable period of training involved in learning the fingerprint specialty. Individuals may be hired in some agencies, primarily as AFIS technicians, to maintain systems, scan images into the system, and so forth. They do not need the level of training required to be an independent latent print examiner. The IAI has a certification program for latent fingerprint examiners, and more and more employers are requiring that certification as a condition of employment. The certification requires passing a challenging written test, demonstrating competency in latent print comparisons, and then maintaining a record of activity and continuing education during the certification period.

Although fingerprints may be the oldest of the forensic science disciplines, it is still one of the most interesting and exciting. More cases may be solved by fingerprints than by any other single type of physical evidence.

Disclaimer

The opinions and conclusions offered in this chapter are those of the authors, and do not necessarily represent the views or official positions of the U.S. Postal Inspection Service or the University of Illinois. Products and product names mentioned in the chapter are for information purposes, and do not constitute endorsements by the authors or by the organizations that employ them.

Acknowledgments

We gratefully acknowledge the assistance of James Snaidauf, Latent Fingerprint Examiner with the Internal Revenue Service National Forensic Chemistry Center in Chicago and adjunct lecturer in the UIC Forensic Science Program, and of Jennifer Bean, a graduate student in the M.S. program, for help with the 2008 revision.

Questions

1. What is the primary value of fingerprints as evidence?

2. What are biometric identifiers? How are fingerprints related to biometric identifiers?

3. What is friction ridge skin? Where is it found on the human body?

4. Who were Herschel, Faulds, Vucetich, and Henry? Describe their contributions to fingerprint science.

5. Who was Bertillon? Why is he an important figure in the history of fingerprint science?

6. What is AFIS? Indicate what the letters stand for and describe the system and how it helps in fingerprint identification.

7. What are the three main types of fingerprints that can be found at a scene?

8. What is meant by "development" or "enhancement" of a latent fingerprint?

9. What is an example of a physical method for enhancing latent fingerprints?

10. What is an example of a chemical method for enhancing latent fingerprints?

11. Explain how you could use a laser in latent fingerprint enhancement.

12. How can a bloody fingerprint be enhanced?

13. What are systematic approaches to latent fingerprint enhancement?

14. What steps and principles are involved in fingerprint identification?

15. What does "A.C.E.V." mean? What are level I, level II, and level III features? How are they used in fingerprint identification?

References and Further Reading

Alphonse Bertillon and the Anthropometric Method for Criminal Identification: http://en.wikipedia.org/wiki/Alphonse_Bertillon, visited 4/28.2008.

Ashbaugh, D. R., Quantitative-qualitative friction ridge analysis, *An Introduction to Basic and Advanced Ridgeology*, CRC Press, Boca Raton FL, 1999.

Almog, J., Fingerprint development by ninhydrin and its analogues, in Lee, H. C. and Gaensslen, R. E. (Eds.), *Advances in Fingerprint Technology*, 2nd ed., CRC Press, Boca Raton, FL 2001, pp. 177–209.

Berry, J. and Stoney, D. A., History and development of fingerprinting, in Lee, H. C. and Gaensslen, R. E. (Eds.), *Advances in Fingerprint Technology*, 2nd ed., CRC Press, Boca Raton, FL, 2001, pp. 1–40.

Cantu, A. A. and Johnson, J. L., Silver physical development of latent prints, in Lee, H. C. and Gaensslen, R. E. (Eds.), *Advances in Fingerprint Technology*, 2nd ed., CRC Press, Boca Raton, FL, 2001, pp. 241–274.

Cowger, J. F., Friction ridge skin, *Comparison and Identification of Fingerprints*, CRC Press, Boca Raton, FL, 1993.

Federal Bureau of Investigation, *The Science of Fingerprints. Classification and Uses*, U.S. Government Printing Office, Washington, D.C., 1998 (Stock number 027-001-00033-5).

Fingerprint science Web site: http://onin.com/fp/ Maintained by Edward German as a service to the community, visited 4/28/2008.

http://www.biometricgroup.com/, International Biometrics Group homepage, visited 4/28/2008.

http://www.biometrics.org/, The Biometric Consortium home page, visited 4/28/2008.

http://www.biometrics.gov/, A companion web site to www.biometrics.org, provides public information on the biometric activities of the U.S. federal government, visited 4/28/2008.

http://www.biometricscatalog.org/, A U.S. Government-sponsored database of public information about biometric technologies, visited 4/28/2008.

Jain, A. K., Bolle, R., and Pankanti, S. (Eds.), *Biometrics—Personal Identification in Networked Society*, Kluwer Academic Publishers, Boston, 1999.

Jain, A. and Pankanti, S., Automated fingerprint identification and imaging systems, in Lee, H. C. and Gaensslen, R. E. (Eds.), *Advances in Fingerprint Technology*, 2nd ed., CRC Press, Boca Raton, FL, 2001, pp. 275–326.

Lee, H. C. and Gaensslen, R. E., Methods of latent fingerprint development, in Lee, H. C. and Gaensslen, R. E. (Eds.), *Advances in Fingerprint Technology*, 2nd ed., CRC Press, Boca Raton, FL, 2001, pp. 105–175.

Maltoni, D., Maio, D., Jain, A. K., and Prabhakar, S., *Handbook of Fingerprint Recognition*, Springer, New York, 2003.

McRoberts, F. and Possley, M., Report Blasts FBI Lab. Peer Pressure Led to False ID of Madrid Fingerprint, *Chicago Tribune*, November 14, 2004.

Menzel, E. R., Applications of laser technology in latent fingerprint enhancement, in Lee, H. C. and Gaensslen, R. E. (Eds.), *Advances in Fingerprint Technology*, 1st ed., Elsevier, New York, 1991, pp. 135–162.

Menzel, E. R., Fingerprint detection with photoluminescent nanoparticles, in Lee, H. C. and Gaensslen, R. E. (Eds.), *Advances in Fingerprint Technology*, 2nd ed., CRC Press, Boca Raton, FL, 2001, pp. 211–240.

Olsen, R. D. and Lee, H. C., Identification of latent prints, in Lee, H. C. and Gaensslen, R. E. (Eds.), *Advances in Fingerprint Technology*, 2nd ed., CRC Press, Boca Raton, FL, 2001, pp. 41–61.

Ramotowski, R. S., Composition of latent print residue, in Lee, H. C. and Gaensslen, R. E. (Eds.), *Advances in Fingerprint Technology*, 2nd ed., CRC Press, Boca Raton, FL, 2001, pp. 63–104.

Forensic Footwear Evidence

William J. Bodziak

19

Introduction

As persons walk about, their shoes track over a large variety of surfaces, constantly acquiring dust, dirt, residue, grease, oils, blood, and moisture. The shoes then deposit these acquired materials back onto other surfaces they subsequently track over. As a result, they leave a variety of both patent (visible) and latent (invisible) two-dimensional shoe impressions. On softer surfaces, such as sand, soil, or snow, they often will cause a permanent deformation of that surface in the form of a three-dimensional shoe impression.

The direct physical contact between the shoe and the substrate results in a transfer of class and individual characteristics from the shoe to the impressions it leaves. The forensic footwear examiner can examine these class and individual characteristics to determine if a specific item of suspect footwear made the questioned crime scene impression, or if that item of footwear can be eliminated. This process begins with the detection and recovery of the footwear evidence from the scene of the crime, enhancing that evidence if appropriate, producing known impressions of the shoes being examined, and finally comparing the crime scene impressions with the footwear. The final result may necessitate the footwear examiner to produce this evidence and provide his or her opinion in a court proceeding.

Footwear impressions are routinely used to prove a suspect was present at the crime scene. This type of evidence is very valuable and most frequently used in homicides, assaults, robberies, rapes, burglaries, and similar crimes where the proof of an individual's presence is, in itself, incriminating.

Forms of Footwear Impressions

Three-Dimensional Impressions

Three-dimensional impressions are those that remain after a shoe has permanently deformed a surface. This type of impression is typically found on exterior surfaces, such as sand, soil, or snow. Some of these impressions are very shallow while others may be deep. Depending on the composition of the substrate, the amount of moisture, and the presence of contaminants such as sticks, stones, and other debris, the resultant quality of the impression can range from those having great detail to those having little or no value. For instance, an impression in a clay-based soil will normally retain greater detail than an impression in a mixture of coarse sand and small rocks. Likewise, an impression in fresh snow will normally retain greater detail than an impression in wet or old refrozen snow. Three-dimensional impressions that retain sufficient detail can be identified with a specific item of footwear.

Two-Dimensional Impressions

Two-dimensional impressions are those made on nongiving surfaces, such as tile, linoleum, or wood flooring, and also include those made on paper, plastics, doors, carpet, clothing, broken glass, countertops, etc. Any surface that can be stepped on or kicked by an item of footwear is capable of retaining a footwear impression. These impressions vary considerably because of the many combinations of dusts, dirt, soil residues, grime, oily materials, or blood acquired on the shoe's sole, making the methods of recovering and enhancing those impressions more numerous and complex.

Some impressions are highly visible, and others are latent. Shoes may track across a surface that contains dust or residue only to track later across a cleaner surface, depositing those trace materials in the form of an impression. Shoes that are wet or muddy, or that have tracked through blood or other opaque materials, can leave a variety of impressions on most surfaces. Even shoes that are relatively clean and dry can leave their impressions on paper or other surfaces such as glass or countertops that may be coated with polish, wax, grease, film, or grime. In each of these cases, the amount and type of material deposited by the shoe and how that material contrasts with the receiving surface determine how visible the impression is. It is interesting to note that often the less visible impressions actually retain greater detail than impressions that result from heavier deposits of residue, dust,

or blood. Regardless of whether the impressions are full or partial, or heavy or light, the examination results depend on the detail retained in the impression.

Information from Footwear Impressions

Footwear impressions located at a crime scene can provide a variety of information that assists in the investigation of a crime.

Identification of Footwear

Based on the agreement of both class and individual characteristics, a suspect's shoe may be positively identified as the exact shoe that left one or more impressions at the scene of the crime, thus proving the suspect's presence at the crime scene.

Elimination of Footwear

Based on confirmable differences of class characteristics, shoes may also be eliminated as the cause of an impression at the crime scene. Elimination of a suspect's footwear may be useful in accounting for all footwear impressions at the scene, and in some cases may constitute exculpatory evidence. Random individual characteristics on a shoe can change with wear as old ones are worn away and new ones are acquired. Not all individual characteristics reproduce all of the time in every crime scene impression, and it is normal to find individual characteristics on a shoe that have not been retained in the impressions it leaves. For that reason, changes in a shoe or the absence of random individual characteristics are normally not used to eliminate a shoe.

Participation in the Crime

Footwear impressions found at the crime scene and identified with the shoes of persons who had no legal authority to be there are highly significant. For instance, footwear impressions left on objects such as broken glass inside the point of break-in, on paper items that were removed from a burglarized safe, on items that may have been knocked to the ground during an assault, or in the blood of the victim, not only contribute significantly toward the proof of a suspect's presence at the scene but also in evaluating his or her participation in the crime.

Location of Impressions

Impressions at the point of entry and exit and at other significant locations within the crime scene may provide a link or relationship to the location of other impressions or additional physical evidence.

Rebuttal or Confirmation of Suspects' Alibis

Suspects often admit their presence at a crime scene. The exact location and diligent documentation of shoe impressions may help to prove that the suspects are lying or being truthful about where they have or have not walked.

Determination of Shoe Brand

The brand name and description of the footwear that left the crime scene impression can often be determined through **footwear databases** or by other means. In the United States, the FBI laboratory maintains a footwear database that includes thousands of shoe designs. At no charge to law enforcement agencies the FBI will search crime scene impressions in an attempt to identify the manufacturer or brand of the footwear. This information may contribute toward the identification of the suspect or may be otherwise useful in the investigation.

Linking Scenes of Crime

Databases in some laboratories can store the footwear impressions recovered from various crime scenes, often linking different crime scenes to one another. This is a particularly useful tool in investigations of repetitive crimes such as burglaries.

Determination of Shoe Size

In many cases, if the manufacturer of the footwear is known, an accurate determination of the size of the shoe that made a full or partial footwear impression is possible. In other cases, the dimensions of full or nearly full impressions can allow for a general estimate of the shoe size.

Number of Perpetrators

The location of more than one suspect shoe design at the scene of the crime may provide important information about the number of persons that committed or were present during that crime. Likewise, the absence of more than one set of footwear impressions, under certain circumstances, may indicate only one individual committed the crime.

Association with Other Evidence

The backtracking of footwear impressions from the point of entry, or tracking impressions exiting the scene, can assist in locating discarded weapons or other evidence, and in associating the footwear impressions with tire impressions or other physical evidence.

Gait Characteristics

Gait analysis is used primarily for medical evaluation of persons with walking problems. The measurements of a person's stride, step length, and step width change as he or she walks more slowly or quickly, and as he or she walks over different surfaces. These variables also exist when known standards of a suspect's gait are obtained. Because of these variations, gait characteristics cannot be reliably used as a means of personal identification.

Tracking

Tracking involves following the path of an individual by observing evidence that person has created as he or she passes over various surfaces. That evidence—referred to by trackers as "sign"—includes shoe prints, bare footprints, crushed debris or displaced rocks, and sticks or leaves that may have been stepped on. It is most commonly used for tracking illegal aliens and searching for missing children, but in some instances has been used to track criminals from the scene of the crime. Most trackers in the United States originate from agencies like the U.S. Border Patrol, where they routinely gain training and experience in tracking methods.

Location and Recovery of Footwear Impressions

It is critical that the proper techniques and materials be used to locate, document, and recover footwear evidence from crime scenes. Unfortunately, some investigators still overlook this evidence, failing to look for it aggressively at the crime scene or failing to recover it properly. Success in locating footwear impressions and then recovering the maximum detail from each impression has a direct impact on the usefulness of this evidence and the results of any subsequent forensic examination.

Impressions that shoes leave may be full, but in most cases they are partial in that they do not represent the entire surface of the shoe's outer sole. Some partial impressions represent only a small percentage of the shoe sole that created it. Regardless, and even in the case of small partial impressions, all impressions can potentially contain sufficient detail for a meaningful examination result. It is not possible simply to look at a crime scene impression and determine its value. That impression's value will not be known until it is fully recovered, enhanced, examined, and compared with shoes. Therefore, all questioned footwear impressions should always be recovered from a crime scene.

Most impressions are on floor or other walked upon surfaces, and if the scene is not properly controlled, the shoes or equipment of other individuals can track over this evidence. To prevent this, the scene should be secured as soon as possible. This should include both the interior and exterior perimeters, because this evidence can be both inside and outside of the scene.

Some impressions are obvious and can be seen immediately upon entering the scene. For instance, bloody shoe prints on a light-colored surface next to a homicide victim are hard to overlook. Locating most footwear impressions, however, requires a more deliberate and aggressive effort. Making a slow visual search, followed by darkening the room and searching for impressions with a high-intensity oblique light source, often reveals many impressions that could not otherwise be easily seen. More aggressive techniques, such as searching for impressions with an **electrostatic lifting device**, may also be appropriate in certain areas, depending on the conditions. In addition, any items that have potentially been stepped on, such as pieces of paper on the floor, broken glass, or other surfaces that may have been walked over or kicked, should be closely examined. In most cases, paper items that are stepped on do not reflect visible footwear impressions, but they almost always contain highly detailed latent dust impressions that can be recovered electrostatically. Likewise, broken glass that falls inside of the point of forced entry is often stepped on and normally retains good to excellent impression detail, yet these footwear impressions can be seen only with proper lighting. Exterior surfaces, including the areas near any forced point of entry or any logical exit path, should also be thoroughly searched.

Notes should provide information about the location of all impressions as well as a brief description of each. These notes should be prepared in a way so they, along with the **general crime scene photographs**, can be used to document and reconstruct the scene and the relevance of the evidence. To document footwear impressions, numbered identifiers, such as those depicted in Figure 19.1, should be placed alongside each impression or other items of evidence. Whenever possible, general crime scene photographs should be taken with the numbered or alpha markers in place. The investigator's notes and photographs should reflect these evidence assignments to the respective impressions in any subsequent lifts, casts, or examination photography. In

Figure 19.1 A general crime scene photograph, taken from a medium range, to document the position of shoe prints and other evidence.

Figure 19.2 A general crime scene photograph, taken from close range, to document the evidence items such as this footwear impression next to identifier #3. This type of photograph is for documentation only and is not intended nor satisfactory for a comparison with a suspect's shoes.

this way, a cast #3 can be directly linked to both general crime scene and examination quality photographs of these impressions, as well as the notes regarding those impressions. An example of this is depicted in Figures 19.1 and 19.2. Figure 19.1 is a medium-range photograph of a burglary crime scene at the exterior forced point of entry. It depicts three pieces of evidence: two shoe impressions (3 and 4), and a discarded pry bar (5) and shows their location and relationship to the overall side of the residence and the point of entry at the door. Figure 19.2 is a close-range photograph depicting shoe impression 3. It is not intended for an examination, but simply to show more closely the position and orientation of that impression.

Any impressions at the scene that can be carefully recovered should be taken to the laboratory. This includes paper and broken glass, area rugs and flooring with bloody shoe prints, and similar types of evidence that can be safely removed. If the impression is on an item that cannot be removed from the scene, it should be recovered in a prescribed and proper manner as described below. All impressions, both partial and full, are of potential value and should be recovered.

Footwear impressions that cannot be removed and taken to the laboratory such as on concrete or asphalt, must first be

photographed in a special manner to provide high-quality photographs for a forensic examination. This type of photography is known as **examination quality photography**. Examination quality photographs of impression evidence are taken strictly for a forensic examination with suspected footwear. The impression in the general crime scene photograph in Figure 19.2 was subsequently photographed in this way (Figure 19.3). To take examination quality photographs, the camera should be a professional film or digital SLR camera. It should have a good lens, be capable of manual focus, and be equipped with a detachable flash unit. If using a digital camera it should be one rated minimally at 8 mega pixels. A ruler, used as a scale, should be positioned along side of the impression and should be used in every photograph. It is also important to place the ruler on the same plane (level) as the bottom of three-dimensional impressions. In this way, the ruler will provide a way to enlarge the photograph accurately to a natural size for examination. If the ruler is not used, or if it has not been placed on the same plane as the impression, it will significantly reduce the ability to use the photograph for a forensic examination. In order to hold the camera in the proper position and to maintain focus, the

Figure 19.3   An examination quality photograph of the footwear impression #3, featured in Figures 19.1 and 19.2. This photograph was taken with a ruler placed along side of the impression and with the camera positioned on a tripod directly over top of the impression. By using the scale of the ruler, this type of photograph can be enlarged to a natural size and then compared with suspect's shoes.

camera must be placed on a tripod. The camera and tripod should be positioned directly over the impression so that the impression and ruler fill the entire frame. The camera should be manually focused on the bottom of the impression and not on the ruler™.

Certain three-dimensional impressions, such as those in white sand or snow, are very difficult to photograph with contrast. For impressions in snow, Snow Print Wax™ or a dark colored aerosol paint can be carefully applied lightly at an oblique angle to highlight the ridges or high spots of the impression, thus adding some contrast in the photographs. The left side of Figure 19.4 depicts an impression in snow. On the right is a photograph of the same impression after it was highlighted with the red-colored Snow Print Wax.

Two-Dimensional Impressions

Once the impression has been photographed, two-dimensional impressions, in many cases, can be lifted. Lifting of two-dimensional impressions can improve the visibility and detail of the impression by transferring it to a surface that provides better contrast. It also enables the removal of the impression from the crime scene to the laboratory. There are many methods and materials for lifting. Electrostatic lifting is a method that utilizes a high-voltage power source to create a static electrical charge that enables the transfer of a **dry origin impression** from the surface to a special black lifting film. A person walking across dirty or dusty surfaces who then steps on a cleaner surface, such as paper or a tile floor, will create dry origin impressions on those surfaces. These dry origin impressions do not have a tight bond with the substrate and can be electrostatically lifted. The method of electrostatic lifting is normally used first, particularly when it is not known if the impression was one of dry or **wet origin**. Should the electrostatic lift be unsuccessful, the evidence will not be harmed, and other lifting or enhancement methods can still be used. Figure 19.5 depicts the value of electrostatic lifting. On the top of Figure 19.5 is a paper folder containing dry origin dust impressions. On the bottom is

Figure 19.4 Impressions in snow are difficult to photograph. On the left is an impression as it would appear in a normal examination quality photograph. The impression was then highlighted with a light spray of Snow Print Wax™ and rephotographed. The highlighted impression provides much better contrast and detail for examination. This is an excellent way to document impressions in snow.

Figure 19.5 The top photograph is a high-contrast photograph of shoe impressions on a folder. The bottom is a photograph of the impression on the black lifting film after it has been electrostatically lifted. Electrostatic lifting not only increases contrast but reveals dry origin latent shoe impressions on evidence and floor surfaces that may have been walked over.

the electrostatic lift of those impressions on the black lifting film. The increased contrast provided by the electrostatic lift allows for the visualization of much greater detail, and thus a more productive examination. Electrostatic lifts are fragile. They must be stored properly and must be photographed before they can be fully utilized in an examination.

If an impression will not lift with the electrostatic method, the impression is either a wet origin impression or is composed of other materials that have bonded to the surface. In some cases, fingerprint powder can be used to enhance impressions. If this is successful, the impression should be rephotographed first and then lifted with a gelatin or adhesive lifter or with Mikrosil® casting material. Unlike adhesive and gelatin lifters, Mikrosil will conform to textured or uneven surfaces. It also will often lift the entire powdered impression, making it a very good choice for impressions that have

been developed with fingerprint powder but are still very faint. Whatever lifting material is used, it should be of a color that provides good contrast with the color of the fingerprint powder. Transparent lifters are not recommended because they do not provide sufficient contrast with either an original impression or powdered impression. A white adhesive lift of a black-powdered impression, such as that pictured in Figure 19.6, is an example of a lift that provides excellent contrast. If the powder were gray or silver, a black gelatin lift would provide the best contrast.

Three-Dimensional Impressions

All three-dimensional impressions should be cast with **dental stone**. Dental stone, like plaster, is a gypsum product. But dental stones, unlike softer plasters, set much harder, have a higher compressive strength, retain

Figure 19.6 A white adhesive lift is one method of lifting a shoe impression that has been enhanced with black fingerprint powder on a nonporous surface.

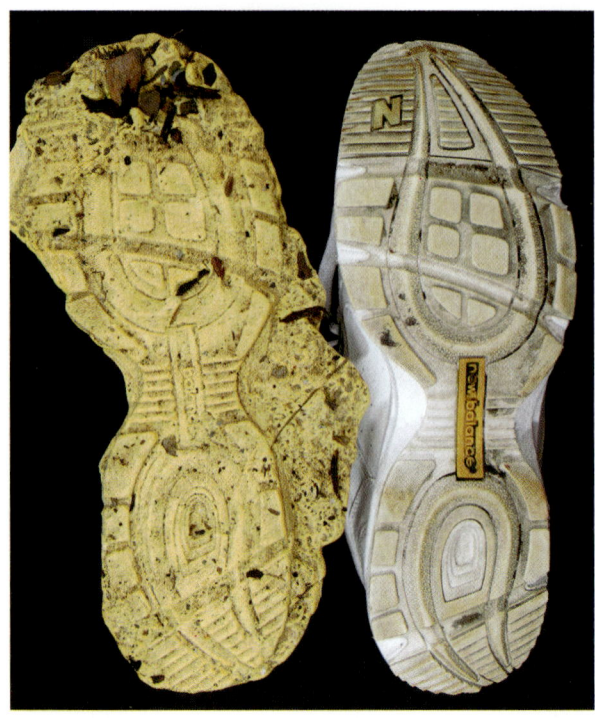

Figure 19.7 On the left is a cast of impression #3 in Figure 19.3 that recovers additional detail and provides a more accurate piece of evidence with regard to size. A cast should always be made of all three dimensional impressions. Alongside of the cast is the suspect's shoe to which it will be compared.

greater detail, and provide a quick and easy way to recover three-dimensional impressions. Dental stones having a compressive strength of around 8000–9000 pounds per square inch are sufficiently hard to be cleaned without loss of detail. Although dental stone can be mixed in a bucket, the more popular and common use for casting footwear impressions involves placing a 2-pound portion into several zip-lock bags to have on hand when needed at a crime scene. The proper amount of water for the 2-pound portion can then be added to the bag at the crime scene and combined within the bag. The exact amount of water will depend on the powder-to-water ratio for each particular dental stone product and can be found on the dental stone box or accompanying information sheet. It is extremely important to mix the water and powder in the zip-lock bag for a minimum of 3 minutes to assure the powder has had ample time to absorb the water. The casting material should not be poured directly into the impression but carefully poured next to the impression in a way so it will naturally flow into the impression. Once certain areas

of the impression are covered with the dental stone, the bag can be moved to facilitate an even distribution of casting material over the entire impression. The dental stone material will harden in approximately 20 to 30 minutes. It will take another 24 to 48 hours for all of the water inside the cast to evaporate after which the cast will become fully hardened. Only then is it safe to clean the cast by immersing it in water and using a soft brush to remove any sand or soil. Figure 19.7 depicts a dental stone cast of the impression featured in Figure 19.3 and a picture of the shoe suspected of making the impression. Impressions in snow can also be cast with special methods and materials.

Enhancement Methods

Many crime scene impressions are indistinct, have poor contrast with the surfaces they are

Figure 19.8   On the left, a faint bloody footwear impression on a piece of fabric provides very little contrast but is an important piece of evidence. This impression was enhanced chemically with Leuco crystal violet (LCV), as shown on the right, which provided excellent enhancement. LCV is used extensively for the enhancement of bloody impressions both at crime scenes and in the laboratory.

on, or are altogether invisible. There are several ways to develop and enhance these impressions, thus providing more detail for later examination. Some of these methods reveal latent footwear impressions that would otherwise go undetected. In other instances, visible, yet faint or indistinct, footwear impressions are improved visually after enhancement.

Forensic Photographic Methods

Specialized lighting and photographic techniques are nondestructive and, therefore, a good first method of enhancement. The use of oblique light, high contrast, ultraviolet, infrared, and other special photographic methods, as well as equipment, such as alternate light sources, can provide increased contrast and visibility of many impressions.

Physical Methods

These primarily include various methods of lifting impressions, powdering impressions with fingerprint powder, and recovering indented footwear impressions on paper items.

Chemical Methods

There are many commonly used methods of chemically treating both bloody footwear impressions and nonbloody impressions. Reagents used to enhance blood impressions include, but are not limited to, Leuco crystal violet, Amido black, Diaminobenzidine (DAB), Luminol, and Fuchsin acid. Some reagents commonly used to enhance residue impressions include, but are not limited to, physical developer (PD) and potassium thiocyanate. The left side of Figure 19.8 depicts a piece of clothing bearing a faint footwear impression in blood. The right side depicts the same impression after enhancement with Leuco crystal violet. Figure 19.9 depicts faint footwear impressions in blood on a magazine. Figure 19.10 depicts the same impression after enhancement with diaminobenzidine.

Digital Methods

Impressions that are photographed or scanned can be digitally enhanced with computer software like Adobe Photoshop®. The software can be used to increase contrast and brightness and can improve the visualization of the

Figure 19.9 A bloody impression on a magazine is hardly recognizable as a shoe print.

Figure 19.10 The same blood impression depicted in Figure 19.9, after it has been chemically enhanced with diaminobenzidine. The impression now reveals excellent detail for comparison, including portions of the name of the shoe brand.

impressions in a number of other ways. Digital methods of enhancement are often used to further enhance impressions that have already been enhanced with photographic, physical, or chemical methods.

Known Shoes and Preparation of Exemplars

Footwear impressions recovered from the crime scene area are intended to be those impressions made by the shoes of the perpetrator or perpetrators. However, impressions left by other persons at the crime scene, or additional impressions of unknown origin, may also be among those recovered. For elimination purposes, it is often necessary to document the designs of the shoes of victims, or of police officers, medical personnel, or other persons who were also present at the scene and whose shoes may have left impressions.

Footwear from Suspected Persons

Shoes seized from a suspect within minutes to an hour after the crime might appropriately include only the shoes that the individual is wearing at the time of arrest. However, most persons own more than one pair of shoes and often possess more than one pair of shoes of their favorite design. For this reason, shoes obtained from a suspect many hours or days after the crime should sensibly include all footwear that a suspect owns and not just the footwear the suspect happens to be wearing at the time of questioning or arrest to avoid missing that evidence. The actual shoes are needed for a forensic examination in order to make known **test impressions** and to allow for the proper evaluation of class and individual characteristics. Photographs of the shoes or test impressions of the shoes made by a third party, in lieu of seizing the actual shoes, are not acceptable for a full and thorough examination.

Elimination Footwear

Elimination footwear includes those worn by medical personnel, police officers, or other innocent persons, who, in addition to the suspect, could possibly have left the recovered impressions. Due to the thousands of footwear designs that exist, crime scene impressions left by the perpetrator will rarely, if ever, be the same design as those worn by other individuals at the crime scene. Because the shoe design(s) worn by the perpetrator(s) is rarely known when the crime scene is processed, it is important to attempt to account for all impressions recovered from the crime scene. By documenting the shoes of others who walked through the crime scene area, their shoe designs can be eliminated as those of the perpetrator(s). Consideration should be given to obtaining a digital photograph or a test impression of the footwear of those other individuals so their shoes can be accounted for. The shoes of such individuals could then be obtained later in the rare event that their shoes were of a similar design as the suspect's shoes.

Known Test Impressions of Footwear

During the examination made between a crime scene impression and suspect's shoes, the examiner makes known impressions of the suspect's footwear to assist in the evaluation and comparison of this evidence. Figure 19.11 depicts a known test impression of a shoe. Methods of making known impressions utilize inks, fingerprint powders, special

Figure 19.11 Several known shoe impressions of the suspect shoe, like the one depicted here, are used to assist in the comparison of the questioned impression with the suspect shoe.

paper, casting materials, and other products that reproduce the characteristics of the shoes being examined. The purpose of making known impressions is not to attempt to recreate the exact circumstances that were present at the scene of the crime, but to produce highly detailed replications of the class and individual characteristics of the shoes. These known impressions are then used to assist the examiner in comparing the shoe directly with the crime scene impressions.

The Examination Process and Conclusions

When shoe soles come into contact with a receiving surface, such as a floor, kicked door, or piece of paper, that contact results in the direct physical transfer of the class and individual characteristics of that footwear. These characteristics can be compared with the respective characteristics of the footwear alleged to have made them. Footwear examinations involve comparisons of both full and partial crime scene impressions with known shoes. The comparison process utilizes side-by-side and superimposition methods, assisted by low magnification, specialized lighting, and the known impression exemplars. The examination of the class characteristics of design, physical size, and general **wear**, and the presence of any individual characteristics, form the basis for the examiner's resulting opinion.

Design

The most obvious class characteristic of any impression is its design or pattern. Thus, it is normally the first area to be compared. There are many thousands of shoe designs, with new ones arriving on the market as older designs are discontinued. Depending on how the shoe has been made, the specific features of that design can even vary slightly among different shoes of that general design. Any design features that are clearly evident in the questioned impression must correspond sufficiently with the suspect shoe in order for the examiner to conclude that the shoe design corresponds. During the examination, if the general design is clearly different, then the suspect shoe can easily be eliminated. If the design of the impression corresponds with the respective area of the suspect shoe, then the examination process continues. Because there are so many design choices the suspect could be wearing, a conclusion that the suspect shoe design corresponds with the crime scene impression design does have some significance.

Physical Size and Shape

The physical size and shape features pertain to the actual dimensional features of the impression and are not a reference to the manufacturer's shoe size. Each shoe design manufactured comes in many different sizes. The soles of these different-sized shoes have different dimensions or proportions of their design throughout the size range. Consequently, the physical size and shape feature of soles is of comparative value during examination of the crime scene impressions. During the examination, the evaluation of the shoe design and of the physical size and shape features of that design are evaluated together. When a crime scene impression corresponds in both design and physical size of that design with the suspect's shoe, the association has a high evidentiary value. This is so because there is an extremely large number of possible design and size combinations, and any particular design–size combination represents an extremely small fraction of one percent of the total overall shoe population.

Wear

When shoes are worn, their sole designs become altered by the abrasive forces created as they make repetitive contact with the ground. Some areas of the sole receive more wear than other areas. As the wear progresses, noticeable areas of greater wear develop. Over time, as long as the shoe continues to be worn, the degree of wear will increase, and the areas of wear will enlarge. The degree of correspondence in wear between a crime scene impression and a perpetrator's shoe recovered

soon after the crime is highly significant. If the suspect's shoes are not recovered until some time later, the extent to which wear can be factored into the examination may be affected. As a routine matter, the examiner should know the dates on which the crime scene impression was made and the suspect's shoes were seized. During the examination, the wear will be examined to assess the *position of wear* and the *degree of wear*. The position of wear is the area or areas on the sole that reflect visible wear. Not all persons wear their shoes the same and the wear areas on the shoes of one person may differ from the wear areas of others. The degree of wear is the extent to which those areas are worn. Some shoes may be virtually unworn, while the design may be completely worn away in certain areas of other shoes. As shoes continue to be worn, the degree of wear will increase. Correspondence of the position and degree of wear of shoes that are obtained shortly after the crime is very significant because it offers an additional means of eliminating other shoes of the same size and design that would not have the same general condition of wear. In some cases, shoes may be new or nearly new and exhibit no visible wear characteristics. In other cases, the detail retained in the questioned impression may not be sufficiently clear to enable an accurate assessment of the condition of wear of the shoe that made it. With these limitations, the relevance of wear may be minimized or incapable of accurate evaluation and comparison. Although general wear can provide a significant contribution to the overall basis and results of the examination, general wear features of footwear, even if matched precisely with the scene impression, are not a basis alone for the identification of the shoe with the impression. In some cases, it is possible to use wear to eliminate a shoe from having made an impression; however, this must be done with caution and is only possible when the questioned impression reflects the wear characteristics clearly.

Individual Characteristics

Individual characteristics are characteristics that have been randomly added to or removed

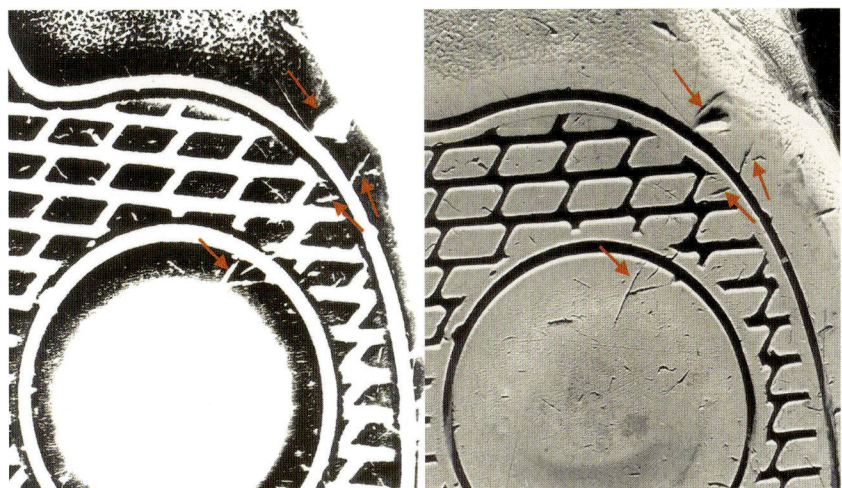

Figure 19.12 A portion of a shoe on the right, with arrows pointing to a few of the many individual identifying characteristics. These individual cuts or tears in the shoe sole are acquired during its use and make this shoe different from all others. The same characteristics can be seen in the test impression that shoe made, pictured on the left. Characteristics like these enable a conclusion that a particular shoe made a particular impression.

from a shoe sole after the shoes are worn. They most commonly include cuts, scratches, gouges, tears, and other physical damage that occurs to the surface of a shoe sole during its use, but they also include materials that may be randomly attached to the shoe sole, such as tar, gum, nails, tacks, and stones wedged between the tread design. Figure 19.12 depicts an impression of an athletic shoe on the left side, with the respective portion of that athletic shoe depicted on the right side. The shoe contains many highly individual cuts and tears. A few of these are indicated with arrows. The randomness of the manner in which each of these features are acquired means none are likely to have the same shape or size nor be in the same position and orientation on any other shoe, much less another shoe of the same design and physical size. It is likely you would not be able to find even one similar-looking individual characteristic in the same position on another shoe of the same design. These randomly acquired features provide a high degree of individuality to an item of footwear and serve to distinguish the soles of one item of footwear from all others.

When a crime scene impression and a shoe sole share sufficient individual characteristics, a positive identification can be made. A positive identification is a conclusion that a particular shoe, and no other shoe, made the crime scene impression. Although positive identifications are normally a result of finding two, three, or many more individual identifying characteristics on a shoe sole, a single characteristic may be all that is needed to identify a shoe with a crime scene impression, provided that the single characteristic was sufficiently clear and reflected sufficient features in common with the scene impression. The number of random individual characteristics is not nearly as important as the specific features, the clarity of those characteristics, and the ability to associate the characteristic's features in the shoe with those in the crime scene impression.

Case Studies

Case 1

During a snowstorm in a large city in the Northeast, a convenience store owner was robbed and murdered in a similar manner to several prior robbery homicides. The assailant quickly entered the store, pointed the gun at the victim's head, and shot him. The assailant then jumped over the counter and emptied the cash register. As the assailant left the store, responding police pursued him on foot. The long chase continued through the blizzard, through buildings and alleys and over flat-topped roofs. Eventually the suspect was captured. He did not have the murder weapon in his possession. Cleverly, the police back-tracked the perpetrator's shoe prints in the snow to an alley where he apparently had hidden momentarily beneath a small covered area. In this covered area they found the gun. Next to the gun were several shoe prints of the same design. All of the shoe prints were photographed and submitted to the laboratory for examination with the suspect's shoes. Examination of the shoe prints resulted in an identification of the shoe prints with the shoes the suspect was wearing when apprehended. This evidence was important because it linked the gun used in the homicide with the suspect. Testimony was given at the trial for this crime and the suspect was found guilty. The gun was subsequently identified with over a dozen other similar robberies and homicides in the same city. The suspect subsequently entered pleas to the other crimes.

Case 2

In a West Coast city, four persons pulled up in front of a bank in a van. One of the individuals remained in the van. The other three entered the bank to commit the robbery. Once inside the bank, all three jumped onto the bank counter, brandishing their guns and demanding that the bank employees give them the bank money. They exited the bank not only with money, but also with the explosive dye pack that was among the stolen money. The dye pack exploded seconds after they entered the van. The van came to a quick stop, and all four bank robbers quickly exited and fled on foot. Two responding police officers gave successful pursuit to two of the suspects. Back at the crime scene, investigators recovered numerous shoe impressions on the bank counter. These were enhanced with black fingerprint powder and lifted with white adhesive lifters. The impressions were searched through the FBI's footwear database. The impressions were all made by the same brand of athletic footwear, although each was a differently designed style of that brand name. Later, these impressions were compared with

the shoes of the two apprehended suspects. One suspect wore a size 5 shoe and the other suspect wore a size 13 shoe. Both the left and right shoes of each suspect were identified with several of the impressions on the top of the bank counter. One suspect's clothing also contained the red dye from the exploded dye pack. Testimony was provided in the bank robbery trial and was instrumental in the jury verdicts of guilty.

Case 3

In one of the islands in the Caribbean, a young man reported that two unknown assailants attacked him and his father at 2 a.m. in his father's office. He advised that he escaped after receiving only minor cuts, but feared for his father's life. The police met him at his father's office and found that the father had been stabbed to death. A pool of blood was present where the father had been stabbed near the front door. Bloody drag marks indicated the father's body had been dragged across the light-colored carpet, down a hallway, and into a back room. The investigators initially thought they could see bloody shoe prints, but were not really sure. They suspected the son whose story was uncertain and constantly changing, and whose wounds were very minor and more than likely self-inflicted. He was also unable to provide any real description of the alleged two assailants that he said had attacked both him and his father. The investigators seized the shoes of the son as well as his father's shoes. The son's shoes were dress shoes with a leather heel and sole. The heel was characterized by its shape and size, and by the cut corner from the inner side of each heel. The sole also had a row of stitching around its perimeter. The carpeting from the entire office was seized and examined approximately two months later. There was no visible evidence of any footwear impressions on the light-colored carpeting, although the pools of blood and drag marks were still visible. The carpeting was chemically treated with Luminol to enhance any faint traces of blood. This enhancement resulted in the detection of 41 footwear impressions. All of these corresponded with either the left or right shoe that the son was wearing. The son's alibi had never included his return to the office after he claimed he fled, so it would not account for any possibility of his shoes acquiring the blood and subsequently depositing all of these impressions. All of the impressions were either within or alongside of the drag marks, linking them to the activity involving the father's body. No other footwear impressions were detected. According to the son, two assailants had killed his father. Yet the absence of other footwear impressions in blood in the office did not support the presence of other assailants and, thus, further challenged the son's alibi. If two other assailants had killed the father and dragged his body though the office, either their shoes or their feet would have left bloody impressions. These did not exist. Faced with this evidence, the son entered a plea of guilty.

Questions

1. How should a scale be properly used in photographing a shoe impression and why is it used?

2. What material is used to cast footwear impressions?

3. Name and briefly describe four methods of enhancing footwear impressions.

4. What areas are examined in a footwear impression comparison?

5. What is the purpose of making known impressions of shoes during an examination?

6. Name three materials that can be used to lift a shoe impression that has been treated and enhanced with fingerprint powder on a nonporous tile floor, and indicate which one will make the most complete lift.

7. Some footwear impressions are latent or hardly visible. Do these impressions contain sufficient detail for examination?

8. What information can a shoe print provide in an investigation? Name at least three.

9. State if a partial footwear impression can be examined and identified. Explain.

10. Are class characteristics or individual characteristics normally used to eliminate a shoe?

Suggested Readings

Abbott, J. R., *Footwear Evidence*, Charles C Thomas, Springfield, IL, 1964.

Bodziak, W. J., *Footwear Impression Evidence: Detection, Recovery and Examination*, 2nd ed., CRC Press, Boca Raton, FL, 2000.

Bodziak, W. J., *Tire Tread and Tire Track Evidence: Recovery and Examination*, CRC Press, Boca Raton, FL 2008.

Cassidy, M. J., *Footwear Identification*, Canadian Government Publishing Center, Ottawa, 1980. (Reprinted by Lightning Powder Company, Jacksonville, FL.).

Giles, E. and Vallandigham, P. H., Height estimation from foot and shoeprint length, *J. Forens. Sci.*, 36, 1134–1151, 1991.

Paine, N., Use of cyanoacrylate fuming and related enhancement techniques to develop shoe impressions on various surfaces, *J. Forens. Ident.*, 48, 585–601, 1998.

Theeuwen, A. B. E. and Limborgh, J. C. M., Enhancement of footwear impressions in blood: comparison of chemical methods for the visualization and enhancement of footwear impressions in blood, *Forens. Sci. Int.*, 95, 133–151, 1998.

Forensic Tire Tread and Tire Track Evidence

William J. Bodziak

20

Introduction

Whatever the illegal activity of today's criminal, it is far more likely than not that a vehicle is used either during the commission of a crime or in traveling to and from the scene of the crime. Although the majority of a vehicle's travel passes over asphalt and concrete surfaces, a vast number of unpaved roads, road shoulders, driveways, off-road surfaces, snow-covered roads, and other run-over objects retain tire impression and tire track information. Investigators should be conscious of the possible presence of this evidence, and when found, it should be properly recovered.

Tire tread impressions reflect the tread design and dimensional features of the individual tires on a vehicle. A tire tread impression is depicted in Figure 20.1. These tread impressions can be compared directly with the tread design and dimension of the tires from a suspect vehicle. If the impression produced sufficient detail, this comparison can result in highly significant conclusions, including positive identifications. With less detail, the comparison can still provide good evidence that the suspect vehicle possessed the same tread design, dimension, and possibly other corresponding features.

Tire tracks are the relative dimensions between two or more tires of a vehicle. Tire tracks reflect general information about the vehicle that left the impressions. By measuring the dimensions of tire tracks at the crime scene, it may be possible to determine or approximate the track width, wheelbase, or turning diameter of the vehicle that created those impressions. Tire tracks can be sometimes be used to profile the type or size of vehicle used and provide other information that can help to include or exclude a suspect vehicle. Some tire tracks are depicted in Figure 20.2.

Figure 20.1 An examination quality photograph of a tire impression, pictured here, depicts the tire tread design. This photograph, taken with a scale placed on the same level as the bottom of the impression, can be enlarged to natural size and compared with the suspect tire and tire impression exemplars from the suspect tire.

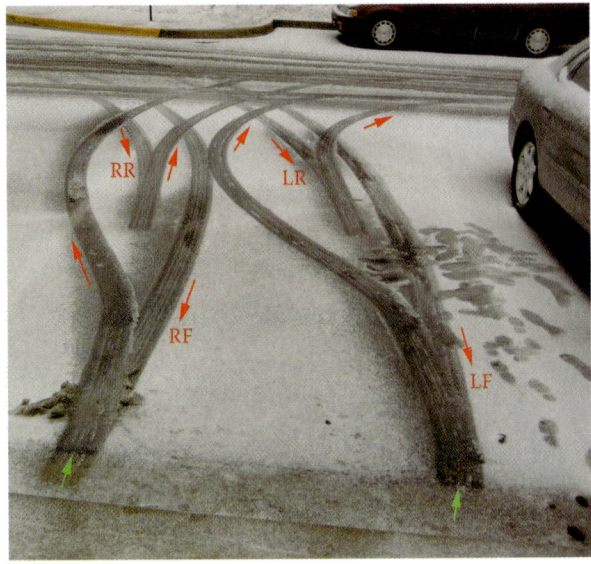

Figure 20.2 A general crime scene photograph of tire tracks in the snow is an important phase of the recovery of tire impressions. The relative position of the impressions, their directionality, and other track features and dimensions must be documented at the scene.

Original Equipment Tires, Replacement Tires, and Tire Construction

There are thousands of tire designs in the world, and most of them come in numerous sizes. Tires that are sold as equipment on new vehicles are known in the automotive industry as **original equipment manufacturer (OEM) tires**; however, most refer to these tires simply as **OE tires**. OE tires of the same size and brand are used on thousands of the same make and model vehicle.

Replacement tires are those purchased to replace worn or damaged tires. They are often not the same design as the OE equipment. The choice of a particular replacement tire design is made by the vehicle's owner and is influenced by factors, such as availability, price, and personal preference. Multiple vehicles with three OE tires of one design that have one replacement tire of a different design occur when a single tire must be replaced and the OE brand and style is either unavailable or not chosen. Instances of multiple vehicles possessing two replacement tires, each of a

different design, is a far more rare occurrence, but a significant one because of the large number of combinations possible. A vehicle with three or even four replacement tires, each of a different design, constitutes a highly unique situation.

Most passenger tires today are of radial ply construction, although bias belted truck tires are still made. Bias belted tires have plies running beneath the tread at an angle (bias) across the tire but with the addition of belts beneath the tread surface. In radial tires, the plies run straight across the tire, from bead to bead. Because of the direction of the radial plies, radial tires are more efficient at reducing the amount of squirm (contraction and expansion) of the tire's surface during use.

Tires are built from a variety of components, including the liner, sidewall rubber, plies, beads, belts, and tread rubber. These components are made of various compounds of unvulcanized rubber, steel, and fabric. They are assembled on a rotating and collapsible drum. The tread rubber, a thick layer of rubber without any design on it, is applied last. The tire at this point, known as a **green tire**, contains no tread design or sidewall information. The green tire is then placed in the tire mold where the tire's components are vulcanized and the tread and sidewall designs are molded into the finished tire.

Tread Nomenclature and Sidewall Information

Figure 20.3 depicts some general tire tread nomenclature. Tire treads are composed of many tread blocks. Some are arranged in ribs around the circumference of the tire, and others are formed to create patterns. These tread blocks are separated by grooves. Some of the grooves may also run circumferentially around the tire, but many grooves run across the tire design. Grooves that run across a tread are called *slots* or *transverse grooves*. If you look closely into the grooves of a tire, you will see some raised areas known as **tread**

Figure 20.3 Some basic terminology of a tire tread.

wear indicators or *wear bars*. Most passenger tire tread blocks also contain very small grooves that are called "sipes."

In addition to the visual information of the tread design, which can be associated with a specific brand name and manufacturer, much information is also molded into the sidewalls of the tire. Each tire has two sidewalls. The outer sidewall or *label side* is the side that a whitewall or raised white lettering is on. This is the side intended to face outward on the car. The inner sidewall, known as the *serial side* of the tire, is the side that is normally not visible. Some of the sidewall information is of importance to the investigator or examiner and should always be noted when examining a tire. The tire brand and style name, such as Michelin XM+S 244, and the size of the tire, such as P 195 75/R15, are normally on both sides of the tire and should be noted first. The serial side traditionally contained the Department of Transportation (DOT) number and the mold numbers; however, tires are now required to have the DOT information on both sidewalls. The **DOT number** begins with the letters "DOT" and contains important information. A DOT number is depicted in Figure 20.4. The two digits (letters or letter and number) that follow the DOT prefix, such as AP, are symbols for the manufacturer and plant code. The plant code indicates the exact location where the tire was made. The Tire Guides, Inc., publication *Who Makes It and Where* lists plant codes and shows that a tire with a plant code of AP was manufactured at the Uniroyal Goodrich Tire Manufacture in Ardmore, Oklahoma. The final three or four digits of the DOT number, which are numbers, are also important since they provide the week and year during which the specific tire

Figure 20.4 The Department of Transportation number, known as the DOT number, includes information that tells where and when that tire was manufactured.

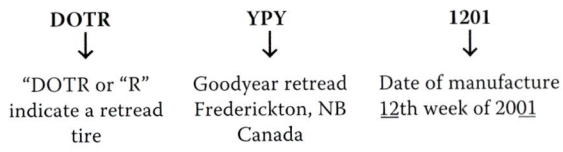

DOTR	YPY	1201
↓	↓	↓
"DOTR or "R" indicate a retread tire	Goodyear retread Frederickton, NB Canada	Date of manufacture 12th week of 2001

Figure 20.5 The Department of Transportation requires a different number arrangement on retread tires. It may or may not begin with the letters "DOT" or "DOTR." The letters and numbers indicate the date and the specific location where that tire was manufactured.

was manufactured. In Figure 20.4, the numbers "3704" indicate the tire was made during the 37th week (37) of 2004 (04). Tires made after January 1, 2000, have four numbers in this position, whereas tires made in 1999 or earlier have three numbers. For instance, a tire made during the 49th week of 1999 or 1989 has the numbers 499.

Retread tires have slightly different DOT numbers, such as DOTR YPY 1201. An example of a retread number is featured in Figure 20.5. The retread DOT number should begin with the letters "DOTR" or "R," but in reality this prefix may not always be present. The original DOT number will still appear on the sidewall of the tire, as required in order to know the date of the original tire carcass. DOTR numbers beginning in the year 2000 consist of three letters and four numbers as in the Figure 20.5 example. The three letters "YPY" identify the location and facility where the tire was retread. The Tire Guides, Inc., publication *Who Retreads Tires* lists, by their three-letter code, thousands of facilities that retread tires. The final four numbers identify the week and year the tire was retread. Tires retread in 1999 or earlier have a three-number designation here.

Numerous tire-size designations have been used over the years. A commonly used size designation for passenger tires is the P metric system. A P metric size designation, such as P 195/65 R 15, can be broken down as follows:

P   = passenger tire
195 = approximate section width in millimeters
65 = aspect ratio
R   = radial tire
15 = rim diameter in inches

Included in the tire size is the aspect ratio, which reflects the ratio of the height of a tire to its width. Tires with low aspect ratios, such as 45 or 55, look flatter and proportionally wider. A tire with a higher aspect ratio, such as 70 or 75, looks taller and narrower.

Noise Treatment

As a tire turns under load, the tire tread blocks vibrate and produce harmonics or noise. If a tire's tread blocks were all one size and pitch, the noise would be more than desirable. To reduce this type of noise emitted by the tires, the tire industry has created tire designs that change the size (pitch) of the tread blocks around the tire, thus creating a variety of pitches. This and other engineered factors that help reduce the noise a tire generates are referred to in the industry as **noise treatment**.

Years ago, noise treatment of a tire was simpler and might only have involved the creation of three sizes of tread blocks, i.e., small (S), medium (M), and large (L). Their sequence around the circumference of the tire was simply S, M, L, S, M. L, S, M, L, and so forth. Later the sequences became more random such as S, L, M, S, S, L, M, M, L, M, L, S, and continuing with a similarly random order of the three sizes around the full circumference of the tire. Figure 20.6 depicts a basic example of noise treatment using a mixture of S, M, and L tread block sizes. The difference in size is visually obvious. You can look at virtually any passenger or light truck tire made today and see how the tread block sizes vary around the circumference of a tire. The modern tire is developed with sophisticated computers that are capable of creating complex tire designs and noise treatments. The mixed arrangement and sequence of varied tread block sizes is far more complex, to the degree that a sizeable segment of a tire impression will only represent one nonrepeating segment of this tire. An example of a more complex design is represented in Figure 20.7. This diagram depicts the varied arrangement of four sizes of tread blocks around the circumference of the tire.

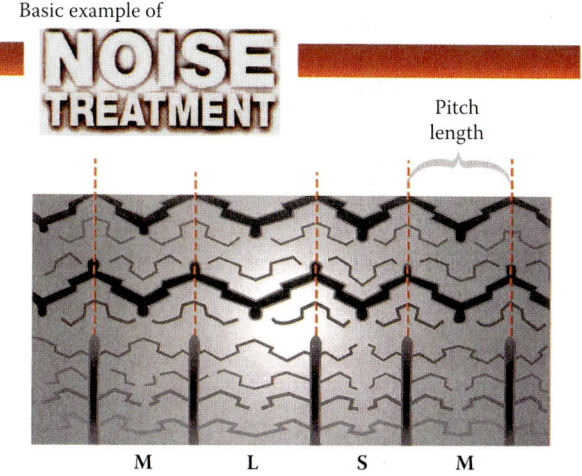

Basic example of

Pitch length

M   L   S   M

Figure 20.6 The tread blocks around most tires are a mixed arrangement of various pitch sizes. This helps reduce tire noise at higher speeds. The industry's method of doing this is known as noise treatment. In this example, the tire has a mixture of tread blocks that vary in their pitch length, and are either small (S), medium (M), or large (L).

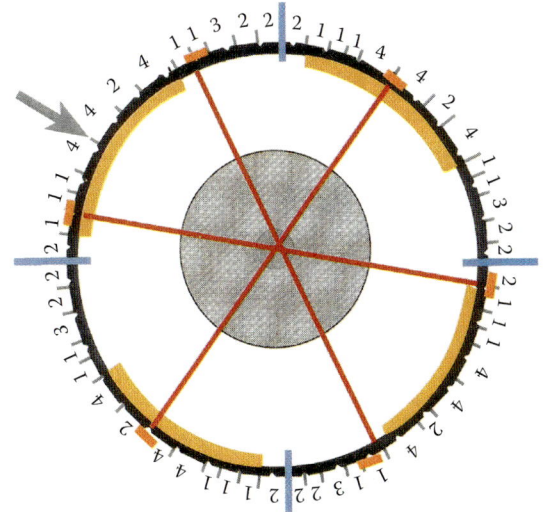

Figure 20.7 A diagram that shows a more complex arrangement of four tread block pitch sizes. In this example, the arrangement repeats four times (in between the blue lines). The red lines indicate the position of the tread wear indicators. Note that even though the noise treatment repeats four times, in each case, the tread wear indicators intersect that noise treatment at a different location. This is extremely important to understand during examination.

Forensic tire examiners must understand the concept of noise treatment and include it in the evaluation of the dimensional aspects of a tire impression as compared to a tire. The noise treatment permits the examiner to find the exact location(s) on a suspected tire or tires of the same design and general tread dimension that potentially made the crime scene impression, improving the quality and accuracy of the full-tread dimensional analysis. Noise treatment is usually different on opposite sides of the tire, which can be helpful in certain cases where the direction of the vehicle is at issue. Off-the-road tires and retread designs often do not have noise treatment.

Tread Wear Indicators (Wear Bars)

A tread wear indicator, also known as a wear bar, is a raised rubber bar that is 2/32 inch above the base of the tire grooves. A tread wear indicator is depicted in Figure 20.3. The Department of Transportation requires that all tires over 12 inches in diameter contain a minimum of six tread wear indicators around the circumference of a tire. As the tire tread wears down to the height of 2/32 inch, the wear bars become very noticeable. The purpose is to indicate to the car's owner that the tire should be replaced. Tread wear indicators will only record in a two-dimensional impression after the tire tread has worn down to the remaining 2/32-inch depth. Tread wear indicators can be retained in three-dimensional impressions, regardless of the condition of the tire, as long as the impression is sufficiently deep to record them.

Tread wear indicators are added to the tire design in different locations relative to the complex and varied noise treatment. Because of this, when tread wear indicators are present in a tire impression, the position of that tread wear indicator, combined with the position of the noise treatment, can offer valuable information during any subsequent examination of tires.

To illustrate the importance of this, Figure 20.7 depicts the positions where the six tread wear indicators occur in the tread design. The four orange sections of Figure 20.7 represent four examples of portions of the tire that might leave an impression or that might be recovered at the crime scene. For illustrative purposes, these four areas contain the same sequence of noise treatment. One of these areas does not contain any tread wear indicator. Three of the sections do contain a tread wear indicator; however, each of those three tread wear indicators is in a different portion of the noise treatment. Thus tread wear indicators can be used in combination with the noise treatment to locate or eliminate the area(s) of the tire that may have made the scene impression.

Retread Tires

In the United States, few passenger car tires are replaced with retread tires. Instead, most retread tires are produced for fleet vehicles such as medium and short haul trucks and school buses, for which the cost difference between new tires and retread tires is substantial.

Two processes are used in the production of retread tires: the mold cure process and the precure process. The mold cure process utilizes strips of raw rubber that are applied to the used tire carcass. The tire carcass is then placed into a mold where the tread design is molded into the new rubber. The precure process utilizes tread rubber already containing the tread design with a system of bonding that attaches it to the original carcass.

Some retread tires reflect valuable individual characteristics that are a product of the retread process and offer the examiner additional information during the examination. The many designs and lower quantities of retread tires may attach additional significance to the association of a retread tire impression with a particular vehicle.

Tire Reference Material and Databases

Thousands of tire designs are made for a variety of vehicles and in a variety of sizes. Since the 1960s, the annually published *Tread Design Guide* by Tire Guides, Inc., has provided photographs of most tires' designs. The publication *Who Makes it and Where* lists where tires are manufactured. Both guides are published yearly in hardcover form. Also available is a more comprehensive version of both guides on a CD-ROM (www.tireguides.com) that includes over 18,000 tire designs. Investigators can use either one of these reference sources to link a tread design from a crime scene to a particular design and brand of tire.

Tire Track Evidence

Tire track width, wheelbase, **turning diameter**, and the relative positions of multiple turning tracks are collectively referred to as *tire track evidence*. Many crime scenes offer little or no tire track information due to limited quantity or quality of that evidence. For instance, at one particular scene, only one tire may leave its impression, and this impression may be relatively straight. At other scenes, a full set of four tire tracks both entering and exiting the crime scene may be retained by the soil. In any case, documenting and measuring the relation of tire tracks to one another can yield valuable information about the vehicle that made the tracks.

Track Width (Stance)

Track width is the measurement made from the center of one wheel or impression to the opposite wheel or impression. Figure 20.8 depicts the measurement of track width relative to a vehicle's tires. In most cases, a vehicle's front and rear track widths are engineered to be slightly different. As a vehicle travels forward in a straight line, the rear tire tracks will track over top of all or most of the tracks left by the front tires. The front wheel measurement is therefore not often clearly present at a crime scene for measurement. If the vehicle is in a turn and the front and rear tracks are separated, the front track width can be measured but will not be accurate. This is because

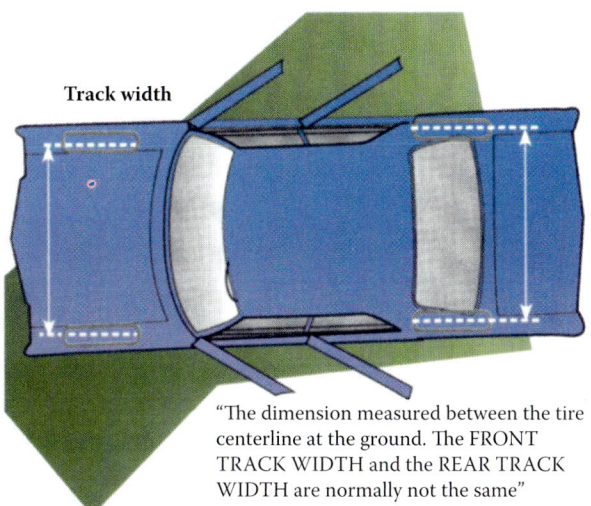

Track width

"The dimension measured between the tire centerline at the ground. The FRONT TRACK WIDTH and the REAR TRACK WIDTH are normally not the same"

Figure 20.8 Track width is the distance between the center of the tire on one side to the center of the tire on the other side. This measurement should be carefully made at the crime scene and cannot be later determined from photographs.

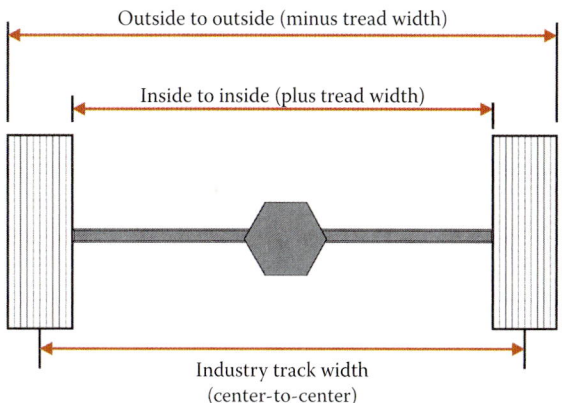

Outside to outside (minus tread width)

Inside to inside (plus tread width)

Industry track width
(center-to-center)

Figure 20.9 Although the industry measurement for track width is from center to center, it is possible to make this measurement in other ways, as illustrated here. The other measurements are a good way to verify the center-to-center measurement or, in some cases, may be the only method to use if the tire impressions are not complete.

as the vehicle enters a turn, the track width between the front wheels will become narrower and cease to be a reliable measurement. Regardless of traveling straight or in a turn, the rear track width will always record accurately so the most reliable crime scene track width will always be obtained from the rear tire tracks. Track measurement must be made perpendicular to the track. Figure 20.9 depicts the way to record a track width measurement. The industry standard is a center-to-center measurement, as depicted in Figure 20.9, and is the best and most accurate measurement to make. However, in some instances, the left or right tire tracks may be partial or limited, so Figure 20.9 also depicts other ways to make this measurement. With outside-to-outside measurement, the width of the tread must be subtracted from the measurement to result in a center-to-center measurement. With inside-to-inside measurement, the width of the tread must be added to the measurement to equal the center-to-center measurement.

Wheelbase

The **wheelbase** of a vehicle is the measurement between the centers of the hubs of the front wheels to the centers of the hubs of the rear wheels. At a crime scene different points

needed to make this measurement are rarely present. Even under optimal conditions, sufficient detail rarely remains in both the front and rear tire tracks that are necessary for this measurement. In some instances, a vehicle parked in snow, or a vehicle parked briefly during a light rain, will leave four patches that mark the bottoms of the four tires. In other instances, tracks in snow or soil that involve turns and changes in direction may enable an accurate measurement.

Turning Diameter

The **turning diameter** of a vehicle is the diameter of the circle a vehicle makes when its steering wheel is fully turned. The measurement pertains to the tracks of the front wheels only, since they are the turning wheels. Some vehicles are capable of smaller turning diameters while others are limited to a larger turning diameter. In general, smaller cars have a much smaller turning diameter than larger cars. The turning diameter and formula for calculating that diameter is illustrated in Figure 20.10. The measurements are made from two selective points, x and x′, on the outer margin of the tire track of the outside front tire and the distance (A) between the outer margin and a point midway between x and

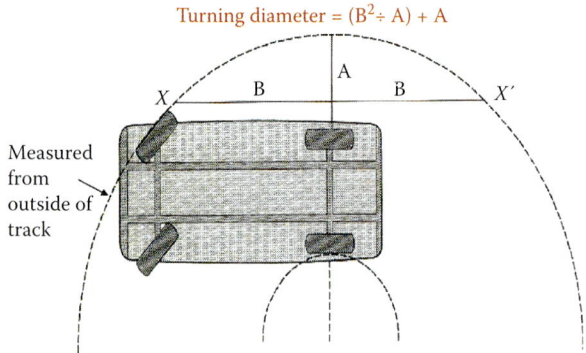

Turning diameter = $(B^2 \div A) + A$

Figure 20.10 Every vehicle has a minimum turning diameter. Some vehicles can turn very sharply in small circles, while other vehicles make much wider turns. This diagram depicts how to measure and calculate the turning diameter if turning tire tracks are present at the crime scene.

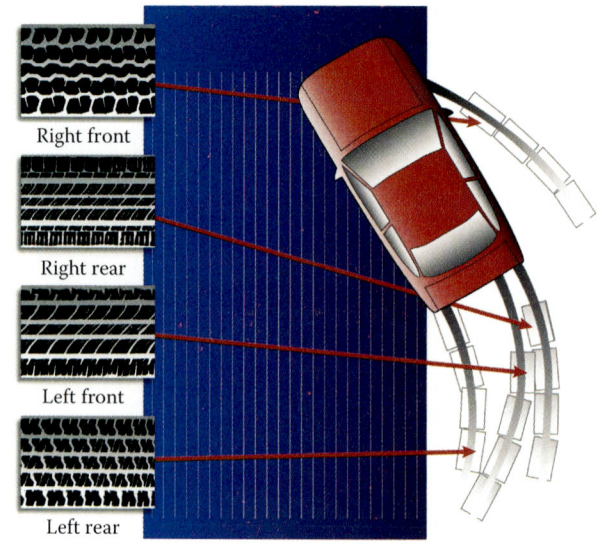

Right front

Right rear

Left front

Left rear

Figure 20.11 Vehicles sometimes have more than one tire tread design, or the tires are in different stages of wear. This information needs to be accurately documented at the crime scene.

x′. The formula can then be used to calculate the diameter of the turn. Vehicles normally increase their amount of turn while entering a turn and decrease their amount of turn while leaving a turn. In addition, vehicles often turn at less than their minimum turning diameter. So any turning diameter measured at a crime scene does not necessarily represent a vehicle's minimum turning diameter and can only be used to eliminate any other vehicle that is not capable of turning at least that sharply. Whenever selecting the points of the turning tracks to measure, remember to select those portions that appear to represent the sharpest turn. It is always a good idea to make more than one measurement from different places in the turn to assure the sharpest point is included.

Tire Positions in a Turn

When a vehicle travels in a straight line, the rear tire tracks run almost directly over the tracks of the front tires. For that reason, there are normally only two tracks to measure in a straight-traveling vehicle, i.e., those of the rear tires. When a vehicle is turning, the front and rear tires track separately and the rear tires will track to the inside of the path of the front tires. This is shown in Figure 20.11. This important knowledge is useful in documenting the relative positions of tire impressions at

a crime scene, particularly if the vehicle had different tires of more than one design or tires that were in different conditions of wear.

Too often, tire impressions are photographed, cast, or measured at a crime scene only for it to be discovered later that no one had documented where they came from. Scenes that contain various tread designs are often complicated and sometimes difficult to evaluate at the scene. It is wise to take your time, approach the area cautiously, and take general scene photographs from many angles. After this has been done, additional examination-quality photographs should be taken, followed by the casting of the tread impressions.

General crime scene photographs of the tire tracks and tire impressions, such as those in Figure 20.2, should be made from various angles to document all tire evidence. Diagrams and other written documentation to describe the number of tracks, their relationship to one another and the surrounding area, the track width, the direction of travel, and other pertinent information should supplement the photographs. Additional information, such as the relationship of the tire impressions to shoe impressions or other tire impressions crossing through the scene, should also be documented, both in writing and through photography.

Then, examination-quality photographs of the tire impressions, such as those in Figure 20.1, should be made. The examination-quality photographs are those taken for later detailed examination of the tire tread. These photographs are made in nearly the same manner as for the recovery of footwear impressions. (See Chapter 19 for more on recovery methods.) One difference lies in the occasional need to document a longer tire impression (over 12 inches in length) by making sequentially overlapping photographs. For instance, with longer impressions, several examination-quality photographs may have to be taken in a sequence that not only records each segment of the tire impression, but also allows the overlapping segments to be spliced together later in order to reconstruct the longer impression for examination. One method that can be used to take these sequenced photographs is illustrated in Figure 20.12. These photographs depict three segments photographed in a sequence that can later be enlarged to natural size and pieced together to reconstruct the full tire impression. A long tape measure or yardstick can be placed alongside of the full length of the impression as a reference to help to splice together the series of overlapping photographs of the long impression later. A scale, such as a flat 12-inch ruler, should be placed alongside the length of each segment

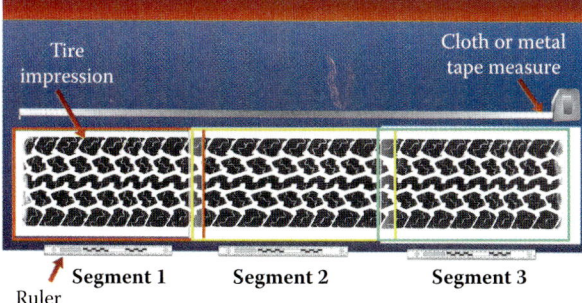

How to photographically document a *Long* tire impression

Segment 1   Segment 2   Segment 3

Figure 20.12 To photographically document long tire impressions, a series of overlapping photographs must be taken. This picture depicts how a 3-foot impression could be photographed with three overlapping photographs. The photographs, once enlarged to natural size, could then be spliced together to recreate the crime scene impression.

of the impression as it is photographed. It is very important that the ruler be placed on the same plane as the bottom of each segment of the impression. After the segment is photographed, the 12-inch ruler can be moved to the next segment being photographed. No ruler or scale should be placed within the impression or across the impression. The camera should be placed on a tripod and adjusted so that an approximately 14–16-inch segment of the impression is photographed each time. Only the center 12 inches will be used. After each segment is photographed, the camera and tripod should be advanced approximately 12 inches along the length of the impression where the next photograph will be taken. The second photograph will overlap the first photograph, and the third photograph will overlap the second, and so forth. This process should be repeated until the full length of the impression is photographed. With this method, the sequential segments can each be enlarged to natural size and then can be assembled to recreate the full impression.

Casting a long tire impression is extremely important. A cast will provide an examiner with more information and will likely result in a far better examination result. Casting a long tire impression may present some challenges not associated with the casting of smaller footwear impressions. To allow the examination to utilize the noise treatment of a tire fully, it is essential to cast long sections of each tire impression being recovered. Although it may not be possible to cast the full length of every tire impression at a scene, any impression that is 4 feet in length or smaller should always be cast in its entirety. Making casts of larger impressions are less practical as they are very heavy, difficult to manage and require excessive amounts of casting materials. For instance, a tire impression that is 10 feet in length cannot be cast in one piece. However, at least one 3- to 4-foot section of the portion of that impression that appears to contain the best detail should be cast. In addition, a 3- to 4-foot section of the best-detailed portion of any other impressions at the scene should also be made. A 3-foot cast of an average tire impression will require approximately 15 to 25 pounds of dental stone depending on the tire tread width and the depth of the impression. This can be easily mixed in a bucket after adding the dental stone to the proper quantity of water. The mixture then can be poured carefully into the impression. After the cast has been poured and begins to harden, information about the impression, such as the impression number and the direction of travel if known, should be written on the back of the cast. Before the casts of the tire impressions are lifted, they should all be photographed in order to verify their relative position to one another and to the overall scene.

Figure 20.13 illustrates a cast of a tire impression. A transparent known impression of the suspect tire has been placed over it to show the correlation of the design and dimensional features. More often than not, a cast of a tire impression will provide the best physical evidence for later comparison with a tire. Casts produce a more accurately scaled representation of the impression than photographs. Casts also recover certain features that a photograph cannot recover, such as the three-dimensional aspects of the impression, its contours, and its uneven qualities and depth. These are very important when making a comparison with the tires. Some deeper impressions may even retain portions of the sidewall detail, so the dental stone should

Figure 20.13 This photograph illustrates a portion of a cast of a tire impression. A transparent inked tire exemplar, taken from a suspect vehicle tire, has been placed over the cast to illustrate how the design, dimension, and noise treatment of the tire correspond with the crime scene cast.

always fill the entire impression in order to recover that detail.

Occasionally a crime scene will involve a vehicle equipped with a **dual tire assembly,** i.e., two tires mounted side by side. When this occurs, both tire impressions should be photographed and cast as a single unit because the relative position of one tire's noise treatment to the noise treatment of the other tire is of extreme importance.

Known Tires and Tire Exemplars

A tire tread examination involves the comparison of a tire impression recovered from the crime scene with tires taken from a known vehicle. Tires from known vehicles can be divided into two categories. First are the tires on a vehicle of a suspect that is believed to have left the impression at the crime scene. Second are those tires from other vehicles, know as elimination vehicles, which include emergency vehicles that may have inadvertently left tire impressions in the crime scene area.

Because there are so many different tire designs, the crime scene impressions will almost always be a different design than any tires from police, emergency, or other elimination vehicles. For this reason, it is normally not necessary to go to great extremes obtaining elimination prints or photographs of these vehicles. A digital photograph or small impression of the tires combined with information identifying the tire brand, the vehicle it belongs to, and the position on the vehicle it came from, are all that is needed. Once documented, these elimination vehicles can always be accessed later should full test impressions of their tires be needed.

If the tire designs of the suspect vehicle are similar to the crime scene impressions, it is necessary to seize those tires for a full examination. When the tires are taken, the position of each tire (left front, right rear, etc.) and the side of the tire that is facing outward on the vehicle should be noted and should be marked directly on the tires before they are removed from the vehicle.

Full circumference test impressions of the tires are required to provide a comparable standard for comparison with a crime scene impression. These are normally made in a prescribed manner by the examiner with inks or powders on long pieces of solid core chart board or on clear film taped over the chart board. The impressions may be taken with the tires mounted on the suspect vehicle, or alternately, on a vehicle that will accommodate that tire size. The tires should always be seized because the actual tires are required in order to perform a full examination.

Tires should not be removed from trucks or trailer rigs that have dual wheel assemblies because of the relationships between the tires mounted on the dual wheel. Demounting those tires would lose their noise treatment position relationship, which, in turn, would lose that important aspect of any dual tire impressions recovered from the crime scene. In order to preserve the position relationship, it is necessary to obtain test impressions directly from any suspect vehicle on which the dual tire assembly is mounted.

Tire Impressions, the Examination Process, and Conclusions

Tire impressions are the resulting transfer of the tread detail of a tire against the substrate. The impressions are a direct one-to-one transfer of their design, size, and other acquired features. These impressions can be three-dimensional impressions in soil, sand, or snow, or can be two-dimensional impressions made after a tire tracks through dust, blood, or other materials. Their features can be physically compared with the tires from a suspected vehicle, and also, if necessary, with the tires from emergency or police vehicles for the purpose of elimination. The examinations routinely allow for elimination of tires as well as positive identification of tires.

The forensic examination of a questioned crime scene tire tread impression begins with a visual comparison of that impression with the tread design of each known suspect or elimination tire. If the design is different, an elimination of that particular tire is made. Should the design be similar, the examination process continues and normally requires at least two staggered full circumference test impressions of each tire being examined. Figure 20.13 depicts a segment of a full circumference inked impression exemplar of a suspect tire that has been found to correspond in tread design and dimension and noise treatment with a cast impression. The known impression of the tire is on clear material, allowing for a direct one-on-one superimposition of the known impression over the questioned cast impression. During this type of comparison, the inked impression is used as a tool to help compare the actual tire with the crime scene impression. The following areas are examined.

Tread Design

Tread designs contain a very specific and detailed arrangement of tread blocks, grooves, and sipes. Some tread designs have similar appearances to others, but to conclude that a design corresponds means that the design is the same and that all of the design portions visible in the questioned impression are also present in the known tire.

Tread Dimension and Noise Treatment

Tread dimension, or size, refers to the specific physical tire tread size, whereas noise treatment is the variance of the pitch (size) of the tread blocks as they are arranged around the circumference of the tire. Most tires come in several sizes. The tread dimension and noise treatment features in the questioned impression should correspond with the known tire. Note that the dimensional features that a tire leaves in some crime scene impressions, particularly those over uneven surfaces, may vary slightly from a known exemplar of that tire taken on a perfectly flat surface. Understanding the noise treatment sequence and other comparison factors enables the proper interpretation of this data during examination.

Wear Features

As tire treads wear, the frictional forces cause erosion of the rubber and ultimately change the visible features of the tread blocks, sipes, and some grooves. For instance, some sipes and even some lateral grooves will partially disappear as the tire tread is worn away. Many of these wear features may be noticeable in the questioned impression and can be compared with the appropriate section(s) of the known tire. Although not in itself a basis for identification, some tires contain highly significant wear features that are recorded in the impressions they leave. The wear features serve to reduce significantly the possible number of other tires of that same design and dimension that would be in the same general condition of wear and that could have potentially left the questioned tire impression.

Random Individual Characteristics

Random, individual characteristics include scratches, cuts, tears, and abrasions that have occurred to a tire in a random manner during its use. They also include the acquisition of stones, glass, nails, and other artifacts that have been either temporarily or permanently embedded themselves in the tread surface in a random manner. These characteristics, if present in both the questioned impression and the known tire, make that tire unique and allow for positive identification of that tire as having made the questioned impression. Figure 20.14 depicts an individual characteristic in the cast crime scene impression and the corresponding individual characteristic on the tire. The characteristic is a small cut that has randomly occurred on the tire during its use. Its size, shape, and orientation, as well as its precise position on the tire, as confirmed by the corresponding noise treatment and other features, make it highly valuable

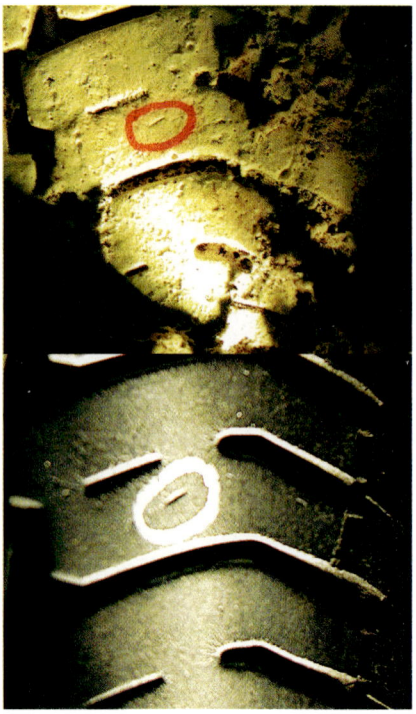

Figure 20.14 In some cases, a tire impression can be positively identified as having been made by a particular tire. The identification requires the presence of individual characteristics that are present both in the crime scene impression and the proper location on the tire. This photograph illustrates an individual characteristic that is visible in the cast above and in the respective area of the tire below.

for purposes of identifying this tire as having made this impression.

Multiple Tires of Different Designs

Whereas most vehicles are equipped with four tires of the same size and design, occasionally a vehicle will have a mixture of tire tread designs. When this occurs, the likelihood of finding another vehicle with the same combination of tire designs is small or nonexistent.

Case Study

In a rural residential neighborhood, a teen-aged girl rode her bicycle home from a friend's house after school. During her trip home, she rode down a dirt road in an isolated area. Along this road she was intentionally struck and forced off the road by a car and knocked off her bike. The assailant placed the stunned girl on the passenger floorboard of his small lime-green vehicle. As he drove away through the adjacent community in which the victim lived, he leaned over to hold her down. Later, some persons in that community recalled seeing a lime-green car with a man leaning over toward the passenger side. They independently provided sketches of the assailant. Many weeks later, the remains of the homicide victim were found tied to a small tree.

At the abduction scene, tire tracks were present where the perpetrator veered off the main path of the dirt road over to a softer shoulder next to the area where the car struck the victim and knocked her off her bicycle. The shoulder was a mixture of soil and rocks, but the detail of a full tire impression was present for several feet. Investigators photographically recovered the impression.

In time, the suspect was identified when someone recognized the artist sketches produced from the witnesses that lived in the victim's neighborhood. His car contained four Yokahama tires and a spare. Some of the tires, including the spare tire in the trunk and the right front tire, reflected some very unusual wear characteristics due to mechanical problems with his vehicle. The right side of the vehicle also contained some rubber scrape marks similar in height to the rubber handle grips of the victim's bicycle. Comparison of the tires with the tire tread impression from the scene resulted in correspondence of the tire design, dimension, and noise treatment. The tire design and dimension was one that had been used in very limited numbers of vehicles imported several years prior. Normally most of those tires would be worn out, but the suspect's vehicle had not been used for an extended period of that time due to his incarceration in a mental facility. His vehicle, therefore, was still equipped with the original tires, which were no longer commonly found. The position and degree of wear also corresponded. Even more significant was the fact that the wear extended around the edge of the shoulder of the tire, beyond where it would normally occur. This unusual wear was found to be a result of a severe mechanical problem with the front right wheel assembly of the suspect's car. The examination results were critical to proving the case at trial.

Questions

1. What four areas of the tread are involved in a forensic comparison?

2. What is "track width"? Explain how it is measured.

3. Differentiate between a "tire impression" and "tire track."

4. What does the term "OE" tires mean, and what are "replacement tires"?

5. What is the "DOT" number on a tire, and what are two important pieces of information that it contains?

6. What does the term "noise treatment" refer to?

7. What are "tread wear indicators," and how often must they appear around a tire?

8. What is the Tread Design Guide, and how might it assist in an investigation with tire tread evidence?

9. Should a cast be made of a three-dimensional tire impression, and if so, why?

10. Explain why it is important to document the positions of tire impressions at a crime scene, both photographically and otherwise.

References and Suggested Readings

Bessman, C. W. and Schmeiser, A., Survey of tire tread design and tire size as mounted on vehicles in central Iowa, *J. Forens. Ident.*, 51, 587–596, 2001.

Bodziak, W. J. *Tire Tread and Tire Track Evidence: Recovery and Examination*, CRC Press, Boca Raton, FL, 2008.

Bodziak, W. J., Shoe and tire impression evidence, *FBI Law Enforce. Bull.*, 53, 2–12, 1984.

Bodziak, W. J., Some methods for taking two-dimensional comparison standards of tires, *Forens. Ident.*, 46, 689–701, 1996.

Bodziak, W. J., *Footwear Impression Evidence*, 2nd ed., CRC Press, Boca Raton, FL, 2000.

Kovac, F. J., *Tire Technology*, The Goodyear Tire and Rubber Company, Akron, OH, 1978.

McDonald, P., *Tire Imprint Evidence*, CRC Press, Boca Raton, FL, 1989.

Nause, L., The science of tire impression identification, *R.C.M.P. Gaz.*, 49, 1–25, 1987.

Nause, L., Forensic Tire Impression Identification, Canadian Police Research Centre, Ottawa, Ontario, 2002.

Pearl Communications, Tread Assistant 2004 (CD-ROM), Carlsbad, CA, updated annually, 760-804-7103.

Tire Guides, Inc., Who Retreads Tires, Boca Raton, FL, published annually (www.tireguides.com).

Tire Guides, Inc., Who Makes it and Where, Boca Raton, FL, published annually (www.tireguides.com).

Tire Guides, Inc., Tread Design Guide, Boca Raton, FL, published annually (www.tireguides.com).

Firearm and Tool Mark Examinations

Walter F. Rowe

21

Introduction

Both firearm and tool mark examiners use microscopic comparisons of markings to associate an item of evidence with a particular source. In the case of firearms examinations, the marks made by a firearm on a fired bullet or cartridge casing are used to determine whether the bullet or casing was fired in a particular weapon. In tool mark examinations, the microscopic features of a tool mark are used to determine whether a particular tool made the mark. In addition to microscopic comparisons, firearms examiners conduct other examinations to further the investigation and prosecution of a crime. They examine bullets and cartridge cases to determine the make and model of the firearm that fired them; they test the functioning of firearms submitted to them for examination; they carry out serial-number restorations; and they reconstruct the circumstances of shooting incidents.

Types of Modern Firearms

Before surveying the types of firearms currently being manufactured, some comments on the mechanisms of automatic and semiautomatic firearms are in order. Automatic and **semiautomatic weapons** function in one of three ways: blowback, recoil, or gas piston. In blowback weapons the fired cartridge pushes the **breechblock** backward against a spring; the expended cartridge is extracted by the moving breechblock and ejected from the weapon. The compressed spring then pushes the breechblock forward, removing a cartridge on the magazine and inserting it into the firing chamber. Most **submachine guns** use the blowback operation. In the recoil operating system, the barrel of the weapon and its breechblock recoil a short distance together; the breechblock then unlocks from the barrel and continues to recoil rearward against a spring. The compressed spring returns the breechblock to its original position, loading a fresh cartridge into the firing chamber. In gas-operated weapons, a small amount of the propellant gases passes through a small hole in the barrel into a gas piston. The expanding gas forces the piston to the rear; a rod connects the piston to the breechblock so that the breechblock is also pushed to the rear.

The types of firearms currently available are summarized below.

Handguns

These are weapons designed to be held in and fired with one hand (although accurate shooting requires the use of two hands).

Revolvers

The cartridges are held in firing chambers in a rotating cylinder. Single-action **revolvers** are fired by manually cocking the hammer and then pulling the trigger. Cocking the hammer both rotates the cylinder to place one of the chambers under the hammer and cocks the firing mechanism. Double-action revolvers are fired by a long **trigger pull** that raises the hammer, indexes a firing chamber under the hammer, and then allows the hammer to drop, firing the cartridge.

Semiautomatic Pistols

These are also called autoloaders or self-loaders. Laypersons sometimes incorrectly refer to these weapons as automatic **pistols**. (Figure 21.1 shows a Beretta semiautomatic pistol.) In semiautomatic pistols, the cartridges are

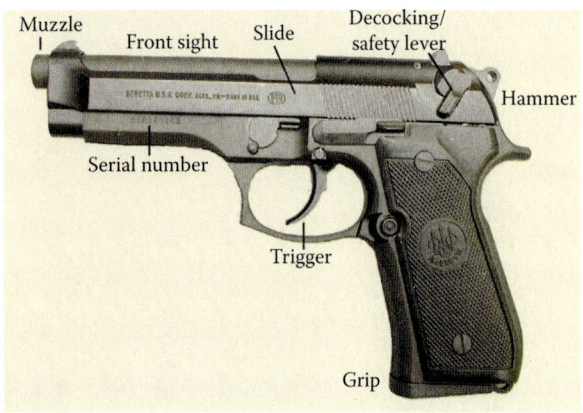

Figure 21.1 Beretta semiautomatic pistol.

held in a magazine (usually removable) and are loaded sequentially into the pistol from the top of the magazine. The typical firing sequence is (1) a cartridge is chambered in the pistol (usually by manually moving the slide or breechblock to the rear, then allowing it to move forward and strip the top cartridge from the magazine) and simultaneously the firing mechanism is cocked; (2) the shooter pulls the trigger, firing the cartridge; (3) some of the energy produced by the fired cartridge is used to move the slide or breechblock to the rear, extracting the spent cartridge from the chamber, ejecting it from the weapon, and cocking the firing mechanism; and (4) a spring compressed by the rearward travel of the slide or breechblock pushes the slide or breechblock forward, stripping an unfired cartridge from the magazine and loading it into the firing chamber.

At the end of this cycle, the pistol is left loaded and cocked; it requires only the pull of the trigger to fire another shot. A true automatic pistol would continue to fire as long as its trigger was depressed and its magazine contained ammunition; however, only one truly automatic pistol, the Mauser Snellfeuerpistole, has been mass-produced. The various "**machine pistols**" manufactured in Nazi Germany during World War II (discussed below) were actually submachine guns.

Rifles

These are rifled firearms designed to be held in two hands when being fired from the shoulder. The different types of **rifles** currently available are listed below.

Lever and Slide Action Magazine Repeaters

In lever action rifles, a lever below the weapon's receiver is dropped to move the breechblock to the rear and cock the firing mechanism. Raising the lever chambers a fresh round in the rifle. In slide (or pump) action rifles, a slide under the barrel is moved to the rear to extract and eject the expended cartridge and cock the firing mechanism; pushing the slide forward chambers a fresh round. The magazine is a tube under the barrel.

Bolt-Action Magazine Repeaters

The two types of bolt actions are straight pull and turn-bolt. In a straight-pull, bolt-action rifle, the expended cartridge is extracted and ejected by drawing the bolt to the rear; moving the bolt forward strips a fresh cartridge from a magazine and inserts it into the rifle's chamber. Movement of the bolt also cocks the rifle. The turn-bolt action operates in a similar fashion except that the handle of the bolt is turned downward to lock the bolt closed. The turn-bolt action was developed to permit the firing of more powerful cartridges. The U.S. Army Springfield 1903 rifle uses the Mauser turn-bolt system.

Single Shot Rifles

These are rifles into which rounds must be loaded one at a time; there is no magazine. Single shot rifles use rolling block, lever action, or bolt action.

Semiautomatic Rifles

Squeezing the trigger of a semiautomatic rifle fires one round. Some of the energy of the discharge is used to extract and eject the expended cartridge, load a fresh cartridge into the firing chamber, and cock the firing mechanism. Box magazines (either built into the weapon or detachable) are most common, although some semiautomatic rifles have drum and helical magazines. Most semiautomatic rifles operate on the gas piston principle, although blowback- and **recoil-operated** semiautomatic rifles are also encountered. The U.S. Army M1 Garand semiautomatic rifle was the first semiautomatic rifle adopted as the service rifle of a major army.

Other Rifles

Double rifles, or combination rifles, are sporting rifles with two rifled barrels or one rifled barrel and one shotgun barrel. The barrels may be over and under or side-by-side. Break-top actions are the most common.

Shotguns

These are smoothbore firearms designed like rifles to be held in two hands when fired from the shoulder. As will be discussed below, shotguns are commonly used to fire pellet loads rather than single projectiles. The different types of shotguns currently available are discussed below.

Slide and Lever Action

The shotgun is loaded and cocked by pumping a slide or lowering and raising a lever. The shotshells are loaded into a tubular magazine under the barrel.

Double-Barreled Shotguns

These are available in two configurations: over/under shotguns, in which the two barrels are aligned one above the other, and side-by-side shotguns. Double-barreled shotguns break open to eject fired shotshells and each barrel is loaded separately. In sporting double-action shotguns, the two barrels often have different **chokes**.

Bolt-Action and Single Shot Shotguns

Bolt-action shotguns use the turn-bolt action described above. Single shot shotguns may use rolling block or break-open actions.

Semiautomatic Shotguns

These use the same mechanisms as semiautomatic rifles.

Automatic Weapons

These are weapons that will continue to fire as long as the trigger is depressed and ammunition is available. A wide variety of **automatic weapon**s are currently available.

Machine Guns

These are military weapons that fire from a mount (e.g., a tripod) or have a bipod attached to the barrel to support the weapon when it is fired. **Machine guns** may also be mounted on vehicles (e.g., tanks and armored personnel carriers) or aircraft (e.g., military helicopters and fighter planes). The ammunition may be fed from a detachable drum or box magazine; however, machine guns are more commonly fed by a fabric belt or one composed of metal links. During World War I, ground forces used heavy machine guns and light machine guns. Heavy machine guns were mounted on tripods, wheels, or sleds; these weapons typically fire belted ammunition. Light machine guns were developed to give assault troops automatic supporting fire. These were usually equipped with bipod mounts and box magazines. Between the World Wars the German army rethought its tactics and developed the concept of the general purpose machine gun. Equipped with a bipod, it would be light enough to be carried and fired by one man. Mounted on a tripod, it also could be used in the role of a heavy machine gun. The German MG42 was the most widely used general purpose machine gun in World War II. After that war, the other major armies of the world adopted the same concept. The M60 is the U.S. Army's general purpose machine gun.

Automatic Rifles

These are actually light machine guns that feed ammunition from a detachable box or drum magazine. The Browning **automatic rifle** (BAR), used by the U.S. Army before and during World War II, is an example of a military automatic rifle. Automatic rifles have been replaced in military use by general purpose machine guns.

Submachine Guns

These are automatic weapons that fire pistol ammunition fed from detachable drum or box magazines. Submachine guns were developed toward the end of World War I to give assaulting infantry close-support automatic weapons. The famous Thompson .45-caliber submachine gun was developed at the end of World War I, but is better known through its use by 1930s gangsters in the United States. Submachine gun designs proliferated during World War II when both Germany and the Soviet Union tried to put as much firepower in the hands of the infantry as possible. Germany produced a whole family of so-called machine pistols, while the Soviet Union armed large infantry formations solely with submachine guns. Submachine guns are accurate only at close range.

Assault Rifles

These are automatic weapons that fire a reduced-charge rifle cartridge. They were first developed by the Germans during World War II to provide airborne troops with an automatic weapon that had better range and accuracy than the submachine gun. Modern armies have embraced the **assault rifle** as the principle weapon of the infantryman. The U.S. Army's M16 and the Soviet Kalashnikov AK47 are familiar examples of assault rifles.

Manufacture of Firearms

Firearm barrels can be rifled in a number of ways. A hook or **scrape cutter** can be used to shave away metal from the interior of the barrel to create grooves. The barrel is prepared for **rifling** by drilling out a metal bar and then reaming the interior of the barrel to smooth it. The barrel is placed on a frame, called a rifling bench, that rotates the barrel as the cutting tool is drawn through it. Hook cutters cut one groove of the rifling at a time; a scrape cutter that has cutting surfaces on both sides can cut two grooves at a time. After the rifling has been cut, the interior of the barrel may be polished by lapping: a soft metal plug is cast inside the barrel and then pushed up and down with rouge (a powder used to polish metal) to remove imperfections. Hook and scrape cutters are not the most efficient rifling methods for the mass production of firearms.

Most mass-produced firearms are rifled by broaching, **swaging**, hammer-forging, or **electrochemical etching**. When a barrel is broached, it is first drilled out to approximately the desired finished bore diameter and reamed to remove drill marks; then a gang **broach** (a long rod having a series of circular cutting tools on it) is forced by hydraulic pressure through the barrel. Each successive cutter on the broach cuts each grove of the rifling a little deeper until the last cutter reaches the desired groove depth. A single pass of the gang broach is sufficient to rifle a gun barrel.

When a barrel is rifled by swaging, it is drilled out to a diameter less than the desired final bore diameter, reamed to remove the drill marks, and then swaged. A rifling button is forced through the barrel, simultaneously engraving the desired rifling pattern on the inside of the barrel, expanding the barrel to the desired bore diameter, and cold-working the interior of the barrel. Cold-working hardens the interior of the barrel and permits the use of cheaper mild steel. Many inexpensive small-bore weapons are rifled by swaging.

A hammer-forged barrel is produced by first drilling out steel bar stock to a diameter larger than the desired finished bore diameter. The barrel is slipped over a hardened steel mandrel on whose surface is a negative impression of the desired rifling pattern. The barrel is hammered down on the mandrel; the mandrel is then knocked out of the barrel. **Hammer forging** is used to produce barrels with polygonal rifling (e.g., those of Glock semiautomatic pistols).

Electrochemical etching is the newest method used to rifle barrels. In this process a piece of steel bar stock is drilled out, and the interior surface of the barrel is coated with strips of chemically resistant polymer. The polymer layers correspond to the lands of the rifling. A chemical etching solution is placed in the barrel and an electrode is inserted. A voltage is applied between the barrel and the electrode so that the barrel becomes the anode in an electrochemical cell. The passage of an electrical current through the cell eats away the metal on the interior of the barrel where there is no protective polymer layer, forming the grooves of the rifling.

Rifling produced by **hook cutting**, scrape cutting, and broaching is often referred to as **cut rifling**. Cut rifling tends to leave more forensically useful marking inside the rifled barrel than do the other rifling methods.

The size of a firearm barrel may be designated in different ways. The most common is to specify its caliber. Caliber is defined as the diameter (in hundredths or thousandths of an inch or in millimeters) of a circle that is tangent with the top of the lands of the rifling. Calibers are merely nominal; the actual bore diameter may be different. For example, a weapon designated as .45 caliber may have an actual bore diameter at the muzzle of .44 inches. The caliber of a weapon can be determined with taper gauges that are inserted into the muzzle or from a sulfur cast of the interior of the barrel. Caliber determinations should be a routine part of the examination of a firearm because the barrel may have been altered from its original caliber (by being bored out), or a barrel with a different caliber may have been fitted to a firearm (such as a .45 caliber barrel mounted on a .38 Smith and Wesson Police Special frame).

A brand new firearm may or may not leave a unique pattern of markings on a fired bullet. Carefully lapped barrels may have too few microscopic imperfections to produce useful patterns of markings on fired bullets. In other cases, a new barrel may have a set of unique irregularities that mark fired bullets. In the 1920s and 1930s, several studies were done on the marks produced by sequentially rifled barrels. These were found to produce unique sets of marks on fired bullets. The best explanation for this observation is that rifling tools wear during the rifling operation so that in effect each barrel is rifled with a "different" tool. A study has also examined the markings made by electrochemically etched barrels and found them to be unique. Of course, once a barrel has been used extensively it will develop its own suite of unique irregularities: in some areas metal will have worn away (erosion) and in others, minute pitting will appear due to propellant combustion residues eating away at the metal of the gun barrel (corrosion).

Cartridge cases will be marked by a number of firearm components: firing pins will mark the primer cap of centerfire cartridges or the rim of the cartridge itself in **rimfire cartridges**; the breechblock or backing plate (in revolvers) may mark the primer cap or even the metal of the casing itself (if the cartridge is powerful enough); in self-loading or automatic weapons the extractors and ejectors will mark the base of the cartridge; and in weapons using magazines, the lips of the magazine may also mark cartridges. Firearms examiners consequently need an understanding of how components, such as firing pins, breechblocks, extractors, and ejectors, are made. Firing pins are finished by hand filing or by turning on a lathe. Breechblocks are hand filed, milled, or turned on a lathe. Extractors and ejectors are finished by hand filing. Wherever finishing operations involve a human operator (filing, milling, or turning on a lathe), the resulting patterns of marks are unique. In some makes of firearms the breechblock is painted so that it will not transfer markings to the primer cap.

The barrels of shotguns may be bored out of bar stock or drawn like tubing. A unique feature of shotgun barrels is the choke in the muzzle. This is commonly a constriction in the muzzle that serves to concentrate the shot pattern. Shotguns may have four degrees of choke: full choke (greatest choke), modified choke, improved cylinder, and cylinder-bore (no choke at all). A few shotguns have been produced with a reverse choke, which is a flaring at the muzzle intended to cause the pellet pattern to spread. The choke of a shotgun barrel may be changed using inserts or adjustable compensators. As a general rule, shotgun barrels do not leave identifiable markings on shot, **wads**, or shot columns. However, if a criminal has sawed off a shotgun's barrel (usually to facilitate concealment), metal burrs may be left at the muzzle; these may mark shot, wads, or shot columns in a unique fashion.

The diameters of shotgun barrels are customarily designated by the gauge of the shotgun. Gauge is defined as the number of spherical lead balls having the diameter of the interior of the barrel that weigh one pound. For example, 12 lead balls having the diameter of the interior of a 12-gauge shotgun barrel (roughly .75 inches) would weigh 1 pound. The only exception to this rule is the .410 gauge, whose actual bore diameter is .410 inches. Rifle calibers were at one time designated using a similar system. Thus, one can still read of 4-bore or 12-bore gauge rifles. These were rifles produced in the 19th century for hunting dangerous game such as elephants.

Firearm Ammunition

Figure 21.2 shows the disassembled components of a typical rifle cartridge and a shotshell. Firearms can fire a variety of projectiles. Bullets may be lead, lead-alloy, semi-jacketed, or full-metal jacket. Lead bullets are soft and readily deformable. Because the surface of a lead bullet can be easily stripped off by the rifling in the gun barrel, lead bullets are used in low-velocity firearms. Because lead is a ductile and malleable metal, lead bullets are easily marked by the rifling; such bullets may also undergo extreme deformation or even fragmentation when they strike a target. Lead alloy bullets contain a small percentage of an alloying element, such as antimony in commercially manufactured bullets or tin in home-made bullets. Lead alloy bullets are harder than lead bullets and are consequently used in weapons having higher muzzle velocities. Lead alloy bullets are sufficiently ductile and malleable to be easily marked by the weapon's rifling; they also readily deform on impact. Lead and lead alloy bullets may be plated with a thin layer of copper (e.g., Luballoy bullets). The copper layer reduces the friction between the bullet and the gun barrel. This layer readily flakes off the surfaces of fired bullets, taking with it the marks made by the rifling.

Semi-jacketed bullets commonly consist of a lead core covered with a thin jacket of brass. The brass typically covers the sides of the bullet, leaving the lead core exposed at the nose (as in hollow point bullets). Soft point bullets are semi-jacketed bullets in which a soft

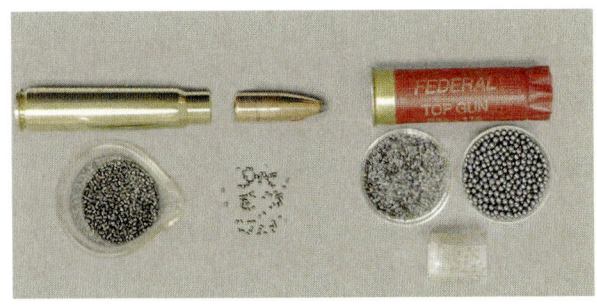

Figure 21.2 Disassembled rifle cartridge and shotshell.

metal plug has been inserted in the nose of the bullet. Soft point bullets are used to hunt large game; the soft metal insert promotes the expansion of the bullet when it enters the target animal. Explosive bullets are produced by inserting percussion caps in the hollowed out noses of hollow point bullets. The jackets of hollow point and soft point bullets may separate from the lead cores when the bullets enter the body of a shooting victim. The treating medical personnel or the forensic pathologist must make every effort to recover the jacket because only the jacket bears markings made by the weapon's rifling. Full metal jacket bullets consist of a lead core covered with a brass jacket. The jacket covers the nose and side of the bullet and occasionally the base as well. Because the jacket of a semi-jacketed or full metal jacket bullet is harder than lead, these bullets are not as well marked by the rifling of the gun barrel as are lead or lead alloy bullets. Moreover, semi-jacketed and full metal jacket bullets are slightly smaller in diameter than lead or lead alloy bullets of the same caliber; they also expand less upon entering the rifling. As a consequence, semijacketed and full metal jacket bullets may not be marked by contact with the bottoms of the grooves of the rifling.

Plastic can also be used to jacket bullets. Teflon coated bullets were developed for police use in barricade situations. The self-lubricating property of Teflon reduces the friction between the bullet and the barrel of the weapon and allows the bullet to reach a very high muzzle velocity. Because of their high velocities, Teflon-jacketed bullets can penetrate automobile engine blocks and also multiple plies of body armor. Nylon-clad bullets are also being marketed. Some European military forces have also produced training ammunition with plastic bullets.

Firearms examiners may also encounter a number of unusual or special purpose bullets. These include frangible bullets, open tubular rounds, and discarding sabot rounds. Frangible bullets consist of iron or copper particles for training, particles simply pressed together or held together with an organic binder. Frangible bullets are used in shooting galleries and also for killing livestock in slaughterhouses. An open tubular round consists of an open tube of copper or brass with a plastic cap at the base to serve as a gas check. Open tubular rounds produce unusual wounding effects and, because of their low mass, can achieve much higher muzzle velocities than conventional bullets. Brass open tubular rounds can pierce bulletproof vests; the Bureau of Alcohol, Tobacco, and Firearms (BATF) was able to prevent the sale of brass open tubular rounds in the United States. Discarding sabot rounds have been used in certain rifle ammunition (e.g., the Remington Accelerator .30-06 cartridge). The discarding sabot projectile consists of a .223-caliber (5.56-millimeter) soft point bullet inserted into a .30-caliber plastic cup or sabot. When this round is fired in a .30-caliber rifle, the plastic sabot acts as a gas check while the projectile is in the gun barrel; then it separates from the .223-caliber bullet. The sabot bears the rifling marks from the gun barrel, while the .223-caliber bullet (which would be recovered from the target) never comes into contact with the rifling. Prefragmented bullets contain metal pellets embedded in plastic: when the bullet hits its target the thin outer jacket disintegrates and the pellets disperse throughout the victim's body. The Glaser Safety Slug and the MagSafe Safety Bullet are examples of prefragmented bullets. These bullets are intended for use in situations where over-penetration of the target cannot be risked (airline hijackings). A variety of solid copper bullets have recently come on the market. They are available in a number of designs: solid pointed nose and round nose bullets, hollow point bullets and hollow point bullets with a central post.

Bullets are produced in a variety of shapes. Wadcutter bullets are flat-nosed, cylindrical bullets used for target shooting. Bullets may have round noses or may be pointed (spitzer bullets) to reduce aerodynamic drag or to facilitate target penetration. Hollow-point bullets have depressions in their noses: hollow points facilitate the expansion of the bullets when they enter tissue. The bases of bullets may be flat or they may be boat-tailed to reduce the turbulent wake of the bullet that contributes to drag. Bullets may have knurled grooves called cannelures engraved around their circumferences. Cannelures may contain lubricant, and the mouth of the cartridge may also be crimped into the bullet's cannelure to hold the bullet in the cartridge. Cartridge casings may also have cannelures. The cannelures in cartridges prevent the bullet from being accidentally pushed into the casing.

Cartridges may be either rimfire or centerfire. Rimfire cartridges have the primer composition in the rolled rim of the cartridge. These cartridges are fired by the weapon's firing pin striking the cartridge rim. A **centerfire cartridge** has a primer cap placed in the center of the cartridge base. In the United States, the Boxer cartridge is popular. This type of cartridge has a central vent hole, which communicates the flash of the exploding primer to the propellant inside the cartridge. The Boxer primer is a small cap made of gilding metal. Inside is a small metal tripod oriented with its apex toward the metal surface of the primer cap. A small speck of primer mixture is placed between the apex of the tripod and the primer cap so that, when the firing pin strikes the primer, the primer mixture is crushed between the primer cap and the tripod. The open side of the primer cap is covered with a thin sheet of treated paper or foil to keep out moisture. The Berdan cartridge and primer are used in Europe. The principal difference between the Boxer and Berdan primer systems is that the Berdan primer does not contain a metal tripod; instead, in the recess for the primer cap, the cartridge has a cone with three vents communicating with the interior of the cartridge.

Centerfire cartridges are produced in a wide range of shapes. These cartridges may be straight-sided or bottle-necked. The cartridges may be belted, rimmed, semirimmed, rimless, or rebated (with the cartridge base slightly recessed from the sides). Cartridge shapes may be divided into 12 types:

Rimmed bottleneck
Rimmed straight
Rimless bottleneck
Rimless straight
Belted bottleneck
Belted straight
Semirimmed bottleneck
Semirimmed straight
Rebated bottleneck
Rebated straight
Rebated belted bottleneck
Rebated belted straight

The shape of the cartridge reflects the shape of the weapon's firing chamber.

Shotguns can fire a variety of projectiles. The most common type of projectile is shot. Shot may be made of lead, lead alloy (chilled shot), or steel. The smaller shot sizes are produced by allowing molten metal to fall through the air down a shot tower. The molten metal separates into spheres of varying diameter that cool and harden as they fall. The hardened shot are then sorted according to size and loaded into shotshells. The larger shot sizes (e.g., 00 buckshot) are individually molded. Modern shotshells are plastic with brass bases. A plastic or fiber disk, called a *wad*, separates the shot in the shotshell from the powder. A second wad may be placed over the shot and the mouth of the shotshell crimped over it to seal the shotshell. The shot may be enclosed in a plastic cup that protects the shot from deformation due to contact with the inside of the shotgun barrel. In some shotshells the over-powder wad and the cup may be combined into a one-piece shot column. Wads and shot columns can act as secondary projectiles and may be recovered from within close-range shotgun wounds; they may also inflict superficial injuries adjacent to shotgun wounds. Shotguns can also fire loads containing a single large-caliber round ball or a single rifled slug. Foster-rifled shotgun slugs are round-nosed conical lead bullets with hollow bases and projecting fins on their sides. The fins are set at an angle so that when a rifled slug is fired it acquires a spin. If the shotgun barrel is choke-bored, the fins are forced into the hollow base as the slug exits the shotgun muzzle. Brenneke rifled slugs also have projecting fins; they are attached to wads or shot columns. The wounding effects of rifled shotgun slugs are similar to those of the minié bullet used in the American Civil War.

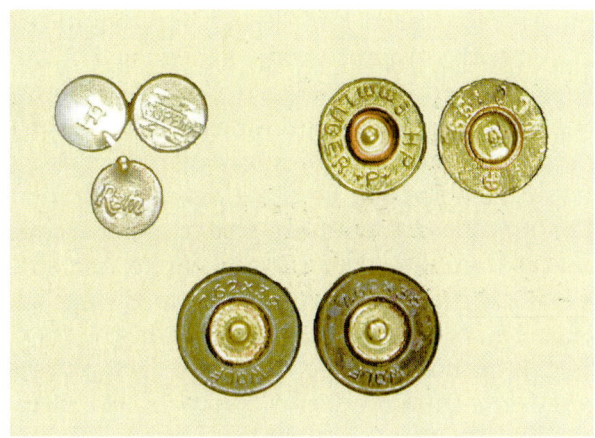

Figure 21.3 Examples of cartridge headstamps.

The base of a cartridge and shotshell (called the *headstamp*) bears information such as the vendor of the ammunition (usually abbreviated, as F for Federal, Rem for Remington, R-P for Remington-Peters, or WIN for Winchester) and the gauge or caliber. The caliber or cartridge type may also be indicated in the headstamp: for example, ".30–06" means that the cartridge is a .30-caliber rimless, bottleneck cartridge introduced in 1906. The headstamp may also consist of a logo (e.g., a diamond on .22-caliber Winchester rimfire cartridges). Figure 21.3 shows a selection of cartridge headstamps.

Smokeless Powder

Modern **smokeless powders** may be single base, double base, or triple base. In single base smokeless powders, nitrocellulose is the only energetic material. Double-base powders contain nitrocellulose and nitroglycerin, while triple base powders contain nitrocellulose, nitroglycerine, and nitroguanidine. Triple-base smokeless powders are used only as propellants in artillery ammunition.

The nitrocellulose used in smokeless powders is produced by the nitration of wood pulp or cotton lint. Gun propellant grades of nitrocellulose have degrees of nitration between 13.15 and 13.25 percent. During the production of nitrocellulose, other characteristics, such as viscosity, particle size, purity, and stability, are carefully controlled. Scrap nitrocellulose and obsolete propellants are also used as sources of nitrocellulose. Obsolete propellant is recycled in this fashion because it is virtually impossible to dispose of such material. It cannot be burned without producing air pollutants, and few landfills will accept quantities of smokeless powder.

Nitroglycerine is manufactured by nitration of natural or synthetic glycerin. Nitroglycerin is considered an energetic plasticizer. It softens the propellant granules, raises their energy content, and reduces their absorption of moisture. Smokeless powders used in small caliber ammunition typically contain a variety of other ingredients: stabilizers, plasticizers, flash suppressants, deterrents, dyes, opacifiers, graphite glaze, and ignition aid coatings. **Stabilizers** react with the acidic breakdown products of nitrocellulose and nitroglycerin (primarily the nitrogen oxides NO and NO_2). Diphenylamine is used as a stabilizer in single-base powder and in double-base powder with a nitroglycerin content of 20 percent or less. Ethyl centralite is used as the stabilizer in double base powders with nitroglycerin contents greater than 20 percent. Typical stabilizer concentrations range from 0.5 to 1.5 percent. Plasticizers soften the propellant granules and reduce the absorption of moisture.

Other plasticizers, such as ethyl centralite, dibutyl phthalate, dinitrotoluene, and triacetin, may also be used. Flash suppressants interrupt the free-radical chain reactions in the muzzle gases. Muzzle gases contain high concentrations of carbon monoxide, which can react with atmospheric oxygen. Low concentrations (0.5 to 2.5 percent) of alkali salts, such as potassium nitrate and potassium sulfate, are used as flash suppressants. Flash suppressants may be present within the powder particles or coated on them; flash suppressants may also be added to the smokeless powder as separate particles.

Deterrents are coatings on propellant grains that reduce their initial burning rate. The reduction in the initial burning rate broadens the pressure peak and increases the muzzle velocity. Deterrent coatings are either of the penetrating type or of the inhibitor ("candy shell") type. Penetrating deterrents include Herkote (Paraplex® G-54 polyester), ethyl or methyl centralite, dibutyl phthalate, and dinitrotoluene. The main inhibiting ("candy shell") deterrent is Vinsol® resin. Deterrent concentrations range from 1 to 10 percent. Dyes are added to powder grains to facilitate brand identification (e.g., Red Dot and Blue Dot smokeless powders). Opacifiers prevent radiant energy from penetrating the surface of the powder grains (and initiating burning within the grains). Carbon black is the most commonly used opacifier. Graphite glaze reduces sensitivity to static electricity, improves the flow of powder grains, and improves the packing density of the powder. An ignition aid coating improves the oxygen balance of the surfaces of the powder grains. Potassium nitrate is the most commonly used ignition aid coating.

The manufacturer of smokeless powder controls a number of characteristics of the powder. The energy of the powder is determined by the formulation of powder (i.e., whether the powder is single base or double base); it is primarily a function of the flame temperature and the average molecular weight of the muzzle gases. The linear burning rate depends on the composition and on the porosity of the powder grains. The surface area of the powder grain is determined by its geometry. Powder grains may be spheres, cylinders, disks, or flakes; the grains may also be perforated. The surface area of the powder grains is the major control factor that the manufacturer uses to adjust the performance of a smokeless-powder product.

Smokeless powders are produced by one of two processes: the **extruded powder** process and the **ball powder** process. In the extruded powder process, the nitrocellulose and other ingredients are kneaded together with an organic solvent to form a dough-like mass. The dough is extruded through small openings in a steel die. A rotating blade cuts off lengths of extruded dough. The dimensions of the extruded grains are controlled by the size of the openings in the die, by the rate of extrusion, and by the speed of the cutting blade. The extruded powder grains are then coated and glazed with graphite. After drying to remove the solvent, the powder grains are screened to remove grains that are too large or too small. In the ball powder process, the nitrocellulose and other ingredients are mixed with solvent to make a lacquer. The lacquer is extruded through a steel die into hot water, where the nitrocellulose forms spherical grains. The solvent is removed by the hot water, and the grains are allowed to harden. The spherical grains are sorted by size, and particles in the desired size range are subjected to further processing. The grains are coated and may be passed between rollers to flatten them into disks or flakes. In both processes the final product is a mixture of several batches of smokeless powder. The batches are blended to create a powder with a specified burning rate.

Collection of Firearms-Related Evidence

Fired bullets may be difficult to recover at the scenes of crimes. They may be embedded in walls, ceilings, door frames, window frames, and the like. Only rubber-coated or heavily taped tools should be used either to probe for bullets or to extract them. It is generally best for the investigator to remove the section of the building structure that contains the bullet so that the forensic firearms examiner can carefully remove the bullet in the laboratory. Some police agencies recommend that bullets be marked for identification by investigators on the nose or on the base. Under no circumstances should such identification marks be placed on top of potentially useful markings on the bullet. Care must also be taken not to dislodge any trace evidence that may be on the bullet surface. As an alternative to marking the bullet for identification, some law enforcement agencies recommend placing the bullet in a sealed pillbox or plastic vial; the container and its seal are then marked for identification. An expended cartridge can be marked for identification inside its mouth. Some agencies recommend not marking expended cartridges but rather placing them in sealed and marked containers.

The serial numbers of all firearms seized should be recorded by investigators.

Investigators will generally mark firearms for identification only if they are being collected as evidence. Even then some common sense is in order; antique or highly engraved firearms should be carefully marked so that their value is not diminished. All removable parts of a firearm that can mark a fired bullet or cartridge should be marked for identification. Thus, a 9-millimeter semiautomatic pistol should be marked on the barrel, the slide, the receiver, and the magazine. A revolver having an interchangeable cylinder and removable barrel should be marked on the cylinder, barrel, and frame. Weapons with removable bolts should be marked on the bolt, barrel, and frame.

For safety, loaded weapons should be unloaded before they are transported to the firearms laboratory. Magazines or clips should be carefully removed and marked for identification; live or expended cartridges should be removed from the firing chamber. Revolvers present a special problem. The investigator should scratch a small arrow on the rear face of the cylinder to indicate the chamber that was under the hammer when the revolver was collected as evidence. The chambers in the cylinder are numbered in a clockwise manner (when viewed from the rear) with chamber #1 being the chamber under the hammer. Each chamber is then emptied, with the live or expended cartridge or its container marked with the number of the chamber.

In general, firearms should not be cleaned before they are shipped to the forensic science laboratory. However, if there is a lot of moisture in the weapon's barrel, it should be removed by passing a dry patch through the barrel. The investigator should note the cleaning in his or her notes and in the letter of transmittal to the forensic science laboratory. He or she should submit the patch as evidence along with the firearm. For shipment to the forensic science laboratory the weapon should be wrapped in a clean protective covering to protect it from contamination. If the firearm is to be processed for latent fingerprints, it should not be covered, but should be placed in a box or other container so that any latent fingerprints are not rubbed off by contact with the packaging.

All ammunition in the possession of a suspect in a shooting case should also be seized as evidence. The firearms examiner may have to estimate the range of fire from a powder pattern or shotgun pellet pattern. Such estimation requires ammunition from the same batch as that which produced the questioned powder or pellet pattern. Live ammunition cannot be sent through the United States mail. Therefore, the investigator may have to hand-carry live ammunition unloaded from a weapon or seized from a suspect to the forensic science laboratory or ship it through a private parcel service.

Laboratory Examinations

Microscopic Examinations

Figure 21.4 shows examples of fired bullets that might be submitted as evidence. Several of them show patterned markings or trace evidence. Bullets should always be examined for the presence of trace evidence as well as patterned markings. Bullets may pick up textile fibers, traces of paint, or bits of concrete and brick from intermediate targets. Bullets may

Figure 21.4 Expended bullets that might be submitted for firearms examination. Upper left: deformed full metal jacket bullets. Upper right: core and detached jacket of hollow-point semijacketed bullet. Lower left: lead-alloy bullet that struck body armor. Lower middle: lead-alloy bullet that passed through window. (Note embedded glass particles.) Lower right: lead-alloy bullets that passed through plywood. (Note embedded wood fibers.)

acquire patterned markings from clothing or window screens. In one case, a state police officer was shot during a traffic stop. One bullet passed through the badge device on his uniform cap. When recovered, the bullet had the pattern of the badge embossed on its nose. In another case, a man was accused of shooting his neighbor to death. The defendant claimed that he had shot the victim during a quarrel in the victim's kitchen during which the victim had attacked him. The discovery of an impression on the nose of the fatal bullet that matched the window screen in the victim's kitchen window (along with an apparent bullet hole in the screen) refuted the self-defense claim.

The goal of the initial examination of a fired bullet is the determination of general rifling characteristics of the firearm that fired it. This information allows the forensic firearms examiner to narrow the possible makes and models of firearm that could have fired the bullet. The general rifling characteristics to be determined are:

- Caliber
- Number of lands and grooves
- Direction of twist of the rifling
- Degree of twist of the rifling
- Widths of lands and grooves

The caliber of a fired bullet can be determined in a variety of ways. If the bullet is not deformed, its diameter can be measured with a micrometer, or it can be compared with bullets fired from weapons whose caliber is known (i.e., so-called fired standards). With either method, allowance must be made for the fact that lead and lead-alloy bullets are larger in diameter than full-metal jacket or semijacketed bullets of the same nominal caliber. If a bullet is severely deformed, its possible caliber(s) may be determined by weighing it. The weight of the bullet will rarely pinpoint its caliber but will serve to eliminate a number of calibers from consideration. For example, a 72-grain bullet may be .32 caliber but it cannot be .22 caliber. If the bullet is fragmented, its possible caliber(s) can be determined by combining the measured width of a land marking with that of an adjacent groove. Tables of combined land

and groove widths have been prepared for a number of makes and models of firearms.

The number of lands and grooves is determined by inspection, as is the direction of twist of the rifling. Rifled firearms have either right twist (Smith and Wesson type) rifling or left twist (Colt type) rifling. The degree of twist of the rifling and the widths of the land and grooves can be determined by microscopic measurements of the rifling marks on the bullet. However, it is generally simpler to compare the questioned bullet to a set of fired standards—bullets fired from firearms having known rifling characteristics. The general rifling characteristics determined from a fired bullet can be used to search a database such as that prepared by the FBI. Both a text form and a computer-searchable form of the database can be downloaded from the Internet.

Class characteristics of firearms can also be determined from expended cartridges. The significant class characteristics are:

- Caliber
- Shape of firing chamber
- Location of the firing pin
- Size and shape of the firing pin
- Size of extractors and ejectors (if any)
- Geometrical relationship of the extractor and ejector

For the most part, these characteristics can be determined by inspection. Determination of the caliber and the size of the firing pin and extractor requires the use of a micrometer or a microscope equipped with a reticule for measurements. The class characteristics can be used to search the FBI database for matching makes and models of firearms.

Some care should be exercised in inferring the general class characteristics of a weapon from fired bullet or cartridge cases. A number of devices are available that permit the firing of types of ammunition other than that for which the weapon was designed. For example, crescent-shaped metal clips can hold rimless pistol cartridges in the cylinder of a revolver. A barrel insert can convert a shotgun into a rifle. Subcaliber cartridges can be wrapped in layers of paper to facilitate loading into the cylinder of a revolver. It may even be possible to chamber and fire a larger caliber cartridge in a weapon. Rifling characteristics of barrels can be altered by reboring them.

If a questioned firearm has also been submitted in a case, the forensic firearms examiner should conduct an initial examination for trace evidence that might be destroyed in the course of his or her testing. For example, blood or other tissue may have been blown back from a contact gunshot wound onto the exterior of the weapon or inside the barrel. There may be textile fibers on the weapon from the shooter's pants or jacket pocket. The weapon should also be processed for latent fingerprints before additional tests are undertaken. Next, the firearms examiner determines if the weapon's class characteristics are consistent with those found on the fired bullets or cartridges. Taper gauges can be used to verify the weapon's caliber; the number and lands and grooves and their direction and degree of twist can be determined with a helixometer. A sulfur cast of the interior of the barrel may also be used to determine all general rifling characteristics.

Once it has been determined that the class characteristics of the firearm and those of the fired bullets or cartridges are consistent, the weapon must be test fired to obtain bullets and cartridges for comparative microscopic examination. Before the weapon is test fired the examiner conducts examinations to determine if the firearm's mechanism operates properly and if the weapon can be safely fired. These examinations are discussed below. If the weapon cannot be safely fired, the examiner may be able to replace damaged or missing parts from the laboratory's firearm collection.

To obtain test-fired bullets, the firearm is fired into a bullet trap. Most firearms laboratories use some type of water trap to catch fired bullets. A horizontal bullet trap consists of a steel tank filled with water. The firearms examiner fires the test shots through a self-sealing rubber membrane. After the test shots have been fired, the tank is opened and the fired bullets removed for subsequent microscopic analysis. A vertical bullet trap does not require a self-sealing membrane. A basket at the bottom of the tank is used to recover the test fired bullets.

Bullets are mounted for microscopic comparison on a comparison microscope. A comparison microscope consists of two compound microscopes connected by an optical bridge. It allows two specimens to be viewed side by side. Some examiners recommend first comparing the test-fired bullets with each other to verify that the weapon's barrel consistently marks the bullets fired through it. If consistent marking does not occur (e.g., if the barrel is severely rusted), further comparison of the questioned bullet(s) and the test-fired bullets would be a waste of time. The comparison process is straightforward, albeit tedious. The examiner searches the surface of one bullet for a distinctive pattern of parallel striations and then rotates the other bullet slowly in an attempt to find a matching pattern. If a matching pattern is found, the bullets will be slowly rotated together to see if additional matching patterns of striations can be found. If no matching pattern is found, the examiner tries to find another distinctive pattern of striations on the first bullet and again tries to find a matching pattern on the second bullet. This process continues until matching striation patterns are found or the examiner exhausts all the distinctive striation patterns on the first bullet. Before the two bullets can be said to match, most firearms examiners require that identical patterns of three or more consecutive striations be found on each bullet. Not all striations on the two bullets must match. Particles of dirt in the weapon's barrel may have made stray markings on the bullets, or the two bullets may have entered the rifling in slightly different ways. Figure 21.5 shows matching striation patterns within the land impressions of two bullets fired from the same semiautomatic pistol. Striation patterns in the groove impressions can also be examined; however, jacketed and semijacketed bullets may not expand sufficiently to fill the groove of the rifling.

Other marking on bullets have also been compared from time to time. **Skid marks** are marks parallel to the axis of the bullet made when the bullet initially enters the rifling; the edges and surfaces of the lands will scrape along the bullet surface before the bullet is fully gripped by the rifling. The term skid mark is

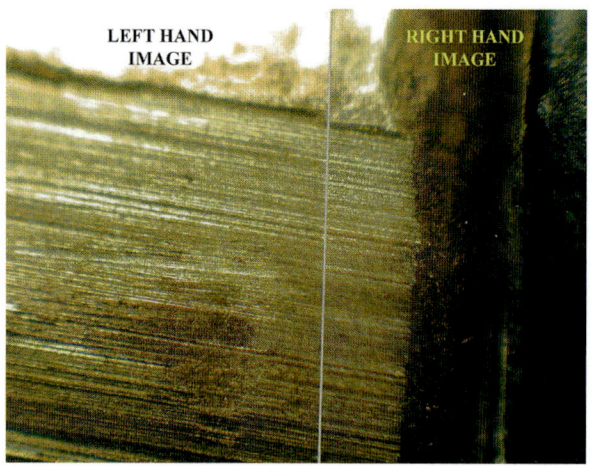

Figure 21.5 Microscopic comparison of striations in the land markings of two bullets fired from the same semiautomatic pistol.

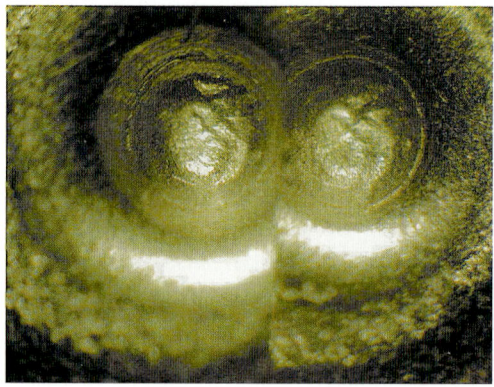

Figure 21.6 Microscopic comparison of firing pin impressions on two cartridges fired from the same semiautomatic pistol.

also applied to marks near the nose of a bullet caused by contact with the **forcing cone** in the barrel of a revolver. The forcing cone is a flare at the breech end of a revolver barrel that is intended to guide the bullet into the rifling. Skid marks made by the forcing cone of a revolver are hard to reproduce; a number of bullets might have to be test fired before one strikes the same spot on the forcing cone as the questioned bullet. **Slippage** marks are made on a bullet when it slips along the tops of the lands without being gripped by the rifling. Slippage marks are the result of the barrel being worn or having been bored out. Slippage marks may also result if a subcaliber bullet is fired in a weapon. Slippage marks are hard to replicate in test firings.

A variety of markings on cartridges can be compared microscopically. There is a logical hierarchy to follow in conducting such comparisons. Firing pin impressions and breechblock markings (also called **bolt-face signatures**) should be compared first because they can only be produced by firing a cartridge in a firearm. Figure 21.6 shows the microscopic comparison of firing pin impressions on two cartridges fired from the same semiautomatic pistol. Chambering, extractor, ejector, and magazine marks may be made by loading a cartridge in a weapon without firing it.

The following conclusions may be reached as a result of the comparison of bullets or cartridges:

- A *positive identification*. The class characteristics are consistent and individual characteristics match. This means that the likelihood that a firearm other than the firearm submitted for examination fired the questioned bullet or cartridge is so remote that it must be regarded as a practical impossibility.
- *Negative identification*. The questioned bullet or cartridge was not fired in the submitted weapon. This means that the class characteristics did not match. In this instance no microscopic comparison will have been undertaken.
- *Inconclusive*. This means either that there was too little information on which to base a conclusion or that while the class characteristics matched, sufficient individual characteristics to declare a match could not be found. The firearms examiner does not render a negative identification in this last instance because the individual characteristics of the firearm may have been altered by further use. For example, there may have been bore erosion due to firing a large number of rounds through the weapon. The weapon may also have been deliberately altered by the shooter. Some weapons (e.g., Glock pistols) do not make usable striation patterns on fired bullets.

The microscopic comparison of bullets and cartridges is a very tedious and time-consuming process. Most law enforcement agencies keep open bullet and cartridge casing files where evidence from open criminal cases is kept for later comparison with test-fired bullets and cartridges from weapons recovered in subsequent investigations. However, the time required for such comparisons is such that they are attempted only when information developed by investigators provides a link between a new case and an open case. The National Integrated Ballistic Information Network (NIBIN) was developed to facilitate linking firearms evidence in cases in different jurisdictions. The system consists of data-acquisition computer workstations and data-analysis computer workstations, all linked in a nationwide computer network. The data-acquisition computer workstation is used to scan and digitally capture images of bullets and cartridge cases. Each captured image has a mathematical signature that is placed in a database. Once a bullet or cartridge case is scanned and its image captured, the database can be searched for possible matches. The search of the database results in a list of potential candidate hits, and the NIBIN system operator can retrieve digitized images for visual comparison using the data-analysis computer workstation. Likely candidates for matches are referred to a firearms examiner for manual comparison of the actual items of evidence. The NIBIN system is designed to be operated by a technician with minimal skills, freeing the more highly trained firearms examiner for other duties. The image databases can be shared among firearms laboratories in different jurisdictions. The United States and Canada have agreed to share their image databases. In 2008 the United Kingdom launched its own version of NIBIN, the National Ballistics Intelligence Programme (NaBIP). NaBIP uses IBIS hardware and software, so that its database is compatible with those of the United States and Canada.

Two states, New York and Maryland, have legally mandated the ballistic "fingerprinting" of new firearms. Before a semiautomatic pistol can be sold in these states, a cartridge case fired in it is scanned into a special database. Theoretically, if the weapon is subsequently used in the commission of a crime, an expended cartridge case from the crime scene can be scanned by a data-acquisition computer work station and matched to the image in the database. Police investigators would then be led to the original purchaser of the handgun. After the expenditure of several million dollars, the Maryland ballistic fingerprinting program has only solved one case. In 2005 an estranged husband was convicted of the murder of his wife with a .40 caliber handgun. The gun was never recovered; however, expended cartridges recovered at the murder scene were matched to a semiautomatic handgun purchased by the defendant. The New York ballistic fingerprinting program has shown a similar record of limited success in solving shooting cases. A low "hit rate" for ballistic fingerprinting it to be expected: most firearms are never used in the commission of crimes. There are also technical difficulties with the imaging and databasing systems. When California was considering a ballistic fingerprinting program similar to the Maryland and New York programs the state attorney general commissioned a pilot feasibility study. This study revealed a high rate of misses (38%). The hardness or softness of the brass in the percussion caps of the centerfire pistol cartridges was identified as the likely culprit: if the brass is too hard, poor firing pin and breechblock impressions with limited detail result, and if it is too soft, the large amount of detail in the impressions confuses the image matching software algorithm. In 2008 the National Research Council (NRC) of the National Academy of Sciences released a study of the feasibility of a national database of images of markings made by new and imported weapons sold in the United States. This study concluded that given the currently available technology such a database would produce too many "matches" to be of helpful to police investigators. This NRC report also recommended further study of "microstamping," a process in which identifying information is engraved on firearm components such as firing pins and breechblocks. This information would be stamped on fired cartridge cases by the weapon's action; it could be read at the scene of a shooting by a police investigator with a magnifying glass or low-power microscope. California has passed a law that would require microstamping on internal parts of new semiautomatic handguns sold in the state by 2010. The NRC report concluded that further research needs to be done on microstamping to determine the durability of the engraved marks as well as their susceptibility to tampering. The cost impact of microstamping for manufacturers and consumers also needs to be examined.

Microscopic comparisons of bullets do not make use of the three-dimensional profiles of the bullet surfaces. In the 1950s, firearm examiner John Davis developed the striagraph, which used a system of levers to amplify the movement of a stylus as it was drawn over the surface of a fired bullet. The striagraph produced a circular graph that reflected the profile of the bullet surface. The striagraph was never widely used by firearms examiners. Recently, researchers have used laser surface profilers to capture the three-dimensional profiles of the surfaces of fired bullets. The profiles are digitized and placed in a computer, where some measure of similarity between two bullets (e.g., a correlation coefficient) can be calculated. In this way the matching of bullets might be made more objective. Methods for obtaining three-dimensional profiles from cartridges are also being explored.

Other Firearms Examinations

In addition to the microscopic examination and comparison of bullets and cartridges, firearms examiners must carry out other examinations on firearms. Before a firearm can be test fired, the examiner must verify that the weapon's action and safety devices work properly. If the weapon does not function properly, the examiner should note that fact and determine the reason for the malfunction. Critical components may be damaged or missing. Because meaningful microscopic comparisons of bullet and cartridges require that the weapon be fired, the examiner may have to replace damaged or missing components with parts taken from the forensic laboratory's firearms collection. Many firearms have both passive and active safety devices; some firearms have

no safety devices. Passive safety devices are features that the shooter does not have to set. Such devices include a half-cock position for the hammers of weapons with exposed hammers, a grip safety, and a firing pin catch or block. A half-cock safety is intended to prevent the hammer from accidentally falling far enough to discharge the weapon. If the hammer is released before it is fully cocked, it will fall to the half-cock position and be stopped. A grip safety is located on the back of the grip of a handgun. It prevents the trigger of the weapon from being squeezed unless the heel of the hand has depressed the safety. A firing pin catch or block prevents the firing pin from moving forward to strike the cartridge primer until the trigger has been squeezed. This prevents the weapon from accidentally discharging if it is dropped. Active safety devices are set by the shooter; they include trigger blocks, slide blocks, and safety decocking levers.

Trigger blocks prevent the trigger from being squeezed. Slide blocks prevent the slide of a semiautomatic handgun from being drawn to the rear and also block the trigger mechanism. A safety decocking lever prevents the trigger from being squeezed and rotates the rear section of a two-part firing pin so that the hammer cannot strike it. When the safety decocking lever is set, the cocked hammer of the weapon can be lowered without risk of an accidental discharge. The safety decocking lever prevents the weapon from being fired normally and also if it is dropped. The firearms examiner must verify that the weapon's safety devices work properly for his or her own safety when test firing the weapon, and because the defendant in a shooting case may allege that the weapon discharged accidentally.

Steel

40 milliliters of concentrated hydrochloric acid, 30 milliliters of distilled water, 25 milliliters of ethanol, and 5 grams of cupric chloride; or

120 milliliters of concentrated hydrochloric acid, 100 milliliters of distilled water, and 90 grams of cupric chloride

Aluminum

20 percent (weight per volume) aqueous solution of sodium hydroxide; or

1 normal aqueous sodium hydroxide rinse, followed by 0.1 molar mercuric chloride in 0.1 normal hydrochloric acid

Zinc alloys

Alternate applications of phosphoric acid–nitric acid (98:2) solution and 5 to 10 percent nitric acid

Ultrasonic cavitation works on a principle similar to chemical and electrochemical etching. The surface where the serial number has been obliterated is polished as described above and then the weapon is placed in an ultrasonic bath where it is subjected to high-frequency vibrations. Minute bubbles form on the metal surface (a process called *cavitation*). Cavitation causes etching of the metal surface. This same phenomenon causes pitting of high-speed marine propellers. In serial number restoration, the etching proceeds fastest where the metal structure has been disordered by stamping the serial number.

Firearm and tool mark examiners may be called upon to restore stamped serial numbers on other items, such as automobile engine blocks. A method has also been developed for the restoration of stamped serial numbers on plastic items (camera bodies, stereos, and computer equipment). Organic solvents are used to swell the plastic. Different types of plastics require the use of different organic solvents.

A serial number may be unrecoverable for a number of reasons. The filing or grinding may have removed too much metal. The criminal may overstamp the serial number with new numbers or letters or obliterate it by attacking the metal surface with a metal punch. Methods have been proposed to prevent the removal of serial numbers from weapons. According to one such idea, a laser would be used to drill holes in the weapon's frame following a grid pattern. Each digit of the serial number would be represented by a hole drilled at a specific location in the grid; the serial number could be removed only by completely removing a section of the frame. No such scheme has ever been implemented.

Range-of-Fire Estimation from Powder and Pellet Patterns

Detailed reconstruction of shooting incidents requires knowledge of ballistics, the branch of physics that studies projectile motion. A bullet encounters four ballistic regimes on its journey from the firing chamber of the weapon to its final resting place in its target. The first regime is that of interior ballistics, where the chemical energy stored in the propellant is converted into the kinetic energy of the projectile. The second ballistic regime is that of transitional ballistics, through which the projectile passes as it moves from interior ballistics to exterior ballistics. In the transitional ballistic regime, a spinning projectile experiences both lateral and vertical jumps. In small arms these lateral and vertical jumps are negligible. In the regime of exterior ballistics, the projectile moves under the combined effects of a number of forces: gravity, frame-of-reference forces (Coriolis and the centrifugal forces), and aerodynamic forces (drag, lift, the Magnus force, pitch damping, and the transverse Magnus force). Generally, the effects of the Coriolis force, pitch damping, and the transverse Magnus force are negligible for small arms. The centrifugal force is included as part of the net gravitational force at a given location.

The center of mass of a projectile will follow an approximately parabolic trajectory under the net effect of the gravitational force and aerodynamic drag. The combined effect of the lift force and the Magnus force is to make the **yaw angle** of the projectile (which is the angle between the axis of symmetry of the projectile and the direction of flight) execute a complicated precessional and nutational motion. With a properly designed projectile, the yaw angle is initially very small; as the projectile moves farther from the weapon, the yaw angle may increase, so that at great distances the projectile may strike its target sideways. If the bullet encounters an intermediate target, the yaw angle may be abruptly changed.

In the reconstruction of shooting incidents, the range and direction of fire are of paramount importance. Firearms examiners are frequently called upon to estimate the range from which a gunshot was fired by examining gunshot residue patterns on the victim's skin or clothing. Gunshot residue consists of particles from the gun barrel, particles from the bullet surface (lead, lead alloy, or brass), particles originating from the propellant (unburned or partially burned particles of smokeless powder as well as soot), and particles originating from the primer (lead, or lead, antimony, and barium). This residue is projected in a roughly conical cloud in the direction of the target. Because the gunshot residue particles are traveling through the air, they experience aerodynamic drag forces. The smaller particles are slowed more rapidly and consequently travel shorter distances than the larger particles. Some gunshot residue may also leak out of the weapon to be deposited on the shooter's hands, face, and hair. Gunshot residue may be seen on the hands of a shooting victim who is grappling for a weapon when it is fired. Revolvers in particular produce distinctive powder-burn patterns on the hands when they are gripped with the hand around the front of the cylinder.

The appearance of a gunshot wound may hold a clue to the range from which it was inflicted. Forensic pathologists usually place the range from which a gunshot wound was inflicted into one of several categories: distant shots, close-range shots, near-contact shots, and contact shots. Distant shots are fired from such a range that no detectable gunshot residue reaches the skin or clothing of the victim. The gunshot wounds inflicted by distant shots consist of a circular or elliptical defect in the skin, which is surrounded by a **marginal abrasion** or contusion ring where the skin has been stretched and torn by the entry of the bullet. The marginal abrasion may be overlaid by a **gray ring**; a ring of propellant combustion products, bullet lubricant, and metal from the bullet surface that has been wiped off onto the skin. The material comprising the gray ring is also termed **bullet wipe**. Bullet wipe may be found on any solid materials through which bullets pass, such as clothing, doors, and walls.

Close-range gunshot wounds are inflicted at ranges short enough for gunshot residue to reach the skin or clothing of the victim. Two types of gunshot residue deposition are seen with close-range shots: stippling (tattooing) or soot (smudging). The large propellant particles that produce stippling travel farther than the finer particles that comprise the soot. Consequently, as the range of fire for close-range shots decreases, the resulting gunshot residue patterns go from widely dispersed stippling to more concentrated stippling plus soot.

At a near-contact range, stippling and smudging are concentrated in a tight circle. The weapon's muzzle flash may tear clothing and char or melt clothing fibers. Woven fabrics typically tear apart along the warp and weft directions, producing a cruciate (cross-like) defect; knit fabrics usually show a circular area of damage. Natural fibers, such as cotton and wool, are charred by the muzzle flash, whereas manmade fibers (being composed of thermoplastic materials) are usually melted. If the gunshot wound is inflicted on a part of the body that is covered with hair, the hair will show characteristic singeing.

Loose-contact gunshots are fired with the weapon's muzzle just touching the target surface. Gunshot residue may be blown outward between layers of clothing. The muzzle flash will produce similar effects to those observed in near-contact gunshots. Tight-contact gunshots over bony plates, such as the vault of the skull or the sternum, produce a characteristic **stellate defect**: an irregular, blown-out entrance wound. This type of wound is caused by the propellant gases separating the soft tissue from the bone and creating a temporary pocket of hot gas between the bone and the muzzle of the weapon. If the gas pressure is high enough, the soft tissue and skin will tear, creating a jagged entrance wound. Blood and other tissue may be blown back into the muzzle of the weapon and onto the hand and forearm of the shooter. Firearms examiners should always check weapons for such trace evidence. The soft tissue may be forced back against the muzzle of the weapon hard enough to receive a muzzle impression. Carbon monoxide in the muzzle gases may also react with blood in the wound area to produce carboxyhemoglobin.

Determination of the range of fire from a gunshot residue pattern requires the original powder pattern, the firearm used to fire the pattern, ammunition from the same lot as that used to fire the pattern, and a knowledge of the weather conditions at the time of the shooting. If the original powder pattern is on a garment, its preservation for examination is straightforward. The powder pattern may require chemical treatment or special imaging techniques to render it visible, particularly if the pattern is on dark-colored or blood-soaked clothing. If the powder pattern is on the victim's skin, scaled photographs may be the best method for the preservation of the original powder pattern. When the victim is deceased, the skin bearing the powder pattern may be excised and treated with preservatives. However, skin treated in this way may shrink or stretch, altering the dimensions of the powder pattern. The court may regard excised tissue as unnecessarily inflammatory evidence and consequently exclude it at trial. Some religious groups would regard the removal and preservation of a portion of the victim's body as extremely sacrilegious.

Determination of the range of fire from a powder pattern requires that the firearms examiner test fire powder patterns at various ranges until he or she is able to reproduce the original pattern in size and density of gunshot residue deposition. For the results of the tests to be meaningful, the firearms examiner must use the same weapon and ammunition as that used to fire the original pattern. The same weapon must be used because weapons vary in their leakage of gunshot residue. A weapon with a worn barrel, for example, might produce more gunshot residue than a weapon of the same type with a pristine barrel. Using ammunition from the same lot is likewise important. Ammunition manufacturers commonly change lot numbers when they exhaust a lot of one of the components (bullets, casings, primers, or propellant). Using ammunition from the same lot, therefore, ensures that critical components, such as primers and propellant, are the same for the questioned powder patterns and the test fired patterns. If police investigators seized any ammunition with the firearm, it may be used to fire the test patterns.

Weather conditions also affect the dispersal of gunshot residue. Wind and rain may disperse the plume of gunshot residue so that little or no residue reaches the surface of the target, even at very close range. Ambient temperature also affects the rate of burning of smokeless powder. While the effect of temperature on powder patterns has not been demonstrated, research has shown that temperature does affect the dispersal of shotgun pellets. Duplication of the weather conditions may not be feasible; in such a case, it may not be possible to estimate the range of fire.

Firearms examiners may need to use special techniques to visualize gunshot residue patterns on dark or bloodstained clothing. Infrared imaging technologies have proven to be useful for visualizing gunshot residue patterns obscured by blood. The hemoglobin and other proteins in blood are relatively transparent to certain wavelengths of infrared radiation, while graphite-coated propellant particles and soot strongly absorb infrared radiation. Chemical methods for visualizing gunshot residue patterns are also widely used. The original Walker test used undeveloped photographic paper impregnated with sulfanilic acid and alpha-naphthylamine, which react with inorganic nitrites in gunshot residue. The Walker test for nitrites has been superceded by the Griess and Maiti tests. To perform the Griess test, the gunshot residue is transferred to a sheet of desensitized photographic paper or filter paper. The paper is dampened with dilute acetic acid and the gunshot residue is transferred to the paper by ironing or pressing the paper and article of clothing together. The photographic paper or filter paper is then immersed in the Greiss reagent. The powder pattern may also be lifted from a garment using a "peelable" low adhesive lifter; the use of a lifter avoids damage to heat-sensitive fabrics and the transfer of clothing dyes whose color may mask that produced by the reaction between nitrites and the Griess reagent. Several recipes for the Griess reagent are currently in use:

1. 3 percent sulfanilamide and 0.3 percent N-(1-naphthyl)ethylenediamine dihydrochloride are dissolved in 5 percent phosphoric acid.
2. 5 grams sulfanilic acid are dissolved in 1000 milliliters of 30 percent (volume per volume) acetic acid. 6 grams of alpha-naphthylamine are dissolved in 1000 milliliters of 30 percent (volume per volume) acetic acid. Fresh Griess reagent is prepared by mixing equal volumes of each of these solutions.
3. 80 grams sulfanilamide are dissolved in 1000 milliliters of 10 percent (volume per volume) phosphoric acid. 4 grams N-(1-naphthyl)-ethylenediamine are dissolved in 1000 milliliters of 10 percent (volume per volume) phosphoric acid. Fresh Griess reagent is prepared by mixing equal volumes of each of these solutions.

The Griess reagent reacts with inorganic nitrites in gunshot residue patterns to form an azo dye through a diazotization reaction. The Griess reagent will not react with inorganic nitrates or with the cellulose nitrate in the partially burned or completely unburned smokeless powder particles in the powder pattern. Alkaline hydrolysis of cellulose nitrate causes a disproportionate reaction that releases nitrites, enhancing the Griess reaction. The powder pattern is sprayed with a 2 percent alcoholic solution of potassium hydroxide and then heated in a laboratory oven at 100 degrees Celsius for 1 hour. Bloodstained clothing may develop unpleasant odors when treated in this fashion. Therefore, a "peelable" lifter should be used to remove the powder pattern for subsequent testing.

The Maiti test is an alternative test for the presence of nitrites in a gunshot residue pattern. When the original Walker test was applied to bloodstained articles of clothing, traces of blood were transferred to the test paper and obscured the orange-red color produced by nitrites. The Maiti test paper is prepared in the following way.

Glossy photographic paper is soaked in photographic fixer, then washed and dried. The dried paper is then soaked in a solution of p-nitroaniline, alpha-naphthol, and magnesium sulfate (each 0.25 percent in 1:1 aqueous alcohol). The photographic paper is dried and stored until it is needed. The test is performed by placing the treated photographic paper on the laboratory bench with the emulsion side up. The article of clothing to be tested is placed with its front side in contact with the paper. A towel soaked in 10 percent acetic acid is placed over the article being tested and pressed with a hot iron. The photographic paper is removed and treated with 10 percent aqueous sodium hydroxide. The gunshot residue pattern appears as blue specks on a yellow background.

Other color tests are commonly used in conjunction with the Griess or Maiti tests. The sodium rhodizonate test is used to visualize the dispersal of lead (primarily from the cartridge primer) in the gunshot residue pattern. Lead residues are transferred from a garment to filter paper by pressing. The filter paper is then sprayed either with an aqueous tartaric acid and monosodium tartrate buffer solution (pH 2.8) followed by a saturated aqueous sodium rhodizonate solution (0.4 grams per 100 milliliters) or with a solution of sodium rhodizonate saturated in pH 2.8 aqueous tartaric acid and monosodium tartrate buffer. A scarlet color appears immediately if lead is present. To avoid false positives from other nonlead inorganic materials, the filter paper should be blotted with clean filter paper and dried with a hair drier. The filter paper is then sprayed with 5 percent aqueous hydrochloric acid until the scarlet color is converted to blue. If the filter paper is again blotted with clean filter paper and dried with a hair drier, the blue color may be preserved indefinitely. If the garment is light colored, the sodium rhodizonate reagent can be sprayed directly on it. Figure 21.7 shows powder patterns that have been enhanced by sodium rhodizonate and also by the Griess test.

Cartridges containing lead-free primers have begun to reach the market. The gunshot residue produced by such ammunition does not contain either lead or barium.

Figure 21.7 Powder patterns treated with sodium rhodizonate (upper patterns) and modified Griess reagent (lower patterns). Patterns on the left are as originally test fired; patterns on the right have been laundered.

Sintox primers, for example, contain zinc peroxide and titanium metal. Consequently, the sodium rhodizonate test cannot be used in the visualization of gunshot residue patterns produced by lead-free cartridges. If the cartridge contains a Sintox primer, zincon reagent can be used to visualize the gunshot residue pattern. Zinc and titanium produce blue-colored compounds with the zincon reagent. Rubeanic acid has also proven useful in the visualization of gunshot residue patterns through its color reaction with copper. Copper in gunshot residue is derived from the jackets of full-metal jacket or semijacketed bullets and from the copper alloy comprising the primer cap. Rubeanic acid has proven to produce useful gunshot residue patterns even when lead-free, full-metal jacket ammunition has been used.

Firearms examiners also estimate ranges of fire from shotgun pellet patterns. When a shotgun is fired, the pellet mass spreads laterally. Thus, pellet patterns fired at different ranges have different sizes and different shot densities. A widely used rule-of-thumb is that the pellets spread 1 inch laterally for each yard down range. In order to determine the range of fire from a shotgun pellet pattern, the firearms examiner needs the original pellet pattern, the shotgun used to fire the pattern, ammunition from the same lot as that used to fire the pellet pattern, and a knowledge of the weather conditions at the time of the shooting. Typically, the firearms examiner then test fires the shotgun at various ranges until he or she obtains a pellet pattern similar in size and pellet density to the questioned

pattern. It is also possible to use regression analysis to determine the range of fire and its confidence limits. If regression analysis is used, the firearms examiner must use some measure of the size of the pellet pattern. The following have all been used as measures of the sizes of pellet patterns:

- The radius of the smallest circle that will just enclose the pellet pattern
- The area of the smallest rectangle that will just enclose the pattern
- The arithmetic mean of the lengths of the sides of the rectangle
- The geometric mean of the lengths of the sides of the rectangle

The firearms examiner fires a series of pellet patterns at selected ranges, measures their sizes, and then calculates the equation for the size of the pattern as a function of range of fire that best fits the data for the test-fired pellet patterns. The equation is then used to calculate the range of fire from the size of the questioned pellet patterns. Confidence limits for the range of fire estimate can be calculated by standard statistical methods.

The sizes of shotgun pellet patterns are affected by the choke of the shotgun barrel. Full-choke barrels fire smaller, denser pellet patterns than cylinder-bored barrels (i.e., barrels without any choke). Double-barreled shotguns often have barrels with different chokes. A number of barrel inserts and adjustable compensators can also be placed in a shotgun barrel to change its choke. Thus, a firearms examiner may have to test fire pellet patterns using both barrels of a double-barreled shotgun and make range of fire estimates for both barrels. Likewise, if the shotgun has an adjustable compensator, the firearms examiner should test fire pellet patterns for the extreme settings of the compensator (greatest choke and least choke) and make range-of-fire determinations for both settings.

Sawing off the barrel of a shotgun may increase the size of the pellet patterns it will fire. However, this phenomenon depends on the type of shotshells fired. With some types of shotshell, little or no spreading is seen, even when the barrel is shortened to 6 inches.

The size of a shotgun pellet pattern can also be affected by the presence of an intermediate target (such as a window, window screen, or door). If the mass of shotgun pellets encounters a target that slows down the leading pellets in the mass, the trailing pellets will overtake and collide with them. The leading pellets will be deflected on new trajectories so that the pellet pattern produced in the final target will be larger than would be the case in the absence of the intermediate target. The shotgun pellets may be marked by the intermediate target or pick up trace evidence from it.

The direction of fire can be determined from a variety of phenomena. It can obviously be determined from the locations of entrance and exit wounds (with due allowance for the possible deflection of the bullet by bone). If the bullet passes through a bony plate, such as the vault of the skull, it will punch out a cone-shaped hole. The wider end of the cone indicates the direction the bullet was traveling. Coning or beveling is observed not only with bone but also with glass, wallboard, and wood. If a bullet strikes a long bone, such as a femur, it will punch out a wedge-shaped segment of bone; the wedge will be displaced in the direction the bullet traveled.

The paths of bullets are also frequently reconstructed at crime scenes. Because most shootings take place at short ranges, the paths of the bullets can be approximated by straight lines. An accurate reconstruction of the path of a bullet requires that the bullet mark two fixed objects within the crime scene. For example, a bullet might pass through (or perforate) the wall of a room, leaving an entrance hole (sometimes referred to as the in-shoot defect) in the wallboard on one side and an exit hole (or out-shoot defect) on the other. Rods, strings, or laser beams can be centered in each of the two holes to approximate the bullet's path. The farther apart the in-shoot and out-shoot defects are, the more accurate the reconstruction will be. If the bullet makes only a single hole at the crime scene, the bullet's angle of impact can be estimated from the shape of the bullet hole. For an elliptical bullet, the ratio of the width of the hole to its length is approximately the sine of the angle of impact. This is the same method used to determine the angle of impact of a blood drop and is subject to the same uncertainties. Even if the path of the bullet cannot be accurately determined, it may be possible to define the general area that the bullet came from and limit the area that must be searched for expended cartridge cases and other evidence of the shooter. If a bullet came through an open window to strike a victim all locations from which a fired shot would not reach the window can be eliminated as locations of the shooter. Suppliers of crime scene processing equipment now often also sell materials for bullet trajectory reconstruction.

Ricochet

Bullets may be deflected from their trajectories by ricochet. When a bullet strikes a target it either penetrates the target, disintegrates upon impact, or ricochets. Ricochets can occur from any surface, including water. For any particular target surface, there is a critical angle of incidence below which bullets will ricochet from the surface. For most surfaces this angle is around 5 or 6 degrees. The angle of ricochet is generally less than the angle of impact. Bullet shape and composition also affect whether ricochet will occur. Round-nosed bullets are more likely to ricochet than pointed bullets, and full metal jacket bullets are more likely to ricochet than semijacketed bullets or lead and lead alloy bullets. A ricochet causes a bullet to lose much of its kinetic energy so that it is more likely to inflict a penetrating rather than a perforating wound. Ricochet also upsets the yaw angle of the bullet so that it may produce a key-hole wound. A ricocheting bullet may pick up trace evidence, such as particles of concrete or brick, from the surface from which it ricochets.

Silencers

The acoustic phenomena associated with the discharge of a firearm also merit discussion. When the hot propellant gases encounter the atmosphere outside the firearm, they generate sound waves. Carbon monoxide in the propellant gases may also be ignited explosively. The muzzle of a firearm, therefore, acts as a point sound source from which sound waves radiate spherically. The bullet itself may also function as a sound generator. If its velocity exceeds the speed of sound, it will produce a bow shock wave as it travels through the air. Firearms may be equipped with silencers to reduce the noise level of their discharges. No firearm can be completely silenced; sound can only be reduced. Two methods for silencing a firearm may be used: external and internal. An external silencer is a device that fits on the muzzle of the weapon. A commercial external silencer consists of a pressed steel shell that contains a series of baffles to break up and diffuse the muzzle gases. Steel wool packing can also be used. Internal silencing involves drilling ports in the weapon's barrel to vent the propellant gases. A steel sleeve packed with steel wool may surround the barrel. Criminals may improvise external silencers (from stacks of washers, for example) or modify a weapon to provide internal silencing. External silencers may produce markings on the surfaces of fired bullets. Misaligned baffles may score the bullets' surfaces, even to the extent of effacing rifling marks. Internal silencing may leave burrs inside the barrel that will mark any bullets fired through it. The markings made by an external silencer or those produced by the burrs left inside the barrel when a weapon is internally silenced may be sufficiently reproducible to permit microscopic comparisons.

The sound waves generated by bullets can be eliminated by using ammunition with a muzzle velocity below the speed of sound. If the silenced firearm is commercially produced, it may be designed around a cartridge with a subsonic muzzle velocity, or reduced velocity ammunition may be produced for it. However, the use of subsonic ammunition may require too much modification of the sighting of a weapon (e.g., a sniper rifle) to be feasible.

In World War I the sound and flash of artillery pieces were used to pinpoint their location for retaliatory counterbattery artillery fire. The observations of multiple observers were pooled to triangulate the position of the artillery pieces. This concept has been updated for law enforcement use. Several major cities (including Washington, D.C.) have deployed arrays of microphones in neighborhoods with high levels of gun violence. Computer software can identify sounds as gunshots and provide responding officers with their approximate locations. The software has the capability of differentiating the sounds of different types of firearms so that it may be possible to determine if multiple weapons were fired in a particular shooting incident.

Firearms Examinations in the Investigation of War Crimes and Human Rights Violations

In the 20th and 21st centuries firearms have been used in the commission of a vast number of atrocities: the 1940 Katyn Forest massacre of 4000 Polish prisoners of war by the Soviet Union; the murders of over a million Jews by Nazi *Einsatzgruppen,* the 1944 massacres of Belgian civilians and American military personnel during the Battle of the Bulge, the 1968 My Lai Massacre (in which 300 to 500 Vietnamese civilians were killed by U.S. Army troops) and the ethnic "cleansings" in the Balkans in the 1990s. In the investigation of such incidents, firearms examination can provide a wealth of useful information. The bullets and cartridges recovered at the scenes of the atrocities may of course be linked to weapons belonging or issued to specific individuals. Bullets and cartridges may also provide evidence as to who committed the atrocities. For example, an archaeological excavation of a mass grave in the Ukraine recovered numerous cartridges for German machine pistols. These artifacts confirmed that the dead were Jewish victims of one of the German execution groups that operated in the Ukraine during World War II (rather than victims of Soviet terror). The cartridges bore the dates 1939, 1940, and 1941, showing that the mass execution occurred no earlier than 1941 (when that part of the Ukraine was occupied by the German army). In this instance the dating of the incident was of critical importance because the archaeological excavations were undertaken in support of Australian prosecutions of accused Nazi war criminals.

Firearms Examinations and Battlefield Archaeology

Battlefield archaeology uses the methods of archaeology to reconstruct human behavior on battlefields. The distribution of artifacts such as bullets, percussion caps, and cartridges can be used to locate battlefields (if the precise location of an engagement is unknown or disputed), to locate troop positions on the field, to provide estimates of the numbers of weapons of a particular type used in the battle and to chart the movements of weapons across the battlefield. Comparisons of firing pin impressions on cartridges excavated on the Little Bighorn battlefield (scene of Custer's Last Stand) allowed archaeologists to track the movement of a particular .45 caliber 1873 Springfield cavalry carbine from the hill where the initial onslaught of the Sioux and Cheyenne warriors overwhelmed several detached companies of Custer's force to Last Stand Hill, where Custer and the remnants of his command were wiped out. Microscopic comparisons of firing pin impressions on the large number of .44 caliber Henry cartridges recovered from the field enabled archaeologists to estimate how many Henry and Winchester repeating rifles were used by the Native American participants in the battle. The Sioux and Cheyenne warriors had more repeating rifles than was previously believed, and these weapons gave them a significant firepower advantage over the troopers of the 7th Cavalry, armed with single shot carbines and Colt revolvers.

The semi-arid environmental conditions of the American West seem to be the most conducive to the preservation of markings on soft copper percussion caps and cartridges. The most significant applications of microscopic firearms examinations have been to 19th century Indian Wars battlefields: Cieneguilla (1854), the Fetterman Massacre (1866), and Hembrillo Canyon (1880). Bullets and cartridges recovered from battlefields sometimes provide a vivid glimpse of the last moments of a soldier's life. An expended .52 caliber cartridge excavated at the scene of the Fetterman Massacre (in which several thousand Sioux, Cheyenne, and Arapahoe warriors ambushed and killed 79 U.S. Army troops and two civilians) had two firing pin impressions made by different Spencer repeating carbines: the cartridge had misfired in one carbine and a desperate soldier had picked up the ejected cartridge and manually loaded it into another carbine.

Types of Tool Marks

The three categories of tool marks are compression (or indented) tool marks, sliding tool marks, and cutting tool marks. Compression tool marks result when a tool is pressed into a softer material. Such marks often show the outline of the working surface of the tool, so that class characteristics of the tool (such as dimensions) can be determined. The individual characteristics of the tool may be more difficult to discern in compression tool marks. Sliding tool marks are created when a tool slides along a surface; such marks usually consist of a pattern of parallel striations. Class characteristics are more difficult to determine from sliding tool marks. For example, screwdrivers, chisels, and pry bars could all make very similar sliding marks. Cutting tool marks are a combination of compression and sliding tool marks. The cutting tool indents the material being cut and, as it does so, the working surfaces of the tool slide over the cut surface.

The quality of a tool mark is very much affected by its substrate, the material on which the tool mark is made. In general, soft metals such as lead, copper, and brass are excellent recipients of tool marks. Many plastics are good surfaces for the retention of tool marks. Painted surfaces are also excellent substrates for tool marks because paint layers consist of a plastic vehicle or binder in which pigment particles are dispersed. Other surfaces, such as raw wood and hard metal, are poor substrates for tool marks. Raw wood is a poor substrate for sliding tool marks because its grain structure has the same dimensions as the striations in the typical tool mark. Hard metal surfaces are poor recipients of tool marks because their hardness prevents them from being marked by tools.

Processing of Tool Marks at Crime Scenes

Tool marks may be found on a variety of surfaces. Points of entry such as doors and windows should be examined for pry marks. Doors of safes and cabinets may also show pry marks. Cutting tool marks may be found on lock hasps, chains, and chain link fences. Once a tool mark is found, the crime scene technician must take care to prevent alteration of the mark. In particular, he or she should make sure that no one attempts to fit a suspect tool into a tool mark; to do so risks alteration of the mark and would vitiate the value of any transferred trace evidence on the tool.

Safe burglaries in particular provide a wealth of tool marks. Safes may be broken into in a variety of ways. For example, safes may be broken open with a cutting torch. While the cutting torch itself leaves no tool marks on the safe, there may be cutting tool marks on the hoses used to connect the oxygen and acetylene tanks to the cutting torch. Lightweight safes used in the home or by small businesses can be opened by prying open the door. This will leave tool marks on the edge of the door and the adjacent frame. The front plate of the safe door can be peeled away by insertion of a pry bar or similar tool between the front plate and the door's frame. Once the front plate is peeled away from the frame, the locking bars in the safe door may be forced back with the pry bar. In this case, tool marks will be found on the front plate and the locking mechanism. A hole may be pounded or chopped through the cladding of the safe. Two areas are usually attacked: the bottom or the front plate adjacent to the lock. If the bottom is the site of attack, the hole will be pounded through into the safe and the valuables extracted through the hole. If the front plate is the site of the attack, the locking mechanism will be forced open to allow access to the interior of the safe. Such "pound" jobs leave indented tool marks at the point of attack. If a small hole can be made in the outer cladding of the safe, it may be enlarged by ripping the cladding. Tool marks will be found at the point where the hole was started in the cladding and on the edges of the rip where the cladding was pulled away with pliers.

Electric drills can also be used to attack safes. A high-torque drill may be used to drill through the locking bars in the safe door. A jig or metal frame will be attached to the front plate of the safe door to position the drill properly. The safebreaker usually draws layout lines on the front plate to ensure that the jig is attached in the right place. These layout lines are important evidence and should be carefully photographed. A "drill" job leaves drill marks, both those made in the locking mechanism as well as those made in attaching the jig to the front of the safe. A core drill may also be used to drill a large hole in the side of the safe through which the safe contents can be removed. A jig is attached to the side of the safe to guide the core drill.

Most of the methods used to attack safes release safe insulation, which can be transferred to the perpetrator's clothing, footwear, or tools. Safe insulation may contain Portland cement, vermiculite, gypsum, diatomaceous earth, or sawdust. Microscopic examinations may associate the trace evidence removed from a suspect's clothes, shoes, or tools with the insulation in the safe. For such examinations to have value, the investigator must ensure that no cross contamination between the safe insulation and the evidence taken from the suspect can occur. The investigator must be sure to collect a representative sample of the safe insulation for comparison.

Once tool marks have been identified at the scene of a crime they should be documented in the usual manner with notes, sketches, and photographs. The photographs should show the locations of the tool marks; however, even macrophotography will rarely show sufficient detail to allow the photographs to be used for laboratory comparisons. Photographs are often useful to the laboratory examiner as clues to how the crime scene marks were made. Once the documentation of the marks is completed, the tool mark evidence should be collected for subsequent laboratory examination. If the object bearing the tool mark can be transported, it should be collected and sent to the forensic science laboratory. The operative rule is that the thing itself is the best evidence. However, if the object cannot be transported, a cast of the tool mark should be made. The ideal casting medium for tool marks reproduces the microscopic detail of the tool mark and is dimensionally stable, easy to use at crime scenes, and inexpensive. A variety of casting materials have been used to make casts of tool marks: negative moulage, low-melting metal alloys (e.g., Wood's metal), and silicone rubber. The material that most closely meets the criteria of an ideal casting material is silicone rubber. It faithfully replicates the microscopic detail of a tool mark. It is dimensionally stable if is stored at room temperature (prolonged storage at 100 degrees Celsius may cause anisotropic expansion of the cast), and it is relatively inexpensive. Silicone rubber casting material is sold as a partially polymerized base with which a catalyst must be mixed to complete the polymerization. The base material is filled with gray or red pigment particles. Silicone rubber casting material is available in two different viscosities: one has the consistency of thick cream, the other the consistency of putty. The thicker material is especially useful for casting tool marks on vertical surfaces. After the catalyst has been kneaded into the mass of base material, the mixture can simply be pressed into the tool mark.

Laboratory Examinations of Tool Marks

In the forensic science laboratory, the tool mark examiner carefully examines the questioned tool mark along with any tools that have been submitted for examination. If the tool mark examiner concludes that the tool mark could have been made by a submitted tool, he or she will use the tool to prepare test tool marks for microscopic comparison. The making of such marks is the most time-consuming part of tool mark examination. For microscopic comparisons of the crime scene tool marks and the test marks to be successful, the test tool marks must be made using the same tool surface as that used to make the crime scene mark; moreover, the test tool mark must be made in the same way (particularly using the same angle of attack). The test tool marks are made using a malleable and ductile material such as lead, tin, or aluminum. These materials accurately reproduce the individual characteristics of the tool surface, but are soft enough that they will not damage the tool surface.

The tool mark examiner makes microscopic comparisons of the test tool mark and the questioned tool mark in order to match markings made by individual characteristics of the tool. Individual characteristics of tools result from manufacturing processes, from wear, and from damage due to misuse. For example, the cutting edges of bolt cutters and wire cutters are usually hand-ground on a grinding wheel. This process results in each cutting edge having a unique pattern of striations. This pattern of striations will in turn produce a unique pattern of markings in a tool mark. However, many modern manufacturing techniques can produce tools that when new do not possess sufficient individual characteristics to permit the tool marks they make to be distinguished from one another. Occasionally what appear to be individual characteristics on a tool are found to be class characteristics. For example, a screwdriver used in a series of break-ins was found to have a pattern of striations on the side of the blade and ripple marks on the face of the blade. These features matched marks left at the crime scenes. However, the forensic tool mark examiner discovered that screwdrivers of the same brand and size purchased at local hardware stores exhibited the same features. The striation pattern on the side of the blade and the ripple marks were produced by the die used to stamp

Figure 21.8 Microscopic comparison of sliding tool marks made with the same screwdriver.

Figure 21.9 Microscopic comparison of cutting tool mark made with the same pair of shears.

out the blades of the screwdrivers. This case emphasizes the need for tool mark examiners to be familiar with the various techniques in the manufacture of tools. Confronted by tool marks made by an unfamiliar tool, the tool mark examiner may have to contact the tool manufacturer for information regarding production methods and even acquire examples of the different stages of tool manufacture. Figure 21.8 and Figure 21.9 show microscopic comparisons of sliding tool marks and cutting tool marks, respectively.

As is the case with firearms examinations, tool mark examiners can reach one of three conclusions:

- *Positive identification.* The class characteristics (as far as they can be determined) are consistent and that individual characteristics match. This means that the likelihood that a tool other than the tool submitted for examination made the tool mark is so remote that it must be regarded as a practical impossibility.
- *Negative identification.* The tool mark was not made by the submitted tool. This means that the class characteristics did not match. In this instance no microscopic comparison will have been undertaken.
- *Inconclusive.* This means that the class characteristics match but that sufficient individual characteristics to declare a match could not be found. The tool mark examiner does not render a negative identification in this instance because the individual characteristics of the tool may have been altered by further use between the time it left the marks at the crime scene and the time it was collected as evidence by investigators.

Tool Marks on Manufactured Items

Tool marks on manufactured items may also be useful in an investigation. For example, in the case of the kidnapping and murder of Charles Lindbergh's son, the planing marks on pieces of lumber used to fabricate the ladder used by the kidnapper led the investigation to the lumber mill in North Carolina where the boards had been cut; the lumber yard where the kidnapper bought the lumber was also identified. Unfortunately, the lumber yard did not keep records of cash transactions, so the identification of the kidnapper had to wait until he spent some of the ransom money.

The following examples of tool marks made during manufacturing have proven to be forensically useful:

1. Hammer marks on the heads of nails and brads
2. Extrusion marks on pipe
3. Machining marks on metal shavings
4. Extrusion marks (and manufacturing defects) in plastic film, plastic cling wrap, and plastic bags
5. Ream marks on sheets of flat glass
6. Punch defect marks on illicitly manufactured drug tablets

Plastic sheet may be produced by extrusion through a die followed by "calendaring" between large rollers. Plastic film and plastic bags are melt-extruded through a circular die (blown-bubble extrusion) and maintained in the form of an air-filled balloon as the hot plastic cools. When plastic film is produced, the balloon is pressed into a sheet between "nip" rolls. When plastic bags are produced, the balloon is allowed to collapse and a heat sealer seals the plastic layers together at specific points rather than throughout their lengths. Both calendaring and extrusion produce patterns of parallel striations on the plastic layers. The calendaring and extrusion processes, as well as the wind-up process, also cause the polymer molecules in the plastic layers to be oriented parallel to the direction of calendaring or extrusion, making the plastic layers **birefringent**. The calendaring and extrusion marks on plastic sheets, plastic film, and plastic bags can be visualized as shadowgraphs or by means of **Schlieren optics**. A polarization table can also be used to compare plastic sheets, plastic film, and plastic bags.

Illicitly manufactured drug tablets can be examined for tool marks left by the pressing process. Samples of seized drug tablets can be compared using a number of features: nature and quantity of the active ingredient (usually a hallucinogen such as LSD or a central nervous system stimulant such as amphetamine); the nature of any **diluent** or excipient (inactive substances such as lactose, maltose, corn starch, or calcium sulfate used to give the dosage bulk); the presence of impurities from the method used to synthesize the active component; physical features such as size and shape, color, and weight; and marks made by punch defects on pressed tablets.

Molded tablets encountered by law enforcement agencies are almost invariably illicitly manufactured. Molded drug tablets are used in the pharmaceutical industry for research purposes only. Molded tables are produced by making the diluent and active ingredient into a paste, pressing the paste into a mold, and allowing it to dry. Unless the molds used to produce the tablets have defects, molded tablets cannot be meaningfully compared, microscopically. Pressed tablets are produced by pressing a mixture of active component and diluent in a die between two punches. The faces of the punches are heat treated to harden them; however, careless handling can damage them so that tablets pressed in them bear microscopically comparable tool marks. The interpretation of such microscopic comparisons must take into account marks on tablets that are not the result of punch defects. "Picking" is caused by the adherence of the drug mixture to the punch face (usually where the punch has a logo or letter on its face). Picking causes small cavities in the surface of the pressed tablets. "**Sticking**" is the result of fine particles sticking to the walls of the die; the particles score the edges of the pressed tablets as they are ejected from the die. The test tool marks used in the microscopic comparisons of pressed tablets are made not by pressing test tablets (a process likely to damage the punch faces if conducted by inexperienced persons), but by making impressions of the punch faces in a semiplastic material such as modeling clay. These microscopic comparisons can determine if a seized punch press made a particular tablet or batch of tablets, if a single punch press or a multistation punch press was used to make a batch of tablets, and if more than one **clandestine drug laboratory** made a particular type of illicit tablet.

Striation Matching for Personal Identification

Striation matching has also been used in personal identification. Human finger and toe nails have striation patterns on their upper and lower surfaces. The size and spacing of these striations is determined by the dermal ridges of the nail bed; the dermal ridges are comprised of dermal papillae, the same structures that also comprise the dermal ridges of fingerprints. It seems reasonable that the pattern of striations on the surfaces of a fingernail or toenail should be unique to a particular nail of a particular person. The limited number of studies of the striation patterns on the fingernails of identical twins have supported this hypothesis. Nails are prepared for examination by pressing them between two flat surfaces, such as microscope slides. The ridges on the under surface of the nail are examined; the ridges on the upper surface of the nail are usually worn smooth and show little detail. The ridges on the nail surface are best observed when the nail is rendered opaque by sputter coating the nail (a procedure commonly used to prepare specimens for scanning electron microscopy). Silicone rubber casts of the nail ridges can also be made and compared microscopically.

Challenges to Firearm and Tool Mark Examinations

In the wake of the 1993 Daubert decision and the extended struggle over the admissibility of DNA evidence, the so-called identification sciences (fingerprint identification, questioned document examination, shoe and tire track examination, and firearm and tool mark examinations) have come under increased scrutiny by the courts. Critics of firearm and tool mark examinations have claimed that these examinations are wholly subjective. Firearm and tool mark examinations are partially objective and partially subjective. The microscopic examinations carried out by examiners are objective; the examiners' interpretations of those examinations in terms of whether a particular firearm or tool made the marks examined microscopically do have a subjective component. However, courts have held that subjectivity does not render tests unscientific and inadmissible. Critics of firearm and tool mark examinations have also argued that markings on bullets and cartridges and in tool marks lack sufficient individuality for the examiner to assert that a questioned bullet or cartridge was fired in a particular firearm or that a particular tool made the questioned tool mark. And they have objected to firearms examiners testifying that their examinations absolutely identify the firearm or tool that made the evidentiary marks that have been microscopically examined. Some courts have agreed that the conclusions firearm and tool mark examiners should be limited to reasonable scientific certainty, rather than absolute certainty. Both the Association of Firearm and Toolmark Examiners (AFTE) and the Scientific Working Group for Firearms and Toolmarks (SWGGUN) have adopted interpretive guidelines that permit examiners only to state that the likelihood that another firearm or tool made the marks is so remote as to be considered a practical impossibility. Critics have also asserted that the methods of firearm and tool mark used in the microscopic comparisons of bullets, cartridges, and tool marks have not been adequately validated. While the number of validation studies is not large, some large-scale studies have been performed (without error, it is important to note). Finally, critics have asserted that proficiency testing of firearm and tool mark examiners is inadequate. This charge has some merit. However, those forensic science laboratories that are accredited by the American Society of Crime Laboratory Directors–Laboratory Accreditation Board (ASCLD–LAB) are required to have all their laboratory examiners participate in frequent proficiency tests. Moreover, the examiners' reports and courtroom testimony must also be monitored, to insure that the examiners' conclusions are well-grounded scientifically. The results of firearm and tool mark proficiency tests conducted by Collaborative Testing Service of McLean, Virginia (currently the only source of firearm and tool mark proficiency tests), have been examined to estimate the error rate for these examinations: for the two time periods whose test results were examined, the false identification rate ranged between 1.0 and 1.3 percent. To date, no challenge to the scientific basis of firearm and tool mark examination has been successful.

Case Study

Police were summoned to the apartment of two women where a shooting had occurred. One of the women was found dead of a perforating (through and through) gunshot wound of the chest. The decedent was wearing a short dressing gown at the time of the shooting. The other woman claimed that she and the decedent had had a lovers' quarrel and that she (the survivor) had attempted to commit suicide with a .22-caliber pistol. A struggle for the gun had ensued, during which the pistol accidentally discharged, killing the decedent instantly. An examination of the garment worn by the decedent revealed two bullet holes: one with bullet wipe under the left arm and a second in the right chest area. No gunshot residue was found either by a visual examination or by chemical tests. Test firings of the .22-caliber pistol with ammunition from the same lot as that with which the pistol was loaded at the time of the fatal shooting showed that the combination of pistol and ammunition produced visible gunshot residue deposits out to a range of 25 inches. Several conclusions could be drawn from these observations. First, the fatal bullet traveled in a left-to-right direction through the victim's chest. Second, the shot was fired from a range greater than 25 inches—a distance greater than the average adult woman's arm length. It was, therefore, highly unlikely that the decedent was struggling for control of the pistol. The surviving woman was charged with murder and subsequently convicted.

Questions

1. How should firearms be marked as evidence?
2. What general rifling characteristics of a firearm can be determined from a fired bullet?
3. In what ways can the caliber of a firearm be determined from a fired bullet?
4. What class characteristics of a firearm can be determined from an expended cartridge?
5. What markings on fired bullets are compared microscopically?
6. What markings on fired cartridges are compared microscopically?
7. What conclusions can a firearms examiner reach as result of a microscopic comparison of bullets or cartridges?
8. How is the range of fire estimated from a powder pattern?
9. How can the range of fire be estimated from a shotgun pellet pattern?
10. What are the three types of tool marks?
11. How should tool marks be processed at the scene of a crime?
12. What conclusions can a firearms examiner reach as result of a microscopic comparison of tool marks?
13. The energetic material in single-base smokeless powder is _____; the energetic materials in double-base smokeless powders are _____ and _____.
14. In a rifled gun barrel the raised areas are called _____ and the recessed areas are called _____.
15. Cut rifling methods are _____, _____, and _____.
16. Barrels with polygonal rifling are made by _____.
17. The two types of metallic cartridge priming systems are _____ and _____.

18. Markings made on the bases of cartridges by breech blocks are also called _____.

19. Three commonly used chemical methods for the visualization of powder patterns produced by ammunition having lead-based primers are _____, _____, and _____.

20. _____ reagent is used for the visualization of powder patterns fired with ammunition containing lead-free primers.

21. An irregular, blown-out entrance wound in the skull is termed a _____ defect.

22. Gunshots that are fired from so far away from the target that no gunshot residue reaches the target surface are termed _____ shots.

23. The ring of bullet lubricant, gunpowder combustion products, and metal from the bullet surface surrounding a gunshot wound is called _____.

24. Two types of firearm silencers are _____ and _____.

25. _____, _____, and _____ have been used to make casts of tool marks.

Suggested Readings

Alakija, P., Dowling, G. P., and Gunn, B., Stellate clothing defects with different firearms, projectiles, ranges and fabrics, *J. Forens. Sci.*, 43, 1148, 1998.

Burke, T. W. and Rowe, W. F., Bullet ricochet: a comprehensive review, *J. Forens. Sci.*, 37, 1254, 1992.

Cork, D. L., Rolph, J. E., Meieran, E. S., and Petrie, C. V., *Ballistic Imaging*, National Academies Press, Washington, D.C., 2008.

Davis, J. E., *An Introduction to Tool Marks, Firearms and the Striagraph*, Charles C. Thomas, Springfield, IL, 1958.

DiMaio, V. J. M. and DiMaio, D., *Gunshot Wounds: Practical Aspects of Firearms, Ballistics, and Forensic Techniques*, 2nd ed., CRC Press, Boca Raton, FL, 1999.

Du Pasquier, E. et al., Evaluation and comparison of casting materials in forensic sciences—application to tool marks and foot/shoe impressions, *Forens. Sci. Int.*, 82, 33, 1996.

Garrison, D., *Practical Shooting Scene Investigation: The Investigation and Reconstruction of Crime Scenes Involving Gunfire*, Universal Publishers, Boca Raton, FL, 2003.

Geradts, Z., Keijzer, J., and Keereweer, I., A new approach to automatic comparison of striation marks, *J. Forens. Sci.*, 39, 974, 1994.

Glattstein, B. et al., Improved method for shooting distance estimation. I. Bullet holes in clothing items, *J. Forens. Sci.*, 45, 801, 2000.

Haag, L. C., *Shooting Incident Reconstruction*, Academic Press, Burlington, MA, 2005.

Hamby, J. E., Firearms reference collections—their size, composition and use, *J. Forens. Sci.*, 42, 461, 1997.

Hatcher, J. S., Jury, F. J., and Weller, J., *Firearms Investigation, Identification, and Evidence*, Stackpole Books, Harrisburg, PA, 1977.

Heaney, K. D. and Rowe, W. F., The application of linear regression to range-of-fire estimates based on the spread of shotgun pellet patterns, *J. Forens. Sci.*, 28, 433, 1983.

Hogg, I. A., *The Encyclopedia of Infantry Weapons of World War II*, Thomas Y. Crowell, New York, 1977.

Hogg, I. A. and Adam, R., *Jane's Guns Recognition Guide*, HarperCollins, London, 1996.

Maehly, A. and Stromberg, L., *Chemical Criminalistics*, Springer-Verlag, Heidelberg, 1981.

Maiti, P. C., Powder patterns around bullet holes in bloodstained articles, *J. Forens. Sci. Soc.*, 13, 197, 1973.

Mathews, J. H., *Firearms Identification*, Vols. I–III, Charles C. Thomas, Springfield, IL, 1962.

Meng, H. and Caddy, B., Gunshot residue analysis—a review, *J. Forens. Sci.*, 42, 553, 1997.

Millard, J. T., *A Handbook on the Primary Identification of Revolvers and Semiautomatic Pistols*, Charles C. Thomas, Springfield, IL, 1974.

Nichols, R. G., Firearm and toolmark identification criteria: A review of the literature, *J. Forens. Sci.*, 42, 466–474, 1997.

Nichols, R. G., Firearm and toolmark identification criteria: a review of the literature, Part II, *J. Forens. Sci.*, 48, 318–327, 2003.

Nichols, R. G., Defending the scientific foundations of the firearms and tool mark identification discipline: Responding to recent challenges, *J. Forens. Sci.*, 52, 586–594, 2007.

Rinker, R. A., *Understanding Firearm Ballistics*, 6th ed., Mulberry House Publishing Company, Corydon, IN, 2005.

Rowe, W. F., Statistics in forensic ballistics, in *The Use of Statistics in Forensic Science*, Aitken, C. G. G. and Stoney, D. A., Eds., Ellis Horwood Ltd., New York, 1991.

Rowe, W. F. and Hanson, S. R., Range-of-fire estimates from regression analysis applied to the spreads of shotgun pellet patterns: results of a blind study, *Forens. Sci. Int.*, 28, 239, 1985.

Scott, D. D. and Fox, R. A., *Archaeological Insights into the Custer Battle: An Assessment of the 1984 Field Season*, University of Oklahoma Press, Norman, OK, 1987.

Starrs, J. E., Once more unto the breech: the firearms evidence in the Sacco and Vanzetti case revisited: Part I, *J. Forens. Sci.*, 31, 630, 1986.

Starrs, J. E., Once more unto the breech: the firearms evidence in the Sacco and Vanzetti case revisited: Part II, *J. Forens. Sci.*, 31, 1050, 1986.

Stone, I. C. and Petty, C. S., Examination of gunshot residues, *J. Forens. Sci.*, 19, 784, 1974.

Warlow, T., *Firearms, the Law, and Forensic Ballistics*, 2nd ed., CRC Press, Boca Raton, FL, 2004.

Questioned Documents

22

Howard Seiden and Frank H. Norwitch

Introduction

In the broadest terms, a **document** is any fixed method of communication between one individual and another. A questioned document is one that in its entirety or in part is suspect as to authenticity or origin. Questioned documents include more than checks, wills, and contracts. A typewritten letter, dollar bill, postage stamp, gas station receipt, or concert ticket can be a questioned document.

The field of questioned documents is one of the older disciplines in the forensic sciences. According to J. Newton Baker's book, *The Law of Disputed and Forged Documents*,[1] "forgery was practiced from the earliest times in every country where writing was the medium of communication." The rule for the identification and comparison of handwriting can be traced back to Roman law under the Code of Justinian in 539 A.D.

Today, the field of questioned documents is credited to the pioneering work of Albert S. Osborn, who was born on March 26, 1858. His publication *Questioned Documents* set forth the basic principles that document examiners still utilize. This book is considered the bible of questioned documents. The significance of the document examiner was brought to the attention of the legal system in the Charles Lindbergh baby kidnapping trial, which became known as the "Trial of the Century." Osborn's testimony was crucial during the trial, demonstrating that the ransom note was written by the suspect Richard Hauptmann.

In the early 1930s and 1940s, the FBI and the U.S. Postal Department, respectively, began their questioned document laboratories. Today many states and county law enforcement agencies employ document examiners.

Functions of a Forensic Document Examiner

Who wrote the threatening letter to the politician? Whose signature is this on the dotted line of the contract? Is the endorsement signature on the back of the check genuine? When was this document typed? What does the faded writing say? Is this a genuine or counterfeit document? An individual who is trained in the field of questioned documents is the person who can answer these questions.

A forensic document examination can involve the comparison of handwriting and signatures, typewriters and printing devices, **alterations** and obliterations of documents, counterfeiting, photocopy manipulation, rubber-stamp impressions, inks, paper, and much more. The document examiner may be involved in criminal cases involving written threats, anonymous letters, extortion, fraud, identity theft, elderly abuse, white-collar crime, and contract disputes. Many examiners are employed by government agencies. Other examiners in private practice are involved in civil casework, such as medical malpractice and **insurance fraud**.

A document examiner's expertise may be needed for the verification of a person's signature on a sign-in sheet. This may be used to prove or disprove the alibi of an individual who claims to have been at a particular location or present at a certain time. The identification of an individual's writing on checks, credit card invoices, or contracts may reveal those who are involved in fraud. The writing-on-demand notes or threatening letters can serve to identify the individual.

From the examiner's analysis, an opinion is rendered in the form of a report that the attorney can use to assist his or her case. If the case goes to trial, the document examiner should be prepared to state his or her opinion clearly to the judge or jury and demonstrate to the court why such an opinion was formed. Such a demonstration usually takes the form of an enlarged photograph of the known and questioned material. The examiner can then point out significant similarities or dissimilarities and discuss their merits before the court.

The document analyst examines only the physical characteristics of the signature or writing. The examiner cannot tell the person's personality from handwriting. (This is known as **graphology**, the role of which in the criminal justice system has been debated.) The document examiner cannot tell the sex, age, race, or educational level of a person from his or her writing.

Becoming a Document Examiner

Okalahoma State University offers online coursework leading to a graduate certificate in forensic examination of questioned documents (QD). This was established in cooperation with the American Board of Forensic Document Examiners. This course of study supports programs that prepare document examiners, trainees, for certification. OSU also has a Master's of Forensic Science Administration Program, (M.F.S.A.) providing an alternate track to the QD graduate certificate. Other schools that offer a degree in forensic science have coursework dealing with the study of questioned documents. George Washington University in the District of Columbia offers a course in questioned documents in its graduate program. Virginia Commonwealth University in Richmond, Virginia has a criminalistics course, which includes questioned documents. Marshall University in Huntington, West Virginia, offers a forensic comparative science class that lists QD as one topic. These are introductory courses with a limited amount of time to deal with this study.

A single course taken at a college does not prepare a person to become a qualified document examiner. Trainees learn the discipline by on-the-job training within a government crime laboratory. Usually a training program consists of a 24-month internship. The trainee works and learns from a senior document examiner. There the trainee has practical problems to work, literature to read, and research to do. There are lectures, tests, actual casework with the trainee giving reasons for arriving at an opinion, and mock trials. The FBI and Secret Service offer schools in questioned documents that the individual should attend for additional training. Examiners should also continuously avail themselves of the various QD seminars and workshops presented by professional affiliations and law enforcement agencies.

Theory of Handwriting

Think back to your early days in school when you spent a period learning how to write. There in class, above the blackboard, were charts demonstrating a particular writing style. Instruction was given on the correct printing of letters, and later on how to write cursively. Depending on your geographic location, you were taught one of various **copybook** styles of **penmanship**, such as the **Palmer** and **Zaner–Bloser** methods. During this early learning period, attention was focused on how to write the letters of the alphabet correctly. As the individual matures, his or her writing tends to deviate from the copybook style of writing, and the emphasis shifts to what is being written rather than how the letters are formed. Neuromuscular coordination and visual perception differ from one individual to another. Individuals may incorporate shortcuts from the copybook style or add an extra flair to their writing, perhaps because they saw it in someone else's writing. By the late teenage years, a person's writing has matured to the point where his or her writing style is unique. Handwriting is an acquired skill that becomes ingrained; it is habitual as well as individualized. This individualization is a basic principle in document examinations.

Collection of Writing Standards

Before an examination can be made, known standard writing must be submitted for comparison. Usually the investigator submits standards for comparison. If the investigator does not submit proper standards, an examination may be limited in scope or not occur. It is crucial for the investigator and document examiner to have a good working liaison.

When submitting written standards, one should realize that like must be compared to like. For example, if the questioned material is printed, the known standards must be printed. If the questioned document is written cursively, it must be compared to cursive writing. If the writing is in pencil, the standards should be written in pencil. A general rule is to duplicate as much as possible the same conditions that occurred when the questioned material was written. Items such as writing instrument, writing position (if known), and type of paper (ruled or not) may be important conditions of the writing act.

Two classes of writing standards are utilized for comparison purposes. These are nonrequest writing, also known as spontaneous or undictated writing, and requested writing or dictated exemplars. Both types of standards have benefits and disadvantages. The nonrequest or undictated writing—material written by the individual during the everyday course of business—is likely to reveal the normal writing **habits** of the individual. No circumstances call attention to or provide undue emphasis to the act of writing. The writer, unaware that the writing will be used as a standard for comparison, is not likely to alter his or her handwriting for the purpose of disguise. The disadvantage is having the nonrequest written material authenticated for court, and obtaining enough comparable letters and words can be difficult.

Requested exemplars are standards written at the request, and usually in the presence, of the investigator or examiner. Their advantage is that they provide writing that is comparable to the questioned material, and authentication is easily accomplished. The inherent problem with requested standards is that they call attention to the writing process. This may inhibit the writer because of nervousness, or may allow the writer to attempt to distort his or her writing for the purpose of disguise. From the examiner's perspective, a combination of both requested and nonrequest writing standards serves as the best material for comparison to the questioned document. The addition of **nonrequest standards** serves as a check against the individual who may attempt some form of disguise.

Normal course-of-business writing standards can be obtained for handwriting examination from:

- Applications (credit, employment, insurance, loan, rental)
- Bank records (deposit slips, cancelled checks, safe deposit record, signature cards)
- Birth certificates
- Business contracts and agreements
- Employment records
- Letters of correspondence
- Real estate (contracts, listings, warranty deeds)
- Receipts (credit card, cash, delivery)
- Registers (attendance, motel, visitor)
- School records
- Tax returns
- Time sheets
- Wills

Process of Comparison

A document examiner compares questioned handwriting or signatures side by side to the known standards. Handwriting attributes are examined both visually and microscopically. Everyone who looks at writing and signatures notices the most conspicuous features first, such as the **slant** of the writing and how the letters are formed. An examiner will look beyond the obvious features and study the subtle, inconspicuous aspects of the questioned signature or writing. By applying basic rules in document analysis, combined with experience observing thousands of letter formations and words, an expert examiner is able to determine if writing is genuine or is not.

A good analogy to handwriting identification taught to beginners is that you have been given a general description of a person. He is male, 30 years old, with dark hair and eyes, 170 pounds, 6 feet tall, with a scar on his forehead. He walks with a permanent limp and has a tattoo of a rose on his left arm.[2] You must find this individual among a group of passengers who are coming off a plane at the airport. The first five characteristics are common; many men fit that general description. With the addition of the next three uncommon characteristics, the field narrows significantly. With all the traits combined, when you see this individual and your brain has processed the description, you will recognize him in the crowd. If the individual differed in weight by a few pounds or in age by a few years, would not be significant. The general description could be off slightly without changing the identification. However, if one of the last three traits were missing, that would be significant, and you may not have the right individual.

The analogy applies to handwriting. Some writing features are common, and some handwriting characteristics are considered uncommon or even rare. The common features are referred to as class characteristics. These are writing attributes observed in a group of writers that are probably derived from a penmanship system they learned. The uncommon handwritten characteristics, known as individual characteristics, are considered distinctive, personal, or peculiar to the handwriting of one person. An experienced document examiner is able to recognize class characteristics and avoid identifying an individual's writing solely on the basis of these common handwriting features. If the writing is naturally executed, and a combination of similarities between the questioned material and known standards is significant and individual, the examiner renders an opinion that the questioned and known material were written by the same individual. If the questioned writing or signature contains a combination of significant dissimilarities or indications of forgery, the examiner may proffer an opinion of not genuine. In doing a comparison, an examiner studies characteristics, such as how letters are constructed, how they are connected, the **beginning** and **ending strokes** of letters, the relative **height ratio** of letters, the spacing between letters and words, the **skill level**, speed, size, and **shading**.

In order to account for the variation in a person's writing, an examiner needs an adequate number of writing or signature standards to compare. Writing variation represents the alternate forms of a single handwritten characteristic found in a person's writing. One principle in document examination is that no two individuals write exactly alike, and another principle is that no one person writes exactly the same way twice. An individual has a repetitive range to his or her writing. Not every letter will be exactly the same or every beginning or terminal stroke of a letter the same. Every time a person writes, the pen or pencil may start at a slightly different speed or point on the paper. However, a basic pattern or habitual style is still inherent within a person's writing. An examiner looks for this pattern in the standards. He or she can then determine whether the questioned writing is within the range of a person's variation.

An illustration of this concept is the "th" height ratio in the word "the." Is the "t" higher than the "h," lower than the "h," or the same height? An examiner would study all the words that a person writes that contain a "th." Let us suppose that one individual habitually writes the "t" higher than the "h." At times, the "t" may be much higher than the "h" and sometimes just a little bit higher. Any slight change with respect to the height of the "t" would be considered variation within his or her writing, as long as the basic habit of making the beginning "t" higher than the "h" remains. A study of two hundred individuals by Muehlberger[3] showed that 5.5 percent made the "t" taller than the "h," 78 percent made the "t" shorter than the "h," 15 percent showed no pattern, and 1.5 percent made the "t" even with the "h."

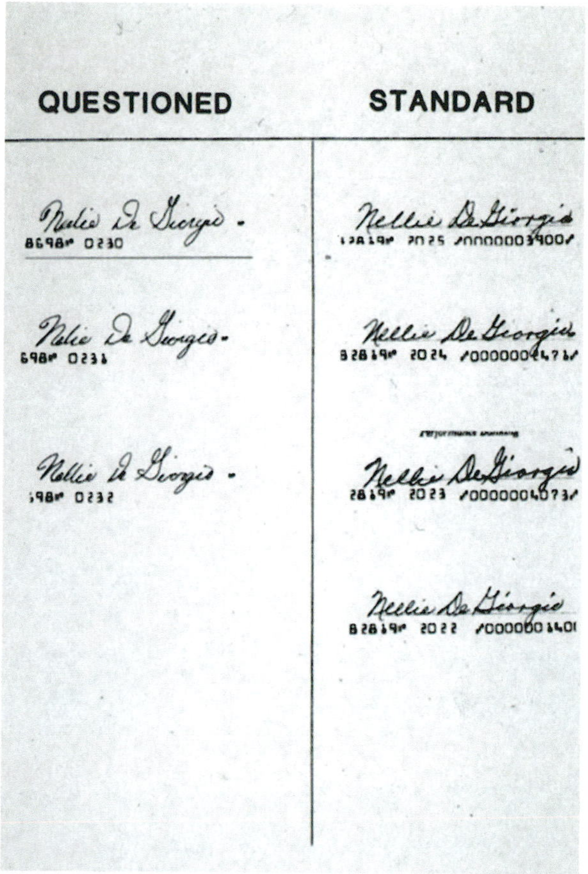

Figure 22.1 A comparison chart used in court to demonstrate a freehand simulation.

Types of Fraudulent Writing

Types of fraudulent writing include **freehand simulations**, tracings, and normal hand forgeries. A freehand simulation is an attempt to draw the signature or writing of another person (Figures 22.1 and 22.2), usually when working with a model signature. This type of forgery requires that the individual maintain the same speed as the original writing and imitate the correct letter formations, height ratio, and **pen pressure** at the same time. This presents the forger with quite a challenge. If his or her attention is directed at maintaining the speed of the writing, as in a stylized signature, then less concentration can be focused on the replication of correct letter formations and **connecting strokes**, and the chances are greater that these characteristics will be wrong. If the forger attempts to make the letters and connecting strokes correct, then he or she will pay less attention to speed of the signature. The **line quality** will suffer—the signature will display an awkward uneven line that appears hesitant and drawn. Adding to this challenge, the forger must continuously suppress his or her own writing habits so they will not be revealed in the simulation. Upon closer microscopic examination, evidence may be observed

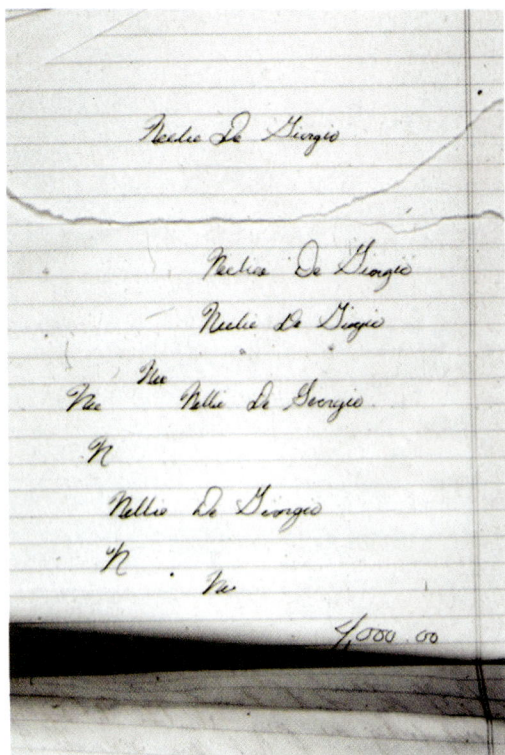

Figure 22.2 A practice sheet revealing the forger's attempt to write the victim's signature.

of the forger's **patching** or **retouching** of the written line to make it appear closer to the genuine signature. Additional characteristics of **blunt starts** and stops, and **pen lifts**, may also be revealed. A competent examiner should have little problem in discerning a freehand simulation.

Tracing, another common type of simulation, involves using an original signature or writing as a guide to produce a fraudulent document. Indications of a tracing may be the presence of guidelines around the questioned signature, such as graphite from a pencil or remnants of a line from carbon paper. Indented impressions around the questioned writing may signify that an impression of the original signature was made with a pointed object and then a pen used to go over the impression. A more direct approach is to take the original signature or writing, place the fraudulent document over it, and trace it. A tracing will normally reveal poor line quality. Instead of the written line appearing smooth and free flowing, it will be uneven and wavy, and appear to have been drawn slowly. The

final product will not reveal the shading differences in a freely and naturally executed signature that is written with a fair amount of speed.

In a **normal hand forgery**, the individual does not attempt to copy the victim's signature or writing. The forger either writes the name in his or her own writing style or tries to distort it. **Disguised writing** is another form of fraudulent writing. The individual attempts to alter his or her writing to be able to deny later that he or she was the author. Methods of disguise vary from altering the slant of writing to changing the size of writing. Some individuals alternate between upper case and lower case letters or printed and cursive forms of letters. Others add additional strokes to letters. The more a person writes, the more difficult it is to suppress habitual writing characteristics. Since the act of writing is an acquired skill that has become habitual, a person will find it challenging to maintain an effective disguise. The smaller, inconspicuous writing characteristics of the individual are revealed the more a person writes. It is difficult to write effectively and concentrate at the same time on altering learned handwriting habits.

Factors That Can Affect Handwriting

Many factors can affect handwriting. For a more detailed overview, the reader should consult the chapter references. The health of the writer is a consideration. Various disorders can produce muscular weakness that prevents proper writing control or reduces the ability to hold a writing instrument correctly. For example, an individual who normally has legible handwriting may suddenly be overtaken by arthritis. The resulting change in handwriting may be abrupt and cause the writing to become unrecognizable as that individual's writing. A stroke may change a person's handwriting severely. The resulting product may be illegible and totally different from the writing before the stroke.

Parkinson's, usually a disease of the elderly, can also affect the middle-aged. It causes the destruction of brain cells, resulting in muscular **tremor** and weakness that have a noticeable effect on handwriting; the hand tremor of Parkinson's results in the wavering of the written line. Essential tremor, another neurological condition commonly observed in the elderly, causes the arm to shake and leads to difficulty in handwriting.

As individuals age, their handwriting may undergo noticeable deterioration. A loss of pen control and smoothness of the line can be observed. Senility can cause changes in handwriting. It is important for the examiner to receive written standards that are contemporaneous with the time the questioned document was written. For example, a will that was written 5 years ago now bears a questioned signature. This signature is naturally executed and stylized. Signatures of the deceased are provided that date back to only a year ago. These standard signatures exhibit a loss of pen control and deterioration in letter formations. A likely explanation is that the individual suffered some recent debilitating condition that affected his or her writing capability. In order to facilitate an examination, the examiner should request signature specimens that were written at about the same time as the disputed signature.

Alcohol's effect on writing has been studied extensively. With increasing blood alcohol levels, the quality of writing starts to degrade. As muscular coordination becomes poorer, the writing becomes less legible and the size of the writing may increase. The writer becomes careless and introduces errors. These effects may not be the same for each individual.

Challenge to the Document Field

The testimony of document examiners has been favorably accepted by the courts for many decades. However, a 1989 article by Risinger, Denbeaux, and Saks[4] criticized the use of expert handwriting testimony and its use in the courts. The authors state that there is no analysis of evidence that document examiners can perform better than nonexperts, and that the error rate in document examiners' investigations is high.

The courts have begun to scrutinize expert witness testimony. In the 1995 case *U.S. v. Starzecpyzel,* SSOF. Supp. 1027 (S.D.N.Y. 1995) the court heard testimony from Saks critical of handwriting examinations. The court ruled that testimony provided by the document examiner qualifies as technical, but not scientific, knowledge—that document examiners provide practical rather than scientific expertise.

In a study by Kam, Fielding, and Conn,[5] experts and nonexperts were tested for proficiency in handwriting identification. The error rate was 6.5 percent for the experts and 38.3 percent for nonexperts. The nonexperts erroneously matched documents that were created by different writers at a rate almost six times greater than the experts.

Kam et al.[6] also conducted research on signature examination between experts and nonexperts. Kam reported an error rate of less than 1 percent (.49 percent) for document examiners, compared to the nonexpert error rate of 6.47 percent, when false signatures were declared authentic. The error rate for document examiners was 7.05 percent, and 26.1 percent for the nonexperts, when authentic signatures were declared not genuine. The results of these tests illustrate that the professional document examiner possesses skills that exceed those of the nonexpert and refute the claims by Risinger, Denbeaux, and Saks.

In an affirmation of the testimony of document examiners, on May 13, 1999, the U.S. Court of Appeals for the 11th Circuit upheld Judge Tidwell's ruling in *U.S. v. Paul,* 175 F3d 906. The district court not only properly admitted expert testimony involving handwriting evidence, but also excluded Denbeaux from testifying. The court stated that his education and training as a lawyer did not qualify him to testify as to the reliability of handwriting examinations because he had no training or education in this field.

A study has been published to validate a basic premise in handwriting that no two individuals write exactly alike. Dr. S. N. Srihari of the Center of Excellence for Document Analysis and Recognition (CEDAR) collected samples of handwriting from 1500 individuals. These samples were examined for such attributes as darkness, slant, height, structure, and concavity of characteristics. Based upon computer algorithms, an extremely high degree of confidence level was established for writer identification. If the results were expanded to the entire U.S. population, then the individuality of handwriting can be validated to a 95 percent confidence. If the finer characteristics were considered that a document examiner looks for, then a confidence level would near 100 percent.[7] This study assists in scientifically supporting handwriting evidence for court.

Future Trends

Researchers are creating software programs that will assist in the analysis of handwriting comparisons. A computer vision system is being developed that may assist the examiner in the style of handwriting and perhaps the nationality or origin of an individual. The author has collected 33 Roman alphabet copybook styles from various countries. The characters were segmented and matched against the copybook styles with a higher success rate when more characters are present in the questioned material.[8]

Alteration, Obliteration, and Ink Differentiation

Often a document examiner discovers and deciphers an alteration to or obliteration of a document, or determines an ink differentiation. The obvious and discernible evidence of a second writing instrument may well prove that an addition has been made after the fact. Occasionally, an examination of what appears to be a rather straightforward medical record reveals a trail of deception and attempts to mask evidence of wrongdoing.*

In addition to a naked-eye or low-magnification visual examination of the questioned material, an examination may also incorporate the use of selective color filters or color filter combinations. This could be the method of choice, for instance, in the examination of obscured, blood-stained documents. Although at times this visual examination reveals evidence leading to a definitive opinion, usually a more thorough examination is necessary. Viewing by microscope may reveal subtle differences, such as a slight change in shading or hue, within the questioned material. Secondary lines, indicative of alteration when present, are usually discovered during this process. More in-depth examinations of the questioned document using selective portions of the **light spectrum** can be accomplished either photographically or by a process generically known as **video spectral comparison** (VSC) (Figures 22.3 and 22.4) or by chemical analysis of the ink, known as thin layer chromatography (TLC).

Combinations of two or more of the following techniques can allow for extremely definitive opinions concerning alterations and obliterations, or ink and pen differentiation.

Video Spectral Comparison (VSC)

While infrared imaging equipment and infrared photography have opened up a new world for the questioned document examiner, the concept is not new. Frequently, we read about the deployment of a new "spy" satellite that employs infrared imaging. "Snooper" scopes and infrared binoculars have been in use for decades.

The human eye perceives color as a reflection of a portion of the light spectrum. When we observe somebody wearing a red shirt, the chemical composition of the dyes in the shirt material has absorbed the entire visible light spectrum with the exception of that

---

* The procedures outlined here may help determine that there was no alteration or addition to a record. The majority of medical records that are examined during the course of a malpractice suit display no evidence of anything extraordinary, but you may not know that just from looking at them.

Figure 22.3 Video spectral comparator (VSC) revealing security features using ultraviolet light.

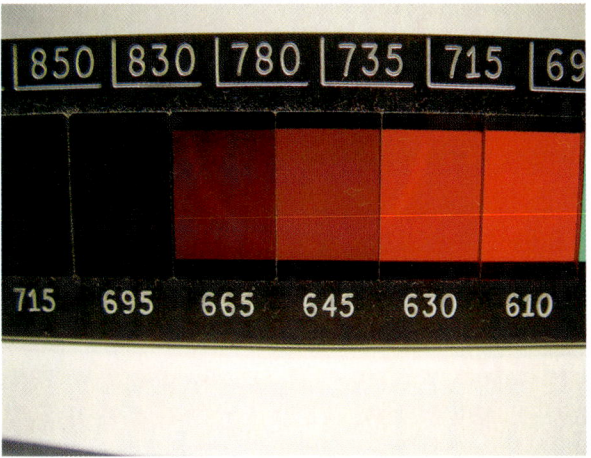

Figure 22.4 Infrared filters used to discriminate inks.

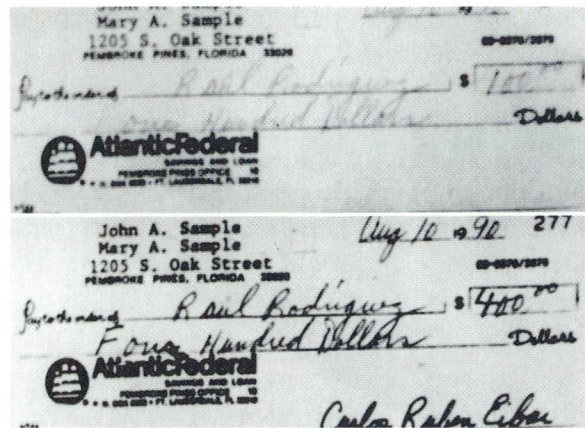

Figure 22.5 An example of check alteration using VSC **infrared reflectance**. The bottom portion of the photograph reveals the altered check in the amount of $400. The top part of photograph displays the original entries.

portion that we have come to interpret as red. However, just as dogs have the ability to hear a much wider spectrum of frequencies than the human ear is capable of hearing, there is much more to the color spectrum than what we are capable of perceiving with the unaided eye. Every time we see a rainbow we observe a graphic display of the visual portion of the light spectrum. The rainbow of colors that we observe, from red to blue, does not, however, define the limits of that portion of the electromagnetic spectrum that is known as color. We cannot observe an area on each side of that rainbow; these areas are ultraviolet (UV) and infrared (IR).

Although we are not able to see in the infrared or ultraviolet regions of the light spectrum with the naked eye, we can use instrumentation to convert those wavelengths into visible images when examining a questioned document. Different inks that appear similar to our unaided eyes may react quite differently when viewed under ultraviolet light or with the use of infrared imaging techniques.

When examined using infrared, an ink can be observed to luminesce or glow, be transparent, or appear unchanged, depending upon its chemical properties (Figure 22.5). The infrared examination may use specialized light filters and films for photographic imaging or equipment specifically designed for infrared imaging. This process is referred to as video-spectral comparison, and the equipment is usually referred to as a video spectral comparator.*

In addition to the VSC being used on ink examinations, this instrument is being routinely employed to detect forgeries in passports,

---

* While there is some conjecture as to the origin of the VSC technology, the acknowledged standards of equipment are the VSC-1, VSC-4, and VSC-2000, developed by Foster and Freeman of the UK.

visas, identification cards, immigration documents, etc. Many security documents utilize luminescent features, such as holograms and watermarks. The VSC can easily check for authenticity. Additional features include an embedded information decoder. This is where personal information is invisible on passports. The individual simply selects the country and the appropriate decoding system is applied to detect forgeries.[9]

Thin Layer Chromatography (TLC)

Another highly selective process that is occasionally employed is thin layer chromatography (TLC). The TLC process is usually used for ink comparisons rather than obliterations. Although by its nature a destructive test, TLC is a most definitive process.[*] When used in conjunction with the other methods described above, TLC affords the examiner the most frequent opportunity to issue conclusive opinions concerning two or more inks on one document. Small, almost microscopic "punches" of the ink are taken from a portion of the written line that is least likely to play a prominent part in any subsequent handwriting examination. Places where the moving pen changed direction or where ink lines intersect are avoided. Punches are made by using a blunted hypodermic needle, preferably one with a small diameter such as those used for insulin injections. These work admirably and do little damage to the document. These punches (some three or four or more) containing portions of the suspect ink are placed into a small test tube, and the ink is separated from the paper portion of the punch by the introduction of a solvent such as pyridine. The resultant ink and solvent solution is spotted onto paper or glass plates specifically made for chromatography and dried.

The paper strip or glass plate is then placed into a beaker or similar glass container containing a small quantity of an alcohol solvent that covers the bottom to a depth of perhaps one half inch, and covered. The solvent diffuses up the chromatographic medium, reaches the ink spots, and continues its upward movement. Portions of the ink spot separate into bands of color and migrate upward along with the solvent. Each of these (usually three or four) bands reaches a stopping place on the strip. A comparison of bands created by different suspect ink areas may at times allow for conclusive opinions of difference. However, even if the TLC bands created from different ink spottings appear in the same pattern, a conclusive opinion that both the questioned inks are from a common source (or writing instrument) is not possible. Each ink manufacturer fills thousands of writing instruments, sometimes for different pen distributors, with ink of the same formulation. These formulations are changed infrequently over the years. The best that can be said is that the inks appear consistent and could have come from a common source. Although most TLC processes have the best results when traditional ballpoint pen inks are involved, other solvents can be employed to use TLC for other inks, such as those found in rollerball or plastic tip pens.

Ink Line Striae

Another method for differentiating between different writing implements (usually limited to ballpoint pens) is microscopic observation of the striae (lines of noninked areas), or lack thereof, left in the ink line by imperfections in the ball or ball housing of the writing pen.

Indented Writing

Indented writing or second page writing is the impression from the writing instrument captured on the second sheet of paper below the one that contains the original writing. This most often manifests itself on pads of paper. Indented writing can be a source of identification in anonymous note cases and is an invaluable investigative procedure when medical records are suspected of containing alterations.

Often, a writing addition to a record or file can be revealed by an impression that has been transferred to the page below. Indented writing on subsequent pages may not be in

---

[*] The term "destructive," while technically correct, is an overstatement. If the TLC test is performed with caution, the "destruction" may require microscopic observation to be seen.

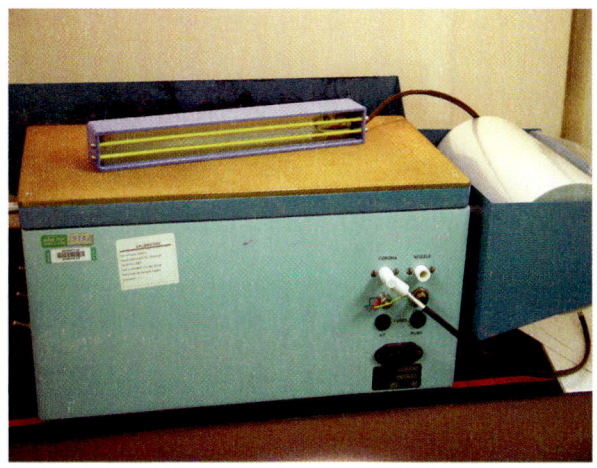

Figure 22.6 The ESDA and the electrically charged wand.

agreement with what appears on the surface of the document. Writing found to be out of position, missing, or added after the fact can often be demonstrated by recovering and preserving indented writing from other pages.

Mystery novels and television and movie plots sometimes depict recovery of indented writing as part of a clue. The method they usually show for reading indented writing from suspect pages is to rub a soft lead pencil or carbon paper over the surface of the document, causing the indentations to be highlighted in relief. Although entertaining, this technique is one way to destroy what might be valuable evidence and should serve as a warning against amateur examinations. Indented writing is normally recovered either photographically using oblique (glancing) light or by use of electrostatic detection apparatus (**ESDA**) (Figure 22.6).

Photography

Until recently, the forensic document examiner applied oblique, or glancing, light to the furrows of indented writing. Photography was then employed to preserve the shadowed indentation. A combination of multiple exposures taken while moving the light source fills in the available indentations with shadow and effectively reproduces the indented writing. While such techniques are often acceptable, they lack the ability to recover invisible microscopic indentations (those occurring

three or four pages down) and have an inherently lengthy processing time.

Electrostatic Detection

The modern well-equipped forensic laboratory employs electrostatic detection to recover indented writing. With ESDA, indented writing can be recovered three, four, or even more pages below the original writing.

A preliminary examination eliminates documents or cases in which the material to be examined is unsuitable for the detection and recovery of indented writing by electrostatic detection. Documents that have been previously processed for latent fingerprints with ninhydrin, or have been saturated with fluids, normally fall into this classification. Thick cardboard mediums are usually incompatible with ESDA.

The document to be processed may need to be humidified slightly if it has been kept in, or had as its source, an arid environment, such as an interior page from a pad of paper. This helps the **electrostatic charge** develop. In more humid climates, several hours, exposure to normal room air serves the same purpose for most documents.

The page suspected of bearing indentations is covered with a cellophane (mylar) material, which is then pulled into firm contact with the paper by a vacuum drawn through a porous bronze plate. This fastens the document and cellophane covering to the plate. The cellophane covering prevents damage to the original document. The examiner then subjects the document and cellophane to a repeated high-voltage static charge by waving an electrically charged wand over the document's surface.

This results in a variably charged surface with the heavier static charge remaining within any impressions, even those that are microscopic in depth. Black toner (similar to that used in dry-process photocopy machines) is then cascaded over the cellophane surface using microscopically sized glass beads as a carrier, or by "misting," i.e., spraying the toner over the paper within a chamber placed over the questioned document. The toner is strongly attracted to static electricity and is retained on the mylar surface in accordance with the amount of residual static charge

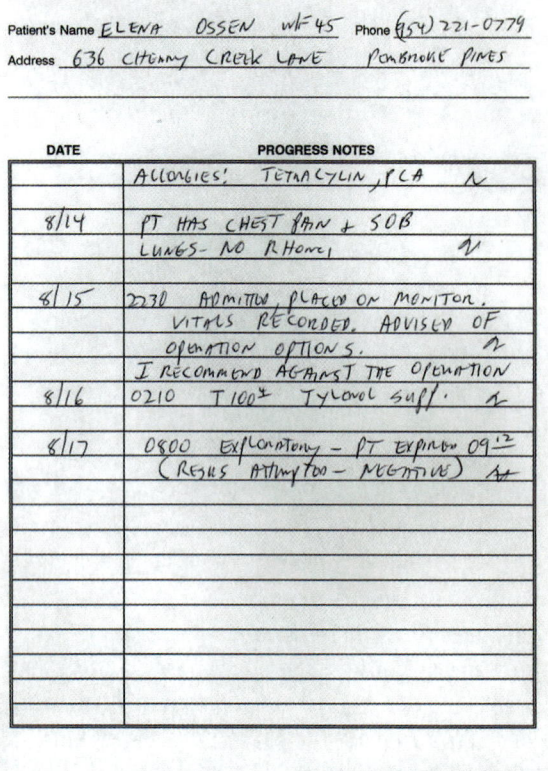

FBN MEDICAL CENTER   Patient Number 96-0424

Patient's Name ELENA OSSEN WF 45   Phone (954) 221-0779

Address 636 Cherry Creek Lane   Pembroke Pines

DATE	PROGRESS NOTES
	ALLERGIES! TETRACYLIN, PCA
8/14	PT HAS CHEST PAIN + SOB
	LUNGS- NO RHONCI
8/15	2230 ADMITTED, PLACED ON MONITOR.
	VITALS RECORDED. ADVISED OF
	OPERATION OPTIONS.
	I RECOMMEND AGAINST THE OPERATION
8/16	0210 T 100° TYLONOL SUPP.
8/17	0800 EXPLORATORY - PT EXPIRED 09:12
	(RESULTS ARRYTOO - NEGATIVE)

Figure 22.7 An original page of a questioned document medical record suspected of being altered. (Note that there are no initials after the questioned entry.)

present at any given surface point. The areas of the document containing the higher static electric charge retain greater portions of the black toner, resulting in a deposit of toner aligned with the indentations in the paper.

The developed indentations may be photographed. They are preserved by placing an adhesive-backed clear plastic sheet over the cellophane while it is still being held in place by the vacuum of the ESDA (Figures 22.7 and 22.8).

The advantages of electrostatic detection are twofold. First, ESDA is nondestructive. The indentations are revealed on the protective cellophane surface and are fixed by applying pressure-sensitive adhesive plastic over the cellophane. The original document remains unharmed throughout the process. Second, ESDA is extremely sensitive, allowing indentations that are not revealed by any other method to be readily observed and recovered. If the recovered indented writing is of a high enough quality, the handwriting in the

indentations, used in comparison with standard material, may even serve as a method to associate somebody to the questioned document.

Photocopy and Photocopier Examination

Like typewritten documents, photocopied documents may contain elements likely to identify the source of the original document, the model of machine that was employed in its production, or the specific machine that was used.

Photocopier Identification

Most modern day photocopying machines operate in a more or less similar fashion. The image of the document to be copied is captured by a camera lens and transferred to a cylindrical drum usually coated with a light-sensitive substance, such as **selenium**. This drum has been charged with a static electric charge, which dissipates when exposed to light. The image of the document transferred to the drum is made of light and dark areas that create similar or corresponding areas of more or less static electric charge on the drum. The drum then is bathed with toner, which is either in a dry or wet form depending upon the specific machine. Toner has an inherent affinity for static electricity and clings to those areas of the drum in quantities proportional to the electrostatic charge. The toner in turn is transferred to a piece of paper that is then subjected to a fixing process, usually in the form of heat, that fuses and attaches the toner to the paper. Paper is pulled through the photocopying machine by **"grabbers."** Marks made by the grabbers are transferred to the paper. These may be small depressions at the edge of the paper, areas of toner, or tonerless spots on the finished photocopied document.

A machine may leave on the photocopy characteristics that are specific to a particular make and model within a manufacturer's line. Grabber marks, paper edge depressions,

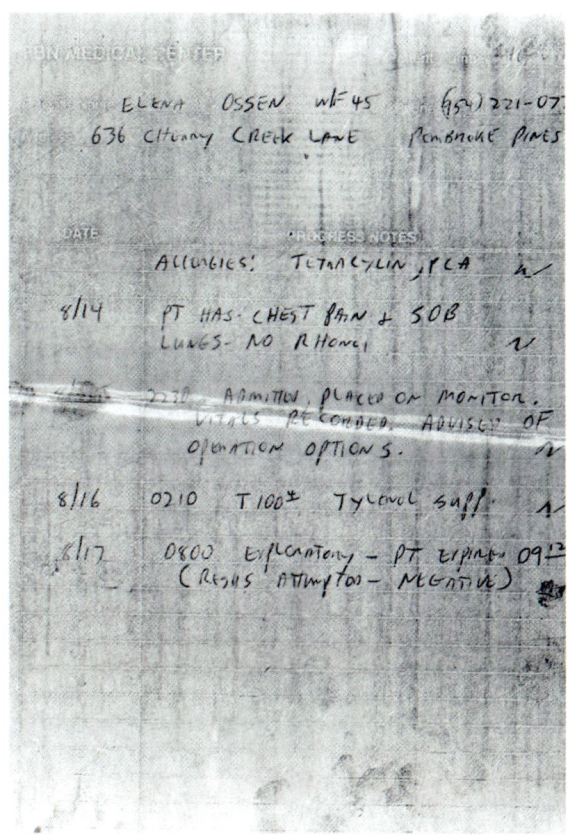

Figure 22.8 The resultant ESDA impressions raised from the next page down in the medical record file reveals a missing line, making it obvious that this entry was written subsequent to the original entries.

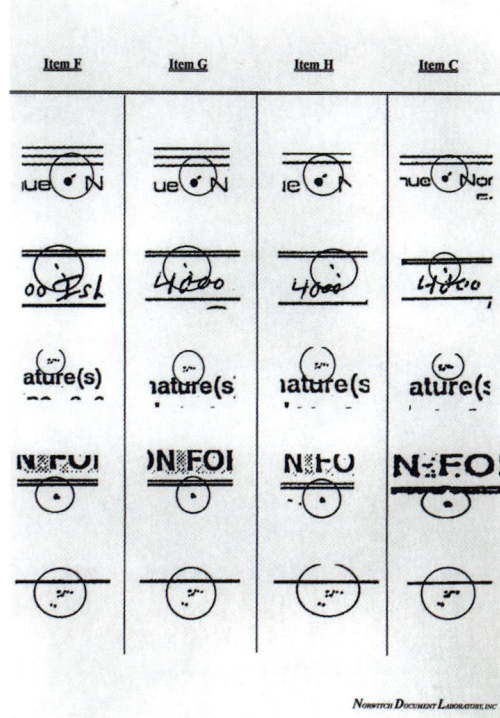

Figure 22.9 A demonstration chart showing the same photocopier "trash" marks on different documents. This proves these four documents were copied on the same photocopier

designs incorporated into specialized paper for specific machines, paper type, and toner type may be such characteristics. The examiner notes characteristics and then searches through reference files. At times highly definitive opinions as to make and model are possible. (If the questioned photocopy is a second, third, or more generation removed from the original document, and if two or more different photocopy machines have been used in its lineage, the examination is more difficult and the likelihood of a definitive opinion is greatly decreased.) The database of reference material must be updated continually because of the short life of most photocopier models and almost daily introduction of new models and technology.

Photocopy Examination

Questioned photocopies can be examined visually for individual characteristic "trash" marks

that may be made because of dirt, scratches, and other extraneous marks on the surfaces of the drum, cover, glass plate, or camera lens of a photocopy machine (Figure 22.9). A comparison of these marks on the questioned document with those marks made by a specific machine can identify or eliminate that particular machine as the source of the document. Similarly, a side-by-side comparison of two or more questioned photocopies may reveal if they are the product of a common photocopy machine. As in photocopier identification, multiple generations of copies and more than one photocopy machine involved severely limit the conclusiveness of the resulting opinions.

Photocopy Forgery

When the document examiner examines a photocopy to determine the genuineness of the original signature as represented by the photocopy, the examination must take into account the possibility that a genuine signature was affixed to a fraudulent document and the composite, or paste-up, photocopied. This may result in what would appear to be a photocopy of an original document bearing a genuine signature. The same may be true of any other portion of a photocopy. Photocopies can be prepared from a composite of parts of two or more documents which, when copied, can appear to be a reproduction of a single document. The resultant copy, made from composites, may or may not display characteristics indicative of its production from two or more document sources.

Indications of spuriousness include misaligned typing; different **fonts** and font sizes; misaligned preprinted matter; incorrect vertical, horizontal, and margin spacing; "shadowing" in the joined areas; disproportionate area sizes; different preprinted material and ink densities; and missing portions of writing or printing (covered by the paste-up, too closely trimmed, or masked by an opaque fluid). The **"trash marks"** surrounding the signature may be of greater or lesser quantity than those on the remainder of the document. This is especially true if either the model signature or document to be used in the paste-up was itself a photocopy. The best indication of a possibly fraudulent photocopy is a claim that the original document has "disappeared" or has been "misplaced."

Even when none of these indications of photocopy forgery is present, the prudent document examiner who issues an opinion about the authenticity of a signature or an entire disputed document, when the submitted evidence is a photocopy, will qualify his or her opinion. The qualifier is a statement that the opinion is predicated upon the questioned document being a true and accurate reproduction of the original document. Many examiners go even farther by including a statement in the Report of Findings that finds the accuracy of opinions involving photocopies upon viewing the original document prior to any court testimony. Photocopies that display prohibitively poor quality may be precluded from examination, but those displaying adequate line quality are deserving of some degree of qualified opinion.

Typewriter and Typewriting Examination

Frequently, typewritten anonymous notes, threatening letters, extortion requests, and altered contracts are the active portion of a questioned document case. While typewriters are often thought of as a single piece of equipment, many machines are often a combination of machine, type element (ball or printwheel), and single- or multiple-pass ribbon (Figures 22.10 and 22.11). This class of machine may be more appropriately referred to as a **typewriting system.** While each major component may have its own identifying characteristics, such as defects in the type element from misuse or the manufacturing process, the possibility of interchanging components between similar models exists. This renders the likelihood of an identification of a specific typewriting system to a suspect typed letter rather remote.

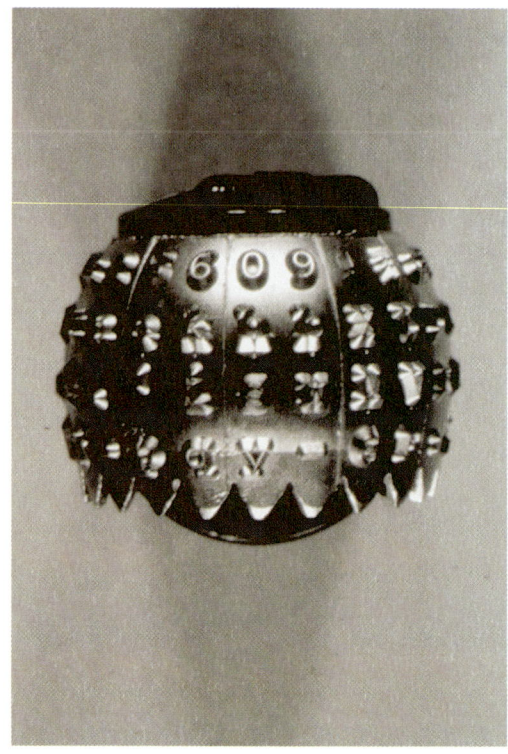

Figure 22.10 An IBM type-font ball. On occasion, defects to the typeface may allow for identification of a specific ball to a questioned page of typewriting.

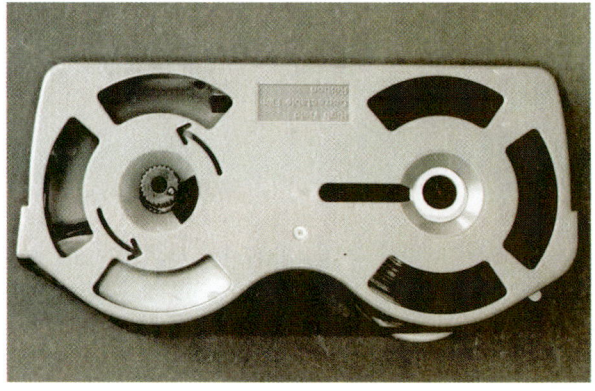

Figure 22.11 A replaceable carbon film typewriter ribbon cartridge. The used portion of this ribbon can be "read."

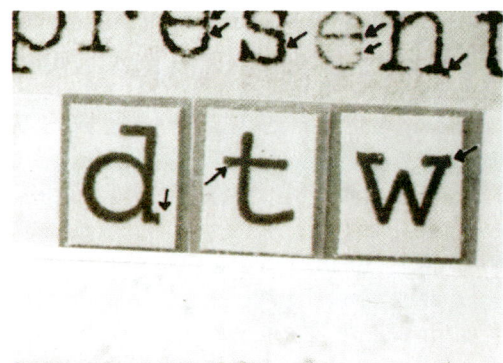

Figure 22.12 Obvious type defects on an older "upstrike" typewriter. These defects are the result of individual letters jamming into one another if more than one typewriter key is hit at the same time.

Evidence Submission

In the case of typewritten material, such as a typewritten letter, the original of the questioned document is almost a necessity. Machine-copied documents are usually unsuitable for a typeprint examination except when differences in type fonts are obvious. Opinions derived from such documents, if any, are generally much less definitive.

The submission of the suspect typewriter is the best standard for the document examiner; however, on some occasions this may not be practical. In these instances, the submitter can place a clean sheet of paper in the suspect machine(s) and type the material in question. If the questioned letter is several pages long, it may not be possible to type all of it. Likewise, if the questioned material is only a few lines, it may be necessary to type it several times.

Typewriting and Typewriter Examinations

At times, the identification of a typewriter make and model from a questioned typewritten document can be accomplished. The typewritten material in question is first classified, then searched for in reference files. Highly definitive opinions are rare due to the limitations of the reference files and the thousands of new machines, printers, systems, and interchangeable fonts available today. Generally, the older the machine, the better the possibility of identification.

Other typewriting examinations consist of visual and microscopic comparison of questioned typewriting and standard typewriting in an attempt to identify a typewritten document with a specific machine. Individual characteristics that include type alignment, spacing, broken or damaged type font (Figure 22.12), ink density, and pressure are considered. Comparisons involving documents typed on later model element machines are rarely definitive.

While generally a lengthy process, the examination of one-pass plastic (carbon) ribbons can also be undertaken to determine if the questioned text is present. During the production of a typewritten document on a machine employing such a ribbon, the carbon film image of each letter or punctuation mark is transferred to the document, leaving a letter-shaped void in the ribbon. These voids, although backward, can be read easily. This process does not work on older typewriters that use fabric ribbons.

Paper and Watermark Examination

Paper examinations usually are necessary when there is some question as to whether one or more pages have been added to a multipage document, or whether a document was created at the time that it purports to have been created. A last will may be one such document. There may be a suspicion that one or more pages have been added or replaced subsequent to the original execution of the document.

The simple examination of the staple and staple hole can sometimes shed light on the authenticity of the document. A will having only one staple in the top corner may have evidence of previous staples (holes) in subsequent pages. This would indicate that one or more pages have been added or replaced and the will restapled.

Paper is commonly made of wood or cotton materials. During the production, various sizings, **fillers**, and coatings are added. Sizings, such as **rosin**, enable the paper to resist ink penetration. Fillers, such as clay, calcium carbonate, and titanium dioxide, improve the surface and color of the paper. Various coatings are added to the paper to improve its appearance and printing properties. These additives vary from one paper type or paper manufacturer to another.

Chemical testing can determine which of these materials is present, and even the type of wood that was used in the paper's manufacture. A comparison of the results of such testing can associate or dissociate a questioned page with a known standard. Unfortunately, these processes are destructive in nature and require sampling of both the questioned and standard paper. This is most often not convenient or allowed by the courts. Many other properties of the paper, however, can be investigated and compared.

Paper Size and Thickness

Although there are standard sizes for paper, such as "letter" (8.5 × 11 in.) or "legal" (8.5 × 14 in.), very small differences in lengths and widths exist between different manufacturers' products and even between different papers in a specific manufacturer's paper line, or even between different runs of the same paper. While these small differences can be measured, the simple process of stacking the questioned and standard paper readily displays differences in size. Minute differences in paper thickness can also be detectable. This

determination requires instrumentation with a paper micrometer. Most micrometers display differences in thousandths of an inch.

Paper Opacity, Color, and Brightness

Paper opacity, color, and brightness are directly related to the chemical additives that were put into the paper during its manufacture. Differences between two papers in these areas may, at times, be easily observed with the naked eye. When held up to a light source, one paper may transmit more light than another. Obvious differences in color or shading between papers can likewise be an unaided observation. Often two papers with brightness that appears similar to the unaided eye display differences when subjected to a short-wave or long-wave UV light source. While one paper may remain dull in appearance, the other may almost glow.

Watermarks

Some papers, when held up to the light, display an area of translucent design—the watermark—incorporated into the paper by one of several different methods during the paper manufacturing process. These designs contain "hidden" clues to the paper manufacturer, the entity for whom the paper was produced, and first date of that paper's production. For example, a questioned document that purports to have been executed in 1975 may contain a watermark that, after research, proves that the paper was not made until 1983.

Miscellaneous Examinations

Rubber Stamps

Rubber stamps are used to affix an impression on a document and can be the subject of a questioned document examination. Some examples are hand stamps and self-inking stamps. Stamps may be individualized through the production process, wear, misuse, and damage. Rubber stamp production today consists of laser engraving or ultraviolet sensitive photopolymers. These techniques eliminate some of the classical individualizing traits commonly found in type that is assembled on a composing rack, such as misalignment, spacing, or flaws in type.

Stamp production using a photopolymer is a liquid plastic that hardens when exposed to ultraviolet light. This process utilizes desktop publishing, to scan the image, and this serves as a negative. In a series of steps, the polymer is exposed to the ultraviolet light through the clear areas of the negative. The plastic hardens forming a stamp impression. Incorrect technique in the production can results in air bubbles in the final die, small pieces tearing away from the harden plastic, low spots on the stamp die.

Laser engraving eliminates the photopolymer and several steps in its production. Natural rubber that has been hardened for use as a stamp, (vulcanized), is cut by a laser engraving machine. A layout file of the stamp is sent to the engraver, and a laser beam burns off the background leaving an impression that is the stamp.

An examiner conducting a comparison between a questioned stamped impression and the rubber stamp must be aware of the possibility of a counterfeit stamp. The question remains how well will physical defects, or individualizing characteristics unique to a particular stamp be reproduced by the production of a second stamp utilizing modern desktop publishing?[10]

Miscellaneous Examinations

Questioned document examiners have occasion to undertake examinations other than those discussed in this chapter. Many of these examinations are specialized forms, or combinations, of those other procedures. They may include, but are not limited to, counterfeiting, charred documents, industrial printing, check writers, **embossing**, and **sequence of strokes**. Readers requiring in-depth information about these areas should refer to the Suggested Readings.

Questions

1. Is it possible for a document examiner to tell the personality of an individual from his or her handwriting?

2. Can you determine the sex and age of the writer from his or her handwriting?

3. What is the difference between requested handwriting standards and nonrequested standards?

4. Explain the meaning of the term "class characteristics" in relation to handwriting.

5. Can a document examiner identify all types of writing?

6. Name one of the methods of ink differentiation for similar-appearing inks.

7. Name one of the methods for the recovery of indented writing.

8. What is a photocopy trash mark and how does it occur?

9. What are the elements of a typewriting system?

10. What instrumentation is used to test paper thickness?

References

1. Baker, J., *Law of Disputed and Forged Documents*, Michie Co., Charlottesville, VA, 1955.
2. Dick, R. M., The identification of handwriting, brief discussion presented at the FDIAI Semi-Annual Conference, Sanford, FL, May 26, 1978.

3. Muehlberger, R. J. et al., A statistical examination of selected handwriting characteristics, *J. Forens. Sci.,* 22, 206, 1977.
4. Risinger, D. M., Denbeaux, M. P., and Saks, M. J., Exorcism of ignorance as a proxy for rational knowledge: the lessons of handwriting identification expertise, *Univ. Pa. Law Review*, 137, 731, 1989.
5. Kam, M., Fielding, G., and Conn, R., Writer identification by professional document examiners, *J. Forens. Sci.* 42, 778, 1997.
6. Kam, M. et al., Signature authentication by forensic document examiners, *J. Forens. Sci.,* 46, 884, 2001.
7. Sihari, S. N. et al., Individuality of Handwriting, *J. Forens. Sci.,* 47, 856, 2002.
8. Cha, Sung-Hyuk et al., Computer-assisted handwriting style identification system for questioned document examination, Image and Video Communications and Proceedings, *Proceedings of the SPIE*, 5685, 220, 2005.
9. Prospectus, VSC600, Foster & Freeman.
10. Seiden, H., Rubber Stamps: Revisited, *Int. J. Forens. Document Examininers.* 4, 58, 1998.

Suggested Readings

Brunelle, R. and Reed, R., *Forensic Examination of Ink and Paper*, Charles C. Thomas, Springfield, IL, 1984.

Conway, J. V. P., *Evidential Documents*, Charles C. Thomas, Springfield, IL, 1959.

Ellen, D., *The Scientific Examination of Documents: Methods and Techniques,* John Wiley and Sons, New York, 1989.

Haring, J. V., *The Hand of Hauptmann*, Hamer Publishing, Plainfield, NJ, 1937.

Harrison, W., *Suspect Documents*, Eastern Press, London, 1958.

Hilton, O., *Scientific Examination of Questioned Documents*, Revised edition, CRC Press, Boca Raton, FL, 1993.

Hilton, O., *Detecting and Deciphering Erased Pencil Writing*, Charles C. Thomas, Springfield, IL, 1991.

Hilton, O., *Scientific Examination of Questioned Documents*, 2nd ed., Elsevier, New York, 1982.

Huber, R. A. and Headrick, A. M., *Handwriting Identification Facts and Fundamentals*, CRC Press, Boca Raton, FL, 1999.

Levinson, J., *Questioned Documents: A Lawyer's Handbook*, Academic Press, New York, 2001.

Morris, R., *Forensic Handwriting Identification, Fundamental Concepts and Principles*, Academic Press, New York, 2000.

Osborn, A. S., *Questioned Documents*, 2nd ed., Boyd Printing Company, Albany, New York, 1929.

Osborn, A. S. and Osborn, A. D., *Questioned Document Problems*, Boyd Printing Company, Albany, New York, 1944.

Osborn, A. S., *The Problem of Proof*, Essex Press, Newark, NJ, 1926.

Seaman-Kelly, J., *Forensic Examination of Rubber Stamps A Practical Guide*, Charles C. Thomas Publisher, Springfield, Illinois, 2002. The procedures outlined here may help determine that there was no alteration or addition to a record. The majority of medical records that are examined during the course of a malpractice suit display no evidence of anything extraordinary, but you may not know that just from looking at them.

Analysis of Controlled Substances

Donnell R. Christian, Jr.

Introduction

Controlled substances are substances (usually drugs) whose possession or use is regulated by the government. Title 21 of the United States Code (21 USC) defines these. It recognizes that many of the substances have a useful, legitimate medical purpose and are necessary to maintain the health and general welfare of the American people. Therefore, it has divided the drugs into regulated schedules based upon their medical use and potential for abuse. Many state and local laws concerning the possession of controlled substances are based on these federal regulations. Others have taken the list of controlled substances from 21 USC and simply made their possession illegal without regard to legitimate use or abuse potential.

The analysis of controlled substances is a basic function of the forensic laboratory. The section that performs this function is known by a variety of names: the drug section, narcotic analysis, and forensic chemistry are just a few. No matter what the name is, the goal is the same: to confirm the presence of a substance that is either statutorily regulated or is illegal to possess.

Many forensic scientists begin their careers in this section of the laboratory. They learn the forensic applications of their academic scientific training. They are exposed to proper evidence handling procedures, including chain-of-custody issues, case documentation, and proper forensic analytical techniques. The examiner learns to consider the long-term ramifications of the analysis (i.e., Can he or she justify an examination and defend the results in an adversarial situation?). The amount and variety of cases provide the fledgling forensic scientist a great deal of experience in a short amount of time. The lessons learned while analyzing evidence in this section serve as the foundation for working in other sections of a forensic laboratory.

Working in the controlled-substances section also cultivates the thought processes. Many disciplines of forensic science require **deductive reasoning**, which is not often emphasized in academic environments. The controlled substance examiner is taught how to use a series of nonspecific examinations to produce a conclusive identification. This thought process is extended to other areas of forensic science in which the identification and comparison of

23

Topics

Standards of Analysis

Scope of Analysis

Chemical Examinations

Instrumental Examinations

Quantitation

Quality Assurance (QA)/ Quality Control (QC)

Clandestine Drug Laboratories

Conclusion

evidentiary items may not be clearly defined. With the introduction of automated instrumentation, this analytical approach is taught less often.

This chapter addresses the basic analytical schemes used to identify controlled substances. The goal is to provide a basic understanding of the analytical process by briefly discussing each step in the sequence along with tables containing generic step-by-step procedures. The last section briefly discusses the controlled substance section's role in the investigation of clandestine laboratories and the prosecution of the persons who manufacture controlled substances.

Standards of Analysis

In October 2000, the Scientific Working Group for the Analysis of Seized Drugs (SWGDRUG) met in Vienna, Austria to finalize its recommendations concerning the examination and identification of controlled substances. Some of these contained proposals for the minimum examination requirements for the identification of controlled substances. Although these recommendations do not hold any statutory authority, they do represent the accepted analytical standards established by a consensus of the scientific community engaged in the analysis of drugs of abuse.

The E-30 Committee of the American Society of Testing Materials (ASTM) has reviewed recommendations of the SWGDRUG analytical protocols and has initiated the process of adopting them as ASTM standards. If the SWGDRUG protocols are adopted as a "guide" they will serve as a best practices reference that does not recommend a specific course of action. If adopted as a "practice" or a "test method" the described analytical protocols would effect how a forensic laboratory examines controlled substance exhibits.

ASTM standards are considered the analytical benchmark. Standard practices are a definitive set of instructions for performing specific operations that does not produce a test result. A standard "test method" is a definitive procedure that produces a result. These standards should be considered when developing a laboratory's analytical protocols. If they are not used as the foundation of the examination procedure the analytical results may not be readily accepted in court.

The standard practices currently under development by the ASTM E-30 committee are:

- WK389: Terminology Relating to Forensic Seized Drug Analysis
- WK390: Practice for Identification of Seized Drugs
- WK2238: Standard Terminology Relating to Seized Drug Analysis
- WK2239: Standard Practice for Education and Training of Seized Drug Analysts
- WK2240: Standard Practice for Quality Assurance of Laboratories Performing Seized Drug Analysis
- WK2241: Standard Practice for Identification of Seized Drugs

ASTM standards reach beyond the specific application used for an examination. Some of the standards address the use, care, and maintenance for the instrumentation used to perform the examination. Others address functions within the forensic laboratory. Each should be addressed when developing an examination protocol, quality assurance program or laboratory standard operating procedures. Some of the auxiliary standards that directly effect the examination of controlled substances include, but are not limited to:

- Chromatography Analysis Standards
  E260: Practice for Packed Column Gas Chromatography
  E355: Practice for Gas Chromatography Terms and Relationships
  E516: Practice for Testing Thermal Conductivity Detectors Used in Gas Chromatography
  E594: Practice for Testing Flame Ionization Detectors Used in Gas Chromatography

E697: Practice for Use of Electron Capture Detectors in Gas Chromatography

E1140: Practice for Testing Nitrogen/Phosphorus Thermionic Ionization Detectors for Use in Gas Chromatography

E1510: Practice for Installing Fused Silica Open Tubular Capillary Columns in Gas Chromatographs

E1642: Practice for General Techniques of Gas Chromatography Infrared (GC/IR) Analysis

D3016: Practice for Use of Liquid Exclusion Chromatography Terms and Relationships

E1151: Practice for Ion Chromatography Terms and Relationships

- Infrared Spectroscopy Analysis Standards

E131: Terminology Relating to Molecular Spectroscopy

E168: Practices for General Techniques of Infrared Quantitative Analysis

E334: Practice for General Techniques of Infrared Microanalysis

E573: Practices for Internal Reflection Spectroscopy

E932: Practice for Describing and Measuring Performance of Dispersive Infrared Spectrophotometers

E1252: Practice for General Techniques for Obtaining Infrared Spectra for Qualitative Analysis

E1866: Guide for Establishing Spectrophotometer Performance Tests

E1944: Practice for Describing and Measuring Performance of Fourier Transform Near-Infrared (FT-NIR) Spectrometers: Level Zero and Level One Tests

E1421: Practice for Describing and Measuring Performance of Fourier Transform Infrared (FT-IR) Spectrometers: Level Zero and Level One Tests

E1642: Practice for General Techniques of Gas Chromatography Infrared (GC/IR) Analysis

- Mass Spectrometry Analysis Standards

E1504: Standard Practice for Reporting Mass Spectral Data in Secondary Ion Mass Spectrometry (SIMS)

As previously stated, SWGDRUG has established three categories of analytical techniques that can be used for the identification of controlled substances. The groupings are based upon the technique's discriminating power. Below are listed the categories and associated analytical techniques:

Category C
Nonspecific Techniques

Chemical color tests	A nondocumentable technique that uses the colors produced by chemical reactions to provide information regarding the structure of the substance being tested.
Fluorescence spectroscopy	A documentable analytical technique that uses the release characteristic wavelengths of radiation following the absorption of electromagnetic radiation (fluorescence) to establish a compound's potential identity.
Immunoassay	A documentable laboratory technique that uses the binding between an antigen and its homologous antibody to identify and quantify the specific antigen or antibody in a sample.
Melting point	The temperature at which a solid becomes a liquid at standard atmospheric pressure. The documentability of this technique depends upon the instrument used.
Ultraviolet (UV) spectroscopy	A documentable technique that uses the absorption of ultraviolet radiation to classify a substance.

Category B
Moderately Specific Techniques

Capillary electrophoresis (CE)	A documentable separation technique using the differential movement or migration of ions by attraction or repulsion in an electric field through buffer-filled narrow-bore capillary columns as an identification tool.
Gas chromatography (GC)	A documentable separation technique that uses gas flowing through a coated tube to separate compounds by their size, weight, and chemical reactivity with the column coating.
Liquid chromatography (LC)	A documentable separation technique that uses liquid flowing through a coated tube to separate compounds by their size, weight, and chemical reactivity with the column coating.
Microcrystalline tests	A technique that uses the microscopic crystals produced by chemical reactions to provide information regarding the identity of the substance being tested. A series of positive microcrystalline tests can be considered to be a conclusive test. This technique can be considered documentable if photomicrographs of the crystals used for identification are taken at the time of the examination.
Pharmaceutical identifiers	Comparing the physical characteristics of a commercially produced pharmaceutical product to known reference material to tentatively establish the composition of the preparation.
Thin layer chromatography (TLC)	A traditionally non-documentable technique that uses solvent(s) traveling through a porous medium to separate compounds by their chemical reactivity. This technique can be documented through photographing or photocopying the developed thin layer plate.

Category A
Specific Examinations

Infrared spectroscopy (IR)	A specific documentable technique that uses the absorption of infrared radiation to produce a chemical fingerprint of a substance. This technique can be used in conjunction with gas chromatography.
Mass spectroscopy (MS)	A specific documentable technique that uses molecular fragment (ion) patterns to produce a chemical fingerprint of a substance. This technique can be used in conjunction with gas and liquid chromatography.
Nuclear magnetic resonance spectroscopy (NMR)	A specific documentable technique that monitors the splitting of nuclear energy levels within a molecule when it is exposed to oscillating magnetic fields.
Raman spectroscopy	A specific documentable technique that uses the inelastic scattering of light by matter to produce a chemical fingerprint of a substance.

The SWGDRUG guidelines provide recommendations for the types and minimum number of tests required to identify seized drugs. A validated Category A technique, with documentable data, supported by one Category A, B, or C technique is the suggested minimum examination criteria. A combination of three different Category B and C techniques can be used if a Category A technique is unavailable. The Category B techniques utilized must produce reviewable data.

Scope of Analysis

Local laws and criminal procedures will be the driving force behind the scope of the analytical process. The laboratory's mission within its agency will also weigh heavily into the depth of analysis each exhibit will receive. For example, the amount of analytical effort involved in the identification of a controlled substance for criminal prosecution purposes is significantly less than that required for intelligence gathering and investigative purposes. The only information required in a criminal prosecution is the identity and amount of controlled substance contained in an exhibit. Laboratories responsible for intelligence gathering will also identify the types and quantity of the exhibit's diluents and adulterants.

The level of analytical detail required not only affects the time involved but the type of instrumentation required. Most forensic chemistry sections can provide a complete range of analytical service wet chemical techniques and a basic mass spectroscopy or infrared spectroscopy. As the level of information detail increases, so does the type and sensitivity of the instrumentation required. For example, the equipment and procedures required to confirm the presence of heroin in a street sample is far less sophisticated than the one needed to identify the region of the world in which the opium used to produce the heroin was grown.

The final issue that determines the depth of analysis is laboratory policy. The laboratory's policy is generally developed through collaboration between the laboratory's management and a peer group consisting of the examiners who perform the examinations on a daily basis. This represents a balance between the need to produce timely results that meet the applicable legal criteria while at the same time not compromising the scientific integrity of the examination.

An example of this collaboration is the need for quantitative analysis. Unless mandated by statute, the amount of a controlled substance in an exhibit is not an element of the crime. However, this information may have investigative significance and can also be used as part of an internal quality control procedure. It may not be realistic to quantitate every exhibit submitted for analysis. Therefore, laboratory and investigators work together to establish a quantitation policy that satisfies the needs of both parties.

The identification of controlled substances is divided into botanical and chemical examinations. Botanical examinations identify physical characteristics specific to plants that are considered controlled substances. Chemical examinations use wet chemical or instrumental techniques to identify specific substances that are controlled by statute. The analytical testing sequence is represented by a simple flow chart (Figure 23.1). For each examination, a series of tests is administered to the sample. Each test is more specific than the last. At the end of the sequence the examiner is able to determine if there is a controlled substance in the sample and to identify it.

Botanical examinations are the most common examinations performed in the controlled substance section of a forensics laboratory. The plants that require botanical examinations include marijuana, **peyote**, mushrooms, and opium. Marijuana is by far the most common botanical examination. It is not uncommon for marijuana examinations to exceed 50 percent of the caseload, although case management techniques utilized by some agencies have dramatically reduced the number of marijuana examinations required. The examination of mushrooms, peyote, and opium poppy samples are rare. However, the examiner must know the physical characteristics of these plants to be able to recognize them when they are presented in case samples.

The controlled substance examiner walks a tightrope when performing botanical examinations. He or she is identifying plants and plant material, not the specific psychoactive ingredients. As a rule, by education and training the examiner is a chemist, not a biologist or a botanist. However, he or she has been trained in the identification of specific types of plants or plant parts and can identify whether plant material is or is not marijuana, peyote, or opium. Beyond that the examiner should not render an opinion as to the identity of the substance.

Botanical examinations are the only area of controlled substance examination in which DNA analysis has a potential application. However, no genetic markers are forensically accepted for use in identifying specific types of plants. Time and financial constraints

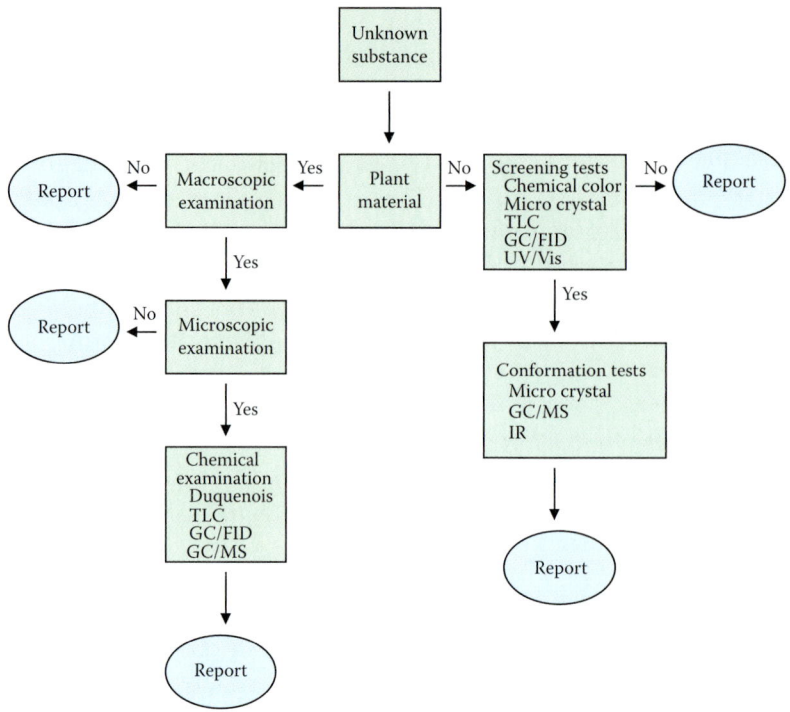

Figure 23.1 Controlled substance analysis flow chart.

also discourage the use of DNA for plant identification.

Marijuana

Marijuana is the common name for the plant *Cannabis sativa l.* Numerous treatises concern themselves with the number of species of cannabis, but it is generally accepted that there is only one species. The different varieties of marijuana—indica, rhutamalus, etc.—are simply variations of the sativa species. To avoid this ongoing debate, some jurisdictions have opted to control all varieties of marijuana by defining the group cannabis as the controlled substance.

The identification of marijuana is a two-step process. The first step establishes the plant or plant material as marijuana through its physical characteristics. The second step is to establish the presence of the plant resin that contains the psychoactive components.

The identification process begins with a **macroscopic examination** of the plant material to establish if the plant material has the class characteristics of marijuana. The marijuana plant structure (Figure 23.2)

Figure 23.2 Marijuana plant (*Cannabis sativa l.*).

has palmate leaf configuration and pinnate leaf structure with serrated edges. The plant stems have a fluted structure. These class characteristics may not be readily observed in samples of crushed plant material that has been submitted for analysis. However, with experience the trained eye can recognize plant material with the macroscopic consistency of marijuana.

A microscopic examination is used to identify the individual characteristics that are unique to marijuana. This examination includes the identification of cystolithic (bear claw shaped) hairs on the top surface of the leaf and finer clothing or guard hairs on the underside of the leaf. The bud material of the plant may have the presence of a red "thread" entwined in it.

The identification of cannabis resin, which contains the psychoactive components, is the second phase of marijuana examination. The **Duquenois–Levine test** is the chemical color test used to confirm the presence of compounds called cannabinoids. Delta 9 tetrahydrocannabinol (THC) is the primary psychoactive compound in this class. The exact mechanism of the reaction of this chemical color test has not been established. However, it has been accepted as specific for the identification of marijuana resin.

Additional chemical tests for marijuana resin include chromatographic examination to establish the presence of specific cannabinoids. Thin layer chromatography (TLC) has been the traditional method of choice. It separates the cannabinoids in the resin and provides a chemical color test to identify their location on the thin layer plate. The questioned sample's pattern of colored spots is compared to a known sample of cannabis resin that is examined at the same time. It is considered a match if the patterns of the known and the questioned samples have the same sequence of colored spots.

The use of gas chromatography (GC) to identify cannabis resin has increased in popularity. GC provides the examiner a documentable method of establishing the presence of various cannabinoids. It also provides a means of establishing the relative amount of each compound in the sample by comparing the size of each peak in the chromatogram. It is considered a match if the retention time of a predetermined number of cannabinoids appear in the chromatogram. The relative size of the peaks is not an issue in the identification process.

Mass spectroscopy (MS) can be used to positively identify individual components of the cannabis resin, specifically THC. One school of thought considers this testing excessive because the law governing the possession of marijuana states the plant material only needs to contain the cannabis resin, not any specific component. Since the Duquenois–Levine test is considered specific for cannabis resin, the mass spectrum of any or all of the cannabinoids is unnecessary. Another school of thought considers the MS examination necessary to positively establish the presence of one or more of the cannabinoids, specifically the psychoactive ingredient THC.

Hashish

Contrary to common belief, hashish is not a potent form of marijuana. **Hashish** is the resin from marijuana that has been isolated from the plant material. It can be found as an oil or in cake form. The oil is added to other substances and smoked. The cake can be smoked separately or added to other materials and smoked.

The analysis of hashish depends upon the statute regulating its possession. The federal law does not distinguish between marijuana and hashish. Some state and local jurisdictions define hashish separately. If the hashish statutes mirror marijuana's, a Duquenois–Levine test may be all that is required to establish the presence of cannabis resin. If the statute identifies specific compounds that must be present (e.g., THC), a confirmatory test such as MS should be performed. Internal laboratory protocols also assist the examiner in establishing the requisite analytical scheme.

Peyote

Peyote is the common name for the small Mexican cactus *Lophophora williamsii*. The indigenous people of Mexico and the southwestern United States have used it in religious ceremonies for centuries. *Lophophora diffusa* is a rare species of peyote that is occasionally encountered. Each variety contains mescaline (3,4,5-trimethoxyphenethlamine), which produces hallucinogenic effects.

The identification of peyote begins with a macroscopic examination of the plant material. The peyote "button" (Figure 23.3) is approximately 1 inch in diameter. It can be divided into 5 to 10 orange-like segments or has a soccer

Figure 23.3   Peyote cactus (*Lophophora williamsii*).

Figure 23.4   Psilocybe mushrooms.

ball-like appearance. Each segment contains a small white tuft of material similar to a cotton ball. Peyote does not have specific microscopic characteristics that can be used for identification purposes. The presence of the cotton-like tuft is generally used as a predecessor to the chemical examination steps.

A chromatographic examination is used to confirm the identity of peyote. The examiner uses this examination to identify a pattern of **alkaloids** characteristic of peyote in a manner similar to the comparison of marijuana resin to known cannabis resin. The chromatographic pattern can also be used to identify the specific peyote species. Thin layer chromatography and gas chromatography techniques work equally well.

The identification of mescaline is not an essential element of the peyote identification process. However, it should be in the chromatographic pattern that is used to identify peyote. A confirmatory test for mescaline is not required since the examiner is identifying the plant, not the psychoactive components.

Mushrooms

The analytical approach to mushrooms is different from the approach to marijuana and peyote, since the possession of mushrooms is not illegal per se. The components within the mushrooms (i.e., **psilocin** and **psilocybin**) are the items that are controlled. Therefore, the addition of a step to confirm the presence of psilocin or psilocybin is required.

The physical identification of mushrooms that potentially contain psilocin or psilocybin is the initial step in the identification process. Over a dozen species of mushrooms contain these compounds. Figure 23.4 is a photograph of a variety of commonly encountered psilocybe mushrooms. The stems of the species most commonly encountered are off-white in color with a blue-gray staining throughout. The color of the mushroom caps ranges from off-white to light brown or tan.

The next step in the screening process is testing for the presence of psilocin and psilocybin. These tests include **chemical color tests**, TLC, and examination of extracts using ultraviolet light (UV). Two chemical color tests are useful in the screening process. Color test reagents, such as Van Urk's, that contain para-dimethylaminabenzaldehyde (pDMBA) turn purple in the presence of psilocin and psilocybin. An aqueous solution of Fast Blue B turns red when exposed to mushrooms containing psilocin and psilocybin. This solution turns blue with the addition of concentrated hydrochloric acid (HCl). For UV and TLC analysis, methanol can be used to extract psilocin and psilocybin from mushrooms. Psilocin and psilocybin absorb UV light, producing a characteristic UV spectrum. They can be separated chromatographically using TLC. Visualization can be achieved using UV light or by spraying the TLC plate with Van Urk's reagent. Psilocin and psilocybin spots on the thin layer plate glow when exposed to UV light and turn violet when over-sprayed with Van Urk's reagent.

The preliminary identification of psilocin and psilocybin is critical in determining the confirmatory test and the sample preparation technique that will be used. Psilocybin is a fragile molecule. Certain sample preparation techniques convert the psilocybin into psilocin. If the preliminary identification is not done prior to the confirmatory test, the examiner cannot definitively say that the psilocin identified in the confirmatory test was originally in the sample or was a result of the conversion of psilocybin during the extraction. Therefore, the determination of the presence of the compounds prior to the confirmatory test is necessary to evaluate the results properly.

Wet chemical extraction techniques must be used for infrared (IR) spectroscopy confirmation of psilocin. Two facts should be considered when using this technique. First, wet chemical extractions cannot separate psilocin from psilocybin. Second, psilocybin may be converted into psilocin during the extraction process. Therefore, if the examiner has not predetermined the presence of psilocin in the sample, it may be erroneously identified.

Gas chromatography–mass spectroscopy (GC–MS) can be used to identify both psilocin and psilocybin. As in IR analysis, the determination of whether psilocin or psilocybin is present prior to analysis is an issue. Direct injection of an extract containing both psilocin and psilocybin into a gas chromatograph results in the detection of only psilocin. The psilocybin does not chromatograph well and can decompose into psilocin. This raises the question of whether the psilocin was there prior to the injection. Chemical reagents can be added to extracts containing psilocybin. These reagents react with the psilocybin to create a stable molecule that can be analyzed using GC–MS. The resulting molecule will produce a mass spectrum that is considered specific to psilocybin.

Documentation

In general, no supporting documentation is generated during the identification of marijuana or peyote, other than the examiner's case notes. The case notes do not independently demonstrate that the examination occurred. Even so, they should describe the visual observations the examiner used in making the conclusions. Drawings of the observed characteristic structures can be made to support such conclusions. Descriptions of the color changes during the Duquenois–Levine test are equally important. A positive identification is dependent upon positive results in both steps of the test.

The use of TLC is not a documentable analytical technique. Unless photographs are taken of the thin layer plate after it is developed, there is no way of verifying the examination took place. A sketch of the developed thin layer plate aids in documenting the test results. However, sketches do not allow for an independent interpretation of the plate by a case reviewer or independent examiner. On the other hand, GC is a documentable technique. To document that the examination occurred, the examiner should mark all paperwork related to GC analysis with the information outlined in the instrumental analysis portion of the chemical analysis section.

The documentation requirements for mushroom examinations are different from those for marijuana and peyote because of the need to identify the specific substance that is the subject of the control. Screening tests should be documented in the examiner's case notes in the same manner marijuana and peyote are documented. However, because the examiner is identifying a specific controlled substance, he or she should use a documentable confirmatory test whose data is included in the case notes. This documentation should include the information outlined in the instrumental examination section of this chapter. This information allows for an independent interpretation of the test by a case reviewer or an independent examiner.

Chemical Examinations

The balance of the samples encountered by the controlled substance section requires the identification of specific compounds within a mixture. The composition of the samples may vary, but the procedure remains the same. Each sample requires a screening step, an extraction or sample preparation step, and a confirmatory step. Chemical examinations can be subdivided into wet chemical and instrumental procedures. Wet chemical procedures are used as a screening method or for sample preparation. Instrumental procedures are used for screening or as a confirmation tool.

Wet Chemical Procedures

Wet chemical procedures are used in the initial stages of the controlled substance identification process. These nonspecific tests provide a method to indicate quickly whether a controlled substance is or is not present within a sample. These procedures can be used to isolate controlled substances for confirmatory testing using instrumental techniques. Wet chemical procedures consist of chemical color tests, microcrystalline tests, thin layer chromatography, and liquid extraction techniques. A series of these tests can be used to identify a controlled substance deductively.

Chemical Color Tests

Chemical color tests are chemical reactions that provide information regarding the structure of the substance being tested. Certain compounds or classes of compounds produce distinct colors when brought into contact with various chemical reagents. These simple reactions can indicate the presence of a generic molecular structure.

Chemical color tests are generally conducted by transferring a small amount of the substance being tested to the well of a spot plate or into a test tube. The test reagent is added to the substance. Some tests may be conducted in a sequential fashion utilizing multiple reagents. The results of each step in the sequence are observed and noted. Positive and negative controls should be run on a regular basis to ensure the reliability of the testing reagents.

A certain amount of subjectivity is involved when a color is reported. It is not uncommon for two people to describe the same color differently. Colors can also be influenced by the concentration of the sample, the presence of diluents and **adulterants**, and the age of the reagent. The length of time the reaction is observed may also influence the color reported. Color transitions and instabilities are not unusual. Allowances should be made for these differences.

Microcrystal Tests

Microcrystal tests are used as a screening tool to confirm a diagnosis made with other testing methods. They are fast and simple to administer, and can be highly specific (although whether they are specific enough to be used as a confirmatory test has been debated).

In microcrystal tests, the test sample is dissolved in a solution. A test reagent is either added to the solution or is already present in the solution in which the sample is dissolved. A reaction between the compound of interest and the test reagent forms a solid compound that is not soluble in the test drop. The solid forms uniquely shaped crystals that can be observed with a microscope. Table 23.1 lists some of the common methods for developing crystals.

Microcrystal identification relies upon the comparison of the crystals formed by the unknown with those formed by a reference

Table 23.1
Crystal Development Techniques

- A drop of a reagent solution is caused to flow into the test drop.
- A drop of the reagent is added directly to the test drop at the center (or vice versa).
- The reagent and test drop mixture is scratched or mixed to induce crystal formation.
- Reactions take place in a capillary tube.
- A fragment of solid reagent is added to a test drop.
- A drop of the reagent is suspended over a test drop (or vice versa).
- A drop of acid, base, or solvent may be added to the test drop to assist in volatilization.

standard using the same reagent. Difficulties obtaining an exact match between the crystals of the unknown and those of the reference sample may arise. Impurities in the unknown may lead to the formation of deformed, irregular, or unusual crystals. This can be overcome by utilizing a cleanup procedure such as TLC, extractions, or particle picking prior to microcrystal analysis.

Other differences in crystal appearance can arise from the concentration of the solution. The crystals in highly concentrated test drops develop rapidly, resulting in a distortion of the classic crystal shapes. Concentrated test drops should be diluted to a concentration that produces classic crystal forms that are conducive to comparison and identification. Reagent age can also affect crystal development. Unknown and reference samples should be run using the same reagents, under the same conditions, and at approximately the same concentration. Polymorphism is occasionally a source of trouble. Sample concentration and reagent age can lead to the creation of different microcrystalline forms. This reemphasizes the comparative nature of microcrystal identification. The comparison should be done using the same sample concentration with the same crystal reagent.

Microcrystal tests can also be used to determine the optical isomer of a compound. Single isomer compounds (*d* or *l*) produce a different crystal form from a **racemic mixture** (*d* and *l*) of the same compound. Single isomer crystals will form if a substance with the same isomer is added to the test solution prior the test reagent. Racemic crystals will form if the opposite isomer configuration is added to the test solution prior to analysis.

The microscopic crystalline structures of a compound can be used to tentatively identify components within a mixture. The examiner can obtain a profile of the various components within the mixture by placing a sample into a liquid test drop in which most, if not all, of the components are insoluble (mineral oil works well for this type of analysis). The component's physical and optical characteristics are then observed under polarized light microscopy or the polarized microscope.

Thin Layer Chromatography (TLC)

TLC is the third wet chemical test used to screen for the presence of controlled substances. It is a separation technique that utilizes molecular mobility and solvent compatibility to separate and distinguish compounds within a mixture. Compounds are separated by their size, shape, and reactivity with the solvent, like rocks flowing down a river. Small compact molecules travel across the TLC plate at a different rate from that of large rambling molecules.

The typical TLC procedure places a sample of the unknown toward the bottom of a glass plate containing a thin layer of silica gel. A sample of a reference compound is placed the same distance from the bottom of the plate. The TLC plate is placed into a tank containing a solvent (or mixture of solvents). As the solvent moves up the TLC plate, the various components within the sample are separated. When the solvent migration is stopped, the TLC plate is removed from the tank and the solvent is allowed to evaporate. The compound movement is then visualized through observation under UV light or through development with a chemical color reagent designed to react with various compounds.

The Rf value is used to establish the identity of the spots on the TLC plate. The Rf value is the ratio of the distance the solvent travels to the distance the compound travels (see Table 23.2). The distance the sample travels is measured from the center of the original sample spot to the middle of the densest portion of the spot after the solvent has traveled across the thin layer plate. The distance the solvent travels is measured from the same origin to the leading edge of the solvent front when the thin layer plate is removed from the developing tank.

Many compounds can have the same Rf value with a given solvent system. Multiple solvent systems are necessary when utilizing TLC for identification purposes. Each solvent system should have differing chemical properties to be able to separate compounds with similar Rf values. The solvent choice for the mobile phase will vary, depending upon the

Table 23.2
Table of Calculation Equations

TLC Rf value	$Rf = (\text{Distance traveled}_{\text{Compound}}/\text{distance traveled}_{\text{Solvent}}) \times 100$
GC relative retention time	$RRt = Rt_{\text{Compound}}/Rt_{\text{Internal standard}}$
Gravimetric percentage calculation	$\% = (\text{Weight}_{\text{Postextraction}}/\text{Weight}_{\text{Preextraction}}) \times 100$
Serial dilution concentration calculation	$\text{Concentration}_{\text{Unknown}} = [(\text{Concentration}_{\text{Std 2}} - \text{concentration}_{\text{Std 1}})/(\text{Data}_{\text{Std 2}} - \text{Data}_{\text{Std 1}})] \times \text{Data}_{\text{Unknown}}$
Serial dilution percentage calculation	$\% = [\text{Concentration}_{\text{Calculated}}/(\text{weight}_{\text{Unknown}}/\text{Volume}_{\text{Test Solution}})] \times 100$
Relative response percentage quantitation	$\% = [(\text{Area}_{\text{Unknown}} \times \text{Area}_{\text{Is std}} \times \text{weight}_{\text{Std}})/(\text{area}_{\text{Is unknown}} \times \text{area}_{\text{Std}} \times \text{weight}_{\text{Unknown}})] \times 100$

compound of interest. If the target compound or group of compounds is known, reference material can be used to select a solvent system that works for that compound.

The use of Rf values for a known solvent system provides only a general insight into the identity of the unknown spot. They should not be relied upon for confirmation of unknowns. A known reference sample, run on the same TLC plate, should be used for comparison.

Many factors can affect Rf values. The adsorbent uniformity on the thin layer plate, sample concentration (spotting is too weak or strong), room temperature during the mobile phase, and development distance of the solvent during the mobile phase can all affect the results. Care should be taken to eliminate variances in the method caused by any of these factors. Placing a reference sample containing the suspected compound on the TLC plate with the questioned sample reduces the variables involved in TLC comparisons.

Extractions

Extractions are used to separate the compound of interest from the rest of the sample. The type of extraction used depends upon the compound of interest and the matrix in which the compound is located. In some cases, multiple extraction techniques are necessary to separate the substance of interest from the remainder of the sample. In other instances instrumental analysis is the only way to separate compounds with similar chemical properties for confirmation.

Extractions are not screening tests per se. However, the fact that the compound is isolated as a result of the extraction indicates that the compound has certain chemical characteristics. These are class characteristics that can be used to support the confirmatory test deductively.

The screening techniques used should be designed to identify as many of the components of the sample matrix as possible. This allows the examiner to select the extraction technique that efficiently and effectively isolates the component of interest from the rest of the compounds. Missing or failing to identify the components within a sample mixture may lead to the selection of an inappropriate extraction technique, which in turn may affect the results of the confirmatory test.

The basic types of extractions include physical extractions, dry washing, dry extractions, and liquid/liquid extractions.

Physical extraction—Physical extractions are the simplest. They involve physically removing the particles of interest from the balance of the sample for later analysis. Physical extraction is appropriate when the examiner observes particles of different size, shades, and consistency within the sample. The particles are separated from the bulk sample by the use of stereomicroscopes, tweezers, sieves, or other devices designed to physically isolate particles of different sizes.

Dry wash and dry extraction—Dry washes and dry extractions are different versions of the same process. The only difference is the substance that is removed from the sample matrix. A dry wash uses a solvent to dissolve and remove adulterants and diluents from the sample matrix, leaving the compound of interest. A dry extraction uses a solvent to dissolve and remove the compound of interest from the sample matrix.

Liquid/liquid extractions—The ability of a substance to dissolve in a liquid can change with the liquid environment. Liquid extractions utilize these solubility characteristics to separate a substance from a mixture. Table 23.3 lists the general solubility guidelines used for liquid/liquid extractions.

During a liquid/liquid extraction (Table 23.4), the sample is initially dissolved into a water solution in which the compound of interest is soluble. This liquid is washed with an organic liquid in which the compound of interest is not soluble, but the diluents and adulterants are. Once the organic liquid is separated, the pH of the water is changed to make the compound of interest insoluble in the water solution. An organic liquid is used to separate the purified substance from the water. Care must be taken when selecting the acidic environment and the organic solvent used in liquid/liquid extractions. Some drugs are subject to **ion pairing**. This means that hydrochloride salt of the drug is soluble in chlorinated solvents (e.g., chloroform) and will choose the chlorinated solvent over an acidic environment with a high chloride concentration (e.g., HCl).

Ion pairing can be used to the examiner's advantage when multiple basic drugs within a matrix need to be isolated. If one of those drugs is subject to ion pairing, it can be isolated from the other drugs that under normal circumstances could not be separated. Table 23.5 outlines a generic ion pairing extraction sequence.

In some instances the compound of interest cannot be isolated because the sample matrix contains multiple drugs of the same salt type. In these instances a combination of techniques may be necessary to isolate the component of interest. An example of a

Table 23.3
Extraction Solubility Guidelines

- Basic drugs are soluble in acidic (pH < 7) water solutions. Acidic drugs are not.
- Acidic drugs are soluble in basic (pH > 7) water solutions. Basic drugs are not.
- Neutral drugs can be soluble in both.
- Free base, free acid, and neutral drugs are soluble in organic solvents.
- Free base and free acid drugs are insoluble in water.
- To determine if a drug is acidic or basic, look at its salt form. Basic drugs have an acid as part of their salt form (e.g., hydrochloride [HCl], sulfate [H_2SO_4], acetate [CH_3COOH], etc.). Acidic drugs have an alkali metal as part of their salt form (e.g., sodium [Na], or potassium [K]). Neutral drugs do not have an associated salt form.

Table 23.4
General Liquid/Liquid Extraction Procedure

- Dissolve the sample in an acidic aqueous solution. (The basic drug[s] will dissolve into the acidic aqueous solution.)
- Add an organic solvent and agitate. (The acidic and neutral drugs will dissolve into the organic solvent.)
- Allow the liquids to separate.
- Remove and discard the organic solvent.
- Make the aqueous liquid basic. (The basic drug[s] will come out of solution.)
- Add an organic solvent and agitate. (The basic drug[s] will dissolve in the organic solvent.)
- Allow the liquids to separate.
- Remove and retain the organic solvent. (The organic solvent can be analyzed by GC–MS or the solvent can be evaporated. IR can be used to analyze the residual free base drug.)

Table 23.5
General Ion Pairing Extraction Technique

- Dissolve the sample in an acidic aqueous solution with a high concentration of chloride ions. (The use of HCl or the addition of a chloride salt will provide the needed environment. The basic drug[s] will dissolve into the acidic aqueous solution.)
- Add a chlorinated solvent (chloroform) and agitate. (The ion-pairing drug will extract into the organic solvent along with the acidic and neutral drugs.)
- Allow the liquids to separate.
- Remove and save the organic solvent.
- Add an acidic solution that is void of chloride ions to the organic solution and agitate. (The ion-pairing drug[s] will dissolve into the acidic aqueous solution.)
- Allow the liquids to separate.
- Remove and discard the organic solvent.
- Make the aqueous liquid basic. (The ion-pairing drug[s] will not be soluble in solution.)
- Add an organic solvent and agitate. (The ion-pairing drug[s] will dissolve into the organic solvent.)
- Allow the liquids to separate.
- Remove and retain the organic solvent. (The organic solvent can be analyzed by GC–MS. The solvent can be evaporated and IR can be used to analyze the residual free base drug.)

Table 23.6
Preparatory TLC Extraction

- Perform a liquid/liquid extraction. (See Table 22.4)
- Prepare a TLC plate using the final organic extract from the liquid/liquid extraction.
- Develop the plate using a solvent system that will provide a distinct separation of the compounds in the mixture.
- Perform a dry extraction of the individual spot areas of the separated compounds.

combination extraction is a TLC separation of the final extract of a liquid/liquid extraction (Table 23.6). The silica gel around the spot corresponding to the compound of interest is physically removed from the TLC plate. A dry extraction or another liquid/liquid extraction is performed to isolate the substance from the silica gel.

Documentation

Chemical color tests are a nondocumentable technique. There is no independent record of the performance of the test. The test documentation solely rests on the examiners handwritten notes, so the examiner should describe as completely as possible the colors or transitions of colors that were observed during the course of the test. A plus (+) or minus (–) notation next to a test name does not provide a peer reviewer insight into what the examiner saw during the performance of the test.

Photographing a chemical color test may or may not be a solution to the documentation issue. Photography does demonstrate the color that was observed during the examination. However, it preserves only a portion of the test. Many chemical color tests have a transition of colors from the beginning of the test to the end, so photographs do not adequately reflect the examiner's observations.

As with color tests, no supporting documentation is generally generated with microcrystal examinations. The examiner's description of his or her observations should be as complete and accurate as possible. When definite crystals are formed, their form and habit should be noted (described, sketched, or photographed). Table 23.7 is a list of descriptive terms that can be used to describe the observed crystals.

Lack of supporting documentation may be less significant if microcrystal tests are used as a screening tool. However, if they

Table 23.7
Crystal Descriptions

Crystal	Description
Blade	Broad needle
Bunch/bundle	Cluster with the majority of crystals lying in one direction
Burr/hedgehog	Rosette, which is so dense that only the tops of the needles show
Cluster	Loose complex of crystals
Cross	Single cruciform crystal
Dendrites	Multibrachiate branching crystals
Grains	Small lenticular crystals
Needle	Long, thin crystals with pointed ends
Plate	Crystals with the length and width that are of the same order of magnitude
Prism	Thick tablet
Rod	Long, thin crystals with square-cut ends
Rosette	Collection of crystals radiating from a single point
Sheaf	Double tuft
Splinters	Small irregular rods and needles
Star	Rosette with only four to six components
Tablet	Plates with appreciable thickness
Tuft or fan	Sector of a rosette

are to be used as a tool to specifically identify a compound or isomer, configuration steps should be taken to provide reliable documentation concerning the examiner's observations. Photomicrographs should be taken of the microcrystals that were used to make the identification. The photomicrographs should be included in the examiner's notes for peer review when necessary.

As with color and microcrystal examinations, no supporting documentation is usually generated with the use of TLC. Accurate notes regarding the type of solvent system used should be included in case notes, along with the Rf calculations for the spots used for compound identification. Any deviations from the referenced method or unusual occurrences noted should also be documented. The examiner should thoroughly describe the observations used to make his or her conclusions, including the colors and patterns observed on the TLC plates as well as any observations made under UV light.

Photography of TLC plates is an option. Photographs can document the examiner's observations of the color and position of the sample spots. If the photograph is scaled properly, a peer reviewer or independent examiner can calculate Rf values.

The extraction phase of the analysis is not used for preliminary or confirmatory identification purposes. However, it is a means to those ends, and it should be documented. Peer reviewers should be able to evaluate the extraction technique used to prepare the sample for any subsequent testing.

Instrumental Examinations

Instrumental examinations are documentable testing methods. This point is key to the confirmation process. It is not enough for the examiner to be able to say the compound had the same chemical fingerprint as a controlled substance. He or she has to be able to prove it beyond a reasonable doubt. That includes subjecting the test to peer review. Instrumental examinations provide the vehicle for this review.

Four basic instruments are routinely utilized in the controlled substances section. This section describes how each of the instruments can be used to identify controlled substances and drugs of abuse. The UV spectrophotometer and the gas chromatograph are used as screening and quantitative tools. The infrared spectrometer and the mass spectrometer are instruments used to confirm the identity unknowns.

Ultraviolet Spectroscopy

Ultraviolet (UV) spectroscopy is an instrumental technique that provides compound classification. It is a screening tool, not a confirmatory test. Although some compounds exhibit unique UV spectra, the spectra are considered class characteristics and do not contain sufficient detail (individual characteristics) to be considered a compound's chemical fingerprint. The two general uses for UV

spectroscopy in the controlled substances unit are general screening and quantitation. The shape of the spectrum provides insight into the identity of the compound. The amount of UV light absorbed can correlate to the amount of substance in the sample.

As a screening tool, UV spectroscopy can identify a class or group of compounds in a sample. Many drug groups produce characteristic UV spectra. Figure 23.5 is an example of the UV spectra of a phenethylamine class of drugs. This class includes amphetamine, methamphetamine, ephedrine, pseudoephedrine, and phenylpropanolamine. Amphetamine and methamphetamine are considered controlled substances, and ephedrine and phenylpropanolamine are not. Nevertheless, the five different compounds exhibit similar UV spectra.

Ultraviolet spectroscopy is a useful tool for single component analysis of samples with a known or suspected composition, such as pharmaceuticals. The UV spectrum can confirm or refute the composition of the preparation under examination. However, if compound identification is required, it should be done using a specific test such as IR or MS.

Mixtures of compounds capable of absorbing UV energy can present an analytical

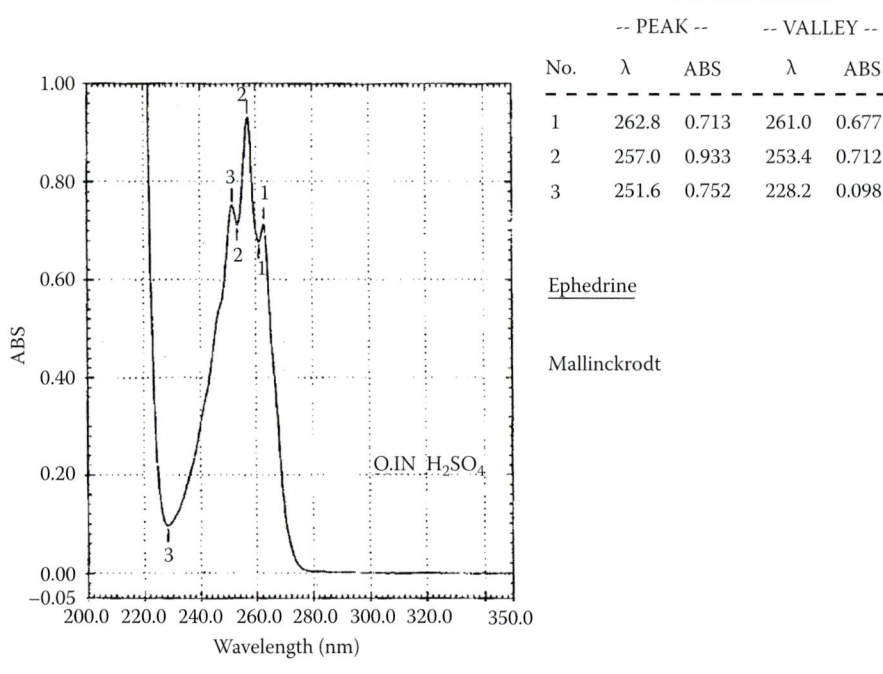

xxx PEAK-PICK xxx				
	-- PEAK --		-- VALLEY --	
No.	λ	ABS	λ	ABS
1	262.8	0.713	261.0	0.677
2	257.0	0.933	253.4	0.712
3	251.6	0.752	228.2	0.098

Ephedrine

Mallinckrodt

Figure 23.5 UV spectrum.

problem. Compounds have differing capacities to absorb UV light. If a noncontrolled substance that is a strong UV absorber is mixed in with a controlled substance that is a weak UV absorber, the resulting UV spectrum may not reflect the presence of the controlled substance.

Quantitation is another venue in which UV spectroscopy is useful. To be effective the sample should contain a single UV-absorbing component. If the sample contains multiple UV absorbers, the component of interest should have distinct resolvable absorption bands. The quantitation procedures can be as simple as comparing the concentration of the suspected tampered sample with that of a known unaltered sample. The UV absorbances should be the same if the concentrations and compositions are identical. A detailed analysis can determine the concentration of the substance in question. The absorbance value of the test sample is compared to the absorbances of a series of known solutions. The concentration of the test sample can be taken from the graph of concentration versus absorbance values of the reference samples.

Gas Chromatography

Gas chromatography (GC) is a documentable chromatography form that can be used in lieu of TLC. The gas chromatograph separates compounds by their size, shape, and reactivity with the chemical coating of the GC column. It is not a specific confirmatory test for controlled substances. However, dual column techniques and the evaluation of alkaloid peak patterns can be used for identification purposes.

Chromatograms from GCs (Figure 23.6) are used to identify unknowns on the basis of the retention time or **relative retention time** of a peak under certain operating conditions. The retention time (Rt) is the time it takes a compound to travel from the injection port of the GC to the detector. The relative retention time (RRt) is the ratio of the retention time of the substance to the retention time of an internal standard that is placed into the sample (see Table 23.2). The RRt is considered a more reliable value. The internal standard provides a reference point to calculate RRt values. It

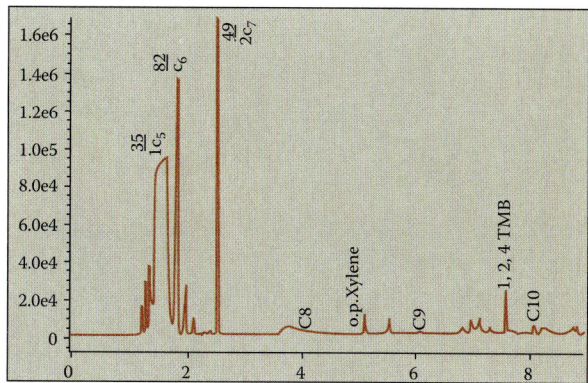

Figure 23.6 GC chromatogram.

also can be used to demonstrate the precision and accuracy of the instrument. The internal standard eluting at the proper time indicates the gas flow and oven conditions are operating properly. The size of its peak of a known concentration of internal standard indicates proper operation of the detector.

Analysis by GC alone is not generally considered confirmation of a controlled substance. More than one compound could possibly have a given Rt or RRt. With conventional detectors (i.e., flame ionization, electron capture, nitrogen/phosphorus, etc.), the examiner cannot definitively tell what compound elutes at a given Rt or RRt. Dual column GC has been used as a confirmatory test. A single sample is injected into a GC that divides the sample into two chromatographic columns. Each column contains a different liquid phase (the interior coating that causes compound separation). A compound is considered identified if the compound has the proper Rt or RRt elutes on both columns.

Commonly, GCs are used as the separation tool for the confirmatory tests of mass spectroscopy (MS) and infrared spectroscopy (IR). The GC separates the compounds, and the MS or the IR provides information concerning the chemical properties of each of the compounds as it elutes from the chromatographic column.

Quantitation is another use for GC. This can be accomplished through a serial dilution method similar to that used in UV analysis, or with a relative response technique. As a quantitation tool, GC has an advantage over UV. The effects of multiple components within the sample are reduced or eliminated because it separates the components of the sample during the analysis.

Mass Spectroscopy

Mass spectroscopy (MS) uses the pattern of molecular pieces (ions) that is produced when a molecule is exposed to a beam of electrons. This characteristic pattern is called the *mass spectrum*. It is considered one of a compound's chemical fingerprints. Figure 23.7 is an example of the mass spectrum of heroin.

Mass spectroscopy cannot differentiate between certain types of isomers. **Stereoisomers** (molecules that are mirror images of each other) have identical mass spectra. The chromatographic retention times of these compounds are also the same. Ephedrine and pseudoephedrine are examples of

stereoisomers that cannot be differentiated by mass spectroscopy. Geometric or positional isomers also produce similar, if not the same, mass spectra. Many times they can be differentiated by the chromatographic retention time of the compound. Other times one or two clusters of ions have ions ratios that are specific to a particular isomer. Methamphetamine and phentermine are two geometric isomers that can be differentiated through the use of mass spectroscopy.

The mass spectrometer cannot distinguish between the salt and free base form of a drug. The salt portion of the compound is below the detection range of the MS, and the detector sees only the free base portion of the compound.

A number of mass spectra libraries are available to assist in the identification of unknowns. However, final confirmation is

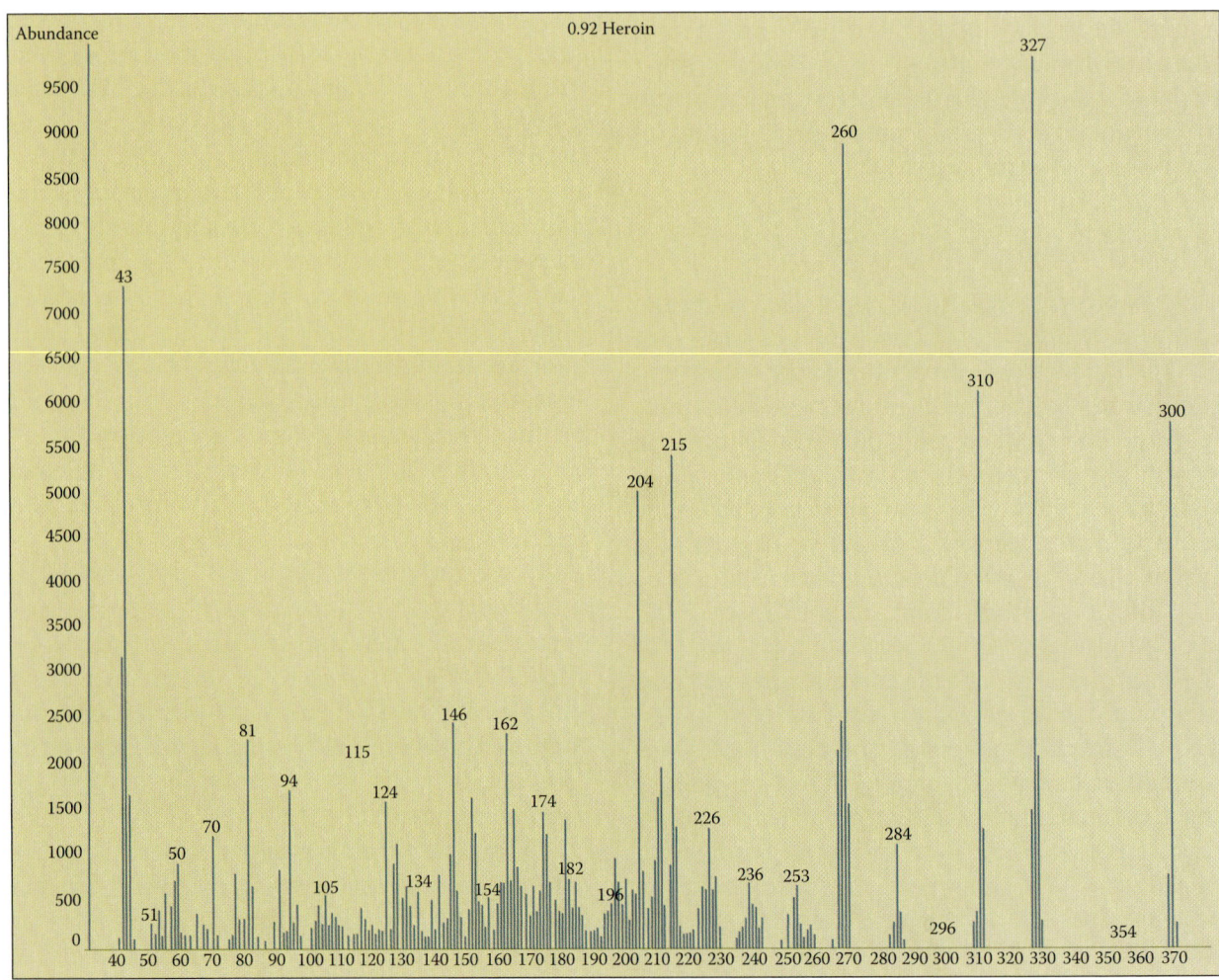

Figure 23.7   Mass spectrum of heroin.

accomplished only by comparing the mass spectra of the unknown to the mass spectrum of a known reference standard. The reference spectra should be obtained on the same instrument, under the same operating conditions.

The operating conditions of mass spectrometers vary slightly among instruments. This variation can cause shifts in ion ratios in various parts of the mass spectrum. Confirmation of a compound's identity must be done by comparing the spectrum of the unknown to the spectrum of a known reference sample that was analyzed on the same instrument under the same operating conditions.

Large concentration differences commonly lead to differences in ion ratios in the mass spectrum. These differences may be evident across the chromatographic peak. The spectrum at the apex may be different from the spectrum on the leading or tailing ends of the peak. Diluting the concentration of the sample can remove these differences, resulting in the same spectra produced across the chromatographic peak.

Infrared Spectroscopy

Infrared (IR) spectroscopy has been the traditional method of confirming the identity of a controlled substance. Traditionally, the sample was subjected to a series of screening tests to establish the compound's suspected identity. The identity of any adulterants and diluents were determined. The controlled substance was then extracted and purified. Finally, an IR spectrum was obtained. Modern technology has introduced instrumentation that can obtain an IR spectrum from a single particle or from a peak in a GC, eliminating the need for complicated procedures.

IR spectroscopy uses a compound's ability to absorb IR light as a means of identification. Organic compounds absorb different portions of the IR spectrum. The pattern that results from charting the absorbance and transmittance of IR light that is passed through (or reflected from) a sample is considered a chemical fingerprint.

Isomer determination is a benefit of using IR spectroscopy as a confirmation tool. Compounds with isomers that are indistinguishable by MS may be differentiated through the use of IR spectroscopy. Figure 23.8 is a comparison of the IR spectra of ephedrine and pseudoephedrine. The salt form of a compound also affects its IR spectra. Each salt

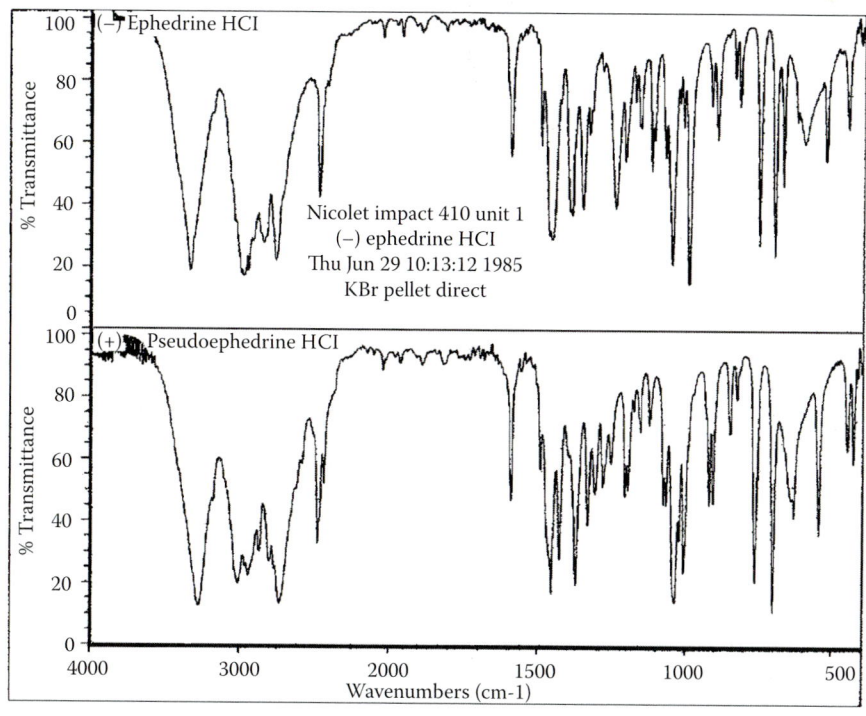

Figure 23.8 Ephedrine/pseudoephedrine FT-IR comparison.

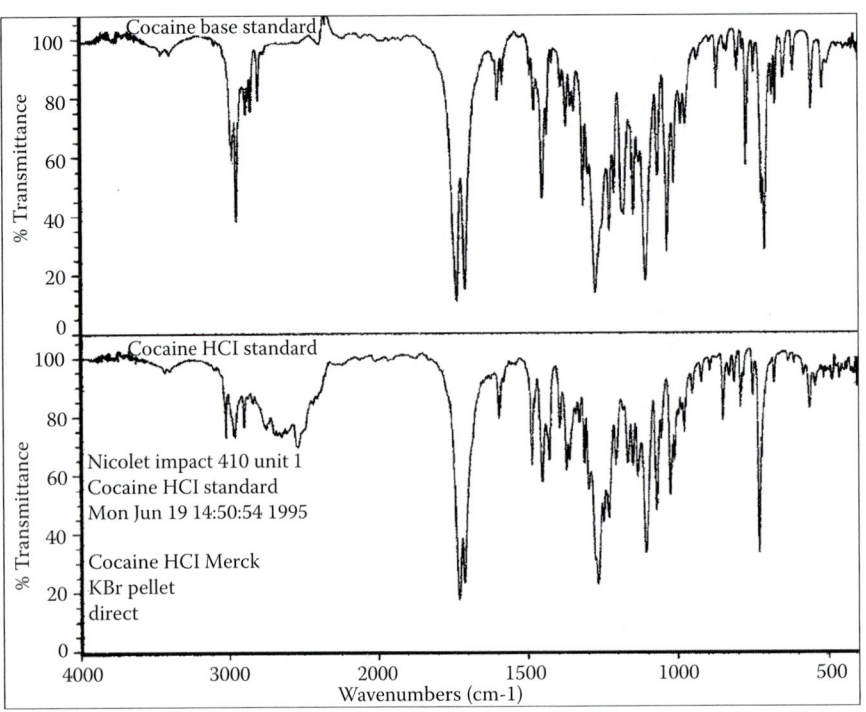

Figure 23.9   Cocaine base/cocaine HCl FT-IR comparison.

type (hydrochloride, sulfate, etc.) contributes to the spectra differently. This provides the examiner with a tool to identify and document the salt form of a compound. Figure 23.9 is a comparison of free base cocaine (crack) and cocaine hydrochloride.

Advances in technology have reduced the time required for sample preparation and analysis. Extractions can be done using a GC. The Fourier transform infrared (FT-IR) spectrophotometer can obtain an IR spectrum of individual peaks traveling through the chromatographic column. Micro FT-IRs can isolate and obtain IR spectra of individual particles within a mixture.

Infrared spectroscopy analyzes the vibrations of different parts of a molecule when it is exposed to IR light. Changing the sample method may affect the way different parts of the molecule can vibrate, which will cause shifts in the peak intensities in the resulting IR spectra. The way the compound crystallizes (or does not crystallize) within the sample matrix that is presented to the instrument will affect the resulting IR spectrum. Transmittance spectra differ from reflectance spectra. The IR spectrum of a vapor phase sample is different from that of a liquid sample, which is different from a sample pressed into a KBr pellet or recrystallized on a salt plate.

Polymorphism can affect a compound's IR spectrum. A single compound may have more than one crystalline form, along with an amorphous form. The way a compound crystallizes affects the vibration within the molecule, which in turn affects the resulting IR spectrum. These variations can occur in the same sampling technique and will have a slightly different IR spectrum.

Because of the variation of IR spectra among sampling techniques, a library of known spectra from traceable sources should be maintained for compound confirmation purposes. The various IR spectra libraries that are available should be used as a screening tool, not as a reference for confirmation. As with MS, final confirmation is accomplished only by comparing the IR spectra of the unknown to the IR spectrum of a known reference standard. The spectra should be produced on the same instrument, under the same conditions.

Documentation

Instrumental techniques are documentable because they generate analytical data in a form that demonstrates the analysis was performed. The data itself is objective and can be subjected to peer review as part of a quality assurance program or independent evaluation at a later date. The interpretation of the data is less subjective than in other areas of the forensic laboratory, although it is still subject to interpretation.

For peer review purposes, case notes or instrument printouts should include the operating conditions of the instrument during the analysis. This allows the reviewer to evaluate whether instrumental results are consistent with the analytical conditions. If necessary, an independent examiner should be able to achieve the same results under the same test conditions. All data should contain, at a minimum, the examiner's initials, case number, solvent information, and date of the analysis. The examiner should have the instrument print this information on the spectra at the time of analysis if the instrument has the capacity to do so. For GC analysis the calculated RRt value should be on the chromatogram or on the printout of the peak retention times. The divisions of the mass value axis on MS data should be such that the examiner can easily determine the mass value of each of the ions of the spectra. The wave number of the significant peaks of an IR spectrum should be labeled or should be easily determinable by a peer reviewer. The examiner should have the instrument print this information at the time of analysis if the instrument has the capacity to do so.

Quantitation

As a general rule there is no statutory requirement to perform a quantitative examination on controlled substance samples. Quantitation is used as an investigative tool or is done as part of a laboratory's internal security policy. With few exceptions, criminal statutes regulate only the possession of a given substance. The concentration does not affect guilt or innocence.

The concentration of a sample may become an issue during the sentencing phase of a trial. Some statutes provide enhanced penalties for possession of a substance over a given concentration. The words "possession of X grams of compound Y," versus "X grams of substance containing compound Y," may affect whether a quantitative exam is required to establish a sentence of 1 year or 10 years.

The four basic methods of quantitating the amount of controlled substance in a sample are microscopic examination, **gravimetric** comparison, UV analysis, and GC analysis.

Microscopic Examination

The quickest and most subjective quantitative method is through microscopic examination. In this technique a sample is placed on a microscope slide and diluted with a solvent in which the components are insoluble. The examiner estimates the percentage of crystals of controlled substance in the sample under observation.

This is the least precise and least accurate method. The method is subject to the examiner's ability to recognize the microscopic crystalline form of the controlled substance under consideration. The uniformity of the bulk sample will affect the accuracy and reproducibility of the results.

Gravimetric Techniques

Gravimetric techniques are methods that can be performed in conjunction with the extraction phase of an analysis. The examiner weighs the sample to be extracted prior to the extraction process. A weight is obtained of the extracted substance prior to the performance of any confirmatory tests. The ratio of the postextraction weight to the preextraction weight provides the percentage of the item that is the controlled substance (see Table 23.2).

An advantage of gravimetric techniques is that the identity of the final extract can be confirmed. All the diluents and adulterants have been removed from the matrix. The resulting residue can be analyzed for purity and identity. A limiting factor in the precision and accuracy of this technique is the efficiency of the extraction solvents. If they do not effectively remove the diluents and adulterants, the calculated controlled substance percentage will be high. If the solvents do not efficiently and completely isolate the controlled substance, the percentage will be low.

The examiner must be aware of the salt form the controlled substance is in before and after the extraction process. This will affect the calculated percentage because the molecular weights of the salt form differ from the molecular weight of its free base form. For example, a 100 percent pure sample of cocaine hydrochloride contains 89.38 percent by weight free base cocaine. The examiner must take into account the mass of the salt when calculating the percentage of controlled substance in the sample, or qualify the conclusion by stating the salt form of the substance identified.

Ultraviolet Techniques

The use of UV light provides an effective method to quantitate a sample if it has a single UV absorber. If the sample has components with overlapping UV absorbances, the instrument cannot determine which compound is contributing to the absorbance. Compounds also absorb UV radiation at different rates. Therefore, UV is not conducive to quantitating mixtures of unknown composition.

Simple yes-or-no concentration comparisons can be accomplished with the use of UV. These comparisons are conducted in association with product-tampering cases in which the product in question may have been diluted or altered. A comparison of the UV spectra of the item in question to a known reference sample can indicate if the unknown has been diluted or altered. The composition of both samples should also be confirmed through a separate examination.

A detailed examination can determine the concentration of the substance in question. To accomplish this, the examiner obtains the UV spectra for a series of solutions with a

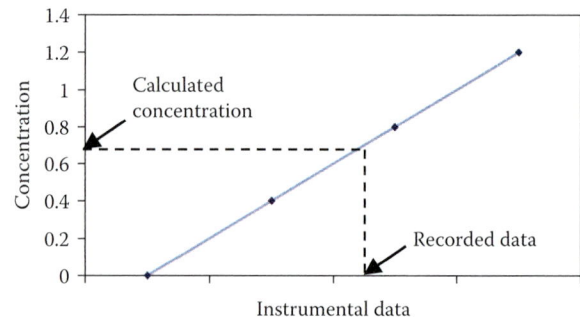

Figure 23.10 Data versus concentration graph.

known concentration of the substance in question. The absorbance values are placed on a concentration versus absorbance graph. A solution of the unknown is prepared and analyzed. The absorbance value is placed on the graph to determine the concentration of the substance in the solution. This value is then used to calculate the percentage of substance in the unknown. Figure 23.10 is an example of a graph used to calculate the concentration of the unknown solution.

UV techniques done properly are precise. However, the accuracy of the results of multi-component mixtures may be in question because of the interference of the UV absorbance of other compounds in the sample.

Gas Chromatography Techniques

The use of GC for quantitation is the most accurate and precise analytical technique. This technique gives the examiner the ability to isolate and quantitate a specific compound in a single test. The identity of the chromatographic peak can be confirmed at the time of the analysis or by analyzing the test solution with a GC–MS using the same chromatographic conditions used during the quantitation.

The serial dilution method uses the relationship between the concentration of a sample and the instrumental data. Doubling the sample concentration doubles the GC peak area or reduces the percentage of UV light that is transmitted. Table 23.8 outlines the serial dilution method. If the analytical data of the serial dilutions of a substance are charted, the

Table 23.8
Serial Dilution Quantitation Method

- A set of three or more standard solutions of the substance under investigation is prepared. (Each solution has a different concentration.)
- A solution of the unknown is prepared using the same solvents.
- Each solution is injected into the GC and the peak areas are recorded.
- The concentrations and corresponding peak areas are plotted on a graph with the peak area on the x-axis and the concentration on the y-axis.
- The concentration of the unknown in the test solution is determined from correlating the peak area of the unknown solution to the corresponding concentration on the graph.
- This concentration is then used to calculate the percentage of compound in the original sample (see Table 22.2).

Table 23.9
Relative Response Quantitation Method

- Prepare an internal standard solution.
- Prepare the reference sample to a known concentration using the internal standard solution.
- Prepare the unknown sample to a similar concentration using the same internal standard solution.
- Inject each sample into the GC and record the peak areas.
- Calculate the concentration of the controlled substance in the test solution (see Table 22.2).
- Calculate the percentage of the sample that contains the controlled substance (see Table 22.2).

concentration of an unknown solution can be determined from its instrumental responses.

The relative response GC method of determining sample concentration uses the ratio of the compound's and internal standard's peak areas, known sample concentrations, and algebra (Table 23.9). Two GC injections using this method can provide the same results as four or more injections using the serial dilution method. This procedure is based on the predictable relationship between sample concentration and the peak area of a chromatogram; i.e., doubling the sample concentration doubles the resulting peak area of the chromatogram.

Quality Assurance (QA)/ Quality Control (QC)

There was a time that the phrase "Trust me. I am from the government and here to help" had a positive connotation. The forensic laboratory used to be considered a source of unquestionable truth. Unfortunately, that trust has been compromised through sloppy laboratory practices, unscrupulous examiners, and mismanagement. Additionally, the United States' adversarial legal system has placed additional scrutiny upon the analytical results generated by the forensic laboratory. As a result forensic laboratories have adopted systematic checks and balances that had been common practice in clinical and industrial laboratories.

A documented quality assurance (QA) and quality control (QC) program is just as important as documenting the results of individual examinations. The examination results may not be accepted by the court if the reliability of the instruments, protocols and chemicals used to perform the examinations cannot be established. Individual quality control programs demonstrate the reliability of the examination process. It is the combination of both which displays the reliability of the whole process (quality assurance).

Quality assurance is a documented system of protocols used to assure the accuracy and reliability of analytical results, and as such, it consists of a variety of components. Proficiency testing and employee qualifications and training standards are directed at the forensic chemist's performing the examinations. Documented evidence collection and handling procedures as well as documented, standardized, and validated analytical protocols are used to ensure the analytical methods used to meet an accepted scientific standard. Instrument maintenance logs, reagent preparation records, and the use of traceable chemicals used in preparation (to testing reagents) document the reliability of the chemicals and equipment used in the examination process. The use of traceable reference material ensures the material being used to generate the data being used for identification purposes is from a known source.

A quality assurance program produces intangible affects that may be difficult to assess. Service is improved through streamlining operation. The numbers of challenges to the analytical results are reduced because of the documented reliability of the equipment, chemicals, and protocols used. The need for reanalysis is reduced, saving the laboratory time and money. The laboratory's image and credibility is improved, which leads to fewer court appearances from testimony stipulation. Each of these effects enhances those of the others. Quality work leads to credibility which enhances staff morale and ultimately producing a more productive work environment.

Clandestine Drug Laboratories

Clandestine drug laboratories are illicit locations that manufacture controlled substances. The types and numbers of labs seized reflect national and regional trends concerning the types and amounts of illicit substances that are being manufactured, trafficked, and abused. They have been found in remote locations, in urban and suburban neighborhoods, hotels and motels, industrial complexes, and in academic and industrial laboratories. They range in size from table top setups used to produce gram quantities to large multi location operations that generate kilograms of final product. In each of these, toxic and explosive

Figure 23.11 Stove top clandestine lab used to extract precursor chemicals.

fumes can pose a significant threat to the health and safety of local residents.

The sophistication of clandestine labs varies widely. The production of substances such as methamphetamine, PCP, MDMA, and methcathinone requires little sophisticated equipment or knowledge of chemistry. However, the synthesis of drugs such as fentanyl and LSD requires much higher levels of expertise and equipment (Figure 23.11 and Figure 23.12).

The investigation of clandestine labs is one of the most challenging efforts of law enforcement. No other law enforcement activity relies on forensic science as heavily. The controlled substance section's involvement commences with the drafting of an affidavit used to obtain a search warrant. Their expertise is needed to process the crime scene. Forensic chemists analyze the evidence in a laboratory and render opinions in a written report or in courtroom testimony. Occasionally, they are called upon to testify on auxiliary issues concerning the clandestine lab investigation that occur after the case has been adjudicated.

Clandestine lab investigation is one of the most demanding tasks of the controlled substances section. It is a roller coaster ride of activity that requires every tool at its disposal. Traditional analytic techniques are used to develop information concerning the type and location of the clandestine lab as well as the identity of the operators. Chemists act as crime scene advisors used to identify significant

Figure 23.12 Clandestine reflux apparatus used to manufacture methamphetamine.

physical evidence as well as potentially hazardous chemicals and situations. Their analysis is also used to corroborate investigator information and establish the identity of the final products as well as the manufacturing methods used to produce them.

The forensic chemists who deal with clandestine labs should specialize in these issues. In bookkeeping, all CPAs are accountants but not all accountants are CPAs. The same is true with forensic chemists. All clandestine lab chemists are forensic chemists, but not all forensic chemists are clandestine lab chemists. The clandestine lab chemist has additional training in clandestine manufacturing techniques as well as in inorganic analysis. This knowledge allows them to expand their analytical scheme to identify the chemicals used in the manufacturing process. Their goal is to identify the manufacturing process, not just the controlled substance final product.

The identification, investigation and prosecution of a clandestine lab is a team effort. It is a collaboration of the efforts of law enforcement, forensic experts, scientists and criminal prosecutors to present a case that definitively demonstrates how a group of items with legitimate uses are being used to manufacture an illegal controlled substance. The goal of the alliance is establishing the existence of a clandestine lab beyond a reasonable doubt. *The Forensic Investigation of Clandestine Labs* (CRC Press) provides the general information needed to understand how the different pieces of the clandestine lab puzzle fit together.

The investigation of clandestine lab activity can be divided into five sections, all of which should involve a forensic chemist. The first section involves recognizing a clandestine lab. The second section deals with processing the clandestine lab site. The third section covers the laboratory analysis of the evidence. The fourth section encompasses generating opinions from the physical evidence. The fifth and final section covers presenting the evidence in court.

Recognition of clandestine lab activity is the first step in the process. The forensic chemist is a subject matter expert who can articulate the common elements encountered in clandestine labs. He can provide a profile of a clandestine lab operator and identify the chemical and equipment requirements, as well as the basic manufacturing techniques utilized with a given set of chemicals and equipment. He is able to describe why a clandestine lab exists and subsequently assists the investigators in securing a search warrant to proceed to the next phase of the process.

Knowing what a clandestine lab is and proving one exists are separate issues. Steps two and three of the clandestine lab investigative process deal with collecting and identifying the pieces of the clandestine lab puzzle. The information gathered from the crime scene must be evaluated. The forensic chemist acts as a technical advisor who assists investigators process clandestine lab sites for physical evidence. He or another chemist subsequently performs laboratory analysis on the exhibits.

Processing a clandestine lab scene is more complicated than the traditional crime scene

Figure 23.13 Personal protective equipment used to process clandestine lab scenes.

search associated with a narcotics investigation. The site of a clandestine lab is, because of the chemicals involved, a "hazardous materials incident" and necessitates invoking different protocols for crime scene processing. There are also a number of preliminary opinions that should be made by evaluating the physical evidence observed at the scene. These issues necessitate an on-scene forensic chemist (Figure 23.13).

A complete forensic laboratory analysis is a critical element of a clandestine lab investigation. The analysis of a reaction mixture is more complex than simply identifying the controlled substance it contains. The identification of precursor and reagent chemical as well as reaction by-products is necessary to establish the manufacturing method used. The identity of unique chemical components within a sample can be used as an investigative tool to connect the clandestine lab under investigation to other illegal activity. These analytical requirements necessitate training and experience beyond that of a traditional forensic chemist who examines drug samples.

Opinions or "What does it all mean?" is the next phase. A large amount of information is collected during a clandestine lab investigation. The forensic chemist collates the information from various sources and creates a profile of the clandestine lab under investigation. He addresses the questions: What type of operation existed?, What was it making?, How was it being made?, and How much could it make?

All the work to this point may be useless if the information cannot be relayed effectively to a jury. The forensic chemist provides expert testimony. He educates the prosecutor, deals with defense attorneys, and presents technical information to nontechnical jurors.

The use of forensic evidence is essential to the successful investigation and prosecution of a clandestine lab. The proper collection and preservation of the physical evidence followed by the complete analysis of the evidentiary samples are key elements. Their information is the cornerstone on which the forensic chemist's opinion is based. If forensic evidence is properly handled the court will have all of the information it needs to make a fully informed decision.

Conclusion

The analysis of controlled substances is a major component of a forensic laboratory. Its analysts perform a wide variety of examinations using a plethora of analytical methods. In a single case an examiner can analyze plant material in one exhibit, a powder-controlled substance in the next, and a liquid reaction mixture from a clandestine laboratory in the final exhibit.

The analytical tools and techniques available to the controlled substance section are as varied as the analysis they perform. The tools can be as simple as a stereomicroscope or as complex as GC–MS. The techniques can be as simple as visual observation with the unaided eye, or as complex as the evaluation of the mass spectrum of a designer drug to determine which controlled substance the unknown resembles.

Questions

1. List two plants that are considered controlled substances that require a botanical examination as part of the identification process.

2. When is it necessary to confirm the identity of the controlled substance in plant material? Give an example.

3. List four wet chemical techniques that can be used in the analysis of controlled substances.

4. List two wet chemical techniques that can be used as both screening tools and sample preparation techniques.

5. List two disadvantages to wet chemical techniques.

6. List two specific and two nonspecific instrumental techniques.

7. What information should accompany instrumental data?

8. When is a library search considered a confirmation and why?

9. Which instrumental technique's spectra are most subject to variations due to sample preparation techniques. Why?

10. List three quantitation techniques in order from most specific to least specific.

11. What are the minimum qualifications for a clandestine lab chemist?
    a. A baccalaureate degree in chemistry
    b. Training in the analysis of controlled substances
    c. Specialized training in the clandestine manufacture of controlled substances.
    d. All the above

12. What is the forensic chemist's objective during the analysis of evidence seized in a clandestine lab investigation?
    a. Identify only the controlled substance
    b. Identify every chemical in the sample matrix
    c. Establish the manufacturing route
    d. None of the above

13. When does the controlled substance section's participation in a clandestine lab investigation begin?
    a. The drafting of the search warrant affidavit
    b. The search of the crime scene
    c. The laboratory analysis of the physical evidence
    d. Court testimony concerning the clandestine lab seized

14. The guidelines of which professional group should be considered when developing examination techniques used to examine controlled substances?
    a. Working Group for the Analysis of Seized Drugs (SWGDRUG)
    b. American Society of Testing Materials (ASTM)
    c. American Society of Crime Laboratory Directors (ASCLD)
    d. A and B
    e. None of the above

15. The guidelines of which professional group hold the weight of law concerning the examination techniques used to examine controlled substances?
    a. Scientific Working Group for the Analysis of Seized Drugs (SWGDRUG)
    b. American Society of Testing Materials (ASTM)
    c. American Society of Crime Laboratory Directors (ASCLD)
    d. All the above
    e. None of the above

Suggested Readings

Alm, S. et al, Simultaneous gas chromatographic analysis of drugs of abuse on two fused silica columns of different polarity, *J. Chromatogr.*, 254, 179, 1983.

Bailey, M. A., The value of the Duquenois test for cannabis—a survey, *J. Forens. Sci.*, 24, 817, 1979.

Baker, P. B. and Phillips, C. F., The forensic analysis of drugs of abuse, *Analyst*, 108, 777, 1983.

Balinger, J. T. and Shugar, G. J., *Chemical Technician's Ready Reference Handbook*, 3rd ed., McGraw-Hill, New York, 1990.

Bartle, K. D., Lee, M. L., and Yang, F. J., *Open Tubular Column Gas Chromatography,* John Wiley and Sons, New York, 1984.

Boke, N. H. and Anderson, E. F., Structure, development and taxonomy in the genus Lophophora, *Am. J. Bot.*, 57, 569, 1970.

Brown, J. K., Shapazian, L., and Griffin G. D., A rapid screening procedure for some street drugs by thin-layer chromatography, *J. Chromatogr.*, 64, 129, 1972.

Budavari, S., Ed., *The Merck Index*, 11th ed., Merck and Company, Inc., Whitehouse Station, NJ, 1989.

Butler, W. P., *Methods of Analysis*, IRS publication 341, 1967.

Christian, D. R., Deviation of cast film heroin spectra, *SW Assoc. Forens. Sci. J.,* 7, 59, 1986.

Christian, D. R., Cast films as an alternative to pellets for solid sample IR, *SW Assoc. Forens. Sci. J.,* 9, 14, 1987.

Christian, D. R., Clandestine drug laboratories, *DRE*, 3, 3, 1991.

Churchill, K. T., Synthetic tetrahydrocannabinol, *J. Forens. Sci.*, 24, 762, 1983.

Clarke, E. G. C., Ed., *Isolation and Identification of Drugs,* Vol. I, Pharmaceutical Press, London 1969.

Clarke, E. G. C., Ed., *Isolation and Identification of Drugs,* Vol. II, Pharmaceutical Press, London, 1986.

Concise Encyclopedia of Chemical Technology, 3rd ed., Wiley Interscience, New York, 1985.

Engel, R. G. et al., *Introduction to Organic Laboratory Techniques: A Microscale Approach,* Saunders College Publishing, Philadelphia, 1990.

Fiegl, F., *Spot Tests for Inorganic Compounds*, 2nd ed., Elsevier, New York, 1987.

Fiegl, F., *Spot Tests for Organic Compounds*, 7th ed., Elsevier, New York, 1989.

Fulton, C. C., *Modern Microcrystal Tests for Drugs*, Wiley Interscience, New York, 1969.

Gill, R., Bal, T. S., and Moffat, A. C., The application of derivative UV-visible spectroscopy in forensic toxicology, *J. Forens. Sci. Soc.*, 22, 165, 1982.

Gough, T. A. and Baker, P. B., Identification of major drugs of abuse using chromatography—an update, *J. Chromatogr. Sci.*, 21, 145, 1983.

Gough T. A. and Baker, P. B., Identification of major drugs of abuse using chromatography, *J. Chromatogr. Sci.*, 20, 289, 1982.

Griffiths, P. D. and de Haseth, J. A., *Fourier Transform Infrared Spectrometry,* John Wiley and Sons, New York, 1986.

Heagy, J. A., Infrared method for distinguishing optical isomers of amphetamine, *Anal. Chem.*, 42, 1459, 1970.

Hughes, R. B. and Kessler, R. R., Increased safety and specificity in the thin-layer chromatographic identification of marihuana, *J. Forens. Sci.*, 24, 842, 1983.

Hughes, R. B. and Warner, V. J., Jr., A study of false positives in the chemical identification of marihuana, *J. Forens. Sci.*, 23, 304, 1978.

Johns, S. H., Wist, A. A., and Najam, A. R., Spot tests: a color chart reference for forensic chemists, *J. Forens. Sci.*, 24, 631, 1979.

Kriz, G. S., Lampman, G. M., and Pavia, D. L., *Introduction to Spectroscopy*, Saunders College Publishing, Philadelphia, 1979.

Liu, J. H. et al., Approaches to drug sample differentiation. III: A comparative study of the use of chiral and achiral capillary column gas chromatography/mass spectrometry for the determination of methamphetamine enantiomers and possible impurities, *J. Forens. Sci.*, 27, 39, 1982.

Mahmoud, A. E. et al., Constituents of *Cannabis sativa* l. XXIV: the potency of confiscated marijuana, hashish and hash oil over a ten year period, *J. Forens. Sci.*, 29, 500, 1984.

Maher, J. T., Narcotics and Other Substances Subject to the Controlled Substance Act of 1970, Drug Enforcement Administration, Public Law 91–513.

Marihuana: Its Identification, U.S. Treasury Department, 1948.

Marnell, T., *Drug Identification Bible*, 2nd ed., Amera-Chem. Inc., Grand Junction, CO, 1997.

McLafferty, F. W., *Interpretation of Mass Spectra*, 3rd ed., University Science Books, New York, 1980.

McLinden, V. J. and Stenhouse, A. M., A chromatography system for drug identification, *Forens. Sci. Int.*, 13, 71, 1979.

Microgram, Drug Enforcement Administration, U.S. Department of Justice.

Mills, T., III and Roberson, J.C., *Instrumental Data for Drug Analysis*, Vols. 1–7, CRC Press, Boca Raton, FL, 1992–1996.

Moss, W. W., Posey, F. T., and Peterson, P. C., A multivariate analysis of the infrared spectra of drugs of abuse, *J. Forens. Sci.*, 25, 304, 1980.

Nakamura, G. R., Forensic aspects of cystolith hairs of cannabis and other plants, *J. Assoc. Off. Anal. Chem.*, 52, 5, 1969.

Nakamura, G. R. and Thornton, J. I., The identification of marihuana, *J. Forens. Sci. Soc.*, 479, 1972.

Nakamura, G. R. and Thornton, J. I., The forensic identification of marijuana: some questions and answers, *J. Police Sci. Adm.*, 1, 102, 1977.

Perrigo, B. J. and Peel, H. W., The use of retention indices and temperature-programmed gas chromatography in analytical toxicology, *J. Chromatogr. Sci.*, 19, 219, 1981.

Pettitt, B. C., Rapid screening for drugs of abuse with short glass capillaries and a nitrogen selective detector, *J. High Res. Chromatogr. Chromatogr. Comm.*, 5, 45, 1982.

Physician's Desk Reference, 52nd ed., Medical Economics, Montvale, NJ, 1998.

Plotczyk, L. L., Application of fused-silica capillary gas chromatography to the analysis of underivatized drugs, *J. Chromatogr.*, 240, 349, 1982.

Ravreby, M. D. and Gorski, A., Effects of crystal habits in heroin on the infrared spectra, in *Proceedings of the International Symposium on the Forensic Aspects of Controlled Substances*, 1988, p. 165.

Saferstein, R., *Criminalistics: An Introduction to Forensic Science*, 7th ed., Prentice-Hall, New York, 2000.

Schepers, P. et al., Applicability of capillary gas chromatography to substance identification in toxicology by means of retention indices, *J. Forens. Sci.*, 27, 49, 1982.

Shugar, G. J. and Ballinger, J. T., *Chemical Technician's Ready Reference Handbook*, 4th ed., McGraw-Hill, New York, 1996.

Stead, A. H. et al., Standardized thin-layer chromatographic systems for the identification of drugs and poisons, *Analyst*, 107, 1106, 1982.

Sundholm, E. G., More economical use of high performance thin-layer plates for chromatographic screening of illicit drug samples, *J. Chromatogr.*, 265, 293, 1983.

Sunshine, I., Ed., *Handbook of Analytical Therapeutic Drug Monitoring and Toxicology*, CRC Press, Boca Raton, FL, 1996.

Velapoldi, R. A. and Wicks, S. A., The use of chemical spot test kits for the presumptive identification of narcotics and drugs of abuse, *J. Forens. Sci.*, 19, 636, 1974.

Vinson, J. A., Hooyman, J. E., and Ward, C. E., Identification of street drugs by thin-layer chromatography and a single visualization reagent, *J. Forens. Sci.*, 20, 552, 1975.

SECTION IV

Forensic Engineering

Structural Failures

Randall K. Noon

24

Introduction

Buildings and structures are the most obvious hallmarks of a civilization. Particular structures, such as the Great Pyramid of Cheops, the Taj Mahal, or the Golden Gate Bridge, symbolize aspects of a culture. Building methods, materials, and architectural style can even be used to broadly characterize a civilization. Distinguishable architectural details among buildings and structures can then be used to separate eras within a particular civilization. For example, the New York skyscrapers built in the 1930s are readily distinguishable from the New York skyscrapers built in the 1970s, despite the fact that both groups characterize the skyline of modern New York.

Babylon's King Hammurabi, ruler of one of the first great metropolises in the history of the world, was one of the first rulers to provide a written, standardized legal code. The Code of Hammurabi includes regulations that constitute the building code of that era. The underlying message of Hammurabi's building code is that a builder has a serious responsibility to ensure that buildings are properly constructed and do not fall down. Similar admonitions to builders are also contained in the Old Testament.

> If a builder makes a house and does not properly construct it, and the house falls down and kills the owner, the builder shall be put to death. If it kills the son of the owner, the son of the builder shall be put to death. If it kills the owner's slave, the builder shall pay for the slave. If it ruins goods, the builder shall pay for all that was ruined. Since he did not construct it properly, the builder shall rebuild the house at his own expense. If a builder makes a house and the walls are shaky, the builder must strengthen the walls at his own expense.

Code of Hammurabi, circa 1780 B.C.

Although these ancient building regulations are nearly 4000 years old, the basic requirement that engineers, architects and builders exercise care and diligence is still the fundamental tenet in modern building codes, such as the Unified Building Code.

Once a structure is erected, factors such as corrosion, weather, various aging effects inherent in the choice of materials, original design mistakes, abuse, unexpected loads and external forces, all work together to bring a building down. These items can be divided into two fundamental categories: static load support deficiencies and dynamic load deficiencies.

Topics

Static Loads

Dynamic Loads

Case Studies

Static Loads

Static loads include the basic weight of the building itself and its contents. A building has to be strong enough to resist gravity and hold itself up. It should do so without excessive deflections and movements that might scare the occupants, make the occupants uncomfortable, or make the building difficult to use. For example, a large foundation settlement in a frame building may allow the floors to sag significantly, which then allows wheeled furniture to slowly slide to the low point, causes doors and windows to jam, and makes walking around an uncomfortable experience.

The static loads of a building are often subdivided into two categories: dead loads and live loads. Dead loads are loads that never seem to change in a building, such as the weight of the floors, walls, supports, and roof. Live loads are loads that can sometimes change due to weather, occupancy, or building use. They include things like the temporary weight of snow or ice on the roof, the weight of the people in the building and where they are congregated at various times of the day, and the weight of furniture, machinery, and equipment in the building and how they are distributed.

A building can collapse when its primary structural components do not have sufficient strength to support the applied static loads. This can happen because of an error in original design; an omission or mistake during construction; abuse or neglect of the building; sabotage; external forces such as earthquakes, storms or floods; use of the building for unintended purposes; or perhaps because of degradation over time by corrosion, wear, or weathering. To compensate for degradation over time, wear, possible minor design mistakes, minor construction mistakes, and certain types of abuse or neglect that can be reasonably anticipated, buildings are designed to support static loads that are several times stronger than what the designer anticipates would typically be needed. This is called the building's *margin of safety*.

Dynamic Loads

Dynamic loads are loads on a building that change during a relatively short period of time. They are repeatedly applied and released. They add to the static loads that a building must be able to handle, which means that a building, or perhaps certain parts of a building, must be made even stronger than is required to handle its static loads. Unexpected dynamic loads eat into a building's margin of safety.

Dynamic loads typically include forces due to strong winds, gusting, or winds from varying directions; machinery inside the building or nearby that pounds or shakes the floors and walls; and ground motion such as earthquakes, heavy traffic, or nearby construction work. Dynamic loads, when sufficiently strong and when applied often enough, can cause some materials to fail due to material fatigue. Fatigue is the premature fracture and failure of material due to the repeated application and release of loads. Fatigue can occur even when there is margin between the intrinsic static strength of the material and the sum of the applied forces. In other words, a varying load can sometimes cause failure even when the varying load is less than the static load strength of the material.

A famous example of excessive dynamic loading is "Galloping Gertie," the nickname given to the Tacoma Narrows Suspension Bridge built across the Tacoma Narrows in Puget Sound in Washington in the late 1930s. It collapsed on November 7, 1940, soon after the bridge had been opened to the public. Gertie was a cable suspension bridge, similar in design to the Golden Gate Bridge. When it was built, it was considered to be a notable achievement due to its relatively light weight, great structural flexibility, and architectural grace. However, it tended to sway and "wave" excessively on windy days.

The waves in Gertie were caused by aerodynamically induced dynamic forces applied to the bridge decking that were not anticipated by the designer. Crosswinds under and over the bridge decking caused the decking to alternately lift like an airplane wing, and then drop. On November 7, 1940, the crosswinds were sufficiently strong to induce the formation of very large waves in the bridge decking. The alternate lifting and falling eventually tore the decking apart and the decking failed catastrophically. Once failure initiated, the decking fell in a sequential pattern as the deck support connections gave way in order, beginning at one spot about one third of the way across the bridge and then working its way back to the beginning of the decking. When a failure proceeds in an orderly sequence like this, the result is called a *domino effect*. If a domino effect failure can be interrupted early, the consequential damage can be significantly mitigated.

Fortunately for engineers, Gertie's failure was filmed (see references). The film has been useful to engineers and bridge architects in their study of the dynamic loads induced in suspension bridges by aerodynamic lift. Since the collapse of the Tacoma Narrows Bridge, bridge architects and engineers have included aerodynamic and vibrational considerations in their bridge and structure designs. In some instances, bridge models are tested in wind tunnels.

Three case studies of building collapses follow. The first describes how a typical older wall-bearing commercial building with brick and mortar walls can collapse due to something as simple as a leaky roof. The second describes how a simple design mistake led to the deaths of 114 people. The third describes how the second-tallest buildings in the United States at the time, the World Trade Center Towers, were brought down by sabotage.

Case Studies

Case 1: Building Collapse Due to Roof Leakage

In the period from post-Civil War Reconstruction to about World War I, many of the small and medium-sized towns that now dot the Midwest were settled and built. Homesteads were built first, followed by the construction of wood frame commercial buildings. When business activity was sufficient, the wood frame buildings were often replaced by more substantial masonry buildings. Many of these masonry buildings still populate the original downtown areas of these towns and cities.

In general, these masonry commercial buildings followed a similar structural design. A typical front elevation is shown in Figure 24.1. Most were made of locally quarried stone or locally fired brick. The mortar was usually made of quicklime from local kilns and sand from local riverbeds. The front width of the building was usually one fourth to one half the length of the building. The long side walls carried most of the structural load.

The roof was sloped with the high side of the roof at the front and the low side at the back, as shown in Figure 24.2. Usually, the front of the building had a large facade

Figure 24.1 Typical two-story commercial building with load-bearing masonry side walls.

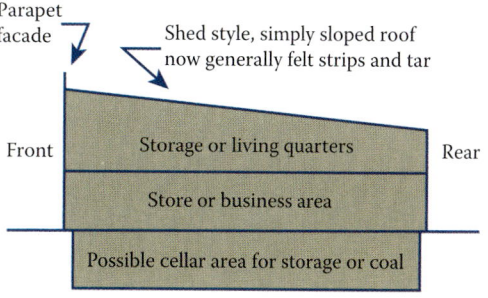

Figure 24.2 Cutaway view showing roof slope and floor usage.

parapet wall for advertising. The sides also usually had parapet walls. The rear portion of the roof usually did not have a parapet wall. Rainwater drained from the front and sides of the building to the rear. When the buildings were first built, rain barrels at the rear corners of the building would often catch the runoff.

The original roof was usually a wood board deck over simple wood beam roof joists. Over the deck were various layers of felt and bitumen, roofing rolls, or, in some cases, overlapping galvanized metal sheeting. In recent years, many of these roofs have been converted to conventional tar and gravel built-up roofs (BURs), or rubber membrane roofs.

The floor and roof decks were usually supported by simple wood joist beams. When it was necessary to splice joists together to span the distance across the side walls, the joists would be supported in the middle by a beam-and-post combination. Splices were often accomplished by overlapping the two pieces where they had been laid over the support post, and nailing or bolting them together. In many cases, however, the joists were simply overlapped and set side by side on top of the post with no substantial fasteners connecting them. It was presumed that the decking or flooring nailed to their upper surfaces would hold them in place. The joists were usually, but not always, side-braced to ensure they would stay vertical.

The floor and roof joists were supported at the ends by the side walls. A bearing pocket was created in the side wall and the end of the wood joist was set into the bearing pocket. Often, the end of the joist was mortared into the bearing pocket so that it would be rigid and vertical.

Sometimes the bearing pocket only extended about halfway through the thickness of the wall. In some buildings, however, the bearing pocket would go all the way, or nearly all the way, through the side wall so that the ends of the floor and roof joists could be seen from the outside. To keep the ends of the joists from weathering, the butts were covered with mortar, tar, or paint. Roof decking

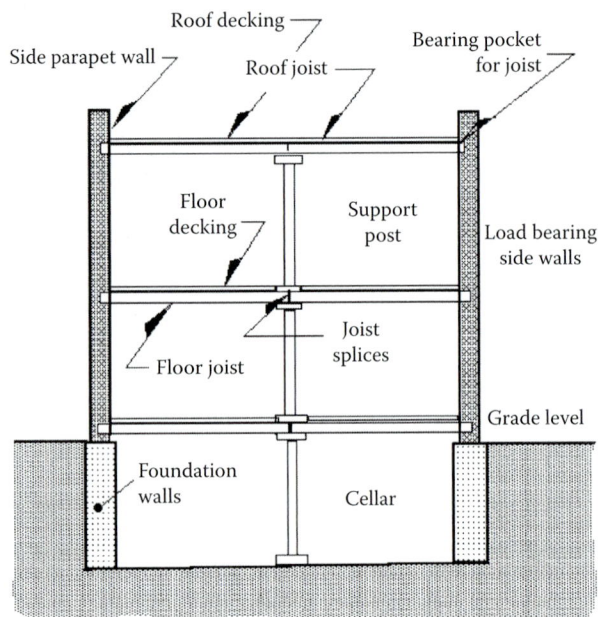

Figure 24.3 Basic structural support system.

and floor decking were nailed directly on top of the joists. Figure 24.3 shows the basic structural support system.

Lime or quicklime for the mortar in these buildings was often obtained by roasting calcium carbonate in kilns. The calcium carbonate may have come from local limestone deposits, chalk deposits, or even marble deposits. The fuel for the roasting process was usually wood and charcoal, although coal and peat were occasionally used when it was convenient.

The usual recipe for making quicklime mortar in that era consisted of two to three volumes of sand to one volume of lime paste. Lime paste was usually two parts by weight of water to one part weight of quicklime. Sometimes the lime mortar would be strengthened by the addition of cement, especially Portland. These mixing ratios and recipes were simply common rules of thumb. Since there were no quality control standards or building codes, conditions and mixture compositions varied greatly. Incomplete decomposition of the limestone in the kiln, or contamination of the quicklime by combustion materials, could affect mortar strength greatly. Most masons added water as they saw fit for "workability."

The sand used in the mixture was usually not sieved for size consistency. Thus, the sand could be well graded, gap graded, or uniformly graded and have significant amounts of silts and fines mixed in. Mortars containing sand of which 48 percent is able to pass through a #100 sieve, for example, have only about one tenth the compressive strength of one in which only 5 percent passes through a #100 sieve.

Even when good practice is adhered to, lime mortar is inferior to Portland cement mortar mixtures. A lime mortar with no cement additives will develop a tensile strength at most of only 26 pounds per square inch after 84 days of curing. This is about one tenth the tensile strength that a modern concrete mixture will develop after 28 days. Because of its slow hardening rate and poor strength, lime-based mortars are not used in standard United States construction anymore except for interior, nonload-bearing walls.

Water Damage

While some of these buildings have been well maintained over the years by their owners, many of them have not, especially their roofs. Water leakage through the roof membrane is a common maintenance problem. The most common point of leakage is the parapet wall flashing where the roof decking abuts the parapet walls. Often, the flashing along the parapet wall dries out and cracks due to ultraviolet light exposure, weathering, and age. Sometimes the membrane is damaged by foot traffic on the roof. Sometimes the flashing membrane pulls away from the wall due to a combination of age-related material shrinkage and weather-deteriorated tar, mastic, or sealing compound. Whichever is the case, deficient flashing allows rainwater to enter the building next to a load-bearing wall and run down the interior side of the wall.

Because there is often a stud wall cavity between the interior side of the masonry wall and the interior finish of the building, the leakage can go unnoticed for years if the water "hugs" the interior side of the masonry wall. The water may not cause any noticeable staining of the interior walls or ceilings because the interior finish may not become wet despite extensive leaking. Because many of the buildings have been outfitted with drop ceilings, wall paneling, and fixtures that obscure the original interior finish, any stains on the original walls and ceilings that may have occurred may no longer be observable. If regular inspections of the roof are not done to check for roof membrane integrity, leakage can go undetected for years.

In this type of building, water leakage through the roof causes two main types of structural damage:

- Weakening of wooden roof and floor joists, and
- Weakening of the load-bearing walls

Leaking water damages wood by providing needed moisture for colonies of bacteria and fungi. The bacteria and fungi establish themselves on the surface of the wetted areas, and then go on to digest the organic materials within the wood itself. The resulting wood rot, or rust, is often black, brown, or white in color. Water leakage can also cause structural wood timbers to swell and soften. If structural timbers are carrying load when they absorb excess moisture, they can permanently deform in response to the loads.

Deformed timbers can cause structural load and deflection problems due to second moments associated with eccentric loading. Second moments are bending moments created when a beam or column is misshapen and the applied forces are no longer centered or symmetric. The eccentrically placed loads cause the beam or column to twist or flex. Trusses with significant second moments, for example, might no longer form a plane but instead flex into a saddle-shaped curve.

Furthermore, wood that has absorbed excess moisture and has become softened may allow nails and bolts that are carrying load to loosen, corrode in place, and pull out. The loss of fastener integrity can cause a significant share of structural headaches.

Both wood rot and distortion are readily observable by inspection. They are expected to occur where the water directly impinges on the wood over a period of time. These areas will also likely be stained by minerals and dissolved materials carried by the water. These minerals and dissolved materials are picked up by the water as it percolates through the building, and are deposited on the wood during the various wetting and drying cycles.

Chemical Attack

A less familiar way that water leakage damages roof and floor joists in these older buildings is by chemical attack. Water that makes its way to the bearing pockets reacts with the lime mortar surrounding the wood. The calcium hydroxide in the mortar is soluble in water and forms an aqueous, caustic solution. The pH of the solution at room temperature may be as high as 12.4, which is more than sufficient to attack the wood embedded in the bearing pocket. When this occurs, the affected surfaces become discolored, lose mass, and appear dimensionally reduced. Material around the bearing surface, usually on the bottom side of the joist, loses strength and flattens out due to compressive failure. Because of this, the joist will often drop down in the bearing box or become loose within it. This loosening may allow the beam to twist or tip.

Structural weakening of load-bearing walls by leakage from the roof is accomplished in several ways. The primary way is the leaching of calcium hydroxide, the main binding ingredient, from the mortar by water that percolates down the wall from the roof. With the calcium hydroxide dissolved from the mortar, the mortar becomes weak and porous. If leaching damage is sufficiently severe, the wall becomes a pile of loose bricks or stones held together by a slightly sticky sand with a high voids ratio (Figure 24.4).

It is easy to verify if a wall has been damaged in this way. First, the water stains on the wall are readily apparent. Secondly, mortar that has been damaged in this way can be easily removed from between stones or bricks with a penknife or sometimes with a person's

Figure 24.4 Mortar leached out from around the bricks. The concrete was applied at the bottom in an attempt to fix the problem.

Figure 24.5 Bricks popping out at base of wall due to "missing" mortar.

fingers. The mortar will flake and crumble easily, like porous sandstone. In severe cases, it is even possible for a person to remove whole stones or bricks from the wall by digging them out with his or her fingernails (Figure 24.5).

The deleterious effect that calcium hydroxide leaching from the mortar has on the structural stability of the wall can be assessed mathematically. This is done by considering the equation that describes the elastic stability of a thin plate under compression, and presuming that it is analogous to the masonry wall under consideration. According to Roark (1938), the critical stress at which buckling occurs when a thin plate is uniformly loaded in compression along two parallel edges is as shown in Equation 24.1.

$$\sigma_{crit} = K[E/(1 - v^2)][t/b]^2 \qquad (24.1)$$

where

t   = thickness of wall,
b   = length of wall
E   = Young's modulus
ν = Poisson's ratio
K   = factor to account for the ratio of height to length and end conditions of the plate (~3.4 to 3.3 in most buildings)
σ_{crit} = compressive stress at which instability occurs

The thin plate modeled by Equation 24.1 is presumed to be homogenous and isotropic throughout, while the load-bearing walls under consideration are composed of discrete units of brick, stone, and mortar. Despite these differences, Equation 24.1 suffices to show the overarching principle that applies in this case, especially if the weakest material values in the wall are presumed to apply in the formula.

It is apparent by inspection of Equation 24.1 that significant changes in Young's modulus, E, greatly affect the critical stress at which buckling in the thin plate occurs. If we compare the condition of the wall before leaching occurs to that after leaching occurs and presume that the dimensions and conditions are all the same except for the condition of the mortar, then Equation 24.2 is true.

$$\sigma_{crit}/\sigma_{crit}' = E/E' \qquad (24.2)$$

Table 24.1 shows some representative values of Young's modulus for brick, stone, and soil.

In relating the values in Table 24.1 to Equation 24.2, it is obvious that if the mortar loses significant stiffness by leaching and

Table 24.1
Young's Moduli for Brick, Stone, and Soil

Material	Typical "E" Value
Limestone	7 to 8 × 10^6 pounds per square inch
Sandstone	~3 × 10^6 pounds per square inch
Brick	3 to 4 × 10^6 pounds per square inch
Concrete	~2 × 10^6 pounds per square inch
Soil (unconfined)	0 to 10 × 10^3 pounds per square inch

essentially becomes equivalent to something between a soil and a porous sandstone, the wall will be structurally weakened, perhaps becoming sufficiently unstable for buckling to initiate.

This structural problem with the walls is further exacerbated if there has been chemical attack on the wooden joists in the bearing pockets. Since most of these walls were constructed without formal tie-ins, the wooden roof and floor joists act as tie-ins to the rest of the structure and help stabilize the walls. If the wall is considered analogous to an Euler column with respect to buckling, tie-ins divide the wall into smaller columnar lengths. This strengthens the wall against buckling. The joist tie-ins also improve the end conditions of the column sections, which further strengthens the wall against buckling (Figure 24.6).

Euler's formula for column buckling is given in Equation 24.3. Note the basic similarity of Equation 24.3 to Equation 24.1, which was taken from Roark's text.

$$\sigma_{crit} = \pi^2 E\ I\ C/L^2 \qquad (24.3)$$

where

σ_{crit} = Euler buckling stress (stress at which buckling could occur)
E   = Young's modulus
C   = factor for end conditions of column
L   = effective length of column
I   = moment of inertia

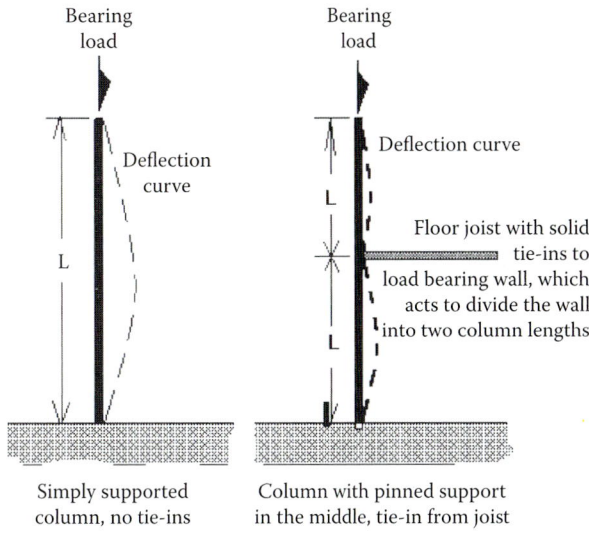

Figure 24.6 Effect of joist tie-in on wall.

When one end of a column is rounded and the other end is fixed, C = 2. When the column is pivoted at both ends, C = 1. When both ends are fixed, C = 4. When one end is fixed and the other free, C = 0.25.

With no effective tie-ins, the wall can be modeled as a column of length L, with a free end and a fixed end. However, if a rigid tie-in is in the middle of the wall, the lower portion becomes a column with a fixed end and a pivoted end, and the upper portion of the wall becomes a column with a free end and a pivoted end. The effective tie-in afforded by the joist not only shortens the length of the column, which reduces the critical stress point, but also improves the end conditions, which further reduces the critical stress point.

Some care should be taken in applying the assumption that the bottom of the wall is a fixed end. Often when water leakage is sufficient, water will run down the wall and pool at the base of the wall (Figure 24.7). If significant damage has occurred at the bottom, it should be considered a rounded-end condition.

When water leakage from the roof has both leached significant amounts of calcium hydroxide from the mortar and caused rot in the bearing boxes, the combined structural effects can be very pronounced. The effective value for Young's modulus is reduced, the overall length of the wall as a column can be doubled, and the applicable factors for column-end conditions are lowered.

Some of the buildings in the category under discussion may have become icons of the city or town in which they were constructed. Because they were built when the town was first established, they have historical and sentimental value to the residents. Unfortunately, the repair of such buildings is neither easy nor cheap. If the damage to the wall is localized and confined to a small area, repairs may entail removal of the stone or brick in the leached areas with complete resetting with a modern cement-based mortar. Since removal of such brick or stone may leave a large hole in the load-bearing wall, such repairs may require significant structural shoring. To do this safely requires the expertise of experienced rehabilitation contractors.

Sometimes when such water damage is finally discovered in a turn-of-the-century building, leaching of the mortar has progressed to the point at which entire sections of load-bearing walls are affected and are at risk (Figure 24.8). Simple tuck-pointing repairs or remortaring in situ are the usual first attempts to deal with such damage because they are the cheapest options. These efforts can hide the damage temporarily, but they cannot reverse it. Such repairs often will not even be cosmetically effective for long because

Figure 24.7 Wall collapse due to roof leak. Leak trickled down inside of wall and leached away the lime-based mortar.

Figure 24.8 Brick wall where face of bricks crack and separate. This is a freeze–thaw effect related to the grade or porosity of the brick.

the new mortar will not be able to adhere to the crumbling old mortar.

Meaningful structural repair and salvage of the wall in such cases requires more than superficial tuck pointing. It may require more resources than are available, or more than the owners are willing to spend. This creates a dilemma: the building is too historically significant to the town to be torn down, but proper repairs would cost the owners too much. Because of the indecision that results, the building usually sits for a long time, decays more, and is eventually razed when decay is obvious and collapse is imminent.

Case 2: Building Collapse Due to Simple Design Error

On Friday evening, July 17, 1981, two of the "skywalks" in the one-year-old, 40-story Hyatt Regency Hotel in downtown Kansas City, Missouri, collapsed during a popular weekly Tea Dance. One hundred and fourteen people were killed, over two hundred people were injured, and claims that were paid out exceeded 120 million dollars.

Just before the collapse, perhaps 1,500 people were enjoying the sounds of "Satin Doll" played by a formally attired, 15-piece dance band in the hotel's atrium. Many patrons were listening to the music while standing in the packed atrium. Many were also on the suspended skywalks on the second, third, and fourth floors. The skywalks hung over the atrium and provided a bird's eye view of the affair.

The skywalks were steel-reinforced, concrete walkways suspended from the ceiling by steel rods. They spanned the open atrium of the hotel like small suspension foot bridges. They provided not only a shortcut for guests across the open atrium, but also provided an entertaining view of the ground floor below.

At about 7:11 p.m., witnesses variously reported that a snap, a pop, or an explosion was heard. The connections between the supporting steel rods and the fourth floor skywalk gave way. The fourth floor skywalk, which was suspended 45 feet above the atrium, then collapsed onto the second floor skywalk, which

was about 30 feet directly below it. Both skywalks fell together to the ground floor, which was 15 feet below the second floor skywalk. The third floor skywalk did not collapse. It was not in line with the other two. As the two skywalks fell, they severed various pipes that then poured water into the resulting heap of twisted steel, broken glass, and rubble.

Some patrons were killed outright by the collapse. Some survivors were buried under the rubble for hours as emergency crews worked to free them. Others escaped direct harm, but will remember always the horror they witnessed that night.

Almost as soon as the event had occurred, and certainly well before evidence could be systematically gathered and assessed, many theories were discussed by news organizations to explain the accident. One popular theory was that people dancing on the skywalks excited the resonant frequency of skywalks and caused it to fall like the Tacoma Narrows Bridge. Others suggested that dancing on the skywalks caused vibrations that caused the connections to fatigue.

The investigation as to what occurred was conducted by several organizations and consulting companies, including the National Bureau of Standards, which is now called the National Institute of Standards and Technology. The main finding is as follows.

The original design of the skywalks was something like that depicted in schematic 1. In the schematic, the bottom skywalk is supported by two rods, and each rod roughly supports half the load of the bottom skywalk. The nut and washer under the skywalk at each end are attached to the rod, and each nut and washer supports about half the load of the skywalk. The nut and washer transfer load from the skywalk to the rod.

The upper portion of the rod supports both the lower and upper skywalks. For simplicity, since we are allowing the two skywalk loads to be roughly equal in load, then the support rod at the left end must support half of the total load of both skywalks, and the rod at the right end must support half of the total load of both skywalks. However, the nut and washer under the upper skywalks at each end only

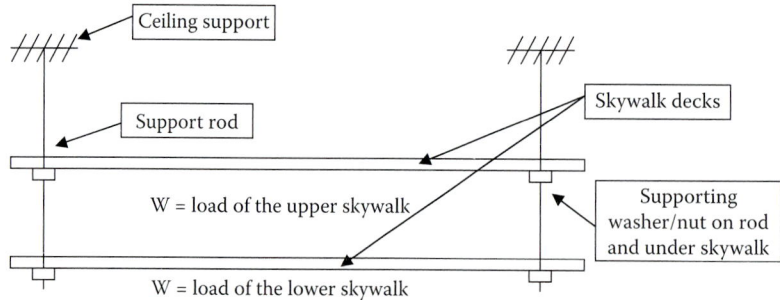

Schematic 1   Basic original design.

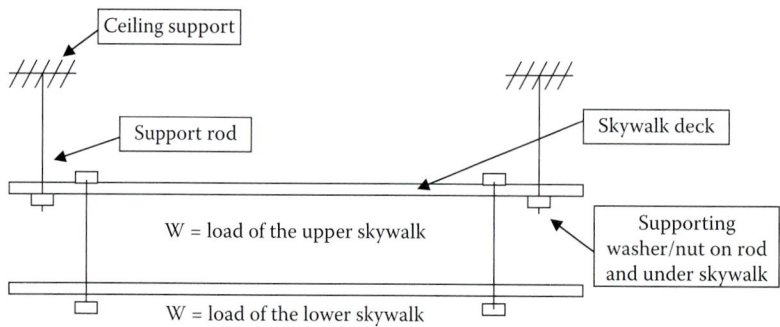

Schematic 2   Changed design.

support half the load of the upper skywalk. In short, all the nuts and washers under the skywalks only support half the load of a skywalk. So far, so good.

Having a single, long rod with a nut and washer in the middle was a problem for the contractor to fabricate, however. Consequently, the contractor requested that he be allowed to cut the rod, and suspend the lower skywalk from the upper one, similar to that shown in schematic 2. Despite a number of oversight checks intended to protect against this sort of error, the change was allowed and this is how the skywalks were built.

The problem with the arrangement in schematic 2 is this: The left side nut and washer supporting the upper skywalk now have to hold twice as much load as before. They have to hold the left half of the upper skywalk and the left half of the lower skywalk. And the same thing occurs on the right side. Unfortunately, while the load on the support nut and washer have now doubled, the design of nut and washer was not changed accordingly; it stayed the same. In short, the loads on the rods stayed the same, but the load on

the nut and washers under the upper skywalk doubled.

By itself, this flaw would not have caused a collapse. The skywalks had for some time been in place and performed satisfactorily. As with most structures, there was sufficient margin built into the structure to forgive this error. However, during the last Tea Dance, the number of people on the skywalks was much greater than was ever anticipated in the original design. The skywalks were intended to be walkways, not densely populated dance floors. By some accounts, the live load of the skywalks were double, or more than double, what the designers had presumed would be maximum.

By doubling the expected live load on the skywalks and by having changed the original design such that the load on the support nuts and washers on the upper skywalk was double, the net result was that the support nut and washer on the upper skywalk were holding four or more times more load than they were intended to hold. This exceeded its design margin. Consequently, the overloaded nut and washer simply pulled through the welded box beam that supported the skywalk and allowed the skywalk to drop. It is quite possible that dynamic loads imparted by dancers moving about the skywalk were the last load increment needed to cause the box beam to give way. Of course, since the lower skywalk was tied to the upper one, when the upper one fell, the lower one fell.

Other items that contributed to the accident were the following: (1) the rod design did not meet building code requirements; (2) the original design was not clear as to how the long support rods were to be threaded in the middle to accept a nut and washer; and (3) some welds in the fabricated box beams that supported the skywalks were deemed poor, and the various administrative inspections and checks, which were intended to prevent such simple mistakes from occurring, failed.

Case 3: Building Collapse Due to Impact

Although several buildings were associated with the World Trade Center (WTC), it conspicuously consisted of two very large buildings: One World Trade Center (the north tower) and Two World Trade Center (the south tower). One WTC, completed in 1972, was 1368 feet tall. Two WTC was completed a year later and was 6 feet shorter. Both buildings were 110 stories high and were structurally more or less the same. For a short time they were the tallest buildings in the world. The foundation of each building was set about 70 feet into the ground. Each floor enclosed almost an acre of office space.

Structurally, each building was a vertical, hollow, rectangular tube within another vertical, hollow, rectangular tube. The outer rectangular tube consisted of 244 14-inch steel box columns spaced 39 inches apart. The inner rectangular tube was 90 feet long and was composed of tightly spaced steel girders around a central core of elevator shafts and stairways. The inner tube supported much of the weight of the building.

The inner tube and the outer tube were connected by steel spandrel members overlaid with steel decking and 4 inches of concrete.

The floor decking system supported the 40,000–square-foot floor and also acted as a structural stiffener between the inner and outer rectangular tubes. Each floor system by itself weighed perhaps 3 to 3.5 million pounds. At the foundation of each building, the total bearing load was approximately 1 billion pounds.

At the time the building complex was designed, it occurred to the designers that accidental impact by an airplane was a possibility. The Empire State Building, which is 102 stories, 1250 feet high, and in the same neighborhood as the WTC, was struck by an errant U.S. Army B-25 bomber during a fog in 1945, just 14 years after the building was completed. In that accident 14 people were killed.

When the WTC was on the drawing boards, the most probable sort of air accident imagined was the accidental impact by a 707 Boeing jet aircraft when it was either landing or taking off from one of the nearby airports. At the time, the Boeing 707 was the largest commercial aircraft in use. A Boeing 707 has a maximum takeoff weight of about 336,000 pounds, a wingspan of 146 feet, a length of 153 feet, a tail height of 52 feet, and a cruising speed of 607 miles per hour; when fully fueled it contains about 23,000 gallons of fuel. It was presumed that if an impact were to occur, the velocity at impact would be similar to landing or takeoff velocity, perhaps 180 mph.

At 8:46 a.m. local New York time on September 11, 2001, the north tower of the WTC was struck by a Boeing 767 that was deliberately steered into the building. A second Boeing 767 struck the south tower at 9:03 am; it also was deliberately steered into the building. Both aircraft apparently impacted the towers at cruising speed in an effort to maximize damage to the buildings (Figure 24.9).

A Boeing 767 has a maximum takeoff weight of about 395,000 pounds, a wingspan of 156 feet, a length of 159 feet, and a cruising speed of 530 miles per hour; when fully fueled it carries about 24,000 gallons of jet fuel. A 767 aircraft is about 18 percent heavier than a 707 aircraft, and cruising speed is almost three times faster than landing or takeoff speed.

Figure 24.9 The south and north towers of the World Trade Center prior to collapse. (Photo courtesy of Michael Rieger, Federal Emergency Management Agency [FEMA], Washington, D.C.)

The impact occurred on the north tower between the 90th and 96th floors. Seismometers located in Palisades, NY, about 21 miles north of the building, recorded a 12-second ground shock with a 0.8-second dominant period that had an equivalent earthquake magnitude of 0.9 on the **Richter scale**. The impact to the south tower occurred between the 75th and 84th floors. It generated the equivalent of a 0.7 magnitude earthquake and had a 6-second ground shock with a 0.7-second dominant period.

Photographs taken of the second aircraft approaching the building indicate that just prior to impact the aircraft was rolled about 45 degrees with the left wing tip down and the right wing tip up. It was also pitched with the nose downward perhaps 10 degrees and was dropping in altitude as it approached the building.

In the north tower, the initial impact severed about two thirds of the steel supports on the tower's north side. Despite this severe structural damage, however, the floors above the impact area did not collapse. After impact, fire immediately broke out in the affected floors and rapidly spread through the crash-affected area, feeding upon the spilled fuel

from the decimated aircraft. The north tower eventually succumbed to the fire and collapsed after 102 minutes at 10:28 a.m. When it fell, it generated the equivalent of a 2.3 magnitude earthquake.

Impact to the south tower caused similar structural damage to the steel supports. In the resulting fire, which immediately ensued after impact, the south tower collapsed, 56 minutes after impact. This is about half the time that the north tower stood before fully collapsing. When the south tower fell, it generated the equivalent of a 2.1 magnitude earthquake.

The kinetic energy at impact of each aircraft that struck the towers is estimated by:

$$KE = (1/2)mv^2 \qquad (24.4)$$

where

m = mass of aircraft
v = velocity of aircraft
v = 530 miles per hour = 777 feet per second
KE = (1/2)(395,000 pounds force/ 32.17 feet per second^2) (777 feet per second)^2
KE = 3,710,000,000 foot pounds force
   = 4,771,000 British thermal units
   = 1.40 megawatt hours

A compact car that weighs 2500 pounds force and is traveling at 100 miles per hour has a kinetic energy of 836,000 pounds force feet. The impact energy of one of the aircraft was equivalent to about 4439 compact cars all traveling at 100 miles per hour.

The momentum of each aircraft at impact is estimated by:

$$P = mv \qquad (24.5)$$

where

P = momentum
m = mass
v = velocity
P = (395,000 pounds force/32.17 feet per second^2) (777 feet per second) = 9,540,000 pounds force seconds

The aircraft did not penetrate through the building and come out the other side, although some parts and some fuel did. Both aircraft buried themselves into the buildings' interiors. On the basis of this fact, it is estimated from kinematic considerations that the impact time was about 0.4 second. This amount of time is consistent with the length of the aircraft, 159 feet, the average speed during impact, ~389 feet per second, and the depth dimensions of the building.

An estimate of the force applied to the WTC towers by the impacting aircraft can be computed using Newton's second law:

$$F = \Delta P\_\Delta t \qquad (24.6)$$

F = 9,540,000 pounds force second/
      0.4 second = 23,850,000 pounds force

In comparison, the horizontal force that either building would have to resist at 126-mile-per-hour hurricane level winds blowing on one face is estimated to be 11 million pounds force. It appears then that the concentrated force on each tower applied by the impact of 767 aircraft well exceeded the expected hurricane wind forces spread over the face of a tower.

Given that the impact to the north tower was approximately centered at the 93rd floor and the impact to the south tower was approximately centered at the 80th floor, the simple moment applied to the foundation of the north tower was about 27,590,000,000 foot pounds force. Similarly, the simple moment applied to the foundation of the south tower by the second impact was about 23,736,000,000 foot pounds force. The moment transmitted to the ground by the north tower was 16 percent greater than that transmitted to the ground by the south tower. This is why the first impact registered higher on the seismometer at Palisades than the second impact.

With respect to the aircraft impacts themselves, both the north and south towers performed admirably. Both towers absorbed remarkable amounts of impact energy, sustained significant structural damage, and endured remarkable applied forces and moments and still stood upright and did not collapse. What the two towers could not endure was the ensuing fire.

Upon impact, both aircraft were wholly destroyed as they penetrated into the towers.

Consequently, both aircraft released all their fuel into the buildings' interiors. Since both aircraft had just taken off and both were bound for the West Coast, both had nearly full fuel tanks. The impact to the north tower was spread over six floors and the impact to the south tower was spread over nine floors. If 24,000 gallons of jet fuel are spread evenly over nine floors, each with an area of about 40,000 square feet, this amounts to 0.067 gallons for every square foot, or 9 ounces of jet airplane fuel for every square foot of space. This is a significant **fire load** per square foot. Since both planes had flown for a while, the actual amount would have been somewhat less.

The **adiabatic flame temperature** of jet fuel is about 3140 degrees Fahrenheit, give or take a few degrees depending upon fuel additives. The adiabatic flame temperature of a fuel is a theoretical calculation of the maximum temperature at which a fuel burns. Even in the laboratory, flame temperatures do not reach this temperature because of chemical disassociation effects at high temperatures that tend to cool the burning process slightly. Actual flame temperatures in an uncontrolled building interior environment, such as the interior of one of the WTC towers, were likely several hundred degrees less than the theoretical adiabatic flame temperature.

Structural steel loses approximately half of its tensile strength at 1000 degrees Fahrenheit. At 1300 degrees Fahrenheit and higher, it loses most of its strength and stiffness and ceases to be a viable structural component. Likewise, steel-reinforced concrete degrades and cracks at temperatures of 1200 degrees Fahrenheit or more. After the fire initiated, temperatures built up significantly within the building interior. With jet fuel as the initial primary fire load, temperatures in excess of 1300 degrees Fahrenheit certainly occurred within the impact areas. As the temperature increased, the strength of structural components within the fire-affected areas diminished. When the temperature of the structural components approached and perhaps exceeded 1200 degrees Fahrenheit, the components could no longer carry their loads and failed.

Because the second impact was spread between the 75th and 84th floors of the south tower, the remaining columns, connections, and supports in the area of the second impact were supporting about 26 floors' worth of weight above the damaged area. Similarly, because the first impact was spread between the 90th and 96th floors of the north tower, the remaining columns, connections, and supports in the area of the first impact were supporting only about 14 floors' worth of weight above the damaged area. As fire in the area of impact increased the temperature of the structural components located there, those components that carried proportionally more load failed first. This is why the south tower collapsed after 57 minutes of burn time and the north tower collapsed after 102 minutes of burn time.

When collapse of the south tower was initiated, all the floors above the impact area dropped onto the floor just below where failure was initiated. Thus, one floor was called upon to hold the weight of the 26 or so floors above it. Since one average story weighed perhaps 9 million pounds, 26 stories weighed about 236 million pounds. Dropping this group of floors one story converts all the potential energy of this mass at a height of one story into impact kinetic energy. A one-story drop of 26 stories, as a group, is about 2,940,000,000 foot pounds force of energy. This amount of kinetic energy, by itself, is about equal to 80 percent of that imparted by the aircraft at impact, and was more than sufficient to initiate collapse of that floor onto the next lower one, and so on, until the building "pancaked" into a heap. This is also a type of domino effect.

This is the principle that building demolition experts use when buildings are deliberately brought down using explosives. Small explosive charges are used to take out well-chosen structural members so that the resulting falling mass from an upper portion of the building drops onto and crushes a lower portion of the building, which in turn falls upon even lower portions of the building. It is a sort of controlled avalanche. In a sense, the inherent potential energy of the building due to gravity is harnessed to destroy itself.

The north tower collapsed in a similar fashion to the south tower. Since the columns, connections, and supports held significantly less weight in the area where the first aircraft

impact occurred—14 stories instead of 26 stories—the columns, connections, and supports in the north tower had to weaken more before failure occurred. The fire had to heat things up to a higher temperature, and this took more time.

When the columns, connections, and supports in the impact area of the north tower weakened sufficiently, the upper 14 or so floors dropped onto the next lower floor and imparted about 1,583,000,000 foot pounds force of kinetic energy to it. Since the single floor could not hold that amount of weight or absorb that amount of kinetic energy without sustaining significant structural damage, it collapsed onto the next lower floor, and so on until it had also "pancaked" into a heap.

The collapse of the two towers occurred half an hour apart. Some have presumed that the collapse of one tower somehow precipitated the collapse of the other—that one tower somehow dragged the other one down or shook the other one down. The 29-minute difference between collapses argues that this was not the case.

In computing the collapse energy of each building by summing the potential energy floor by floor, it was found that the north tower had about 664 billion foot pounds force of collapse energy, and the south tower had 632 billion foot pounds force of collapse energy. The north tower had a little more collapse energy because the "pancake" effect started at about the 93rd floor. Floors above the 93rd floor dropped down at the same time. In the south tower, however, the "pancake" effect began at about the 80th floor. Thus, all the floors above the 80th floor started their drop from a significantly lower point (Figure 24.10).

The equations used to estimate the collapse energy of the two towers are

$$(24.7)$$

$$E_{North} = \sum_{i=2}^{n=93} (W)(H)(n) + (16W)(94H) = 5874(WH)$$

$$= 664 \times 10^9 \text{ foot pounds force}$$

$$E_{South} = \sum_{i=2}^{n=80} (W)(H)(n) + (29W)(81H) = 5588(WH)$$

$$= 632 \times 10^9 \text{ foot pounds force}$$

Figure 24.10 View of the World Trade Center after collapse. Note the exterior outer support columns, which are about the only structural components still standing. (Photo courtesy of Michael Rieger, Federal Emergency Management Agency [FEMA], Washington, D.C.)

where

W = average weight of one floor
H = average height of one story
n = number of building stories

This is why the north tower generated a larger Richter magnitude ground vibration than the south tower, 2.3 versus 2.1 (Figure 24.11).

The 9–11 event caused many engineers, architects, and planners to consider whether building codes for skyscrapers should incorporate defensive, and some have even suggested offensive, measures to deal with the possibility of airplane crashes and other direct acts of sabotage. Some defensive architectural measures were evaluated and incorporated into some types of buildings after the April 19, 1995, Murrah Federal Building bombing event in Oklahoma City where 168 people were killed. Likewise, the U.S. Embassy bombing attack in Lebanon in 1983 that killed 60 people, and the bombing of the U.S. Embassy in Kenya in 1998 where 212 were killed and perhaps 4000 were injured have caused people to think more defensively about building design. The topic of deliberate or accidental airplane impact, however, is still being debated. As the public memory of the event has somewhat faded, new plans for even taller skyscrapers are being planned.

As a footnote, 7.5 metric tons of steel taken from the rubble Twin Towers has been incorporated into the USS New York, a new amphibious assault ship as a memorial to that event.

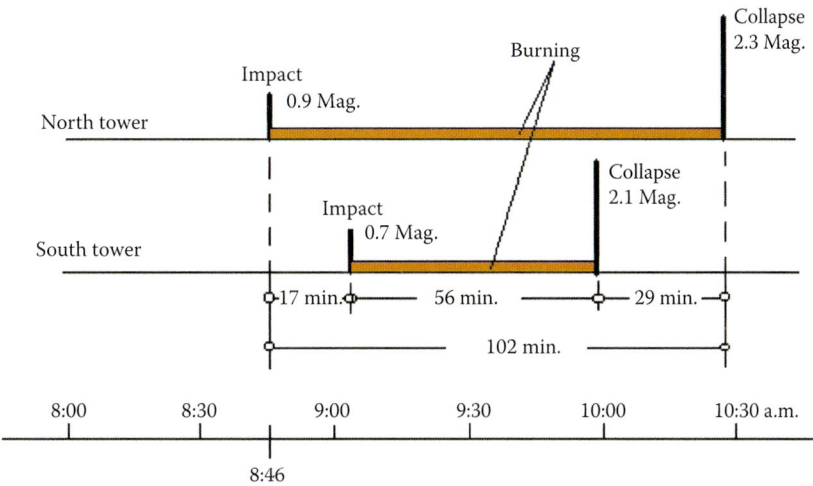

Time line - September 11, 2001

Figure 24.11 Comparison of north and south tower timelines.

Questions

1. What is the difference between the dead load and the live load of a building?

2. What are dynamic loads in a building?

3. What caused the Tacoma Narrows Bridge to collapse?

4. Why are lime-based mortars no longer used for load-bearing walls?

5. When the WTC was designed, what kind of possible aircraft accident was envisioned?

6. If a Boeing 707 instead of a Boeing 767 had struck a WTC tower under the conditions assumed for an accidental impact during take-off, how would its kinetic energy compare to that of the aircraft that struck it on September 11, 2001?

7. When the two WTC towers were initially impacted by aircraft, seismometers located 21 miles away indicated that impact to the north tower caused greater surface motion, 0.9 Richter magnitude versus 0.7 Richter magnitude. Why?

8. Did the impacts of the aircraft into the two towers of the WTC directly cause the towers to collapse? Why or why not?

9. What happens to steel when it is heated in an intense fire?

10. How do demolition experts use small explosive charges to collapse large buildings deliberately?

11. Is it possible to thread a rod in the middle for use with a nut and washer as was required in the original design for the Hyatt Regency Skywalks?

12. If a contractor makes a request to change something in the design, who is ultimately responsible for approving the change?

13. In most jurisdictions, building inspectors are not trained engineers or architects. What do inspectors do when they inspect a building?

14. Search the web for photographs of the Hyatt Regency skywalk collapse and print them out. Be sure to print out photographs of the box beam joint where the support rod pulled through.
    a. Were the nut and washer stripped from the rod?
    b. Was the bottom of the box beam bent inwards?
    c. How was the box beam fabricated?

15. What are the procedures in your jurisdiction for approving a building design, inspecting it, and gaining approval for use?

Suggested Readings

Carper, K., *Why Buildings Fail*, National Council of Architectural Registration Boards, Washington, D.C., 2001.

Carper, K., Ed., *Forensic Engineering*, 2nd ed., CRC Press, Boca Raton, FL, 2001.

Field, J. and Carper, K., *Construction Failure*, 2nd ed., John Wiley and Sons, New York, 1996.

Hrenchir, T., The night the skywalks fell, *Topeka Capitol Journal*, July 15, 1981.

Kent, W., *Mechanical Engineer's Pocket Book*, 9th ed., John Wiley and Sons, New York, 1915, pp. 372, 370.

Mark Ketchum's Bridge Collapse Page, The Tacoma Narrows Bridge, http://www.ketchum.org/bridgecollapse.html.

Noon, R., *Forensic Engineering Investigation*, CRC Press, Boca Raton, FL, 2001.

Petroski, H., *To Engineer Is Human*, Vintage Books, New York, 1992.

Petroski, H., *Design Paradigms: Case Histories of Error and Judgement in Engineering*, Cambridge University Press, London, 1994.

Ratay, R., Ed., *Forensic Structural Engineering Handbook*, McGraw-Hill, New York, 2000.

Roark, R., *Formulas for Stress and Strain*, 2nd ed., McGraw-Hill, New York, 1938, p. 300.

World Trade Center Building Performance Study, team leader Gene Corley, FEMA 403, May 2002.

Basic Fire and Explosion Investigation

David R. Redsicker

Introduction

The investigation of fires and explosions seeks their origins and causes through applications of basic scientific method. The investigation of fire involves applications of both physics and chemistry. Fire science is the knowledge of fire, including the chemistry and behavior of fire or explosions in given circumstances. The basic method is the systematic approach that addresses the following criteria:

1. *Recognition of the need.* One must first determine that a problem exists. In this case, a fire or explosion has occurred and its cause must be determined and listed so that similar incidents can be prevented in the future.
2. *Definition of the problem.* Having determined that a problem exists, an investigator or analyst must define how the problem can be solved. In this case, proper origin and cause investigation must be conducted. This is done by an examination of the scene, by a combination of other data collection methods such as the review of previously conducted investigations of the incident, interviews with the witnesses or other knowledgeable persons, and the results of scientific testing.
3. *Collection of the data.* Facts about the fire incident are now collected. This is done by observation, experiment, or other direct means of gathering data or experience and can be verified.
4. *Analysis of the data.* All of the collected and observed information is analyzed in the light of the investigator's knowledge, training, and experience. Subjective or speculative information cannot be included in the analysis, only facts that can be clearly proven by observation or experiment.

Chemistry and Behavior of Fire or Explosion

Fire has frequently been defined as the rapid oxidation process with the evolution of heat and light. An explosion is the sudden conversion of potential energy (chemical or mechanical) into kinetic energy with a production and release of gases under pressure. These may be divided into two subdefinitions: high order explosion (a rapid pressure rise or high force explosion characterized by shattering the confining structure or container) and a low-order explosion (a slow rate of pressurization or low force explosion characterized by a pushing or dislodging the confining structure or container). The components necessary for a fire are best described by the fire tetrahedron (a four-sided solid geometric form) made up of the four components: fuel, heat, oxygen, and uninhibited chemical chain reactions. The behavior of a fire can be affected by the addition or subtraction of any of the components during the progression or suppression of a fire. The fuel is any substance that will burn or support combustion. Fuels are found in three basic states: solid, liquid, or gaseous vapor. An example of solid fuel is wood, a liquid (gasoline), and a gaseous vapor (natural gas).

It should be understood that although the fuels may exist in three different states, they can only be volatilized and consumed in the vapor state. The burning of solid and liquid fuels takes place at the surface of the fuel in the area where the vapors have been created by heat and subsequently ignited. An oxidizing agent is required to support combustion. The most common is the oxygen in the Earth's atmosphere. Normal atmospheric air contains 21 percent oxygen. Typically flaming combustion occurs above 15 percent oxygen in the atmosphere. A smoldering fire may continue at much lower percentages. Generally, the higher the ambient temperature, the less oxygen is required for combustion.

Obviously, oxygen is necessary for the normal combustion process. If the oxygen in the environment is limited, it is consumed during the fire process. This results in the production of carbon monoxide, which is very common at most fire scenes. Carbon monoxide asphyxiation is the primary cause of death in fatal fires. Carbon monoxide is also a fuel with an ignition temperature of 1128 degrees Fahrenheit and is the likely cause of most backdrafts or smoke explosions. Fuel is matter that exists in three physical states: solid, liquid, and gas. Solids decompose with heat, vaporize, and become gases. Liquids do not burn; however, the vapors rising from the liquid surface do burn similar to wood vaporization. Gaseous fuels are those in which molecules are in rapid movement and random motion. They have no definite shape or volume and assume the shape and volume of the container. How a gas defuses in air depends on its vapor density relative to air. The nearer this is to the vapor density of air (a value of 1.0), the greater the ability of the gas to mix with air. Various gaseous fuels have different vapor densities. Methane equals 0.6 and is lighter than air. Propane is 2.5 and is heavier than air.

When a gaseous fuel diffuses into air, the mixture may ignite or explode. The percentage of gas to air at which this occurs is the lower limit of the flammable (explosive) range of that gas. The upper limit is the percentage at which the mixture is too concentrated to ignite (e.g., natural gas range is 5 to 15 percent).

Liquid fuels also assume the shape of their container. Because a liquid's boiling point indicates imminent vaporization, it is one measure of the volatility of a liquid fuel. Two other indicators are flashpoint and fire point. The flashpoint is the temperature at which a liquid gives off sufficient vapors to form an ignitable mixture at its surface (gasoline is 45 degrees Fahrenheit and kerosene is 100 degrees Fahrenheit). The fire point is the temperature at which a liquid produces vapors that will sustain combustion. This is generally several degrees higher than the flashpoint (e.g., gasoline is 495 degrees Fahrenheit and kerosene is 110 degrees Fahrenheit).

The generally accepted temperature used to distinguish flammable from combustible liquids is defined as 100 degrees Fahrenheit; temperatures below 100 degrees Fahrenheit classify flammable liquids, and temperatures greater than 100 degrees Fahrenheit classify combustible liquids.

Heat is the tetrahedron component necessary to increase the temperature of the fuel in the presence of oxygen and cause ignition. One must understand how heat is produced and transferred as it relates to a fire. Heat is the energy possessed by a material or substance due to molecular activity. There are five basic methods of heat production.

Chemical. Chemically produced heat is the result of rapid oxidation. The speed of the oxidative reaction is an important factor. Rust is a product of oxidation, but a very slow one.

Mechanical. Mechanical heat is the product of friction. Internal metal components of machinery can overheat due to lubricant breakdown or ball bearing failure and cause ignition of available combustibles.

Electrical. Electrical heat is the product of arcing, shorting, or other electrical malfunction. Poor wire connections, too much resistance, a loose ground, and too much current flow through an improperly sized wire are other sources of electrical heat.

Compressed gas. When a gas is compressed, its molecular activity is greatly increased. Consider the operation of a diesel engine. The gaseous fuel is compressed within the cylinder, increasing its molecular activity. The heat generated by this activity eventually reaches the ignition temperature of the fuel itself. The resulting contained explosion forces the piston back to the bottom of the cylinder, and the process repeats over and over again.

Nuclear. Nuclear energy is the product or splitting of atomic particles. The tremendous heat energy in a nuclear power plant produces steam to turn steam turbines. Once a fire initiates, it can only continue through a transfer of the heat by molecular activity.

Three of the most common forms of heat transfer are conduction, convection, and radiation.

Conduction. The transfer of heat through direct contact is called conduction. If you touch a hot stove, the pain you first feel is the result of conducted heat passing from the stove directly to your hand. In a structural fire, superheated pipes, steel girders, and other structural members, such as walls and floors, may conduct enough heat to initiate fires in other areas of the structure. Heat transfer by convection is responsible for the spread of fire in almost every structure.

Convection. Entails the transfer of heat by a circulating medium, usually air or liquid. The superheated gases evolved from a fire are lighter than air and consequently rise. As they travel and collect in the upper levels of the structure, they can and do initiate additional damage.

Radiation. Radiated heat moves in invisible waves and rises much the same as sunlight or x-rays. Radiated heat travels at the same speed as visible light (186 miles per second). It is primarily responsible for the exposure hazards that develop and exist during a fire. Radiant heat travels in a direct or straight line from the source until it strikes an object.

Chemical chain reaction is a complex series of events that must be continuously and precisely reproduced to maintain flaming combustion. Two of the events required are (1) the oxidation reaction produces sufficient heat to maintain continued oxidation, and (2) the fuel mass must be broken down into similar compounds and liberated (vaporized) from the mass itself; in turn, these unburned vapors must combine with available oxygen and be continuously drawn up into the flame.

Behavior of Fire

There are many factors that affect the progression of fire. Initially, fuel supply, oxygen, and an available heat source are necessary. However, as the fire evolves and spreads to other combustibles or flammables in proximity, the fire progression may increase rapidly. As long as there is a sufficient supply of these three basic components, the fire will continue until either structural components of a building or fire suppression efforts change the conditions.

Basically a fire extends horizontally and vertically from its **point of origin**. It follows the path of least resistance through ceilings, doorway and window openings, and stairwells as it progresses unimpeded. Other factors that may affect the progression and growth of the fire may be an environmental element (heavy winds and direction may increase the intensity of the fire as well as the path).

During a fire's progression, it normally transcends four phases.

1. *Incipient.* This is the earliest phase of fire, which may last anywhere from a fraction of a second to several hours or days depending on the fuel or **ignition source**.

2. *Emergent smoldering.* In this phase, the products of combustion become increasingly pronounced. Some fires, such as smoldering mattress fires, may pass directly from this second phase to oxygen-related smoldering (phase 4).

3. *Free burning.* During this phase of the fire, the rate and intensity of open burning increases. The intensity of the fire doubles with each 18 degrees Fahrenheit (10 degrees Celsius) increase in temperature. Heat rapidly evolving from the original point of the fire is convected and collects at the upper areas of the structure or room. Additional heat is transferred through conduction and radiation. During this phase, a flashover may occur when the temperature reaches the ignition temperature of all the combustible items in the room.

4. *Oxygen-regulated smoldering.* This final phase occurs when the oxygen-enriched air in the room during phase 3 (free burning) is depleted. The depletion of oxygen supply causes the flaming combustion to end, and it is replaced to a large extent with glowing combustion. This produces heavy dense smoke and gases, which are forced from the room

under pressure. The fire continues to smolder and the temperatures exceed 1000 degrees Fahrenheit. The resulting superheated mixture of gases needs only a fresh supply of oxygen to resume free burning at an explosive rate. This type of explosive ignition is referred to as a *back draft*. One of the superheated gases produced by fire is carbon monoxide (CO). This odorless, colorless gas collects and mixes with oxygen to within its explosive or flammable limits, 12.5 to 74 percent of the atmosphere by volume. Carbon monoxide is highly flammable and has an ignition temperature of 1128 degrees Fahrenheit. When the ignition occurs, the entire cloud of smoke within the area or room literally explodes or bursts into flames.

Origin and Cause Analysis

The methodology behind determining the origin and cause of a fire is, for the most part, the guideline to proper investigative procedures. To determine where a fire started (origin), one must first evaluate those areas of the structure that were not damaged or were less affected by fire than the area of heaviest fire damage. Typically, the heaviest fire damage occurs at or near the **area of origin**. Once the area of origin has been identified through the evaluation of **fire patterns** and physical evidence, including charring, melting, and distortion of components or items in a fire, the point of origin is the point or spot within the area that the fire initiated. To identify the point of origin, one must evaluate the effects of the fire within the area of origin to identify the most severely damaged lowest point of origin.

Once the point of origin has been identified, then the cause of the fire can be identified through the careful inspection and analysis of potential heat sources identified at or near point of origin. The cause of a fire can usually be categorized as one of four classifications:

1. Accidental, and explainable (may include negligent acts)
2. Natural, act of nature (lightning)
3. Incendiary, intentional act of setting a fire
4. Undetermined, cause unknown, unable to be identified

Determination of the cause and ultimately the responsibility for a fire involves the recognition of the degree of human intervention (factor). Almost all fires have some type of human involvement whether intentional or accidental. The degree of involvement often involves the culpability of a person or persons involved. Important questions to ask and answer are: What did they do? When did they do it? Where was the work performed? Why did they do the work? How did they do the work? Ultimately, the determination of the cause and responsibility for the fire may identify circumstances and factors that were necessary for the fire to occur. These circumstances and factors may be certain equipment involved in the ignition of the fire, the presence or absence of combustible or flammable material, and the circumstances of the human factor that brought the ignition source and fuel together to cause the fire. This, in turn, may bring about certain legal actions, either criminal or civil. In the criminal justice system, an individual may be charged with the intentional starting of the fire for various motives ranging from financial to revenge. Criminal charges may include arson, insurance fraud, or murder. In the civil arena, the interest is in obtaining some type of monetary compensation in the form of subrogation. Subrogation is defined as the legal action of substituting one creditor for another, as when an insurance company seeks to recover the costs it paid out due to a manufacturer's defective product or negligence on the part of a service provider.

One other area of concern and consideration before we get into the actual physical examination of the fire scene is the topic of spoliation. **Spoliation** is the intentional or negligent destruction or alteration of evidence. The physical examination of the fire scene is approached in the systematic method addressed in the introduction. With this in mind, the investigation takes on an increasingly focused analysis commencing with the exterior of the structure and then the interior, including the room of origin, the point of origin, and the determination of the cause. The reconstruction and examination of the fire scene can be seriously impeded by indiscriminate or haphazard handling of the routine firefighting operation known as overhaul. This involves the inspection of, and when necessary, the movement or removal of debris in an effort to discover concealed embers or flames that might rekindle the fire. If circumstances permit, the room of origin should not be overhauled before an investigator is on the scene. Otherwise, evidence may be destroyed.

The purpose of the exterior examination is to document the fire conditions and patterns, which help to document the fire spread and direction of fire progression. Also included in this examination is the documentation of the utilities including the electric, gas, or other fuel service into the structure. The purpose of this evaluation is to either eliminate or document the involvement of the utilities in the cause of the fire. During the next phase of the investigation, the examination of the interior of the structure progresses from the least amount of fire damage to the heaviest fire damage, which ultimately identifies the area or areas of fire origin. Remember that more than one area or point of origin may be an indicator of an intentionally set fire (incendiary). The absence or presence of contents, the condition of the content, and the fire spread as determined by various char, heat, and smoke patterns further help to evaluate the ultimate point of origin of the fire.

Fire safety systems should be evaluated, including smoke detectors, heat detectors, fire escapes, and fire equipment to improve fire safety and fire prevention. To this end, it is important to document the location and integrity of the fire safety systems as they relate to the fire origin and the ultimate factor in the injury or death of the occupants of the structure. Local, state, and federal fire safety codes should be reviewed and addressed in the report along with the origin and cause analysis. Once the area of fire origin has been determined, whether it is on the exterior or interior of the structure, it is important to evaluate the fire patterns in formulating an accurate conclusion to the point of origin of the fire. To properly identify the cause of the fire, it is necessary to examine all potential ignition sources at the point of origin. This would include the electrical utilities in the room including electrical outlets, lighting equipment, and appliances. Heating equipment should be evaluated, including the normal domestic heat source for the room and any supplemental heating equipment such as space heaters or secondary heating equipment (fireplaces, wood stoves, gas-fired appliances, kerosene heaters, etc.).

In addition to the normal potential heat sources, others that may be necessary to identify and evaluate may include candles and smoking material, as well as certain chemicals that may be subject to spontaneous heating (stains, paints, or other natural oil products).

As one identifies and systematically eliminates potential heat sources to the point where all natural or accidental heat sources have been eliminated and the fire patterns indicate the potential use of an **accelerant**, the possibility of an **incendiary fire** cannot be eliminated. Further investigation includes securing samples from the point of origin to determine or identify a possible flammable or combustible **liquid accelerant**. It is necessary to take additional **comparison samples** along with char samples. Comparison samples would include carpeting, wood flooring, linoleum, or certain furniture upholstery, which may have been involved at the point of origin of the fire. The samples must be properly secured and identified as well as documented with photographs for future reference. As part of the origin and cause analysis, certain items may have to be secured and retained for other parties to examine as well. For example, if the fire is clearly accidental and an appliance has been identified as the cause of the fire, it is necessary to identify the manufacturer of the item so they may have an opportunity to examine the fire scene as well as the evidence. All parties involved in litigation may perform additional laboratory analysis or testing under a protocol agreed upon in advance.

Fatal Fire Investigation

It goes without saying that the investigation of fires is a challenge because of the effects of the fire on the evidence. It is typical for a fire scene to be altered by the fire itself and, to some degree, the fire suppression efforts. To further complicate a scene, an incident with fatalities only highlights the importance of cooperation and thoroughness in the investigation. Not only is the investigator trying to investigate the origin and cause of a fire, he or she is potentially involved in a crime scene investigation and must determine not only the cause of death, but who or what was responsible for the victim's demise. To this extent, a thorough and time-consuming investigation must be pursued to separate those incidents that are tragic accidents from those that are criminal in nature.

Of utmost concern in a fatal fire is the potential loss of physical evidence. A fire will obviously alter evidence to some degree. How the evidence is identified, documented, and preserved is of paramount importance to the ultimate success of the case resolution. It is important and critical to document the scene thoroughly, including the relationship of the origin and cause of the fire to the victims. The victim's condition and position is important in later evaluation of cause of death. The coordination of efforts at the fire/crime scene with the medical and postmortem investigation is equally important. Many times the local medical examiner or coroner will prefer to conduct his or her own on-scene examination of the victim's body before removal from the scene. Ultimately, the cause of death and cause of fire may be closely related if the victim was the perpetrator of the fire and succumbed to his or her own acts of negligence or intentional incendiary activities.

As for identifying fire victims, many times the fire is so intense, the only method of identification may be through medical or physical examination and autopsy. The most basic identification is known as "gross identification" whereby a relative or friend identifies the victim by visual identification. Additional identification may be by fingerprint comparison, forensic odontology (dental comparison), and ultimately medical or physical examination for tattoos or scars, evidence of surgical procedures, unique or unusual deformity, gender, race, build, features and approximate age, personal papers, jewelry, and clothing.

Motor Vehicle Fires

This section is a brief overview of the investigation of the origin and cause of motor vehicle fires. Modern vehicles are changing drastically. The most common internal combustion engine can vary from unleaded gasoline to diesel, hydrogen, natural gas (methane), and propane. New vehicles are electrically charged; some combine an internal combustion engine and electrical charge. Knowledge of the fuel source for the operation of the vehicle is obviously essential. Motor vehicle fires can be typically characterized as noncombustible small containers and frames with a concentration and wide range of combustible or flammable materials. The fire patterns produced on all of these materials are an important part of the investigation. This documented physical evidence is compared and contrasted with a pre-fire vehicular history, statements made by the owner or driver of the vehicle, mechanical history, prior accidents, witness information, and research and problematic conditions with a particular vehicle, such as technical service bulletins and recalls. The best practical experience is to observe vehicle fires. Obtaining permission in salvage yards and other areas that retain vehicles that have been involved in fire occurrences is greatly beneficial to developing investigative expertise. When examining salvage yard vehicles, begin with vehicles with small fires so you can observe how the fire patterns migrate and travel to different compartments and how they react with certain materials in the vehicle. Then you eventually gain a greater knowledge in analyzing larger vehicle fire investigations.

Recording the information of your investigation, whether through notes, dictation, or photographs, are obvious starting points. When starting an examination of a motor vehicle, begin with the exterior and move to the interior compartments. When performing various investigations, you may have information regarding the fire occurrence, such as witness statements or prior repairs. The pre-fire history of the vehicle can greatly assist in the investigation. Circumstances must be compatible with your conclusions. A vehicle parked for hours away from other structures and vehicles has only one of two possible causes: a fire starting within the vehicle, either of electrical origination or incendiary. The other fuels will most likely become involved only after the initiating failure.

Exterior Examination

An investigator should first document missing components and damages (these may be previous repairs), then record the extent of fire damage to the engine, passenger compartment, and trunk/payload if applicable. Note the indicators left (driver) and right (passenger) for each compartment. Document and consider exterior fuel loads and their contribution to the fire: the fuel tank, tires, or fiberglass/plastic body panels. Document and consider the position and condition of the windows, doors, and, if possible, the locking mechanisms.

Ventilation through an open window or door will produce distinct patterns. Determining if a fire originates inside or outside the vehicle will require a detailed examination of the interior compartment in comparison to the patterns on the exterior.

Interior Examination

Typically, all vehicles have two or three compartments with separation barriers. Documenting these physical characteristics of the vehicle is the next step in the origin and cause determination. Accounting for missing or present components is foremost in determining an explanation for missing components. Examples of these items include the radiator core, air bag assemblies, radio/stereo equipment, and seats. The interior compartments can be evaluated at opening areas where there are barriers to determine the path the fire traveled, such as the bulkhead openings that separate the engine compartment from the passenger compartment.

The wire bundles within the dashboard may not have much, if any, insulation material remaining on them. Generally, if a fire originates in the engine compartment and travels to the passenger compartment, the areas of wire bundles near or at the bulkhead openings will have sustained damage or melting. If a fire originates in the passenger compartment with extension into the engine compartment, the affected materials will also include hoses and other components besides electrical wiring. Passenger compartment fuel loads are to be taken into account for evaluating the fire patterns that are observed. Documenting these additional fuel loads may include items such as seat cushions, dashboard plastics, or personal items including tissues, papers, refuse, and food. There also may be applied or stored containers of flammable and combustible materials that will produce fire patterns.

Accidental ignition sources within the passenger compartment most frequently are electrical in origin. Mechanical failures are the next most common, such as those from catalytic converters. These are actually origins outside of the passenger compartment. The fuel loads are within the passenger compartment. Electrical failure sources are from wiring connections, switches, or motors (seat motors, window motors, etc.). Of course, the introduction of an outside heat source, such as smoking materials or incendiary objects, can point to the origin. Although the vehicle manufacturer's materials are not readily conducive to ignition and open flame by the accidental contact with smoking materials, they are additional possible sources of fuel. The trunk, if equipped, has very similar ignition sources as the passenger compartment. It contains electrical components, such as lights, stereo equipment, and wiring. The fuel loads can vary greatly. A spare tire is normal, but there can also be a vast assortment of stored vehicle and personal items, such as flammable/combustible materials, gasoline containers, transmission and power steering fluids, aerosol cans, clothes, etc. Fire extension to the trunk area is usually easy to discern considering the burn patterns on the trunk lid and side panels as well as items lying on the trunk floor.

The most condensed compartment of potential fire origin is the engine compartment, which contains fluids and electrical and mechanical components in one small combined area. Once again, document how the vehicle is equipped and what may be missing. Determine whether it is missing as the result of fire damage, was removed before the fire, or failed before the fire and became dislodged from the vehicle. An example of this situation would be an oil filter unthreading and coming dislodged. The sources of fluid-related failures may occur at many potential areas, such as hoses, connections, reservoirs, fuel lines, radiator, air conditioning, oil lubrication, and transmission—all have the potential for ignition when these fluids are released in various forms with heat and ignition sources. To determine ignition, the heat source must be sufficient to ignite these materials.

Determining the flashpoint is much less necessary when considering that many of these fluids are carried in high-pressure systems and can be sprayed at high pressure and atomized against heat/ignition sources. Within an engine compartment, the most common ignition source in relation to the ignition of expelled fluids is from the exhaust manifolds. Contact of these fluids with such a heat source is often sufficient in the ignition of the vapors of these fluids or providing ignition from open electrical sources to ignite the vapors. The temperatures of exhaust systems can be up to 650 degrees Fahrenheit (343 degrees Celsius) during normal operation. A poorly operating engine can greatly increase the temperatures of the operating systems. The hottest temperatures are typically in the area located in front of the catalytic converter.

Electrical System

As for the electrical system, one must consider the battery and whether it was connected. Some models of vehicles do not have the battery in the engine compartment, but in the trunk compartment or underneath the rear passenger compartment seats. Documenting the primary and secondary wiring within the vehicle is also essential in determining its potential involvement in the initiation of the fire. A short circuit condition in the primary wire will often leave arcing signatures or fusing of the metal. One must also consider if this electrical failure was the cause of the fire occurrence or a victim in the fire path.

Overloaded wiring will cause wires to heat up severely and can cause insulation materials to reach their melting or ignition temperatures. Surrounding components can further insulate the overheated wires, such as other wires within the harness, seats, or carpeting, which will not allow heat to dissipate adequately. The compounding heat may not open the fuse or links. Overloaded electrical wiring heats uniformly through its length between the connections or between the point of short circuit and the energy source. Failures at connections or damaged areas may not overload the wire and cause the uniform insulation melting. Examining the fuse panel and the relay panel is important to determine if tampering with these safety areas has been bypassed. Many times these items are greatly affected during the fire and cannot be fully documented.

Mechanical faults, such as overheating due to failed mechanical components, may occur. A failed exhaust system can produce sparks when it contacts the pavement as the vehicle is being driven. A flat tire can produce sparks to ignite gases, vapors, and liquids that are in a sprayed or atomized form.

Obtaining information regarding the vehicle's maintenance or history as well as statements from the driver and owner and any other witnesses is important to compare it to observations the investigator has made of the vehicle. A basic list of questions is provided that can be expanded, depending on the information gathered.

- Was the vehicle being driven at the time of the fire occurrence, and if so, for how long?
- If the vehicle was being operated, was there any prior indication of problematic conditions?

- What was the first indication that the problem existed?
- What were the last services performed on the vehicle?
- Can a vehicle maintenance history be obtained for review?
- When was the last time the vehicle was fueled up and what was the total amount of fuel?
- If the vehicle was not being operated and was parked, when was the last time the vehicle was driven? By whom?
- Can other keys be accounted for?
- If the vehicle was parked, where was it parked?
- What personal items were in the vehicle (radar detectors, cellular telephones, stereo equipment, etc.)?
- What was the last known mileage on the vehicle?

Checklists may be useful for vehicle fire investigations to be certain that nothing is overlooked. A checklist for most vehicles would include the particular case information that has been obtained for the investigative personnel. The particulars with the vehicle should include the make, model, year, vehicle identification number, and odometer reading. Exterior examinations should include details about the tires, wheels, rims, and their conditions. The status of the windows and doors of the vehicle should be documented, including any evidence of force or damage before the fire occurred. The exterior body panels should be examined starting from the front and going 360 degrees around the vehicle to detail the different body parts. Documentation of the engine compartment should detail the engine, type of transmission, braking system, fuel system, belts, accessories, alarms, battery, fuses, oil, transmission fluid, radiator fluid, brake fluid, power steering fluid, and windshield fluid. Passenger compartment examinations should include personal contents, safety devices (such as airbags), seats, stereo equipment, ignition cylinder, glove box, dashboard components, interior panels and the carpeting, padding, and mats on the passenger compartment floor. Also, one can use an enlarged version of the vehicle diagram to show the areas of the highest and lowest fire intensity.

Variables to consider are wind direction, location of the vehicle burn or recovery, and possible contents or items surrounding the vehicle. By examining vehicles on a regular basis, an investigator can pick up on consistencies. Starting with small fire losses will help one visualize how fire patterns progress.

Collection of Fire Evidence

During the course of a fire investigation, it may become necessary to collect and secure evidence for further testing or analysis to confirm or eliminate a potential fire cause. In most cases, the collection of evidence is critical in determining the cause of a fire loss (Figure 25.1). Evidence is usually defined as any finite or tangible material that is legally obtained in an effort to prove the cause of a fire (Figure 25.2). For the purpose of this discussion, we will focus on evidence securing, preservation, and testing as it relates to the public sector; that is, police agencies and fire municipalities. Many of these principles, however, also apply to the private sector, which include investigators for the insurance industry.

As indicated earlier, the investigation of fire losses have become more organized and thorough over the past 25 years. Many municipalities, even small volunteer fire departments, have their own fire investigation team.

Figure 25.1   Overall view of the bedroom where the third point of origin was located on the bed.

Figure 25.2   Closer view of the bed showing the area where the local authorities cut the burn pattern from the bed. The arrow indicates the location of the ignition device that was not recovered by the local authorities at the time of their inspection. It was later identified as a partially burned diaper.

These teams of investigators are usually the first to investigate a fire scene after a fire is extinguished. Depending on the size of the loss, an arson task force may also be brought in to investigate the loss. Usually, if there is a fatality involved, the task force or Fire Prevention Bureau will be summoned as a matter of standard operating procedures. In some instances when the dollar loss amount is exceedingly high, or if significant loss of life occurs, the local or state authorities may approach the Federal Bureau of Alcohol, Tobacco, and Firearms.

Because the field of fire investigation has grown, the scrutiny of evidence collection, preservation, and testing has also increased. Criminal and civil cases depend on how evidence is collected and removed from the initial inspection date to the day on which it is presented in a court of law.

There are four recognized methods of removing evidence from a scene that are accepted in the court system today. They include exigent circumstances, consent of property owner, administrative search warrant, and criminal search warrant. All of these methods have been upheld in the judicial system. Exigent circumstances are generally recognized when fire department personnel enter a scene to extinguish the fire. In this particular instance, no authorization from the property owner is required because the purpose of the fire department is to extinguish the fire and investigate the loss. Consequently, after a fire is extinguished, the fire investigation team, local fire marshal, or other agencies brought in to investigate the loss can legally remove any property they believe confirms or refutes potential fire causes. Once the fire agency releases the scene, however, they cannot remove any additional property without a search warrant issued by a court or consent from the property owner.

Property owner consent is the usual method of entry for private sector investigative firms. Either the individual homeowner, tenant, or management association grants permission for the investigator to enter the property and investigate the fire. Private sector investigators normally use this method. The public sector or municipality investigators can reenter a released scene if the property owner grants permission.

Administrative search warrants are normally granted when an agency has the responsibility to investigate the fire and permission by the property owner has been denied. Under normal situations, the investigation is limited to the documentation of the origin and cause of the fire occurrence. If, at any time during the investigation, potential criminal evidence is uncovered, the investigation must stop and a criminal search warrant must be obtained. A criminal search warrant is requested when evidence suggests a crime at the location has been committed. The facts are presented in front of a judge and "probable cause" of a crime must be presented. If the evidence of a crime is not convincing to the judge, the judge will deny the request of a criminal search warrant.

Evidence Collection

When the investigator believes that the cause of the fire could potentially be incendiary, samples of the burn materials around the point(s) of origin should be removed and tested to identify the potential traces of accelerants. Some municipalities have accelerant detection dogs that are specifically trained to detect trace odors of accelerants. Many of the municipalities that handle or use accelerant detection dogs will remove samples of the flooring or other materials where the dog detects the odor of a potential accelerant.

When removing samples for testing, it is imperative that a comparison or control sample also be submitted. A control or comparison sample is the same material removed from a different room or different area of the room than where the fire originated. For example, if a sample of carpeting is removed from the center of a specific room for testing, a comparison sample of the same carpeting removed from a different room or possibly from a protected area, such as under a cabinet or furniture item, should be secured if possible. Laboratory tests will show if a potential flammable component is inherent in both samples. If possible, only one type of material should be tested per sample. For instance, if a fire is determined to originate in the center of a room and the floor is a hardwood floor with carpeting, the investigator should have at least three samples to test to determine the presence of accelerants, as well as three control samples. The samples would include the carpet remnants in one sample, the carpet padding in a second sample, and the wood flooring in a third sample. If it is not possible and if the investigator removes only the carpeting and padding and places it in one sample container, then the control sample should also contain the carpeting and padding (Figure 25.3). Likewise, if a sample of flooring material consisted of vinyl

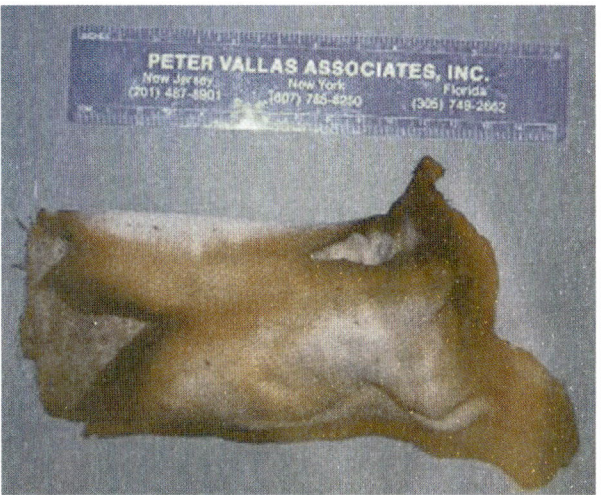

Figure 25.3 Closer view of rolled up diaper that had been ignited and thrown on the bed. It was found to contain paint thinner.

Table 25.1
Flammable and combustible liquid classification system

Class Number (class name)	Peak Spread Based on N-Alkane Carton Numbers	Examples
1 Light petroleum distillates	C4–C8	Petroleum ethers; pocket lighter fuels; some rubber cement solvents; Skelly solvents; Varnish Makers and Painters (VM&P) naphtha
2 Gasoline	C4–C12	All brands of automotive gasoline, including gasohol; some camping fuels
3 Medium petroleum distillates (MPD)	C8–12	Paint thinners, mineral spirits; some charcoal starters, dry cleaning solvents; some torch fuels
4 Kerosene	C9–C16	No. 1 fuel oil, Jet-A (aviation) fuel; insect sprays; some charcoal starters; some torch fuels
5 Heavy petroleum distillates	C10–C23	No. 2 fuel oil; diesel fuel 0 variable single compounds such as alcohols, acetone, toluene, xylenes; some lamp oils, camping fuels, lacquer thinners, turpentine, and others

tile and wood, a control sample of the same vinyl tile and wood flooring should be submitted for comparison purposes.

Table 25.1 highlights the various classifications of accelerants, with examples for each classification.

In all cases, each step of the debris removal and sample collection should be documented either with photographs, or by videotaping the entire process, or both. This will eliminate questions regarding potential cross contamination of samples or improper sample removal.

Evidence Containers

For accelerant detection, there are three acknowledged containers for placement of evidence once it is secured. They consist of metal cans, glass jars, or special evidence bags. All of the containers have advantages and disadvantages. Metal cans are considered the recommended container for solid and liquid evidence, which is to be tested for accelerants. The metal cans should be unlined and clean. Lined cans can reduce the ability to properly detect trace elements of accelerants in the sample or will alter the results, and therefore, should not be used under any circumstances.

The advantages of unlined metal cans are that they are inexpensive and come in several different sizes. The major disadvantage of the metal cans is that over time, the metal can rust and corrode, and if the sample had not been tested, accelerant vapors could escape through the deteriorated openings in the can. Glass jars are also acceptable to secure samples. If using glass jars, the caps of the jars should not have glued liners or rubber seals. When the samples are heated during the gas chromatograph testing, the gluing agent, which is normally comprised of some solvents, can alter the results of the test. Rubber seals can deteriorate or dissolve in the presence of vapors from accelerants. This allows leakage of the vapors and loss of the sample. Most glass jars will have a threaded cap that can be secured to the top of the jar with no glued liner or rubber seal attached.

Some of the advantages of using glass jars to collect samples are the availability of jars, as well as the different sizes of the jars. The sample can also be viewed without opening the jar and disturbing any vapors unlike a metal can. The obvious disadvantage is that the glass jar can be broken if dropped or shipped improperly.

Special evidence bags are increasingly used by municipalities primarily for the variety of the sizes and shapes, and the cost factor.

They are also easily stored, and, as with the glass jars, the evidence can be viewed without disturbing the seal. These special evidence bags have a seal that has to be properly closed; otherwise vapors will escape. Another disadvantage of the bags is that they are also susceptible to damage and can tear easily. These are not common plastic bags, but are bags specifically designed for the collection and preservation of samples.

Most of this section has concentrated on evidence collection and preservation with regard to an incendiary fire. When a fire is ruled accidental by a fire department or municipality, their investigation stops, and no evidence is collected. Some municipalities do secure evidence in fires that are ruled accidental. Investigators for the private sectors, such as the insurance company, will, in all probability, investigate the loss as well. Therefore, any evidence secured by the marshal or the fire department should be released to the investigator handling the investigation for the property owner. This sometimes is difficult to determine if a fire occurs in a tenant-occupied building, or if a fire develops in a condominium because there are several parties affected and often each party has a separate insurance carrier and, therefore, will have an independent investigation. When the property is released into another party's possession, a receipt form should be completed and a copy maintained by the fire investigator or fire marshal so he or she can direct the calls to the individual who retained possession of the property. The private sector investigators have the ability to further examine any appliance or items removed from an accidental fire scene through failure analysis or engineering evaluations of the equipment.

Case Studies

Case 1

This case involved a residence that was equipped with a lightning protection system. It was installed by a licensed contractor specializing in the installation of this type of equipment. The fire occurred during the lightning season.

The fire report as well as a follow-up weather report confirmed lightning on the date of loss. During the investigation, it was learned that the extent of fire damage was affected by the delayed response of the local fire department. The home was located in a rural area on a remote private road on top of a hill. The fire department was a volunteer organization that required a response of local fire fighters from either their homes or places of business to the fire department and ultimately to the fire scene. As a result, the home was totally destroyed.

The scene examination involved the normal origin and cause analysis with exterior scene examination as well as interior inspection of the remains of the property. During the inspection, the electrical system was identified on the west side of the residence. The point of origin was on the southwest corner of the residence approximately 75 feet from the incoming electric service. The service entrance was underground. The area of origin was a utility room that housed the water pump, water storage tank, and electric water heater. The water well system was a submersible pump located near the southwest corner of the residence. The water line fed from the well casing underground to the storage tank in the utility room. The water line was standard ⅝-inch copper. The electrical systems grounding for the incoming service to the panel was completed with stranded copper cable to the water line.

During the evaluation of fire patterns on the structural wood components (walls and remaining ceiling structure), a very low burn pattern was noted at the corner of the utility room where the water line entered through the sill plate. Upon closer inspection of the water line leading to the water storage tank, an unusual scorch and arc mark were discovered spiraling around the water line and terminating at the pump pressure switch. To further evaluate this unusual evidence, the buried water line out to the well pump head was dug out and exposed. During this excavation, which was approximately 24 inches below the surface of the ground, a second grounding cable was uncovered affixed to the water line. This was traced to the corner of the building and up to the eave level where it dropped into the fire debris of the house because the roof was completely gone. Inspection of this grounding cable continued to the first lightning rod on the south end of the residence. Subsequently, it was tied into additional lightning rods that were on the residence. After completion of the excavation, it was determined that the lightning rod system had been grounded to the same water line as the grounding for the electric service in the residence.

As a result of the lightning strike to the lightning arresting equipment, it grounded to the water line, which continued into the residence via the copper water line to the water storage tank. As it passed through the sill plate, a smoldering fire ignited in the wall and subsequently involved the entire structure.

Case 2

The residence is a three-story townhouse in a complex of six identical units. This particular townhouse was occupied by a young couple. The fire occurred on a Sunday morning at around 10 a.m. The husband was an avid dirt bike/motorcycle enthusiast who raced on Saturdays. The motorcycle was stored in the garage of the townhouse along with his other vehicle. The garage was separated from the main living area by a fire-rated sheetrock wall on wood studs. The door leading from the garage into the entry foyer was a steel fire-rated door off of the back of the entry foyer on the same level as the utility room where the natural gas-fired water heater, furnace, and laundry area were located. The second level consisted of a kitchen, living room, dining room, and a den. The third level contained three bedrooms and a bathroom.

On the date of loss, the homeowner had been in the garage cleaning his motorcycle and decided to run down to the local store to get the newspaper. When he pulled out of the garage, he left the overhead door open. While he was away from the residence, his wife was working out on her exercise treadmill and decided to take a shower. After her shower she was in the kitchen on the second floor, which is

in the front of the residence. She was alerted to the fire by an explosion and heavy smoke conditions coming from the ground level of the residence. She ultimately evacuated the second floor level via a deck off the living room.

Upon the arrival of the fire department, they were faced by heavy fire in the garage and utility room that was ultimately suppressed. The insurance carrier for the homeowner requested an independent investigation of the origin and cause of the fire. During this investigation, it was found that the dirt bike motorcycle was lying on its side in the garage adjacent to the wall that separated the garage from the utility/laundry room area (Figure 25.4). The fire department had already breached the wall during their fire suppression efforts. During examination and reconstruction of the scene, it was noted that although the wall was constructed with the proper fire-rated sheetrock, there was a void at the base of the wall. The void was due to the lack of a wooden sill plate adjacent to where

Figure 25.4 Overall view of the point of origin from the garage side showing the location of the motorcycle in relation to the hole in the wall.

Figure 25.5 Close-up view of the hole in the wall made by the fire department as viewed from the garage side.

the motorcycle was lying (Figure 25.5). The reason the sill plate was missing was because the contractor failed to use the proper length 2 × 4 wood framing and chose to use whatever piece was left over, creating a 6- to 8-inch void at the base of the wall (Figure 25.6). On the opposite side of the wall in the utility/laundry room were the water heater and furnace (Figure 25.7). There was carpeting on the utility room floor with a burn pattern extending from the void in the wall over to the area of the water heater and furnace.

An inspection of the furnace and water heater was conducted to determine which might have been involved in the ignition of this fire. It was obvious from the inverted **"V" patterns** on both sides of the wall between the garage and utility/laundry room that there had been a flammable/combustible liquid burning on the floor. This was further confirmed by the patterns on the motorcycle as it lay on the floor, particularly around the fuel tank cap, which was still in place. Closer inspection of the patterns in the utility/laundry room revealed soot patterns around the opening to the burner area of the water heater (Figure 25.8). This was not present on the furnace.

As a result of the investigation, the following facts were considered in the conclusion as

Figure 25.6 View of the same hole in the wall as viewed from the utility room side. Note the defect in the wall where the sill plate is missing.

Figure 25.7 View into the utility room showing the water heater and furnace on the left and the hole in the wall on the right.

Figure 25.8 Close-up view of the base of the water heater. The arrow indicates the soot mark caused by the ignition of the gasoline fumes by the burner of the water heater.

to the origin and cause of the fire. The husband confirmed that the motorcycle still contained approximately 1 gallon of gasoline in the tank after he completed racing. It did not have a kickstand and was leaning up against the wall. After he left to go to the store, the garage door was left in the open position allowing air movement into the garage from the exterior. When the bike fell over, the gas leaked from around the gas cap, which was not sealed tightly because he had been working on the bike earlier. As the liquid gas spilled from the tank, it flowed under the wall through the void. The vapors were eventually ignited by the water heater and flashed back along the carpet under the wall through the void to the puddle of gasoline in the garage.

The ignition sequence from the water heater was due to the water heater's recycling to heat water after the wife had taken a shower. As a result, the potential **subrogation** in this case involved the general contractor for the townhouse complex as well as the individual subcontractor for negligence in their workmanship in constructing the wall between the garage and the utility/laundry room.

Case 3

Although the incendiary fire investigation may initially commence on a similar path as the two previous examples, the fact remains that once this fire has been identified as a crime,

the scene investigation takes on a dual role for the local authorities. In this example, we encountered not only an intentionally set fire, but also the motive was to cover up a homicide. The unfortunate victim was a kind and generous grandmother who was dealt an untimely and unjustified death by her conspirators, a greedy daughter-in-law, grandson, and foster child.

The scene is a rural farm where the victim lived with her son and his family. To avoid any burden on them, she chose to live by herself in a mobile home on the property. Unfortunately, the son was a long-haul truck driver who was frequently absent, which left the burden of care to the daughter-in-law. Thus, the plot was hatched to dispose of the lady and inherit the entire estate with "no strings attached."

The plan commenced with the 14-year-old foster child distracting the victim in the living room area of the mobile home while the 27-year-old grandson prepared his "accidental" fire setup under the bedroom at the opposite end of the mobile home. In the early morning hours of the following day, they locked the door to the mobile home from the exterior by placing a screwdriver in the lock hasp of the door so that the grandmother would be unable to exit the home once the fire began. The fire setup failed and the grandson started the fire with a propane torch flame applied to the combustible underside of the bedroom floor. Smoke began to fill the mobile home and the grandmother was alerted to the impending danger of the fire. However, her initial attempts to exit were prevented by the blocked door. Later, there was evidence that she attempted to break open the door with a hammer, which was found on the floor just inside the door as well as impact marks on the aluminum edge of the doorframe. This did not prevent her from continuing her attempts to free herself from the fire. She calmly removed the interior screen of the small window above the kitchen sink climbed up onto the sink, through the window, and to safety.

This incident occurred 1 week before Christmas. There was heavy snow on the ground and extremely cold conditions. Unfortunately, when she landed in the snowdrift outside the window, her worst fears were not over. The grandson and foster child were on the lookout for any problems with their plan. When they heard her attempting to break the door down, they were waiting for her outside in the hedgerow. As she fell to the ground, they approached and struck her numerous times in the head with a steel pipe.

The daughter-in-law now thought that the fire had progressed sufficiently to call the fire department and allow the plan to unfold. When the fire department arrived with emergency medical services (EMS) close behind, they found the victim unconscious and bleeding in the snow outside the window. As the fire department proceeded to suppress the fire, which did not involve the opposite end of the mobile home, the EMS personnel treated the victim. They quickly stabilized the grandmother and began transporting her to the nearest hospital. En route to the hospital, they radioed to the emergency room and gave their initial assessment of the victim's injuries. The assessment included numerous and apparent skull fractures. This radio transmission was also relayed to the local authorities who responded to the fire scene to assist in the investigation. From the onset, the daughter-in-law, grandson, and foster child appeared remarkably calm throughout the ordeal.

As the investigation unfolded, it was apparent that the victim's injuries were not consistent with the physical evidence at the fire scene. The three remaining witnesses were separated and interviews commenced while the on-scene investigation continued. The initial physical evidence regarding the fire quickly focused on the lack of an accidental ignition source under the bedroom area of the mobile home, such as an electrical ignition source or something related to the heating system.

In addition, the unconventional method of exiting the mobile home through the window was puzzling because access to the mobile home was through the front door, which could not be locked from the inside or outside at first appearance. Upon closer inspection, the presence of the hammer on the floor just inside the door along with the unusual impact marks on

the doorframe raised suspicion. Closer inspection of the exterior hasp revealed unique indentations in the aluminum siding above and below the hasp.

Further inspection of the area of fire origin in the bedroom of the mobile home confirmed suspicions about the fire cause. There was no accidental or natural ignition source in the bedroom since the fire vented through the floor under the bedroom from beneath the mobile home.

During the interviews with the only witnesses, the first break came during the interview of the 14-year-old foster child. During his recitation of the events leading up to the discovery of the fire, he commented about getting a pony as his reward. Further inquiry into the "reward" revealed the entire story and conspiracy with the other members of the family. The plan was to make the death of the grandmother appear to be accidental while her son was away on a long-distance road trip. They carefully planned her demise. Initially when the fire set did not go off as planned, the grandson set the fire with the propane torch by igniting combustibles under the bedroom area of the mobile home. He apparently assumed that the fire would destroy any evidence if the mobile home was extensively damaged by fire. The foster child was in charge of making sure that she could not escape from the mobile home by placing a screwdriver through the hasp to secure the door. Eventually, they saw her exit through the kitchen window, falling into the snowdrift below. At this point, they began beating her with the steel pipe. Once the fire was sufficiently in progress, the daughter-in-law called the fire department while the two young men hid the screwdriver and steel pipe in the barn. What was more disturbing about the case was that the 27-year-old grandson worked for the insurance company that carried the grandmother's life insurance policy. He apparently had devised the plan with his mother after reviewing the policy language.

Because the other parties to the conspiracy refused to cooperate and their stories were inconsistent with the information developed from the juvenile, it was necessary to obtain a search warrant. In the search warrant, the specific language was developed from the interview with the juvenile and identified specifically where the screwdriver and steel pipe were hidden. The screwdriver was recovered from an eaves trough on the edge of the main house roof. The steel pipe was recovered from two mattresses hidden in the hayloft of the barn. The screwdriver was compared to the marks on the outside of the mobile home door in the lock hasp and were determined to match. After photographing and documenting the measurements along with careful securing of the evidence for further laboratory testing and analysis, the three conspirators were arrested and charged with arson and homicide.

In the end, the juvenile was a state's witness against the two adults and served 2 years in a juvenile detention facility. The 27-year-old grandson was tried separately from his mother and sentenced to 25 years to life. As for the daughter-in-law, she had been incarcerated in a local county jail pending her trial; however, she passed away from a terminal illness approximately 1 month before her trial was to commence.

Questions

1. What is fire?
   a. Rapid oxidation (i.e., similar to rusting process)
   b. An open flame that produces heat, smoke, and light
   c. A rapid oxidation process with the evolution of light and heat in varying intensities
   d. All of the above

2. An incendiary fire is
   a. A fire involving melted plastic residue
   b. A fire that starts accidentally
   c. A fire that starts by an unknown cause
   d. A fire that is never investigated
   e. A fire that is intentionally set

3. A "V" pattern on a vertical surface indicates
   a. High temperatures from a fire
   b. That an accelerant was used
   c. A flame occurred close to the surface
   d. The approximate ignition temperature of the fire was greater than 100 degrees Fahrenheit

4. The area of origin is best described as
   a. The general area in which the fire began
   b. The exact spot where the fire started
   c. The search area of a fire scene
   d. The crime scene search area of an incendiary fire
   e. None of the above

5. The point of origin is
   a. The exact physical location where a heat source and a fuel come in contact with each other and a fire begins
   b. The precise point where an ignition source, fuel, and an oxidizer come together to create a fire
   c. The point within an area of origin where the fire originated
   d. All of the above

6. The cause of a fire is
   a. The person responsible for setting the fire
   b. A material that yields heat through combustion
   c. The process of initiating self-sustained combustion
   d. The luminous portion of burning gases or vapors
   e. The circumstances and conditions that bring together a fuel, an ignition source, and an oxidizer that result in a fire

7. Which of the following should be noted when forming an examination of a motor vehicle at a fire scene?
   a. Location of the vehicle with regard to the point of origin
   b. The VIN (vehicle identification number)
   c. Fire patterns on the vehicle
   d. The location of tire pads, if any are found
   e. All of the above

8. Which of the following procedures should be observed when collecting evidence that might contain accelerant for laboratory testing?
   a. The samples should be sealed in large plastic bags
   b. A fresh pair of disposable latex gloves should be worn for the taking of each sample
   c. As soon as you get back to your office, you should write down where you collected the samples
   d. The location of each sample should be photographed
   e. b and d above

9. Which of the following is an accepted method of fire scene documentation?
   a. Photography
   b. Sketching
   c. Videography
   d. All of the above

10. Which of the following can be used in the identification of a fatal fire victim?
    a. Gross identification (visual)
    b. Dental
    c. Radiographs (x-rays) of old injuries
    d. Fingerprints
    e. All of the above

11. The generally accepted temperature used to distinguish flammable from combustible liquids is defined as
    a. 100° Fahrenheit
    b. 200° Fahrenheit
    c. 300° Fahrenheit
    d. 451° Fahrenheit

12. The incipient phase of a fire is the
    a. Last stage
    b. Middle stage
    c. Earliest stage

13. Spoliation is the intentional or negligent destruction or alteration of evidence.
    a. True
    b. False

14. Name four recognized methods of removing evidence from a scene that are accepted in the court system.
    a. Exigent circumstances
    b. Consent of property owner
    c. Administrative search warrant
    d. Criminal search warrant
    e. All of the above

15. Ultimately the cause of death and cause of fire may be closely related if the victim was the perpetrator of the fire and succumbed to his or her own acts of negligence or intentional incendiary activities.
    a. True
    b. False

Suggested Readings

Lee, H. C., Palmbach, T., and Miller, M. T., *Henry Lee's Crime Scene Handbook*, Academic Press, New York, 2001.

NFPA 921, *Guide for Fire and Explosion Investigations*, NFPA, 2001.

Redsicker, D. R., *Fire and Arson Investigation Techniques*, Practical Video, LLC, Ithaca, New York, 2001.

Redsicker, D. R. and O'Connor, J. J., *Practical Fire and Arson Investigation*, 2nd ed., CRC Press, Boca Raton, FL, 1997.

Redsicker, D. R., *Practical Methodology of Forensic Photography*, 2nd ed., CRC Press, Boca Raton, FL, 2000.

Vehicular Accident Reconstruction

Randall K. Noon

26

> Personally, I'd rather be thrown clear than be trapped in a car by a seat belt. Actually, one drink tends to relax me, and I really drive better.
>
> **Often heard said by many misinformed dead people**

Introduction

Unlike arson, poisoning, stabbing, beating, siege machines, and other methods of mayhem whose history is as old as human civilization, motor vehicle accidents are a relatively new instrument of mayhem. Obviously, there were no vehicular accidents before the invention of self-propelled vehicles.

In 1900, there were only 8000 cars registered in the United States. Amazingly, by 1915 there were 2,332,426 registered cars in the United States. The rapid adoption of motor vehicles by the public is similar to the social change stories associated with the telegraph, the telephone, radio, television, personal computers, and cellular telephones. All of these marvels compressed time and distance, and promoted personal independence. From the beginning, motor vehicles had sufficient mass and velocity to cause serious accidents when they ran into buildings, objects, animals, and people. By 1915 there were sufficient numbers of them to frequently crash into one another as well.

Except for the years during World War II when civilian vehicle production was halted, the number of registered cars in the United States has steadily increased with population. In 2005, there were 136,568,083 registered passenger cars, and approximately 8 million new passenger cars were bought or leased. Of course, these figures do not include trucks of all sorts, including SUVs (sport utility vehicles), motorcycles, mopeds, buses, public vehicles, military vehicles, and nonregistered vehicles such as all-terrain vehicles, tractors, and go-carts. Typically, there are about half as many trucks in use as there are registered cars, and there are over 6 million registered motorcycles.

Topics

Primary Causes of Vehicular Accidents

Analytical Tools Used to Evaluate Accidents

Converting Scene Data into an Event Sequence

Basic Energy Method

Basic Momentum Method

Tips and Solution Strategies

Although the statistics of deaths and injuries on our roadways are sobering, there is an optimistic side to them. Thanks to improvements in safety, the death rate has declined significantly in the past three decades. In 2004, for example, the death rate was 1.48 deaths per million miles. This is the lowest it has been in the 40 years since such records have been kept. In 1977, the death rate was 3.3 deaths per million miles. Amazingly, the death rate has been reduced more than half in less than three decades.

How has this been accomplished? There are three primary reasons. First, and most important, more people are regularly wearing seat belts. Survey figures released in 2001 from the National Highway Traffic Safety Administration (NHTSA) indicate that 73% of people observed in the front seat of vehicles during the summer of 2001 were wearing seat belts. In a more recent NHTSA survey, 80% of those surveyed claim that they wear seat belts whenever they drive or ride as a passenger. In 1970, however, hardly anyone wore them. This change in cultural attitude toward seat belts has been accomplished by three decades of persistent public service advertising, driver training classes in secondary schools, and state and local law enforcement. It is no small accomplishment.

Second, there has been a significant reduction in drunk driving. Drunk driving is no longer winked at as it once was, and state and local drunk driving laws are significantly tougher and more consistently enforced. In 1987, more than half of all traffic fatalities were related to drunk driving. By 2005, this figure had dropped to 39%. The credit for this reduction goes to organizations such as Mothers Against Drunk Driving (MADD) and Students Against Drunk Driving (SADD) for vigorously pursuing this issue, as well as state and local law enforcement.

Last but not least, the crashworthiness of cars and vehicles has improved. Safety seats for children are now mandatory. Vehicle interiors have been redesigned to reduce impalement injuries. Air bags and side air bags are now standard features in new cars, as well as crumple zones and impact-absorbing bumpers. Side doors are stronger. Engines have been designed to slide under occupants rather than impale them. Seats have better anchors. Seat belts have evolved from simple two-point lap belts to three-point harness restraints with backlash neck supports. Steering wheel columns are collapsible. Gas tanks and fuel lines have improved. Braking systems have improved. Headlights and taillights have improved. Tires have improved, and so on.

Despite these gains, however, vehicular accidents still have a major economic impact in the United States. The economic losses due to vehicular accidents were about $230 billion in 2000. This includes lost wages, legal expenses, medical expenses, funeral expenses, insurance administrative costs, and property damage. A fatality accident currently costs perhaps just over $1 million computed on the basis of an average discounted lifetime, and alcohol-related accidents are still responsible for approximately one-third of all economic losses.

Figure 26.1 Rollover damage. Note crumpling of roof, windshield, top of hood, broken side mirror, and rear fender.

Primary Causes of Vehicular Accidents

Table 26.1 is a list of the reported causes of vehicular accidents in 1999. The first column

Table 26.1
Causes of Vehicular Accidents—1999

Reported Causes of Vehicular Accidents	Injury Accidents (%)	Fatal Accidents (%)	All Accidents (%)
Improper driving	67.2	72.6	62.2
Excessive speed	13.0	23.0	10.6
Right of way	25.8	20.1	22.9
Failure to yield	19.2	10.8	13.8
Failed to stop at sign	1.7	4.6	3.2
Disregard of signal	4.9	4.7	5.9
Driving left of center	1.7	9.6	1.3
Improper passing	0.9	1.1	1.2
Improper turn	2.4	1.2	3.0
Following too close	3.4	0.5	6.3
Other imprudent driving	20.3	17.1	16.9
No improper driving stated[a]	32.8	27.4	37.8

[a] States include driving under the influence of alcohol or drugs in this category. The rationale is that the accident was due to the physical condition of the driver and not because of a specific driving error.

Figure 26.2 Fire damage to hood from engine compartment. Less than 1% of accidents involve fires. Most fires are caused by fuel leaks in the engine compartment.

reports the causes of accidents in which people were hurt. The second column reports the causes in which people were killed. The last column reports all accidents, including fender benders, in which no one was hurt or killed (Figure 26.2).

The reader may note that the category "equipment failure" is not listed. This is because it is not as significant as the other factors. Consistently through the years, the three most significant causes of improper driving have been excessive speed, right of way, and failure to yield. These three categories combined represent the causes for half of all vehicular accidents. The other significant factor is alcohol and drugs. Even though there have been significant reductions in the number of accidents caused by alcohol and drugs, this category still constitutes the underlying cause for approximately one-third of all accidents.

Table 26.2 summarizes what vehicles most commonly collide with when there is a fatality. Not unexpectedly, nearly half of all fatal accidents involve collisions with other vehicles. Approximately one-fourth of all fatalities involve collisions with fixed objects, such as telephone poles, bridge abutments, concrete walls, overpass piers, and trees. The category "noncollisions" includes accidents such as driving off a cliff or into a lake.

From these statistics, it follows that most vehicular accident evaluations include determining the speed of the vehicles at the time of

Table 26.2
Fatal Accidents in 1999

Category	Percentage
Collisions between vehicles	48
Collisions with fixed objects	26
Collisions with pedestrians	12
Noncollisions	11
Collisions with bicycles and other nonmotorized vehicles	2
Collisions with trains	1
Collisions with animals or horse-drawn vehicles	0.2

the accident, the relative positions of the vehicles with respect to yielding or right-of-way requirements, adherence by the drivers and pedestrians to traffic signs or controls in the area of the accident, whether a driver is sober, and so on. In short, most accident evaluations attempt to determine which of the "cause" categories in Table 26.1 might be applicable.

Analytical Tools Used to Evaluate Accidents

The two most important analytical tools used by engineers in vehicular accident evaluations are the laws of conservation of momentum and conservation of energy. The application of these principles usually feasible, given the facts and information usually contained in police accident reports. A well-prepared police accident report usually notes skid mark lengths, final vehicle positions, initial travel directions, and the point of impact, either directly or implied. This constitutes most of the usual data needed to solve momentum and energy equations.

In any scientific cause analysis, reliance on verifiable physical evidence should be primary. Verifiable physical evidence in vehicular accidents usually includes tire and skid marks, a representational plan of the accident area, the location and depth of impact damage on the vehicles and other damaged items observed at the scene, the mechanical condition of the vehicles, the type and location of impact debris noted at the scene, paint transfer marks and other vehicle-to-vehicle and vehicle-to-object contact marks, road conditions, weather conditions, and the physical and mental conditions of the drivers. Other information that could bear upon the causation of the accident includes specific vehicle model performance, roadway specifications, type and placement of traffic control devices and signs, and date and time of the accident.

Although eyewitness accounts of the accident are important sources of information, they must be carefully scrutinized. Distance, lapsed time, and speed—the key elements of most vehicular accident evaluations—are difficult to judge accurately even for an experienced observer. Further, drivers involved in the accident are notorious for underestimating their own speed and overestimating the speed of the other driver.

Sometimes, eyewitnesses to the same accident disagree on the most fundamental aspects of the accident. They may intertwine their own conclusions and ideas about the accident into their narrative of what occurred. The eyewitness may not be impartial. He or she may be the wife, ex-wife, boyfriend, ex-boyfriend, cousin, close friend, or sworn enemy of one of the parties involved in the accident, especially in a small town. A proper scientific accident reconstruction can often provide sufficient information to discriminate between who is telling the truth, who is embellishing the truth, and who is not telling the truth.

The engineering disciplines commonly applied in accident reconstruction include kinematics, statics, dynamics, mechanics of materials, machine design, ergonometrics, and traffic engineering. Skill in college-level mathematics is required to manipulate and solve the various vector equations, integrals, and complex algebraic expressions that are often involved. A familiarity with other sciences, such as chemistry and metallurgy, is also useful. All these disciplines are normally included in traditional mechanical and civil engineering curricula, although the specific applications are perhaps different from those demonstrated in the classroom.

For example, a typical undergraduate civil engineering course in mechanics of materials concentrates on the factors that contribute to the initiation of failure. Generally, the elastic properties of a material are the most studied. In an accident reconstruction, however, the structure being studied, the vehicle, often has already failed in a material sense; it has been crushed, twisted, and distended by impact. The more important material properties in this case are the plastic properties of the material.

Converting Scene Data into an Event Sequence

An engineer or investigator often does not have the opportunity to examine an accident scene before it is cleaned up and the vehicles are towed away. In fact, it is typical for an engineer or investigator to be assigned the analysis of an accident days, weeks, or perhaps even years after the actual event. For this reason, detailed scene documentation is very important. Facts missed during the initial information-gathering stage at the time of the accident can make or break a court case years later. It is difficult, however, to unerringly determine at the time of the accident which facts will be contested and which facts will be accepted without a problem at a trial many months or even years later.

In general, the following information is typically available to assist in analyzing a vehicular accident:

- Police accident report
- Photographs of the accident scene and accident vehicles taken by the police or insurance adjusters
- The accident vehicles that have been towed to a salvage lot
- Statements of the involved parties and witnesses

A good police accident report is generally fundamental to a good accident reconstruction analysis. In addition to providing the basis information of when, where, and who, a good police report will:

- Diagram the position of the vehicles as found after the accident
- Diagram the tire marks, impact marks, impact debris, and other items found at the scene
- Contain statements made by the involved parties or witnesses directly after the accident
- Contain photographs taken at the time the scene was being "worked" by the authorities

After an accident, the vehicles may be impounded by the local authorities in a salvage yard. This may provide an opportunity to make measurements, perform tests, photograph, and otherwise document the condition of the vehicles. It is a good idea to relate marks or debris noted by the police at the accident scene in the accident report to damage observed and measured on the vehicles. Whenever possible, it is best to view firsthand the vehicles involved in the accident. A person never knows what he or she will readily see that was missed by everyone else (Figure 26.3).

In transporting the vehicles from the accident scene to the salvage yard, additional damages or modifications may occur that are unrelated to the accident itself. For example, doors may have been cut off to extricate victims, or flat tires may have been fixed so that the vehicles could be towed. The vehicles may have been roughly handled, dragged, or dropped to be placed in rows or particular parking slots in the salvage yard. Sometimes, parts are stolen along the way or after the vehicle has been in storage for a time, especially radios, CD players, headlights, grilles, hubcaps, tires, and easily sold items. Such postaccident changes to the vehicles should be noted and segregated from the analysis of the actual accident damage (Figure 26.4).

Some accident reconstructions are done long after the vehicles have been scrapped. Most cities or states retain an accident vehicle only for a short time. In lieu of the actual vehicles, the engineer may have to rely on photographs taken by the police or an insurance

Figure 26.3 Skid marks at accident scene. Barber pole pattern in skid mark indicates tire was rolling forward at same time it was skidding sideways.

Figure 26.4 Distended lamp filament shows that light was on at time of impact; a very useful piece of information.

adjuster. By applying reverse descriptive geometry drafting techniques, it is sometimes possible to determine the dimensions of crush, impact area, and other accident parameters from photographs.

After all the aforementioned data are examined and assessed, it is a good idea to outline the accident scenario chronologically in general terms. Many engineers and investigators do this in reverse because the place where the vehicle or vehicles came to rest is often the most well-established fact. Alternately, some investigators initially write down both the first and last events because the beginning point is often well established also, and then begin to list what happened between the two endpoints.

In any case, a basic accident scenario, or reconstruction, is outlined in general terms, and then details are "worked" into it until the scenario contains all the known facts in a logical sequence. This is done before any engineering analysis is done. Later, when the engineering analysis is done to determine speeds, directions, and other computed data, the computed data are added to the scenario as details. Of course, the computed details must be consistent with the known physical facts as well as the known physical laws and constraints. If not, further modifications may be needed until it is. If the scenario is laid out logically, a major reconstruction faux pas or inaccurate witness statement is relatively easy to detect. The end product of this process is the accident reconstruction scenario.

Basic Energy Method

The basic energy method is simply the application of the conservation of energy law to vehicular accidents. When there are sufficient data, the application of this method makes it possible to mathematically determine the speed of a vehicle just before the accident event.

The conservation of energy law states that in any physical process, the total energy of the system at the beginning of the process is equal to the total energy of the system at the end of the process. The total energy at the end of the process includes any irreversible work done during the accident event. Irreversible work simply means that once some energy is used for a particular process, it cannot be converted back into the kinetic energy it once was. In an irreversible process, the change from one energy type to the other is one way. An example of irreversible work is skidding. The initial kinetic energy of the vehicle is converted to irreversible work as tires rub against pavement and brake pads rub against brake disks. The energy expended during skidding cannot be retrieved to make the car go at its original speed. Equation 26.1 mathematically summarizes the conservation of energy principle in three different formats:

$$E_{start} = E_{end} \qquad (26.1)$$

$$0 = E_{end} - E_{start}$$
$$0 = \Delta E/\Delta t$$

where
   E = Total energy of the system
   t = Time increment
   Δ = Mathematical symbol of an operator that means "the change of …"

The third equation shows that the total change energy during the time that the accident occurred is zero. No energy was added; no energy was removed.

When a vehicle is moving, it possesses kinetic energy. The amount of kinetic energy it has depends on its mass and how fast it is going. The faster it moves, the more kinetic energy it has. Conversely, when a vehicle is not moving, it has no kinetic energy. It follows then that bringing a vehicle to a stop requires that a vehicle's kinetic energy be reduced to zero (Figure 26.5).

Because the total energy at a given speed is constant, reducing a vehicle's kinetic energy to zero requires that it be converted into another type of energy, something other than kinetic. In most accidents, this is done by using the kinetic energy to do some type of irreversible work, such as braking, skidding, or various types of crushing, twisting, or bending.

The kinetic energy of a vehicle, or any object for that matter, is given by the following formula:

$$KE = (1/2)mv^2 \qquad (26.2)$$

where
   KE = Kinetic energy
   m = Mass of the vehicle
   v = Velocity of the vehicle

Because velocity is squared in Equation 26.2, when the velocity doubles, the kinetic energy increases four times. This is why the braking distance for a vehicle traveling at 60 mph (miles per hour) is approximately four times more than the braking distance of the same vehicle traveling at 30 mph. Damage severity and injury severity, like braking distance, also increase in direct proportion to the square of the velocity.

When a vehicle comes to a complete stop in the normal way, its kinetic energy is reduced to zero by applying the vehicle's brakes. The brake pads rub against a disk or drum, and the friction between the two surfaces converts the kinetic energy of the vehicle into irreversible work. Some of the energy is used to abrade away material on the two rubbing surfaces; some of it is converted into heat. As the vehicle's kinetic energy is dissipated, the vehicle slows. When all the energy has been dissipated, the vehicle stops.

In an accident, the initial kinetic energy of a vehicle can be dissipated in many ways. One of the most common ways is skidding. To see how this works, consider the following example of a simple skid.

A car skids 100 ft to a stop on dry concrete pavement. Halfway through the skid, the car

Figure 26.5 Typical truck "duals" skid mark. Straight lines in skid marks indicate tires not rotating much, or rotating and skidding in the same forward direction.

strikes a pedestrian. How fast was the car going just before the driver applied his brakes and initiated skidding, and how fast was the car going when the pedestrian was struck?

Since the definition of work is force applied through a distance, the irreversible frictional work done in skidding 100 ft on dry, concrete pavement is given by the following formula:

$$E_{work} = (mg)fd \qquad (26.3)$$

where

E_{work}	=	Work done by skidding
m	=	Mass of the vehicle
g	=	The acceleration of gravity, 32.17 ft/s^2
(mg)	=	Weight of the vehicle and its contents, mass × acceleration of gravity
f	=	Frictional coefficient between the tires and the pavement, which is about 0.75 in this case
d	=	Distance skidded

If Equation 26.3 is set equal to Equation 26.2, that is, if the kinetic energy of the car before skidding is set equal to the energy dissipated by skidding and then we apply algebra to simplify the results, the following is obtained:

$$\text{Energy at start} = \text{Energy at end} \qquad (26.4)$$

$$
\begin{aligned}
KE &= E_{work} \\
(1/2)mv^2 &= mgfd \\
v &= [2gfd]^{1/2}
\end{aligned}
$$

Thus, the initial speed of the vehicle just before skidding began can be computed by measuring the skid mark. The other values in the formula, g for the acceleration of gravity and f for the tire-to-pavement frictional coefficient, are already known. This formula, or variations of it, is widely known as the **skid formula**. Most police officers who have had some formal accident reconstruction training at an academy are familiar with the use of this formula.

In our example, where the skid mark was measured to be 100 ft long, the initial speed of the vehicle is calculated to have been 69 ft/s or approximately 47 mph just before the skid began.

To determine the speed of the vehicle when the pedestrian was struck, a person can simply work backward from where the vehicle stopped to where the pedestrian was struck and then reapply the skid formula. In this case, the car skidded 50 ft after striking the pedestrian. Thus, the speed of the car when it struck the pedestrian is equivalent to a skid of 50 ft. This computes to a speed of 49 ft/s or approximately 33 mph.

This result is notable. If the driver had been traveling at the posted speed limit of 35 mph or less, it is possible that he might have been able to come to a stop, or almost come to a stop, in 50 ft. Fifty feet is the distance in which the driver responded after recognizing the situation. Because the driver was not traveling at or less than the posted speed limit, but was traveling at 47 mph or perhaps more, he could not stop within 50 ft. Consequently, excessive speed is an important factor in both the cause and severity of the accident. Being struck by a vehicle at 33 mph is certainly very different in severity from being struck at 2 mph (Figure 26.6).

If desired, foot-by-foot velocity figures can be computed concerning the speed of the vehicle during the course of the skid. The graph formed by plotting velocity versus skid distance follows a parabola. For example, at 25 ft from the end of the skid, the speed of the

Figure 26.6 Skipping truck skid marks are usually the result of light trailer load or when trailer brakes alone are used. Measure the whole length; do not omit "skips."

vehicle was 24 mph. At 75 ft from the end of the skid, the speed of the vehicle was 41 mph. In short, at any given point during the skid, the speed of the vehicle can be computed from the skid formula.

Another way that kinetic energy can be dissipated in an accident is by crushing the front end. It takes work to push in the front end of a vehicle. Most people know by experience that if a car impacts an unforgiving, massive brick wall, the faster the car impacts the wall, the deeper the front end is crushed. Although each make and model of vehicle has its own specific crush versus speed relationship, vehicles within the commonly accepted vehicle categories tend to have similar crush versus velocity relationships. The commonly accepted categories for cars in this regard are minicompact, subcompact, compact, intermediate, full size, largest size, and miscellaneous, which is a category for very large passenger cars. Each year, both government and private organizations test vehicles to determine how much they crush for various speeds and conditions. Some agencies process this raw data into a composite rating for crashworthiness, as is done each year by *Consumer Reports*.

Although graphs of crush versus impact speed into a brick wall, or fixed barrier, for each vehicle are sometimes compiled and used directly, it is often more useful to plot crush depth versus the kinetic energy of the vehicle. When this is done, the graph tends to be a flat straight line constant in the usual range of accident speeds.

To show how crush considerations fit into the energy method, let us reconsider the simple skid example. However, this time assume that the vehicle skids 100 ft on dry concrete, strikes a pedestrian at the midpoint of the skid, and then crashes into an unyielding brick wall such that the front end of the compact car is more or less evenly crushed 14 in. (inches) deep across the front.

It is obvious that if the compact car was still moving forward after 100 ft of skidding and then crushed the front end to a depth of 14 in., it must have been going faster that the skid formula alone would indicate for a skid 100 ft long. In this case, the "after energy"

Figure 26.7 Side crush in car. Because car with crush was sitting still, all the crush energy is credited to the other car's kinetic energy.

terms must include both the energy dissipated in skidding and the energy dissipated in crushing the front end of the car (Figure 26.7). The energy expended by hitting the pedestrian is trivial and is ignored.

In this example accident, the car is a subcompact car that has a kinetic energy-to-crush depth coefficient of 5000 pound force feet per inch, and the car has gross weight of 3000 pound force. This means that it takes about 5000 pound force feet of kinetic energy to cause 1 in. of crush across the front of the car.

Thus, the energy balance in this example is as follows:

Kinetic energy = Skid work _+ Crush work

$$(1/2)mv^2 = mgcdf + Kx \qquad (26.5)$$

where
  m = Mass of vehicle (not weight)
  g = Gravitational constant
  f = Frictional coefficient of tire on pavement
  K = Crush coefficient
  x = Average crush depth across front of vehicle

Solving the preceding equation algebraically for v gives the following:

$$v^2 = 2gdf + 2(K/m)x \qquad (26.6)$$

$$3000 \text{ lbf} = mg$$
$$93.25 \ s^2 \text{ lbf/ft} = m$$
$$v^2 = (2)(32.17 \text{ ft/s}^2)(100 \text{ ft})(0.75) +$$
$$2(5000 \text{ lbf-ft/in.})$$
$$(14 \text{ in.})/(93.25 \ s^2 \text{ lbf/ft})$$
$$v = 80 \text{ ft/s or } 54 \text{ mph}$$

From Equation 26.6, it is known that 225,000 foot-pound force of energy was expended by braking and 70,000 foot-pound force of energy was expended smashing into the wall. The total energy "account" is then 295,000 foot-pound force. By reapplying Equation 26.4, it is possible to figure out how much energy is expended when a car skids just 50 ft.

Skid energy expended by skidding 50 ft

$$mgdf = (3000 \text{ lbf})(50 \text{ ft})(0.75)$$
$$= 112,500 \text{ ft/lbf} \quad (26.7)$$

If this amount is then subtracted from the total amount of energy expended in the accident, 295,000 foot-pound force, then the remainder is the amount of energy left after 50 ft of skidding. Setting this equal to the kinetic energy formula allows computation of the speed when the car struck the pedestrian.

Kinetic energy left after 50 ft = total energy – energy expended by skidding 50 ft

$$(1/2)mv^2 = 295,000 \text{ ft/lbf} - 112,500 \text{ ft/lbf}$$
$$v = 63 \text{ ft/s or } 43 \text{ mph} \quad (26.8)$$

Summarizing the second example, the car was initially traveling at 54 mph. After it skidded 50 ft, the speed reduced to 43 mph. This is the speed at which the pedestrian was struck. The car then skidded 50 ft more, collided with a brick wall, and came to a stop. Just before it struck the wall, and after it had skidded a total of 100 ft, the car was traveling at 39 ft/s, or 26 mph. When it struck the brick wall, its speed went from 26 to 0 mph in 14 in.

As an aside, let us estimate the deceleration rate during the collision with the wall. When the car first contacted the wall, it was traveling at 39 ft/s. After 14 in. of crushing the front end of the car, the car came to a stop. Using some basic kinematics, the deceleration is computed as follows:

$$v^2/2x = a \quad (26.9)$$

where
v = Velocity
x = Crush distance
a = Deceleration

$$(39 \text{ ft/s})^2/(2)(1.17 \text{ ft}) =$$
$$a = 650 \text{ ft/s}^2 \text{ or about } 20 \ g\text{'s}$$

Without a seat belt or air bag to provide cushion and absorb the kinetic energy of the driver's body, the driver would otherwise be thrown into the steering wheel, dashboard, and windshield area at a speed of 26 mph. This is the equivalent of a fall from about 24 ft. The driver would then impact the steering wheel and dashboard, and decelerate at $20g$'s (gravitational acceleration) to a stop in 14 in. A couple of additional inches might be gained due to interior crush of the steering wheel. This would be very tough on a person's body. It would likely result in serious injuries or death.

The preceding two simple examples demonstrate the usefulness of the energy method. Because of the variety of damages that can be done by a vehicle in an accident, it is not possible to give examples of all the ways energy dissipation can be calculated in a single chapter. Some of the techniques require a higher level of mathematics than basic algebra. However, the basic essence of the method is this: the initial speed of a vehicle can be computed by accounting for all the irreversible work done by that vehicle to bring itself to a stop. Accurate results are obtained when all the ways in which the vehicle dissipated energy are accounted for.

In cases where not all the work terms can be accounted for, the energy method can still be useful. The method can provide a lower bound for the speed of the vehicle. In other words, if an irreversible work term is left out, the computed speed of the vehicle will be less than the actual speed. Sometimes this is enough. For example, if the lower bound estimate of the speed of a vehicle found that it was going at least 70 mph in a 20 mph zone, knowing precisely the actual speed of 76 mph may be moot.

Basic Momentum Method

Newton's second law defines *momentum* as the velocity of an object times its mass. Typically, the algebraic term for momentum is written as follows:

$$p = mv \qquad (26.10)$$

where

p = Linear momentum
m = Mass of vehicle (not weight)
v = Velocity of vehicle

Force is related to momentum in the following way:

$$F = \Delta p / \Delta t \qquad (26.11)$$

where

F = Force
Δ = An operator which means "change of"
p = Momentum
t = Time

That is, force is equal to the change in momentum divided by the corresponding change in time.

When vehicles collide, as they contact each other they exert equal but opposite forces on the other. This is a consequence of Newton's third law. Because of this, the combination of two vehicles colliding does not cause any change in total momentum. The sum of the net forces between the two vehicles is zero because they are equal and opposite to each other. Thus, as long as there are no external forces being applied on the vehicles, during a collision the net momentum of the vehicles just prior to the collision is equal to the net momentum just after the collision. The preceding is a definition of the conservation of momentum principle.

There are two basic types of collisions: elastic and plastic. A fully **elastic collision** between two bodies is one in which the deformation of each body obeys Hooke's law. Hooke's law states that, within certain limits, the deformation of a material is directly proportional to the applied force causing the deformation. A more practical working definition of an elastic collision is one where the two bodies return to the same shape after the collision that they had before the collision.

With respect to passenger cars, elastic collisions occur only at very low speeds, 2.5 mph or less. Because injuries and damages at this level of speed are relatively small, the analysis of fully elastic collisions between passenger cars is more of academic than practical significance.

In a **plastic** or **nonelastic collision**, the deformation in the contact zone of each body does not follow Hooke's law. A significant portion of the deformation is permanent. Once dented and smashed, it stays dented and smashed. Nearly all vehicular accidents of significance are plastic collisions.

Consider the following simple example. The driver of a 2500-lb car is sitting at a light waiting for it to turn green. The driver has his foot firmly on the brake pedal so that the brakes are fully engaged. A very unobservant driver in a 4000-lb SUV approaches the first driver from the rear and is in the same lane. The driver of the SUV does not see the driver of the car and drives right into him. After impact with the car, the driver of the SUV quickly applies his brakes. The car leaves a skid mark 25 ft long. The SUV leaves a skid mark 13 ft long. How fast was the SUV traveling when it initially struck the car?

In qualitative momentum terms, the accident scenario is as follows. The SUV had an initial speed and momentum. When it struck the car, some of its momentum was imparted to the car, which caused it to lurch forward. The remainder of the momentum stayed with the SUV, but its speed now was slower because it has lost some of its momentum.

Using the skid formula, Equation 26.4, and assuming that the coefficient of friction between the car's tires and the pavement is 0.7, the speed of the car after impact is computed to be 33.6 ft/s, or 23 mph.

Using the same skid formula and the same coefficient of friction, the speed of the SUV after impact is computed to be 24.2 ft/s, or 16.5 mph. Because the 2500-lb car was a little faster than the SUV, the vehicles separated after impact.

The momentum of the car after collision was as follows:

$$p_c = m_c v = (W_c/g)v = (2500 \text{ lbf}/32.17 \text{ ft/s}^2)$$
$$(33.6 \text{ ft/s}) = 2611 \text{ lbf/s} \qquad (26.12)$$

where

p_c = Momentum of car after collision
m_c = Mass of car
W_c = Weight of car
g = Acceleration of gravity
v = Velocity after collision, at the beginning of skidding

Similarly, the momentum of the SUV after collision was as follows:

$$P_{SUV} = m_{SUV} v = (W_{SUV}/g)v = (4000 \text{ lbf}/32.17 \text{ ft/s}^2)$$
$$(24.2 \text{ ft/s}) = 3009 \text{ lbf/s}$$

Prior to the collision, the car had no momentum because it had no velocity. All the momentum was supplied by the SUV. Consequently, the combined momentum of the car and SUV after the collision is equal to the momentum of the SUV alone before the collision. Applying this allows us to algebraically solve for the velocity of the SUV before the collision:

$$\text{Total momentum} = \text{Momentum before} = \text{Momentum after}$$

$$P_{total} = P_{before} = P_{after} \qquad (26.13)$$

$$m_{SUV} v = 2611 \text{ lbf/s} + 3009 \text{ lbf/s} = 5620 \text{ lbf/s}$$
$$m_{SUV} = 124.3 \text{ lbf-s}^2/\text{ft}$$
$$v = 45.2 \text{ ft/s or 31 mph}$$

In short, the SUV was initially going at 31 mph when it struck the rear of the car. The car was initially pushed forward at 23 mph and then came to a stop by skidding, and the SUV slowed down to 16.5 mph due to the impact and also came to a stop by skidding. The car came out of the collision traveling a little faster than the SUV, which allowed the two vehicles to separate.

This example demonstrates not only the use of the principle of conservation of momentum to determine the speed of a vehicle prior to impact, but it also shows how the energy method and the momentum method are often used together to obtain a complete solution. The "after impact" speeds of the vehicles were computed from the skid marks using the energy method, whereas the "before impact" speed of the SUV was computed using the momentum method. One method supplied information for the other.

It is also possible to determine the "before impact" speed of the SUV by measuring the crush depth at the front of the SUV and at the rear of the car, and equating the work needed to create the crush in both vehicles to the "before impact" kinetic energy of the SUV. By doing this, the "before impact" speed of the SUV would then have been computed using two independent methods.

In cases where there is sufficient information to compute the speed of a vehicle using two independent methods, and the two solutions reasonably converge, this tends to affirm that the solution is correct. It is an excellent way of checking the accident reconstruction.

As with the energy method, the preceding simple example of the momentum method does not exhaust the possibilities; there is much more to it than this. Because momentum is conserved in three linear directions and three rotational directions, it is not possible in this short chapter to give examples of all the ways and combinations the momentum method can be applied. Furthermore, the application of the momentum method in more than one dimension requires the use of vectors, centroids, moments of inertia, and vector computational techniques. This cannot all be explained in a brief chapter, and is a course of study by itself. However, the basic essence of the method is this: the initial speed of a vehicle can be computed by accounting for all the momentum transfers from that vehicle. Accurate results are obtained when all the ways in which the vehicle transferred momentum are accounted for.

Tips and Solution Strategies

Vehicular accident reconstruction is not a "clean" discipline. In some cases, people just like you or your loved ones have been killed or horribly maimed. It can be unpleasant to examine the interior of a car involved in such an accident. There may be civil or criminal litigation pending, there are angry or grieving people involved, and no matter what the investigator finds or concludes, there is likely to be an expert from the "other side" who disputes the findings or even disparages the investigator personally. It is a "messy" endeavor. An accident reconstruction is not a simple textbook problem; it is a serious matter involving people's lives, reputations, and fortunes.

It is advisable that a person not reach into places in a car before visually checking them. An extendable machinist's mirror is often handy for this. People who use drugs will often hide their syringes under seats, under mats, in car pockets, or other places where the paraphernalia cannot be seen easily by the cursory look of an officer. It would be unfortunate to be accidentally stuck with a contaminated syringe needle. When discovered, items such as these should be reported to the police agency investigating the accident. The police usually thoroughly check a vehicle for such things, but occasionally they might miss an item.

When drug- or alcohol-related paraphernalia are found, additional testing may be required to determine if the driver was under the influence at the time of the accident. Even if the driver had been tested for drugs after the accident, because there are so many varieties of illegal and legal drugs, it is possible that the driver may not have been given the correct test.

In cases where prescription drug containers are found in the vehicle, they may indicate that the driver was taking medication at the time of the accident. Check the label for warnings and admonitions about drowsiness or operating machinery while under the influence of the medication. A count of the remaining pills compared to the date of issue of the prescription may indicate whether the medication was being taken at the time. The name of the pharmacy is useful; it may be contacted by authorities to develop a medical profile of the driver before the accident.

Some vehicles have weapons hidden in them, especially around the driver's area. Guns, electric stun guns, crowbars, knives, chains, long metal flashlights, and baseball bats are the usual items. These items are mentioned because some of them can harm the person who inadvertently finds them. Any items removed from a vehicle should be documented. When criminal proceedings are pending, the appropriate law enforcement authorities must be notified if any items are removed, and what will be done with them. Usually, evidence chain-of-custody-type documentation will be required.

Other items in the driver's area may indicate whether the driver was fully attentive at the time of the accident. Open maps, open books, open newspapers, cellular telephones in the open position, CB radios with the microphone off the hook, food scattered and splattered on the top of the dashboard and windshield, and makeup items in an open position and scattered on the top of the dashboard may indicate that the driver's attentions were elsewhere. Some cellular telephones indicate the time and telephone number of the last call. This can be useful in determining if the driver was talking on the telephone when the accident occurred. When the physical evidence indicates that there were things going in the driver's field of view, but the driver indicates that the other car "came out of nowhere" or "I never saw him," this is suggestive that the driver's attention was focused on other tasks.

In an experiment conducted by Daniel Simmons of the University of Illinois and Christopher Chabris of Harvard University, two teams of three players in a team tossed around two basketballs for 1 min in a videotape. One team was dressed in white shirts, the other in black shirts. Viewers were asked to count the number of passes made by the white team. After 35 s, a person in a gorilla suit walked into the room, beat his chest, and 9 s later, exited. Incredibly, only 50% of the viewers tasked with counting passes noticed the gorilla. They were focused on counting passes. The effect is called **inattentional blindness**.

Sound level is also an important indicator of attention. Warning signals, such as the horn on a locomotive, train crossing bells, emergency vehicle sirens, and the horns of other vehicles cannot be heard by the driver if the noise level inside the vehicle is too loud. A rule of thumb to use is that for an external noise to be perceived, it should be at least three decibels higher in intensity than the ambient noise level. For this reason, it is useful to check the position of volume control knobs on radios, CD players, and similar, and verify whether the side windows were up, the air conditioner or heater was on, and so on.

In intersection-type accidents, the width of a person's peripheral vision can be important. A person who regularly only perceives objects 50° on either side of dead ahead may not be cognizant of a vehicle that is 60° from the center of his vision and is approaching fast. It is useful sometimes to chart the visual angles of the vehicles involved in an accident to determine who could see whom, and when.

The vehicle identification number (VIN) is usually located at several points in a car, most notably at the base of the windshield, left-hand corner of the driver's side, and also on the inside of the driver's side door. With a vehicle's VIN it is possible to (1) determine whether it has been brought back to the dealership to have a recall item fixed, (2) determine the maintenance that has been done on the vehicle at authorized dealerships, (3) determine at which dealerships the maintenance was done, or (4) track whether the vehicle has been stolen or involved in previous accidents.

Coded into the VIN is the vehicle type, the factory where it was made, and the year and week it was built. Most car companies encode the VIN in a bar code so that it can be read by a scanner from the outside of the windshield.

A common question that is posed to an accident investigator is, who was driving at the time of the accident? The question can be especially important when there is only one survivor. The estates and families of the deceased occupants may wish to sue the driver of the vehicle for wrongful death, negligence, or something similar. It may be that the survivor was found after the accident to be very drunk and is subject to criminal prosecution if he actually was the driver.

When the occupants in a car do not wear seat belts, the identification of the driver is generally easy. Often at impact, the unbelted driver will be hurled against or through the windshield in the direction of impact. If the driver was hurled against the windshield but the windshield did not break, it is probable that hair and tissue of the driver will have been left in the "spider" crack pattern in the windshield where his head or face impacted. If the driver was the only one in the car with blond hair, for example, it can be as easy as examining the hair found in the spider cracks under low magnification to verify color.

Because the driver has the steering wheel in front of him, if he is not seat-belted, the steering wheel will often cause easily identifiable injuries. Bruises in the shape of a steering wheel are an obvious giveaway. Where the steering wheel has been bent or damaged, the driver will often have corresponding injuries.

If a driver is injured and bleeds in the car, the blood DNA of the stains and a profile can be determined and matched to the occupants. This type of chemical analysis, however, requires the services of a specialized laboratory, which may not be readily available in some areas. It is important not to inadvertently cross-contaminate the sample locations so that this evidence is not disputable.

The position of the other occupants at the time of the accident may be determined by reviewing each person's injuries with respect to his or her trajectory through the vehicle after impact. For example, a person in the back seat may be thrown forward between the two front seats headfirst into the dashboard. This person would have head and shoulder injuries consistent with this trajectory.

Sometimes, the cause of a crash or accident is blamed on the failure of an important metal part. It may be claimed that the part failed first, which then precipitated the accident.

In such cases, it is important to examine the fracture and determine its metallurgical characteristics. A hardness test will determine its approximate strength and provide information as to whether it has been heat-treated. The type of fracture is also very important. A fatigue-type fracture indicates that the part has been partially cracked for some time before the accident. Thus, it can be argued that the part was weak at the

Figure 26.8 Distended lamp filament used to prove that left side, front signal light was on at time of accident.

time of the accident and susceptible to failure (Figure 26.8).

A shear fracture, tensile fracture, or bending fracture with no indication of fatigue usually indicates that the part failed at the time of the accident, perhaps being damaged as part of the accident process. This is especially true when the fracture surface is new and no evidence is present that the fracture existed prior to the accident, such as rust, corrosion, or extensive discoloration due to exposure to exhaust fumes or oil.

Sometimes, when a car has undergone previous repair work, the mechanic may have used an inferior replacement part. Nuts and bolts, for example, are not all the same simply because they fit. Some nuts and bolts are made of stronger material than others. If a high-strength nut or bolt is replaced with a common grade item, it is possible that the replacement will not perform properly. It may back off, rust, or corrode prematurely, or simply fail due to weakness. There are known cases of "bootleg" nuts and bolts. These are components that display the markings of a high-strength component, but are actually low-strength, cheap substitutes designed to bilk the buyer.

In a similar vein, improper weld repairs may sometimes precipitate mechanical failure. Many vehicles have frames made of high-strength alloys that have been heat-treated. If a body shop attempts to weld these alloys in the usual way, the weldment will not have the same characteristics as the original metal. There will also be metallurgical changes in the base metal immediately next to the weldment, the heat-affected zone (HAZ). Usually, the material in the HAZ is weaker and more susceptible to fatigue than the original base material. This can often be detected by a hardness test profile across the suspect weldment.

In some older foreign cars, using the wrong kind of brake fluid can precipitate an accident. Older European cars often used natural rubber for seals and o-rings in brake systems. U.S. cars and vehicles use neoprene-type synthetic rubber for the corresponding parts. The brake fluid used in most U.S. cars can sometimes dissolve the natural rubber-type parts, causing them to leak and fail. Thus, in some cases, the wrong kind of brake fluid could cause an accident.

Tires are often blamed for causing accidents. The most common failure occurs when a tire is underinflated. When this occurs, the bending stresses in the tire wall are greatly increased, and the tire carcass heats up. Because the strength of the rubber and carcass is sensitive to temperature, the tire wall will have less mechanical strength at the elevated temperature. The combination of high temperature and increased bending stresses causes the materials in the tire to fatigue. Eventually the tire fails, often by a blowout.

A tire that blows out due to underinflation will usually have a blowout in the tire wall. It often will be located at the point of maximum bending in the tire wall. If a tire has been run underinflated for a time, the point of maximum bending can often be clearly observed by discoloration of the tire wall in that area. There will be a boundary line that runs around the tire at a more or less constant radius from the center. It will be dark on one side of the line and normally colored on the other.

The "hardness" of a tire or other rubber-type product, including many plastics, is measured with a durometer. If a tire has been run underinflated, the durometer number of the heat-affected tire will be measurably lower than that in areas that were not subject to the excessive bending stress and heat buildup. A durometer profile of the tire can be taken across the tire to map out the extent of softening that occurred before the blowout.

When a tire blows out due to underinflation, the point where the blowout occurs will exhibit material **fatigue**. The cord fibers that first failed will be frayed unevenly and will have a disheveled, even dirty appearance. This will be in contrast to fibers that simply snapped in tension later on when the tire failed generally. Where the bending stresses are highest in the tire wall, usually at a point closest to the exterior surface, the fibers will have come loose from the rubber matrix and may have even "wallowed" around in the matrix. This is in contrast to the fibers in the rest of the tire, which may still adhere fast to the matrix.

When a tire fails some time before the accident, it will often be damaged by having been run flat. There may be cuts in the tire from the rim, and even holes where the pavement abraded away material if the brakes were locked. If the wheel continued to roll some time after the tire lost air, the rims may be damaged and rim marks may be visible in the pavement. The presence of rim marks along the accident pathway of a vehicle is often used to determine when a tire went flat.

Impacts to tires by potholes, curbs, and roadway edges can be severe enough to cause a tire to lose air and deflate. Often, the rim will be bent where the impact initially occurred with the curb or whatever, and this will also correspond to tire damage at the bead. Because most late model cars and trucks use tubeless tires, damage to the tire bead by impact will usually cause the tire to lose seal and deflate.

Examination of the tire tread can also provide significant clues as to whether the tire was consistently run with low or high pressure. A consistently underinflated tire will have greater tread depth in the middle than at the edges. Conversely, a consistently overinflated tire will have less tread depth in the middle than at the edges.

If a tubeless tire is not mounted properly, it is possible for the tire to have a "hung bead." This is where the bead of the tire is not properly seated on the bead seat of the wheel. However, it may be sufficiently seated to hold air during initial mounting and low-speed driving. The tire may suddenly fail during cornering, making a curve at higher highway speeds, or simply encountering typical bumps in the road. An excellent study of this mode of failure is contained in the publication, *Research Dynamics of Vehicle Tires*, Vol. 4, by Andrew White, which is listed in the reference section.

Naturally enough, when a tire fails due to punctures from something on the road like a nail or sharp object, the puncture will occur in the tread portion of the tire. Dismounting of the tire and inspecting the tire interior may occasionally find the article responsible for the puncture embedded within the tire itself. Damage from a sliding contact with a curb or curb offset will usually cause damage to the outer sidewall of the tire.

Point of impact is often the most contested issue with respect to the scene examination of an accident. In some states, it is not even permitted to discuss in court the issue of point of impact or to even use the term. This is because of questionable past practices in accident reconstruction that used what is often called the debris method.

The debris method is simply an examination of the scene for debris on the ground at the accident scene, for things like headlight glass, turn signal lens parts, and other parts that were damaged on the vehicle. The general area where these items are found is mapped out and the center of the area is sometimes considered the point of impact (POI) (Figure 26.9).

Figure 26.9 Skid pattern showing car was rotating counterclockwise as it crossed over to the oncoming side and struck a tree.

This is a poor method because it does not take into account the difference in speeds between an impacting vehicle and the times separated from the vehicle by impact. The debris items may actually travel and come to rest several yards from where actual impact occurred. Because of ricochets and glancing impacts, debris can be hurled in directions that do not correspond to any vehicle's direction of travel. However, as practiced, the method often assumes that accident events occurred more or less in the same area that the debris was found.

Sometimes, when a collision has damaged the radiator or engine, the position of spilled liquid would be called the POI. In itself, this is also poor practice. It may take some time for fluid to drip down onto the pavement. In the time it takes to begin flowing and drop down, the car may have moved significantly away from the actual POI. The position of pooled fluids does, however, mark the location of the vehicle after it has come to rest and can be a useful reference point after the vehicles have been removed from the scene.

A more accurate method for determining the POI is to look for gouge marks or telltale tire marks. A skid, rim mark, or other tire mark that suddenly deviates in a new direction can be a good indicator of the POI. A severe head-on collision will sometimes cause the front tires to become flat due to downward forces. A skid mark that terminates at a short rim mark or rim gouge will often denote the POI in such cases.

When impact occurs, other parts of the car may be damaged and be pushed down to the pavement, making gouges at or leading away from the POI. Tie rod gouges and bumper gouges are common in severe collisions in which the parts were immediately detached and forcefully pushed down into the pavement (Figure 26.10).

When assessing accidents involving trucks, weigh stations can sometimes provide useful data. When a commercial truck is checked at a weigh station, not only is its weight checked but also the time this is done is registered on the weigh ticket. This places the vehicle at a known location at a known time.

Figure 26.10 Paint transfer marks in contact are "tag" marks, and can be chemically compared to the other vehicle to confirm who hit whom.

This is useful if the truck is then involved in an accident after leaving the weigh station, and it is suspected that the truck was speeding. When an accident is reported, the time is recorded by the dispatcher. Furthermore, when an officer arrives at the accident scene, he or she usually checks and determines how long before the report of the accident was made the accident occurred.

Thus, combining the time and location from the accident report and a weigh station ticket provides an accurate base for determining the average rate of speed of the truck between the two points. Obviously, if the average speed is above the speed limit, there must have been more distance driven above the speed limit than below it. This is especially useful if the truck driver has claimed during the accident interview that he drove the whole distance from the weigh station to the accident scene at or under the posted speed limit.

Since 1983, car bumpers have been designed to prevent physical damage to the front and rear ends of vehicles in low-speed collisions. The design requirement per 49CFR Part 581 is that it protect the vehicle from damage in a fixed-barrier impact of 2.5 mph across the length of the bumper, or 1.5 mph at the corners. This is assumed to be the equivalent of a 5 mph impact with a parked car of about the same mass. This requirement only applies to passenger cars. It surprises many people to learn that this does not apply to SUVs, pickups, or minivans.

The standard first came into effect in 1972. The original standard required cars to withstand a 5 mph front and 2.5 mph rear impact across the bumper with a fixed barrier, without significant damage occurring to the vehicle. This was changed in 1979, and was changed again in 1983. Bumpers made in the years 1979 to 1982 are sometimes called Phase I bumpers, because the regulation was called Phase I. Similarly, bumpers made since 1983 are called Phase II bumpers for the same reason. Similar Canadian and European bumper standards are slightly different.

This tidbit of information concerning U.S. bumper standards can be useful in diagnosing low-speed collisions. The 2.5 mph fixed-barrier impact requirement is equivalent to stating that when impact occurs across the length of the bumper at the front of the car, it can elastically absorb the kinetic energy the car possesses at 2.5 mph without significant damage occurring. Although the National Highway Traffic Safety Administration does not conduct low-speed tests on vehicles, the Insurance Institute for Highway Safety conducts low-speed crash tests on selected models each year.

Accident avoidance strategy is where it is assumed that certain parties to the accident traveled at the legal speed and performed reasonably in all their actions including braking, turning, and accelerating. This presumption is then compared to the actual events that occurred. If there is a significant difference between the two scenarios, the difference lends itself to an explanation of what occurred.

Recent car models are now being equipped with "black box" recorders.

When air bags began to be installed in cars in the late 1970s, a "trigger" computer chip was also installed to tell the air bag equipment when to deploy. Initially, this was a simple accelerometer connected to a logic circuit. When the acceleration, or change of speed per change of time, exceeded a certain value, as might occur if a car were impacting a telephone pole, the chip would send a signal that caused the air bag to deploy. Later, to ensure that the bag would not deploy inadvertently, additional parameters were added, e.g., bumper movement or brake pedal actuation, to ensure that the car was really involved in an accident.

This basic technology has rapidly evolved, however, as manufacturers such as Ford and GM have realized the possibility of obtaining real-world performance information about their safety equipment directly from the field. Most current air bag computer chips can record the performance of a vehicle during the last 5 seconds before a crash. Parameters commonly recorded include vehicle velocity, engine RPM, throttle position, seat belt engagement, and brake actuation. All these parameters can be downloaded through the vehicle's diagnostic patch and plotted against time. The information is not only useful for automobile designers to improve the performance of their products, but it is also very useful in court cases.

In August 2003, for example, despite various legal objections, police in South Dakota obtained and used black box data from a Cadillac as evidence in the hit-and-run trial of Congressman Bill Janlow. Owing to the publicity this case and other similar high-profile black box cases have received, black box data is now often subpoenaed and is used in lieu of engineering analysis, or as a supplement to it. The black box is perceived by many as an impartial third party witness to the accident.

Many new cars and trucks also come equipped with combined global positioning system (GPS) and automobile cellular telephone devices, such as GM's "On-Star," that automatically indicate to a central office when and where an accident has occurred. When such an accident occurs, a dispatcher is alerted by a signal and calls the vehicle, makes inquiries, and summons the appropriate help. The conversations of the dispatcher with the driver and the authorities that are summoned is recorded. This conversation may be subpoenaed as evidence in a later trial.

Similarly, many new and old vehicles are now equipped with GPS direction finder devices. New vehicles may have built-in units, but older vehicles may be equipped with after-market units made by the same companies that perform the same function. Many of these devices are essentially GPS data loggers. Depending on how they are programmed or configured by their owners, these devices can preserve and provide plots of the vehicle's speed and position versus clock time during the trip. Rather than recording just 5 seconds of dynamic data prior to an accident, these devices can record a vehicle's speed and position versus time for an entire trip. There have been recent court cases where such GPS recorded data has been used to disprove speeding tickets. In one case, the driver disputed a speeding citation based on a radar gun measurement by a police officer. The driver claimed that the radar gun was out of calibration and measured incorrectly. The driver won his case by offering alternative data: his own GPS device's record of his vehicle's speed at the time and location the officer was using the radar gun.

In short, vehicular accident analysis is rapidly becoming automated. The ability to record on-the-spot statements about an accident from the drivers and passengers; the ability to exactly note the location and time of an accident; and the ability to apparently impartially provide speed and location information about both vehicles seconds even hours before an accident occurred, before the police or emergency personnel even arrive at the scene, not only currently exists, but is becoming routine.

Questions

1. In 2000, the death rate was 1.6 deaths per million miles. In 1977, the death rate was 3.3 deaths per million miles. The death rate has been halved in almost three decades. How was this done?

2. A car skidded on a roadway that has a coefficient of friction with the road of 0.7. A police officer measured the skid marks and noted them to be 100 ft long. Two years later at a trial, the lawyer for the driver argues that the officer did not measure the skid marks correctly. He asserts, based on photographs, that the skid marks were no more than 92 ft long. Is the difference significant?

3. Improper driving is consistently responsible for approximately two-thirds of all vehicular accidents. What are the three primary categories of improper driving that cause accidents?

4. If improper driving is responsible for approximately two-thirds of all vehicular accidents, what is responsible for the other one-third?

5. What are the two most important analytical tools that an engineer uses to evaluate a vehicular accident?

6. In analyzing a vehicular accident, reliance on verifiable physical evidence should be primary. Give examples of verifiable physical evidence.

7. In most vehicular accident cases, eyewitness testimony is usually gathered and recorded. Explain why this type of evidence should be evaluated with caution.

8. Apply the energy method to solve the following. On low beams, a car has headlights that clearly illuminate the roadway 155 ft ahead but no farther. If the coefficient of friction between the road surface and the car's tires is 0.65, how fast can the driver go and still be able to brake to a complete stop when his lights first show a deer in the roadway?

9. Apply the momentum method to solve the following problem. A fully loaded semitractor trailer that weighs 84,000 lb collided head-on with a compact car that weighs 2500 lb. Both vehicles were going 60 mph when the collision occurred. What was the speed of the truck and car right after the collision?

10. Car A, weighing 3217 lb and traveling at 30 mph, impacts the rear end of car B, which also weighs exactly 3217 lb. Car B was not moving; it was stationary. The two cars "stick" together after impact. During the impact, the driver of car B mashed on his brakes and left a skid mark. The driver of car A allowed his car to roll freely after impact; the engine was no longer powering his car, and his car did not leave any skid mark. How long was the skid mark of car B? The coefficient of friction is 0.7. Ignore energy dissipation that might occur in crushing damage to the two cars.

11. Alcohol decreases in a man's blood at a rate of about 0.017 percent per hour. Bill has an accident, and his blood alcohol level is checked about two hours after the accident at a local hospital. During that time he is in police custody and does not drink anything. The analysis determines that his blood alcohol at the time of testing is 0.05%. If the legal limit is 0.08%, was he legally drunk at the time of the accident?

12. Alcohol decreases in a woman's blood at a rate of about 0.015 percent per hour. Susan has been drinking and has a blood alcohol level of 0.10 percent, as determined by a breath analyzer administered by a barkeeper. Being a responsible driver, Susan does not want to drive while legally drunk. How long does she have to wait to have a blood alcohol of less than 0.08 percent?

13. At time A, a car is 300 ft from an ungated but marked railroad crossing and is traveling at 30 mph (44 ft/s) along a gravel road that is perpendicular to the train tracks. Because the area is sparsely populated, a train is approaching from the driver's left side at 60 mph (88 ft/s). At time A, the train is 600 ft from the railroad crossing.
    a. If the driver has a peripheral vision of 50° on either side of dead ahead, and looks straight ahead, will he see the train at time A?
    b. At time A, how long will it take the train and car to meet?
    c. Under these conditions, will the driver ever see the train if he continues to look straight ahead?
    d. If the driver hears the train's horn, how far to the left does he have to turn his head to see the train?
    e. If the driver perceives the danger and begins braking after 0.5 s, which is a typical time for middle-aged men, in what amount of time can he stop safely? Assume a coefficient of friction with the road of 0.6.
    f. If the sound level of the train's horn is 95 decibels at 25 ft, and drops 6 decibels each time the distance is doubled, can the driver hear the train's horn just prior to reaching the minimum distance for safe braking?
    g. What could the driver do to avoid the accident?

14. Data taken from a vehicle's black box are as follows. At 0 seconds, the air bag was deployed upon impact with another vehicle. Assume that the coefficient of friction with the pavement is 0.72.

Item	4 s	3 s	2 s	1 s	0 s
Speed	71 mph	70 mph	69 mph	54 mph	39 mph
Brakes	Off	Off	Off	On	On
Throttle	100%	100%	100%	60%	5%
RPMs	3000	2960	2935	1750	750

    a. Was the driver braking hard just prior to impact?
    b. When he started braking, did he take his foot off the gas?
    c. When do you think he actually perceived there was a danger?

15. A car reportedly traveling at 70 mph is struck by another vehicle and immediately begins leaking radiator fluid. The spot where radiator fluid is first noted on the highway is called the point of impact by the officer at the scene. Prepare a scientific case explaining why this is a bogus call by the officer.

Suggested Readings

Baker, J. and Fricke, L., *Traffic Accident Investigation Manual,* 9th ed., Northwestern University Traffic Institute, Evanston, IL, 1986.

Consumer Reports, www.consumerreports.org.

DUI Statistics, www.geocities.com/Heartland Plains/3121/statistics.html.

Green, P. E., *Bicycle Accident Reconstruction: A Guide for the Attorney and Forensic Engineer,* 2nd ed., Lawyers and Judges Publishing, Tucson, AZ, 1999.

Langhaar, H.,*Energy Methods in Applied Mechanics,* John Wiley and Sons, New York, 1962.

Limpert, R., *Motor Vehicle Accident Reconstruction and Cause Analysis,* 2nd ed., Michie Company, Charlotteville, VA, 1984.

Lind, V. G., *Principles of Physics,* Saunders College Publishing, Philadelphia, 1994.

National Joint Committee on Uniform Traffic Control Devices, Manual on Uniform Traffic Control Devices for Streets and Highways, U.S. Department of Commerce, Bureau of Public Roads, Washington, D.C.

National Safety Council, *Accident Facts,* Itasca, IL, published annually.

Noon, R., *Engineering Analysis of Vehicular Accidents,* CRC Press, Boca Raton, FL, 1994.

SAE, *Accident Reconstruction: Automobiles, Tractor-Semitrailers, Motorcycles, and Pedestrians*, published periodically by the Society of Automotive Engineers.

SAE, *Field Accidents: Data Collection, Analysis, Methodologies, and Crash Injury Reconstructions,* Society of Automotive Engineers, published periodically by the Society of Automotive Engineers, Collection of papers presented at symposium, Detroit, MI.

SAE, *SAE Handbook,* Vol. 4, On-Highway Vehicles and Off-Highway Machinery, Society of Automotive Engineers, 2000.

Shermer, M., None so blind (introductory article about inattention blindness), *Scientific American,* March 2004, p. 42.

The Economic Cost of Motor Vehicle Crashes, Blincoe, National Highway Traffic Safety Administration Technical Report, NAD-40, 1994.

The World Almanac and Book of Facts, World Almanac Books, New York, 2008, published annually.

Tong, N. and Lantz, S., Eds., *Symposium on Vehicle Crashworthiness Including Impact Biomechanics*, ASME, AMD Vol. 79, BED-Vol. 1, 1986.

U.S. Department of Transportation, Work Zone Traffic Control, Standards and Guidelines, Federal Highway Administration, 1995.

White, A., *Research Dynamics of Vehicle Tires,* Vol. 4, Research Center of Motor Vehicle Research of New Hampshire, 1965.

www.nhtsa.dot.gov/cars/problems, Web page for the National Highway Traffic Safety Administration.

www.autosafety.org, Web page for the Center for Auto Safety.

www.hwysafety.org, Web page for the Insurance Institute for Highway Safety.

SECTION V

Cybertechnology and Forensic Science

27

Informatics in Forensic Science

Zeno Geradts

Introduction

In forensic science, the influence of informatics is more evident then ever before. The use of the Internet is integrated in society, and databases of forensic evidence (DNA, fingerprints, cartridge cases, etc.) are linked for solving crimes—even international crimes—and for scientific research.

This chapter provides more background information on the ways the world of forensic informatics has evolved, and because we are in the middle of a rapid process, there will be an overview of the future expectations for forensic informatics, including the use of data mining in forensic science. This chapter will not cover **digital evidence,** because that is discussed in Chapter 28.

Forensic Databases

Forensic databases have been known for years in linking and solving crimes. In the past, most of these databases were in files with an appropriate classification; however, in the past decades, many of these files have been computerized. The fingerprint databases are well known for their effectiveness. However, there are numerous other databases such as handwriting, firearms, cartridge cases, bullets, DNA, paint, etc. Some examples of databases are discussed in this chapter.

Databases for handling evidence as Laboratory Information Management Systems (**LIMS**) are used in forensic laboratories and should be structured correctly to gain the maximum benefit from these.

Image Databases

In this chapter, we will focus on image databases of:

- Fingerprints
- Handwriting
- Cartridge cases
- Toolmarks
- Shoeprints
- Drugs and tablets

In all these databases, the image has to be searched through the Query By Example (QBE) framework for formulating similarity queries over the images.[1] In QBE, a user formulates a query by providing an example of the object that is similar to the object the user would like to use. The oldest database in forensic science is the fingerprints database. Although there is a lot of information on these databases, in this chapter we discuss some background of the fingerprint work before continuing with the separate databases.

Fingerprints

The fingerprint databases are the most well known and have been used for many years. The techniques are similar to those of other image database search methods. First, the image acquisition is done, then the features are extracted from the digitized pattern and stored in a database. A matching algorithm is used in the database; an authentication decision is made, depending on similarity.

In the commercial systems, four major approaches are being used:[2]

- *Syntactic approach:* The ridge patterns and minutiae are approximated by a string of primitives.
- *Structural approach*: Features based on minutiae are extracted and then represented using a graph data structure. Use of the topology of the features does the matching.

- *Neural networks approach*: A feature vector is constructed and classified by a neural network classifier.
- *Statistical classifier approach*: Statistical classifiers are used instead of neural network classifiers.

For fingerprint, digital image processing is used, for example, for filtering out regular patterns from a fingerprint by using the fast Fourier transform (FFT).

Handwriting

For forensic handwriting comparison, different systems[3] exist on the market. The oldest system is the **Fish system**, which was developed by the Bundes Kriminal Amt (BKA) in Germany.

Another system is Script, developed by TNO in the Netherlands. In both systems, handwriting is digitized with a flatbed scanner, and the strokes of certain letters are analyzed with user interaction. The system uses contents semantics of the handwriting for the features. For each image, the **script system** measures interline distance and the interword distance. For each word, all letter heights, the slant, and the word width are measured. For Fish, the black/white statistics are computed, and upper and lower loops are analyzed, if available. These systems are used for threatening letters and analyzing who might have written them (Figure 27.1).

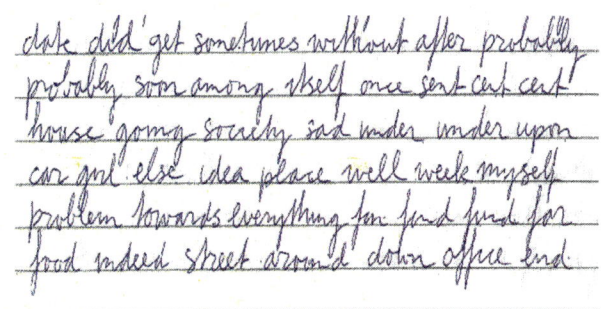

Figure 27.1 Example of handwriting.

Cartridge Cases

When a firearm is loaded and fired, the mechanisms and parts in the firearm that come into contact with the cartridge case cause impressions and striations, which can be characteristic for the firearm being used. The striation marks on bullets are caused by the irregularities in the firearm barrel as well as larger and more distinct lands and grooves of the rifling.

The ejected cartridge case shows marks (Figure 27.2) that are caused by the firing pin and the breech face as the cartridge is repelled back in the breach by the force of rifling. The feeding, extraction, and ejection mechanisms of the firearm will also leave characteristic marks. These marks on cartridge cases and bullets are compared with the test-fired bullets. Often, the cartridge case is the most important forensic specimen in the identification of weapons, because bullets are commonly deformed by the impact. The examiner can also determine, using class characteristics, what kinds of firearm or which make and model have been used.

There is a commercial system with databases that can be used for acquiring, storing, and analyzing images of bullets and cartridge cases. In the United States, the systems are in a network named the National Integrated Ballistic Information Network (NIBIN), which is based on the IBIS software.

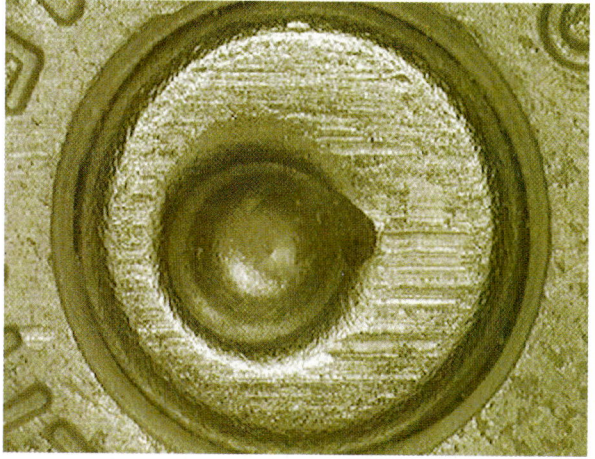

Figure 27.2 Cartridge case with breech face and firing pin marks.

The systems capture video images of bullet striations and of the markings left on cartridge cases in a standardized way. These images are used to produce an electronic signature that is stored in a database. The system then compares this signature to that of another fired bullet or cartridge case—or to an entire database of fired bullets and cartridge cases.

Both systems have image matching algorithms. The methods of image matching applied in these systems are not known.

For evaluation of these systems, it is important that the user have a hit list (the order of the most relevant image matches that are displayed) that is reliable. For Bullets, BulletTrax 3D http://www.forensictechnologyinc.com/p2-1.html has been developed, which compares the three-dimensional (3-D) image of the bullet to a database.

Toolmarks

In toolmark examination, it is possible to determine if a tool has made a toolmark on the basis of striation marks or impression marks. Often, the tool (e.g., a screwdriver) has characteristic damage and manufacturing marks. These characteristics allow a comparison of a test mark made with the tool and the toolmark found at the crime scene. Often, toolmarks are found at burglaries. For this reason, the police will have a database of toolmarks to link the various burglaries that have been committed by a suspect.

In this part of the research, we at the Netherlands Forensic Institute handled striation marks. For comparing the striation mark, we take a signature of the mark itself. This signature can be made by taking 50 vertical lines and following the striation mark of a screwdriver itself with an algorithm (Figure 27.3).

In this way, the user has to determine if the striation mark has been sampled correctly. Furthermore, tests are made with structured light[4] to get a 3-D image of the striation mark. It appears that the shape of the tool can be digitized, which might be easier for faster comparison because otherwise all striation marks

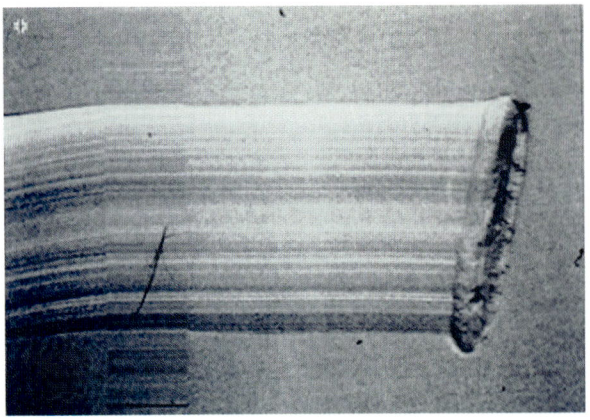

Figure 27.3 Sampling a signature of the striation mark of a screwdriver.

should be compared to one another, where the shift and partial zooming has to be taken into account.

In the structured light approach, a sequence of different projection structures is created with a computer-controlled micromirror projection. The entire procedure of projecting and grabbing the sequence of images takes a few seconds. The resulting images (object views) are processed and interpreted finally as one object view. In this resulting view, the stripes carry the desired code numbers. The border lines in the image and calculation of the X, Y, and Z coordinates of the object surface points are defined. The result can be viewed as a depth map (or image); it represents a 3-D sampling of the surface. The accuracy and density of this method—known as *Gray Code Projection*—depend on the largest number of stripes in the grids that could be handled by the camera system and by the accuracy of the geometric calibration. This information can be processed and visualized with 3-D data. This kind of database is not often used in practice because commercial efforts were stopped owing to the expenses (labor cost at the laboratory) and the market situation.

Shoeprints

Shoeprints are frequently found at the crime scene. It is hard for a suspect to hide shoeprints because often they will be latent in dust and can be visualized with various techniques. For shoeprints, several databases might be useful in forensic investigation:

- Database of shoeprints found at the crime scene
- Database of shoes of suspects
- Database of shoes on the market

The database of shoeprints found at the scene of the crime can be compared with the database of shoes of suspects. Because people change their shoes (because of wear), this database could be limited to 6 months to be useful in most cases. The database of the shoes that are on the market can be useful for finding which brand and model the perpetrator might have worn. This database should be built centrally because of the time it takes to construct.

In Figure 27.4, the data entry screen is shown for the system for shoes. For ease of processing and standardization, it is necessary to have a test print of the shoe in the database. Furthermore, we developed a standard classification system with the police, containing all shapes that can be found on shoes, such as circles, triangles, and the kind of classes. In each class, there is a subclass to specify the shape of the subclass.

The classification takes time, and examiners differ in their method of classifying shoeprints; attempts have been made to automate this process. The Forensic Institute's staff has tested several algorithms for automation.

First, the shapes should be segmented into different visible shapes, and then classified automatically. For basic forms (triangles, circles, etc.), this method appeared to work with Sammon Plots of Fourier Descriptors.[5] The decision for the shape recognition is done with a neural network. There currently is an interesting research project on the classification of complete shapes by fractals.[6]

To automate this kind of search for real shoeprints, much more work has to be done. Often, real shoeprints are faint in their appearance. Most often, these searches will result in a list of interesting shoes, which might be compared manually for damage and other features that could result in identification. The shoeprint database SICAR[7] is used in practice in the United Kingdom. Efforts in the Netherlands, Germany, and Switzerland have stopped due to expenses involved with these projects. Because shoeprints can be

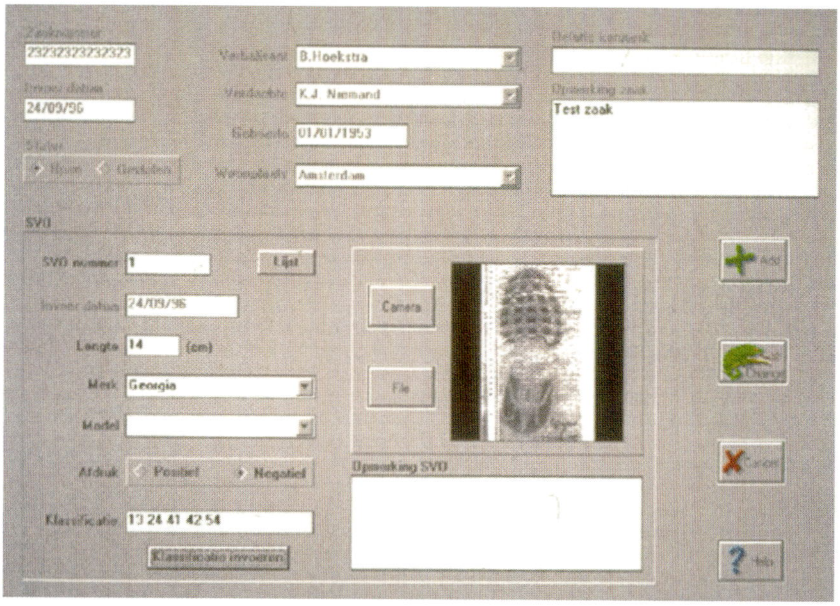

Figure 27.4 Data entry screen for suspects' shoes.

important as evidence in forensic science, it is expected that more databases will be available in future.

Database of Drug Tablets

Drug tablet (e.g., XTC) cases are submitted to the laboratory for forensic investigation. For information about manufacturers of illegal drugs, a database is available with images of the drug tablets and information about the chemical composition. Often, the manufacturer of the tablets will have stamps with logos on them (Figure 27.5).

The correlation method for this database should be easy to use, and invariant to rotation of tablets in the database. As tablets are depicted with a standard camera with a sidelight source, two images of the same pill may differ due to light variations. Another problem is that the tablet or the logo can be damaged, and the logo is not visible anymore. Because the logo is three dimensional and is captured with a regular camera, the resulting two-dimensional (2-D) image has to be compared. The correlation method should be insensitive to the preceding factors. Currently, several efforts are under way to build such a database.

Discussion

There are many different approaches to searching in image databases. For each kind of forensic image database, an appropriate method should be selected. A database for faces (e.g., L1ID[8]) is commercially available. The development of new algorithms that are faster and the implementation of search algorithms in hardware might provide new developments in these systems.

Biometric systems will provide more information for forensic science. An example of biometry is fingerprints; however, there are many other kinds of research projects going on, for example, gait, earprints, faces, and irises. For forensic investigation, databases can be important to validate the conclusions that are being drawn. In court, forensic evidence is sometimes challenged, and there should be more research to have a larger statistical background on uniqueness.

DNA Databases

DNA databases have been very successful. In the United Kingdom, there are large

Figure 27.5 Database of drug tablets.

databases for solving burglaries and other crimes.[9] The available database depends on the laws in a country that dictate the types of crimes for which DNA can be stored in a database (e.g., CODIS [Combined DNA Index System][10]). Databases for known criminals of sexual crimes are effective. Furthermore, many homicides and missing persons cases have been solved with these databases, which will become more important in the coming years as they are being filled with new data. Efforts have begun to link these databases between countries and states.

Paint Databases

Paint databases of cars can be used to determine the brand and model of car involved in an accident. Because the composition of the paint used by manufacturers differs, this information has been used for many years. These databases are exchanged internationally, and the Bundes Kriminal Amt in Wiesbaden has the largest known collection of car paint in their database. For comparison, it is important to have a standardized procedure for the analysis.

Forensic Archiving of X-Ray Spectra

For the forensic archiving of x-ray spectra, the Spectra Library for Identification and Classification Engine (**SLICE**) was developed under contract with the Federal Bureau of Investigation.

SLICE was designed to provide several utilities not available on commercial EDS (energy dispersive spectroscopy) suites. Most distinctive is the ability to archive spectra and narrative information in a true database architecture. Additionally, spectra can be displayed in nested fashion, permitting critical visual comparison. From SLICE, the future spectral archive of Combined Storage of Energy Dispersive X-ray Spectra (CODEX) will be accessed. CODEX is the working name of the database distributed to participating laboratories, including a master database, associated infrastructure, protocols, etc. The concept is similar to the existing utility CODIS for DNA.

SLICE is the user utility installed on a personal computer associated with the EDS analyzer on the scanning electron microscope.

Video Image Processing and Animation Software

Because of the number of installed camera surveillance systems, it is common for a crime to be recorded on a CCTV system. If this occurs, the CCTV images can be used as evidence in court. The court often asks for image processing to get a clearer image of the video, and of comparison of the video images with a suspect. Furthermore, the court sometimes asks if an image has been manipulated. In the past, the systems where analog; since the move to digital systems, where files are stored on a hard disk, the number of CCTV systems on videotapes is declining rapidly.

Image Enhancement

Surveillance video images often are of poor quality. This is caused by light conditions, the resolution of the system, and videotape wear. In practice, it is rare for image processing (apart from contrast stretching) to improve the quality with these kinds of images on a single image.

We investigated the different kinds of image restoration based on Wiener restoration, maximum entropy, and PDEs for still images.[11] For artificial images, this works; however, for real images, it does not often work. For a sequence of images, the contents of these images can be combined in a final image, which is called

Figure 27.6 Example of a surveillance image.

the *super resolution method*. There are several commercial products for working with the video files and doing the analysis.

In Figure 27.6, an example of a surveillance image is shown. In this image, there were questions as to whether to visualize the number plates and to identify the persons, or to magnify the faces. It is important to validate the methods, before giving the results to court.

Surveillance Systems

In the past, time-lapse videotapes were submitted to the laboratory. The reason for this is that the videotape should handle more than 2 hours, otherwise the handling of the videotapes becomes inconvenient for the operators in shops and banks. Often, the tape will cover 48 hours; however, this means that fewer images per second are recorded than the normal 25 for PAL (European analog video format). For a 48-hour videotape, 1 frame per second is recorded.

Nowadays, we mostly see digital systems, which often have less quality, depending on the settings. Because there are no standards yet for digital systems, they can be hard to handle at the forensic laboratory. Often, the system itself should be seen before the optimal images are acquired. Digital systems can either be a hard disk with JPEG compressed images, MPEG streams, or digital tapes. Often, the format of storage is proprietary on the hard disks, and there is no standardization in this field yet. These systems have many options for the operators, where the number of images per second and different actions for the alarms can be set. It is important to have a database of players to view the different formats.

Noise Removal

Many types of noise exist in a surveillance system; however, because the following types are distinguished for the video system if noise is introduced by playing and digitization, it might be worthwhile to play the recording more often and average the final result. One danger with this method is that the tape can be damaged if it is played too often.

Averaging over multiple frames can also reduce the noise. This can be valuable if there is a night recording with a scene that does not move. One problem is that if people move, as seen in Figure 27.7, they are not visible anymore. The value of this method depends on the part of the image that has to be visualized.

Image Restoration

In the past, research has been published on image restoration of a video sequence by calculating the movement or the blur of a lens system. These approaches often do not work because the blur function is not known.

Tracking

Other research involves tracking a person in a video sequence.[5] This might be useful if the court asks for magnification of a sequence of images; otherwise, it takes too much time to work on the case manually. There are large efforts going on for tracking persons and combining the different cameras in court.

Commercial Products

There are several commercial products for forensic video image processing, including Avid (http://www.avid.com/products/forensic/), Video Investigator™,[13] and Impress.[14] Other products,

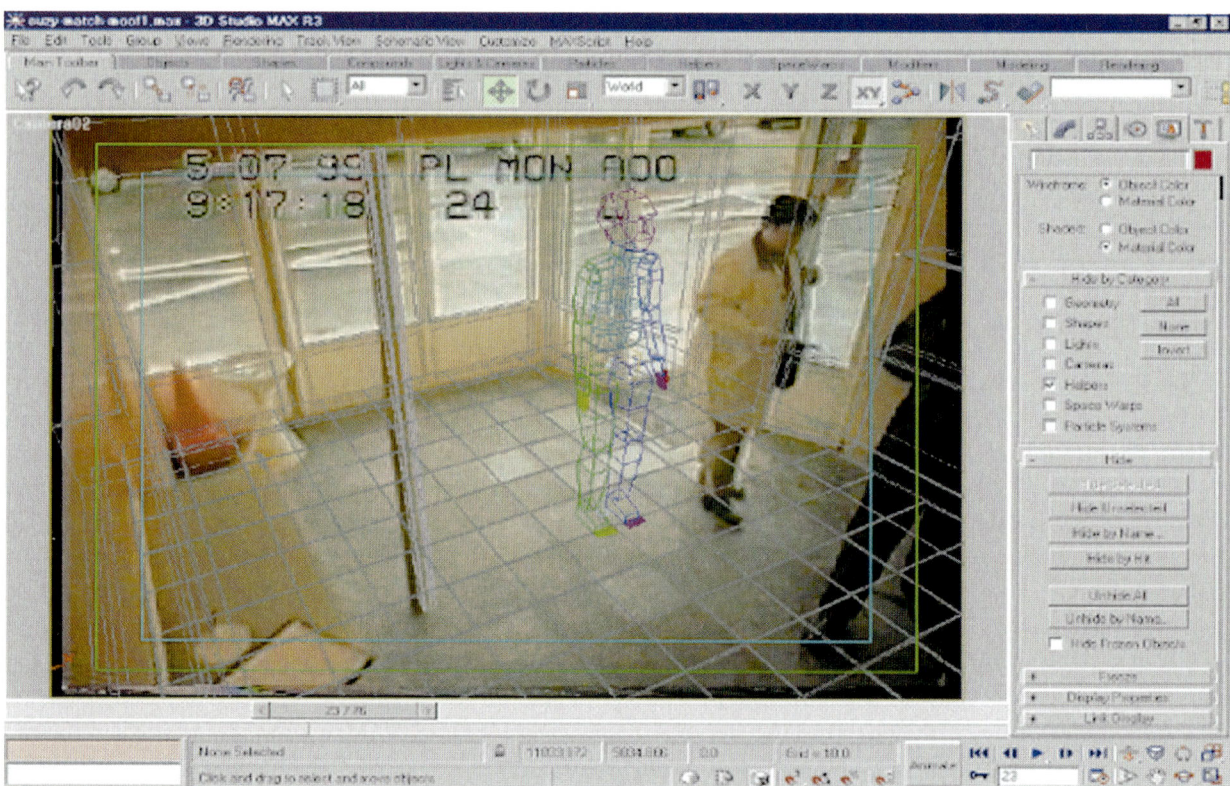

Figure 27.7   Example of human model in three dimensions.

such as Adobe® Premiere and Photoshop®, are also used.

Three-Dimensional Reconstruction

For the comparison of a suspect with a video image, there are several implementations.

One approach that we use is bringing the suspect to the same video recording system and asking the individual to stand in the same position as the people that were visible on the video recording. The problem with this approach is that the place should be available (often, they have to rebuild the place or the shop owners do not cooperate) and the suspect must cooperate as well. Furthermore, there should be people available who look similar to the suspect because one can draw a conclusion about the uniqueness of features that are combined.

The other method is photogrammetry. This means that many parts of the scene have to be measured, and the length and size of a suspect can be compared. We also used a human model that was available in a commercial software package to determine and compare the size of the person.

In Figure 27.7, an example of such a human model (in 3-D Studio MAX) is shown with the real human being. In this way, the length is measured. However, it can be difficult to determine the real length because there might be errors in the measurements or there might not be enough reference points in the image. Furthermore, if someone bows, it is not known how much it differs from the real length, and the influence of shoe heel height and head size should be taken into account.

For this reason, although the 3-D images have a certain appeal to people's imagination, these kinds of models should be validated for each image before they can be used in court, otherwise the evidence may not be accepted.

Output to Court

The court should use the output to interpret what happened at the scene of the crime. Often, we will submit a written report with video prints to the court; however, because they should watch the complete sequence, we will sometimes send a composite videotape. The problem with this is that the lawyer might ask if the digital evidence has been tampered with. We prefer to send a CD-ROM or DVD-ROM to the court with a hash-code based on, for example, SHA-1 and MD5 of the movie files that are on these discs.

Image Integrity

Images can be manipulated. Compared to other evidence (e.g., DNA or fingerprints), the question often arises if images have been tampered with. For the analog videotapes, we might investigate the signals, and we will say that the videotape seems to be an original. If the videotape has been altered professionally, we may not be able to detect this.

One other problem that arises is with child pornography cases. In the Netherlands, the possession of virtual child pornography is not illegal. Virtual child pornography means that there is a composition of a child and another occurrence in such a way that the child seems to be abused; however, it never happened in reality. For this reason, we had a case in which the lawyer claimed that his client's images were virtual. The court asked us to investigate for manipulation.

We proved that some of the images were manipulated by examining the edges and the light effects. Because all images were compressed with a poor compression scheme, it was hard to find the fine details in the image. It is difficult to prove that an image has not been altered. This can only be claimed by testimony and other types of investigations.

Another issue is whether a certain videotape has been recorded with a certain VCR used as evidence. Often, there are signals in the videotape where this information might be checked. However, it is weak evidence.

A stronger type of evidence is available with cameras. If the question arises whether a camera has recorded a video image or a still image, there might be information in the image itself. One of the strongest kinds of evidence was

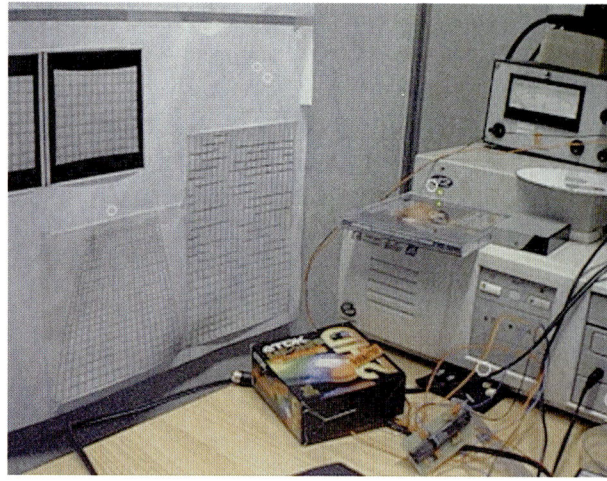 

Figure 27.8 Camera identification based on pixel defects.

found with pixel defects in a camera. We tested certain inexpensive digital cameras, and it appeared that there were enough pixel defects to identify the camera. An example is shown in Figure 27.8. Cameras might hide serial numbers in files or by stenography. These ideas are under research at the moment. For this reason, digital systems might provide good methods for authentication. Because charge-coupled device (CCD) and CMOS sensors become better in quality, the pixel defects are often not seen anymore. A powerful alternative is to employ pixel response nonuniformity (ref. Lukas J., Fridrich J. and Goljan M.: "Digital camera identification from sensor noise," *IEEE Transactions on Information Forensics and Security*, vol. 1(2), June 2006), which also can identify a certain sensor based on slight variations between the pixels. Validation of results with different cameras of the same model and brand is strongly recommended when using these methods in court.

Discussion

Video image processing is important in the court; however, often expectations are too high. The responsibility of the forensic examiner is to inform the court on the possibilities, limitations, and the artifacts that might arise from image processing. Many image processing methods should be validated before being used in court because the risk exists that if a certain method is not accepted in the court, it might be very hard to use the method again in other cases, although it may have been a method that could be useful in forensic investigations.

Use of Networks in Forensic Science

In forensic science, the use of networks is important for communication between laboratories, and makes it possible to exchange large quantities of data. The use of networks makes it possible to exchange visual information and advice on crime scenes. Furthermore, networks also can be a source for investigation themselves because attacks on networks by **hackers** can be a forensic investigation.

Mobile networks are becoming more standardized and, for these networks, communication at remote locations or at crime scenes is appropriate. Third-generation mobile communication provides high bandwidths and makes it possible to communicate in video streams.

One problem with using these types of networks is that there is a security issue. If a DNA database is on an unprotected network, data could be stolen and this would not be acceptable to the general public because of privacy issues.

For this reason, it is important to secure the networks by whatever means. This can be done by encryption if there is a possibility for interception of the signal between the systems. Furthermore, there should be firewall software for protecting the data sources. If the data are very sensitive, as with DNA, the protection should be higher, and a closed network should be considered.

Virtual private networking is being used more often. With these types of network systems, one can communicate over an insecure line with encryption. If the signal is intercepted (as can be done easily by unknowns on the Internet) and strong encryption is used, it should not be possible to read the information that has been sent.

Often, forensic scientists spend a lot of time in court as well as in waiting. The travel time and the time for waiting can be reduced by testifying online. In some countries and states, this is permitted. Good training is required to use this solution, and more experience is needed before this process is used in complex cases.

If a crime officer is at a crime scene, he or she might provide advice as to how the evidence should be collected. This might be a challenging way of improving the quality of the forensic evidence in some cases.

An overview of the different aspects of using the Internet with online literature follows. On the Internet, different methods exist to find information. The different search strategies and indexes that are available on the Internet are also explained, and an overview is given of the literature databases that are available, and how to access them.

There are different literature sources on the Web. Medline is a database that has become available for free on the Internet; however, there are many other sources that could be of interest. We will also look back at how these services started and try and predict how these services will develop.

The security issues of using the Internet and ways to prevent security risks are often considered a challenge. There are different ways of reducing these risks. One of the methods is using encryption programs and communicating between parties in a secure way.

Because bandwidth is growing very rapidly, the communication of real-time video and images is a reality. Furthermore, other techniques of imaging (three dimensional) will become widely available on the Internet and will need other standards for handling the information. Mobile high-speed connections are available. This means that the forensic scientist can give advice to crime officers by watching the crime scene in a committee.

More efficient search techniques (data mining) of huge collections of images and text are required, and the development of new implementations on the Internet will have an impact on forensic science in general.

Literature Resources

Many new databases are becoming available on the Internet. The oldest databases are literature references and bibliographic information. Use of these databases previously required a subscription, and usually, the librarian accessed them because the charges were high and the interfaces not very user-friendly.

These old databases are going to change and will offer Web access services. There are other services with full text and images on the Internet. Many databases with links to this information exist.

The next revolution will be searching for image content on the Internet. Several approaches seem to be successful in demo databases. In the following, examples of these databases are discussed.

Literature Databases

Medline

Medline[15] is very useful for finding information on forensic medicine. It contains over 11 million records of literature and covers 3800 biomedical journals from 1966 onwards. Furthermore, it is also possible with PubMed to search full text in 400 journals.

STN

STN[16] has an extensive resource of over 200 literature databases. Depending on the search question, different databases can be used. Access to the database requires a subscription.

The Chemical Abstracts (CAS) database is the first database to use for chemical questions. It covers over 8000 journals that are monitored and contains over 15 million records. The Science Citation Index covers 5600 leading scientific, technical, and medical journals and starts with the year 1970. Currently, there are more than 17 million records in the database.

The Inspec and Compendex database can be useful for engineering and technology. Compendex covers over 4 million records, and 4500 journals have been monitored since 1970. The Inspec database covers physics, electronics, and computing and has 6 million records.

The various databases can be searched with a simple Web interface, separately or in combination. With the more sophisticated interface, the user can combine databases and get statistics on the search.

Some of these databases might be freely available on the Internet via universities or libraries with a user identification number and password.

FORS

The FORS database is available at http://www.forensic.gov.uk/forensic/products/fors/. It provides a paid service for references.

Infotrieve

This database http://www4.infotrieve.com has a wide variety of journals online.

Google Scholar

scholar.google.com is a very good source, with over 600,000 hits on the word *forensic*.

NCJRS

Most current forensic journals are searchable in this database, for example, *Journal of Forensic Sciences;* also, smaller journals such as the *Journal of Forensic Identification* and the *AFTE Journal* (for firearms and toolmarks examiners) are covered in this database. For retrieval of older literature (before 1997), the paid databases should also be considered.

Patents

Because more companies are getting involved in forensic science, many patents are applied for. In the database Delphion,[19] European, American, Japanese abstracts, and world patents can be searched. There are 1114 U.S. patents with the word *forensic* from the year 1996 to 2001. The U.S. patents are available for free in this database.

The U.S. Patent and Trademark Office has full-text patents that are searchable and retrievable at http://www.uspto.gov/patft/. For patent searches in other countries, charges apply. The information in patents can be valuable when examining products. The European Patent Office (Figure 27.9) has a database at http://ep.espacenet.com/, where all American, European, world patents, and Japanese abstracts can be searched for free.

Forensic Groups and Societies

In each field, there are several active forensic groups. Some were established several decades ago and have generated a huge amount of literature.

The American Society of Questioned Documents[20] provides an extensive source of databases of literature. A database that is online is a computer printer database in which 1600 printers are stored. Information on subscribing to this information is available online.

The Association of Firearms and Tool Marks Examiners (AFTE)[21] has indexed its literature references for firearms and toolmark examination.

Some working groups are established under scientific working groups of the Federal Bureau of Investigation[23] and working groups of ENFSI www.enfsi.eu (European Network of Forensic Science Institutes).[24] They sometimes make literature listings available online. Many mailing groups on forensic science are online.

Libraries

Many libraries have their complete collection online. However, bookshops that are online often

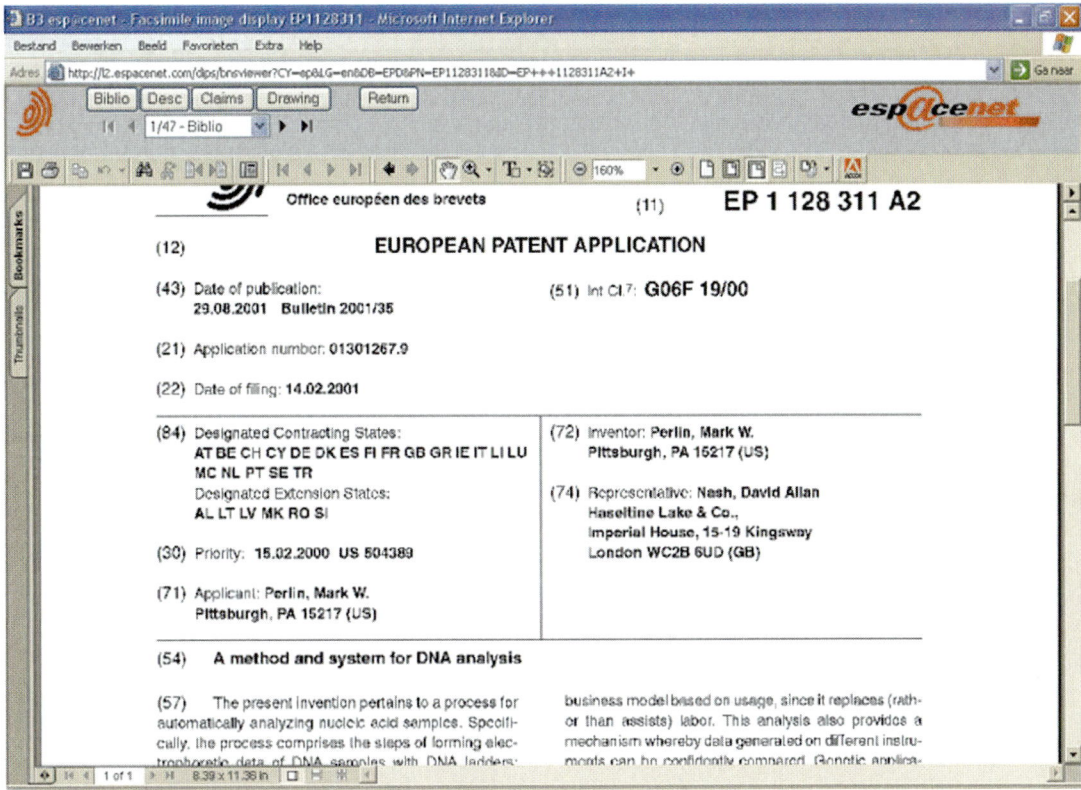

Figure 27.9 Screen of www.espacenet.com.

have more recent books (e.g., www.amazon. com). Furthermore, there are several literature sources online, such as Worldcat, which is a combination of libraries in the world.[25]

Forensic Journals

An example of an online journal is *The Journal of Forensic Sciences*,[26] the full text of which is accessible online for members in the form of PDF files. It also provides a search engine for the journal abstracts. Others can pay for each article they view by means of a credit card. Often, the contents of a journal are listed on the Internet (e.g., *Forensic Science International*). There are also combinations of journals found at http://www.forensicsource. org/, where five journals are accessible online with a subscription.

There are many journals that have their information online. Even journals with high impact factors will have articles online that are older than 1 year. A listing of these journals

is available at www.freemedicaljournals.com. There are also several forensic journals listed here.

Databases of Links of Forensic Information

There are several listings of links to forensic information on the Internet. Zeno's Forensic Site[29] is an example of this. It was started in 1994 and now has a forum with thousands of postings and a search engine that can be useful in finding relevant information. Furthermore, the links are rated by the users, and overviews are given of how many times the information was requested. Users can also add their own forensic Web space on this site.

The power of discussion forums is shown in several instances. Often, the forensic groups mentioned earlier have discussion groups and mailing lists for questions. An overview of forensic discussion lists is available.[32]

Databases of Images and Patterns

Firearm Image Library

This image library shows images of firearms online.[34] Users can also submit their images of firearms. Furthermore, a database of firearm sounds is provided.

Drug Database

This site shows legal drug databases and the retrieving symbols that are given on the drugs.[35]

Ignitable Liquids Reference Collection

The Ignitable Liquids Reference Collection (ILRC) (Figure 27.10) is a compilation of reference materials used by forensic analysts to conduct fire debris analysis.[36] It consists of a comprehensive set of physical specimens of ignitable liquids and accompanying characterization data used in the analysis of fire debris samples in accordance with the American Society for Testing and Materials (ASTM) E-1387 and E-1618 standard test methods. Access to the ILRC database requires a Java-enabled browser.

Security

Nowadays, with the number of viruses and trojan horses expanding, keeping updated with the latest information is important, and awareness and protection against viruses and other threats is necessary.

E-mail messages can also be encrypted. In Microsoft Outlook, a standard encryption is included; however, at the time of writing, it is not used very often. Another software package that is used more often is PGP.[37] A government-approved encryption method is recommended, for example, AES.

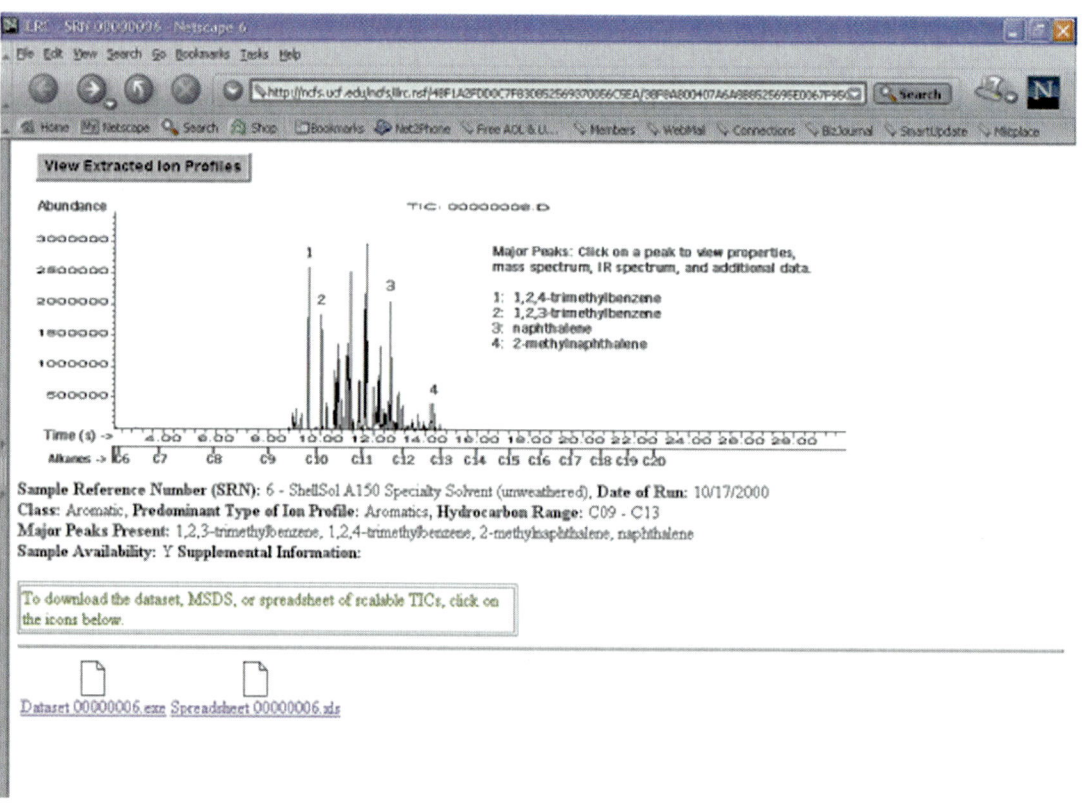

Figure 27.10 Ignitable liquids reference collection.

Expectations

Future implementations of databases will combine video, audio, and text. One of these developments is seen in the proposed MPEG standards, which are available.

We see that bandwidth is growing in households and companies with the appearance of ADSL and cable modems. Also, developments in mobile combination will make it possible to have more mobile stations at the crime scene. In this way, it will become easier for the forensic laboratories to provide advice via remote location.

There are also implementations of remote systems for pathology (telepathology). Another issue is forensic profiling. In theory, one can link all databases with one another and extract data from them, based on profiles. Of course, one has to be aware of accidental matches. In D6.7 Forensic Profiling www.fidis.net, an overview is given of the possible risks.

Risks and Protection

When using software in forensic casework, validation of the software and hardware is recommended. Furthermore, the manipulation of data is a risk if the system has been hacked. By computing hashes of files, it is possible to testify that the file is the same as it was at the time of examination.

The abuse of data is another risk. If the forensic databases are copied by third parties, huge privacy issues could result, for instance, if a country's DNA database became available on the Internet. A high priority should be placed on protection of these systems and the procedures around them.

Court Presentation

In court, the use of a computer can be useful for presentation of the evidence. The use of 3-D animation can also clarify the forensic examination in court.

The use of these software tools should be acceptable to the court. This means that one should take care of colors, symbols, and other attributes (such as sounds) that are used in court. If there is more than one possibility for an animation, these different options should also be visualized or it should be clarified that the animation chosen is an example of what might have happened.

It is also a possibility that the use of specific software may be challenged in court. If the software is commonly used and is the de facto standard, it will be accepted as evidence. The court may ask for a proof that the software is doing what is says. Using different packages and comparing the results can validate the software. Furthermore, it is also possible to write one's own programs and compare the results with commercial products. If this is not acceptable, a source code review should be considered. Often, it is difficult to acquire the source codes of commercial software packages; one way is to sign nondisclosure statements with a company. Source code reviews of commercial packages are very time consuming, especially, if they contain many lines of source code. Another problem is that the source code of libraries may not be available, and the source code of the operating system may not be widely distributed (expect for Linux). For that reason, a partial review is feasible if requested in court. In practice, this does not happen very often.

Future Expectations

In the future, more data sources will be online and available for use within law enforcement. There should be a standardized communication because profiling in forensic databases will be easier to use. LIMS, image databases, and mobile communication also will be integrated. This might help to solve more crimes. Furthermore, wireless applications and in future high-speed wireless connection—for example, ultra wide band (UWB)—will contribute to more mobile applications for forensic science. Validation of results remains important in all these developments.

Questions

1. Which kinds of image databases are mostly used in forensic laboratories?

2. What is important when filling image databases?

3. What are the new developments in these databases?

4. What are the risks of image processing in court?

5. Which precautions should be taken for 3-D reconstructions?

6. Which database has been used for years with good results?

8. Discuss the process of photogrammetry.

9. What are the risks of using networks in forensic science?

10. Which kind of encryption is used often for e-mail communication?

11. How can one validate information that has been found online?

References and Suggested Reading

1. Huijsmans, D. P. et al., *Visual Information and Information Systems*, 3rd International Conference, VISUAL '99, Amsterdam, The Netherlands, June 1999, Proceedings Lecture Notes in Computer Science 1614, Springer-Verlag, Heidelberg, 1999.

2. Jain, L. C. and Smeulders, A. W. M., *Intelligent Biometric Techniques in Fingerprint and Face Recognition,* CRC Press, Boca Ration, FL, 1999.

3. Franke, S., Schomaker, L. R. B., Vuurpijl, L. G., and Giesler, S., A common ground for computer-based forensic writer identification. *Forensic Science International,* 136(Suppl.), 84., 2003.

4. van Beest, M., Zaal, D., and Hardy, H., The forensic application of the Mikrocad 3D Imaging System, in *Proceedings of European Meeting for Shoeprint/Toolmark Examiners 2000*, Vantaa, Finland, p. 77, 2000.

5. Geradts, Z., Keijzer, J., and Keereweer, I., Automatic comparison of striation marks and automatic classification of shoe prints, in *Proceedings, SPIE Investigative and Trial Image Processing*, Vol. 2567, p. 151, 1995.

6. Alexander, A., Bouridane, B., and Crookes, D., Automatic classification and recognition of shoe profiles, in *Information Bulletin for Shoeprint/Toolmark Examiners, Proceedings*, European Meeting for Shoeprint/Toolmark Examiners, National Bureau of Investigation, Vantaa, Finland, p. 91, 1999.

7. http://www.fosterfreeman.co.uk/solemate. html; accessed April 20, 2002.

8. http://www.l1id.com/; accessed April 20, 2002.

9. http://www.forensic.gov.uk/forensic/ entry.htm; accessed April 20, 2002.

10. http://www.fbi.gov/congress/congress00/dadams.htm; accessed April 20, 2002.

11. Bijhold, J., Kuijper, A., and Westhuis, J., Comparative study of image restoration techniques in forensic image processing, in *Proceedings, SPIE Investigative Image Processing*, Vol. 2942, p. 10, 1999.

12. Geradts, Z. and Bijhold, J., Forensic video investigation with real-time digitized uncompressed video image sequences, in *Proceedings, SPIE Investigation and Forensic Science Technologies*, Vol. 3576, p. 154, 1999.

13. http://www.cognitech.com; accessed April 20, 2002.

14. http://www.imix.nl/impress/; accessed April 20, 2002.

15. http://www.nlm.nih.gov/databases/ freemedl.html; accessed April 20, 2002.

16. http://www.cas.org/stn.html; accessed April 20, 2002.

17. http://www.dialog.com; http://openaccess.dialog.com/tech/; accessed April 20, 2002.

18. http://www.ingenta.com; accessed April 20, 2002.

19. http://www.delphion.com/ibm.html; accessed April 20, 2002.

20. http://www.asqde.org; accessed April 20, 2002.

21. http://www.afte.org; accessed April 20, 2002.
22. http://www.for-swg.org/; accessed April 20, 2002.
23. http://www.enfsi.eu; accessed April 20, 2002.
24. http://www.oclc.org/worldcat/ accessed April 17, 2008.
25. http://www.aafs.org; accessed April 20, 2002.
26. http://marygoldgupta.freeyellow.com/index. html; accessed April 20, 2002.
27. http://haven.ios.com/~nyrc/homepage.html; accessed April 20, 2002.
28. http://forensic.to/forensic.html; accessed April 20, 2002.
29. http://www.google.com; accessed April 20, 2002.
30. http://www.mamma.com; accessed April 20, 2002.
31. http://forensic.to/links/pages/General_information_resources/Mailing_Lists/; accessed April 20, 2002.
32. http://www.mssm.edu/genetics/dna_ dbase.html; accessed April 20, 2002.
33. http://www.recguns.com/gunPictures.html; accessed April 20, 2002.
34. http://pharmacology.miningco.com/health/ pharmacology/mlibrary.htm; accessed April 20, 2002.
35. http://ncfs.ucf.edu/ncfs/ilrc.nsf; accessed April 20, 2002.
36. http://www.pgp.com; accessed April 20, 2002.
37. http://www.cselt.it/mpeg/; accessed April 20, 2002.

Computer Crime and the Electronic Crime Scene

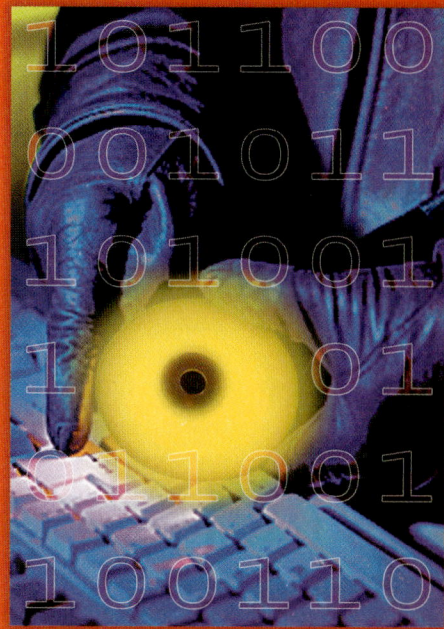

Thomas A. Johnson

In the mid-1960s, our nation experienced its first series of criminal activity in which a computer was used as an instrument to perpetrate an economic crime. In his book, *Fighting Computer Crime*, Donn B. Parker reports that the first federally prosecuted case of a computer crime occurred in 1966 and involved a consultant working under contract with a Minneapolis bank to program and maintain its computer system. This case was unique: The individual was prosecuted for embezzlement of bank funds; he changed the checking account program in the bank's computer so that it would not identify and automatically notify bank officials of overdraft charges in his personal checking account (Parker 1997, 8).

By 1973, the largest recorded and prosecuted computer crime had occurred in Los Angeles and resulted in the destruction of the Equity Funding Insurance Company, with a loss of $2 billion. Twenty-two executives and two auditors were convicted for creating 64,000 fake people, insuring them, and then selling those policies to reinsurers (Parker 1997, 65). Law enforcement agencies were not prepared for the use of sophisticated computers in these economic crimes. In fact, the first federal agencies to participate in these criminal investigations were the Internal Revenue Service (IRS) Criminal Investigation Division, the U.S. Secret Service, and the Federal Bureau of Investigation (FBI). On examining the training provided by those agencies to their personnel, it is clear that there was little or no instruction offered in terms of computers and their use in criminal acts. Agents who were assigned to these cases had to develop and refine their individual skills to address the challenges they were encountering in the field.

Introduction and Historical Developments

The IRS Criminal Investigation Division (IRS-CID) was the first federal investigative agency to contract with a university to develop and refine the skills of an elite group of special agents to confront this new and emerging trend in criminal activity. Michael Anderson and Robert Kelso were among the first group of IRS-CID agents to receive this training in computers and to play a leadership role within their agency. Another pioneer in this newly emerging field was Howard Schmidt, who would eventually be called on to serve as vice chairman of the President's Critical Infrastructure Group. Howard's career began in a small municipal police agency in Arizona, and he eventually served in several important federal agencies where, through his vision and encouragement, he created programs to train other law enforcement personnel at the local, state, and federal levels of government. Howard Schmidt's skills did not go unnoticed by the corporate community, and, as computer crime was increasing, the corporate community turned to him and a select few others for assistance in combating these new developments in corporate criminal activity.

Universities also were not prepared for how computers might be used in the commission of criminal activity. As a result, law enforcement had to rely on the insights of such leaders as Howard Schmidt and Michael Anderson, who were both instrumental in developing training seminars for their colleagues. Indeed, the very beginning efforts of organizations such as the International Association of Crime Investigative Specialists (IACIS), and the High Technology Criminal Investigation Association (HTCIA) were specifically developed to offer training, instruction, and sharing of information in this important area. Eventually the HTCIA began developing chapters in various states and regions and, to this day, is one of the most respected organizations for professional, in-service training of law enforcement officials interested in computers and their role in criminal activity.

If law enforcement agencies were ill-prepared for the challenges they would confront in computer crime and economic crime cases, our prosecutorial agencies were even less prepared for this growing criminal activity. One only has to examine the dearth of statutory law in each of our states to realize that we were not prepared to prosecute these cases. Once again, our nation had to rely on a small cadre of people who saw these challenges and played a most formidable role in providing their colleagues with the training in this area. Leaders such as Kevin Manson, Tony Whitledge, Ken Rosenblatt, Gail Thackeray, and Abigail Abraham provided enormous assistance not only to their colleagues but also to state legislators in the framing of a new statutory law to address this new criminal activity.

In the early 1980s, the SEARCH Group, Inc., under the leadership of Steve Kolodney (and afterwards, Gary Cooper), perceived a need for training law enforcement managers in information management systems. Fortunately, the SEARCH Group also had two outstanding pioneers in the field of training police officers in computers—Fred Cotton and Bill Spernow, who began one of our nation's first outreach efforts in training municipal and state police in this important area. The contributions that both Fred Cotton and Bill Spernow have made in this field are measured by the esteem in which their professional colleagues held them. The contribution of the SEARCH Group is also evident in that during the entire decade of 1980 to 1990 they provided the only Peace Officer Standards and Training (POST) instruction to law enforcement officers in the state of California. Indeed, another major deficit of our nation's ability to address computer crime centered on the fact that virtually every one of our states' training agencies provided no training at all to their law enforcement agencies in computer crime. In fact, until the early 1990s, state POST agencies were not offering even occasional training courses or instruction in this area.

In the mid-1990s, our nation experienced a greater collaboration between federal, state, and local law enforcement agencies in addressing mutual training strategies. The Information Technology Working Group was an important step forward, as the then U.S. Attorney General Janet Reno appointed a small group of approximately 40 people from agencies within the federal, state, and local communities to join together in developing a cooperative blueprint for how our nation might best confront the growing problem of individuals using computers as an instrument for committing crime. After a series of meetings, they decided on a strategy of "Training the Trainers" so that a new and larger population of officers could reach out to their colleagues and provide instruction in this new area of criminal activity. Accordingly, a training curriculum had to be developed, and the U.S. Department of Justice funded several meetings of the nation's leading experts in an effort to develop a series of courses that would be provided for state, federal, and local law enforcement personnel. After 2 years of course development, the National White Collar Crime Center was allocated the responsibility for delivering these courses to law enforcement personnel at the local and state levels. The federal effort of training new agents and in-service agents was allocated to the FBI, U.S. Secret Service, IRS-CID, U.S. Customs Agency, U.S. Postal Inspectors Division, and Federal Law Enforcement Training Center.

Having had the privilege of serving as a member of the Information Technology Working Group, as well as having been active in our higher education community, I saw a critical need to begin to mobilize our university community to address the unique needs of our law enforcement and prosecutorial agencies in addressing this growing problem of computer crime. Ironically, our nation's universities had numerous computer science departments and over 1,000 criminal justice programs, but there existed no coherent educational strategy to provide the theoretical and pragmatic skill sets that were required if our justice community was to seriously make inroads into this growing problem. Computer science departments were focused on educating their students in programming languages, database skills, and a number of other areas that provided assistance only to a small subset of our justice communities' need. At the same time, most, if not all but a few, educational institutions with criminal justice departments simply were not equipped with the faculty to address the problem of computer crime.

As a result of working in the area of computer crime since 1980, coupled with the knowledge of universities' computer science and criminal justice departments, in 1996 the University of New Haven formulated both a graduate and undergraduate certificate in forensic computer investigation. This certificate program includes a sequence of courses that address three target discipline areas: computer science, law, and forensic investigation. These course offerings were initiated in 1997 at both the main campus in Connecticut and the branch campus in Sacramento, California. As we have had the privilege of working with our nation's leaders in this field, we have utilized over 21 outstanding experts who have joined us in the capacity of practitioners-in-residence or distinguished special lecturers to offer this program. In 1998, we responded to the need for providing online educational courses and began offering both a graduate and undergraduate certificate in information protection and security at both campus locations. In 2001, we began offering a master's of science in criminal justice with a concentration in forensic computer investigation at our main campus. Finally, in 2002, we began offering the nation's first master's of science degree in national security with a concentration in information protection and security. This graduate degree is offered both at the main Connecticut campus and the California campus at Sandia National Laboratory in Livermore, California. These programs developed at the University of New Haven serve as a model in our attempt for universities to play a larger role in providing both the training and educational courses to the men and women of our justice community.

Several of our nation's universities, aside from the efforts of the University of New Haven, have made notable contributions in this area. Among these are Carnegie-Mellon Institute, with its formidable efforts in computer emergency response teams (CERT); Purdue University, led by the pioneering efforts of Eugene Spafford; the University of California at Davis, led by Matt Bishop's work in computer security; the Naval Postgraduate School Campus at Monterey, with its outstanding computer science department; and Dartmouth University's new program in research led by Michael Vattis. These are only a small section of the outstanding contributions being made by our academic community today.

Crime Scenes with Digital and Electronic Evidence

The electronic crime scene with digital and electronic evidence poses new challenges for the investigator. This new environment is unique not only because the evidence may be difficult to detect but also because of how its evidentiary value may be hidden through steganography and encryption. Furthermore, there is a degree of anonymity in which perpetrators can hide their true identity in the forging of certain criminal acts and endeavors. Therefore, the rapid technological advancements occurring in our society through the digitalization of data and information are presenting new challenges to investigators. This electronic evidence is both difficult to detect and quite fragile; therefore, the latent nature of electronic evidence requires very skilled investigators.

Additional challenges that continue to confront the investigator encountering an electronic crime scene center on the global nature of the evidence. In many criminal cases involving computers and electronic technology, we encounter multijurisdictional issues that challenge the very legal structure of all nations' legal and statutory codes. For example, today we find criminal enterprises being initiated from different nations throughout the world, and to effectively investigate, apprehend, prosecute, and convict these individuals we must utilize appropriate judicial search warrants. It is also necessary that the penal codes of the respective nations have statutory authority for legal action to be pursued.

The "I Love You" virus in 2000, which caused an estimated $10 billion in damages, was released by an individual in the Philippines and created havoc to computer systems throughout the world. Despite the extensive damage, this case was not prosecutable because the Philippines did not have legal restrictions against behavior of this type when this virus was released.

Also, the attack on Citibank in New York by Vladimir Levin and members of a mafia group in St. Petersburg, Russia, created an enormous legal problem for the FBI because their investigator had to examine banking systems in over seven different nations where the electronic transfer of money was deposited. The application for search warrants and the timely tracking of this event was a challenge to even the most skilled set of investigators. Levin was arrested and sentenced to 3 years in prison and ordered to repay Citibank $240,000.

An additional problem with this new-age criminal activity, which relies on technology and electronics, is the ease with which one person can impersonate another through elaborate spoofing schemes. A related activity that has cost our nation's businesses enormous financial losses is identity theft. With this crime of identity theft, it generally takes the victim approximately 6 to 9 months of work with credit agencies, bill collectors, and other credit entities before they can have any semblance of restoring their good name and credit standing.

Because personal computers can store the equivalent of several million pages of information, and networks can store many times more this amount of data, the location and recovery of evidence by a trained computer forensic specialist working in a forensic laboratory may take several days or weeks. As mentioned earlier, searching computer files is an extraordinarily difficult process, because files can be moved from one computer to another throughout the world in a matter of milliseconds. Files can also be hidden in slack space of the computer hard drive or stored on a remote server located in other geographic jurisdictions. Files can also be encrypted, misleadingly titled, or commingled with thousands of unrelated, innocuous, or statutorily protected files. It is to address these challenges that the FBI has developed a Computer Analysis Response Team (CART Team); the IRS has a Seized Computer Evidence Recovery Team (SCER Team); and the Secret Service has an Electronic

Crime Special Agent Program (ECSAP) (U.S. Department of Justice 2002, 35).

It is evident that these new technologies are requiring more skills for our investigators, prosecutors, and judges. Accordingly, the role of our educational institutions in preparing current and next-generation criminal justice personnel to address these challenges is becoming more critical as each new technology is developed and introduced to our society.

Computers, Electronic Equipment, Devices, and Information Repositories

In July 2001 the U.S. Department of Justice, through the Office of Justice Programs in the National Institute of Justice, released the Technical Working Group for Electronic Crime Scene Investigation's (TWGECSI) report, *Electronic Crime Scene Investigation: A Guide for First Responders*. The gathering of our nation's experts to organize their advice to assist law enforcement personnel and agencies in preparing to address this new paradigm change in crime was one of our nation's first important efforts to address this problem. The identification of the types of electronic equipment and its purpose was to inform law enforcement personnel of the potential use and value of such equipment.

Both first responders to crime scenes and investigative personnel must appreciate the unique attributes of electronic equipment and be prepared to identify and assess its importance at a crime scene. This suggests the types and purposes of electronic equipment should be well understood as to their functionality and value to their owner. Also, from the viewpoint of assessing the potential impact on the victim, a thorough knowledge of this new environment will prove most useful and beneficial to law enforcement because the crime scene must be protected and processed in accordance with forensic science principles. Because electronic evidence is so fragile, we must train officers in the preservation and collection of electronic evidentiary materials.

Digital evidence can easily go unrecognized, or be lost, if not properly processed. We must also ensure the integrity of digital evidence, because it is easily alterable. Therefore, the importance of training first responding officers to an electronic crime scene is an extremely critical function, and one that must be addressed by state and local law enforcement agencies throughout our nation.

Today, given the ubiquitous presence of computers, answering machines, handheld personal digital assistants, facsimile machines, and other electronic equipment, almost any crime scene may conceal information of value in a digital format. The acquisition of this information is totally dependent on the actions of the first responding officer, who must have the ability to visualize and perceive the presence of such evidentiary material.

The Value of Equipment and Information

The type of computer system or electronic environment the investigator may encounter at a crime scene has a certain tangible and intangible value to the owner, victim, suspect, or witness. Because this value is measured not only in financial terms but also in terms of informational value, there are numerous perspectives that the investigator must be prepared to analyze. It is possible that the owner of a computer system may have become a victim or may become a suspect in a case involving criminal activity. For example, the computer system can be the target of criminal activity, or it can be an instrument used to commit criminal activity. Data residing on the hard drive will provide the answer and appropriate documentation as to each possibility. More often than not, the information that resides within these computer and electronic systems is of greater value than the systems themselves. The proliferation of new technologies at extremely economical prices will continue to make the investigator's job more difficult. We are now in an era where computer communications can occur by using **RAM CACHE**, thus avoiding writing to the hard drive, and this can occur in a networked environment from any point to any other point within our world. Also, the development of encrypted hard drives will make the investigator's job both more difficult and more expensive. As RAM CACHE communications become used by those attempting to commit criminal activity, the impact will be felt by law enforcement, homeland security, national security, and intelligence agencies.

Information Repositories: Informational Value

Just as information residing within electronic systems has value to the owner, victim, or suspect, it also has value to law enforcement, prosecution, defense, and the judiciary as they engage their respective roles in the full investigative and judicial process.

The valuable information residing within these computers and electronic systems will permit our judicial system to weigh the accuracy of allegations, establish the circumstances and truth as to the purported criminal activity, and demonstrate with documented digital evidence the nature of the criminal activity or violation. This, of course, is totally dependent on the correct processing of the electronic crime scene, both technically and legally. The search and seizure of any electronic systems must withstand the scrutiny of the Fourth Amendment and all appropriate case and statutory law.

It is incumbent on our law enforcement agencies to provide the technical competence to evaluate this new form of criminal activity, while being fully compliant with all appropriate legal mandates.

Information Collection

The investigator may look for information on a suspect or criminal by searching for electronic data that may reside in four specific locations:

1. Computer hard drive
2. File servers (computer)
3. Databases from governmental agencies, as well as private and corporate databases
4. Electronic record systems from governmental to private and commercial sectors

The first responding officers to a crime scene in which electronic equipment is present must recognize the presence and potential value of this electronic equipment. They also must provide the necessary security to ensure protection of potential evidence located on hard drives and file servers as the case moves from a preliminary investigation to a full investigation.

The searching and seizure of computer hard drives for information must be done within the parameters of a lawful search either incident to arrest or with appropriate judicial search warrants, or both. The investigator performing the search of a computer hard drive must be sufficiently trained and educated in the use of appropriate software utilities used in scanning hard drives. Furthermore, the officer must use the department's approved protocol for conducting such a search. This includes creating a disk image on which to perform the search of the targeted hard drive while maintaining the integrity of the original hard drive and ensuring that none of the data residing on the hard drive are modified by the software utilized to search for appropriate information. The imaged hard drive should also be duplicated for eventual defense motions regarding discovery of the data, in the event the defense counsel wishes their forensic computer experts to review or perform independent analysis of the hard drive.

Information on individuals, whether they are suspects, victims, or individuals of particular interest, can be obtained through a wide array of governmental and private electronic record systems. Financial reports and credit histories contain a vast storehouse of data not only on the individual in question but also on spouses, relatives, and friends. Because law enforcement agencies also have the responsibility of protecting the privacy of individuals, great care must be exercised in searching the enormous range of databases that now exist within our society. This implies that legal rules must be vigorously adhered to through use of subpoenas and application for judicial review or search warrants.

Management of the Electronic Crime Scene

Managing an electronic crime scene is quite similar to managing any other crime scene, with the exception that specific skills and training background will be required of the forensic computer investigator. In addition, the type of crime committed will invariably call for an exceptional team effort by the seasoned crime investigator in cooperating with the electronic crime scene investigator. Because most police organizations do not have adequate resources to fully staff their departments with individuals who possess such demanding skill attributes, it is not uncommon to find that regional task forces have been developed to address these issues. However, this can lead to complications regarding jurisdictional issues, command and control, collection of evidence, and sharing of information with other members of the crime scene team. Because most electronic crime scenes are photo-rich environments, all of the traditional crime scene mapping, photographing, and diagramming are essential to the proper investigation. The crime scene may contain computers that may need to be searched not only for information residing on their hard drive but also for fingerprints and DNA from the keyboard, diskettes, and other areas of the computer. Therefore, a protocol for addressing such issues must be planned beforehand and made available to all personnel, should implementation of such requirements become necessary.

Electronic Crime Scene Procedures

The value of the National Institute of Justice's *Electronic Crime Scene Investigation: A Guide for First Responders* centers on the awareness and assistance that the typical first responding officers will need in both identifying and protecting electronic instruments found at the crime scene. Their publication provides brief descriptions, photographs, primary use, and potential evidence for the following:

- Computer systems and their components
- Access control devices such as smart cards, dongles, and biometric scanners
- Answering machines
- Digital cameras
- Hand-held devices such as PDAs and electronic organizers
- Hard drives, both external and removable hard drive trays
- Memory cards
- Modems
- Network components with local area network (LAN) cards, network interface cards (NICs), routers, hubs, and switches
- Servers
- Network cables and connectors
- Pagers
- Printers
- Removable storage devices and media
- Scanners
- Telephones, such as cordless and cell phones
- Miscellaneous electronic items, such as the following:
  - Copiers
  - Credit card skimmers
  - Digital watches
  - Facsimile machines
  - Global positioning systems (GPS)

This booklet for the first responding officer provides a good orientation to the types of devices one might encounter at an electronic crime scene. It also highlights the idea that data of informational value to the crime scene investigator can reside in unusual electronic places. At the same time, the first responder should note that data can be lost by unplugging the power source to an electronic instrument, and great care must be taken to protect the crime scene (National Institute of Justice 2001, 9–22).

There are occasions when the first responding official to a call-for-services event may not be a police officer; the official may, in fact, be responding to a medical emergency or fire assistance call. If these respondents perceive the incident to be a potential crime, they will have to call for police services, in which case there may be a multiagency responsibility for securing the potential or real crime scene. A recent example of this situation occurred in the "Frankel Case" in Stamford, Connecticut, where the first responding personnel to a fire alarm notification were fire personnel. After observing computers throughout the estate, including even in bathroom areas, and what appeared to be a deliberate effort to burn computer components within the kitchen area of the estate, the fire personnel notified the fire arson investigator, who not only notified the local police department but also encouraged the local department to notify the federal authorities. Fortunately, this arson investigator had received educational courses in the area of computer crime and quickly realized the nature of the electronic evidence and took appropriate action.

It is interesting to note in this case that although the local police department had personnel trained in many areas, they did not have any personnel trained in electronic crime scenes. The arson investigator prevailed on them to contact a federal agency, which initially declined to get involved in the case. The arson investigator was familiar with a guest instructor who had lectured in a computer crime course, so he called on her and described the situation. This guest instructor, who was also a federal agent well trained in the area of computer crime, realized the importance and significance of the situation and subsequently notified the original federal agency as to the seriousness of this case. The federal agency reevaluated the situation and joined in a multiagency investigation that resulted in the arrest of the subject by German police authorities. Thus, the perseverance of the first responding personnel, along with their training and education, resulted in an international

investigation of a multimillion dollar fraud and embezzlement case. The scope of the computer involvement in this case can be assessed by the fact that it required 16 federal agents over 3 months to process all of the computer evidence in this case.

In most cases, the first responding officer's initial duty is to provide aid or assistance to a victim or victims, if any. Second, it is incumbent on the responding officer to take into custody any suspect at the crime scene and to identify witnesses or ask them to remain until crime scene investigators arrive at the scene. Finally, the first responding officer must secure the crime scene to prevent contamination of the scene or destruction of materials that may possess evidentiary value. As the preceding case revealed, often it is the education, experience, and initiative of a first responder that can go beyond traditional role expectations and requirements and play an important part in the successful resolution of a case. This suggests that we really need more than technicians who will respond to crime scenes; we need those who have the benefit of a rich education and broad training perspective.

It is generally accepted as good police practice that, when entering an electronic crime scene in which there are no injured parties or suspects in need of detention, the following guidelines be followed:

1. Secure the scene so as to minimize any contamination.
2. Protect the evidence, and, if people are at the scene, do not permit anyone to touch any computers or other electronic instruments. Have all electronic devices capable of infrared connectivity isolated, so as to control for data exchange. This will include cell phones, PDAs, and other similar instruments.
3. Evaluate the electronic and computer equipment at the scene and make a determination as to whether assistance will be required in the processing of the scene. Few officers can be expected to handle the more complex and sophisticated electronic environments. In some cases, the need for a

consultant may be required. Also, personnel with appropriate skills may be located from a regional or federal task force.

4. Observe whether any computers are turned on, and, if so, take the following precautions so as not to inadvertently lose any data on the computers:

   a. Photograph the computer screen if it is left on and it appears useful.

   b. Document the scene through videotape, photography, and crime scene sketches.

   c. Label and photograph all cards and wires running to and from the computer to peripheral devices.

   d. Do not turn off computers in the conventional manner because the computer could be configured to overwrite data. Therefore, in stand-alone computers, it is best to remove the power plug from the wall. Also, if a telephone modem line is in use, disconnect the cable at the wall. It is important when authorities encounter a network as opposed to a stand-alone computer that no one remove the power cord from the server. If the agency does not have personnel who are trained to work in a network environment, other assistance should be requested, and the scene should remain secured until such assistance is available.

   e. Collect any material germane to the electronic or computer environment, including manuals, peripherals, diskettes, and any medium capable of storing data.

5. Inform the crime scene supervisor, in the event the crime scene will require the use of fingerprinting powders to develop potential latent prints on the computers, that no aluminum-based powders should be used to dust for fingerprints on the computer, because it could create electrical interference. In fact, the forensic processing of the computer and its hard drive should occur prior to any dusting for fingerprints. However, the forensic computer investigator or the person who will actually process the computer should also take care to not preclude a subsequent search for traces of DNA evidence and an examination for latent fingerprints.

6. Take care in disassembling and packaging items for transport to either the police evidence and property room or the crime laboratory for the processing of the equipment:

   a. Maintain the chain of custody on all evidence; therefore, follow and document the appropriate protocols.

   b. Package, transport, and store electronic instruments and computers with minimal to no exposure to situations that might compromise the data residing within their storage mechanisms. Electronic instruments and computers are very sensitive to environmental temperatures and conditions and other radio-wave frequencies.

   c. Place a seizure diskette in and evidence tape over drive bays of computers that will be seized prior to removal and transportation.

7. Transport computers and other electronic instruments and evidence with caution so as not to damage or lose the fragile electronic data. It is advisable not to transport this equipment in the trunk of a police car because this is the area where the police unit's two-way radio is located, and the signals may damage the data stored in the computer and other electronic instruments.

8. Store and maintain computers and electronic equipment in an environment that is conducive to preserving the data contained in that equipment and is free from any nearby magnetic fields. In those cases where the forensic computer investigator may participate as a member of a raiding team, there will obviously be time to prepare and plan for appropriate action, as opposed to being called to a crime scene as a result of the first responding

officer's request for assistance. In the case of a preplanned raid, the forensic computer investigator will clearly be aware of the criminal activity and will have the opportunity to engage in presearch intelligence. This will permit the opportunity to engage skilled personnel who will be able to process the scene on arrival. The presence of a network may be determined, and appropriate plans can be developed for processing this environment. Also, it may be possible to gather useful information about the situation from the Internet service provider (ISP). In short, knowledge about the location, equipment, type of criminal activity, and other pertinent facts will enable the forensic computer investigator to assist the prosecuting attorneys in the preparation of search and seizure warrants. Also, involvement as a member of the raiding team will permit a more tailored plan in which minimal loss of data to the computer and electronic environment will occur.

Initiating the Forensic Computer Investigation

Once a forensic computer investigator is called on to initiate a formal assessment of a case involving a computer, either as an instrument of crime, a repository of data, information associated with a crime, or a target of a criminal act, it will be necessary for the forensic computer investigator to prepare an investigative protocol to correctly gather and preserve any appropriate evidentiary material.

In the collection of evidence from a computer hard drive, it is important to make a bitstream copy of the original storage medium and an exact duplicate copy of the original disk. After the evidence has been retrieved and copied, the **bitstream data copy** of the original disk should be copied to a working copy of the disk so that the analysis of the data will not contaminate the evidence. In the analysis of the digital evidence, you may have to recover data, especially if the users have deleted files or overwritten them. Depending on the type of operating system being used by the suspect, the computer investigator will determine the nature of the forensic computer tools that will be applied. For example, in examining Windows, DOS systems, Macintosh, UNIX, or LINUX systems, one has to understand the file systems that determine how data is stored on the disk. When it is necessary to access a suspect's computer and inspect data, one will have to have an appreciation and working knowledge of the aspects of each operating system (Nelson, Phillips, Enfinger, and Steuart 2004, 50, 51, 54). For example, in Windows and DOS systems, one must understand the following:

- Boot sequences and how to access and modify a PC's system (**CMOS** and **BIOS**)
- How to examine registry data for trace evidence in the user account information
- Disk drives and how data is organized, as well as the disk data structure of head, track, cylinder, and sectors
- Microsoft file structure, particularly clusters, file allocation tables (FATs) and the NTFS; because data can be hidden, as well as files, that may suggest a crime has occurred
- Disk partition in which hidden partitions can be created to hide data

An excellent and detailed explanation of the UNIX and LINUX operating systems can also be found in the *Guide to Computer Forensics and Investigations* (Nelson, Phillips, Enfinger, and Steuart 2004, 74–76, 80).

Additional information on initiating a forensic computer investigation will be provided in greater detail in subsequent chapters of this text. In the interim, a brief taxonomy of crimes impacting the forensic computer investigator may be useful to review:

The computer as an instrument in criminal activity

- Child pornography and solicitation
- Stalking and harassment
- Fraud
- Software piracy

- Gambling
- Drugs
- Unauthorized access into other computer systems
- Denial-of-service attacks
- Data modification
- Embezzlement
- Identity theft
- Credit card theft
- Theft of trade secrets and intellectual property
- Extortion
- Terrorism

The computer as a target of criminal activity:

- Theft
- Virus attack
- Malicious code
- Unauthorized access
- Data modification
- Intellectual property and trade secrets
- Espionage to government computer systems
- The computer as a repository of criminal evidence
- Child pornography and child exploitation materials
- Stalking
- Unauthorized access into other computer systems
- Fraud
- Software piracy
- Gambling
- Drugs
- Terrorism-attack plans
- Terrorist organizations' Web site recruiting plans
- Credit card numbers in fraud cases
- Trade secrets
- Governmental classified documents as a result of espionage activities

A most informative and detailed taxonomy that examines 14 criminal activities and directs the forensic computer investigator to assess these criminal activities against 5 categories where general information may be located and 70 categories in which specific information can be considered is provided in the National Institute of Justice's guide,

Electronic Crime Scene Investigation: A Guide for First Responders (National Institute of Justice 2001, 37–45).

Investigative Tools and Electronic Crime Scene Investigation

Forensic computer investigators have a number of software tools and utilities available for use in analyzing a suspect's computer. Some of the tools available are as follows:

- Safeback
- Maresware
- DIBs Mycroft, version 3
- Snap Back Dot Arrest
- Encase
- Ontrack
- Capture It
- DIBS Analyzer
- Data Lifter
- Smart
- Forensic X

Each agency will equip its forensic computer investigators with the hardware tools necessary to disassemble a computer system and remove necessary components. In many cases, the tool kit will also include necessary materials for packaging, transporting, storing, and evidencing materials. Depending on the workload and caseload of each agency, the use of software and tool kits will vary with the agency's needs and policies.

Legal Issues in the Searching and Seizure of Computers

The Fourth Amendment to the U.S. Constitution limits the ability of law enforcement officers to search for evidence without a warrant. The Fourth Amendment specifically states: "The right of the people to be secure in their persons, houses, papers, and effects against unreasonable searches and seizures, shall not be violated, and no warrants shall issue, but upon probable cause, supported by oath or affirmation, and particularly describing the place to be searched, and the persons or things to be seized."

Searching and Seizing Computers without a Warrant

The U.S. Supreme Court has held that a search does not violate the Fourth Amendment if it does not violate a person's reasonable expectation of privacy. The U.S. Department of Justice's Computer Crime and Intellectual Property Section suggests in their July 2002 revised manual that a reasonable expectation of privacy of information stored in a computer is determined by viewing the computer as a closed container such as a file cabinet. The Fourth Amendment generally prohibits law enforcement from accessing and viewing information stored in a computer without a search warrant. However, this reasonable expectation of privacy can be lost if a person relinquishes control to a third party by giving a floppy diskette or CD to a friend, or bringing the computer to a repair shop (U.S. Department of Justice 2002, 8–10).

The Fourth Amendment applies only to law enforcement officers and does not apply to private individuals as long as they are not acting as an agent of the government or with the participation or knowledge of any government official. Therefore, if a private individual acting on his or her own conducts a search of the computer and makes the results available to law enforcement, there is no violation. In *United States v. Hall*, 142 F. 3rd, 988, (7th Cir. 1998), the defendant took his computer to a computer repairman, who, in the process of evaluating the computer, noticed computer files that on examination contained child pornography. The repairman notified the police, who obtained a warrant for the defendant's arrest. The court upheld the action and rejected the defendant's claim that the repairman's search violated his Fourth Amendment rights (U.S. Department of Justice 2002, 13).

There are exceptions to requiring a warrant in computer cases, and these situations involve consent, exigent circumstances, the plain-view doctrine, and searches incident to arrest. The issues that emerge in consent center around parents, roommates, and siblings, and whether they have the authority to consent to a search of another person's computer files. The courts have held that parents can consent to searches of their minor child's room, property, and living space. However, if the child is living with the parents and is a legal adult, pays rent, and has taken affirmative steps to deny access to his parents, the courts have held that parents may not give consent to a search without a warrant (*United States v. Whitfield*, 939 F. 2nd, 1071, 1075 [D.C. Cir. 1991]).

The exception to requiring a search warrant in exigent circumstances applies if it would cause a reasonable person to believe that entry was necessary to prevent physical harm to the officers or other persons or to prevent the destruction of evidence.

The exception to requiring a warrant under the plain-view doctrine permits evidence to be seized if, in the process of conducting a valid search of a computer hard drive, the officer finds evidence of an unrelated crime while conducting the search (*Horton v. California*, 496 U.S. 128 [1990]). However, the exception to a warrant under the plain-view doctrine does not authorize agents to open and view the contents of a computer file that they were not otherwise authorized to open and view. In *United States v. Carey,* 172, F. 3rd 1278, (10th Cir. 1999), a detective, while searching a computer hard drive for drug trafficking evidence, found a JPG file and discovered child pornography. The detective then spent 5 hours and downloaded several hundred JPG files in a search not for drug trafficking, which the original search warrant authorized, but for more child pornography. The defendant argued to exclude the child pornography files on the grounds that they were seized beyond the scope of the warrant. The government argued the detective seized the JPG files because they were in plain view. The Tenth Circuit rejected the government's argument, stating that the first JPG file was appropriate, but they could not rely on the plain-view doctrine to justify the search for additional JPG files containing child pornography evidence beyond the scope of the warrant (U.S. Department of Justice 2002, 21–22).

In the situations of searches incident to an arrest, the courts have permitted a search without a warrant as an exception for electronic pagers. However, the courts have not resolved this issue with reference to electronic storage devices, such as PDAs, cellular phones, laptop computers, or those devices that contain more electronic information than pagers.

Searching and Seizing Computers with a Warrant

To obtain a search warrant from a judicial officer requires the preparation of two important documents. The law enforcement officer must first prepare an affidavit, which is a statement made under oath that describes the basis on which the officer believes the search is justified by probable cause. The second document is the actual search warrant, which must describe the place to be searched and the items or persons to be seized. In federal search warrants it is also recommended that the officer or agent include an explanation of the search plan or strategy.

In criminal investigations involving the use of computers, it is important to describe in the search warrant whether the property to be seized is the computer hardware or the information that the computer contains. If the computer is an instrument of a crime, then the search warrant would specify the computer hardware itself. On the other hand, if the officer's probable cause is based on the information stored in the computer, then the search warrant would focus on the content of the relevant files rather than the storage device (*United States v. Gawrysiak*, 972 F. Supp. 853, 860 [D. N.J. 1997], Aff 'd 178 F. 3d 1281 [3D Cir. 1999; also *Davis v. Gracey*, 111 F. 3D 1472, 1480 [10th Cir. 1997]; U.S. Department of Justice 2002, 50–51).

Although criminal investigations and the requirements for fulfilling search warrant requirements will vary from state to state, as well as from state to federal jurisdiction, under the federal rules of criminal procedure, Rule 41 would be the guiding force in the previously described search warrant preparation and application. Another important consideration in preparation of search warrants will be whether the target of the investigation is a business, because the economic aspect of seizing computers could have devastating consequences for a legitimate business.

In fact, search warrant requirements for business establishments have to address the issue of reasonable expectation of privacy that people have in their office space. The issue of consent by business managers, supervisors, coworkers, and whoever has common authority over an area can be an important aspect if the search were conducted without a warrant. Another aspect of searching workplace environments would be the public workplace as opposed to the private workplace. The reasonable expectation of privacy would be at variance in the public workplace as opposed to the private workplace.

The complexity of forensic computer investigations entails an appreciation and understanding of the legal requirements, both in terms of the elements of an offense and the procedural requirements for effecting a search and seizure of evidentiary material. In addition, the forensic computer investigator is also required to understand the intricacies of the computer itself, and how it might be used either as an instrument to commit a criminal offense or as a repository of information about crimes.

Summary

This chapter has provided an introduction to the paradigm change that is occurring with reference to crime: Today's criminals are using computers as their tools to take advantage of new technological possibilities. The forensic computer investigator has to be prepared to investigate criminal acts in which the computer may be a target of the criminal. This implies that individual, corporate, and government computers are at risk as targets of opportunity. The data that resides in these computers has value and is subject to loss, in some cases at enormous expense. Therefore, the forensic computer investigator must be cognizant of this environment and how to develop systematic plans for investigating those who use computers and sophisticated electronic equipment in the commission of criminal acts. The computer also serves as a repository of data in which the criminal has either stored the fruits of his or her criminal activity, or provides evidence as to the unlawful actions the criminal has utilized in using his or her computer to attack or harm another individual, corporation, or government.

The categorization of an electronic crime scene rich in new technologies that store data and information of potential evidentiary value suggests that we must educate our law enforcement officers to recognize characteristics of this new environment so that they can function effectively in it.

Questions

1. Enumerate the major challenges investigators confront with digital and electronic evidence in the new electronic crime scenes.

2. How does the investigator deal with computer crime cases that involve encryption?

3. The global nature of computer crime involves multi-jurisdictional issues that challenge the very legal structure of all nation's legal and statutory codes; what must the computer crime investigator do to be in compliance with appropriate legal requirements?

4. It is generally accepted as good police practice that, when entering an electronic crime scene, the investigator should follow guidelines, identify and discuss these eight major guidelines.

5. Computer crime cases involve a level of complexity that requires the investigator to possess a knowledge of both legal requirements and procedural requirements for effecting search and seizure of evidentiary material. Identify the most common mistakes made by the investigator.

6. Describe how the forensic computer investigator collects evidence from a hard drive.

7. Why should a bit stream copy of the original storage medium and an exact duplicate copy of the original disk be made?

8. In the analysis of digital evidence, is it possible to recover deleted files? If so, describe the process one would have to consider using.

9. Identify the typical cases with which investigators will be involved when the computer is used as an instrument in criminal activity.

10. When the computer is a target of criminal activity, what types of cases is the investigator most likely to encounter?

References

National Institute of Justice, *Electronic Crime Scene Investigation: A Guide for First Responders*, Washington, D.C.: U.S. Department of Justice, July 2001.

Nelson, B., Phillips, A., Enfinger, F., and Steuart, C., 2004, *Guide to Computer Forensics and Investigations*, Canada: Thomson Course Technology, 25 Thompson Place, Boston, MA 02210.

Parker, D. B., *Fighting Computer Crime: A New Framework for Protecting Information*, New York: John Wiley and Sons, 1997.

U.S. Department of Justice, Computer Crime and Intellectual Property Section, Criminal Division. CCIPS Manual, July 2002.

SECTION VI

Forensic Application of the Social Sciences

Forensic Psychology

Louis B. Schlesinger

Law and the legal system are the formal means by which society attempts to regulate human behavior. Because psychology is the study of behavior, the principles of behavioral science are relevant to every area of law. The relevance of these principles was explicitly recognized in the early 20th century, when researchers found a direct relationship between the vagaries of human memory and perception, and the reliability of eyewitness testimony and other legal processes. Thus, the subspecialty known as **forensic psychology**—the application of psychological findings to legal processes—was born. Since the mid-20th century, this intersection of psychology and law has grown steadily in two general directions: (1) psychological research findings have increasingly been used to inform various legal processes; and (2) the practice of clinical forensic psychology has become an integral part of the overall field of forensic science. This chapter will review these two areas of forensic psychology.

The chapter also presents the following relevant topics: the history of the emergence of forensic psychology as a specialized discipline; the philosophical and procedural differences between psychology and law, and between clinical and forensic assessments; the use of psychological tests, narcoanalysis, and hypnosis in a forensic context; and the types of deception commonly encountered in forensic examinations of offenders.

Psychological Research and Law

Almost a century ago, Hugo Münsterberg (1908) realized that psychological research findings had the potential to inform the criminal justice system about the unreliability of eyewitness accounts. "Every chapter and subchapter of sense psychology may help clear up the chaos and confusion which prevail in the observations of witnesses." But it was not until 65 years later that psychologists developed specific research designs to assess the accuracy of eyewitnesses. In 1974, for example, Buckhout (1974) published the results of an experiment in which subjects witnessed a purse-snatching crime and were asked to identify the culprit. Only 7 of 52 witnesses made the correct identification. Buckhout concluded, after a series of similar findings, that memory is selective and not a copying process. Recollection is falsified with the passage of time and is influenced by beliefs, motives, stress, environmental factors, expectancy, and stereotypes. On the basis of his research, Buckhout asserted that identification by eyewitnesses is faulty approximately 90% of the time.

Additional eyewitness experiments by Loftus and Palmer (1974) were equally compelling, and led these psychologists also to conclude that eyewitness testimony is highly unreliable. Repetitive research studies—over 2000 by 1995 (Cutler and Penrod, 1995)—have again demonstrated the unreliability of eyewitness testimony. These are important findings considering that courts rely heavily on eyewitness identification. If an eyewitness to a crime testifies that he or she saw the defendant commit the act, the chances of conviction are extremely high, despite the presentation of a credible alibi.

Problems in the accuracy of eyewitnesses stem from two sources: the unreliability of the human information processing system, and the procedures used by law enforcement officials to obtain eyewitness accounts of crimes. Kassin et al. (2001) cite the numerous "criminal justice procedures" that can seriously affect an eyewitness's accuracy: "Wording of questions, line-up instructions, confidence malleability, mug shots, induced bias, postevent information, child witness suggestibility, attitudes and expectations, hypnotic suggestibility, alcohol intoxication, cross-race bias, weapon focus, the accuracy–confidence correlation, the forgetting curve, exposure time, presentation format, and unconscious transference."

Notwithstanding a century of psychological research, the courts seemed reluctant to accept the proposition that eyewitness identification is unreliable. That stance changed, however, with the advent of DNA evidence, which proved conclusively that many defendants had been falsely convicted, solely on the basis of uncorroborated eyewitness testimony. As a result of such shocking cases, in 1999 attorney general Janet Reno familiarized herself with the voluminous eyewitness identification research literature in psychology. She subsequently ordered the National Institute of Justice to develop national guidelines for use by law enforcement officials when they collect eyewitness evidence (Wells et al., 2000).

The same type of psychological research is being conducted on earwitness testimony (Wilding et al., 2000; Yarmey, 1995). Some of these studies (Ollson et al., 1998) have found that earwitness accuracy is even poorer than eyewitness accuracy, so that the earwitnesses had even more confidence in their inaccurate identifications than the eyewitnesses did. Thus, psychological research—which began in the laboratory almost a century ago, with studies on human perception and memory—is now helping to change some procedures in the criminal justice system.

Considerable attention also has been given to the psychological study of false confessions (Kassin, 1997, 1998; McCann, 1998), a phenomenon that is much more common than was previously thought. Just as in eyewitness identification, certain characteristics (such as low self-esteem, eagerness to please, anxiety, compliance, and need for notoriety), as well as procedural tactics used by law enforcement officers, can lead individuals to make false confessions.

Freud (1959) believed that many criminals harbor unconscious guilt feelings and, therefore, commit crimes to be punished; some false confessions might be connected to similar psychodynamics. Other psychological factors might also be involved (Gudjonsson, 1999). For example, vulnerable individuals who have become fatigued or frightened during an interrogation may develop a desire to confess, even though they are not guilty of the crime. Kennedy (1961) reported the case of a false confession, later retracted, that resulted in the execution of an innocent man. The accused, who was illiterate and confused, explained that he had confessed because he was upset and fearful of the interrogation. However, just as a subject may temporarily delude himself into believing he is guilty, he may also delude himself into believing he is not guilty, especially when the offense was ego-dystonic and explosive.

The application of psychological research, in general, by the legal system has had an interesting history. In 1954, in the landmark case of *Brown v. Board of Education* (1954), the U.S. Supreme Court justices—for the first time, and only in a brief footnote—cited psychological research that helped bolster their decision. Twenty-five years later, psychological science had become so integrated into the legal system that, in the U.S. Supreme Court case of *Ballew v. Georgia* (1978), 25 research studies (on the effect of group size on a jury's decision-making behavior) were referenced. Justice Blackmun's opinion in this case reads as though it were lifted directly from an article in an experimental psychology journal.

Psychological research in many other areas, such as the ability of mental health professionals to predict violent behavior, has also been cited by the U.S. Supreme Court in several recent decisions. Although courts have been slow in accepting the credibility and usefulness of psychological research, that barrier has long since been removed (Loftus and Monahan, 1980), psychological studies will surely continue to influence the functioning of the legal system.

Clinical Forensic Psychology

In 1909, William Healy established the first court clinic, an adjunct to the Cook County Juvenile Court in Chicago (Witmer, 1940). Because the juvenile justice system has always emphasized rehabilitation, as opposed to punishment, psychological intervention with juveniles was a natural starting point. In 1914, the first adult court clinic was established; and soon mental health practitioners, at that time mostly psychiatrists, were routinely evaluating and treating criminal defendants, either informally or through various legislative acts. Most of the work in the adult clinics was centered on pretrial examinations to determine competency and criminal responsibility, or presentence evaluations to help the judge determine proper disposition for the offender, including the sex offender.

Society has always been in a quandary about how to handle the repeat sex offender, whose conduct obviously stems from psychological sources. The serial rapist, child molester, exhibitionist, voyeur, or sex murderer has been present since premodern times (Schlesinger, 2000), and the serial nature of such offenses, as well as their general repulsiveness, made resolution of this problem foremost in the minds of legislators. Accordingly, mental health professionals offered their expertise in both diagnosis and treatment of such offenders. Beginning in the mid-1930s, many states took psychologists and psychiatrists up on their offer and enacted various **sexual psychopath laws**, which mandated the evaluation and treatment of sex offenders. The expertise of psychiatrists and psychologists was readily accepted, in part, as a way to allay public hysteria resulting from some brutal sex crimes.

By the end of the 1960s, almost every state had developed specialized diagnostic and treatment programs—whether free-standing or a part of a prison—to manage and solve the sex offender problem (Vuocolo, 1969).

By working in sex offender programs, psychologists developed considerable expertise with this population of offenders, whose psychopathology was very different from the type usually encountered in general clinical practice. Research quickly grew (Brancale and Ellis, 1952), and there was great optimism about the ability of scientific psychology to solve a major criminal problem. However, by the early 1980s, optimism faded into embarrassment as many offenders—deemed suitable for release—continued to commit offenses. Despite the noble intentions of both the law and mental health practitioners, many states concluded that sex offenders simply could not be successfully treated. By the late 1980s, therefore, many states closed their programs (in part or whole) and mandated longer periods of incarceration; treatment now played a secondary role, if any role at all.

Forensic psychology was growing on other fronts as well. In the 1960s, as a result of President Kennedy's initiative, community mental health centers were established. Psychologists began working in these new outpatient clinics, treating a variety of conditions. Juvenile delinquents—and adults whose crimes were the result of substance abuse and other psychological problems—were referred for outpatient treatment, either in lieu of more costly incarceration or following release from prison. In some mental health clinics, specialized programs were established to treat offenders. The courts began requesting progress reports, and frequently the legal status of the released offender, such as return to prison or continuation of parole, was based largely on his progress in therapy.

Prior to the 1970s, most forensic evaluations requiring court testimony were provided by psychiatrists rather than psychologists. Clinical psychology, which only began after World War II, did not gain licensure as an independent profession in all states until the late 1970s. Accordingly, the only way a psychologist could become involved in a forensic case, as an independent private practitioner, was under the auspices of a psychiatrist. Occasionally, psychologists were called on to testify in court, but usually the testimony was limited to the results of **psychological testing**. Psychologists had not yet gained public acceptance as members of an independent profession and, therefore, were not permitted to consider specific psycholegal issues such as competency or criminal responsibility; these were considered "medical" issues, under the sole purview of psychiatry.

In several legal cases, beginning with *People v. Hawthorne* (1940), courts were asked to decide whether psychologists should be allowed to give testimony. In *Hawthorne*, the court held that there was "no magic of an MD degree," and allowed properly trained psychologists to testify about a defendant's mental state. In a similarly decided civil case (*Hidden v. Mutual Life*, 1957), the court also permitted psychologists to testify. However, the decision considered landmark is *Jenkins v. United States* (1962), because in this case a federal court allowed psychologists to testify as expert witnesses in criminal matters.

Despite these early court decisions, it was not until the 1980s that a psychologist's testimony was used regularly. During this time, laws were enacted that allowed patients to be reimbursed for psychological services by insurance companies. As a result, mental health care, especially health care provided by psychologists, became available to most of the population, whereas previously it had been provided only to the wealthy or to those who went to publicly funded centers. Thus, psychology gained general acceptance by society as an independent mental health profession. That acceptance was even reflected in all forms of the media, including television and film. Because the court is a reflection of society (and by this time all court participants, such as judge, jury, and attorneys, had become familiar with the works of psychologists), it was natural that psychologists' expertise would be brought into the courts. Many states had to change specific laws, so that instead of allowing forensic examinations to be performed only by psychiatrists, they now included licensed psychologists as well.

By the early 1990s, forensic psychology was well on its way to becoming a major participant in legal decision making. At the beginning of the 21st century, an area of practice that had been previously dominated entirely by psychiatry was now beginning to be greatly influenced by psychology. In fact, psychiatrist Larry R. Faulkner (president of the American Academy of Psychiatry and Law) noted the expanding role of forensic psychologists in his 1999 presidential address (Norko, 2000).

Along the way, forensic psychologists established standards for the profession and appointed credentialing bodies, such as the American Board of Forensic Psychology, to certify that practitioners have expertise in this subspecialty. The path to becoming a forensic psychologist is still typically that of the clinical psychologist, with additional forensic training and experience. Recently, however, several universities have established programs offering doctoral degrees in forensic psychology, specifically designed to provide students with a broad background in basic clinical skills and in relevant areas of law and criminal behavior.

Today, forensic psychologists are involved in assessment, treatment, and provision of testimony in a variety of legal cases in areas such as family law (including custody and visitation matters), civil law (personal injury, workers' compensation, wills, and contracts), and criminal law (including the various types of competencies, responsibility, and sentencing). Many psychologists are even developing subspecialties in forensic psychology, limiting their practice to one or two types of legal issues such as criminal offenses, sexual harassment, or product liability. In the area of treatment, they provide services to offenders in prisons, in lieu of incarceration, or following release, as well as to the victims of crimes.

Psychology and Law: An Uneasy Alliance

Despite the potential value of psychological science in informing every aspect of law, there remains what Melton et al. (1997) refer to as an uneasy alliance between the two disciplines. In many instances, psychologists dread having to go to court, and often shun a case that might eventually involve some type of legal entanglement. The judiciary is equally displeased when mental health professionals come to court. Judge David Bazelon (1974), for example, has expressed his frustration over the frequent inability of mental health professionals to provide clear and helpful testimony. However, the problems that psychologists and other scientifically trained professionals have when they enter the legal arena are much deeper than simply their discomfort with cross-examination or their lack of familiarity with legal standards and procedures. There are basic philosophical differences, or what Melton et al. (1997) refer to as differences in paradigms, between the two disciplines; these differences give rise to major tensions when science enters the courtroom.

Slovenko (1973) has reviewed many of these philosophical differences between the scientific disciplines, such as psychology and law. For example, lawyers use an adversarial approach (or "the fight theory") to arrive at "the truth." Psychologists, on the other hand, emphasize cooperation; they often work together in an attempt to understand a research problem or a difficult patient. Lawyers are concerned with assigning moral responsibility, guilt, and blame. Psychologists are taught (particularly in their clinical training) not to moralize or to make moral judgments about others. If a patient engages in some unethical behavior, the psychologist is trained not to judge the patient, but to accept him and seek to understand him. Time is always a major concern in law, particularly in court schedules. Defendants, for example, have a constitutional right to a speedy trial; this right, as well as a need to keep the jury intact, is always in the mind of the judge. In science, time is not a major issue. It may take years for a psychologist to finish a research project, write a paper, or treat a patient. Any time constraints that psychologists work under are usually self-imposed.

General theories are important in science. Accordingly, scientific psychology is heavily invested in theory; many psychologists even identify themselves professionally as adherents to one theoretical school (psychoanalytic, behavioral, cognitive) or another. In law, however, theory gives way to individual cases and their resolution. In fact, law is often antitheoretical, in the sense that a court will not try a case simply because it raises a theoretically interesting issue.

Thus, when psychologists enter the legal arena, they must realize they are not on their own turf and must play by the rules of the court; they must also fully appreciate the different (and in many ways conflicting or competing) philosophical or paradigmatic approaches between the two disciplines. Table 29.1 lists a number of differences between law and science that often make for an uneasy alliance.

In addition to underlying philosophical or paradigmatic conflicts, differing assumptions about the psychologist's role often create conflict between the two disciplines. For example, psychologists are called to court because they have expertise beyond that of the judge, jury, or layperson. They are not there to answer ultimate legal questions, such as the question of an individual's "sanity," which can only be determined by a jury. Determinations of sanity are essentially moral judgments defined by various legal standards. Psychologists, psychiatrists, and other mental health practitioners have expertise in mental disorders, but

Table 29.1
Philosophical (Paradigmatic) Differences between Law and Science

Law	Science
Process is adversarial	Process is cooperative
Assumes that people have free will	Assumes that people's actions are determined by external and internal forces
Makes moral judgment of responsibility	Does not make moral judgments; science seeks to understand
Proximate (closest) cause is legally responsible cause	All effective causes, no matter how distant, are important
Values are important	Values are less important than formal relationships between events
The practice of law is very much an art	Science attempts to remove tactics
Trials involve ceremony	Very little ceremony is utilized
Law attempts to resolve a specific case	Science attempts to contribute to general principles
Loyalty is primarily to one's client	One's loyalty is given only to the truth
Rules of evidence screen data	Scientific method is utilized
Legal system is frequently pressed by time	Scientists usually are not pressed by time
People decide outcome of case	Objective experiments decide results
Subjectivity is accepted and even considered desirable	Subjectivity must be reduced through objective experiments
Procedure is critical	Procedure is comparatively less important
The average person, layperson, or reasonable person is the standard of behavior	The unique case is important
Attorneys try to develop evidence helpful to their side and hide evidence that is not helpful	Science is interested in evidence that does not support the theory; that is how new understanding emerges
Law will suppress evidence even if it demonstrates guilt (if evidence is not properly obtained)	Science uses evidence from any source

not in legal sanity. Accordingly, many leading forensic psychologists advise their colleagues to refuse to answer ultimate legal questions posed to them because these questions are to be decided by the jury, not by an expert. However, as a practical matter, attorneys and judges routinely ask for the expert's opinion, or they ask the expert to apply or translate psychological findings to legal standards, such as a defendant's state of mind or his competency to stand trial. If the psychologist refuses to answer these questions, his services are deemed not useful and are subsequently not requested. The issue of experts addressing "ultimate issues" continues to be hotly debated within both professions (Champagne, 1991; Shuman, 1996).

Forensic Assessment: Distinctions between the Clinical and Forensic Approaches

Within the profession itself, even among experienced practitioners, the distinction between

Table 29.2
Dimensions Distinguishing Therapeutic from Forensic Assessment

1. ***Scope***. Clinical assessments stress diagnosis or treatment needs, whereas forensic evaluations more commonly address narrowly defined, nonclinical events or interactions. In forensic assessments, diagnosis or treatment needs are in the background.

2. ***Importance of Client's Perspective***. In a clinical setting, the focus is on understanding the client's unique view of the situation or problem. In a forensic setting, a more "objective" appraisal is necessary, because the forensic examiner is concerned primarily with accuracy; the client's view, though important, is secondary.

3. ***Voluntariness***. Persons seeking mental health therapy commonly do so voluntarily. Persons undergoing forensic assessments commonly do so at the behest of a judge or an attorney.

4. ***Autonomy***. In the clinical setting, patients have some say in decisions about the objectives of treatment and the procedures to be followed. The objectives in a forensic evaluation are determined by the relevant statutes or legal "tests" that define the legal dispute.

5. ***Threats to Validity***. Although unconscious distortion of information is a threat to validity in both clinical and forensic contexts, the threat of conscious and intentional distortion is substantially greater in the forensic setting.

6. ***Relationship and Dynamics***. Treatment-oriented interactions emphasize caring, trust, and empathic understanding as building blocks for developing a therapeutic alliance. Forensic examiners may not ethically nurture the client's perception that they are there in a "helping" role. In forensic settings, there are divided loyalties and limits on confidentiality. Concerns about possible manipulation of defendants in the legal context dictate that the forensic examiner maintain an emotional distance from the subject.

7. ***Pace and Setting***. In a clinical-therapeutic setting, evaluations may proceed at a leisurely pace. Diagnoses may be reconsidered over the course of treatment and revised well beyond the initial interviews. In a forensic setting, a variety of factors, including court schedules and limited resources, may limit the opportunities for contact with the client and place time constraints on completing the evaluation or reconsidering opinions.

Source: Melton, G.B. et al., *Psychological Evaluations for the Court*, 2nd ed., Guilford Press, New York, 1997. With permission.

a clinical and a forensic assessment often is poorly understood. This lack of understanding stems from their clinical training. Clinical psychologists (as well as psychiatrists and other mental health professionals) are taught to evaluate their patients through interviews, psychological tests, and, sometimes, additional information from family members, hospital and school records, and reports from previous therapists. Following the evaluation, a diagnosis is made, and a treatment plan is developed. Clinical practitioners are also taught to listen to their patients and to accept their patients' symptom descriptions as valid. In some instances, a patient's account of symptoms or presenting problem may seem questionable, but these instances are rare. Lying, deceit, exaggeration, and **malingering** do occur in daily clinical practice, but they are secondary matters because the emphasis is primarily on clinical diagnosis and implementation of a treatment plan.

In forensic practice, however, the examiner cannot automatically accept the litigant's description of what has occurred, because the litigant has an obvious motive to lie, exaggerate, or distort symptoms and events. For example, he may be seeking to create a psychological defense, or to recover money damages, or to transfer from jail to a hospital. Therefore, the traditional clinical approach, which most mental health professionals have been trained in, cannot be used in forensic assessments. The criminal defendant's version of events, as well as his background and symptom description, must be corroborated (Davidson, 1965; Melton et al., 1997). Unfortunately, the distinction between a clinical and a forensic assessment is not stressed enough in books on forensic psychology or psychiatry. Many well-known textbooks (Blau, 1998; Cooke, 1980; Hess and Weiner, 1999) do not even mention the basics of a forensic assessment, but instead focus on various laws, legal tests, and legal standards that forensic practitioners need to know.

Melton et al. (1997) describe seven distinctions between a forensic and a clinical (or what they refer to as a therapeutic) assessment. They note differences in the scope of evaluation, the client's perspective, voluntariness, autonomy, threats to validity, relationship and dynamics, and pace and setting (Table 29.2). These differences are in addition to the fundamental distinction. In a forensic assessment, there is a high likelihood that the defendant may not be truthful, a circumstance that does not exist in the majority of nonforensic assessments.

Unfortunately, many mental health professionals who practice in a forensic setting do not always fully understand these distinctions. At least, they often do not demonstrate that understanding in the process of arriving at their conclusions. In a study of the use of third-party information in forensic assessments (Heilbrun et al., 1994), for example, a majority of the forensic evaluators said that they did "incorporate" such information, but their reports did not indicate how this information was integrated, if at all. The following case is illustrative of the improper way to conduct a forensic evaluation of a young man charged with murder.

Case Study 1

A 22-year-old male (A) was arrested and subsequently charged with the murder of an 18-year-old associate. He gave the police two different versions of what occurred. First, he said that his cousin (who had been recently released from jail) was the murderer, and that he himself had only acted as an assistant "because I feared for my life and my family's life." Next, A told the police that this statement was untrue and that his cousin was not even present. Instead, he said in the course of an argument he had killed his friend; he then removed his friend's clothing, so that no fingerprints or trace evidence would be found, and buried the body near a local stream. Several days later, A gave a third statement. In this version, he was attacked by five youths, who hit him over the head with a brick and killed his friend; when he awakened, he saw them burying the body close by. He was frightened, left the scene, and did not report it to the police because he was fearful. One week later, A gave yet a fourth version of events: he was with a group of friends, and they killed the victim after he (A) had passed out from alcohol and marijuana.

A was evaluated by a psychologist retained by the defense attorney. The defendant gave the expert another version of the homicide. This time A said he heard voices that commanded him to choke his friend and bury him. This statement, the fifth rendition of what occurred, evidently prompted the defense expert to conclude that the defendant was psychotic, and insane, at the time of the homicide. This expert was aware of the facts of the case; at least, he noted in his report that he had read all the police reports and statements. Apparently, however, he did not even consider the possibility that the defendant was lying and malingering. Here, the expert approached the defendant as he would a patient seen in a hospital or private office setting. In fact, he referred to the defendant in his report as "the patient," which he was not.

The next case raises another important issue frequently encountered in a forensic psychological consultation. An attorney will retain an expert because of his or her extensive knowledge and experience with a particular disorder such as substance abuse, sex offenses, juvenile delinquency, posttraumatic stress, or neuropsychological dysfunctions. Often, however, these experts, who have mainly treated patients with these conditions in private practice, have an insufficient understanding of the basics of conducting a forensic examination. Thus, like the expert in Case 1, they approach the case as if the defendant were a patient seen in consultation in an office setting. In Case 2, a defense-retained psychologist, who had extensive experience in treating victims of battered woman syndrome, evaluated a defendant charged with two homicides.

Case Study 2

A 38-year-old female (B) with a 20-year history of drug abuse and related offenses (including drug selling and prostitution) rented a room in a boarding house occupied mainly by other drug addicts. There, she met a 42-year-old drug dealer (A) and began a 5-month intimate relationship with him. She continued to sell drugs and became known as A's "enforcer." She carried a gun and frequently intimidated drug buyers by pistol-whipping them or hitting them over the head with a pipe for not paying on time, or for no reason at all.

The victims in this case were two drug users who owed A and B $600. According to numerous witnesses, B urged A to track down the victims, who were then abducted by B and A, taken to the boarding house, and—again at B's instigation—killed. Several people, who were in the hallway, overheard what was going on because the walls in the boarding house were "paper thin." These witnesses heard the female victim begging B for her life and struggling as B approached her and strangled her. They heard the male victim choking as B shoved newspaper down his throat while A held a gun to his head. They also heard B, who seemed to be enjoying the intimidation and the killings, suggest that she and A should take the bodies to a local field and set them on fire.

A and B were arrested shortly afterward. A refused to give a statement, but B gave four different versions of what occurred, implicating others and finally admitting that she participated, but only as an assistant to A, who she said was the primary actor. Nevertheless, she was charged with murder and sent to jail, where she was awaiting trial.

After B had been in jail for 3 years (her trial delayed by a number of legal maneuvers), her defense attorney retained a psychologist with expertise in the treatment of battered women. The psychologist began her interview by asking the defendant to describe her relationships with prior boyfriends, including her codefendant. B gave a history typical of battered woman syndrome. During the 3 years that she had been incarcerated, she attended weekly battered woman's workshops (as did all the female inmates), where she heard many lectures on the topic. The psychologist concluded (basing her findings solely on what B told her) that the defendant had battered woman's syndrome, was passive and compliant, had low self-esteem, and killed the victims to prevent herself from being killed by A. "B was intimidated at the time of the murders into helping A, which would be considered reasonable for any person in that situation, even more so for someone who was battered and who was passive and compliant. One can attribute to her the status of being a kidnapping victim or the victim of a terrorist. She was his victim and had to comply in order to save her own life." This expert did not ask B about several witness statements that described her as aggressive, intimidating, and taking the lead role in the murders. She also did not consider B's criminal record, which included a long history of violence, drug selling, prostitution, and several assaults.

When B's case came to trial, the jury concluded that she was responsible for both murders and that the battered woman plea lacked merit. In this case, the defense psychologist was not incompetent or untruthful, but she simply did not know how to conduct a forensic evaluation. She was basically practicing clinical psychology in a forensic setting, approaching the defendant as she would any patient who came to her private office. Although this expert—like the expert in Case 1—reported that she had read the legal discovery, her conclusions demonstrate that she ignored these findings or, at the very least, believed that her own evaluation of the defendant was more valid. Neither expert helped the defendants or provided much psychological understanding of their crimes.

Psychological Testing: Traditional and Specialized Forensic Tests

Psychological testing is a quantitative or quasi-quantitative method of evaluating personality, psychopathology, and mental functioning. For many years, testing has been used to supplement information obtained through clinical interviews and various background records. Testing has always had general appeal because its purpose, in large part, is to reduce the subjectivity of the clinical evaluation, as well as to assess the individual from a different perspective. Because of its emphasis on objectivity (often including norms and standards), psychological testing has been used a great deal in forensic assessments, where the need for accuracy is paramount (Heilbrun, 1992).

Table 29.3
Traditionally Used Psychological Tests

Personality Inventories	**Tests of Memory Functioning**
California Psychological Inventory (CPI)	Wechsler Memory Scale Revised (WMS/WMS-R)
Millon Clinical Multiaxial Inventory (MCMI/MCMI-2)	
Minnesota Multiphasic Personality Inventory (MMPI/MMPI-2)	**Tests for Neuropsychological Impairment**
	Halstead–Reitan Neuropsychological Battery
Projective Personality Tests	Luria–Nebraska Neuropsychological Battery
Bender–Gestalt	
Projective Figure Drawings	**Tests for Specific Disorders**
Rorschach	Beck Depression Inventory (BDI)
Thematic Apperception Test (TAT)	Michigan Alcohol Screening Test (MAST)
	Psychopathy Checklist Revised (PCL-R)
Tests of General Intellectual Functioning	Structured Interview of Reported Symptoms
Wechsler Adult Intelligence Scale Revised (WAIS-R/WAIS-III)	
Wechsler Intelligence Scale for Children (WISC-III)	

Source: Melton, G.B. et al., *Psychological Evaluations for the Court*, 2nd ed., Guilford Press, New York, 1997. With permission.

Table 29.3 lists a number of traditionally used psychological tests.

Melton et al. (1997) offer several criticisms and purported limitations of traditional psychological testing in a forensic context. These authors believe that "most traditional tests have [not] been developed or validated specifically to inform judgments about legally relevant behavior." They also argue that such testing is unnecessary for three reasons: (1) the tests do not directly address specific legal questions that the evaluation is attempting to answer; (2) the tests represent compilations of characteristics found in groups of individuals, so that the results might not apply to a particular individual being assessed; and (3) testing taps current psychological functioning, whereas legal questions frequently relate to a defendant's previous psychological functioning.

These views, in our estimation, are too narrow and somewhat misleading. Testing is not designed, nor should it be used, to address any specific legal standard. Nevertheless, it helps the examiner obtain a more complete understanding of the individual being assessed, and therefore to make an accurate psycholegal determination. As Heilbrun (1992) notes, there is often an underlying psychological construct that psychological testing can help to clarify. For instance, in many cases, an underlying problem not initially recognized through interviewing (such as organicity, deviant sexual preoccupation, or thought disorder) has been detected only as a result of psychological testing.

The notion that testing has little value because of its use of norms and aggregate data makes little sense because the primary purpose of testing is to use established standards to compare one person with others on various characteristics. Finally, although some tests do tap current functioning, some tests also assess more stable traits and dimensions. For example, if an antisocial personality disorder (or schizophrenia) was present at the time of testing, it very likely was present at the time of the offense. In general, the value of psychological testing and its forensic application depends largely on the psychologist's skill in making test findings understandable and relevant to attorneys, judges, jury members, and others with no background in behavioral science.

Projective Tests

Projective tests developed slowly over an extended period (Rabin, 1968). They are based on the notion that if an individual is shown an ambiguous stimulus and asked to respond to it, his or her responses will reveal aspects of his or her personality. In other words, inner thoughts, wishes, conflicts, and feelings will be projected onto the ambiguous stimuli, enabling the psychologist to see the subject's inner life.

Rorschach

Perhaps the most widely used projective test is the Rorschach, developed by the Swiss psychiatrist Hermann Rorschach (1921) and brought to the United States by David Levy around 1924 (Klopfer and Kelly, 1942). In this test, a subject is shown a series of 10 inkblots and is asked to describe what he or she sees. Although this test has had its critics, mostly nonpractitioners, it has sustained continued popularity for more than 80 years (Piotrowski and Keller, 1992). It has been used not only in clinical settings but also in forensic contexts, where (contrary to popular myth) its scientific validity has only rarely been questioned (Meloy et al., 1997; Weiner et al., 1996). The Rorschach's primary use is to assess personality structure, dynamics, presence or absence of a thought disorder, and accuracy of reality testing.

Thematic Apperception Test (TAT)

Another projective test, the Thematic Apperception Test (TAT), has had a long and distinguished history, beginning with the early work of Murray (1938; 1943; 1951); it continues to be a widely used and widely respected projective technique (Geiser and Stein, 1999). In this test, a subject is shown a number of pictures depicting various everyday situations and is asked to create a story based on the picture; in the process, it is assumed, he will reveal his wishes, thoughts, conflicts, feelings, and motives. One of the main attributes of the TAT is its simplicity: "The manifest material yielded by the TAT is not mysterious in appearance. The comparatively untrained person can appreciate fabricated stories. He needs little acquaintance with technical symbols and test scores to obtain at least a superficial feeling for the moods and perspective of the subject who produced the stories .... Even the experienced tester profits from this property, which differentiates the TAT from other tests" (Rosenwald, 1968).

Although the Rorschach has enormous value in elucidating personality structure, level of personality integration, and extent of inner cohesiveness, the TAT is often more helpful in uncovering relevant psychodynamics. Its usefulness is demonstrated by the response of a 22-year-old offender who committed a murder following a sexual assault. Some important psychodynamics were revealed through the story he created for one TAT card: "This is a sex crime. A rape crime. He feels sorry. He forced her. She didn't want to do it and he did. He liked the woman. He then kills the woman so she doesn't go and talk to the police. He chokes her. He thought he'd be in jail for rape, so he decided to choke her. He thought she would tell the cops that he raped her. He also killed her because he didn't want his wife and children to know he raped a woman. He doesn't get caught because the police don't have any proof he killed her. He liked killing her, too." In other instances, a subject may recognize the purpose of the TAT and may intentionally create nonrevealing stories to conceal his or her inner motives and dynamics. Although such a guarded protocol may not be helpful in uncovering the offender's inner life, it may point out a conscious attempt to conceal what is going on internally, also an important finding.

Projective Figure Drawings

Another projective technique, Projective Figure Drawings, has been used as part of the traditional psychological test battery. This test is easy to administer and has a rich clinical yield (Hammer, 1980). The subject is merely asked to draw a picture of a house, a tree, and a person; a person of the opposite sex; the worst thing that he or she can think of; or similar topics, such as a person in the rain, a family, and the like. Projective drawings are perhaps the least standardized of the projective tests; therefore, interpretations should be made only at a level of face validity (i.e., a level that can be easily understood by an average person). As Uhlin (1978) has noted, "In order for projective drawings to function alone in criminal proceedings, they must be powerful and descriptive to the degree that [they] will touch the sensitivity of untrained laymen and overcome what may be strong personal subjectivity." We, therefore, recommend that examiners avoid "wild" interpretations based on clinical lore (a hole in the trunk of a tree supposedly representing the developmental

Figure 29.1 Sketch by an adolescent who committed an unprovoked attack on a female.

Figure 29.2 Sketches by a 13-year-old who sexually assaulted several female classmates.

stage of childhood trauma, absence of smoke from the chimney indicating lack of family warmth, etc.).

Sometimes an offender's drawings will reveal very little, because he or she may be resistant or may exert little effort. Figure 29.1 depicts a sketch done by an adolescent who committed an unprovoked attack on a pregnant woman. Note the line quality of the stick in comparison with the rest of the drawing. A 13-year-old boy who sexually assaulted several female classmates produced the sketches in Figure 29.2. He was preoccupied with an undescended testicle.

In some cases, an offender's random drawings, not done in the context of formal testing, can also reveal inner thoughts and fantasies. Figure 29.3 shows sketches made by individuals who had sadistic fantasies but denied engaging in such activity. They considered their drawings a mere pastime.

The Bender–Gestalt was initially developed as a screening test for organicity (Bender, 1938) and only later came to be used as a projective technique (Hutt, 1968). Here, the subject is asked to copy nine geometric figures on a piece of blank paper. The presence of organicity (subtle forms of brain damage) is determined by perceptual-motor incoordination, distortion, rotation of designs, and other indicators. Several scoring methods have been developed

to help objectify the detection of organic involvement (Hutt, 1968; Koppitz, 1964). When the Bender–Gestalt is used as a projective test, the psychologist interprets the way the subject draws the various geometric designs. As with the interpretation of figure drawings, interpretations of the Bender–Gestalt should be made with caution, common sense, and at the level of face validity; and speculative or "wild" conclusions should be avoided.

Personality Inventories

Personality inventories are psychological tests that, unlike the projective techniques, are highly standardized and have considerable empirical validation for what they are designed to assess. These inventories (e.g., the Minnesota Multiphasic Personality Inventory [MMPI/MMPI-2], the California Psychological Inventory [CPI], and the Millon Clinical Multiaxial Inventory [MCMI/MCMI-2]) typically assess personality traits and characteristics and provide a general personality profile of the subject. Many forensic psychologists use these tests as part of the standard test battery.

Other personality instruments are designed to assess specific disorders, such as psychopathy (the Psychopathy Checklist, [Hare et al., 1990]), or to assess malingering

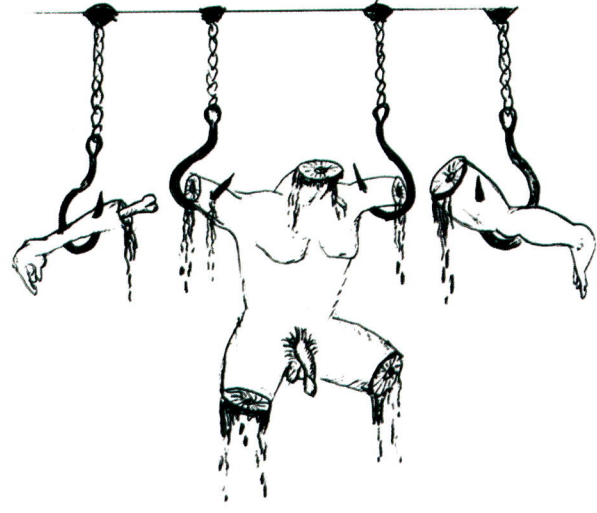

Figure 29.3 Sketches by individuals who had sadistic fantasies but denied engaging in sadistic behavior.

(Structured Interview of Reported Symptoms, [Rogers et al., 1991]). Personality inventories or structured interviews can be especially helpful when they are used in conjunction with clinical findings and projective tests (Ganellen, 1996). Perhaps the greatest value of personality inventories is their capacity to spell out, in an objective way, characteristics of the offender that might not otherwise be discerned, or to corroborate a finding that is detected clinically and on projective tests.

Intellectual and Cognitive Assessment

Intellectual and cognitive evaluation is essential in any forensic assessment. Conclusions about questions such as an offender's criminal responsibility or competency can change dramatically if he or she is found to be intellectually limited or perhaps mentally retarded. Using instruments, such as the Wechsler Adult Intelligence Scale (WAIS), researchers have found differences in overall Intelligence Quotient and in subtest patterning (Dieker, 1973) between various types of offenders and controls. Cognitive style, intellectual strengths and weaknesses, and an offender's ability to make use of his intellectual strengths, all are relevant in forensic assessments. Meaningful information often emerges from an offender's response to a test item or group of items. For example, in a case reported by Schlesinger (2001), a contract murderer who killed more than 100 victims, and remained undetected for 30 years, had overall intelligence (as measured by the WAIS) only within the low-average range. However, he had superior intellect in areas of social comprehension and practical reasoning, and these strengths helped him elude the authorities for as long as he did.

Neuropsychological tests, such as the Halsted–Reitan or the Luria–Nebraska, have been used routinely in forensic assessments when an offender's behavior, especially if he or she suddenly became violent, suggests that he or she might have some neurologic impairment. A number of research studies have attempted to demonstrate various neuropsychological differences between normal control subjects and criminal offenders. For example, Lewis et al. (1986) found neuropsychological and neurologic deficits in the violent offenders they studied. Other researchers (Nestor, 1992; Nestor and Haycock, 1997), however, found less clear-cut results.

Specialized Forensic Tests

Because traditional psychological tests were not developed to address specific forensic issues, a number of researchers have developed specialized tests that attempt to focus directly on specific legal questions. These tests are often used in forensic assessments to supplement the clinical interview and traditional psychological testing. Table 29.4 lists some **specialized forensic assessment techniques** that are currently available, along with the forensic issue they are purported to address.

Although these specialized instruments have merit, they often do not add a great deal beyond what can be obtained in a clinical interview that focuses on a specific issue. For example, it is usually not difficult to question an offender about the elements of *Miranda*, or the various elements of the competency-to-stand-trial standard. In fact, in some instances, the test may not cover every element listed in a particular state statute. Therefore, the psychologist who relies solely on the test may miss some important areas. Moreover, each case rests on its own set of findings, facts, and circumstances. Reliance solely on test results—specialized or traditional—is never recommended and could prove to be self-defeating. Although most specialized forensic assessment instruments appear to have face validity, they should be used with caution, in a supplemental way, and never in place of an evaluation of the specific legal standard as outlined in the statute of a particular jurisdiction. As Melton et al. (1997) have stated, some forensic assessment tests "provide some of the trappings of sophisticated instruments, with multiple scales and apparently quantified measures, but [they] are conceptually flawed or have little empirical research to validate the author's claims about the instrument's capabilities."

Interviews in Altered States of Consciousness

Some psychologists, in collaboration with psychiatrists and other specialists, conduct

Table 29.4
Specialized Forensic Assessment Instruments

Competence to Waive Miranda Rights
Comprehension of *Miranda* rights
Comprehension of *Miranda* vocabulary
Function of rights in interrogation

Competence to Stand Trial
Competency screening test
Competence to stand trial assessment instrument
Interdisciplinary fitness interview

Legal Insanity
Rogers criminal responsibility assessment scales

Child Custody/Parental Fitness
Ackerman–Schoendorf scales for parent evaluation of custody
Guardianship and Conservatorship
Multidimensional functional assessment questionnaire
Community competence scale

Competence for Medical Treatment Decision Making
Hopkins Competence Assessment Test

Source: Melton, G.B. et al., *Psychological Evaluations for the Court,* 2nd ed., Guilford Press, New York, 1997. With permission.

supplemental interviews with criminal defendants after the defendants have received an intravenous injection with substances such as sodium amytal (**narcoanalysis**) or have been placed in a hypnotic trance. Narcoanalysis has mostly been used for therapeutic purposes, such as treatment of conversion disorder, relieving mutism, or more recently to uncover repressed memories of abuse (Piper, 1993). The general theory of narcoanalysis is that with reduced inhibitions and defenses (as a result of the injected drug) the mute patient will talk, the hysterically paralyzed patient will move, and the patient who claims amnesia for events will verbalize what happened (Kavirajan, 1999). However, sodium amytal is not a "truth serum" as is popularly believed. Many experts have concluded (MacDonald, 1954; Redlich et al., 1951) that individuals who are determined to deceive and give false information before a narcoanalytic interview will also deceive under the influence of the drug. However, some offenders have confessed to a crime following a narcoanalytic interview; one such case, involving a sex murderer, was reported by Revitch and Schlesinger (1981). As with any other procedure, the usefulness of narcoanalysis depends largely on the practitioner's skill and experience.

Hypnosis has been used in criminal investigations and in psychological therapy for many years; however, its usefulness (just as in narcoanalysis) is not always clear-cut. In some cases it has helped witnesses recall details of an event or give a detailed description of a crime they could not otherwise recall. Zonana (1979) was unable to get a confession or to retrieve the memory of an event with narcoanalysis, but he succeeded with hypnosis. In many other cases, however, hypnosis has proved unsuccessful, and some offenders have been able to fool even an experienced hypnotist (Revitch and Schlesinger, 1981). To minimize such problems, Relinger and Stern (1983) have developed guidelines designed to help standardize forensic hypnotic procedures and to elicit accurate memories during hypnotic interrogation. Nonetheless, we believe that both hypnotic and narcoanalytic techniques are applicable only in isolated cases

for detection. Both are far more useful for therapeutic purposes or for understanding the complex psychodynamics of the offender, particularly one who is consciously cooperative.

Complicating Factors in Forensic Assessment: Deception Syndromes

In forensic settings, distortion and exaggeration of symptoms commonly occur for a variety of (obvious) reasons. As a result, the forensic practitioner should develop a "low threshold" for suspecting deception in this population. One common type of deception is malingering (also referred to as *simulation*), a conscious attempt to feign a (mental) illness; another is dissimulation (or what Rogers [1988] refers to as defensiveness), a conscious and deliberate attempt to minimize or deny symptoms of a mental disorder. Criminal offenders might feign mental illness to avoid criminal responsibility or they might also deny mental illness to get released from prison or a hospital. In some cases, one can detect malingering by listening with a "third ear." For example, offenders who simulate symptoms of an illness often select a currently "trendy" illness, such as multiple personality disorder or posttraumatic stress disorder; moreover, they often greatly exaggerate the clinical picture. Thus, malingerers "see less than the blind and hear less than the deaf." They apparently believe that the more bizarrely they behave, the more psychotic they will look. However, it is very difficult to maintain malingered psychotic symptoms for extended periods (Enoch et al., 1967).

Schneck (1962) has described a phenomenon that he calls **pseudomalingering**, in which a mentally ill individual feigns the mental illness that he or she actually has. Such behavior, he believes, gives the individual a feeling of control over his illness. In one such case, reported by Arieti and Schreiber (1981), a schizophrenic man, under the delusion that God had ordered him to exterminate the human species, killed one of his sons, two other boys, and several women. In court, the offender attempted to portray the layman's conception of how a mentally ill person would behave: he jumped on the floor, rolled around, and barked like a dog. He was, in essence, attempting to simulate an illness that he actually suffered from.

A somewhat similar phenomenon was observed by Ganser (1898) and is now referred to as **Ganser's syndrome**. Ganser noted that certain prisoners who he was treating would give inaccurate approximate answers to simple questions, such as 2 + 2 = 5, 10 − 1 = 8. He believed that these patients were not malingering, but instead had a true disorder that gave the appearance of malingering. This syndrome has been noted in different cultures and different countries for more than 100 years. If the offenders are malingering, they do a terrible job of it, and it is unclear what mental disorder they are trying to portray. Nevertheless, because it does occur in prison populations, Ganser's syndrome is a disorder that forensic psychologists need to be aware of despite its rarity (Cocores and Cohen, 1996).

Some offenders who commit irrational acts may claim (and actually believe at some level) that there is a rational motive for their conduct. Schlesinger and Revitch (1999) found that some sexual burglars steal insignificant objects to convince themselves that they were there not for pathologic reasons (which they do not fully understand), but to commit a robbery. Other offenders try to convince themselves that their offense was unplanned, an accident, or perhaps provoked by the victim. Sometimes an offender's logical and "common-sense" confession of his alleged motive cannot be taken at face value. For example, a man who killed a woman by inflicting 150 stab wounds claimed that he wanted to prevent her from identifying him; and the prosecution accepted his explanation. Here, the authorities tried to make a pathologic act, with complex levels of motivation, seem simple.

In addition to the previous intricacies of malingering, many offenders simply engage in direct lying. Some criminal defendants lie to avoid responsibility for wrongdoing; others are pathological liars who lie to enhance themselves (Weston, 1996), or perhaps to gain a feeling of power, domination, and control over others (Schlesinger, 2000).Thus, the evaluation of offenders is a complex, multileveled task that requires sophistication, understanding, common sense, and experience with all the different devices that offenders might resort to when undergoing a forensic assessment.

Conclusion

This chapter has reviewed several of the many areas where psychology and law interact. However, in addition to the traditional applications of forensic psychology covered in this chapter, forensic psychology can make a contribution in a number of other ways. For example, forensic psychologists can consult with law enforcement officers in areas such as psychological profiling, a topic covered thoroughly in Chapter 30. Forensic psychologists also are being called on with increasing frequency by nonforensic mental health professionals to help them better understand patients who might constitute a threat to others or who manifest paraphiliac or obsessional or stalking behavior. With their expertise not only in law and legal standards but also in the psychopathology and psychodynamics of crime, forensic psychologists can be of great practical help by providing information on a patient population that is frequently foreign to the nonforensic clinician. Although forensic psychology has certainly come a long way, Münsterberg's century-old vision of using scientific psychology to inform the legal system has yet to be fully realized.

Questions

1. What is forensic psychology?

2. What has psychological research demonstrated about the accuracy of eyewitness identification?

3. What factors contribute to false confessions?

4. Discuss the development of the professional practice of clinical forensic psychology.

5. What court decisions are important to the professional practice of clinical forensic psychology?

6. What philosophical (or paradigmatic) differences between psychology and law make for an uneasy alliance?

7. What is the distinction between a clinical and a forensic assessment?

8. List several traditional psychological tests and several specialized forensic tests.

9. Discuss complicating factors in forensic assessment, specifically with regard to deception.

10. What is Ganser's syndrome?

11. Freud believed that many criminals commit crimes because of _____.

12. The first time psychological research was cited in a Supreme Court decision was _____.

13. The landmark case allowing psychologists to testify as expert witnesses in criminal matters was _____.

14. Feigning an illness is called _____.

15. Psychological tests that attempt to focus directly on specific legal questions are called _____.

References and Suggested Readings

Arieti, A. and Schreiber, R. F., Multiple murder of a schizophrenic patient: a dynamic interpretation; *J. Am. Acad. Psychoanal.*, 9, 501, 1981.

Ballew V. Georgia, 98 S. Ct. 1029, 1978.

Bazelon, D. L., Psychiatrists and the adversary process, *Sci. Am.*, 230, 18, 1974.

Bender, L., *A Visual Motor Gestalt Test and Its Clinical Use*, research monograph No. 3, American Orthopsychiatric Association, 1938.

Blau, T. H., *The Psychologist as Expert Witness*, 2nd ed., John Wiley & Sons, New York, 1998.

Brancale, R. and Ellis, A., Psychiatric and psychological investigation of convicted sex offenders, *Am. J. Psych.*, 102, 17, 1952.

Brown V. Board of Education of Topeka Kansas, 745 S. Ct. 686, 1954.

Buckhout, R., Eyewitness testimony, *Sci. Am.*, 231, 23, 1974.

Champagne, A., An empirical examination of the use of expert witnesses in American courts, *Jurimetrics*, 31, 375, 1991.

Cocores, J. and Cohen, R. S., Ganser's syndrome, prison psychosis and rare dissociative states, in *Explorations in Criminal Psychopathology*, Schlesinger, L. B., Ed., Charles C Thomas, Springfield, IL, 1996.

Cooke, G., *The Role of the Forensic Psychologist*, Charles C Thomas, Springfield, IL, 1980.

Cutler, B. L. and Penrod, S. D., *Mistaken Identification: The Eyewitness, Psychology, and the Law*, Cambridge University Press, New York, 1995.

Davidson, A. H., *Forensic Psychiatry,* 2nd ed., Ronald Press, New York, 1965.

Dieker, T. E., WAIS characteristics of indicted male murderers, *Psychol. Rep.,* 32, 1066, 1973.

Enoch, M. D., Trethowen, W. H., and Barker, J. C., *Some Uncommon Psychiatric Syndromes,* John Wright and Sons, Bristol, U.K., 1967.

Freud, S., Some character types met with in psychoanalytic work, in *Collected Papers,* 3rd ed., Vol. 4, Basic Books, New York, 1959 (originally published 1946).

Ganellen, R. J., *Integrating the Rorschach and the MMPI-2 in Personality Assessment,* Lawrence Erlbaum, Hillsdale, NJ, 1996.

Ganser, S. J., A peculiar hysterical state, *Arch. Psych. Nerven-Krankheiten,* 30, 633, 1898.

Geiser, L. and Stein, M., *Evocative Images: The Thematic Apperception Test and the Art of Projection,* APA Press, Washington, D.C., 1999.

Gudjonsson, G., The making of a serial false confessor: the confession of Henry Lee Lucas, *J. Forens. Psych.,* 10, 416, 1999.

Hammer, E., *The Clinical Application of Projective Drawings,* 6th ed., Charles C. Thomas, Springfield, IL, 1980.

Hare, R. D. et al., The revised psychopathy checklist: descriptive statistics, reliability, and factor structure, *Psychol. Assess.* 2, 338, 1990.

Heilbrun, K., The role of psychological testing in forensic assessment, *Law Human Behav.,* 16, 257, 1992.

Heilbrun, K. et al., The use of third party information in forensic assessment: a two state comparison, *Bull. Am. Acad. Psych. Law,* 22, 399, 1994.

Hess, A. K. and Weiner, I. B., *Handbook of Forensic Psychology,* 2nd ed., John Wiley and Sons, New York, 1999.

Hidden v. Mutual Life Insurance Co., 217 F. 2d 818 4th Cir., 1957.

Hutt, M.L., *The Hutt Adaption of the Bender—Gestalt Test: Revised,* Grune and Stratten, New York, 1968.

Jenkins v. United States, 307 F. 2d 637, D.C. Cir., 1962, en banc.

Kassin, S. M., The psychology of confession evidence, *Am. Psychol.,* 52, 221, 1997.

Kassin, S. M., More on the psychology of false confessions, *Am. Psychol.,* 53, 320, 1998.

Kassin, S. M. et al., On the general acceptance of eyewitness testimony research, *Am. Psychol.,* 56, 405, 2001.

Kavirajan, H., The amobarbital interview revisited: a review of the literature since 1966, *Harvard Rev. Psych.,* 7, 153, 1999.

Kennedy, L., *Ten Rollington Place,* Simon and Schuster, New York, 1961.

Klopfer, B. and Kelly, D. M., *The Rorschach Technique,* World Press, Tarrytown, New York, 1942.

Koppitz, E. M., *The Bender–Gestalt Test for Young Children,* Grune and Stratten, New York, 1964.

Lewis, D. O. et al., Psychiatric, neurological and psychoeducational characteristics of 15 death row inmates in the United States, *Am. J. Psych.,* 143, 838, 1986.

Loftus, E. and Monahan, J., Trial by data: Psychological research as legal evidence, *Am. Psychol.,* 35, 270, 1980.

Loftus, E. and Palmer, J. L., Reconstruction of an automobile destruction: an example of the interaction between language and memory, *J. Verbal Learn. Verbal Behav.,* 13, 585, 1974.

MacDonald, J. M., Narcoanalysis and the criminal law, *Am. J. Psych.,* 111, 283, 1954.

MacDonald, J. M., *Psychiatry and the Criminal: A Guide to the Psychiatric Examination for the Criminal Court,* Charles C Thomas, Springfield, IL, 1969.

McCann, J., A conceptual framework for identifying various types of confessions, *Behav. Sci. Law,* 16, 441, 1998.

Meloy, J. R., Hansen, T. L., and Weiner, I. B., Authority of the Rorschach: legal citations during the past 50 years, *J. Personality Assess.,* 69, 53, 1997.

Melton, G. B. et al., *Psychological Evaluations for the Court,* 2nd ed., Guilford Press, New York, 1997.

Münsterberg, H., *On The Witness Stand,* Doubleday, New York, 1908.

Murray, H. A., *Explorations in Personality,* Oxford University Press, New York, 1938.

Murray, H. A., *Thematic Apperception Test: Pictures and Manual,* Harvard University Press, Cambridge, MA, 1943.

Murray, H. A., Uses of the T.A.T., *Am. J. Psych.,* 107, 577, 1951.

Nestor, P. G., Neuropsychological and clinical correlates of murder and other forms of extreme violence in a forensic psychiatric population, *J. Nerv. Mental Dis.,* 180, 418, 1992.

Nestor, P. G. and Haycock, J., Not guilty by reason of insanity of murder: clinical and neuropsychological characteristics. *J. Am. Acad. Psych. Law,* 25, 161, 1997.

Norko, M. A., 1999 presidential address, Dr. Faulkner: The survival of forensic psychiatry. *Am. Acad. Psych. Law Newsl.,* 25, 1, 2000.

Ollson, N., Juslin, P., and Winman, A., Realism of confidence in earwitness versus eyewitness identification. *J. Exper. Psychol. Applied,* 4, 101, 1998.

People v. Hawthorne, 291 N.W. 205, Mich., 1940.

Piotrowski, C. and Keller, J. W., Psychological testing in applied settings: a literature review from 1982–1992, *J. Training Pract. Prof. Psychol.,* 6, 74, 1992.

Piper, A., Truth serum and recovered memories of sexual abuse: a review of the evidence, *J. Psych. Law,* 21, 447, 1993.

Rabin A. I., Projective methods: an historical introduction, in *Projective Techniques in Personality Assessment,* Rabin, A.I., Ed., Springer, New York, 1968.

Redlich, F., Ravitz, L. J., and Dession, G. H., Narcoanalysis and truth, *Am. J. Psych.,* 107, 586, 1951.

Relinger, H. and Stern, T., Guidelines for forensic hypnosis, *J. Psych. Law,* 11, 69, 1983.

Revitch, E. and Schlesinger, L. B., *Psychopathology of Homicide,* Charles C Thomas, Springfield, IL, 1981.

Rogers, E., Ed., *Clinical Assessment of Malingering and Deception,* Guilford Press, New York, 1988.

Rogers, R. et al., Standardized assessment of malingering: validation of the SIRS, *Psychol. Assess. J. Clin. Conslt. Psychol.,* 3, 89, 1991.

Rorschach, H., *Psychodiagnostics: A Diagnostic Test Based on Perception,* Hans Hüber Verlag, Bern, 1921.

Rosenwald, G. L., The Thematic Apperception Test, In *Projective Techniques in Personality Assessment,* Rabin, A. I., Ed., Springer, New York, 1968.

Schlesinger, L. B., The potential sex murderer: ominous signs, risk assessment, *J. Threat Assess.,* 1, 47, 2000.

Schlesinger, L. B., Ed., *Serial Offenders: Current Thought, Recent Findings,* CRC Press, Boca Raton, FL, 2000.

Schlesinger, L. B., The contract murderer: patterns, characteristics and dynamics, *J. Forens. Sci.,* 46, 108, 2001.

Schlesinger, L. B., Stalking, homicide, and catathymic process: a case study. *Int. J. Offender Therapy Compar. Criminol.,* 46, 64–72, 2002.

Schlesinger, L. B., A case study involving competency to stand trial: Incompetent defendant, incompetent examiner, or malingering by proxy? *Psychology, Public Policy and Law,* 9, 381–399, 2003.

Schlesinger, L. B., *Sexual Murder: Catathymic and Compulsive Homicide,* CRC Press, Boca Raton, FL, 2004.

Schlesinger, L. B. and Revitch, E., Sexual burglaries and sexual homicide: clinical, forensic and investigative considerations, *J. Am. Acad. Psych. Law,* 27, 227, 1999.

Schneck, J. M., Pseudo-malingering, *Dis. Nerv. Syst.,* 23, 396, 1962.

Shuman, D. W., Assessing the believability of expert witnesses: science in the jury box, *Jurimetrics,* 37, 23, 1996.

Slovenko, R., *Psychiatry and Law,* Little, Brown, Boston, 1973.

Uhlin, D. M., The use of drawings for psychiatric evaluation of a defendant in a case of homicide, *Ment. Health Soc.,* 4, 61, 1978.

Vuocolo, A. B., *The Repetitive Sex Offender,* Quality Printing, Roselle, NJ, 1969.

Weiner, I. B., Exner, J. E., and Sciara, A., Is the Rorschach welcome in the courtroom? *J. Person. Assess.,* 67, 422, 1996.

Wells, G. L. et al., From lab to the police station: a successful application of eyewitness research, *Am. Psychol.,* 55, 581, 2000.

Weston, W. A., Pseudologica fantastica and pathological lying, in *Explorations in Criminal Psychopathology,* Schlesinger, L.B., Ed, Charles C. Thomas, Springfield, IL, 1996.

Wilding, J., Cook, S., and Davis, J., Sound familiar? *Psychologist,* 13, 558, 2000.

Witmer, H. L., *Psychiatric Clinics for Children,* Commonwealth Fund, New York, 1940.

Yarmey, A. D., Earwitness speaker identification, *Psychol. Public. Policy Law,* 1, 792, 1995.

Zonana, H. V., Hypnosis, sodium amytal and confession, *Bull. Am. Acad. Psych. Law,* 7, 18, 1979.

Forensic Psychiatry

Robert L. Sadoff

Definition of Forensic Psychiatry

Forensic psychiatry is the subspecialty of psychiatry that deals with people who are involved in legal matters, either criminal or civil. It is primarily the assessment and evaluation of individuals rather than the treatment of psychiatric patients. However, there are areas in which the forensic psychiatrist also may be involved in the treatment of criminal offenders. This area is referred to as **forensic psychotherapy**. In this treatise, forensic psychiatry will be limited to the assessment of individuals involved in legal matters.

History of Forensic Psychiatry

The history of forensic psychiatry in the United States is intimately entwined with general psychiatry and inpatient psychiatry in institutions and asylums, especially during the 19th and 20th centuries. Forensic psychiatry came into its own in America when Isaac Ray published his *Treatise on the Medical Jurisprudence of Insanity* in Boston in 1838. Ray influenced a number of judges and others in criminal cases and also set standards for treatment in his hospital in Rhode Island. Forensic psychiatry proliferated in about the mid-20th century when the general revolution in psychiatry in the United States occurred, primarily in the 1960s and 1970s. Major changes in criminal law and insanity encouraged the use of psychiatrists in various aspects of the criminal justice system. Psychiatrists were also more involved in commitment procedures, changes in patients' rights, and in protecting the public primarily through the *Tarasoff* (1973) rulings in various states.

For psychiatrists to be prepared for the major changes that occurred, they needed to know what the laws were affecting their participation in each jurisdiction. Psychiatrists began working with lawyers and legislators to effect changes in the law that would involve psychiatric testimony.

Scope of Forensic Psychiatry

The field of forensic psychiatry may be divided into three major subgroupings: criminal forensic psychiatry, civil forensic psychiatry, and administrative forensic psychiatry. Administrative issues include matters such as confidentiality, privileged communications, and privacy issues. They also include patients' rights such as the right to treatment, the right to refuse treatment, and newer considerations involving commitment or involuntary hospitalization. Malpractice issues may traverse all three areas of forensic psychiatry: for example, boundary crossings may begin as an administrative matter leading to loss of license, but may also involve a civil lawsuit against the offending psychiatrist. The case may be so serious as to also generate criminal charges against the psychiatrist.

Civil legal matters include primarily personal injury cases, domestic relations matters, and competency issues. There are several areas of competency in civil cases that may be assessed, including competency to write a will (testamentary capacity), competency to manage one's own affairs, competency to vote, competency to enter into commercial transactions, competency to parent one's children (sometimes a consequence of domestic relations cases), and competency to testify (testimonial capacity).

Criminal forensic psychiatry will be the focus of this chapter and will include such issues as competency to stand trial, legal insanity, sentencing issues, and treatment of the mentally ill offender.

The forensic psychiatrist may be called to evaluate a juvenile who has either started in juvenile court to be transferred up to adult court or, in some jurisdictions, started in adult court to be transferred down to juvenile court.

The issues have become more stringent in terms of maintaining an individual in juvenile court or transferring him or her from adult to juvenile court. The concerns of the community have become involved in the decision to transfer. Other considerations are also made by the court, but for the forensic psychiatrist, the major concern is whether the juvenile is amenable to treatment in the juvenile system until the age of either 19 or 21, depending on the jurisdiction. This will, of course, depend on the age of the juvenile for the time required for treatment and the condition to be treated. Character disorders, such as antisocial personality disorder, are rarely treated in a short time. However, an acute psychotic condition may be treated fairly rapidly, and the person may be amenable to such treatment and not be considered a danger to the community when not psychotic. The forensic psychiatrist should confine his or her opinions to the medical issues involved and not be tempted to become more involved in the social/legal issues that present themselves.

Assessment of Competency

In general, the forensic psychiatrist is called by attorneys or judges to assess individuals either for mental state in the past (legal insanity or testamentary capacity), in the present (competency to stand trial or other forms of present competency), or in the future (prediction of dangerousness). Only the assessment of the current state of competency is based primarily on a present examination of the individual in question. The retrospective and prospective assessments require other materials besides a psychiatric examination. One must have records of previous hospitalizations, psychiatric examinations, medical assessments, and other observations of an individual to give testimony "within reasonable medical certainty." Also, to assess a person's later degree of "dangerousness" is a risky thing for a psychiatrist to do. I recommend that psychiatrists not become involved in prediction of dangerousness unless the prediction is based on clinical terms, such as future violent behavior under certain clinical conditions. For example, one can testify appropriately before a judge that James Jones, a diagnosed paranoid schizophrenic, has on several occasions become violent when he stopped taking his medication and went into the local bar to consume alcohol.

One can testify that an individual with such a diagnosis, who is noncompliant with his medication and becomes psychotic and drinks alcohol to excess and has a pattern of becoming violent under those clinical conditions, will likely become violent in the future if those clinical conditions are repeated. Thus, it is important to advise the court that one must prevent such clinical conditions from occurring by making certain that Jones continues to take his medication or to give it in injectable form that will last for several weeks as well as to prohibit him from using alcohol. If he is found using alcohol or is not in compliance with taking his medication, the judge may have the option of hospitalizing or confining the individual to prevent violent behavior.

The comments in this chapter are reserved for forensic psychiatrists and not for other mental health professionals who may come to court to testify. Clearly, there are a number of other mental health professionals, including psychologists, social workers, psychiatric nurses, psychotherapists, and others who are called to testify in various cases. Often, further testing is required before a forensic psychiatrist can testify in a particular case. In those cases, special tests such as psychological tests, neuropsychological tests, physical tests, neurologic tests, and others may be given to further evaluate the defendant in a particular case. Sometimes, hypnosis or sodium amytal is utilized to bring forth memories that have been repressed. Each of these is a special area that needs further elaboration and again is beyond the scope of this chapter.

Phases of Criminal Procedure

To fully understand the role of the forensic psychiatrist in criminal matters, one must have a brief understanding of the criminal procedure in which the psychiatrist works. Basically, there are three phases to the criminal procedure: the pretrial phase, the trial phase, and the posttrial phase. In the pretrial phase, the forensic psychiatrist may be called to assess a defendant for matters of competency either to confess, to give a statement to the police, to

testify, or to stand trial. Each of these issues of competency has a particular set of criteria the assessing psychiatrist must include in his or her questioning to arrive at a valid opinion (within reasonable medical certainty) in order to be of assistance to the court.

The trial phase primarily includes the mental state of the defendant at the time of the alleged offense (legal insanity) or competency at the present time to proceed with the trial. Each of these issues will be discussed more completely in the following text.

In the posttrial phase, the issue is usually one of assessing the information given to the court. Was the defense attorney competent in preparing his or her case and providing the jury or the court sufficient information about the mental state of the defendant? Other posttrial issues include sentencing of the defendant who is found guilty. Should the defendant be sent to a hospital for treatment or to a prison? What type of treatment will the individual require, depending on where he or she is sent? Should medication be a part of the treatment program? Also, if the individual is found not guilty by reason of insanity, is the defendant considered to be "dangerous" such that particular recommendations for commitment are made to the court following the finding of insanity?

Competency to Stand Trial

When considering competency of an individual, the forensic psychiatrist must first ask the question, "Competency to do what?" In criminal matters, the primary concern is that of competency to stand trial. However, another important competency is that of making a statement to the police or giving a confession. Was the individual competent to make the statement? Did he or she understand his or her Miranda warnings and could he or she voluntarily proceed after having been given the warnings? This is, of course, a retrospective assessment that one has to make, putting together all the data available at the time. Often, the case hinges on whether the confession of the defendant is valid. If the confession is deemed to be invalid, or involuntarily given or coerced by the police, there may be little or no other evidence against the defendant to support the prosecution. There is often great difference in the interpretation by different assessors about the validity of a confession. It is important for the assessing psychiatrist to consider all evidence and all data available. This may include talking with the police officers or detectives who were present at the time the confession was made. Sometimes, the defendant will perceive that he or she was coerced by the police officers and made statements that are not corroborated by the evidence of the officers' statements. Sometimes it becomes a matter of "whom do you believe?"

The issue of competency to stand trial, however, is perhaps the most important one facing a number of defendants. Most defendants consider themselves competent to stand trial. However, to be found competent, the defendant must be free of mental illness that impacts his or her ability to know what is happening in his or her case. For example, if a defendant does not know what he or she is being charged with or what the consequences of these charges are, he or she may be found incompetent to stand trial. Many states have issued a list of questions that must be asked of the defendant before the assessing psychiatrist can validly give an opinion about competency to stand trial. These factors include the nature of the charges, the consequences of the charges, and the individual's understanding of the courtroom procedure, including the role of the principals in the courtroom. For example, the defendant should know the role of the judge, the prosecutor, his or her own attorney, and the jury. Some defendants may have an intellectual understanding of the nature of the courtroom proceedings but may have delusions about the principals and their roles. For example, one defendant believed the judge and the prosecutor were in "cahoots" or conspiracy with each other to find the defendant guilty. That alone may not suffice to vitiate competency, but taken in the context of his total mental illness, which was paranoid schizophrenia and which included delusional beliefs about his own attorney, could be a major factor against his competency to stand trial.

Not remembering specific information alone is not a bar to competency. However, if a person had been hit on the head or had a bona fide amnesia for the events of the crime, he or she may not be in a position to effectively assist counsel in preparing his or her defense. The *Dusky* (1960) case insists that the defendant must have a rational understanding as well as a factual understanding of the case against him. He or she must also be able to rationally consult with his or her attorney and not just factually.

In the pretrial phase, there is also the evaluation of the defendant for competency to plead. Rarely, a defendant may be delusional about a particular case that he or she reads in the newspapers and wishes to confess to the killing and to plead guilty. However, there may be insufficient evidence against him or her and the plea may be invalid because it is based on psychotic delusions.

Other forms of competency may arise during the trial phase or posttrial phase. For example, the individual may not be competent to be sentenced to prison because he or she is psychotic and requires mental hospitalization. At a later phase, the individual may not be competent to be executed in the event he or she is given the death sentence.

In the event that a defendant is found incompetent to stand trial, he or she will be hospitalized, usually at a forensic psychiatric hospital where he or she will be treated (or perhaps educated) until he or she is competent. In some rare cases, the defendant, after a period of treatment, will be deemed incompetent to stand trial for the foreseeable future. Under the case of *Jackson v. Indiana* (1972), the defendant will then be committed to a civil hospital after the charges have been suspended, and the case will be treated as a civil case rather than as a criminal case. There is no set time for such assessment, but usually after 1 to 2 years of treatment with no success, the psychiatrist may be asked by the court to give an opinion about whether the defendant will become competent in the foreseeable future.

Legal Insanity

During the trial phase, there are several areas in which the forensic psychiatrist may be called by the court or by either side to give an opinion about the state of mind of the defendant at the time of the committing of the alleged offense. The most written about, but perhaps one of the less common conditions, would be legal insanity. The definition of insanity varies in different jurisdictions. Most states follow the McNaughten rule that was promulgated in England in 1843. McNaughten rules are primarily a cognitive test of insanity. Most jurisdictions indicate that a person would be found not guilty by reason of insanity if at the time of the commission of the crime, the person was suffering from such mental infirmity or disease that he did not know the nature and quality of his or her action. This is purely cognitive in that the individual could not or did not know what he or she was doing or what he or she was doing was wrong.

Modifications to McNaughten have been attempted in various jurisdictions over the past several years. Adding a volitional arm to the test has been attempted in that if a person did know what he or she was doing and did know that it was wrong and yet could not control his or her behavior, the individual would be found legally insane. Perhaps the most ambitious attempt was by the American Law Institute Model Penal Code that was introduced by Judge Bazelon in the *Brawner* case in 1972, replacing the *Durham* decision of 1954 that Judge Bazelon had promulgated in the District of Columbia Court. *Durham* stated that a person would be found not guilty by reason of insanity if the person's behavior was a product of mental illness. That test proved to be too broad and too inclusive and was not helpful in discriminating the insane from the sane. Judge Bazelon, in the *Brawner* case, then adopted the Model Penal Code of the American Law Institute, which stated, "A person would be found not guilty by reason of insanity if at the time of the commission of the crime the person lacked substantial capacity either to appreciate the criminality of the behavior or to conform his conduct to the requirements of the law." In that test, there are three elements of the mental state, including the cognitive, the conative (emotional), and volitional. It was the test used in the *Hinckley* case in 1982 that

led to the finding of not guilty by reason of insanity for John W. Hinckley, Jr.

It was also following that finding of not guilty by reason of insanity that Congress changed the law for the federal jurisdiction in 1984, adopting the Omnibus Crime Code for Insanity, which stated that a person would be found not guilty by reason of insanity if at the time of the commission of the crime the defendant could not appreciate the criminality of his behavior. The use of the word "appreciate" broadens the scope that limits McNaughten to "knowledge." The law also prohibited the psychiatrist from addressing the ultimate question of insanity. He or she could give an opinion about mental illness and its effect on the defendant, but could not give an opinion about insanity, which was the domain of the jury or the trier of fact. The burden of proof also switched from the prosecution proving sanity beyond a reasonable doubt to the defense proving insanity to a clear and convincing degree.

Also following the *Hinckley* decision was the concept of "guilty but mentally ill." It is improper to state "guilty but insane" because insanity indicates a lack of **mens rea**. A crime consists of two parts: the guilty act (*actus reus*) and the guilty intent (*mens rea*). To vitiate one of those, that is, the *mens rea* by insanity, could not then lead to a guilty finding. The concept of "guilty but mentally ill" was adopted in 13 states and, in Pennsylvania, the definition of mental illness is that of the American Law Institute Model Penal Code, which is that the person lacked substantial capacity because of mental illness either to appreciate the criminality of his or her behavior or to conform his or her conduct to the requirements of the law. In Pennsylvania, if a defendant pleads "not guilty by reason of insanity," the judge must also read the charge for "guilty but mentally ill."

Diminished Capacity

Another concept involving the state of mind of the defendant at the time of the commission of the act is that of diminished capacity. Diminished capacity has different meanings in different jurisdictions. For example, in

Pennsylvania, the concept of diminished capacity reduces the degree of homicide from first degree to third degree only. The concept is that a person could not form the specific intent to kill and, therefore, the degree of the homicide is reduced from first to third degree. In New Jersey, however, the concept has a much more profound meaning in that it is a total defense and that if a person is found to have a diminished capacity, that is, the individual could not act purposefully or knowingly, then he may be acquitted of the charges against him. Generally speaking, diminished capacity means that the individual at the time of the commission of the alleged offense had a diminished ability to meet all the criteria for the charges against him, that is, the individual could not or did not form the specific intent or could not act in a knowing and purposeful manner.

Posttrial Sentencing

Following the trial phase, there is the posttrial phase, which involves sentencing. The forensic psychiatrist may be called by the court to give opinions about the sentencing of a particular individual who may require treatment in a psychiatric hospital or may require medication. The psychiatrist will give testimony at the request of the court in such cases. There may be specific needs the defendant would have in a particular institution. The psychiatrist is presumed to be well versed on the treatment facilities in the community and give such testimony for the court's assessment regarding sentencing.

In a specific case of death penalty issues, the psychiatrist may be called to testify whether the defendant is competent to be sentenced to death. He or she may also be called to testify whether the defendant is competent to be executed at a later date. There are also issues where the defendant has asked for all of his or her appeals in the death penalty case to stop and asks that he or she be executed summarily. The forensic psychiatrist may be called to evaluate the competency of the individual at that time to determine whether he or she is competent to waive his or her rights under the Constitution.

Cases Requiring Forensic Psychiatrist Testimony

The forensic psychiatrist is not called in all criminal cases. In fact, the percentage of cases involving mental illness is quite low. Statistics have shown that only in 1 of 100 cases is there an insanity plea and only in 1 in 10 of those cases pled, where there is a contest, is insanity found. Most cases involving theft or burglary do not require a forensic psychiatrist. However, the lawyers and courts will request psychiatric consultation in cases of homicide and in other felonies, such as robbery, arson, or kidnapping, and almost all sexual cases. With the increase in the use of drugs that are involved in criminal matters, the forensic psychiatrist may need to assess the role of medication or illegal drugs on the mental state of the defendant at the time of the commission of the alleged offense. Alcohol has frequently played a role in criminal matters, and the effect of alcohol on the mental state is one that often requires psychiatric consultation. More recently, there have been charges that various medications, such as fluoxetine HCl (Prozac), may have an effect on violent behavior. There have been no scientific correlations with such use of fluoxetine and, as far as I know, no cases have been successfully litigated blaming fluoxetine or other selective serotonin reuptake inhibitors (SRRIs) for violent behavior in either criminal or civil cases.

In the matter of domestic relations crimes, such as spouse abuse, spouse murder, or child abuse, psychiatric consultation is often sought and required. Attorneys or judges will seek psychiatric consultation in such matters because of the increase in the "syndrome diagnoses," such as the "rape trauma syndrome" or "battered spouse syndrome," which are often pled by the defense. Attorneys also seek psychiatric consultation to present information and testimony to the court about the family and its dysfunction. The issue of false memory or repressed or recovered memory also comes into play when dealing with children and criminal behavior. This is a very controversial area in psychiatry, and many will differ on the presence of such memories and their origin. A full discussion of the problem in this area is beyond the scope of this chapter, but suffice it to say that it is a controversial area that requires further research and refinement.

Examination and Report of Findings

The forensic psychiatrist, when called either by defense, prosecution, or the court, has an obligation to pursue the case with as much information and data as is available. It may not be sufficient just to do a psychiatric examination of the defendant. One must have records of treatment, criminal cases, and police investigation reports as well as statements made by the defendant either on tape, videotape, or informally given to police officers. In some cases, several examinations are required after further testing is completed. Further testing may include psychological tests or other physical tests of organic nature (magnetic resonance imaging [MRI], x-rays, blood tests, etc.). It is often important to obtain a full battery of psychological tests and, sometimes, neuropsychological tests to determine various psychodynamics or presence or absence of organic brain conditions in the defendant. It may be helpful in some cases to have either a social worker or a psychologist conduct interviews of family members and others who knew the individual at or around the time of the commission of the offense in question. Once all the data are collected, including the testing and the interviews and several examinations of the defendant, then the forensic psychiatrist is in a position to discuss his or her findings either with the prosecutor or the defense attorney.

It is important for the forensic psychiatrist to have all the data that are available and not to be limited by either the prosecutor or the defense attorney to the information they want the expert to have. That can only backfire and hurt once the expert is in court

giving testimony. Working with an attorney is a very important part of preparation for the case. One must sit down with the attorney, either prosecutor or defense attorney, and go over the data available and the opinions rendered. One must also look to the potential cross-examination or the weaknesses of the case. One must also attempt to help the attorney cross-examine the adversarial witness whose opinion differs and will be given for the other side.

In preparation of the report, the forensic psychiatrist should include all available data, including the examination, in detail. The examiner should indicate that he or she has given the defendant the limitations of confidentiality, and that should be recorded in the report. The dates and times of the examination should also be recorded, so one will know how long the examiner spent with the defendant and also the amount of time one spent in reviewing other information and records. The report should be all-inclusive and give as much support to the conclusions drawn as possible.

Testifying in Court

When testifying in court, the forensic psychiatrist gives all the information that is prepared in advance on direct examination. One should not go into court without adequate preparation. The cross-examination should not be a great surprise to the experienced forensic psychiatrist. One should know what to expect in terms of the kinds of questions to be asked in attempting to weaken the testimony. Generally speaking, the defense looks to the severity of the mental illness in attempting to convince a jury that the defendant is mentally infirm and, therefore, legally insane. The prosecution may agree that there are elements of mental illness, but they do not rise to the level of insanity. The prosecution often utilizes behavior to show that a person knew what he or she was doing at the time and also knew that it was wrong, despite the presence of mental illness. If the expert is well prepared and stays within his or her area of expertise, then there is no need for rancor or arguing with the cross-examining attorney. Most experiences in court are quite pleasant and not disruptive. However, there are times when the cross-examining attorney will consider the expert either an ally of the defendant or an ally of the prosecution and will treat the expert as such. It is best for the expert to maintain a calm and professional attitude and not get caught up in the passion of the cross-examination.

Training in Forensic Psychiatry

Forensic psychiatrists began training one another through fellowship training programs at various universities and, in 1968, several individuals got together at an American Psychiatric Association meeting in Boston to help train one another. Out of this group emerged the American Academy of Psychiatry and the Law (AAPL), initially spearheaded by Dr. Jonas Rappeport from Baltimore. The first meeting was held in Baltimore in October 1969. Eight individuals attended that first meeting, and 100 were recruited for the next year. Currently, there are more than 2200 members. AAPL has maintained its high level of educational function through annual meetings, board review courses, and committee structure to help the continuing education of forensic psychiatrists.

Board certification in forensic psychiatry occurred through the American Board of Forensic Psychiatry (ABFS), spearheaded by AAPL and the American Academy of Forensic Sciences (AAFS) in the early 1970s. More than 200 individuals have been board certified by the American Board of Forensic Psychiatry (ABFP) until it was wound up in 1994, at which time the American Psychiatric Association (APA) and the American Board of Psychiatry and Neurology assumed the function of accrediting and certifying forensic psychiatrists. The new certification is time limited to 10 years, at which time a new examination must be passed for the individual to remain certified.

Training programs in forensic psychiatry have also proliferated to the point where there are now over 25 accredited programs in the United States. At present, an individual wishing to be certified in forensic psychiatry must take a 1-year accredited program of training in forensic psychiatry. Correctional psychiatry is also included in the training to become a forensic psychiatrist. However, the profession branches to either correctional or forensic once the training is completed. Most individuals working in correctional psychiatry do not do forensic work but have a general psychiatric practice or are full time in corrections. Most individuals working primarily in forensic practice do not have correctional appointments.

As forensic psychiatry has become managed care free, a number of individuals have sought economic refuge in the field. As a result, there are a number of people who are poorly trained and inexperienced in forensic work are coming to court to conduct assessments and testify in cases. The American Psychiatric Association and AAPL have set up independent peer review committees to attempt to maintain a high standard of forensic work in the United States. At present, these committees are voluntary, but it appears that in the future there will be a systematic peer review in forensic psychiatry by APA and AAPL.

Case Studies

Case 1

The cases in this field are so numerous that it is difficult to select any one or two that are outstanding. Generally speaking, the forensic psychiatrist involves himself or herself with homicide cases, including serial murderers, mass murderers, and sexual homicides. All sexual cases, such as rape or child sexual assault, may be evaluated and assessed by forensic psychiatrists. Perhaps one case that does stand out is that of a serial murderer who was also suffering from dissociative identity disorder. This young man was accused of killing four prostitutes but had no memory of the killing. He was a mild-mannered, large, ungainly sort who did not socialize well. He had a rich fantasy life, but, indeed, was found to have a number of distinct personalities, one of which was amoral and homicidal. When he learned that he harbored within himself the murderous alter ego, he decided he would never be comfortable outside of confinement and agreed to stay in a hospital or in prison for the rest of his life to prevent any further deaths. The host personality had no awareness of the existence of the alter at the time of the homicides. It was only later that he learned, through treatment, that he harbored such a homicidal alter. Ultimately, other alters, or partial alters, also emerged, confirming the diagnosis. With the help of psychiatric treatment, he was able to integrate the alters into one to keep them from emerging over a long time. He is comfortable in a prison, and the purpose of one of the alters was to have him locked up and, thus, did the killing. Because the host was in prison already, there was no need for the homicidal alter to emerge and to wreak any further havoc.

Case 2

Another case involved an individual who had been given the death penalty for numerous crimes that he had committed. In the course of his appeals, he suddenly said to his attorneys that he wanted to stop all appeals and be put to death forthwith. The federal court judge hearing the case needed an evaluation of his state of mind to determine that he was competent to stop the appeals and that he knew the consequences of his decision making. The judge hired three independent psychiatrists from different parts of the country to assess the prisoner. This is an individual who had a diagnosis of several personality disorders and may have been seen to be psychotic on occasion. However, on examination, he was found to be without psychosis and clearly understood the reasoning for his decision making. His idiosyncratic view of life and religion and himself all intertwined to effect the decision,

which was unusual, but not psychotic. He was found to be competent to proceed and, against the wishes of his attorneys, he was found to be competent. However, he had recanted his wish to be put to death and had filed for further appeals before the judge could make a determination of his competency.

Case 3

Competency to confess has been an issue in a number of cases. The one we will now describe concerns a young man who claimed the police did not read him his Miranda Rights and did not deal with him fairly. He said they threatened him and beat him, although there were no physical marks of beating. The problem was the police did not tape record or videotape the confession and the statement that he made and the proceedings by which the confession was obtained. Many jurisdictions mandate videotaping all statements and confessions, and others do not. From a forensic psychiatric standpoint, it appears to be a good idea for police departments to videotape, or at least audiotape, the statements made by suspects so that when evaluation or assessment is conducted later, there is a record that will support their contention.

In this particular case, the individual was articulate and hired several individuals who substantiated his claim that his will was overridden and that he did not give a truly voluntary statement. This is a man who had a longstanding antisocial personality disorder and who had been involved in a number of different kinds of crimes and had escalated his criminal behavior from fraud to burglary to robbery to homicide. It was clear, from a total assessment of the individual, that he was attempting to have the case thrown out by claiming that his statement was not voluntary and there was very little objective evidence beyond his statement for the prosecution to use in sustaining a conviction. After heated testimony on both sides, the judge determined the statement was competently given and could be admitted into evidence. The defendant then had no choice but to plead guilty and accept a negotiated plea for life imprisonment rather than the death sentence.

Summary

In summary, psychiatrists may testify in a number of different kinds of cases, some of which are criminal. Others include civil cases or administrative matters that come before the court. In the case of criminal matters, the psychiatrist may testify after a thorough evaluation of an individual or an issue to be decided by the court either at pretrial, trial, or posttrial phases of the criminal proceedings. Forensic psychiatrists should be well prepared before testifying and should be thorough and neutral in their assessment of the individual. The ethics in forensic psychiatry mandate striving for neutrality and objectivity as much as possible. Any conflict of interest should be noted early and reported to the court. In the case of any presumed conflict of interest, the forensic psychiatrist should likely withdraw from the case to avoid any appearance of impropriety. The practice of forensic psychiatry is quite satisfying and enjoyable, but also stressful for most psychiatrists. There are particular rules and guidelines that must be adhered to and will be monitored in the future through peer review. The field has grown in terms of training and professional involvement. There are standards that have been set by the various national organizations and board certification for those who wish to pursue the field in a highly qualified manner.

Questions

1. To be found competent to stand trial, what does a person need to be able to do?

2. For a person to be found legally insane, it must be shown that the person had a mental illness that prevented the defendant from doing what?

3. For a defendant's statement to the police to be valid, the statement must be _____ and _____.

4. How frequently is the insanity defense raised in criminal cases?

5. The concept that decreases a defendant's culpability in a criminal case is called _____.

6. In what phases of the criminal procedure may a forensic psychiatrist be called by attorneys or judges to consult or to testify?

7. When a juvenile is transferred from adult court to juvenile court, the procedure is called _____.

8. For what purpose may the forensic psychiatrist be called by the court after a verdict has been given?

9. For what purpose, following a death penalty sentence, may a forensic psychiatrist become involved?

10. Name three areas of civil litigation in which a forensic psychiatrist may become involved.

11. In order to be eligible to take the Board Examination in Forensic Psychiatry, the applicant must complete _____.

12. The ethical forensic psychiatrist strives for
    a. Neutrality
    b. Independence
    c. Thoroughness
    d. All of the above

13. The three phases of the criminal procedure are

    _____

    _____

    _____

14. In the Brawner case, Judge Bazelon replaced the Durham decision for insanity with what test?
    a. McNaughten
    b. American Law Institute Model Penal Code
    c. Omnibus Crime Code Test for Insanity
    d. Volitional Test for Insanity

15. A malpractice case against the psychiatrist may include
    a. Administrative issues
    b. Civil legal matters
    c. Criminal complaints
    d. Loss of license to practice medicine
    e. All of the above

References

Jackson V. Indiana, 406 US 715, 1972.

Dusky V. United States, 362 U.S. 402, 1960.

Durham V. United States, 214F 2d 862, D.C. Cir. 1954.

U.S. V. Brawner, 471F 2d 969, D.C. Cir. 1972.

McNaughten: Clark and Finney, 8 Eng Rep. 718, 1843.

Tarasoff V. Regents of the University of California, 33 Cal. App. 3d, 275, 1973.

Other Selected Readings

APA, *Issues in Forensic Psychiatry*, American Psychiatric Association Press, Washington, D.C., 1984.

Brooks, A., *Law, Psychiatry and the Mental Health System*, Little Brown and Company, Boston, 1974.

Crime and Delinquency Issues, National Institute of Mental Health, *Competency to Stand Trial and Mental Illness*, NIMH, Rockville, M.D., 1973.

Dattilio, Frank M., Ph.D.., ABPP and Sadoff, Robert, L., M.D., *Mental Health Experts: Roles and Qualifications for Court*, 2nd ed., PBI Press, Mechanicsburg, Pennsylvania, 2007.

Group for the Advancement of Psychiatry, *Misuse of Psychiatry in the Criminal Courts: Competency to Stand Trial*, Mental Health Material Center, New York, 1974.

Group for the Advancement of Psychiatry, Committee on Psychiatry and the Law, *The Mental Health Professional and the Legal System*. Brunner/Mazel, New York, 1991.

Gutheil, Thomas G., M.D., *The Psychiatrist as Expert Witness*, American Psychiatric Press, Washington, D.C., 1998.

Gutheil, Thomas G., M.D., *The Psychiatrist in Court: A Survival Guide*, American Psychiatric Press, Washington, D.C., 1998.

Halleck, S. L., *Law in the Practice of Psychiatry: A Handbook for Clinicians*, Plenum Medical Book Company, New York, 1980.

Malmquist, Carl P., M.D., M.S., *Homicide: A Psychiatric Perspective*, 2nd ed., American Psychiatric Publishing, Washington, D.C., 2006.

Perlin, M. L., *The Jurisprudence of the Insanity Defense*, Carolina Academic Press, Durham, NC, 1994.

Sadoff, R. L., *Forensic Psychiatry: A Practical Guide for Lawyers and Psychiatrists*, 2nd ed., Charles C Thomas, Springfield, IL, 1988.

Sadoff, R. L., *Violence and Responsibility*, Spectrum Publications, New York, 1978.

Schetky, Diane H., M.D., and Benedek, Elissa P., M.D., *Principles and Practice of Child and Adolescent Forensic Psychiatry*, American Psychiatric Publishing, Washington, D.C., 2002.

Shapiro, D. L., *Criminal Responsibility Evaluations: A Manual for Practice*, Professional Resource Press, Sarasota, FL, 1999.

Simon, R. I., *Clinical Psychiatry and the Law*, 2nd ed., American Psychiatric Press, Washington, D.C., 1992.

Simon, Robert I., M.D., and Gold, Liza H., M.D.: The American Psychiatric Publishing Textbook of Forensic Psychiatry, American Psychiatric Publishing, Inc., Washington, D.C., 2004.

Stone, A. A., *Mental Health and Law: A System in Transition,* Crime and Delinquency Series, NIMH, Rockville, MD, 1975.

Whitlock, F. A., *Criminal Responsibility and Mental Illness*, Butterworths, London, 1963.

Serial Offenders

Linking Cases by Modus Operandi and Signature

Robert D. Keppel

31

Introduction to Crime Scene Assessment

The method used to determine the signature and modus operandi (MO) of a killer is similar to that for "profiling" a case. The major differences lie in the type of information used and the results. To perform signature analysis and profiling, one must thoroughly examine the police case file.

Throughout history, police investigators, psychologists, and forensic psychiatrists have analyzed cases to determine the "profile" of an unknown offender. This process has been referred to as applied criminology, psychological profiling, **crime scene assessment**, criminal personality profiling, crime scene profiling, and investigative psychology, among others. For the purposes of this chapter, the process will be called crime scene assessment. Several outcomes are possible from crime scene assessment (Keppel and Walter, 1999): (1) determining the physical, behavioral, and demographic characteristics of the unknown offender; (2) developing a profile of postoffense behavior of the offender and strategies for apprehension; (3) developing interviewing strategies once the offender is apprehended; (4) determining the signature of the offender; and (5) determining where evidence may be located.

For any of the preceding outcomes, the type of information used for analysis may differ. Typically, the information comes from the police investigative file, which includes officer's reports, statements, crime laboratory reports, crime scene diagrams, photographs, videotapes of crime scenes, and autopsy reports. This chapter, however, will focus only on the method used to determine the signature of the offender. To accomplish this, it is important to understand how the offender's signature differs from his modus operandi from case to case.

What really confirms that two or more crimes can be linked to the same offender has not been through modus operandi analysis per se, but through signature analysis. It is the purpose of this chapter to explain the differences between modus operandi and signature, and demonstrate the uses of both in linking murders to the same offender.

Historical Perspective

The concept of modus operandi has been subjected to extensive analysis through the criminal appeals of serial offenders, and its definition has remained stable. Generally, the terms *modus operandi, method of operation,* or MO are used interchangeably to describe a certain criminal's way of operating (*State v. Pennell,* 1989; *State v. Prince,* 1992; *State v. Code,* 1994; *State v. Russell,* 1994).

Police investigators and prosecutors need to have cases linked for their own purposes. From an investigative standpoint, the linking of crimes enables investigators to pursue one suspect instead of operating without the knowledge that particular cases are linked. Prosecutors want similar cases linked so the defendant can be tried on multiple charges in the same trial. Of course, it is the defendant's prerogative to request that each charge be heard separately by a different jury.

The threshold for using MO and signature as evidence at trial that cases are linked differs from state to state. For example, the Supreme Court of Virginia held that "evidence of other crimes, to qualify for admission as proof of modus operandi, need not bear such an exact resemblance to the crime on trial as to constitute a 'signature,' but it is sufficient if the other crimes bear a singular strong resemblance to the pattern of the offense charged and the incidents are 'sufficiently idiosyncratic to permit an inference of pattern for purposes of proof,' thus tending to establish the probability of a common perpetrator" (*Timothy Wilson Spencer v. Commonwealth of Virginia,* 1990).

The phrase *modus operandi* first appeared in literature in 1654 in a piece called Zootomia: "Because their causes, or their modus operandi (which is but the Application of Cause and Effect) doth not fall under Demonstration." The term appears to have become popular in the 1800s, with citations in the *Edinburgh Review* in 1835, Mill's *Logic III* in 1843, and in Kenneth Grahame's short story *Justifiable Homicide* in the Pagan Papers in 1898 (Grahame, 1994; *Oxford English Dictionary,* 1933).

The earliest mention of *modus operandi* in the United States is not associated with criminal law or police investigations, but is traced to the law of patents dealing with inventions of machines (Robinson, 1890). In the case of *Whittenmore v. Cutter* (1813), the Massachusetts Circuit Court declared: "By the principles of a machine (as these words are used in the statute) is not meant the original elementary principles of motion, which philosophy and science have discovered, but the modus operandi, the peculiar device or manner of producing any given effect. The expansive powers of steam, and the mechanical powers of wheels, have been understood for many ages; yet a machine may well employ either one or the other, and yet be so entirely new, in its mode of applying these elements, as to entitle the party to a patent for his whole combination."

The first pioneer in the use of the concept of modus operandi in police operations was Major L.W. Atcherley, Chief Constable of the West Riding Yorkshire Constabulary in England. His efforts superseded the establishment of modus operandi files in Scotland Yard in 1896 by 17 years. Major Atcherley devised workable clearinghouses for information on various criminal's methods so they could be tracked from district to district. He constructed 10 categories relating to an offender's modus operandi:

1. *Classword.* Kind of property attacked (whether dwelling house, lodging house, hotel, etc.)
2. *Entry.* The actual point of entry (front window, back window, etc.)
3. *Means.* Whether with implements or tools (such as a ladder, jimmy, etc.)
4. *Object.* Kind of property taken
5. *Time.* Not only the time of day or night but also whether at church time, on market day, during meal hours, etc.

6. *Style.* Whether the criminal describes himself as a mechanic, canvasser, agent, etc., to obtain entrance
7. *Tale.* Any disclosure as to his alleged business or errand that the criminal may make
8. *Pals.* Whether the crime was committed with confederates
9. *Transport.* Whether bicycle or other vehicle was used in connection with the crime
10. *Trademark.* Whether the criminal committed any unusual act in connection with the crime (such as poisoning the dog, changing clothes, leaving a note for the owner, etc.)

These 10 categories related specifically to a criminal's modus operandi; however, Major Atcherley recognized that special individual things or unusual acts occurred at the scene of a crime, and he called these the criminal's *trademark* (Fosdick, 1916). This recognition of a criminal's trademark was the precursor for what would become known as the offender's *signature* today. By the late 1930s, modus operandi identification techniques and procedures became a standard part of criminal investigation literature (Soderman and O'Connell, 1936). Edwin Sutherland (1947) defined modus operandi as the "principle that a criminal is likely to use the same technique repeatedly, and that any analysis and record of the technique used in every serious crime will provide a means of identification in a particular crime."

The investigative use of modus operandi began to change in the late 1980s. It became apparent that the MO of an offender would change slightly from crime to crime. For example, a burglar who normally used a pipe wrench on the front doorknob to gain entry might change his MO after discovering that the front door was unlocked. Thus, the burglar would enter through an unlocked front door without using the pipe wrench and his MO changed to fit the circumstances. Also, it became apparent that a murderer or rapist would have to do the same thing from crime to crime (Douglas and Munn, 1992; Douglas et al., 1992; Geberth, 1996; Hazelwood and Michaud, 1999; Keppel, 1995; Keppel, 2000; Keppel and Birnes, 1997; Keppel and Weis, 1999).

What Is a Killer's Signature?

A killer's **signature**, sometimes referred to as his psychological *calling card*, is left at each crime scene across a spectrum of several murders. Homicide detectives are trained to look for the unusual—for those characteristics that distinguish one murder from all others. Thus, when one sees something rare in one murder and recognizes the same element 1 week later, one sees the personification of a lone killer in those unusual acts. For example, when the killer in one murder intentionally leaves the victim in an open and displayed position, posed physically in a spread-eagle position as if for a bizarre photographic portrait, and when he savagely beats the victim to a point of **overkill** and violently rapes her with an iron rod, you have to consider such behavior as fundamentally unusual. When a second murder is committed in which the killer has done the same things as in the first, even though he may slightly modify one or two of the features, there is little doubt the two murders are related. The crime analyst reacts to the killer's signature in this instance (Keppel and Birnes, 1997).

Some people confuse MO, or modus operandi, with signature as if the two were the same thing. They are not. An MO is simply the way a particular criminal operates. If a criminal commits breaking-and-entry burglaries by using a glass cutter to get through a door and suctions the glass away so it does not fall to the ground and make noise, that's his MO. If the criminal uses flypaper instead of a suction cup to hold the glass fragments together so they do not make noise, that is a different MO. When police find flypaper traces at a crime scene, they go back to their files and look for breaking-and-entry burglars who have used flypaper and use this knowledge to form a list of suspects. For the crime of murder, MO includes only those factors necessary to commit the murder and can change over time as the killer discovers that some things he or she does are more effective (Douglas et al., 1992; Keppel, 1995).

Basically, an MO accounts for the type of crime and property attacked, including the person, the time and place the crime was committed, the tools or implements used, and the way the criminal gained entry or how he or she got his victim, which includes disguises or uniforms, ways he represented himself to a victim, or props such as a bike or crutches.

The problem is that MO investigations can be too much of a good thing. Police detectives rely so much on MO that if it even slightly changes from crime to crime, they begin looking for a different criminal despite other striking similarities from crime to crime. For example, the race of the victim changes in a series of killings in which there are three white female victims and one black woman. Even though the crimes may be similar—all on or near university campuses, all committed in isolated areas within 10 to 15 miles of one another, all cases of rape–murder—the fact that one young victim was black, whereas the others were white usually sends police off on a hunt for two suspects. That is what happened in Rochester, New York, in the Arthur Shawcross cases. When the killer attacked both black and white prostitutes and other women who crossed his path who were not prostitutes, investigators searched for different killers (Keppel and Birnes, 1997).

Signature Versus Modus Operandi

There are crime scene indicators that relate murders even when the MO changes. Many sexually sadistic repetitive killers, for example, go beyond the actions necessary to commit a murder. As discussed previously, the MO of a killer can and does change over time as the killer finds that some things he or she does are more effective. The Federal Bureau of Investigation's (FBI) John Douglas and others (1992) of the Behavioral Sciences Unit in Quantico, Virginia, said that the modus operandi of a killer is only those actions that are necessary to commit the murder.

Beyond the MO, there are many, many killers who are not satisfied with just committing the murder; they have a compulsion to express themselves (or do something that reflects their unique personality). The killer's personal expression is his signature, an imprint he leaves at the scene, an imprint he feels psychologically compelled to leave to satisfy himself sexually. The core of a killer's signature will never change. Unlike the characteristics of an offender's MO, the core remains constant. However, a signature may evolve over time, as in some cases where a necrophilic killer performs more and more postmortem mutilation from one murder to the next. The FBI's Behavioral Sciences Unit defends the premise of a constant signature by saying that the elements of the original personal expression only become more fully developed (Douglas et al., 1992).

Douglas et al. once described the nature of the signature as the person's violent fantasies, which progress in nature and contribute to thoughts of committing extremely violent behavior. As a person fantasizes over time, he develops a need to express those violent fantasies. Most serial killers have been living with their fantasies for years before they finally bubble to the surface and are translated into behavior. When the killer finally acts out, some characteristic of the murder will reflect a unique aspect played over and over in his fantasies. Likewise, retired New York Police Department homicide detective Vernon Geberth (1996) wrote that it is not enough simply to consummate the murder; the killer must act out his fantasies in some manner over and beyond inflicting death-producing injuries. This "acting out" is the signature of the killer.

Detectives who investigate a series of murder scenes look for the same type of extraordinary violence and a bizarre set of similarities, and their gut instinct will tell them there is more here that is alike than different. Another homicide investigator, however, might point out that the killer used a pipe wrench as a blunt instrument here, a hammer there, and in a third crime no one can even figure out what weapon was used. Maybe one time the killer draped a pair of underpants on the victim's left leg, then, at the next crime scene, the underpants were on the victim's right leg or maybe still on the bed. In each case, though, the victim was obviously beaten well beyond the point of death by an assailant whose violence seemed to increase in frenzy while he was attacking her. Also, the killer seemed preoccupied with the victim's clothing and took some time to arrange the crime scene even though there might have been people living just upstairs. These are the psychological calling cards the killer actually needs to leave at each scene. Other examples of signatures are mutilation, overkill, carving on the body, leaving messages, rearranging or positioning the body, engaging in postmortem activity, or making the victim respond verbally in a specified manner. These constitute a signature. What is important about a killer's signature, then, is that killers learn to treat victims the way they do in their fantasies, always attempting to satisfy their fantasies as they move from one victim to the next.

The Shoreline Murders

The Shoreline district lies north of the city limits of Seattle in unincorporated King County, Washington. The area includes a shopping mall to the north, near the King and Snohomish County line, made up of small- to mid-size businesses, convenience shops, and stores. The immediate area of the murders consists of multifamily dwellings, apartments, and some single-family homes. An apartment complex is located on a cul-de-sac in the 19700 block of 22nd Avenue Northeast in an area called Ballinger Terrace. Not many murders occur in the area over a year; in the previous year, the neighborhood experienced the average variety of crimes and one murder of a male victim. However, within 30 days, the locale experi-

enced two separate, atypical murders within the same apartment complex.

Robert Lee Parker was convicted of two counts of aggravated first-degree murder. Before trial, members of the King County Prosecutor's Office in Seattle requested that a signature analysis be completed on the two murders. Their main question was, "Were both murders committed by the same person?" The analysis did not include any information about Parker or evidence about why he was connected to either case.

Renee Powell: The First Victim

Renee Powell relocated from St. Louis to Seattle. As a registered nurse, she was employed by several hospital facilities in the area. Powell was described as a 43-year-old white female, 5 feet 4 inches tall, small build, and weighing 100 pounds. She had no criminal record and had not been a crime victim.

Powell was last seen alive at approximately 7:30 p.m. on February 24, 1995. Previously, she had driven to a nearby Albertson's Supermarket to purchase cigarettes, a newspaper, and some ice. Police investigation revealed that upon returning to her apartment on 197th Place, she had sufficient time to make a jar of iced tea. Also, it was discovered that Renee was doing laundry in the building's laundry room, which was adjacent to her lower apartment unit. The apartment structure was a two-story residence containing four units. No one would see Powell again until firemen discovered her charred remains inside her apartment shortly after midnight on February 25.

At about 11:50 p.m., neighbors reported a fire in Powell's apartment. By 12:40 a.m., the firemen had put out the fire and discovered Powell's body. They discovered that she was bound, gagged, and constrained by a ligature. Homicide detectives from the King County Police were called. Investigation revealed that Powell probably heard a noise at the front door of her apartment because the killer had broken her door open. She probably had no time to respond. Powell was discovered face down on the floor of her bedroom with a bookshelf pulled down and lying on top of her body. The killer appeared to have stripped her naked from the waist down and then tore her shirt from her body. Her bra remained fairly intact, pushed up, exposing her breasts. Her left arm was bound with an electrical cord cut from a study lamp later found in her bedroom.

Arson investigators determined that separate and distinct fires had been started around the residence. They did not communicate with each other. The first fire was started in the master bedroom near the victim's body. The second began in the living room next to the fireplace.

The autopsy examination discovered that Powell suffered two stab wounds. One was in the right abdomen and stomach; the other, in her left back in the parasacral muscle. The gag in her mouth was an elastic bra, tied tightly and fastened in the back with a double overhand knot. The medical examiner removed a segment of electrical cord from around the victim's left forearm and outside of the shirt. The loops were tied with a complicated set of overhand knots. The plug end of the cord was present and the other end appeared cut.

The presence of conjunctival petechiae indicated that there was probable asphyxia. The victim's body was more badly burned in the front than in the back. There was no soot found in the throat; therefore, death occurred before the fires. Powell had been vaginally raped, and semen was preserved as evidence.

Further investigation revealed that the killer had stolen items from the apartment, including an overcoat, a dress, a videocassette recorder, several bottles of wine, a duffel bag, and some frozen meat.

Barbara Walsh: The Second Victim

Barbara Walsh was a 54-year-old white woman who lived alone. She had been widowed more than 20 years previously and never remarried. In the weeks before her death, she had not developed any significant relationships. Like Renee Powell, Walsh lived in a lower unit of the same apartment complex. Walsh worked as a receptionist at Group Health Hospital.

Thirty days after the murder of Renee Powell, at about 10:30 p.m., a neighbor saw Barbara Walsh returning from the laundry room of their fourplex. Walsh also lived on

Figure 31.1 Aerial view of Powell and Walsh's apartments in proximity to Parker's residence.

NE 197th Place, which was about 100 yards northwest of Powell's apartment (Figure 31.1). Unlike Powell's one-door apartment, Walsh's had a front door and a back sliding door that opened to a common patio and a wooded area. Her sliding door could not be unlocked from the outside, so police investigators surmised that during one of Barbara's trips down to the laundry room, the killer slipped inside.

At about 1:06 a.m. the following morning, neighbors reported smelling smoke and discovered a fire in progress in Walsh's apartment unit. Fire personnel extinguished the fire and discovered the body of Barbara Walsh. They could see that she was face down, bound, and gagged with a ligature. She was found on the floor of her bedroom with her head next to the foot of her bed. She was nude except for her shirt, which was shoved up nearly to her neck. Multiple fires had been set within the apartment. In the bathroom, between the sink cabinetry and a throw rug, a Trojan™ condom wrapper was located. Having found a knife in Walsh's kitchen, electrical cords from her lamps, and tights from her drawers and in her laundry, the killer was prepared for his night's work.

The autopsy report stated that Walsh was gagged with three pairs of tights: tan (innermost), white (overlain over the tan), and green (overlain over the white). The tan and white tights circled circumferentially around the back of the head once. The green tight circled once through the mouth and once around the anterior aspect of the neck at approximately the level of the thyroid prominence. All of the crotch regions of the tights were located anteriorly. The white tight was knotted once in the midline posterior region and once around the green tight, slightly left of midline. Approximately 11 inches away was another knot in the green tights through which was threaded a yellow electrical cord. The pathologist found several strands of blue yarn located next to the male adapter of the cord. The knife used in the Walsh murder was found on the couch where one of the fires was started.

Police discovered that similar blue yarn was also tied to the bedstead, as if the victim had been tied to the bed at one point. Of note was a knife found in the kitchen that police believe the killer used on the victim. The knife had a 7-inch blade attached to a 4-inch handle (Figure 31.2). All items and materials used by the killer belonged to the victim. Several items were taken from the victim's residence. Those items included a television set, videocassette player, compact disc player, wicker baskets, a box of silverware, Raggedy Anne and Andy figures, miniature red wire old-fashioned bicycles, a glass prism, and small polished stones.

On the victim's left wrist was a ligature that was extensively burned and consisted of multiple types of wires. There was a 9½-inch length of black insulated wire, incompletely burned through and attached by a few strands to a portion of yellow-tan insulated wire. The

Figure 31.2 Knife used in the Walsh murder, found on the couch where one of the fires was started.

yellow-tan wire, from a lamp in the residence, was wrapped and knotted circumferentially around her wrist. The knot was located on the lateral aspect. One portion of the wire completely encircled the thumb.

Walsh suffered three stab wounds clustered on the right side of her abdomen. The stab wounds were gaping and extensively charred around the edges. Near this cluster of wounds was a solitary stab wound. All of the stab wounds proceeded from right to left without appreciable upward or front-to-back deviation.

In summary, the pathologist stated that Walsh died as a result of ligature strangulation by the stocking, which also served as a gag. Additionally, a ligature was present on her left wrist and multiple abdominal stab wounds were identified, which produced injuries insufficient to account for her death. The thermal injuries were incurred after death, and no soot was present in her throat.

Signature Analysis

In these sexually perverted murders, the killer's approach to the victims and his selection of the location were preparatory, enabling the killer to carry out his highly personalized fantasies. Thus, evidence left as a direct result of carrying out his fantasy was far more revealing of the killer's nature than his MO.

In testimony at a hearing regarding separation of the two charges of first-degree murder, the following characteristics were described as being features of the killer's signature in the Shoreline murders. First, the act of binding was present in both murders. The killer used binding materials found at the scenes. Binding materials were not brought to each scene by the killer. The use of electrical cord and the ligatures exceeded the necessary violence to control the victims for rape-murder. The electrical cord binding and loops around both victims were the specific and necessary control devices that the killer had to use at each crime scene. Typically, these types of arm bindings are used by killers who prance the victim around (much like a dog on a leash) and poke them with a knife, thus evoking terror and satisfying his anger.

Second, the number of stabbing strokes was necessary for this killer and increased from the first murder to the second murder. The killer stabbed Powell, the first victim, twice and inflicted four stab wounds on Walsh, the second victim. Third, the disposition of both victims' bodies reflected this killer's personal feelings. The killer had to leave the victims in a sexually degrading and submissive position. Both were essentially nude from the neck down and intentionally placed face down. The killer purposefully left the victims so they would be found.

Fourth, the taking of **souvenirs** enabled this killer to relive the event at some future time. Such thievery was crucial to this killer's needs. Psychologically, the killer regarded these victims as "bitches," which, in his mind, justified his thefts. Finally, the presence of arson was evidence of another form of violence inflicted by the killer. The arson fires were a product of refinement and learning. There were more fires set at the second murder scene. Setting fires at the crotch of both victims was totally unnecessary, but was an act this killer felt compelled to do.

In summary, as this killer proceeded from one victim to the next, his true signature evolved. More stab wounds, more percussive activity with the body, and additional fires allowed this killer to feel more attached to his victims and vent his anger. These factors led to the conclusion that the two victims were killed by the same person.

Statistical Analysis

The Homicide Investigation and Tracking System **(HITS)** in the Washington State Attorney General's Office is a central repository of murder and sexual assault information in the state of Washington. The HITS program is a database with 227 query capabilities (Keppel and Weis, 1993). Before a hearing on the separation of charges, a statistical analysis was performed to determine the relative frequency of the signature characteristics in the Powell and Walsh murders.

At the beginning of the analysis, there were 5788 murder cases in the HITS program. The first search revealed that there were 1164 cases in which the body recovery site was the victim's home. Of those cases, there were 90 victims that were discovered bound in some way. From those 90 victims, 49 cases were found in which trophies or significant items were removed by the killers). Of those 49 cases, there were 16 victims who received stabbing or cutting wounds. When those 16 victims were checked, only two victims, Powell and Walsh, were found burned. The rarity of these characteristics was significant to the prosecution.

Catching the Killer

Detective follow-up work and crime laboratory analyses further corroborated the opinion that these two murders were committed by the same person. Robert Parker lived across the street from the woods that overlooked the apartment complex where Renee Powell and Barbara Walsh lived. Police detectives contacted Parker's residence during the initial canvas, but the residence was in the name of his girlfriend and Parker provided a false name. Parker's residence was only 130 feet from Walsh's home and 150 yards from the home of Powell. Parker was known to go out for long periods alone at night without explanation.

In late October 1996, detectives were contacted by a therapist. The therapist was treating a woman by the name of Princess Gray. The therapist told police that her client had information about murders that occurred in the Shoreline area. On November 1, 1996, detectives contacted Princess Gray in the King County Jail. Gray had been charged and booked for assault and reckless endangerment and was awaiting trial. Gray told detectives that Parker told her two "white guys" were involved in the first lady's murder and Parker had stolen Powell's property from them.

She related that on the night of the murder, Parker brought home a videocassette recorder, eight or nine bottles of wine, and freezer food that included pork chops. He also had a container with $30 to $40 in change. Gray said that after the second murder, Parker came back with a television, another videocassette recorder, and a compact disc player. Parker also brought some spices. Gray was asked if Parker had also brought any trinkets. She said, "Yes, things you set on your table." Detectives subsequently recovered most items from Gray's residence. They also recovered Powell's duffel bag from Parker's residence at the time of his arrest for the murders. Inside the duffel bag's pocket were polished stones that Walsh was known to collect. Killers frequently remove items belonging to their victims as souvenirs or for monetary gain (Douglas et al., 1992). Parker's retention of stolen items found in the possession of Princess Gray and himself contributed to the evidence against him in these cases.

More specific evidence linking Parker to both murders was discovered. From semen found on Powell's vaginal swabs, investigators requested that DNA analysis be performed and compared to Parker's DNA. It was found to be a match.

In addition, hair was found on Walsh's bathroom counter. The hair was protected from the fire by a towel that lay on top of it (Figure 31.3). DNA analysis was performed on the hair and compared to Parker's DNA, and another match was discovered.

For the experienced homicide detective, linking murder cases by distinguishing between a killer's MO and his signature should not be difficult. What is problematic is that the elements of the signature, at times, can be hidden due to decomposition of the remains or contamination of the crime scene. The two scenes in the Shoreline cases were contaminated by fire, but the early discovery of the fire prevented more destruction of evidence that would have possibly hidden the killer's signature elements.

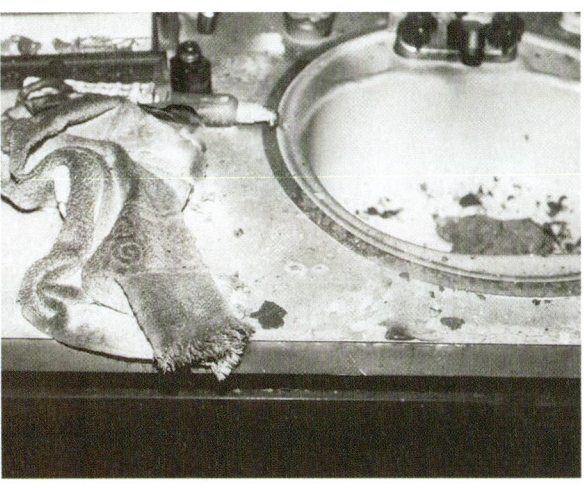

Figure 31.3 Location of the towel in Walsh's bathroom. Hair, linked by DNA profiling to Parker, was found under the towel.

Summary

A killer's method of operation contains those actions that are necessary to commit the murders. They may change from one murder to the next as the killer gains experience and finds more beneficial methods of operation from murder to murder (Keppel, 1995). Whatever the killer does beyond the murder, such as some forms of binding, arson, unnecessary stabbing, and arranging the bodies, should be the major focus of investigators in determining if murders are committed by the same person. It is the signature that remains the same from the first offense through subsequent offenses. The ritual may evolve, but the theme persists (Douglas et al., 1992).

Acknowledgments

This chapter was taken from *Serial Offenders: Current Thought, Recent Findings,* Schlesinger, L.B., Ed., CRC Press, Boca Raton, FL, 2001.

References

Douglas, J. E. and Munn, C., Violent crime scene analysis, *Homic. Invest. J.*, Spring: 63, 69, 1992.
Douglas, J. E., Burgess, A. W., Burgess, A. G., and Ressler, R. K., *Crime Classification Manual*, Lexington Books, New York, 1992.

Fosdick, R. B., *European Police Systems*, The Century Co., New York, 1916.

Geberth, V. J., *Practical Homicide Investigation: Tactics, Procedures, and Forensic Techniques*, 3rd. ed., CRC Press, Boca Raton, FL, 1996.

Grahame, K., *Pagan Papers*, John Lame, The Bodley Head, New York, 1994.

Hanfland, K. J., Keppel, R. D., and Weis, J. P., *Case Management for Missing Children Homicides*, Washington State Attorney General's Office, Seattle, WA, 1997.

Hazelwood, R. and Michaud, S. G., *The Evil That Men Do*, St. Martin's Press, New York, 1999.

Keppel, R. D., Signature murders: a report of several related cases, *J. Forens. Sci.*, 40, 658, 1995.

Keppel, R. D., Signature murders: a report of the 1984 Cranbrook, British Columbia, cases, *J. Forens. Sci.*, 45, 50, 2000.

Keppel, R. D. and Birnes, W. J., *The Riverman: Ted Bundy and I Hunt the Green River Killer*, Pocket Books, New York, 1995.

Keppel, R. D., and Birnes, W. J., *Signature Killers*, Pocket Books, New York, 1997.

Keppel, R. D. and Walter, R. A., Profiling killers: a revised classification model for understanding sexual murder, *J. Offender Ther. Comp. Criminol.*, 43, 417, 1999.

Keppel, R. D. and Weis, J. P., HITS: Catching Criminals in the Northwest, *Federal Bureau of Investigation Law Enforcement Bulletin*, April 14, 1993.

Keppel, R. D. and Weis, J. P., Time and distance as solvability factors in murder cases, *J. Forens. Sci.*, 39, 386, 1994.

Keppel, R. D. and Weis, J. P., *Murder: A Multidisciplinary Anthology of Readings*, Harcourt Brace Custom Publishing, Orlando, FL, 1999.

Oxford English Dictionary, The Clarendon Press, Oxford, 1933.

Robinson, W. C., *The Law of Patents for Useful Inventions*, Little, Brown, and Co., Boston, 1890.

Soderman, H. and O'Connell, J., *Modern Criminal Investigation*, Funk and Wagnalls, New York, 1936.

State of California V. Cleophus Prince, 9 CAL.APP.4th 1176, 10 CAL.RPTR. 2d 855, 1992.

State of Delaware V. Steven B. Pennell, Del. Super., 584 A. 2d 513, 1989.

State of Louisiana V. Nathaniel Code, 627 So. 2d 1373, 1994.

State of Washington V. George W. Russell, 125 Wash. 2d 24, 882 P. 2d 747, 1994.

Sutherland, E. H., *Principles of Criminology*, J. B. Lippincott, New York, 1947.

Timothy Wilson Spencer V. Commonwealth of Virginia, 240 Va. 78, 393 S.E. 2d, 1990.

Whittenmore V. Cutter, 29 F.Cas. 1123, No. 17, 601, 1 Gall. 478, 1813.

Criminal Personality Profiling

Michael R. Napier and Kenneth P. Baker

32

Introduction

This chapter presents the evolution of criminal personality profiling, and how investigators can be assisted in an investigation of violent crime by knowing specific profiling characteristics and their significance. The profiling concepts presented here are the result of extensive experience and some validating criminal research. This chapter describes one approach to profiling that is adopted by the Federal Bureau of Investigation's (FBI) Behavioral Sciences Unit (BSU). The purpose of offender profiling is to supply offender characteristics to help investigators narrow the field of suspects based on the characteristics of the crime scene and initial investigative information.

"Criminal profiling has been used successfully by law enforcement in several areas and is a valued means by which to narrow the field of investigation. Profiling does not provide the specific identity of the offender. Rather, it indicates the kind of person most likely to have committed a crime by focusing on certain behavioral and personality characteristics."[1]

In presenting this chapter, three themes wend their way from start to finish:

1. What began informally within the Behavioral Science Unit at the FBI Academy as an effort to understand violent criminal behavior has now evolved into an accepted investigative tool for officers tasked with solving violent crimes. Through the study and application of profiling, understanding certain aspects of violent criminal behavior has become more widespread among professional investigators.
2. Policing today is performed by professionals possessing better education and training, and who use professional concepts that have developed along with other advancements in forensic science and investigative knowledge. This development guides a professional approach to complex investigations. By applying this understanding of personality profiling together with the fast-developing body of experience from the forensic sciences, police officers are achieving and sustaining their hard-earned reputations as professionals.

3. The profiling process provides an approach to conducting a thorough and competent behaviorally based examination of crime scenes and investigations of violent crimes that display offender characteristics amenable to the profiling process.[2] As with other forensic evidence left at a crime scene, offenders also leave readily retrievable behavioral evidence from which the trained investigator may discern the offender's motivation and some personal traits. Behavior evident at crime scenes serves the investigator by revealing characteristics of the offender both as an individual and as a criminal. This knowledge often reveals characteristics of the perpetrator, making him more identifiable as a potential suspect.

Profiling offers no "recipe" or secret formula for bringing these behavioral perceptions to life in the context of an investigation. There is, however, a need for in-depth experience, a systematic analysis of case materials, and expertise in crime analysis methods. Such expertise applies an understanding of human behavior and follows behaviorally based principles. Some practitioners may imitate profiling expertise, or make zealous claims, but those lacking a background grounded in extensive criminal investigative experience, coupled with proper academic and forensic training as well as a network of qualified professionals (certified by the FBI), often deliver little more than a superficial comprehension of these intricate perspectives. In some cases, self-professed experts may severely harm rather than help an investigation.

When employing a professional consultant for violent crime analysis, the client must thoroughly check credentials and references to determine if the purported expert satisfies each of the following criteria:

1. Has the expert had the opportunity to review criminologic research conducted by the FBI's BSU, prominent academicians, and other law enforcement researchers in the field?

2. Has the expert been trained extensively in the theoretical and practical aspects of criminal behavior, crime scene reconstruction, crime scene assessment, as well as developing a powerful background in criminalistics, forensic pathology, wound ballistics, terminal ballistics, bloodstain pattern analysis, and other forensic fields?

3. Does the expert have extensive investigative experience solving violent crimes, including hands-on experience in processing crime scenes, examining the effects of wounding and injuries, observing autopsy protocols, and experience in applying **criminal investigative analysis** to a wide range of violent crimes?

4. Does the expert approach the investigation of an unsolved violent crime from a law enforcement perspective rather than a clinical psychology one?

History of Profiling and Criminal Investigative Analysis

Behavioral profiling has an ancient history sprinkled with interesting characters that have gained fame as "super sleuths." Years ago, authors and moviemakers in Hollywood fell victim to the allure of the erudite investigator who solved crimes by accurately interpreting minor clues that mixed behavioral and forensic evidence. The characters of Sherlock Holmes and Charlie Chan brought commonsense crime-solving skills into our hearts and minds and had a positive effect on the slowly evolving development of critical, creative police detectives.

Experienced investigators of violent crime developed a cumulative "sense" of offender traits and slowly modified their analyses to each new case. Thus, every investigator worth his salt "profiled" his cases and used that profile to guide his investigative efforts. The profile directed him in the most appropriate use of precious resources and focused his efforts toward the most viable suspect. Talented investigators knew that offender behavior in a crime scene, and the offender's interaction with the victim, often suggested personal qualities of the suspect.

During the 1950s and 1960s, a first-rate cadre of sworn law enforcement officers was selected to attend the FBI's National Academy (NA) at Quantico, VA. There they were exposed to new ideas from colleagues and from FBI instructional staff. As new techniques were developing, experienced investigators in violent crime began to exchange ideas and experiences on various types of violence. They were further trained in scientific investigative techniques, abnormal psychology, and the latest findings in the study of violent criminal behavior.

In these course groupings, they were exposed to the deviant, and often depraved, world of violent criminal behavior. A pattern developed among the officers attending the NA. They were encouraged to bring their most challenging unsolved investigations, and the cases were discussed and "brainstormed." These dedicated officers sought the opportunity to expose their investigative theories to special agents (calling themselves *behavioralists*) in the FBI's esteemed BSU.

The officers brought years of experience and intuition about "how" and "why" an offender acted in a particular way. They brought a collective understanding and wisdom that revealed identity characteristics of certain types of offenders. By combining their experience and investigative talents with the FBI's up-to-date training in abnormal psychology and criminology, an expertise soon developed. This expertise became an art form.

The BSU staff was actively accumulating a growing success rate in solving difficult crimes. Their success was based on informal feedback from the officers who had supplied the cases for examination, and who then applied their investigative recommendations upon returning to their home jurisdictions. As a natural follow-up, the FBI initiated research projects to seek formal validation of certain profiling concepts. From this merger of practice and theory, the concept of criminal personality profiling evolved.

During the same time, the FBI's National Academy was open to more and more officers, and the NA instructor staff continued to grow. This fresh talent supplemented the pioneers and augmented the knowledge that they had accumulated. Formal studies began to examine the crimes of rape, murder, arson, child molestation, bombings, and **equivocal deaths** of all sorts.

These studies involved both quantitative and qualitative research. The qualitative research involved face-to-face interviews with convicted offenders. The interviews often lasted for days. They were made more revealing through the prior collection and mastery of the offender's criminal conduct, the development of extensive protocols to guide and standardize the research interviews, and by the experience levels of the agent-interviewers. It was not usual for agent-interviewers to have better and more accurate recall of the facts of a case than the offender who was directly involved. This approach allowed the researchers to use investigative records and forensic reports to compare and correct, not simply accept the offender's recollection and version of events.

From this research, a wealth of data emerged that led to new findings regarding the criminal behavior of various serial criminals, especially in homicides and rapes. Of considerable significance was the conclusion that in many cases the findings validated the combined anecdotal experiences of the investigators attending the NA, and the agents at the BSU. In fact, many of their ideas were validated, and this theoretical development led to the development of a special "language" of profiling.

In some instances, these data serve as a scientific basis for the projection of offender traits, typologies, and characteristics based on the analyses of crime scenes and of victim statements. This works especially well when applying serial rape typologies.[3] From the core experiences of those behavioral science innovators, many additional law enforcement processes emerged from profiling that form the basis for criminal investigative analysis (CIA). This is a collection of various derivative approaches and techniques for the analysis and investigation of all forms of violent crimes.

The processes falling under the CIA canopy are the following:

- Crime analysis from a behavioral perspective
- Criminal personality profiling
- Crime scene analysis and reconstruction
- Search warrant applications
- Investigative strategies
- Interview and interrogation strategies
- Prosecution strategies
- Case linkages of serial offenses
- Equivocal death analyses
- Threat analyses

General Concepts

Although skepticism remains in academia and some police departments concerning the feasibility of profiling, case submissions continue to flood the FBI's BSU and its operational successor, the National Center for the Analysis of Violent Crime. The success record continues to grow for cases that have previously bewildered experienced and accomplished investigators. CIA processes have been directly credited with the resolution of several hundred violent crimes that would have otherwise been placed in, or remained in, an unsolved status. Many of these cases had languished without the development of a motive or suspect.

The CIA collection of techniques is now applied daily and has been of assistance in cases ranging from political terrorism to threats to use weapons of mass destruction to rape and homicide matters. It has aided in suspect development, the structured interview and interrogation of potential suspects, and to the clearing of innocent suspects.[4]

To further clear up any confusion over the popular, but incorrect, use of the term *profiling*, a more precise definition is necessary. As popularly understood, everything involved in the analysis of crimes by behavioralists was called profiling. This term was commonly used by the media, in books and movies, and by advocates of the new tool as a panacea to solving violent crimes. The term was also loosely applied as a screening tool for airline passengers in an attempt to identify skyjackers. Following the violence of the 1960s, others used the term in an attempt to identify characteristics of a lone assassin.[5] The use of the term profiling has also recently involved drug traffic stops along major interstate corridors. The evolution and refinement of the FBI's process of profiling clearly identified only one use for the term, which was to seek characteristics of an unknown offender involved in an unsolved violent crime.

Profiling has been given many definitions. One helpful view is to define profiling as the identification of certain characteristics of an unknown, unidentified offender based on the way he[6] committed a violent act, and his interactions with the victims. By reading the offender's behavior at the crime scene, certain descriptive traits and characteristics can be attributed to the unknown offender. Some traits are learned from witness descriptions, but the most valuable clues are based on a combination of crime scene examination, investigative experience, an understanding of offender and victim behaviors, knowledge of wound patterns,[7] forensic evidence, and the results of the research conducted by the FBI/BSU.[8]

As the result of such a comprehensive analysis, a *profile* may elevate one suspect from among those developed in an investigation. A helpful profile will describe the offender so that his own associates will recognize him. Categories of descriptors include the following: sex, race, age (approximate), criminal history, residency in relation to the crime scenes, employment history, social adjustment, sexual adjustment or abnormalities, alcohol or drug usage, educational level, and interpersonal skills.

Profiling, as previously stated, is an art (and is not usually acceptable for courtroom presentation). It was not conceived for such an application. However, analysts qualified as expert witnesses frequently present the results from the collection of other CIA processes in courtrooms.

For clarification, the CIA process may be visualized as a tree trunk that supports and nourishes several distinct branches. The branches are seen as independent systems; however, they are closely bound under the CIA umbrella. As previously noted, these processes are routinely and successfully applied to a broad range of criminal investigations involving violent crimes. In addition to the types of cases previously cited, CIA is used in investigations involving arson and bombings, workplace and school violence, extortion, kidnapping, missing persons, and domestic and international terrorism.

Victimology

One good source of information for investigators trying to understand the crime scene involves detailed knowledge of the crime victim's life and lifestyle. The history of a victim in some ways defines the analysis of the crime. From one perspective, unsolved violent crime frequently results from an incomplete or faulty understanding of the victim. So important is this phase of investigation that the study of the victim must be thorough and undertaken immediately.

People have a variety of personality factors, even quirks, which help define them. They appear throughout one's daily life and often relate directly to victimization. The victim's characteristics may also offer insight regarding the personality of the attacker. Why was that person attacked instead of some other person?

The goal is to develop sufficient **victimology** to be able to answer the question: "Was the victim's lifestyle a contributing factor toward victimization?" As the investigator initiates the search for the perpetrator, a full understanding of the crime scene and interactions between the victim and the offender may be elusive. Unfortunately, cases have remained unsolved because key information about the victim and his or her lifestyle has not been uncovered. These intimate details usually explain or clarify why victimization occurred with this victim, at this particular location, at this time, in this particular manner, and by this particular offender.

Victimology data are so crucial to the progress of an investigation, and its ultimate resolution, that special attention must be paid to it at the outset. Simply stated, by learning the personality, attitudes, lifestyle habits, and perspectives of a victim, the crime scene and offense can be analyzed and evaluated for a better behavioral understanding of what transpired and why the crime occurred.

Trophy or Souvenir: What Is Taken from the Victim?

The psychological processes of offenders are varied and personalized. Different offenders may commit the same or similar acts, but their psychological frames of reference may be dichotomous. An excellent example of individual differences in offenders is demonstrated by the significance attached by the offender to the items taken from the victims. These items fall into three general categories: evidentiary, valuables, and psychological. As part of the offender's modus operandi (MO), he may take evidentiary items that tend to identify him, leading to his arrest. The offender may take valuables to fulfill his or her motivation for financial gain, or the offender may take items of minimal value; often, items of little monetary value are not missed. This category of (nonvaluable) items falls into the realm of the offender's psychosexual motivation, drive, and thinking. These thefts have nothing to do with protecting the identify of the offender or fulfilling the purpose of financial gain; rather, they are driven by an inner need of the offender, a drive that takes behavior beyond what is required for the offender to be "successful" in his criminal endeavor.

These same items may be taken by different offenders in different crimes while working with similar MOs, and with similar type victims. However, the reasons for the behavior may be very dissimilar in that the stolen items are closely identified with the victim or the commission of the crime. They are called either a **trophy** or a **souvenir**, depending on the meaning placed on them by the offender's mental process.

In the mind of the offender, the item retained may represent an accomplishment, a victory within the context of his criminality, so it is retained as a trophy. For other offenders, the same item would represent, to him, a fondly remembered occurrence and is retained as a souvenir for inclusion in his masturbatory fantasies.

Fantasy in Criminal Thinking

Fantasies are powerful forces in shaping violent criminal acts. Fantasies play a critical role in many crimes but have their greatest impact in sexually violent crimes. Fantasies do not create criminality, but they do reinforce and deepen criminal thought.[9] Fantasies are daydreams. They may be positive and enriching, or they may be negatively focused on power, control, and domination. Fantasies may be visualized on a continuum, ranging from the low point of repetitive, underdeveloped thoughts, through repetitive, detailed, well-developed, elaborate, and coherent thoughts.

Offenders using the latter type of fantasy are noted for having more "evidence-free" crimes due to the repetitive nature of the thinking pattern wherein the criminal planning is more extensive, more well-rehearsed mentally, and subsequently more effective. Criminal sophistication can often be seen in this type of criminal fantasy. A prime reason for the difference in fantasies and, hence, a difference in offenders is the ability to use imagination. Imagination is a forerunner and indicator of intelligence. Intelligence in fantasy generally leads to more effective planning and corrective thinking that is often seen as criminal sophistication. Hence, the more elaborate the fantasy, the more evident the sophistication, and the more difficult the task of the investigator.

Fantasies may be used to organize a collection of deviant thinking into a criminal thought pattern. This is commonly referred to as *premeditation*. Fantasies may become evident in sexualized violence as those crimes often feature behaviors that go well beyond acts that are strictly necessary to accomplish or complete the intended criminal act.

Requirements for Offender Profiling

Some would equate profiling with crystal ball gazing, but the process is more fundamental and logical than the mysteries of soothsayers and tarot card readers. A few basic tenets provide the foundation for profiling. First, how a person acts helps to define who a person is, to form their personality. A basic axiom follows from this belief: *behavior reflects personality.*

Second, at crime scenes, offenders often leave trace evidence consisting of fingerprints, footprints, tire tracks, DNA materials, hair, threads from clothing, etc. Likewise, an offender leaves behind traces of behavior at the crime scene. This behavior can be inferred logically by an experienced investigator. It is one thing to recognize a bloodstain on an item at the crime scene; it is another, however, to understand the behaviors that likely explain the blood being in a particular location rather than some other. A second axiom follows this understanding: *behavior left at a crime scene by an offender can be discerned.* This knowledge links closely with the notion that how a person behaves in the crime scene may reflect his personal traits.

Third, the cumulative research into human behavior by the FBI and many others has enabled behavioralists to classify offenders into useful typologies. Typologies are broad categories of characteristics that generally describe offenders based on multiple factors. What is contained in a typology may be applicable to some offenders, but not applicable to all offenders in the broad category. These typologies are of value to the development of a useful profile and analysis. However, the real talent lies in adjusting the descriptors to discriminate among the offenders in a particular case from one who may merely be on the periphery of an investigation. It is here that criminal investigative training, experience, and law enforcement, or nonclinical perspective, help determine the usefulness of a proffered profile.

All of this analysis is useful for the direction of an investigation, but it must be more than mere opinion. It must be based on concrete data. Therefore, considerable information is required before a trained criminal investigative analyst can adequately conduct a crime analysis. Special efforts may be required to ensure the requisite data are available. The good news, however, is that most of what is needed to conduct a valid crime analysis is generally found in police investigative files. Anyone offering an opinion regarding an unsolved crime, without access to the necessary materials listed in the following text, is likely premature in his opinions. Such speculation is harmful to the credibility of the crime analysis process, and may be harmful to the ultimate resolution of the case at hand. Furthermore, such speculation lends credibility to those who tend to seize every opportunity to criticize profiling as a version of psychological "witch-doctoring." Officers seeking the assistance of outside consultants must supply the necessary requested data or, with equal candor, indicate its nonavailability. The investigative data used for analysis are equivalent to the data utilized and examined in a scientific study. If the data are flawed or incomplete, the resulting analyses and conclusions will be equally flawed.

Analysts must have access to the following categories of information as well as ready access to the case investigator: all investigative files, crime scene photos, crime scene examination results, autopsy findings and photographs, toxicology reports, victimology findings, demographic and crime data, statements of experts, and results of laboratory examinations.

Results: Type of Analyses

A profile is an outline of characteristics of an individual unknown offender. A CIA report involves much more, and in some ways is more comprehensive and complex. The services offered within CIA concepts are self-explanatory; however, some may require additional comments. The heart of the CIA process involves *crime analysis* from a behavioral perspective.

Crime Analysis

Crime analysis is one of the several processes under the CIA umbrella. In crime analysis, the investigator gathers all available data about the commission of the criminal act, the victim, and the offender. With the information from the crime scene commingled with data obtained in the initial stages of the investigation (victimology, etc.), the investigator begins to assemble a sequencing of the events involved in the violent crime. Each act or behavior resulting in specific crime scene characteristics is examined by asking "why" and "how" the event occurred, and "what" type of person would have acted in the manner that created the scene. The analysis generally results in the ability to reconstruct the interaction between the offender and the victim, and permits an explanation of the individual pieces of the crime scene puzzle. This process allows for a better understanding of the offender's *motivation, criminal sophistication, and possible prior relationship with the victim.* The investigator is also well served in the interview and interrogation process by being able to present his read on the crime, and the offender's role in the criminal act. As new investigative data are obtained, the circumspect investigator rethinks his analysis and adjusts his investigative and interview strategies accordingly.

Analysis involves logical thinking guided by specific parameters provided. Specialized bodies of knowledge peculiar to the field of application best define the remaining parameters. The knowledge and limitations of scientific investigative techniques, the understanding of criminal thinking as presented academically and practically, and the special expertise of criminal investigators across a broad range of violent offenses serve as examples. It is with these special abilities that some CIA analysts are able to identify what others fail to see, connect what is unconnected, dispose of what is extraneous, and correctly synthesize what is missing from a crime scene or a victim, witness, or suspect statement.

Crime Scene Reconstruction

Crime scene reconstruction is a process within CIA and crime analysis that provides the investigator an understanding of how the victim was approached and controlled, as well as the likely interactions between the offender and the victim. With this knowledge, the investigator is able to understand the critical elements of "how" and "why" the violent crime likely occurred. By sorting out the various elements of the crime, he is in a position to see how the offender acted as the crime unfolded, thereby gaining a grasp of the offender's motivation for selecting this victim, at this time and at this location. With this analysis and reconstruction, the investigator is better able to understand how the offender perceived himself as the crime developed, thereby providing clues as to his personal perspectives, personality, character traits, and ultimately his identity.

A special part of crime analysis is the ability to reconstruct and sequence criminal acts as they occurred in the interaction between the victim and offender. By describing the initiating behavior and distinguishing the role of individual "actors" in the crime, the investigator obtains a feel for the crime, that is, an understanding of what occurred, how it occurred, why the crime occurred, why it occurred with these participants, why it occurred at a specific location, and why it occurred at all. The level of planning or spontaneity exhibited by the offender, and the criminal sophistication demonstrated within the crime may also be determined. This reconstruction provides the investigator with a broader foundation from which to conduct an investigation, and provides the ability to interrogate suspects with authority, and thereby gain a genuine confession that outlines how and why the crime occurred.

Investigative and Prosecutorial Strategies

An investigator's analytical knowledge shapes his approach to the investigation as well as to individuals developed as suspects. Avenues for exploration and the psychological strengths or weakness of individuals may be brought into play in the investigative and interview phases. A strategy may be helpful to prepare media statements in a particular manner and

sequence them, or remain silent on specific aspects of a case. Statements can be framed to include insider knowledge about the vulnerabilities of a suspect.

Similar uses have been made of such knowledge in the prosecution of certain cases. Analysts who have developed a thorough understanding of the crime and a personalized perspective of the offender and his motivation can assist the prosecutor in many ways, and in each phase of prosecution. By understanding the offender's behavior, his true motivation and underlying psychological or psychosexual cognitive processes may be revealed. This knowledge may influence the order of proof offered, the structuring of questions, and the ordering of questions asked of witnesses. Specific questions of witnesses may include "trigger" terminology of special significance to the defendant. This insider knowledge may be used to tie together the themes in the prosecution strategy and may be included in arguments before the court and jury. These contributions may be used to "educate" the court, but also have specific impact on the defendant.

Although these efforts may be used in the overall prosecution strategy, they have their greatest impact on the defendant's decision to take the stand on his own behalf. They may also be used to prompt the defendant to take the stand to explain or place his "spin" on his behavior. These strategies may also be used to influence the defendant's decision regarding offering a specific witness or testifying himself.

Interview and Interrogation Strategies

Case resolution often depends on the results of the investigator's interview and interrogation skills. The origin of offender-specific interview and interrogation strategies rests squarely on two prongs.

Prong one involves the behavioral evidence retrieved from the crime scene. This evidence will often reveal information about the offender's criminal makeup, to include his motivation, victim selection process, interpersonal skills, anger control, criminal sophistication, and a variety of additional personal elements. These data are obtained from analyzing the totality of the crime scene and are combined and refined with the second prong of the interview and interrogation strategy.

Prong two involves the specific background and behavioral traits of a suspect. A cause for concern is the propensity on the part of some investigators to engage a suspect prematurely, without dedicating sufficient time to developing critical behavioral data. Often, some of these data are contained in prior police investigative files, but some must be gathered through the investment of shoe leather. A valuable source of information about a suspect can be derived from the impressions of officers who have previously engaged him in prior interviews. Suspects have favorite mechanisms that they use without conscious thought, to protect their self-concept. Although there are a number of such mechanisms available to a suspect, each will have developed and honed his habits over the years. These mechanisms are generally deeply ingrained within their personality and have become second nature.

Most criminal personalities rely heavily on three ego-defense mechanisms. This grouping of defenses is often called *RPMs* and refers to psychological techniques for bending reality. The initials represent *rationalization, projection,* and *minimization.* Through a combination of these mechanisms, a suspect changes objective reality to a personalized view, thereby allowing him to "save face" and feel better. One of the principle goals of an interviewer or interrogator is to make it easy for someone to provide useful information. By tapping into the suspects' protective mental process, the investigator gives the strong impression he understands the suspect's criminal behavior. The investigator is often able to convey that, when all the factors of the crime and interplay with the victim are considered, the offender's behavior is understandable.[10]

Case Linkages for Serial Offenses

Investigators are most concerned with serial offenders. Their concern stems not only from the cumulative injuries and personal devastation by one offender, but because a particular offender is proving a difficult match for their investigative skills. The traditional technique for case linkage was by use of a criminal's method of operation (MO). MO has three general purposes: (1) to enable the successful completion of the criminal act, (2) to prevent the offender's identification, and (3) to ease his escape. The problem with applying MO as a device for case linkage is that MO is learned behavior, making it likely to change over time. Sometimes, however, MO is all that is available to the investigator.

A more reliable method for case linkage is the recognition of aspects within a crime that go beyond the elements strictly necessary to accomplish the purposes of MO. Some offenders, particularly sexual offenders, have an inner or psychological drive that compels them to incorporate aspects beyond the purposes served by MO. These are called psychosexual drives and originate in, are shaped by, and are fueled by fantasy. These aspects become personalized to a particular offender and, therefore, are static and unchanging. These elements often become ritualized and repetitive.

One rapist-murderer investigated by one author established an MO of selecting black, female prostitutes who were all dressed in a similar manner and all worked the same "stroll." All were manually strangled. In the process of committing the sexual homicides, he removed their jeans. At some point he gathered the jeans in his hands, applied great strength, and tore the area of the jeans where the seams are joined at the crotch. Only the offender knew the motive for such behavior, but it was significant enough to him that he repeated it at each crime scene. The finding of such personalized behavior may be sufficient to link cases in the investigative phase. Taking these findings one step further, the question must be asked, "Is this behavior truly unique and sufficient for testimonial linkage of this suspect with that series of cases?"

This prospect raises the standards of certainty because of its startling effect on the guilt determination role of a juror and upon the criminal justice process in general. A testimonial analysis of crime linkage points to the defendant, so the process of arriving at the conclusion must meet high standards. Before such declarations may be made, a thorough search must be undertaken to determine that a wider range of investigators have not encountered that behavior. For testimonial purposes, the crime analyst must be in a position to state that those linking behaviors have not occurred in his experience or in the general experience of any other analyst, and that the behavior cannot be located in published criminology accounts. The behavior must be truly unique to that offender. Because of the high standard applied here, the one-of-a-kind behavior will, in all likelihood, embody a series of behaviors, any one of which could have occurred in prior crimes, but which in combination appear only in this series of crimes.

Equivocal Death Analyses

On occasion, an investigator is faced with the dilemma of knowing the cause (agent) of death, but not the accurate manner of death, that is, homicide, suicide, accidental death, or death due to natural causes. In such cases, crime scene indicators, coupled with victimology and other data uncovered in the initial investigation, are used to provide clear and compelling evidence for determining the proper manner of death.

In these cases, victimology is usually the key ingredient to resolving the equivocal death determination problem. It is helpful for the investigator to create eight columns for listing investigative findings. The eight columns should bear these headings:

1. Factors consistent with homicide
2. Factors inconsistent with homicide
3. Factors consistent with suicide
4. Factors inconsistent with suicide
5. Factors consistent with accidental death
6. Factors inconsistent with accidental death

7. Factors consistent with **natural causes**
8. Factors inconsistent with natural causes

As the investigator collects and reviews each piece of investigative data, he determines if the finding fits as a factor that can be listed under any of the eight columns. It is advisable to cross-check the factors by relisting each of the factors to show they *do not* support the conclusion of each of the four manners of death. For instance, an item that is listed as being indicative of homicide should be relisted as not supporting the manners of death by suicide, accident, or natural causes.

The goal is to arrive at a preponderance of factors listed under one of the headings. From this listing, it becomes evident that some factors weigh more heavily than others (e.g., bloodstain analyses, DNA findings, gunpowder residue analyses, or other scientific evidence). Thus, the investigative conclusions will lead to the accurate manner of death through a thorough and systematic review of all evidence.

Search Warrant Application

Over the years CIA has been successfully used in search warrant applications. It is possible to describe for the court what occurred in the commission of the crime, and what that behavior reveals about an offender. This information can be used to demonstrate how individual factors relate to the body of knowledge concerning offenders, their propensities, and their peculiarities. These elements may be commingled to indicate what special items an offender may retain from his past criminal activities, or in contemplation of additional criminal acts. In the past, this information has included the likelihood that specific items would be kept as well as the type of concealment. This input has occasionally overcome a "staleness" issue and identified crime-linking items for seizure.

Concepts and Applications in Sexual Assault Investigations

Victim Risk

Victim risk involves subjectively classifying victims according to their vulnerability to becoming a victim of violent crime. The evaluation of the victim's likeliness of becoming a victim is utilized in sexual assault and homicide investigations. Generally, victims are classified as high, moderate, and low risk. These classifications are based on the everyday details of their lifestyles. Of particular emphasis in studying the victim is how she handled stress, dealt with security measures or untoward sexual advances, regarded relationships with people, or how she handled her sexuality and that of others. The areas of interest may be expanded to include scenarios she may have become involved in and in what activities she would never have willingly engaged. In brief, the most intimate details about victims regarding sexual practices, involvement in illicit activities, attitudes toward alcohol or drug use, etc., must be determined to understand how she was confronted, and how the offender gained and maintained control over her to accomplish the crime.

A high-risk victim is one who because of individual features, including occupation, sexual or drug history, choice of associates or life patterns is highly vulnerable to becoming a victim. Such victims come into contact with a wide variety of threats, dangers, or deviant personalities, and thereby make crime analysis more difficult and uncertain. A moderate risk rating may be the result of the area where a victim is visiting, the occupation of the victim, or possibly the activity being undertaken at the time of victimization. A person with a low-risk victimology is one who, due to lifestyle traits and habits, was highly unlikely to have become a victim of a violent crime. This type of victim may live alone, always keep her doors locked, and not venture out at night.

Offender Risk Assessment

The significance of this classification is for the officer to appreciate the type and characteristics of the offender who would come into contact with a particular type of victim. For example, for a low-risk victim, the investigator is likely to begin with the assumption that the offender may have specifically targeted the victim, or found her personal vulnerability, or that the offender was personally associated with the victim. Some offenders exhibit concern for personal risks, whereas others do not appear to even consider risk factors, such as time of day, potential for witnesses, need to conceal his identity, the presence of alarms or security cameras, etc.

Victim Selection

The psychological processes involved in the selection of victims range from simplistic to complex, depending on the offender's motivation. The complexity is somewhat commensurate with the richness of the offender's fantasy, criminal sophistication, and constraints of the moment. When a victim is targeted, it is an indication of some level of association between the victim and the offender. Other victims may have merely had the misfortune of being "at the wrong place at the wrong time." Some offenders have idealized their victim type and involved them in their mental playacting; conversely, offenders may act with impulsivity and chose a victim simply based on her availability.

Some of the more elusive serial killers established "rules" for their victims. Such rules are unknown to the victim, and how they respond to gambits thrown their way by the offender often determines their fate. In these situations, the offender creates scenarios and measures the victim's attitudes and responses to determine if they measure up to the standards he has established. The infamous Ted Bundy created an idealized standard for his victims wherein they had to be "worthy" of his selection.[11]

One common finding within the victim selection process involves a conscious evaluation of the victim as to whether the offender can control her or if she will present a special challenge to his criminal planning abilities. As one serial rapist explained to one of the authors, he graded the victim's evident level of confidence and assertiveness before approaching her.[12] If the potential victim was wary, he avoided her. If the potential victim were confused by her surroundings or oblivious to events surrounding her, he would assess these signs of weakness to determine his chances of success with that particular woman.

Offender Typologies

Roy Hazelwood, one of the premier designers of the CIA process, modified the earlier findings of Nicholas Groth, Ann Burgess, and Lyle Holmstrom in typing rapists.[13] Hazelwood proffered six categories of rapists based on their motivation.[14] Two primary divisions of rapists involved their motivation to offend based on a manipulation of power or an expression of anger. These typologies were subdivided to refine and distinguish offenders inside the power and anger motivations. Hazelwood also provided typing for rapists who raped incidental to another crime, such as armed robbery or for the rare rapist who was a "lonely heart" seeking an outlet for sexual arousal.

Such typologies are very useful in analyzing rapes and identifying likely offenders. Based on the extensive research of the BSU, Hazelwood's typologies included the findings that went beyond describing the offender's motivation. Those typologies also provided presumptive clues as to the relationship between the offender and victim, the offender's residency in relation to the crime, prior criminal offending, the offenders' age range and race, etc.

False Allegations

False allegation is a term used by investigators to describe an unfounded claim of having been sexually assaulted, or having received threatening or obscene notes or telephone calls. The real impact of the false allegation is twofold. First, every investigative unit has limited resources, with regard to money and personnel, with which to resolve all reports of crimes against persons. These resources are inevitably spread thin in false allegation cases because they unnecessarily tie up police resources. Second, any real or presumed false allegation impacts the investigative efforts by utilizing resources needed for legitimate victims. Generally, these types of cases result in obtaining a confession from the pseudo-victim that the alleged offense did not in fact occur. In such cases, personal counseling is generally started and prosecution of the pseudo-victim is waived in lieu of therapy.

Applications in Homicide Investigations

Face-to-face interviews with experienced criminals and a review of their criminal acts revealed two distinct patterns in homicidal offenders, with some actors falling in the middle ground. They could be categorized as either organized or disorganized based on the behavioral evidence from their crimes.[15] **Organized offenders** exhibited clear evidence of forethought regarding their approach to and subsequent dealing with a victim. They plan and rehearse the crime in detail, allowing for variances in victims and locations. They conceive, before the crime, what will be necessary to prevent identification and apprehension and what tools or implements will be needed to efficiently deal with the victims in the manner they desire. Within the constraints of their ability to fantasize and imagine, they leave little to chance. They leave few clues as to identity and select victims who cannot be linked to them.

Disorganized criminals are well defined by their impulsiveness and generally thoughtless approach to crime. Their crimes will typically be messy and chaotic, with reliance on using criminal tools readily available at the crime scene. Their lack of planning often results in abundant clues to their identity and a plethora of identity-linking evidence at the crime scene. Cardinal indicators of disorganization are usually found in the evidence surrounding their ability to approach, obtain, and maintain control of the victim throughout the crime.

Sexual Homicide Characteristics

Some homicides have special characteristics that separate them from other homicides and require special investigative techniques. Examples of specialized homicides include those by arson, domestic homicides, erotomania-motivated killings, for-profit homicides, staged crimes,[16] and lust murders,[17] to name a few. Another significant category is that of *sexual homicide*.

Sexual homicides have characteristics requiring the use of specific strategies, knowledge of distinctive traits of the offender, and specific interview and interrogation skills. If a homicide appears random, the investigator should begin by exploring the crime scene and the victim's lifestyle for information that suggests a sexual interest by the offender toward the victim.

Sexual homicides are also characterized by criminal acts performed on the body as evidenced by the exposure of sexual anatomy, the posing of the victim after death in a sexually suggestive or degrading manner, the insertion of foreign objects, or redressing the victim. Each of these crime scene behaviors suggests a sexual interest by the offender, although such behavior may not be considered significant at the outset of the investigation. Some clues may be obliterated or damaged by altering the crime scene through arson or by some form of **staging**, to misdirect the investigation. It is strongly recommended that the investigator begin by noting sexual overtones when no other motive is obvious.

The question of whether or not the offender had a sexual interest in the victim may be answered through a meticulous examination of the victim's lifestyle. Such examination may identify an individual who paid special attention to the victim before the crime, or had made sexually suggestive comments about the victim to his friends. In some cases, the victim's friends may have knowledge of someone having made overtures to the victim regarding a desire for a close personal relationship. Such information may also reveal a person who was previously rebuffed by the victim. Past events of these types may only become available to the investigator who makes a special effort to determine the victim's lifestyle history.

Lust Murder

Lust murder is a term historically used to describe criminal acts involving an assault on sexual parts of the victim's body. The acts involve ways in which the victim's body has been assaulted or mutilated. By definition, the mutilation must have been intentionally inflicted after death, and involves the removal or mutilation of a part of the body. The *lust* terminology used as a classification of homicide may confuse an investigator who equates lust with sexual passion or arousal. As used in this concept, however, the lust refers solely to the sexual portions of the body selected by the offender for abuse. Customarily, these acts are those of the disorganized offender.

Body Disposal Choices

Offenders have choices to make when confronted with the reality that they have murdered another person and now have a body to dispose of. The body disposal *location* and *methodology* are often revealing as to a prior victim and offender relationship, the offender's criminal sophistication level, the degree of planning by the offender in committing the crime, the attitude of the offender toward the victim as an individual, whether the victim represented a class of people (i.e., women in general or a type of woman), and the offender's knowledge of the body disposal site. The location may indicate whether or not the offender resides close to the crime scene and body disposal site. The choices facing the offender are limited. He may simply walk away from the body, abandoning it where it fell, or he may choose to spend a brief or a considerable time to conceal the body. The offender may arrange the body in a manner that, in his thinking, makes a statement about the victim. Furthermore, he may also move the body to another location and dump the remains at the second location.

To walk away from the body, abandoning it where it fell, may signal a lack of planning or forethought. It may also suggest a certain pattern of "disorder" that might lead to the conclusion that the crime was more spontaneous and impulsive, rather than controlled and thought out. The offender who invests time and effort in the concealment of the body suggests he has thought about the crime, and is aware of the need to delay discovery of the body; the offender may need time to establish an alibi because his name will likely surface in the investigation. The concealment of the body also serves the purpose of concealing any linking evidence, thereby denying the investigator vital information on which to establish an investigation based on linkages to known associates or activities. A highly organized offender may choose the disposal site prior to selecting the victim.

The murderer who chooses to pose the body is likely attempting to achieve one of two goals. He may be trying to leave the body in a way that offends the discovering party, or society in general. The alternative is that the offender may be expressing his inner thoughts or anger and hatred toward this victim in particular or others represented by this victim (possibly prostitutes, drug dealers, etc.).

By moving the body to another site, the offender creates multiple crime scenes. The vehicle used for transportation becomes a crime scene, as does the new location. The significance of the acts is related directly to the thought process of the offender. If the movement is part of the process of concealing the body, those acts are likely from an organized offender, who has specific goals in mind to protect his identity or delay the discovery of the crime. The movement of the body may be to place it in an area where it will not be found, or if found, the area itself would be noted for specific activity, that is, a lovers' lane or a trash dump site. The movement of the body may also accomplish the purpose of allowing the offender to simply "dump" the body in a more remote area. Such actions often provide critical hair, fiber, and DNA forensic evidence

that can be found in vehicles; such evidence often links the victim to the specific offender.

Crime Scene Staging

On occasion, investigators will find a crime scene that has been tampered with before the arrival of authorities. This is referred to as *crime scene staging*. Crime scene staging involves acts that are committed to send the investigation off course, and away from the offender.

A crime scene may be staged to mislead the investigator from considering a particular person a logical suspect. An example of this type of staging might involve a man who kills his wife and wants the police to believe the homicide was committed by a burglar. His efforts would be directed toward causing the scene of death to include aspects that suggest the offender came to steal or rob. Staging may be attempted by ransacking a bedroom, a jewelry case, or by removing valuables from the premises.

Innocent family members who discover a body in a position that is personally disturbing to them may also stage a scene. The body may be nude or the manner of death may be unacceptable to the family as it offends their values. A good example of this would be the discovery by the wife of a husband found hanging in the basement, partially clothed. The wife may assume her husband committed suicide. If suicide violates strict religious beliefs or causes concern regarding specific clauses in an insurance policy, the wife may alter the scene. To overcome objectionable aspects, she may rearrange the body, to include redressing the victim. These acts, while possibly criminal, are usually motivated to achieve protective goals and may mislead the investigator.

In both scenarios, the investigator will likely find multiple aspects of the crime scene that are inconsistent. The various details of the crime scene will be *internally inconsistent*, which means they will not fit together logically, nor will they be properly supported by forensic evidence. As an example, the husband who killed his wife may have committed the crime out of uncontrolled anger. If so, the crime scene will usually display facets of that anger within the evidence present. The injuries inflicted on the wife's body may reflect overkill. The forensic and behavioral aspects of her death will reflect anger; however, the staged portions will reflect elements of burglary. These inconsistencies will alert the investigator to look for other incongruent clues. If the crime scene has been staged, a pattern of incompatible features will likely be discovered. The discovery of inconsistent details may include staging of the point of entry, the items that are missing from the crime scene, or the nature of the assault on the victim. The most likely tip-off that the scene has been staged may be the husband's display of emotion or his story about the crime. They generally do not correctly approximate the details reflected by an analysis of the crime scene.

An individual who finds it necessary to stage a crime scene is providing clear signs of a prior relationship with the victim. He may as well be stating, "here I am, look at me." Staged crime scenes alert seasoned investigators to start the investigation at "ground zero" and work outward.

Conclusion

This chapter has sought to provide an overview of the evolution of a distinctive manner of examining and investigating one element of violent crimes. *Profiling* evolved from work initiated by the FBI's BSU. This new approach to solving violent crimes was undertaken in the mid-1970s and was the result of interaction of the BSU staff and experienced police officers from across the United States and the world. These professionals exchanged their cutting-edge understanding of the behavioral sciences, criminal behavior, and anecdotal knowledge derived from their criminal investigative experience. The resulting technique of profiling continued to grow and has now progressed from an informal exchange of ideas, to a series of more complex processes. These processes are now contained within the body of criminal investigative analysis.

Initially, profiling was viewed as a major breakthrough by some but was viewed skeptically by others. The term, still widely used today in law enforcement and the behavioral sciences, saw some success, and as it evolved, several other processes were lumped together under the general term of profiling. For clarity's sake, it must be repeated and understood that *profiling applies, in FBI parlance, only to the interpretation of the offender's behavior in the commission of a crime, to determine the unknown offender's personal traits and characteristics. As noted previously, profiling is an art, not a science.*

Many of the analytical processes regarding the behavior of violent offenders grew from the initial concept of profiling and are again being scientifically validated by behavioral research within the FBI and academic community. However, the evolving CIA procedures are directly anchored within the innovative and groundbreaking research initially conducted by the FBI's BSU, and upon an unquestioned, yet practical, empirical application of these processes. CIA processes have met the toughest test of all; they have been tried and determined to be valuable techniques in solving violent crimes or dealing with violent criminals. These additional processes, although sometimes overlapping, are individually distinctive and are unquestionably differentiated from profiling.

Under the CIA umbrella, the processes flow from conducting crime analysis from a behavioral perspective. Such analysis begins with a close examination of the victim, crime scene, and the forensic findings that are discovered at the scene. Applying innovative investigative techniques, a knowledge of criminal behaviors and personalities, and the results of the studies previously conducted by BSU, the interactions between the offender and the victims can be reconstructed to establish a working understanding of offender motivation, victim selection, the dynamics of the crime, and the offender's preoffense and **postoffense behavior**.

These same proficiencies allow the trained investigator to formulate skillful investigative strategies and more sophisticated interview and interrogation approaches to suspects. In addition, they allow the investigator to lend valuable assistance to the prosecutor in structuring his case in chief. The CIA processes are several and varied and can be applied to all manner of violent crime investigations.

One of the more important contributions of profiling and the processes within CIA is the fact that the large gap that formerly existed between those who investigate crime and those who research and search for criminal etiology has been closed. These processes have come together so that "theory has met practice" in a way that is beneficial to both groups. As members of the FBI's BSU continued to conduct qualitative research with incarcerated criminals, some of the more adventurous academicians were lured from university campuses to participate in research projects. The joint sharing of ideas and perspectives has benefited both. Federal agents have learned that not all academic researchers are "snobbish, pointy-head idealists," and academicians have learned that some agents are very bright, and have the intellectual ability and capacity to conduct behavioral research. They share mutual interests in learning about criminals, especially violent criminals.

Furthermore, a majority of police investigators now use profiling techniques, and have developed a healthy understanding of its limitations. Profiling and the CIA processes now available are applied by law enforcement on a daily basis. CIA processes have earned praise and respect for their effectiveness in conserving and focusing police resources, and in assisting in the successful resolution of violent crime investigations. Another pleasant complement to the CIA process is the continuing evolution of forensic science. Scientific findings are recognized to be part of the solid foundation around which CIA and profiling opinions are formulated. It is reasonable to expect that as new methods are developed and validated scientifically, the field of CIA will continue to evolve and remain a powerful force in the effort to prevent, solve, and hopefully reduce violent crime.

Questions

1. Discuss the differences between profiling and criminal investigative analysis (CIA).

2. Discuss the concept of case linkage analysis.

3. Discuss the concept of "victimology," its appropriate use by an investigator, and its impact on criminal investigations.

4. What is a "profile?" What are its limitations and uses?

5. List the services or techniques included under the concept of criminal investigative analysis.

6. What are trophies and souvenirs? How are they similar and how are they different?

7. How does "fantasy" influence criminal conduct?

8. Discuss the process of victim risk assessment.

9. What is crime scene construction? Discuss its usefulness in a criminal investigation.

10. Describe the CIA technique of providing interview and interrogation strategies to investigators.

Suggested Readings

Depue, R. L. and Depue, J. M., To dream, perchance to kill, *Sec. Manage.,* June, 1999.
Douglas, J. E. et al., Criminal profiling from crime scene analysis, *Behav. Sci. Law, R*, 401, 1986.
Douglas, J. E., An overview of criminal personality profiling, *FBI Law Enforce. Bull.*, November 23, 1981.

Douglas, J. E. et al., *Crime Classification Manual*, Lexington Books, New York, 1992.

FBI, Crime scene and profile characteristics of organized and disorganized murderers, *FBI Law Enforce. Bull.*, August, 1985.

Groth, N., Burgess A., and Holmstrom, L., Rape, power, anger and sexuality. *Am. J. Psych.*, 134, 1239, 2001.

Hassel, C. V., The political assassin, *J. Police Sci. Admin.*, 4, 399, 1974.

Hazelwood, R. R., *Glossary for Criminal Investigative Analysts*, FBI Behavioral Science Unit, FBI Academy, Quantico, VA, 1988.

Hazelwood, R. R. and Burgess, A. W., Eds., *Practical Aspects of Rape Investigation: A Multidisciplinary Approach,* 3rd ed., CRC Press, Boca Raton, FL, 2001.

Hazelwood, R. R. and Douglas, J. E., The lust murderer, *FBI Law Enforce. Bull.*, April, 1980.

Hazelwood, R. R. and Michaud, S. G., *Dark Dreams*, St. Martin's Press, New York, 1998.

Michaud, S. G. and Hazelwood, R. K., *The Evil Men Do*, St. Martin's Press, New York, 1998.

Napier, M. R. and Adams, S. H., Magic words to obtain confessions, *FBI Law Enforce. Bull.*, 67, 1998.

President's Commission on the Assassination of President Kennedy, Hearings on the *Investigation of President John F. Kennedy*, Vols. 1–26, U.S. Government Printing Office, Washington D.C., 1964.

Westveer, A. E., Ed., Managing death investigations, *FBI Law Enforce. Bull.*, 1997.

Articles

Ault, R. L., Hazelwood, R. R., and Reboussin, R., Epistemological status of equivocal death analysis, *Am. Psychol.*, January 1994.

Burgess, A. W. et al., Serial rapists and their victims: reenactment and repetition: Current perspectives in human sexual aggression, *Ann. NY Acad. Sci.,* 528, 277, 1987.

Burgess, A. W. et al., Sexual homicide: a motivational model, *J. Interper. Viol.* 1, 251, 1986.

Crowell, N. and Burgess, A. W., Understanding violence against women, *Nat. Acad. Sci.,* Washington, D.C., 1996.

Dietz, P. E. and Hazelwood, R., Atypical autoerotic fatalities, *Med. Law*, 1, 307, 1982.

Dietz, P. E., Hazelwood, R., and Warren, J. I., The sexually sadistic criminal and his offenses, *Bull. Am. Acad. Psych. Law*, 18, 163, 1990.

Douglas, J. E. et al., Criminal Profiling from crime scene analysis, *Behav. Sci. Law*, 4, 401, 1986.

Ford, C. V., The Munchausen syndrome: a report of four new cases and a review of psychodynamic considerations, *Psych. Med.*, 4, 31, 1973.

Geberth, V., Psychological profiling, *Law Order*, 46, 1981.

Groth, A. N., Burgess, A. W., and Holmstrom, L. L., Rape: power, anger and sexuality, *Am. J. Psychiatry,* 1324, 11, 1239, 1977.

Harpur, T. J. and Hare, R. D., Assessment of psychopathy as a function of age, *J. Abnormal Psychol.*, 103, 604, 1994.

Hazelwood, R. R., The behavioral-oriented interview of rape victims: the key to profiling, *FBI Law Enforcement Bulletin*, September 1983.

Hazelwood, R. R., Equivocal deaths: a case study, *Am. Soc. Criminol.*, November 11, 1983.

Hazelwood, R. R., Dietz, P. E., and Burgess, A. W., Sexual fatalities: behavioral reconstruction in equivocal cases, *J. Forens. Sci.*, 27, 1982.

Hazelwood, R. R. and Burgess, A. W., An introduction to the serial rapist: research by the FBI, *FBI Law Enforcement Bulletin*, 58, 16, 1987.

Hazelwood, R. R. and Warren, J. I., The serial rapist: his characteristics and victims, parts I and II, *FBI Law Enforcement Bulletin*, January and February, 1989.

Hazelwood, R. R., Reboussin, R., and Warren, J. I., Correlates of increased aggression and the relationship of offender pleasure to victim resistance, *J. Interpersonal Violence*, 4, 1989.

Icove, D. J., Principles of Incendiary Crime Analysis, Ph.D. dissertation, University of Tennessee, 1979.

Icove, D. J. and Estepp, M. H., Motive-based offender profiles of arson and fire-related crimes, *FBI Law Enforcement Bulletin*, 49, 18, 1987.

Knight, R. A., Carter, D. L., and Prentky, R. A., A system for the classification of child molesters: reliability and application, *J. Interpersonal Violence*, 4, 3, 1987.

Knight, R. A., Warren, J. I., Reboussin, R. et al. Predicting rapist type by crime scene variables, *Crim. Justice Behav.*, 25, 46, 1998.

Lanning, K. V., *Child Molesters: A Behavioral Analysis for Law Enforcement Officers Investigating Cases of Child Sexual Exploitation*, Monograph, 3rd. ed., National Center for Missing and Exploited Children, Washington, D.C., 1992.

Luke, J. L., The role of forensic pathology in criminal profiling, in *Sexual Homicide*, Ressler R., Burgess, A., and Douglas J., Eds., Lexington Books, Lexington, MA, 1988.

Meloy, J. R., Stalking (obsessional following): a review of some preliminary studies, *Aggression Violent Behav.*, 1, 147, 1996.

Porter, B., Mind hunters: tracking down killers with the FBI's psychological profiling team, *Psychology Today*, April 1983.

Warren, J. I., Reboussin, R., Hazelwood, R. R., Cummings, A., Gibbs, N. A., and Trumbetta, S. L., Crime scene and distance correlates of serial rape, *J. Quantitative Criminol.*, 14, 35–59, 1998.

Books

Abrahamson, D., *Confessions of Son of Sam*, Columbia University Press, New York, 1985.

Allen, W., *Starkweather: The Story of a Mass Murderer*, Houghton-Mifflin, Boston, 1967.

Bevel, T. and Gardiner, R. M., *Bloodstain Pattern Analysis with an Introduction to Crime Scene Reconstruction*, 2nd ed., CRC Press, Boca Raton, FL, 2000.

Bledsoe, J., *Bitter Blood*, E. P. Dutton, New York, 1988.

Breo, D. L., and Martin, W. J., *The Crime of the Century*, Bantam Books, New York, 1993.

Brizer, D. A. and Crowner, M. L., *Current Approaches to the Prediction of Violence,* American Psychiatric Press, Washington, D.C., 1989.

Bugliosi, V. and Gentry, C., *Helter Skelter*, Norton, New York, 1974.

Cartel, M., *Disguise of Sanity: Serial Mass Murderers*, Pepperbox Books, Toluca Lake, CA, 1985.

Clarke, J. W., *American Assassins: The Darker Side of Politics,* Princeton University Press, Princeton, NJ, 1982.

Clarke, J. W., *On Being Mad or Merely Angry*, Princeton University Press, Princeton, NJ, 1990.

Coston, J., *To Kill and Kill Again*, Penguin Books, New York, 1992.

DeForrest, P., Gaensslen, R. E., and Lee, H. C., *Forensic Science: An Introduction to Criminalistics*, McGraw-Hill, New York, 1983.

Douglas, J. E., Burgess, A. W., and Burgess, A. G. et al., *Crime Classification Manual*, Jossey-Bass, San Francisco, CA, 1992.

Douglas, J. and Olshaker, M., *Mind Hunter*, Scribner's, New York, 1995.

Englade, K., *Cellar of Horror,* St. Martin's Press, New York, 1988.

FBI, *Handbook of Forensic Services*, Federal Bureau of Investigation, Washington, D.C.

Fisher, B., *Techniques of Crime Scene Investigation*, 6th ed., CRC Press, Boca Raton, FL, 2000.

Flowers, A., *Bound To Die*, Kensington, New York, 1995.

Geberth, V., *Practical Homicide Investigation*, 3rd ed., CRC Press, Boca Raton, FL, 1996.

Gollmar, R. H., *Edward Gein*, Charles Hallberg, New York, 1981.

Groth, A. N., *Men Who Rape*, Plenum, New York, 1979.

Gudjonsson, G. H., *The Psychology of Interrogations, Confessions and Testimony,* John Wiley & Sons, New York, 1992.

Guttmacher, M. S., *The Mind of the Murderer,* Farrar-Straus, New York, 1960.

Hare, R. D., *Without Conscience: The Disturbing World of the Psychopaths among Us*, Simon and Schuster, New York, 1993.

Harris, T., *The Silence of the Lambs*, St. Martin's Press, New York, 1988.

Hazelwood, R. R. and Michaud, S. G., *Dark Dreams*, St. Martin's Press, New York, 2001.

Jeffers, H. P., *Who Killed Precious?* Pharos Books, New York, 1991.

Kassin, S. M. and Wrightsman, L., Confession evidence, in *The Psychology of Evidence and Trial Procedure*, Kassin, S., and Wrightsman, L., Eds., Sage Publications, Beverly Hills, CA, 1985.

Klausner, L. D., *Son of Sam*, McGraw-Hill, New York, 1981.

Larsen, R. W., *Bundy—The Deliberate Stranger,* Prentice-Hall, Englewood Cliffs, NJ, 1980.

Levin, J. and Fox, J. A., *Mass Murder: America's Growing Menace*, Plenum Books, New York, 1985.

Leyton, E., *Compulsive Killers*, New York University Press, New York, 1986.

Malmquist, C. P., *Homicide: A Psychiatric Perspective*, American Psychiatric Association Press, Washington, D.C., 1996.

Maples, W. R. and Browning, M., *Dead Men Do Tell Tales*, Bantam, Doubleday, Dell, New York, 1994.

Masters, B., *The Shrine of Jeffrey Dahmer*, Hodder and Stoughton, London, 1993.

McGuire, C. and Norton, C., *Perfect Victim,* Dell Publishing, New York, 1988.

Meloy, J. R., *The Psychopathic Mind, Origins, Dynamics, and Treatment*, Jason Aronson, Northvale, NJ, 1988.

Michaud, S. G., *Lethal Shadow*, Onyx Publishing, New York, 1994.

Michaud, S. G. and Aynesworth, H., *The Only Living Witness*, Linden Press/Simon and Schuster, New York, 1983.

Michaud, S. G. and Hazelwood, R. R., *The Evil That Men Do*, St. Martin's Press, New York, 1999.

Moenssens, A. et al., *Scientific Evidence in Civil and Criminal Cases,* 4th ed., Foundation Press, New York, 1994.

Monahan, J., *Predicting Violent Behavior: An Assessment of Clinical Techniques*, Sage, Beverly Hills, CA, 1981.

Monahan, J. and Steadman, H. J., Eds., *Violence and Mental Disorder,* University of Chicago Press, Chicago, 1994.

Norris, J., *Serial Killers: The Growing Menace*, Dolphin-Doubleday, New York, 1988.

O'Brien, D., *Two of a Kind: The Hillside Stranglers*, New American Library, New York, 1985.

Prothrow-Stith, D., *Deadly Consequences*, Harper-Collins, New York, 1991.

Quinsey, V. L. et al., *Violent Offenders—Appraising and Managing Risk*, American Psychological Association, Washington, D.C., 1998.

Rennie, Y., *The Search for Criminal Man: The Dangerous Offender Project,* Lexington Books, Lexington, MA, 1977.

Ressler, R. P., Burgess, A. W., and Douglas, J. E., *Sexual Homicide*, Free Press, New York, 1988.

Rule, A., *Lust Killer*, New American Library, New York, 1983.

Rule, A., *The Stranger Beside Me*, New American Library, New York, 1984.

Saferstein, R., *Criminalistics: An Introduction to Forensic Science*, 7th ed., Prentice-Hall, Englewood Cliffs, NJ, 2001.

Samenow, S. E., *Inside the Criminal Mind,* Time Books, Random House, New York, 1984.

Spitz, W., *Medicolegal Investigation of Death*, Charles C. Thomas, Springfield, IL, 1994.

Wilson, C. and Seaman, D., *Encyclopedia of Modern Murder*, *1962–1982*, Crown, New York, 1985.

Wilson, J. and Herrnstein, R., *Crime and Human Nature*, Simon and Schuster, New York, 1985.

Winn, S. and Merrill, D., *Ted Bundy: The Killer Next Door*, Bantam Books, New York, 1980.

Wolfgang, M. E., *Patterns in Criminal Homicide*, University of Pennsylvania Press, Philadelphia, 1958.

Wolfgang, M. E. and Weiner, N. A., Eds, *Criminal Violence*, Sage, Beverly Hills, CA, 1982.

Endnotes

[1] Douglas, J. E. et al., Criminal profiling from crime scene analysis, *Behav. Sci. Law*, R, 401,1986.

[2] *Editor's note:* Recall that offender profiling is only one of the many characteristics supplying potential evidence under the umbrella of crime scene assessment as a component of crime scene reconstruction. The other elements of crime scene assessment include modus operandi, signatures, hostage negotiation assessments, threat assessments, and solvability factors.

[3] Hazelwood, R. R. and Burgess, A. W., *Practical Aspects of Rape Investigation: A Multidisciplinary Approach*, 3rd ed., CRC Press, Boca Raton, FL, 2001.

[4] Michaud, S. G. and Hazelwood, R. R., *The Evil Men Do*, St. Martin's Press, New York, 1998.

[5] President's Commission on the Assassination of President Kennedy, *Hearings on the Investigation of President John F. Kennedy*, Vols. 1-26, U.S. Government Printing Office, Washington, D.C., 1964. See also Hassel, C. V., The political assassin, *J. Police Sci. Admin.*, 4, 399, 1974.

[6] For simplicity, we will use the male pronoun for offenders and female pronouns for victims, but this usage is not intended to suggest that the roles cannot be reversed.

[7] Westveer, A. E., Ed., *Managing Death Investigations*, FBI, Washington, D.C., 1997.

[8] FBI, Violent crime, *FBI Law Enforc. Bull.*, 54, 1985.

[9] Depue, R. L. and Depue, J. M., To dream, perchance to kill, *Sec. Manage.*, June, 1999.

[10] Napier, M. R. and Adams, S. H., Magic words to obtain confessions, *FBI Law Enforc. Bull.*, 67, 1998.

[11] Michaud, S. G. and Hazelwood, R. R., *The Evil Men Do*, St. Martin's Press, New York, 1998.

[12] Personal interview conducted as part of the FBI's research on serial rapists, 1998.

[13] Groth, N., Burgess, A., and Holmstrom, L., Rape, power, anger and sexuality. *Am. J. Psych.*, 134, 1239, 2001.

[14] Hazelwood, R. R., Wolbert, A., and Burgess, A., Eds., op cit., 2001.

[15] FBI, Crime scene and profile characteristics of organized and disorganized murderers, *FBI Law Enforc. Bull.*, August, 1985.

[16] Douglas, J. E. et al., *Crime Classification Manual*, Lexington Books, New York, 1992.

[17] Hazelwood, R. R. and Douglas, J., The lust murderer, *FBI Law Enforc. Bull.*, April, 1980.

SECTION VII

Legal and Ethical Issues in Forensic Science

Forensic Evidence

Terrence F. Kiely

33

Introduction

The importance of forensic science to criminal law lies in its potential to supply vital information about how a crime was committed and who committed it. If the information survives the screening function of the rules of evidence, it can be accepted as evidence of a material fact in the ensuing trial.

Evidence is simply court-approved information that the trier of fact, typically a jury, is allowed to consider when determining a defendant's guilt or innocence. The admissibility or inadmissibility of trial information—whether eyewitness testimony, photographs, physical objects, or scientifically generated information such as DNA—is determined by the trial court's application of the rules of evidence. This set of evidentiary rules is basically exclusionary in nature; that is, they serve to filter out information presented by either side that may be irrelevant to the factual and legal issues at hand or that violate long-standing prohibitions such as those against the admissibility of hearsay or substantially prejudicial information. The system of rules that constitutes the law of evidence controlling the flow of information in civil and criminal litigation is exclusionary.

A central concept regarding the admissibility of trial information is the prerequisite of a solid supportive foundation for any offer of evidence, especially in instances of scientifically generated data such as ballistics, fingerprints, or fiber or hair analyses. A foundation consists of sufficiently supportive information presented to a judge to convince him or her that the proposed witness or item of information has the potential to be true, and hence a jury could reasonably determine that it is or is not true.

An example of a nonscientific foundation might occur in a trial concerning a fatal automobile crash, where the plaintiff offers a witness who wishes to testify as to the speed of the defendant's vehicle. The foundation here might consist of preliminary testimony that the witness was in an opportune position to see the accident and was a licensed and experienced driver capable of estimating the relative speeds of two automobiles. In instances of forensic or scientifically generated information, such as toxicology or forensic pathology, the required foundation is typically much more complex, to allow an expert to offer an opinion in a case.

Information generated by the forensic sciences is referred to as forensic evidence simply to distinguish it from nonscientifically generated information, such as witness statements and other circumstantial data, addressing the period preceding, during, and following a crime. Prior to allowing an expert forensic scientist or crime scene technician to render an opinion linking a defendant to a crime scene, a court may require a showing by the offering party that the scientific basis underlying a forensic expert's testimony is generally accepted in the scientific community or, per a federal Daubert standard, that it is relevant and reliable. Once the information produced and testified to by expert witnesses successfully survives the evidence rules and foundational process, it becomes circumstantial evidence along with other inference-based information available for jury consumption.

The aspect of the forensic sciences that is of interest to practitioners in the criminal justice system is its potential for the production of forensic evidence, that is, facts that, when combined with probability assessments geared toward a defendant's participation in a crime, typically aid in establishing one or more of essential elements of the crime. Those elements, such as *actus reus* (affirmative act), intent, and causation, must be proved beyond a reasonable doubt.

Forensic Evidence

Discussions of the use of science in criminal law typically revolve around the subject of forensic evidence, that is, facts or opinions generated or supported by the use of one (or typically more than one) of the forensic sciences routinely used in criminal prosecutions. The list of such disciplines is extensive, and their legal ramifications will receive extended attention in this chapter. Much of the rest of this book is devoted to the description and analysis of the individual forensic sciences.

The central concept in the utilization of the findings of forensic science is the crime scene. Although a crime scene can be the basement of a counterfeiter or the broken back door of a supermarket, the term often refers to the scene of a violent crime such as a sexual assault or a homicide. The identification, collection, and testing of crime scene evidence are the focus for the training of forensic scientists, and they are also the central source and reference point for analysis of many legal issues involved directly or indirectly in the field of forensic evidence.

Direct evidence is information that establishes directly, without the need for further inference, the fact for which the information is offered. A clear example would be eyewitness testimony that the defendant fired the fatal shot in a murder prosecution. All forensic evidence is primarily offered as circumstantial evidence of a material fact required for a conviction. Evidence obtained through forensic anthropology, forensic entomology, forensic geology, DNA, fingerprints, hair, fiber, and footwear and tire impressions, as well as numerous other types of information generated by the body of forensic sciences, all serve the vital function of bringing to light important inculpatory or exculpatory facts.

It is important to understand that forensic evidence is subsumed under the general evidence category of circumstantial evidence. Circumstantial evidence, which includes the lion's share of evidentiary offerings in U.S. courts, allows the trier of fact to accept as proven a fact for which direct evidence is unavailable by inference from a fact that is directly proven. Examples would be the linking of crime scene DNA, hairs, fibers, glass, footprints, fingerprints, and bullets or shell casings linked in some fashion to the defendant, which is offered to infer the defendant's presence at the crime scene and thus inferentially connect him to that crime scene.

In many ways, the O.J. Simpson murder trial was a timely catalyst for the current renewed interest in the subject of forensic science. The success of the CSI and related fictional and nonfictional television shows speaks to the strong public interest in this area of crime and science. More to the point here, there has been a noticeable increase in the attention paid by trial counsel and judges to the rights and wrongs of crime scene investigation and testing, including alleged failure to conduct an adequate forensic investigation, contamination of crime scene samples, deficient testing, and a host of other crime-scene-related issues. Law schools and postgraduate legal training courses have recently begun reemphasizing the importance of forensic evidence courses along with the more familiar tools of criminal law, such as criminal procedure and federal courts.

Criminal Law Theory and Evidence

The scientific nature of information generated by one or more forensic sciences, such as hair or fiber evidence, may require a preliminary determination of whether the scientific methodology on which a forensic expert's testimony is based is either generally accepted in the scientific community or, under the federal Daubert standard, is relevant and reliable.[1] If information produced and testified to by expert witnesses successfully survives the evidence rules and foundational processes, it and other items of inference-based information become available for jury consumption.

The aspect of the forensic sciences of interest to practitioners in the criminal justice system is their potential for the production of forensic evidence or facts that, when combined with probability assessments geared to the defendant's participation in a crime, aid in establishing one or more essential elements of the crime. Those elements, such as *actus reus* (affirmative act), intent, and causation must be proved beyond a reasonable doubt.

How does forensic evidence differ from other evidence? Forensic science involves the application of scientific theory accompanied by laboratory techniques involving a wide variety of traditional academic natural sciences, such as anthropology, DNA analysis, and geology. Certain disciplines associated with forensics are nontraditional in nature, such as footwear impression techniques or fingerprint analysis. Many disciplines utilize the comparison microscope and other microscopy instrumentation

with superb results in the investigation and prosecution of crime. It is important to remember that the reason for using the forensic sciences is to generate forensic evidence. That is the forensic part of forensic evidence. The intent is to get to the evidence part. All this carefully gathered information is generated to meet the goal of establishing material facts at or before trial, not to demonstrate the latest technological advances or most recent methodologies.[2]

Class and Individual Characteristics

Forensic evidence comes into court in two basic forms: (1) class-characteristic evidence that does not reference a particular suspect, and (2) individual characteristics that do, inferentially, associate a particular individual with the commission of a crime.[3] Testimony that the pubic hairs found on a rape-homicide victim came from a Caucasian male or that shell casings found at the scene came from a certain make and model of firearm are two typical examples of class-characteristics statements. The second type of potential testimony generated by forensic science is the individual characteristic or matching statement that serves to link data found at the crime scene to a particular defendant. Testimony finding that court-ordered pubic hair exemplars obtained from the defendant are consistent in all respects to the hair found on the victim or that fibers found on a victim's clothing are consistent with fibers from a defendant's jacket are typical examples. DNA "matching" is another obvious example of an individual or matching statement.

Class-characteristics statements garnered from forensic analyses illustrate the great value in a criminal investigation of statements drawing contextual lines for subsequent attempts to link a particular suspect to a crime scene, especially by excluding other potential suspects. The ultimate goal of all forensic science is the linking of a potential offender to a crime scene by way of testimony as to individual characteristics, connecting a physical sample obtained from the suspect with a similar sample from the crime scene. The exclusionary potential of class or individual forensic findings is equally important, as it can eliminate a suspect or void a conviction based on the lack of adequate forensic evidence. Many reported criminal law decisions discuss where and how such linkages have been successfully proved by forensic experts. On the other hand, according to Barry Scheck and Peter Neufeld of the Cardozo School of Law's the Innocence Project, postconviction DNA analyses have resulted in the release of more than 150 prisoners.[4]

Forensic Evidence and the Crime Scene

The value of forensic evidence for police and prosecutors lies in its ability to interpret the physical data found at a crime scene and, hopefully, to link a particular suspect to it.[5] These materials in any actual case could include fingerprints, footprints, blood spatter, semen, toolmarks, soil samples, glass, insect matter, and a host of other items. It is important to recognize that in each case, the idea of a crime scene actually encompasses, for scientific and legal purposes, four crime scenes, each having its own set of rules and guiding principles:

1. *The physical crime scene created and left by the perpetrator.* Armed with a growing array of increasingly sophisticated tools, forensic scientists are able to see more items at a crime scene and to successfully test them for linkage to a defendant.
2. *Crime scene material collected and transported by crime scene personnel.* Once the physical crime scene is released by the forensic staff, it becomes a bad memory in a community, and the crime scene moves to the laboratory.
3. *Crime scene material that can be tested by a crime laboratory and the ensuing results of any such tests,* for example, hair analysis or DNA testing. The laboratory personnel will cull out testable crime scene data and discard or leave unexamined a portion of the crime scene material presented to them from the forensic staff who collected the data.
4. *Crime scene information allowed into evidence by the court* according to the case issues and rules of evidence. This fourth crime scene is the one that counts for prosecutorial or defense purposes during the trial of a suspect. Whereas the first three crime scene settings provide extremely important investigative leads for the police crime scene investigators and detectives and will be the basis for search warrants, the fourth is the eventual goal of forensic science, that is, to produce usable and persuasive evidence for use at trial.[6]

Laboratory Matches and Courtroom Rules of Evidence

It is essential to understand the very limited number of occasions when an expert is allowed to make any *absolute* claims of a match. Most of the forensic sciences, including DNA, do not support any such claims, and the courts have consistently refused to allow such statements in testimony or prosecutorial closing arguments. Francis Bacon's fear in the early period of the development of the scientific method that scientists might "give out a dream of our imagination for a pattern of the world" is still a major concern of criminal defense lawyers in cases involving forensic science experts.[7] According to the defense bar, statements of forensic scientists wrapped in impressive credentials and complex foundational testimony have always put a shine on prosecution witnesses' testimony and glazed the entire case with an aura of certainty that it may not possess. This is especially the case, they argue, where needed forensic financial support for indigents is typically not forthcoming.

The terms allowed by courts to support the "identification," for example, of a crime scene hair with a sample taken from a defendant include the following:

- Match (reversible error in most states)
- Compatible with
- Consistent with
- Similar in all respects
- Not dissimilar
- Same general characteristics
- Identical characteristics
- Could have originated from

These conclusory linkage pronouncements and variations on them are the grit and gristle of forensic testimony in a wide variety of crimes and forensic disciplines. Such testimony is not grossly unfair and certainly does not constitute a fraud on the court. Quite the contrary. A less-than-certain opinion nonetheless has a powerful effect on a jury. These linkage discussions are not framed in general or universal terms, but are grounded in some significant relationship between the items found at the actual crime scene and the defendant in the case. The guilt-oriented inferences rising from such less-than-certain testimony is very powerful evidence in any case, requiring defense counsel to provide alternative inferences or to challenge the credentials or opinion base of the testifying expert or experts. This point will be emphasized in the extensive discussion of the Sutherland murder case to follow in this chapter.

This type of testimony is typically geared to support the basic common sense of the jury, the common sense used by ordinary people in connecting facts to events. It might even be seen as a scientific contribution to the venerable *who is kidding whom* test known to all jurors. It is up to the defense counsel to achieve a sufficient knowledge of the expertise at issue to be able to effectively cross-examine the expert about the basis for his or her conclusions and to elicit the precise characteristics that serve as the basis of the opinion at issue.[8]

Police and prosecutors can use all sorts of aids as investigative tools, including experience, hunches, and informants, but the use of physical data recovered from a crime scene is determined by the "evidentiary" care taken during the crime scene investigation, including the seizure, collection, and protection of physical evidence before and after laboratory analysis. If the authorities do not recognize the need for care and do not collect, store, and transfer crime scene data properly, it may very well be useless as material evidence at trial.

Forensic Science, Forensic Evidence, and Litigation

Forensic and other types of evidence are used to reconstruct the events that encompass the crime being prosecuted. Given the rule for speedy trials and other constitutional protections, not the least of which are the rules of evidence, such reconstructions are often a formidable task for prosecutors and defense counsel. Increasingly, circumstantial proof presented in criminal trials comes in the form of forensic evidence. The long history of proof of crime has always depended more on the experience of juror's lives than any startling analysis developed in a laboratory. Logic and common sense have always had and will continue to have as great, if not greater, force than probabilistically based forensic facts.[9]

In 81 B.C., the famous orator Marcus Tullius Cicero, then the leading defense lawyer in Rome, represented Sextius Roscius of Ameria, who was accused of murdering his father to get possession of the patrimonial estates in the country. In the absence of forensic aid, Cicero relied on the jurors' sense of community mores, experience, common sense, and history:

> ... I won't even ask you why Sextius Roscius killed his father. I only ask how he killed him .... How did he kill his father then? Did he strike the blow himself, or get others to do the job? If you are trying to maintain that he did it himself, let me remind you that he wasn't even in Rome. If you say he got others to do it, then who were they? Were they slaves or free men? If they were free men, identify them. Did they come from Ameria, or were they some of our Roman assassins?... If they were from Rome, on the other hand, how had Roscius got to know them? For after all he himself had not been to Rome for many years,

and had never on any occasion stayed there for more than three days at a time. So where did he meet them? How did he get into conversation with them? What methods did he use to persuade them? He gave them a bribe. Who did he give it to? Who was his intermediary? Where did he get the money from, and how much was it?[10]

The marshaling of facts that comport with the life experience of triers of fact remains the bedrock of any criminal justice system.

Proof of fact in litigation is increasingly focusing on inferences flowing from the application of the findings of one or more of the natural sciences. The methodologies change as science progresses. The legal system has survived many such changes and will survive yet more as the 21st century progresses. For litigants, the important aspect of this increasing dependence on scientific method as a basis for determining dispositive facts is the facts that are generated, not the method used to do it. The existence or nonexistence of a matter of fact depends in large part on the theory of fact finding being used by the fact seekers.

The legal antagonism between forensic scientists and the courts can be encapsulated in two questions. How far can forensic scientists, with their access to advanced technology, go in making definitive statements about a crime scene or in linking a suspect to a scene? And how far do we let them go given the fact that we have a constitution? The importance of these questions reflects the tension between (1) the empirical basis of and credibility ascribed to such "matching" statements and (2) the impact that such "matching" statements may have on a jury, which might be inclined to interpret such testimony, albeit given in a qualified manner, as true. The concern has always been that a criminalist's testimony that a hair or fiber obtained from a suspect is "consistent in all respects" or is "not dissimilar" will be internalized by jurors as a statement of a definite match. Indeed, courts routinely require the opinions of most forensic experts to be couched in such qualified terms, with the possible exception of fingerprint and ballistics testimony.[11]

The "matching" process used by forensic scientists requires them to explain how a physical item from a crime scene was analyzed to provide the purported link between the defendant and the crime scene. Each piece of data recovered from a crime scene—whether from hair, fiber, soil, glass particles, blood products, foot or tire prints, or firearms—can be broken down into a series of subcomponents for purposes of analysis and comparison. These analytical processes and the response of the criminal justice system to them have been discussed in the earlier chapters of this book. It is important that prosecutors and defense counsel make a detailed study of these separate disciplines.

In both civil and criminal cases, the contending parties seek to prove or disprove a connection between a defendant's act or omission and a resultant death or injury. In civil cases, the "science" at issue is often centered on questions of causation, typically consisting of studies that may only be probative by way of extrapolation. Such testimony does not provide the individualizing expert testimony typically provided by forensic scientists in criminal litigation. In criminal cases, the use of forensic science means that some form of laboratory work has been performed to resolve factual matters in the case. In both civil and criminal cases, the information provided from scientific sources must be relevant to one of the issues in the case. In civil cases, this typically involves the question of whether some commercial application of some scientific formulation "caused" the plaintiff's death or injury.

The importance of the testimony of a particular scientist lies in general or class statements about units of crime scene data or an opinion linking the defendant to the crime scene through an individual or "match" opinion. In either case, the scientific foundation or basis for any such testimony, as in civil cases, is of the utmost concern to the law.

The term *forensic evidence* encompasses two distinct ideas and processes. The *forensic* part refers to the scientific processes through which facts are generated. The manner in which DNA is extracted, tested, and subjected to population analyses serves as a major example. The methodologies of hair, fiber, and fingerprint examination are other illustrations. The area of forensic science encompasses a fairly discrete number of well-known disciplines, whereas the "science" addressed in product liability and environmental civil cases does not lend itself to such finite boundaries.

Although there are repetitive areas of scientific focus in civil cases involving chemistry, pharmaceuticals, biology, and mechanical or electrical engineering, there is much less of an opportunity to discuss the general outlines of acceptable methodology in the arena of civil law. In contrast, the criminal courts do require the forensic sciences to provide broad reviews of their methodology. Nonetheless, the legal concerns are basically the same.

The *evidence* part of the concept of forensic evidence refers to a distinct set of procedures that are unique to the litigation process. These legal procedures are separate and distinct from the scientific processes that serve as the bases for the decision to admit or exclude evidence, including forensic evidence in criminal cases.

It is important to recall the fundamentally different reasons for the introduction of scientifically generated information in the civil and criminal litigation systems. The use of the term *litigation* is important here because the process of litigation brings the issues discussed to the fore. This is quite distinct from nonlegal contexts, where the nature or acceptability of scientific methodologies or opinions are of central importance, such as grant requests, patents, contractual disputes, or publication in a scientific peer-reviewed publication.

As noted, forensic information generated by one or more of the forensic sciences comes to the law in one or both of two forms of expert-witness opinion. The first is referred to as a class-characteristics statement that speaks generally to some aspect of the crime scene under examination. Testimony that the pubic hairs found on a rape-homicide victim came from a Caucasian male or that shell casings found at the scene came from a certain make and model of firearm are two typical examples of such types of statements. The second type of potential testimony generated by a forensic science are known as individual or matching statements, that is, those that serve to link some data found at the crime scene

to a particular defendant. Testimony finding that court-ordered pubic hair exemplars obtained from the defendant are consistent in all respects to the hair located on the victim, or that fibers found on the victim's clothing are consistent with fibers from the defendant's jacket will serve as examples. This idea of class-characteristics statements references the reality that many confident general statements can be made under the auspices of an individual forensic discipline.[12]

Litigation as History

The basic circumstantial evidence, inference-based argument used in modern trials, whether aimed at proving a scientific result or the routine establishment of an important fact, has served the law as the primary method for proof of a past event, such as the commission of a crime and identification of a perpetrator. All trials are attempts to establish a version of history as regards a past event such as a sexual assault, robbery, burglary, or homicide. The state has its version of what happened, and the defendant has another. The trial is an effort to convince a jury of the correctness of one or the other versions of the past event at issue, the facts leading up to it, and the identity of important participants. As noted by the famous American historian, Carl Becker:

> Let us admit that there are two histories: the actual series of events that once occurred, and the ideal series that we affirm and hold in memory. The first is absolute and unchanged—it was what it was whatever we do or say about it; the second is relative, always changing in response to the increase or refinement of knowledge. The two series correspond more or less; it is our aim to make the correspondence as exact as possible; but the actual series of events exists for us only in terms of the ideal series we affirm and hold in memory. This is why I am forced to identify history with knowledge of history. For all practical purposes history is, for us and for the time being, what we know it to be.[13]

Becker's observation can apply to any factual search in litigation, including efforts to establish scientific facts that will determine the central issues in environmental, product liability, medical malpractice, and criminal cases.

The ultimate goal of litigation is not to find absolute truth. Any system that allows a jury to reach a verdict of *guilty* or *not guilty* in such important matters would appear to have something else in mind. The hope of the American litigation system is to provide the best, fairest, and optimal context for a jury to find the truth. This goal of providing the best opportunity for a jury to find its version of the truth is especially important to understand before we discuss the court's current preoccupation with forensic and a host of other science questions.

Litigation involving questions of science or the nature of the validity of modes of scientific inquiry has been part of the legal system since the start of our nation, beginning with patent cases in the 18th century. In examining the background of the current preoccupation of legal scholars and courts with the meaning and application of science in civil and criminal cases, one is struck by the absence of argument on that point until fairly recent times.

The real-life context from which the science-based questions addressed in this book arise are based on the proffer of expert testimony in criminal cases. One side, at a pretrial hearing, may seek to challenge the propriety of testimony by the opposing side's experts or, more commonly, may challenge the reliability or acceptability of the methodology used by the expert in forming an opinion. According to established evidence law theory, any witness can be challenged on several grounds:

- A case may not require his or her expertise.
- A jury is capable of deciding the disputed fact without the need for lengthy and potentially prejudicial testimony.
- An expert witness can be challenged on his or her basic qualifications and ability to give an opinion in the field at issue.
- The expert may have insufficient education or experience to have anything of value to offer.
- The methodology utilized by an expert to support his or her opinion may not

be scientifically sound or capable of supporting the proffered opinion.
- The methodology may be sufficiently scientifically sound to support an opinion, but the opinion based on the method is not sufficiently derived from that scientific methodology.

These process-based objections are key factors in the current state and federal controversy over the utilization of expert scientific opinion in America's courts.

Forensic Science and Forensic Evidence: Sutherland Case

We will now present a case study arising from the rape-murder of a 10-year-old child in a rural Illinois community. The Sutherland case is essentially a circumstantial evidence case, that is, one without eyewitness testimony or any other direct evidence of the defendant's participation in the crime. The case is appropriate to our discussion here because virtually all of the facts pointing toward the defendant's guilt were generated by expert testimony based on several of the traditional forensic sciences. Hair analysis, fiber analysis, footwear impressions, and tire-tread impressions were at the center of the state's proof.

Nuclear DNA evidence was not tested or presented because DNA testimony was not utilized in Illinois courts at the time of the murder, which occurred in 1987. There were no witnesses to this crime. The sole evidence linking the defendant, Cecil Sutherland, to the crime was the testimony of a small number of forensic scientists. As will be discussed below, the defendant's conviction was affirmed by the Illinois Supreme Court in 1993. In 2000, the same court granted the defendant a new trial based on incompetence of counsel in his first trial as the result of a new appeal, by new counsel. The Sutherland case was retried in Illinois in 2004, 17 years after the date of the murder; Sutherland was found guilty a second time and was sentenced to death.

The Illinois Supreme Court decided on *People v. Sutherland* in 1993. The defendant was convicted of aggravated kidnapping, aggravated criminal sexual assault, and murder[14] based solely on circumstantial evidence, most of which was generated by forensic science. In 2001 the Illinois Supreme court reversed the conviction, based on a finding of incompetence of counsel.[15] Both Sutherland cases are excellent examples of the interaction of forensic science with the overriding body of considerations that constitute the legal process.

The Sutherland case study serves as a clear example of the ongoing interrelationship between the world of forensic science and the investigation and proof of crime.

Facts of the Case

At 9 a.m. on July 2, 1987, an oil-field worker discovered the nude body of 10-year-old Amy Schultz of Kell, Illinois. The body—lying on its stomach and covered with dirt—was found approximately 100 feet from an oil-lease access road in rural Jefferson County. There were shoeprints on her back, and several hairs were found stuck in her rectal area. A large open wound on the right side of the neck exposed the spinal cord area. A pool of blood around the head indicated that the murderer had killed the girl where she lay.[16] The victim's shirt, shorts, underpants, shoes, and socks were found scattered along an oil-lease road. Automobile tire impressions were found 17 feet from the body, and a shoeprint impression similar in design to that on the body was found near the tire impressions. The police took casts of the tire and shoeprint impressions.

The medical examiner (ME), Dr. Steven Neurenberger, performed an autopsy on July 3, 1987. He observed a 14.5 cm wound running from the middle of Amy's throat to behind her right ear lobe that cut through the neck muscles, severing the carotid artery and jugular vein and cutting into the cartilage between the neck and vertebrae. The right eye was hemorrhaged, and there was a small abrasion near the left eyebrow. The ear was torn from the skin at its base, and both lips were lacerated from being compressed against the underlying teeth. Linear abrasions on the outer lips of the vagina demonstrated that force had been applied to the back, forcing the vagina against the ground.

Searching for internal injuries, the ME found three hemorrhages inside the skull, a fractured rib, a torn liver, and tearing of the rectal mucosa. The victim's vocal cords were also found to be hemorrhaged, and her esophagus was bruised. The ME deduced from these injuries that the killer had strangled the child to unconsciousness or death, anally penetrated her, slit her throat, and stepped on her body to force exsanguination. The ME placed the time of death between 9:30 and 11:00 p.m. on July 1, 1987, based on the contents of her stomach.[17]

Prosecution's Forensic Evidence: Tire Tracks

Several months after the discovery of Amy's body, the police at Glacier National Park in Montana notified Illinois authorities about Sutherland's abandoned car, a 1977 Plymouth Fury. At the time of the murder, Sutherland had been living in Dix, Illinois, in Jefferson County, on the line between Dix and Kell Counties. Illinois police authorities ascertained that the perpetrator's car had a Cooper "Falls Persuader" tire on the right front wheel. Illinois state deputies and David Brundage, a forensic scientist, traveled to Montana, where they made an ink impression of the right front wheel of Sutherland's car. Brundage evaluated the plaster casts of the tire print impressions made at the scene of the crime and eventually testified that the tire impressions left at the scene were consistent in all class characteristics with only two models of tires manufactured in North America, the Cooper "Falls Persuader" and the Cooper "Dean Polaris."[18] After comparing the plaster casts of the tire impression at the scene with the inked impression of the tire from Sutherland's car, Brundage concluded that the tire impression at the scene corresponded with Sutherland's tire and could have been made by that tire. Brundage, however, was unable to exclude all other tires as having made the impressions due to the lack of comparative individual characteristics, such as nicks, cuts, or gouges.[19]

Mark Thomas, the manager of mold operations at the Cooper Tire Company, determined "mal" wear similarity, which meant that Sutherland's tire could have made the impression found at the crime scene. Thomas also compared blueprints of Cooper tires with the plaster casts of the tire impressions and determined that the "probability" was "pretty great" that a size P2175/B15 tire—the same size as Sutherland's "Falls Persuader" tire—had made the impression preserved in the casts. He admitted that there were a great number of such tires on the roads of America.[20]

Prosecution's Forensic Evidence: Hair Evidence

Criminalist Kenneth Knight compared the two pubic hairs recovered from Amy Schulz's rectal area with Sutherland's pubic hair. He also made comparisons with pubic hairs from members of Amy's family as well as pubic hairs from 24 prior offenders, concluding that the pubic hairs found on Amy did not originate from her family or the 24 suspects, but "could have originated" from Sutherland.

Knight also examined 34 dog hairs found on Amy's clothing and concluded that the dog hairs were consistent with and could have originated from Sutherland's black Labrador, Babe. Knight also testified that the dog hairs on Amy's clothes were dissimilar from her family's three dogs, her grandparents' dog, and dogs of three neighbor families. Tina Sutherland, the defendant's sister-in-law, testified that Sutherland usually carried Babe in his car, making it virtually impossible to be in the car without getting covered with dog hair. Multiple dog hairs found in Sutherland's car were found to be consistent with the hairs from his dog, Babe.[21]

Prosecution's Forensic Evidence: Fiber Evidence

Knight also examined the victim's clothing for foreign fibers, finding a total of 29 gold fibers in her socks, shoes, underwear, shorts, and shirt. He testified that all but one of the gold fibers found on Amy's clothes "could have originated" from the defendant's auto carpet, but could not exclude all other auto carpets as possible sources. He also testified that the one remaining gold fiber found on Amy's clothes could have originated from the defendant's car upholstery.

Knight also examined and compared 12 cotton and 4 polyester fibers found on the front passenger-side floor of Sutherland's automobile with cotton and polyester fibers from the victim's shirt. He concluded that the fibers from the car displayed the same size, shape, and color of the fibers from the shirt and thus could have originated from the shirt. He also compared three polyester fibers found on the front passenger seat and floor with fibers from the victim's shorts and found them consistent in diameter, color, shape, and optical properties, and he opined that the fibers from the car could have originated from the shorts.[22]

Forensic defense expert Richard Bibbing agreed with the state's expert's conclusions on all the comparison evidence except as to the cotton fibers found in the defendant's car. He did not agree that the cotton fibers were consistent based on differences in size and color.[23]

Before examining the Illinois Supreme Court's analyses in the two decisions in the Sutherland case, we will raise a series of questions to consider the relationship between forensic evidence and justice. The following questions are a summary of concerns that are continually raised in criminal trials where basic forensic evidence plays an important role.

People v. Sutherland: Forensic Science and Justice Issues

What facts, assumptions, or surmises may be obtained from the examination of hairs or fibers gathered at a crime scene? What could serve as the basis for any such assumptions or projections, or simply guesses? What value should be assigned to any such factual estimations in our criminal justice system, in which life, liberty, and justice to a victim are all in play? What does it mean to say that hairs or fibers or tire tracks are or are not *consistent with* or *dissimilar to* or *substantially similar to* others? What is the basis for such statements, and what value should be allocated to them if one set of samples was taken from a crime scene and the exemplars from a suspected perpetrator?

What is the meaning of the long-held requirement that the elements of a crime must be proved beyond a reasonable doubt? How does a forensically generated circumstantial fact fit in prosecutorial efforts designed to meet such a high bar of proof in cases partially supported by hair or fiber evidence? How much does hair, fiber, or tire-tread evidence depend for its force on other more traditional observation by eyewitnesses?

How much of the testimony in the area of hair or fiber analysis and comparison has to do with scientific theory or recognized scientific methodology? What science, if any, has been traditionally associated with hair, fiber, or tire-tread analyses, and how has that changed as we enter the 21st century? Are hair, fiber, and tire-tread comparisons deemed to be scientific because of the theoretical underpinnings of those who are devoted to its functioning in a criminal investigation and trial? Or do these comparisons become scientific through the use of microscopy and other technologies that make such evidence essentially observational?

Should it make any difference if expert opinions are simply a combination of experience and modern microscopy? What else, from a forensic scientist's standpoint, is there to say about hair, fiber, or tire-tread analyses and the factual assumptions that follow? Should hair, fiber, or tire-tread analysis carry as much weight as fingerprint, impression, ballistics, tool marks, or DNA evidence?

The predictive capabilities vary widely in the trace areas of hair, fiber, soil, paint, and glass evidence, with something less, or much less, than individual identification of a sample exemplar with crime scene data. So, for each separate discipline discussed, the courts must ask what this science can say and what it cannot say. What are the basic methodologies used by practitioners in this field to bring forth "identifying" evidence? How many accepted modes are there of comparing hair, fiber, tire casts, soil samples, DNA, bullets, shell casings, etc.? How have the courts responded to these various techniques and their exclusionary or inclusionary claims? The definitive exclusionary capability of these *trace* sciences is very important. The trick here is trying to figure out how strong is the inclusion.

The Court's Analysis: Hairs and Fibers

The defendant argued that the prosecution's circumstantial hair, fiber, and tire-print comparison evidence was insufficient to prove guilt beyond a reasonable doubt, contending that the probative value of the state's forensic evidence lay merely in establishing that defendant could not be excluded as the possible offender, not that he must be found by a jury to actually be the offender.[24]

The court ruled that the evidence here, when viewed in the light most favorable to the prosecution, established that the defendant was proved guilty beyond a reasonable doubt. The overwhelming and overlapping nature of the circumstantial evidence supported the jury finding that Sutherland kidnapped, sexually assaulted, and murdered 10-year-old Amy Schulz.[25] The court rejected the defendant's claim that the prosecutor had overstepped the bounds in arguing that the forensic testimony had established a series of fiber "matches," when the actual testimony was couched in terms of "consistency." The state argued in its closing that:

> In every single case the fibers found on Amy's sock shoes and underpants and shorts, shirt were consistent with the fibers from the defendant's car carpeting and dissimilar to all the carpets in her home environment and in her grandparents' house and the vehicles that they drive and in the business where her father works, so there can be no doubt that she got them from there. They came from one place. Those fibers on her clothing came from the defendant's car ... The red shorts are a very big part of this case ... Mr. Bibbing [defense expert witness] didn't examine the shorts at all, and we know from Ken Knight's testimony

that fibers from the shorts were found in the passenger side of the car.

… This evidence doesn't stand alone. It can be considered together with the carpet fibers on her clothing, the seat fabric fiber on her shirt, the dog hair all over her clothes, the foam rubber on her clothing, the defendant's tire impressions being the same as that found near Amy, and the clothing fibers from Amy's shirt and shorts which were deposited in the front passenger side area of the car.

… You know, with regard to the evidence in the car that Amy was in there, you know what's uncontradicted in this case? The evidence that the red polyester fibers from her shorts were found in the passenger side area of the defendant's car. That is fibers just like them—uncontradicted because the defense expert didn't look at them .

The defendant argued that these alleged misstatements constituted reversible error, citing the important case of *People v. Linscott*,[26] decided in 1991. In *Linscott*, the state's evidence established that hairs found in the victim's apartment were consistent with the defendant's hairs. As in the Sutherland case, the state's expert could not conclusively identify the hairs as originating from the defendant. Despite the expert witness's testimony to such effect, the prosecutor argued to the jury that 'the rug in the area where Karen was laying [sic] was ripped out sometime later, rolled up, and shipped to the laboratory. And that another group of hairs were obtained. The head hairs of Steven Linscott.'[27] The Linscott court found such overreaching to be reversible error.

The Sutherland Case: The Court's Analyses

In the Sutherland case, the court was also of the opinion that the prosecutor's overstatement of the fiber-comparison evidence was improper. The court ruled that prosecutorial misconduct in closing argument warranted reversal and a new trial, but only if the improper remarks resulted in substantial prejudice to the defendant. In other words, the comments must have constituted a material factor in the conviction—circumstances absent in Sutherland's case:

We do not find that the remarks in this case substantially prejudiced the defendant. Unlike *Linscott*, the evidence in this case was not closely balanced. The State presented an overwhelming volume of circumstantial evidence: the tire print found by the crime scene was consistent with defendant's car's tire; the dog hair on the victim's clothing was consistent in all respects to the defendant's dog's hair and the dog hair found in his car; the foreign fibers found on the victim's clothing were consistent with the carpeting and upholstery in defendant's car; the clothing fibers found in the defendant's car were consistent with the fibers in the victim's clothing; finally, the pubic hair found on the victim were consistent with the pubic hair standards obtained from the defendant. Given the amount of evidence, it is implausible to think that the prosecutor's remarks could have been a material factor in the conviction. In this case, the jury would not have reached a different result, even if the prosecutor had not made the remarks. [Citations omitted.] Accordingly, defendant was not denied a fair trial and we will not disturb the conviction.[28]

Sutherland's Second Appeal

Seven years later, in 2001, the Illinois Supreme Court reversed Sutherland's conviction and granted him a new trial,[29] not on the basis of any perceived weaknesses in the specifics of the forensic case, but on the basis of the incompetence of his counsel at trial. The defendant filed a postconviction petition in the circuit court raising a variety of claims. The court dismissed most of them but granted an evidentiary hearing on the following allegations: (1) that defendant's trial counsel was ineffective in failing to discover and present evidence that defendant's purchase of "Texas Steer" boots[30] and installation of the Cooper "Falls Persuader" tire on his car both occurred after the date of the crime; and (2) that the conviction of Amy's step-grandfather, William Willis, for sexual abuse subsequent to her death constituted evidence of the defendant's actual innocence.

A mental health counselor and sex offender treatment provider then testified that, based on her research, pedophiles attracted to prepubescent females show a 22% crossover in also molesting males, and those attracted to males show a 62% crossover in also molesting females. She testified that she had reviewed William Willis's medical and psychological evaluations and had spoken with him briefly.

In her opinion, there was a high probability that Willis would cross over from sexually abusing young boys to abusing young girls. She also testified that Willis was prone to outbursts of anger and that when a victim resisted, he used more violent physical force, escalating from fondling to anal rape. [31]

Ronald Lawrence, a friend of the defendant, testified that he had changed all of the tires on Sutherland's car two separate times after Amy Schultz's death and before the defendant left for Montana. Lawrence explained that he and the defendant had to change tires frequently because the rock road leading to Lawrence's house contained metal particles and railroad spikes. Lawrence testified that he told police and the public defender after Sutherland's arrest that he had changed the tires on the defendant's car after the date of Amy Schultz's murder.[32] This additional evidence, insufficiently addressed by Sutherland's original counsel, was found to be of great importance by the Illinois Supreme Court.

Testimony presented at the postconviction hearing indicated that, prior to Sutherland's trial, defense counsel was aware of evidence that he did not own a pair of Texas Steer boots at the time of Amy Schultz's murder. Specifically, trial counsel testified at the evidentiary hearing that Sutherland informed him prior to trial that he had purchased his Texas Steer boots two months after the crime occurred. Counsel also testified that he was aware that the defendant's mother had the boots in her possession at the time of trial, but that he did not request to examine them. Additionally, the defendant's mother testified that at the time of the murder, her son typically wore a different kind of boots.[33]

The testimony at the postconviction hearing also indicated that Sutherland's trial counsel was aware prior to trial of evidence that the defendant claimed that he had changed the tires on his car after the time of Amy Schultz's death but before he drove to Montana. Specifically, there was substantial testimony presented at the hearing that counsel had obtained this information from three sources: Sutherland, Sutherland's mother, and Sutherland's friend Ronald Lawrence. Counsel acknowledged at the hearing that he was aware of such evidence, but failed to investigate it or present it at trial. The court found this combination of mishaps adequate to reverse the conviction on grounds of incompetence of counsel:

> We hold that trial counsel was ineffective in failing to investigate and present evidence concerning the boots and tire. Because the State's evidence at trial consisted primarily of a variety of items introduced to associate defendant with the crime scene, an attack on the suggested links between defendant and the boots and tire could have played a prominent role in the defense.

> Trial counsel testified at the evidentiary hearing that his main trial strategy was to discredit the expert testimony purportedly tying defendant to the crime. In light of this strategy, it was incumbent on counsel to utilize available means of casting doubt on the physical evidence which the State relied upon. Although counsel sought to convince the jury that the hair and fiber evidence introduced by the State was not conclusive proof of defendant's guilt, he failed to present the jury with evidence discrediting two of the most salient and significant items in the State's case. Counsel's performance thus fell below a reasonable level of assistance.

> … We also find that counsel's ineffective performance caused substantial prejudice to defendant. Although the State presented numerous items of evidence associating defendant with the crime, none of them was singularly compelling. If counsel had succeeded in raising questions as to whether the boots and tires owned by defendant played any role in the crime committed against Amy Schultz, there is a reasonable probability that the jury also would have doubted at least some of the other physical evidence which the State attempted to link to the crime, and hence quite possibly may have acquitted defendant.[34]

The Sutherland case exemplifies the shifting nature of inferences during the trial process. The seemingly solid forensic case was reversed by a second look at the available evidence without questioning the findings of the forensic experts who testified at trial. The possibility of inferences establishing innocence may always trump seemingly irrefutable forensic evidence. The two Sutherland cases, with a third yet to be tried, illustrate the tremendous impact but lack of absolute certainty in the area of forensically generated circumstantial evidence. The new trial, focusing on the same forensic evidence, may yet carry the day.

Conclusion

The Sutherland case illustrates all of the points discussed in this chapter, which has attempted to provide an overview of forensic evidence. A great deal remains to be said about the court's response to forensic testimony admitted in a host of discrete areas, such as blood spatter analysis, DNA, forensic anthropology, odontology, entomology and fingerprint analysis. The new century will bring rapid and amazing new developments in this vital area of criminal law and science. It is more important than ever for lawyers and courts to increase efforts to understand and responsibly use the awesome potential of the world of forensic science in our criminal justice system. The essential goal of the use of forensic science in the trial of crimes is not the absolute truth of the theory being utilized, but rather that the case facts generated using any such theory reflect basic rightness and common sense.

Theories come and go. The criminal justice system's need to fairly and responsibly search for facts continues into the 21st century. It remains to be seen how the nation's courts will respond to the forensic science of the 21st century. As noted by author John Horgan in his insightful study of the end of 20th-century science:

Science's success stems in large part from its conservatism, its insistence on high standards of effectiveness. Quantum mechanics and general relativity were as new, as surprising, as anyone could ask for. But they were believed ultimately not because they imparted an intellectual thrill, but because they were effective: they accurately predicted the outcome of experiments. Old theories are old for a good reason. They are robust, flexible. They have an uncanny correspondence to reality. They may even be true.[35]

Questions

1. How is forensic evidence the same as any other type of evidence?

2. How does forensic evidence differ from other types of evidence?

3. Why is most forensic evidence, with the possible exception of ballistics and fingerprints, considered to be only circumstantial evidence of a fact?

4. Briefly identify the four stages of a crime scene for legal admissibility purposes.

5. What is the primary purpose of forensic evidence in a criminal trial?

6. In the term *forensic evidence*, explain the differences between *forensic* and *evidence*.

7. What does historian Carl Becker mean when he says there are always two histories of the same event?

8. Name five examples of the use of the forensic sciences in the Sutherland case.

9. What important facts or assumptions may be established from the examination of hairs or fibers gathered at a crime scene?

10. What does it mean to say that one or more hairs, fibers, or tire tracks associated with a suspect are consistent or substantially similar with another found at a crime scene?

11. What does it mean to say that one or more hairs or fibers or tire tracks are or are not consistent or not dissimilar or substantially similar with another? What would be the basis for any such statements, and what value should be allocated to them if one set of exemplars was taken from a crime scene and the others from a suspected perpetrator?

12. How much does hair, fiber, or tire-tread evidence depend for its force on other more traditional observation by eyewitnesses?

13. Why are statements that a hair or fiber is consistent with those associated with the defendant admissible as circumstantial proof of guilt, while a finding of inconsistent excludes the defendant?

14. What does it mean to say that the forensic sciences are used to prove a suspect's presence at a crime scene?

15. What does it mean to say that before a forensic scientist may render an opinion, that the proponent must establish a foundation for any such testimony?

16. The admissibility or inadmissibility of trial information—whether eyewitness testimony, photographs, physical objects, or scientifically generated information such as DNA—is determined by the trial court's application of the rules of evidence. What does it mean to say that these evidentiary rules are exclusionary in nature?

17. A central concept regarding the admissibility of trial information is the prerequisite of a solid supportive foundation for any offer of evidence, especially in instances of scientifically generated data such as ballistics, fingerprints, fiber, or hair analyses. What is a foundation?

18. Direct evidence is information that establishes directly, without the need for further inference, the fact for which the information is offered. A clear example would be eyewitness testimony that the defendant fired the fatal shot in a murder prosecution. Circumstantial evidence allows the trier of fact to accept a fact for which direct evidence is unavailable as a proven fact, by inference from a fact that is directly proven. All forensic evidence is offered as circumstantial evidence. Why is that the case? Provide three examples.

19. What is the purpose of the so-called *Frye* or *Daubert* hearings that often precede the offer of a wide variety of new or expanded applications of some discipline associated with the forensic sciences?

20. What is the difference between class-characteristic evidence and individual-characteristic evidence in the presentation of forensic evidence in a criminal trial?

References and Suggested Readings

Anon., *Reference Manual on Scientific Evidence*, 2nd ed., Federal Judicial Center, Washington, D.C., 2000.

Baden, M. and Roach, M., *Dead Reckoning: The New Science of Catching Killers*, Fireside Press, 2002.

Bodziak, W., *Footwear Impression Evidence*, 2nd ed., CRC Press, Boca Raton, FL, 2000.

Browning, M. and Maples, W. R., *Dead Men Do Tell Tales: The Strange and Fascinating Cases of a Forensic Anthropologist*, Doubleday, New York, 1994.

Burnett, D. G., *A Trial by Jury*, Alfred A. Knopf, New York, 2001.

Carney, T., *Practical Investigation of Sex Crimes: A Strategic and Operational Approach*, CRC Press, Boca Raton, FL, 2003.

Di Maio, V., *Gunshot Wounds: Practical Aspects of Firearms, Ballistics, and Forensic Techniques*, 2nd ed., CRC Press, Boca Raton, FL, 1999.

Di Maio, V. J. and Di Maio, D., *Forensic Pathology*, 2nd ed., CRC Press, Boca Raton, FL, 2001.

Eckert, W., *Introduction to Forensic Sciences*, 2nd ed., CRC Press, Boca Raton, FL, 1995.

Eckert, W. and James, S., *Interpretation of Bloodstain Evidence at Crime Scenes*, 2nd ed., CRC Press, Boca Raton, FL, 1998.

Fisher, B., *Techniques of Crime Scene Investigation*, 6th ed., CRC Press, Boca Raton, FL, 2000.

Fisher, D., *Hard Evidence*, Dell Paperbacks, New York, 1995.

Geberth, V., *Practical Homicide Investigation: Tactics Procedures, and Forensic Techniques*, 3rd ed., CRC Press, Boca Raton, FL, 1996.

Geberth, Vernon, *Sex-Related Homicide and Death Investigation: Practical and Clinical Perspectives*, CRC Press, Boca Raton, FL, 2003.

Hazelwood, R. R. and Burgess, A., eds., *Practical Aspects of Rape Investigation: A Multidisciplinary Approach*, 3rd ed., CRC Press, Boca Raton, FL, 2001.

Hilton, O., *Scientific Examination of Questioned Documents*, CRC Press, Boca Raton, FL, 1993.

Imwinkelried, E. J., *Evidentiary Foundations*, 4th ed., Lexis Law Publishers, New York, 1998.

Kiely, T., *Forensic Evidence: Science and the Criminal Law*, CRC Press, Boca Raton, FL, 2000.

Lee, H., *Famous Crimes Revisited: from Sacco-Vanzetti to O. J. Simpson*, Publishing Directions, 2001.

Lee, H., *Henry Lee's Crime Scene Handbook*, Academic Press, 2001.

Ramsland, K., *The Forensic Science of C.S.I.*, Boulevard Trade, 2001.

Robertson, B. and Vignaux, G. A., *Interpreting Evidence: Evaluating Forensic Science in the Courtroom*, John Wiley & Sons, New York, 1995.

Saferstein, R., *Criminalistics: An Introduction to Forensic Science*, 7th ed., Prentice-Hall, Englewood Cliffs, NJ, 2000.

Wilson, C., *Clues: A History of Forensic Detection*, Warner Books, New York, 1989.

Endnotes

1 See Chapter 33, Current Legal DNA Issues, in this volume. For a more detailed discussion, see Kiely, T. F., *Forensic Evidence: Science and the Criminal Law*, CRC Press, Boca Raton, FL, 2001, chap. 1. Also see, Kiely, T. F., *Science and Litigation: Products Liability in Theory and Practice*, CRC Press, Boca Raton, FL, 2002, chap. 2. Brown, H., Eight gates for expert witnesses, *Houston Law Rev.*, 36, 743, 1999. Brown, H., Procedural issues under Daubert, *Houston Law Rev.*, 36, 1133, 1999. Graham, M.H., The expert witness predicament: determining "reliable" under the gatekeeping test of Daubert, Kumho, and proposed amended rule of the Federal Rules of Evidence, *U. Miami Law Rev.*, 54, 317, 2000.

2 Although proof at trial is the primary purpose of generating forensically based facts, such facts are also routinely used to generate investigative leads and to provide factual support for search warrants and charging instruments, such as indictments and criminal complaints.

3 Division of the information supplied to the criminal justice system into *class* and *individual* is of the utmost importance for both forensic scientists and the criminal bar, and it has received extensive examination in previously cited chapters in this volume. This chapter focuses on the legal acceptance or rejection of these specific offerings by experts in the forensic sciences.

4 See Dwyer, J., Neufeld, P., and Scheck, B., *Actual Innocence*, Doubleday, New York, 2000, for an absorbing and in-depth look at the phenomenon of wrongly convicted prisoners freed by contemporary applications of posttrial forensic science.

5 See Kiely, T. F., *Forensic Evidence: Science and the Criminal Law*, CRC Press, Boca Raton, 2001, for a full study of modern case law addressing utilization of the forensic sciences in contemporary criminal trials.

6 See *People v. Sutherland*, 155 Ill. 2d 1, 610 N.E. 2d 1 (1993); *McGrew v. State*, 682 N.E. 2d 1289 (Ind. Sp. Ct. 1997); *People v. Linscott*, 142 Ill. 2d 22, 566 N.E. 2d 1355 (1991).

7 Bacon, Sir Francis, Novum Organum: *Aphorisms on the Interpretation of Nature and the Empire of Man*, 1620, reprint, Urbach, P. and Gibson, J., trans., Open Court, 1994, pp. 29–30.

8 See Kiely, T. F., *Forensic Evidence: Science and the Criminal Law*, CRC Press, Boca Raton, FL, 2001, chap. 3, for a detailed examination of this issue. Also see Saferstein, R., *Criminalistics: An Introduction to Forensic Science*, 6th ed., Prentice-Hall, Englewood Cliffs, NJ, 1998.

9 The French mathematician Pierre-Simon Laplace observed in 1820 that "[t]he theory of probabilities is at bottom nothing but common sense reduced to calculus." See Laplace, Pierre-Simon, introduction to *Theorie Analytique des Probabilities*, 1820.

10 Cicero, *Murder Trials*, Grant, M., trans., Penguin Books, New York, 1990, p. 67.

11 This is not meant to imply that these forensic sciences (fingerprints and ballistics) have been sufficiently challenged on their basic assumptions to justify all given opinions. See Saks, M.J., Merlin and Solomon: Lessons from the Law's Formative Encounters with Forensic Identification Science, *Hastings Law J.*, 49, 1069–1081, 1998, for an analysis of the heretofore unquestioning acceptance by the courts of most forensic sciences, in particular, the much debated discipline of handwriting analysis.

12 See Saferstein, R., *Criminalistics: An Introduction to Forensic Science*, 6th ed., Prentice Hall, Englewood Cliffs, N.J., 1998; Eckert, W., *Introduction to Forensic Sciences*, 2nd ed., CRC Press, Boca Raton, FL, 1997; Fisher, B., *Techniques of Crime Scene Investigation*, 5th ed., CRC Press, Boca Raton, FL, 1993; Bodziak, W., *Footwear Impression Evidence*, CRC Press, Boca Raton, FL,

1995; Geberth, V., *Practical Homicide Investigation*, 3rd ed., CRC Press, Boca Raton, FL, 1996; Di Maio, V. J. and Di Maio, D., *Forensic Pathology*, CRC Press, Boca Raton, FL, 1993; Pickering and Bachman, *The Use of Forensic Anthropology*, CRC Press, Boca Raton, FL, 1997; Janes, Ed., *Scientific and Legal Applications of Bloodstain Pattern Interpretation*, CRC Press, Boca Raton, FL, 1999; Ogle and Fox, *Atlas of Human Hair: Microscopic Characteristics*, CRC Press, Boca Raton, FL, 1999. See also Wecht, C. H., Ed., *Forensic Sciences*, Matthew Bender Co., New York, 1997 (a five-volume, 90-chapter loose-leaf collection of a wide variety of forensic science subjects, both traditional and contemporary).

[13] Becker, C. L., Every man his own historian, *Am. Hist. Rev.*, 37, 221, 1932.

[14] *People v. Sutherland*, 155 Ill. 2d 1, 610 N.E. 2d 1 (1993). The defendant in this case is currently awaiting a retrial of his case. See discussion *infra*.

[15] *People v. Sutherland*, 194 Ill. 2d 289, 742 N.E. 2d 306 (2001).

[16] At the time of defendant's indictment in connection with Amy Schulz's death, he was serving a 15-year sentence in a federal prison after pleading guilty to shooting at employees of the National Park Service at Glacier National Park, in Montana. Prior to the trial, the defense filed a motion *in limine* to exclude from evidence knives found in his possession at the time of his arrest in Glacier National Park. The trial court denied the motion, ruling that the knives had "some slight probative value" and would not substantially prejudice the defendant by their introduction.

[17] Id.

[18] Id.

[19] *People v. Sutherland*, supra, n. 20, p. 9.

[20] Id., p. 10.

[21] Id.

[22] Id., p 11.

[23] *People v. Sutherland*, p. 17.

[24] Id.

[25] *People v. Sutherland*, supra, note 20, p. 11.

[26] *People v. Linscott*, 142 Ill. 2d 22, 566 N.E. 2d 1355 (1991). See also *People v. Giangrande*, 101 Ill., App. 3d 397, 56 Ill. Dec. 911, 428 N.E. 2d 503 (1981).

[27] *People v. Linscott*, 142 Ill. 2d at 30, 153 Ill. Dec. 249, 566 N.E. 2d 1355. The prosecutor also distorted the mathematical probability regarding the hair-comparison evidence. Despite the lack of a solid foundation, the prosecutor argued that the odds of another individual having hair with the same characteristics as the defendant's hair were about 1 in 3 million (142 Ill. 2d at 33, 153 Ill. Dec. 249, 566 N.E. 2d 1355).

[28] *Sutherland*, supra, note 20, p. 12.

[29] *People v. Sutherland*, 194 Ill. 2d 289, 742 N.E. 2d 306 (2001).

[30] The issue of the boot print was not addressed during the original trial and was not investigated adequately by original defense counsel.

[31] *People v. Sutherland*, 194 Ill. 2d 295 (2001).

[32] Id.

[33] Ibid., p. 298.

[34] *People v. Sutherland*, 194 Ill. 2d 289, 311–312. In the late summer of 2004, Cecil Sutherland was again convicted on his retrial and again given the death penalty. Two areas of new forensic science were utilized that must be noted in this chapter: testimony that the dog hair found on the victim were Black Labrador dog hairs [breed specific] and mitochondrial DNA analysis that allegedly showed that it was Cecil Sutherland's Black Labrador that was the source of the dog hair found on the body [dog specific]. These two very controversial additional evidentiary offerings will no doubt be the subject of Daubert/Frye scrutiny on appeal.

[35] Horgan, J., *The End of Science*, Little, Brown, New York, 1996, p. 136.

Countering Chaos

Logic, Ethics, and the Criminal Justice System

Jon J. Nordby

34

Introduction

In the rational game governed by the rules of the scientific method, the popularity of a belief has no firm logical relationship to its truth.[1] Initially, perhaps, natural philosophers[2] themselves became divided on this issue, but gradually, through rational disagreement,[3] the truth emerged: the evolving conclusion stood the test of assaults by contrary argument. Popular belief remained largely outside developments in science, well beyond this methodological arena, perhaps shielded by ignorance, or perhaps subject to the irrational pressures of special interest and seductive influence.

Under the gaze of a contemporary public sometimes influenced by both of these irrational pressures, the forensic scientist must construct and deliver an expert opinion. In this context, the word *opinion* itself suggests uncertainty. Like *argument*, opinion also has a technical meaning beyond our everyday use. Legal opinions, for example, involve interpretations of and inferences from law, which may be subject to challenge on legal grounds. Similarly, scientific opinions involve interpretations of and inferences from data, which also can be subject to challenge on scientific grounds. This is why two forensic scientists often may interpret the same data differently. However, not just any challenge works; the challenge must remain rational, and the argument must be developed in the spirit of rational disagreement in either law or in science.

In science, such rational disagreements embody our notion of scientific progress. When issues become clarified through rational argumentation, assessments proceed through a sort of open challenge to the relevant scientific communities. With reasons in the form of testing and experiment openly available for scientific examination, progress results either through identifying difficulties and suggesting remedies or through confirming results and supplying additional confirming instances.[4] In the same sense, forensic scientists have a scientific obligation to present their reasoning as clearly as possible, showing how their conclusions follow from the scientific work applied to a given case.

However, in forensic science, the audience does not consist of other scientists working in similar areas: the immediate audience truly represents the public through jurors, attorneys, and judges. Clarity of expression remains a vital element of all such communications.

Good science, and good forensic science, then, produces reasoned opinions. This, of course, may not be true about some of the other communications associated with a case. Although *some* scientific opinions eventually are proved to be true, *other* scientific opinions, with good reasons supporting them, could actually be false, even though they command rationally measured, cautious acceptance in the face of each currently available alternative conclusion.

With this fact presumably in mind, the court often will ask us for some measurement of a conclusion's *degree of scientific certainty.* Although the exact meaning of this certainty criterion remains less than clear, the basis of such a request for a certainty assessment must remain solidly within the methodological realm of forensic science. Scientific certainty must remain independent of extramethodological features, such as popular opinion.

Bad science, pseudoscience, or celebrity opinion overflowing from radio and TV talk shows and network news programs more often than not remains *mere opinion*: opinion entertained, even held strongly and passionately, with *no rational support* in sight. Such passionately held opinions often remain firm, fixed, and ultimately unaffected by any factual or logical assault. As scientists, we realize that our measure of a reasoned opinion can change as the evidence develops. However, using a common example, the Earth has remained the same shape, in the relevant commonsense notion at stake here, regardless of the different opinions enjoying popularity over the centuries. Even if the majority of ancient scientists believed that it was flat, the Earth did not quickly change shape when majority opinion changed in favor of the sphere. Certainly,

the passion attached to such opinions has no rational effect upon the Earth's actual shape.

We also must remember that some of these ancient scientists had excellent, compelling reasons supporting their conclusion that the Earth was flat. It therefore makes perfect sense to say that such scientists were rationally justified in believing that the Earth was flat, given arguments from relevant data, even though their conclusion was, in fact, mistaken. In the same instance, it makes perfect sense to say that some ancients who believed that the Earth was round, based on the revelatory visions of a goddess, the power of poetry, or the necessities of theology, were not rationally justified in their belief. We have learned over the centuries that scientists should not navigate scientific waters with an eye fixed solely on conclusions. Instead, we must navigate with a critical[5] eye focused firmly on the methods dictated by logic.[6]

To review this topic, which was discussed in Chapter 1, forensic scientists must develop reasoned opinions by recognizing evidence, distinguishing it from coincidence, and applying a sound method. The method must allow for happenstance and uncertainty, carefully acknowledging room for relevant future discoveries. Most importantly, as reflected in the history of science, it must leave room for error.

Scientifically reliable methods help forensic scientists develop reasoned opinions, views that may not be proved conclusively true, but views toward which the explanatory patterns emerging from the evidence, together with the evidence itself, most unambiguously point. Reasoned opinions developed from scientifically acceptable methods avoid subjective, unsupported, and untested hunches and guesses. The bloodstain pattern expert must offer principled reconstruction and experimental confirmation, rather than mere observation.

Although the observation may be correct, its truth is merely coincidental. Because truth often hides among the debris of coincidence, a method that reveals the truth more often than not earns the mantle of reliability. Such a method advances mere guessing beyond a

1 in 2 chance of being correct when choosing between some statement and its negation: "O. J. Simpson is guilty of murder" versus "O. J. Simpson is not guilty of murder." Remember what our first physics, chemistry, and math teachers taught us: "Show your work." Even with the *wrong answer* on the page, applying the *right method* counted for something. At the very least, showing the conclusion's development helps us as well as others to discover the source of our error. Careful scientists must also recognize that sometimes, even often, our method falls short of establishing *anything* of scientific reliability. The forensic scientist's opinion then simply must be an explanation of why the available evidence prevents us from reaching any conclusion of value to the issues before the court.

Reliable scientific method in the forensic context emerges by critically considering the *activities* of forensic scientists from legal, ethical, and scientific perspectives.

Legal Components of Reliability

Unlike theoretical scientists confined to research laboratories, forensic scientists apply their skills to data with the legal status of evidence: data relevant for the resolution of legal questions presented in courts of law. As such, forensic scientists, unlike other scientists, incur specific legal obligations to which they must conform. Initially, one might think that forensic scientists could thereby answer all questions about the right thing to do in a specific instance by appeal to legal obligations alone.[7]

Similarly, police detectives, sworn to enforce the law; prosecutors, sworn to protect the rights of society; and defense attorneys, sworn to protect the rights of the accused, all might make identical appeals to settle questions about the right thing to do in all cases. "What's right is what the law says, period."[8] This would both simplify methodological matters and significantly shorten this chapter.

However, the issues are more complex than that.

Although the legal obligations of forensic scientists do provide some necessary foundation for doing both the morally and scientifically right things, they do not offer the practical guidance necessary to navigate the hazardous seas of forensic scientific practice.[9] To understand the weakness[10] of this legalistic approach, consider, briefly, what the law actually says about the obligations of law enforcement, attorneys, and scientists serving courts of law.

Law Enforcement and the Law

Citizens expect police to have the wisdom of Solomon, the courage of David, the strength of Samson, the patience of Job, the leadership of Moses, the kindness of the good Samaritan, the faith of Daniel, the tolerance of Jesus, and, finally, an intimate knowledge of every branch of the natural and social sciences. Anyone with all these might be a good police officer.

August Vollmer, **Police and Modern Society**, *1936*

Civil societies require policing to protect people from being hurt by others. In one place, policing may function cooperatively with the protected citizenry, whereas in another jurisdiction such cooperation may be less evident. Even though much may depend on the skills and personalities of individual officers, in most civil societies, law enforcement maintains order through its legally constituted authority.

In a dissenting opinion, *U.S. v. Wade*, 388 U.S. 218 (1967), Justice Byron White states what he takes to be the legal obligations of law enforcement officers:

Law enforcement officers have the obligation to convict the guilty and to make sure they do not convict the innocent. They must be dedicated to making the criminal trial a procedure for the ascertainment of the true facts surrounding the commission of the crime. To this extent, our so-called adversarial system is not adversary at all; nor should it be.

Yet when criminologists[11] examine and describe the actual actions of law enforcement officers in specific cases under jurisdictions across the United States, they present evidence of, at best, a selective appeal to Justice White's statement. For example, an overly enthusiastic embrace of the "obligation to convict the guilty" might ignore the obligation "not to convict the innocent."

This enthusiasm to convict the guilty may lead law enforcement personnel to view the criminal trial as a mechanism to secure that conviction against their suspect, rather than as a means to "ascertain the true facts surrounding the commission of the crime." It certainly does not take a criminologist's sociological research to convince us that law enforcement often plays an adversarial role in criminal cases.[12] Practical explanations of what it means to execute these obligations and what to do when these obligations conflict are notably absent from the law.

Attorneys and the Law

In *Berger v. U.S.*, 295 U.S. 78, 88 (1935), words found in the majority opinion help delineate the legal role of prosecutors:

> He is the representative … of a sovereignty whose interest … in a criminal prosecution is not that it shall win a case, but that justice shall be done … . He may prosecute with earnestness and vigor—indeed he should do so. But, while he may strike hard blows, he is not at liberty to strike foul ones.

This is the ethical ideal. At the other extreme, sadly, some decisions to prosecute cases are based on *winnability* rather than on the desire to seek justice.[13] Some decisions to prosecute a case may depend on the social status of the victim, the level of public enthusiasm, and political[14] pressures. The reality of burgeoning case loads, shrinking budgets, and scarce resources, common in prosecutors' offices everywhere, dictate severe practical limits to *interest in justice*. Again, practical advice about how to manage any of these conflicts,

legitimate or otherwise, does not appear in legal form.

Defense attorneys are called to the same high ethical standards as prosecutors. The rules of professional conduct for attorneys

> … define the minimum requirements for lawyers with complex competing obligations to their clients, to their adversaries, and to the system of justice. They presume an adversary system that is not an end in itself, but rather a means to justice and that is tempered accordingly. In their preamble, the Rules call on lawyers to hold themselves to a higher standard than the minimum required to avoid disciplinary action.

The defense can fail to meet these standards when obligations to provide clients with spirited defenses conflict with obligations to the system of justice itself. Both prosecution and defense advocates may veer from their ideals without straying from the letter of the law. It is within this setting that forensic scientists are called to uphold their own ethical standards.

Forensic Scientists, Scientific Values, and the Law

> Who can be wise, amazed, temperate, and furious, loyal and neutral, in a moment? No man.
>
> **Shakespeare**, *Macbeth* [115]

Courts, of course, describe the duty of forensic scientists who give testimony; as any witness before the court, forensic scientists must swear to tell the truth. Common sense tells us that truth-telling demands describing what is while avoiding distortions of fact. (A fact usually is defined accurately, if not somewhat deceptively, as "what is so.") Distortions of fact might include, for example, the deliberate omission of relevant facts, or inviting misinterpretations and encouraging incorrect conclusions from such incomplete presentations. Such distortions may even appear in the scientist's own curriculum vitae, or résumé.[15] This is the reason why a court will examine experts' credentials so carefully before allowing their testimony. The duty to tell the truth and to avoid distortions of fact becomes especially difficult when scientists' testimony is limited to answering *yes* or *no* questions asked by relevant counsel.

When scientists qualify as expert witnesses, the court must assess the merit of the expert opinion.[16] Courts have approached "truth-telling while rendering a scientific opinion" by appealing to "generally accepted scientific practices" in the relevant scientific field. One might suggest that *reliable scientific methods* are simply *widely accepted methods*. However, this places the cart before the horse. Let us think a minute. Is a method accepted because it is reliable? Or is a method reliable because it is accepted? Most scientists vote *yes* to answer the prior question, and *no* to the second. Recent court decisions raise questions about the adequacy of the so-called *general acceptance criterion* for scientific truth. In fact, the history of science builds on legitimate challenges to general acceptance that eventually become generally accepted.[17] General acceptance does not usefully ensure reliable method. (See the discussion of *Daubert* and *Frye* rulings in Chapter 32.)

Judging expert methods and opinions to be scientific cannot be taken to mean judging those methods and opinions to be "as in fact practiced by scientists," or even "as actually practiced by the majority of scientists in a given field." The term *scientific* means more than an objective description of sociological fact. It conveys some sense of correctness or scientific virtue. The criterion of *being scientific* excludes the crackpots; so *scientific*, as used in courts, has an intractable value component meant to capture the notion of proper, good, or correct science—science as practiced by good scientists. This is not merely a descriptive term; it must incorporate a prescriptive or value element to point toward correct, proper, or relevant methods. To judge expert testimony, the court does not require some statistical comparison between the testimony and some accepted scientific studies; it requires an assessment of the scientific reliability of the expert's methods applied to reach the stated conclusions.

The value element built into *scientific* is not an exclusively ethical element. There are many nonmoral virtues and values, including the scientific virtues of correctness, completeness, simplicity, and so on. (Moral virtues generally apply to persons or their actions, either individuals or groups of individuals, whereas nonmoral virtues may not be so limited.) However, the value issue for the forensic scientist concerns the moral virtue of truth-telling mixed with the scientific virtue of good science, a central element of reliability. These, in turn, include many complex combinations of virtues, both moral and nonmoral.

Moral issues of autonomy, duty to self and duty to others, face the forensic scientist who develops and presents expert testimony. In short, for the forensic scientist, telling the truth in court involves a complex mix of both being a good scientist and being a good person. So, no law or set of laws define the legal obligations of forensic scientists, let alone explicitly state their ethical duties.

Organizational Codes of Scientific Conduct

Forensic scientists, physicians, and other professionals also may look to professional organizations' codes of conduct to illuminate correct conduct for scientists. Sadly, much of what passes for professional ethics embodied in codes of professional conduct reduces to lists of permissible and prohibited conduct designed to prevent professional heresy, or simply avoid troublesome litigation. Matters become even worse when professional ethics reduces to conduct designed to avoid embarrassing some specific agency or organization that supplies the code of conduct. Even if we developed explicit and robust codes of conduct, their usefulness would remain doubtful at best.[18] The ethical conduct of forensic science involves much more than a list of dos and don'ts. Professional ethics is not some random, extraneous entity that attaches to forensic practices as an afterthought. It must remain an essential element of doing science, or it has simply no value at all.

Scientists have obligations to their science, to professional oaths, to the rules of conduct established by professional organizations, and, like all human beings, to themselves. These, of course, accompany a scientist's legal obligations to the court as well as contractual obligations to employers or clients. This mixture presents the potential for unprecedented conflicts among these many legal, personal, and professional obligations.

Alan R. Moritz, MD, presented a notable address to the 35th Annual Meeting of the American Society of Clinical Pathologists in 1956—*Classical Mistakes in Forensic Pathology*—recounting his experience with these mistakes and recommending how they could be avoided.

> This Sherlock Holmes type of expert may see certain bruises in the skin of the neck and conclude without doubt that they were produced by the thumb and forefinger of the right hand of the strangler. He may see an excoriation of the anus and maintain unequivocally and without benefit of other elements of scientific proof that the assailant was a sodomist. He ignores the essential component for proof of the correctness of any such scientific deduction, namely, the nonoccurrence of such lesions or changes in control cases. Such a pathologist usually has the happy faculty of failing to remember the many similar bruises of necks that were known to have been produced by mechanisms other than pressure by the thumb and fingers. He fails to remember the many anal and rectal excoriations that were caused by injuries other than sodomy. Such a pathologist is a delight to newspaper reporters owing to the fact that he "makes good copy." He may be highly esteemed by the police and by the prosecuting attorney because he is an emphatic and impressive witness.

Moral as well as scientific value conflicts may arise when one meets such experts on the other side of the courtroom. Not only is the science in question, but an innocent person may be convicted because of it. What are our obligations as scientists to point out such errant, overly emphatic testimony? Does such an emphatic expert witness violate any professional codes of conduct, or if they in fact do not, should codes be revised to cover such behavior? What makes such behavior wrong, if indeed it is wrong at all? Deeply held values of truth and good science may conflict with equally deeply held values of scientific independence, which often results in a legitimate difference of opinion derived from examinations of the same data.[19]

This conflict also applies institutionally, for example, to crime laboratories themselves. A natural public skepticism develops whenever any institution remains largely unregulated, or even worse, self-regulated.[20] Alleged scientific and moral problems with the work of the FBI laboratory, one of the most respected laboratories in the world, point to conflicts at institutional levels requiring the expert examination of legal, ethical, and scientific values. They involve making moral judgments of individual actions compared with those of institutional actions.

These conflicts, often aired in the press, contribute to public cynicism and foster a deep distrust of the legal system, often producing great harm. The issues often involve conflicts among hidden social, moral, and scientific values, especially those concerning the right way of conducting the business of forensic science, including cost containment measures, accreditation issues, and competency questions. These represent everyday issues and questions that face anyone engaged in the practice of forensic science or forensic medicine.

Scientifically Reliable Methods

The conduct and nature of reliable scientific methods and investigative techniques appear among the chapters of this text while covering specific forensic sciences and their practice. Each chapter invites the reader to develop insights into the methodological matters considered here and in Chapter 1. To aid the reader in this process, consider the relationship between what a forensic scientist *does* and what a forensic scientist *says*. This relationship helps exhibit the nature of scientific methods applied to a case, often helping us to distinguish between reliable and unreliable approaches to the evidence.

As scientists, we conduct scientific investigations; as forensic scientists, we must write explanatory reports and testify, in both ways explaining our investigations before the court. Although all scientists may write reports in some fashion, forensic scientists must write for a nontechnical audience, and can make no assumption that the reader or listener understands either the language or the activity. Writing and speaking, therefore, also become integral elements in the practice of the forensic sciences. Two necessary virtues that we met before apply here: simplicity and clarity. To explain technical scientific matters to other scientists is one thing; to explain them both accurately and clearly to a lay audience is quite another.

Once you, as a forensic scientist, write something in a report, or say something under oath, you own that forever, good, bad, or indifferent.[21] (The same applies to professional publications such as journal articles and books, as well as to public oral presentations and media interviews.) Navigating these perilous waters involves exercising the logical acumen required to apply reliable scientific method and explaining its application to the case at hand, both clearly and accurately. Welcome to the public sciences.

With this endeavor, two immediate situations often arise: (1) forensic scientists' statements appear overly definitive or precise, and (2) such statements appear overly inconclusive or imprecise. Assuming that we do not deliberately sabotage our own efforts, both issues must result from the forensic scientist's honest attempt to be accurate and truthful while inadvertently failing to maintain logical standards. When a degree of precision expressed by a conclusion fails to mirror the available precision among the data, red flags ought to fly. Opposing counsel should focus on these two possible issues. For example, when the data support a degree of accuracy to within one whole number, a conclusion expressed to within three places of the decimal remains totally unsupported. Similarly, when noting a weapon's distance as measured from a baseline during the investigation of a crime scene, statements such as "the distance from the wall appeared to be 2.25 meters" introduce a totally inappropriate sense of imprecision. A scientist has no logical grounds for waffling on such a definite measurement unless preparing to discuss why 2.25 meters is, in fact, wrong.

There are other logical difficulties with the use of standard phrases that appear throughout transcripts of scientific testimony. Some phrases may express a caveat about the nature or quality of available data, such as "based upon the current data ..." or "in the condition received ..." as well as others. Such expressions remain useful provided the analyst remains aware of their logical implications and intends to communicate these consequences.

However, matters can become complex very quickly when words and phrases such as *likely*, *tends to be*, and so on, are used incautiously by forensic scientists. These words imply that some degree of probability attaches to the surrounding claims. Again, when this is intended, such use may be entirely in order. However, when couching claims in these words, the forensic scientist must be prepared to supply a foundation for the probabilistic nature of the attendant opinion. The absence of such a foundation for probability assessments points toward a lacuna in the analysis, a situation that opposing counsel will undoubtedly probe to expose a scientific weakness, whether real or imagined. The fact remains that the scientist's own words created this avenue of investigation. If such an avenue remains irrelevant to the analysis, the scientist should choose his or her words carefully and not invite the excursion in the first place.

Complex scientific testimony logically requires the same simplicity and clarity applied in the development of the scientific explanations at issue. Misunderstanding often is due to deviations from the proper methods needed to answer scientific questions. When subjects challenge readers or listeners, there is no benefit in presenting such subjects in a confusing or disorganized manner. A disorganized presentation of scientific results usually mirrors a disorganized investigation of scientific data.

Scientific Testimony

The most important thing for a forensic scientist to remember on the stand is to talk only about familiar territory, and to avoid giving opinions on anything beyond these specific areas of expertise. Methodological wisdom as the essential element of reliable scientific method can be gained only through long hours of study, extended patient exercise, and frequent application interrupted with, yet corrected by, unfortunate error. This wisdom develops over a lifetime; it can never be completely attained, and it can never be completely adequate. There remains evermore to learn. This development depends on loving forensic science and forensic medicine so much that the excitement of seeking and learning can never be satisfied.

This lifelong enthusiasm will play an important part for the scientist required to testify. In court, expert witnesses are asked to tailor testimony to jury pools assumed to have the education and attention spans of the very young. Despite this challenge, the obligation of the forensic scientist is often to educate the jury in complex scientific concepts. An honest enthusiasm for the subject will not be lost on the audience and, with the application of simplicity and clarity, will aid the expert in communicating as effectively as possible.

Other factors associated more with appearance than substance also apply on the witness stand. We might think it unfortunate that attention is paid to how we dress or wear our hair, but whether we like it or not, how we are perceived enhances our ability to communicate. A relaxed, yet alert, posture, friendly eye contact with the jury, clear speech, and appropriate gestures all contribute to the effectiveness of our teaching.

Nothing, however, is more important or effective than the rational power of fact, logic, and convincing method. And nothing can displace the forensic scientist's ethical obligation to present that information with as much honesty, clarity, and wit as possible. The effect on the jury and the legal outcome ultimately remain outside any scientist's realm of responsibility.

Selected Readings

Achinstein, P., *The Nature of Explanation*, Oxford, 1983.

Achinstein, P., *The Concept of Evidence*, Oxford, 1983.

Brody, B. A., Confirmation and explanation, *J. Philos.*, 65, 282–299, 1968.

Cartwright, N., *How the Laws of Physics Lie*, Oxford University Press, New York, 1983.

Cartwright, N. and Nordby, J. J., Essay #6, For Phenomenological Laws, in *How the Laws of Physics Lie*, Clarendon Press, Oxford University Press, New York, 1983, pp. 100–127; further acknowledgment of Nordby, J. J., in *Introduction*, pp.1–20.

Corrington, R. S., *An Introduction to C. S. Peirce: Philosopher, Semiotician, and Ecstatic Naturalist*, Rowman and Littlefield, 1993.

Dostoyevsky, F., *Crime and Punishment*, Modern Library College Editions, Random House, New York, 1950.

Dretske, F., Laws of nature, *Philos. Sci.*, 44, 248–268, 1977.

Earman, J. (Ed.), *Testing Scientific Theories*, Minnesota Studies in the Philosophy of Science Vol. X, University of Minnesota Press, MN, 1984.

Eco, U. and Sebeok, T. A., (Eds.), *Dupin, Holmes, Peirce: The Sign of Three*, Indiana University Press, Bloomington, 1983.

Harman, G., The inference to the best explanation, *Philos. Rev.*, 74, 1, January 1965.

Kent, B. and *Charles S. Peirce: Logic and the Classification of the Sciences*, McGill-Queen's University Press, Montreal, 1987.

Kvanvig, J. L., How to be a reliabilist, *Am. Philos. Q.*, 23, 2, April 1986.

Laudan, L., *Science and Values: The Aims of Science and their Scientific Debate*, University of California Press, 1984.

McMullin, E., *The Inference that Makes Science, The Aquinas Lecture, 1992*, Marquette University Press, Milwaukee, WI, 1992.

Mill, J. S., *A System of Logic,* London, 1843.

Nordby, J. J., How approximations take us away from theory and toward the truth, *Pac. Philos. Q.*, July 1983.

Nordby, J. J., Bootstrapping while barefoot (crime models vs. theoretical models in the hunt for serial killers), in *Synthese: The Philosophy of Applied Science*, Vol. 81, Nordby, J. J. and Weil, V., Eds., Kluwer Academic Publishers, Dordrect, Germany, 1989, pp. 373–389.

Nordby, J. J., Can we believe what we see if we see what we believe? Expert disagreement, *J. Forens. Sci.*, 37, 4, July 1992. Reprinted, *Int. Soc. Air Safety Invest. Forum*, 26, 3, September 1993.

Nordby, J. J., Science is as science does: The question of reliable methodologies in real science, *Shepard's Exp. Sci. Evi. Q*, 2, 3, Winter 1995.

Nordby, J. J., A member of the Roy Rogers Riders Club must follow the rules faithfully, *J. Forens. Sci.*, 42, 6, November, 1997.

Nordby, J. J., *Dead Reckoning: The Art of Forensic Detection*, CRC Press, Boca Raton, FL, 1999.

Nordby, J. J., Is forensic taphonomy scientific? in *Forensic Taphonomy: Vol. II,* Haglund, W. D. and Sorg, M. H., (Eds.), CRC Press, Boca Raton, FL, 2002.

Nordby, J. J. and Weil, V., (Eds.), *Synthese: The Philosophy of Applied Science*, Vol. 81, Kluwer Academic Publishers, Dordrecht, Germany, 1989.

van Fraassen, B. C., *The Scientific Image,* Oxford University Press, New York, 1980.

Whewell, W., *The Philosophical Foundations of the Inductive Sciences Founded on their History,* Vol. I and II, 2nd ed., London, 1847.

Endnotes

1 Confidence in a belief's truth accompanies confidence in the *rational methods developing, challenging, and supporting the belief*. Such confidence neither necessarily nor automatically attaches to the belief itself.

2 "Natural philosopher" referred to those we call "scientists" in today's language. It was not until 1846 that a natural philosopher named William Whewell coined the term *scientist*. But for him, we might today be called "forensic natural philosophers." *Forensic scientist* certainly sounds better.

3 *Argument* is a technical term as in "legal argument." In natural philosophy, philosophy, and of course, natural science, an argument is a set of statements consisting of reasons offered in support of some conclusion. Argument evaluation involves an assessment of those reasons and an investigation of their relationship with the conclusion at issue. The area of philosophy focusing on the nature of argument is called *logic*; that focusing specifically on arguments in the sciences is called *philosophy of science*.

4 This birds-eye view is not offered as a comprehensive description of all progress in science. It merely suggests that disagreements are welcomed, not shunned, by scientists who participate in contemporary debates within their research areas.

5 *Critical* in the scientific and philosophical domain means "careful, or rationally cautious."

6 For an extended discussion of forensic scientific method, see Jon J. Nordby, *Dead Reckoning: The Art of Forensic Detection*, CRC Press, December, 1999.

7 Although initially plausible, this view quickly degenerates into absurdity. Similar positions became targets for the astute reasoning of both Plato and Aristotle, as well as many other so-called political philosophers.

8 Such a view is called a *strict liability* interpretation of law, or is sometimes referred to as *legalism*. The relation of law and ethics is addressed in the AMA Code of Medical Ethics: "The following statements are intended to clarify the relationship between law and ethics. Ethical values and legal principles are usually closely related, but ethical obligations typically exceed legal duties. In some cases, the law mandates unethical conduct. In general, when physicians believe a law is unjust, they should work to change the law. In exceptional circumstances of unjust laws, ethical responsibilities should supersede legal obligations." The fact that a physician charged with allegedly illegal conduct is acquitted or exonerated in civil or criminal proceedings does not necessarily mean that the physician acted ethically. This distinction is also supported by the *Ethics Manual of the American College of Physicians*. The manual raises ethical issues and presents general

guidelines. In applying these guidelines, physicians are asked to consider individual circumstances and use their best judgment. Physicians are morally as well as legally accountable, and the two may not be concordant. Segregation and slavery, for example, were once legal in this country, but are never morally defensible. Physicians must keep in mind the distinctions and potential conflicts between legal and ethical obligations when making clinical decisions and must seek counsel when concerned about the potential legal consequences of ethical decisions. We refer to the law in this manual for illustrative purposes only; these references should not be taken as a statement of the law, which can vary from state to state, or of the legal consequences of a physician's actions. "The law does not always establish positive duties (what one should do) to the extent that professional (especially medical) ethics does. Our current understanding of medical ethics is based on the note defined from which these positive duties emerge. The relative value of such principles, and conflicts among them, often account for the ethical dilemmas physicians must face."

9 Certainly, legal obligations cover some significant behavioral territory; however, compared with the behavioral territory that forensic scientists, law enforcement, prosecutors, and defense attorneys must cover, gaps and conflicts within the rule of law substantiate the need for additional considerations.

10 A criminologist is a social scientist, or sociologist who studies the social phenomena of crime and the criminal justice system. Criminology (social science) is frequently confused with criminalistics (physical science) in court and in the media, as well as among members of the general public. Often, this inappropriate adversarial role begins with what I have called *expectation-laden observations*.

  I discuss this problem in a work cited earlier: Can We Believe What We See If We See What We Believe? Expert Disagreement, *J. Forens. Sci.*, 37, 4, July 1992. A contagious condition afflicting law enforcement, lawyers, and forensic scientists, it involves locking into a solution and locking onto it so firmly as to lock out alternatives better suited to the evidence.

11 In a quote attributed to Dallas prosecutor Henry Wade in Earl Morris' docudrama *The Thin Blue Line*, Wade allegedly says, "Any prosecutor can convict the guilty: it takes a great prosecutor to convict the innocent."

12 In this use, the word *politics* carries a contemporary connotation. I do not here invoke the ancient Greek sense of *Polis*.

13 Stated in the "King County Bar Association Guidelines of Professional Courtesy," adopted by the KCBA Board of Trustees, Seattle, WA, January 1989; Revised June 1992, as printed in *The King County Bar Association 1997 Membership Directory*, The Publishing Co. of North America, Inc., Lake Helen, Florida 1997, p. 20.

14 Prof. Carol Henderson of Nova Southeastern University Law School has studied this matter extensively. She concludes that about 30% of all scientific curricula vitae contain falsehoods. Her conclusion was noted in *Scientific Sleuthing*, 25, 2, Summer 2001.

15 This assessment involves, among other things, decisions concerning the testimony's admissibility and sufficiency.

16 The germ theory of disease, the relationship of the planets in our solar system, the existence of atomic and subatomic particles, and many other "advances" provide common, yet useful, examples.

17 It is doubtful that they would apply to specific cases (the problem of covering laws); they would carry with them the issues of any legalistic or strict liability approach to human action; it would remove personal discretion and human judgment, and serve to erode, if not eradicate, the human autonomy necessary for assignments of responsibility for conduct.

18 *Classical Mistakes in Forensic Pathology*, 35th Annual Meeting of the Society of Clinical Pathology, 1956, pp. 1389–1390.

19 Any robust account of scientific method must account for the fact that legitimate disagreements are not only possible but are in fact likely. The phenomena at issue are simply that complex.

  In 1989, Eric Lander, a molecular biologist, wrote an article in *Nature* stating his view of the problem: "At present, forensic science is virtually unregulated—with the paradoxical result that clinical laboratories must meet higher standards to be allowed to diagnose strep throat than what standards forensic labs must meet to put a defendant on death row." Fortunately, matters have somewhat improved since 1989.

[20] Paul Kish, in James, *op. cit.*, page 84.
[21] *Black's Law Dictionary* defines *probable* as "appearing founded in reason, or experience, more evidence for than against, supported by evidence inclining the mind to believe, but leaving room for some doubt, likely, apparently true, yet possibly false."

Biohazard Safety Precautions

Introduction

Attention to biohazard safety all too often is limited to the medical community. Many tasks performed during the investigation and prosecution of criminal activity by police officers, investigators, forensic scientists, and attorneys could involve contact with numerous biohazards. Forensic environments including the crime scene, morgue, and laboratory allow numerous opportunities for exposure to biohazards and pathogens. The ways exposure can occur and the types of exposures are more innocuous than they are in a traditional medical setting, but the personnel who handle forensic evidence must understand the general rules of biosafety and apply them to the diverse situations that can arise in criminal investigations.

This outline will present basic explanations of the occupational statutes, **blood-borne pathogens**, and general guidelines geared toward overall biohazard safety. The best defense against biohazard exposure is to apply this information along with a healthy dose of common sense.

Universal Precautions Policy

The Occupational Safety and Health Act (**OSHA**) defines universal precautions: "All blood and certain other body fluids are considered potentially infectious for hepatitis B virus, hepatitis C virus, human immunodeficiency virus (HIV), and other blood-borne pathogens." Blood-borne pathogens are defined as pathogenic microorganisms that are present in human blood, human blood components, and products made from human blood that can cause disease in humans. This means that blood and certain other body fluids are considered potentially infectious. Body fluids that require the application of universal precautions are as follows:

1. Blood
2. Semen
3. Vaginal fluid
4. Tissue (unfixed)
5. Synovial fluid
6. Amniotic fluid
7. Peritoneal fluid
8. Spinal fluid
9. Pleural fluid
10. Pericardial fluid
11. Any body fluid that is visibly bloody

Topics

Universal Precautions Policy

Basic Rules for Biohazard Safety

Universal precautions are necessary when dealing with a person known to be infected. If an individual is known to be infected with a blood-borne pathogen, such as HIV, the information should certainly be transmitted to all individuals who might be exposed to this individual's body fluids. However, it is not always possible to identify the source individual or to know infectious status. This is particularly true with body fluids found at crime scenes or on items of evidence. Exposure may occur through skin (especially if broken), eyes, mucous membranes including the nose, mouth, and respiratory tract, and via parenteral means (subcutaneous or intravenous contact). Therefore, a policy of following universal precautions is recommended *at all times.*

Human Immunodeficiency Virus

HIV is a retrovirus associated with acquired immunodeficiency syndrome (AIDS). Its core contains RNA and enzymes, and the outer covering consists of lipids and proteins. Viruses cannot reproduce without infecting host cells. HIV binds to the surfaces of helper T cells (white blood cells) of the host in order to reproduce. The RNA and enzymes of the HIV are released into a host cell that begins copying the viral RNA into double-helix DNA. The DNA containing RNA of the HIV is then incorporated into the DNA of the host cell, which is then deceived into producing more and more virus particles. As HIV gradually depletes the number of helper T cells, the immune system of the host is rendered virtually powerless.

The host becomes vulnerable to other viruses, bacteria, fungi, protozoa, and cancers that ordinarily would not infect an individual with an intact immune system. Body fluids that can transmit HIV infection are blood, semen, vaginal secretions, and possibly breast milk. HIV has been isolated also from saliva, tears, urine, cerebrospinal fluid, amniotic fluid, tissues of infected persons, and experimentally infected nonhuman primates. Although the virus has been isolated in all these body fluids, only blood, semen, vaginal secretions, and possibly breast milk have

Table A.1 Survivability of HIV

Condition	Survivability
Liquid at room temperature	15 days (minimum)
Refrigerated liquid	Much longer than 15 days
Dried form at room temperature	3 to 13 days
Dried form, frozen	Much longer than 13 days

been implicated in transmission of the virus. The survivability of HIV has been estimated under the conditions listed in Table A.1.

Hepatitis

Viral hepatitis is a general term for five types of the disease:

1. Hepatitis A (HAV), formerly called *infectious hepatitis*, is contracted through fecal or oral routes.
2. Hepatitis B (HBV), formerly called *serum hepatitis*, is contracted through nonintact skin or punctures such as needle sticks. It is one of the most common laboratory-associated infections.
3. Hepatitis C (HCV), formerly called *nonA nonB hepatitis*, is contracted through nonintact skin and punctures such as needle sticks. It has been detected primarily in blood and serum, less frequently in saliva, and rarely in urine or semen.
4. Hepatitis D (HDV) is a defective virus and requires the presence of hepatitis B for replication.
5. *NonA, nonB, nonC hepatitis* is a collective name for all other hepatitis viruses contracted via nonintact skin or punctures such as needle sticks. A human host is 100 times more likely to contract hepatitis than HIV. Although HIV has received more attention, the risk of contracting HBV from the handling of biological materials is actually much greater.

When source blood is known to be infected with both HBV and HIV, the likelihood of contracting HBV is 100 times greater than the likelihood of contracting HIV. The HBV is known to survive much longer than HIV even in the dried state.

Basic Rules for Biohazard Safety

It is essential that universal precautions be observed at all times and exposure to biohazards be minimized by eliminating the known routes of exposure by minimizing or eliminating the contact route to the body from the scene, autopsy room, laboratory work space, and supplies and equipment whenever possible. Contact routes include the following:

1. Direct oral route via blood or other body fluids entering the mouth
2. Indirect oral route via smoking or placing pencils in the mouth
3. Ocular route via splashing infectious material into the eyes or rubbing the eyes
4. Inoculation route via puncture wounds caused by needles or other sharp objects
5. Respiratory route via direct respiratory contact or aerosol production

Hands should be washed with warm water and a germicidal soap immediately following contact with infectious materials and after completion of all procedures, including the removal of protective gloves.

Minimizing Biohazard Contamination at the Crime Scene

Establish a *clean area* as a working space, and keep it uncontaminated. Remember that gloves, equipment, and supplies taken to a crime scene may become sources of contamination. Steps to take include the following:

1. Leave all supplies and equipment in the clean area until needed.
2. Organize and assemble equipment to minimize contamination. Take only necessary equipment and supplies into the scene because anything that enters the scene becomes potentially contaminated.
3. Determine the type of personal protective equipment (**PPE**) needed based on the types and quantities of biological fluids present and whether they are in the wet or dry state relative to the tasks to be performed.

Each scene should be assessed initially by questioning knowledgeable individuals who have already arrived and performed an evaluation walk-through. During the evaluation walk, minimize the number of personnel entering the scene and wear protective shoe covers and gloves at a minimum.

Guidelines for Selecting PPE

1. Wear protective gloves as a minimal precaution.
2. Wear face or eye shields if the potential for fluid splash exists.
3. Wear a face mask if the potential for aerosolization is present.
4. Wear a protective suit and shoe covers if clothing contamination is possible.
5. Replace gloves and other PPE items when visibly soiled, deteriorated, or torn.
6. Discard surgical or examination gloves. Do not wash or disinfect them for reuse.
7. Remove gloves and other PPE before leaving the work area.
8. Use gloves if you have cuts or dermatitis.
9. Set up a discard location for disposal of contaminated PPE and supplies when the scene is exited.
10. Secure disposal bags to prevent accidental spills.
11. Prepare bleach or other decontamination solution fresh each day. It should be diluted 1:10 with water (approximately ½ cup of bleach in a quart of water). A minimum of 10 minutes contact time is necessary.

Precautions for Evidence Collection

1. Be aware of syringes, needles, knives, razors, nails, broken glass, or sharp metals nearby.
2. Do not reach into confined or blind spaces before checking them with a flashlight or mirror.
3. Use caution when collecting biological specimens to avoid overt contamination.
4. Change gloves between each item of evidence collected to prevent cross-contamination.
5. Use masks and eye protection to avoid aerosol exposure when scraping or cutting dried stains.
6. Use caution with knives and scissors to avoid surface cuts and contamination from body fluids when cutting out stains.
7. Use a syringe to collect a wet stain. Do not use a needle on a syringe unless absolutely necessary, because a needle will make the opening smaller and collection will be more difficult. Needles also present sharp hazards.

Packaging Evidence

1. Package each item separately to prevent cross-contamination.
2. Use puncture-resistant containers when packaging sharp pieces of evidence. Write "SHARP" on the outside of each container.
3. Package wet items in plastic only when necessary.
   a. Keep evidence in plastic only as long as needed for transport.
   b. Air-dry at room temperature as soon as possible. Do not use direct sunlight or a hair dryer to speed drying time.
   c. Repackage dried evidence in a paper container.
4. Use tape when closing a container. Write your initials across the tape to ensure integrity of the package.

5. Do not use staples. They create sharp hazards. It is also difficult to ensure the integrity of a package that has been fastened with staples.
6. Ensure that appropriate labeled containers are prepared for the disposal of contaminated items.
7. Follow all steps of the established procedure at the discard station.

Discard Station Procedure

1. Items should be handled in sequential order.
   a. Disposable equipment and supplies
   b. Nondisposable equipment and supplies
   c. PPE
2. Disposable equipment and supplies should be classified if they represent sharp hazards.
3. Nonsharp disposables such as paper scales, film boxes, and film canisters may be discarded directly into collection containers.
4. Sharp disposables such as scalpels, pens, or pencils, and broken glassware should be first packaged in cardboard, plastic, or another type of rigid container, and discarded into a collection container to prevent punctures.
5. Nondisposable equipment and supplies such as ink pens to be reused, markers, crime scene kits, cameras, safety glasses, reusable rulers, and scales should be decontaminated with diluted bleach or commercial wipes and returned to their proper storage areas. Cleaning supplies should be discarded into a collection container.

Removal of PPE to Minimize Exposure

1. Remove eye covers, and decontaminate them if reusable. Return them to their proper storage location. Place cleaning supplies and disposable PPE items in a discard container.

2. Remove masks and head coverings with caution to prevent indirect exposure, and place them in a discard container.

3. Remove shoe covers via a roll-off technique. Invert the uncontaminated interior over the contaminated outer surface, and pull the cover off. Place it in a discard container.

4. Roll off the protective suit. Invert the uncontaminated interior over the contaminated outer surface, and pull the suit off. Place it in a discard container.

5. Remove gloves last. Start at the wrist, and peel the glove off one hand. Grasp the removed glove in the still-gloved hand. Roll down the top of the remaining glove. Grasp the uncontaminated inner surface of the glove, and peel it off while tucking the first glove inside the second glove. Place the gloves in a discard container.

6. Close the discard container by grasping the bag beneath the rolled-down top. Pull the top of the bag upward to close it. Secure the tops of bags with tape or twist ties to prevent spills.

7. Wash hands with warm water and germicidal soap. Vionex™, Cal-stat™, or Sani-cloth™ wipes may be used if no hand-washing facility is immediately available. Wiping is a temporary alternative. It is not a substitute for hand washing.

The removal and disposal of contaminated materials, supplies, testing items, and PPE are responsibilities of the agency at a crime scene.

Selection of PPE

An important consideration in biohazard safety is selecting the proper PPE to be used. Selection should be based on the best conservative determination of types of exposure that might occur in a given situation. When in doubt, use more PPE than less. The decision tree in Figure A.1 will assist in the selection of the proper PPE.

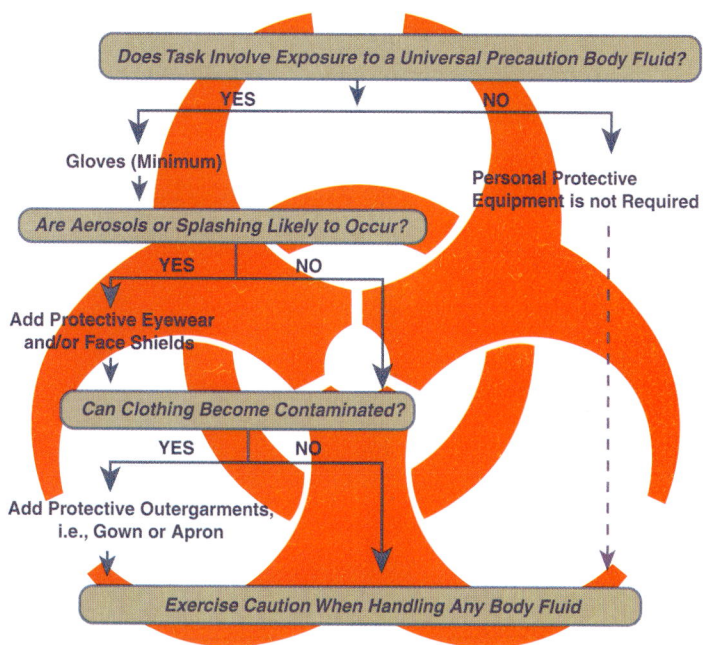

Figure A.1

Suggested Readings

29 C.F.R. 1910.1030, Blood-borne pathogen standard, 56 *Fed. Reg.*, 64179, December 6, 1991.

Bigbee, D., *The Law Enforcement Officer and AIDS*, 5th ed., U.S. Department of Justice, Washington, D.C., 1994.

Centers for Disease Control, 36 *Morb. Mort. Weekly Rep.,* 2S, Atlanta, GA, August 21, 1987.

Centers for Disease Control, Guidelines for Prevention of Transmission of Human Immunodeficiency Virus and Hepatitis B Virus to Health Care and Public Safety Workers, Atlanta, GA, February 1989.

Masters, N.E., *Safety for the Forensic Identification Specialist*, Lightning Powder Co., Salem, OR, 1995.

Forensic Web Sites

General

http://www.forensic.to/forensic.html
Excellent Web site with numerous forensic links.

http://www.crimelynx.com/forensic.html
Excellent for links to forensic megasites, books and journals, crime scene investigation, and forensic disciplines.

http://www.tncrimlaw.com/forensic
Sections on crime scene investigation, criminalistics, trace evidence, DNA analysis, forensic pathology, and more.

http://www.forensicpage.com
Excellent Web site containing a full range of forensic subjects and links to additional sites.

http://www.kruglaw.com/forensic.htm
Web site of law office of Kim Kruglick with numerous links to forensic and law topics.

http://www.forensicportal.co.uk
Guide to Internet forensic resources with primary focus on the United Kingdom.

http://www.trutv.com
Web site of Trutv formerly Court TV. Information on current and past criminal trials and links to other topics of forensic interest.

http://www.law.depaul.edu
Web site of the Center of Law and Science of DePaul University College of Law. Offers comprehensive information on law and forensic science.

http://www.afip.org
Web site of the Armed Forces Institute of Pathology. Provides consultative services in forensic pathology, toxicology, anthropology, and DNA technology related to deaths of military personnel.

http://www.DrugAbuseStatistics.samhsa.gov
Drug Abuse Warning Network provides data on drug-related deaths as reported by medical examiners and coroners throughout the United States.

http://www.forensic-ent.com
Interesting Web site maintained by Dr. Stephen W. Bellington. Provides basic information and techniques in forensic entomology.

http://www.forensic-entomology.com
A good Web site of Dr. J.H. Byrd. Explores the significance of insects in forensic investigations.

http://www.toxlab.co.uk
Web site of the Regional Laboratory for Toxicology in the United Kingdom. Offers a wide variety of toxicological assays.

http://www.criminalprofiling.com
Web site that provides information on offender profiling, behavioral science, crime dynamics, and case studies.

http://www.lawenforcement.com
Good Web site for law enforcement product and technology news.

Education

http://www.ncfs.ucf.edu
The National Center for Forensic Science is a program of the National Institute of Justice. Web site is hosted by the University of Central Florida. Provides information on education, research, training, tools, and technology to meet current and future needs.

http://www.tafns.com
Academy of Forensic Nursing Science

http://www.mdpdtraining.com
Web site of the Miami-Dade Police Department's Metropolitan Police Institute, Miami, Florida. Offers a wide variety of forensic training courses.

http://www.hcleeinstitute.com
Web site of the Henry C. Lee Institute of Forensic Science at the University of New Haven.

http://www.forensic.msu.edu
Web site of Michigan State University's Forensic Science Program. Links to national and international undergraduate and graduate forensic science programs.

http://www.practicalhomicide.com
Web site of Vernon Geberth. Offers seminars on homicide investigation.

http://www.mcri.org
Web site of the McCrone Research Institute. Offers training courses and research in applied microscopy.

http://www.nfstc.org
The National Forensic Science Technology Center provides accreditation to forensic laboratories along with forensic training and educational materials.

http://www.epic-photo.org
Evidence Photographers International Council provides training in forensic photography.

Professional Organizations

http://www.aafs.org
American Academy of Forensic Sciences

http://www.abfp.com
American Academy of Forensic Psychology

http://www.aapl.org/index.html
American Academy of Psychiatry

http://www.criminalistics.com
American Board of Criminalistics

http://www.csuchico.edu/anth/ABFA
American Board of Forensic Anthropology

http://www.abfo.org
American Board of Forensic Odontology

http://www.abfp.com
American Board of Forensic Psychology

http://www.abft.org
American Board of Forensic Toxicology

http://www.slu.edu/organizations/abmdi
American Board of Medicolegal Death
Investigators

http://www.asqde.org
American Board of Questioned Document
Examiners

http://abfde.org
American Board of Forensic Document
Examiners

http://www.acs.org
American Chemical Society

http://www.ascld.org
American Society of Crime Laboratory
Directors

http://www.afte.org
Association of Firearm and Toolmark
Examiners

http://www.physanth.org
Association of Physical Anthropologists

http://www.forensicacademy.org
Australian Academy of Forensic Sciences

http://www.bafm.org
British Association in Forensic Medicine

http://www.cacnews.org
California Association of Criminalists

http://www.csfs.ca
Canadian Society of Forensic Sciences

http://www.cap.org
College of American Pathologists

http://www.forensic-science-society.org.uk
Forensic Sciences Society

http://www.firearson.com
International Association of Arson
Investigators

http://www.iabpa.org
International Association of Bloodstain
Pattern Analysts

http://www.tiaft.org
International Association of Forensic
Toxicologists

http://www.theiai.org
International Association for Identification

http://www.iafn.org
International Association of Forensic Nurses

http://www.iifes.org
International Institute of Forensic Engineering Sciences, Inc.

http://www.mafs.net
Midwestern Association of Forensic Scientists

http://www.nafe.org
National Academy of Forensic Engineers

http://www.criminaljustice.org
National Association of Criminal Defense
Lawyers

http://www.ndaa.org
National District Attorneys' Association

http://www.thename.org
National Association of Medical Examiners

http://www.natari.org
National Association of Traffic Accident
Reconstructionists and Investigators

http://www.njafs.org
New Jersey Association of Forensic Scientists

http://www.neafs.org
Northeastern Association of Forensic
Scientists

http://www.nwafs.org
Northwest Association of Forensic Scientists

http://www.soft-tox.org
Society for Forensic Toxicologists

http://www.southernforensic.org
Southern Association of Forensic Scientists

Government Agencies

http://www.atf.treas.gov
Bureau of Alcohol, Tobacco, and Firearms

http://www.usdoj.gov/dea
Drug Enforcement Administration

http://www.fbi.gov
Federal Bureau of Investigation

http://www.fema.gov
Federal Emergency Management Agency

http://www.ojp.usdoj.gov/nij
National Institute of Justice

http://www.ojp.usdoj.gov
Office of Justice Programs

http://www.ncstl.org
National Clearing House for Science,
Technology and the Law

DNA Testing Laboratories

http://www.bodetech.com
Web site of Bode Technology Group. Offers
DNA testing for general forensic casework,
mass disasters, felon databanking, and forensic research.

http://www.cellmark-labs.com
Web site of Cellmark Diagnostics. Offers DNA
testing for all casework, databasing, and proficiency test needs.

http://www.lifecodes.com
Web site of LifeCodes. Offers DNA testing for
paternity and forensic cases and pretrans-
plant testing.

http://www.mitotyping.com
Web site of Mitotyping Technologies. Offers
mitochrondrial DNA typing.

http://www.hsc.unt.edu/
Web site of the DNA Identity Laboratory at
the University of North Texas Health Science
Center.

http://www.serological.com
Web site of the Serological Research Institute
specializing in the analysis of biological evidence encountered in criminal and civil
matters and providing products for forensic
serological/DNA analysis.

Forensic Equipment and Supplies

http://www.peaveycorp.com
Web site of Lynn Peavey Company. Catalog of
crime scene and fingerprint products.

http://www.sirchie.com
Web site of Sirchie Fingerprint Laboratories,
Inc. Catalog of crime scene and fingerprint
products.

http://forensicsource.com
Web site of Forensic Source, Catalog of crime
scene and fingerprint products.

http://www.forensicbody.com
Web site of Andre Anyon, who constructs body
models for crime scene re-enactment and
bloodstain analysis.

http://www.hematrace.com
Web site of Abacus Diagnostics, Inc. Manufacturer of ABAcard® Hematrace® for the
forensic identification of blood at crime scenes
and SALIgAE® for the forensic identification
of saliva at crime scenes.

http://www.perkinelmer.com
Web site of Perkin Elmer Corporation,
the world's leading provider of infrared,
chromatography (GC, GC-MS, HPLC,
UV/visible spectrophotometers and fluorescence) instrumentation.

http://www.nikonusa.com
Web site of Nikon, Inc. Leading manufacturer and distributor of 35-millimeter and digital cameras and accessories.

http://www.promega.com
Web site of Promega Corporation, a worldwide leader in innovative products for genetic identity analysis in forensics, DNA databasing, and paternity testing.

http://www.varianinc.com
Web site of Varian, Inc., a manufacturer of GC-MS instruments and a full line of sample preparation products.

http://www.fishersci.com
Web site of Fischer Scientific Worldwide, a leading distributor of scientific and laboratory equipment.

http://www.edsci.com
Web site of Edmund Scientific Company. Offers a wide range of scientific products.

http://www.forensicsolutionsinc.com
Web site of Forensic Solutions, Inc. Offers evidence storage and preservation products.

http://www.appliedbiosystems.com
Web site of Applied Biosystems. Offers instrument-based DNA systems, reagents, kits, and software.

Publishers of Forensic Science Texts

http://www.academicpress.com
Academic Press

http://www.ccthomas.com
Charles C Thomas

http://www.taylorandfrancispress.com
Taylor & Francis/CRC Press Publishing Company

http://www.elsevier.com
Elsevier Science Publishing Company

http://www.pearsonhighered.com/
Prentice Hall Publishing Company-Owned by Pearson Higher Education

Editors and Contributing Authors

http://www.bloodevidence.com
Web site of coeditor and contributing author Stuart H. James.

http://www.finalanalysisforensics.com
Web site of coeditor and contributing author Jon J. Nordby.

http://www.rkwrightmd.com
Web site of contributing author Ronald Wright.

http://www.academy-group.com
Web site of contributing authors Michael Napier and Kenneth P. Baker.

http://www.questioneddocuments.com
Web site of contributing author Frank Norwitch.

http://www.forensic.to/forensic.html
Web site of contributing author Zeno Geradts.

http://www.petervallas.com
Web site of contributing author David R. Redsicker.

http://www.iff-murder.com
Web site of contributing author Robert Keppel.

http://www.bodziak.com
Web site of contributing author William J. Bodziak.

http://www.gwu.edu/~wfrowe
Web site of contributing author Walter F. Rowe.

http://www.healthsciences.okstate.edu/ forensic/index.htm
Web site of contributing author R. Tom Glass.

Trigonometric Tables: Sine and Tangent Functions

Degrees	Sine	Tangent	Degrees	Sine	Tangent
0.0	0.0000	0.0000	46.0	0.7193	1.036
1.0	0.0175	0.0175	47.0	0.7314	1.072
2.0	0.0349	0.0349	48.0	0.7431	1.111
3.0	0.0523	0.0524	49.0	0.7547	1.150
4.0	0.0698	0.0699	50.0	0.7660	1.192
5.0	0.0872	0.0875	51.0	0.7771	1.235
6.0	0.1045	0.1051	52.0	0.7880	1.280
7.0	0.1219	0.1228	53.0	0.7986	1.327
8.0	0.1392	0.1405	54.0	0.8090	1.376
9.0	0.1564	0.1584	55.0	0.8192	1.428
10.0	0.1736	0.1763	56.0	0.8290	1.483
11.0	0.1908	0.1944	57.0	0.8387	1.540
12.0	0.2079	0.2126	58.0	0.8480	1.600
13.0	0.2250	0.2309	59.0	0.8572	1.664
14.0	0.2419	0.2493	60.0	0.8660	1.732
15.0	0.2588	0.2679	61.0	0.8746	1.804
16.0	0.2756	0.2867	62.0	0.8829	1.881
17.0	0.2924	0.3057	63.0	0.8910	1.963
18.0	0.3090	0.3249	64.0	0.8988	2.050
19.0	0.3256	0.3443	65.0	0.9063	2.145
20.0	0.3420	0.3640	66.0	0.9135	2.246
21.0	0.3584	0.3839	67.0	0.9205	2.356
22.0	0.3746	0.4040	68.0	0.9272	2.475
23.0	0.3907	0.4245	69.0	0.9336	2.605
24.0	0.4067	0.4452	70.0	0.9397	2.748
25.0	0.4226	0.4663	71.0	0.9455	2.904
26.0	0.4384	0.4877	72.0	0.9511	3.078
27.0	0.4540	0.5095	73.0	0.9563	3.271
28.0	0.4695	0.5317	74.0	0.9613	3.487
29.0	0.4848	0.5543	75.0	0.9659	3.732
30.0	0.5000	0.5774	76.0	0.9703	4.011
31.0	0.5150	0.6009	77.0	0.9744	4.332
32.0	0.5299	0.6249	78.0	0.9781	4.705
33.0	0.5446	0.6494	79.0	0.9816	5.145
34.0	0.5592	0.6745	80.0	0.9848	5.671
35.0	0.5736	0.7002	81.0	0.9877	6.314
36.0	0.5878	0.7265	82.0	0.9903	7.115
37.0	0.6018	0.7536	83.0	0.9925	8.144
38.0	0.6157	0.7813	84.0	0.9945	9.514
39.0	0.6293	0.8098	85.0	0.9962	11.43
40.0	0.6428	0.8391	86.0	0.9976	14.30
41.0	0.6561	0.8693	87.0	0.9986	19.08
42.0	0.6691	0.9004	88.0	0.9994	28.64
43.0	0.6820	0.9325	89.0	0.9998	57.29
44.0	0.6947	0.9657	90.0	1.000	0.000
45.0	0.7071	1.000			

Metric Measurements and Equivalents

Decimal Prefixes and Multiples for Area/Length, Mass Weight, and Volume

Prefix	Symbol	Multiple	Magnitude	Common Name
yotta	Y	10^{24}	1 000 000 000 000 000 000 000 000	heptillion
zetta	Z	10^{21}	1 000 000 000 000 000 000 000	hexillion
exa	E	10^{18}	1 000 000 000 000 000 000	quintillion
peta	P	10^{15}	1 000 000 000 000 000	quadrillion
tera	T	10^{12}	1 000000 000 000	trillion
giga	G	10^{9}	1 000 000 000	billion
mega	M	10^{6}	1 000 000	million
kilo	K	10^{3}	1 000	thousand
hecto	h	10^{2}	1 00	hundred
deca	da	10^{1}	10	ten
unit*	—	—	1	one
deci	d	10^{-1}	0.1	tenth
centi	c	10^{-2}	0.01	hundredth
milli	m	10^{-3}	0.001	thousandth
micro	μ (mu)	10^{-6}	0.000 001	millionth
nano	n	10^{-9}	0.000 000 001	billionth
pico	p	10^{-12}	0.000 000 000 001	trillionth
femto	f	10^{-15}	0.000 000 000 000 001	quadrillionth
atto	a	10^{-18}	0.000 000 000 000 000 001	quintillionth
zepto	z	10^{-21}	0.000 000 000 000 000 000 001	hextillionth
yocto	y	10^{-24}	0.000 000 000 000 000 000 000 001	heptillionth

* unit = meter (m) for area/length, gram (g) for mass/weight, and liter (L) for volume (capacity).

Metric Equivalents

Length Measurements

Some common multiples in powers of 10 include:

kilometer (km) = 10^3 meters

decimeter (dm)= 10^{-1} meters

centimeter (cm) = 10^{-2} meters

millimeter (mm) = 10^{-3} meters

micrometer (μm)* = 10^{-6} meters

nanometer (nm) = 10^{-9} meters

angstrom (Å) = 10^{-10} meters

1 millimeter (mm) = 1000 micrometers (μm)* = 1,000,000 nanometers = 10,000,000 angstroms (Å)

1 angstrom (Å) = 0.1 nanometers (nm) = 0.0001 micrometers (μm) = 0.00000001 centimeters (cm)

10,000 angstroms (Å) = 1000 nanometers (nm) = 1 micrometer (μm) = 0.0001 centimeters (cm)

10,000 micrometers (μm) = 10 millimeters (mm) = 1 centimeter (cm) = 0.01 meters (m)

* Micrometer (μm) may appear as the older designation, micron (μ)

1 millimeter (mm) = 0.1 centimeter (cm)

10 millimeters (mm) = 1 centimeter (cm)

10 centimeters (cm) = 1 decimeter (dm) = 100 millimeters (mm)

10 decimeters (dm) = 1 meter (m) = 1000 millimeters (mm)

10 meters (m) = 1 dekameter (dam)

10 dekameters (dam) = 1 hectometer (hm) = 100 meters (m)

10 hectometers (hm) = 1 kilometer (km) = 1000 meters (m)

Area Measurements

100 square millimeters (mm^2) = 1 square centimeter (cm^2)

10,000 square centimeters (cm2) = 1 square meter (m2) = 1,000,000 square millimeters

100 square meters = 1 are (a)

100 ares = 1 hectare (ha) = 10,000 square meters

100 hectares = 1 square kilometer (km^2) = 1,000,000 square meters

Volume Measurements

Some common multiples in powers of 10 include:

1 liter (L) = 10^3 milliliters (mL)

1 milliliter (mL) = 10^{-3} liters (L)

1 microliter (μL) = 10^{-6} liters (L)

1 microliter (μL) = 0.001 milliliter (mL) = 0.000 001 liter (L)

1000 microliters (μL) = 1 mL = 0.001 liter (L)

1,000,000 microliters (μL) = 1000 milliliters (mL) = 1 liter (L)

10 milliliters (mL) = 1 centiliter (cL) = 0.01 liter (L)

10 centiliters (cL) = 1 deciliter (dL) = 100 milliliters (mL) = 0.1 liter (L)

10 deciliters = 1 liter (L) = 1000 milliliters (mL)

10 liters = 1 dekaliter (daL)

10 dekaliters = 1 hectoliter (hL) = 100 liters (L)

10 hectoliters = 1 kiloliter (kL) = 1000 liters (L)

Cubic Measurements

1000 cubic millimeters (mm^3) = 1 cubic centimeter (cm^3) = 1 milliliter (mL) (liquid)

1000 cubic centimeters (cm^3) = 1 cubic decimeter (dm^3) = 1,000,000 cubic millimeters (mm^3)

1000 cubic decimeters (dm^3) = 1 cubic meter (m^3) = 1 stere = 1,000,000 cubic centimeters (cm^3) = 1,000,000,000 cubic millimeters (mm^3)

Mass/Weight Measurements

Some common multiples in powers of 10 include:

kilogram (kg) = 10^3 grams (g)

centigram (cg) = 10^{-2} grams (g) .

milligram (mg) = 10^{-3} grams (g)

microgram (µg) = 10^{-6} grams (g)

nanogram (ng) = 10^{-9} grams (g)

picogram (pg) = 10^{-12} grams (g)

1 milligram (mg) = 1000 micrograms (µg) = 1,000,000 nanograms (ng) = 1,000,000,000 picograms (pg)

1 picogram (pg) = 0.001 nanograms (ng) = 0.000001 micrograms (µg) = 0.000000001 milligrams (mg)

10 milligrams (mg) = 1 centigram (cg)

10 centigrams (cg) = 1 decigram (dg) = 100 milligrams (mg)

10 decigrams (dg) = 1 gram (g) = 1000 milligrams (mg)

10 grams (g) = 1 dekagram (dag)

10 dekagrams (dag) = 1 hectogram (hg) = 100 grams (g)

10 hectograms (hg) = 1 kilogram (kg) = 1000 grams (g)

1000 kilograms (kg) = 1 metric ton (t)

Useful Metric Measurement Conversions to English System

centimeters × 0.40 = inches

meters × 3.28 = feet

meters × 1.09 = yards

kilometers × 0.62 = miles

grams × 0.035 = ounces

kilograms × 2.20 = pounds

liters × 2.10 = pints

liters × 1.06 = quarts

liters × 0.264 = gallons (U.S.)

milliliters × 0.034 = fluid ounces

Useful English System to Metric Measurement Conversions

inches × 2.54 = centimeters

feet × 0.30 = meters

yards × 0.91 = meters

miles × 1.60 = kilometers

ounces × 28.3 = grams

pounds × 0.45 = kilograms

pints × 0.47 = liters

quarts × 0.95 = liters

gallons (U.S.) × 3.8 = liters

fluid ounces × 29.6 = milliliters (cubic centimeters)

Temperature Conversions
Degrees Celsius (C) = {Degrees Fahrenheit (F) − 32} × 0.555
Degrees Fahrenheit (F) = {Degrees Celsius (C) × 1.8} + 32

Fahrenheit Degrees°	Celsius Degrees°
−40°	−40°
0°	−17°
25°	−4°
32°	0°
50°	10°
75°	24°
100°	38°
125°	52°
150°	66°
175°	79°
200°	93°

AAS Atomic absorption spectroscopy. A technique for metal analysis utilizing the reduction of a metal in solution to an atom, usually by a flame.

ABC American Board of Criminalistics.

ABO Human antigenic system designating the blood groups A, B, AB, and O.

A.C.E.V. Four steps (analysis, comparison, evaluation, and verification) followed to evaluate and identify a latent fingerprint according to Ashbaugh; accepted widely in the latent fingerprint examiner community. Sometimes pronounced *ace vee*.

Acid phosphatase Enzyme group that catalyzes the hydrolysis of certain organic phosphates. Seminal acid phosphatase (SAP) is produced in the prostate gland and serves as a presumptive test for semen.

Accelerant Agent, often an ignitable liquid, that acts to initiate a fire or increase its rate of spread.

Accreditation Endorsement of policies and procedures.

Accuracy Ability of a measurement to match the value of the quantity measured; correctness.

Adenine (A) One of the four nucleotide bases in DNA. The others are cytosine, guanine, and thymine.

Adiabatic flame temperature Theoretically, the highest temperature at which a fuel can burn. It is derived mathematically. Because certain combustion products tend to disassociate at high temperatures, the true maximum burning temperature, even under ideal conditions, is usually slightly lower.

Adjudicated Settled in a court of law.

Adulterant Material used to increase the mass of a controlled substance. Adulterants produce physiological effects and give the illusion that more controlled substance is present than actually is.

AFIS Automated fingerprint identification system.

Agar (agarose) Multiple sugars (polysaccharides) extracted from seaweed used to make the gel that supports DNA fragments during separation.

Agglutination A clumping of red cells by an antibody specific for characteristics on the red cell.

Algor mortis The postmortem cooling of the body.

Alkaloid Substance formed in the plant tissues and in the bodies of animals. Morphine and codeine are alkaloids of opium.

Allele One of two alternate forms of a gene occurring at a locus on homologous chromosomes.

Allele frequency Percentage of total gene copies of one type in a population. For example, a population of 100 people will include 200 alleles. If 80 are of one type, the allele frequency is 80/200 or 40/100 or 40 percent.

Allelic marker Allele form of a gene used to identify chromosomal segments suspected of association with a certain phenotype. For example, allelic markers may be used with a family pedigree in which a phenotype is common to

697

identify chromosomal segments that contain the gene responsible for the phenotype.

Allometry The growth of part of the body in relation to the growth of the whole. The adjective form is *allometric*.

Allozyme Genetically determined allelic form of enzyme.

Alteration Change of a written or printed portion of a document, usually accomplished after obliterating or masking the original information.

Alu repeat Very common (100,000+ copies per genome) short, interspersed nuclear element (SINE) found in humans and other primates.

Alveolus The bony socket in either the maxilla or mandible that holds the tooth.

Amphetamine A controlled substance along with its analogs, such as methamphetamine that creates an excitatory condition (stimulation), state of wakefulness, and euphoria.

Amplicon Product of the amplification of DNA or RNA.

Amylase The proteolytic enzyme found only in salivary and pancreatic secretions.

Anagen stage The growth period of hair.

Angle of acceptance or angle aperture (AA) Maximum angle between light waves that an OBJ can collect.

Angle of impact Acute or internal angle formed by the direction of a blood drop and the plane of the surface it strikes.

Anion A negatively charged ion which will migrate toward the anode in an electric field.

Annealing Pairing of complementary strands of dot-blot DNA analysis (DQ-alpha and amplitype PM).

Anode A positive electrode.

Antemortem Predeath.

Antibody An immunoglobulin molecule with specific receptor sites formed in response to an antigen.

Antigen A substance that can stimulate an immune response when introduced into a host.

Anthropometry Method of identification, devised by Alphonse Bertillon in the late 19th century, consisting of a set of body measurements thought to form a unique profile. The system has been obsolete for a century, but is an important precursor of fingerprint identification.

Aorta Largest artery in the body. It receives blood from the left ventricle of the heart and distributes it to smaller arteries that supply the entire body with blood.

Aperture Opening into a cavity, for example, a nasal aperture or a camera aperture.

Aqueous Made from or containing water.

Area of origin, fire-related General area where a fire started. This term is used when a fire originates in a large area or when the exact point of origin cannot be determined.

Arterial spurting (or gushing) Bloodstain patterns resulting from blood exiting the body under pressure from a breached artery.

Assault rifle Automatic weapon designed to be fired by one man. Ammunition is fed from a magazine.

Association Establishment of a relationship between objects (evidence and other items) through examination.

Associative evidence Evidence that associates individuals or objects to a crime scene.

Atomic absorption spectroscopy Quantitative analysis technique based on the absorption of light by a vaporized and atomized element.

Atomic emission spectroscopy Technique based on the emission of light by excited, vaporized, and atomized elements. Excitation can arise from any of a number of energy sources. The instruments are usually polychromatic devices. The method is most useful for quantitative analysis; qualitative use is also popular.

Atomic mass spectroscopy Technique based on detection of vaporized and atomized elements and their ionized isotopes. The detection and display of the spectra are based on the mass-to-charge ratios of the ions. The method is specific for qualitative analysis and also valuable for quantitative analysis.

Auricular surface Part of the ilium (upper pelvic bone) that forms a joint with the sacrum.

Autoerotic fatality Death occurring during solo sexual activity. Such deaths are accidental and most often involve hanging.

Automatic rifle Light machine gun in which the ammunition is fed from a magazine.

Automatic weapon Firearm that continues to fire as long as its trigger is depressed and ammunition is available.

Autoradiogram X-ray film image of DNA fragments or bands identified by labeled DNA molecules (probes). An autoradiogram is used for visual comparison and computer-assisted sizing of fragments.

Autosomal Refers to all chromosomes except the sex chromosomes.

Autosome Any of the 22 chromosomes (long, independent DNA molecules) not involved in sex determination.

Avulse Expel or remove.

Avulsed Expelled or removed.

Axial illumination Narrow, nonangular illumination surrounding the optical axis of a transmitted light microscope, produced by a low numerical aperture setting of the condenser. It improves contrast and allows more accurate and precise determination of refractive indices by immersion methods and causes a decrease in resolving power.

Azostix™ Commercial test strip for detecting urea in blood by measuring the shift in pH resulting when urea is catalyzed to ammonia and carbon dioxide by urease. Used forensically to detect urine.

Back spatter Blood directed back toward the source of energy or force that caused the spatter. Back spatter is often associated with entrance gunshot wounds.

Ball powder Smokeless powder manufactured by extruding nitrocellulose lacquer into hot water.

Ballistics Branch of physics that deals with the flights of projectiles. Ballistics is divided into four regimes: interior, transitional, exterior, and terminal.

Base line Method for measuring a crime scene and the evidence present; after a fixed line is identified, the items of evidence are measured from that line at right angles.

Base pair Combination of two nucleotides (A and T or G and C) held together by weak hydrogen bonds. The DNA double helix is formed when a base pair nucleotides in the DNA strands are connected by these bonds. The DNA strands are held together by strong chemical bonds. The two halves of the molecule are held by the weaker hydrogen bonds. The double helix may be visualized as two strips of velcro. Weak bonds hold it together and the two strips are difficult to break or rip.

Becké line A method utilized for the determination of the refractive index of materials, such as glass.

Beginning stroke Initial stroke of a letter.

Billiard ball effect Spreading of a shotgun pellet pattern caused by an intermediate target that slows the leading pellets in the shot mass. The trailing pellets overtake them and deflect them on divergent trajectories.

Biological profile A systemized study of the generic characteristics of shape and size of human remains that may allow for an estimation of age, sex, and population ancestry.

BIOS Basic Input-Output System. The name given to the system progam which provides basic input and output functions for the personal computer. BIOS is held in ROM (Read Only Memory) and is nonvolatile memory that cannot be damaged.

Biometrics Science and technology of using individually variable features of the human body for identification.

Bipedal locomotion Pattern of movement characterized by the use of two feet. Humans use bipedal locomotion.

Birefringent Having two or more indices of refraction. When placed between polarizing filters, birefringent materials exhibit bands of color. The specific colors exhibited when white (polychromatic) light is used as the illuminant are determined by the differences in the indices of refraction and the thickness of the birefringent material. The noun form is *birefringence*.

Bite marks   The usually horseshoe-shaped pattern left in inanimate objects or the pattern injury left in the tissue of a victim.

Bitstream Data Copy   Duplicates all data in a cluster, including anything that is in the slack space and unallocated space where digital forensic evidence may be hidden. A copy may not retain digital evidence.

Blood-borne pathogens   Microorganisms originating in human blood, its components, and products derived from it, that can cause disease in humans.

Blood group   A set of chemical characteristics, exhibited in the blood and sometimes other fluids which is genetically controlled and can be identified through analytical methods.

Bloodstain   Transfer resulting when liquid blood comes into contact with a surface or a moist or wet surface comes into contact with dried blood.

Blot   Technique for transferring DNA fragments from a gel to a different support medium such as nylon. It is based on capillary action or electrical or vacuum transfer. Also called Southern blot after its developer, E.M. Southern.

Blowback   An operating principle of automatic and semiautomatic firearms. The fired cartridge blows back against the breechblock, forcing it to the rear, and extracting and ejecting the expended cartridge casing. *Blowback* also describes the blowing back of blood and other tissue onto a firearm or a shooter from a near-contact or contact shot. See *gas operation/recoil operation*.

Blow flies   Large metallic-looking flies belonging to the family Calliphoridea in the order Diptera often attracted to a dead body immediately after death.

Blunt start   Lack of one continuous movement of a writing instrument as it touches paper in the initial writing stroke.

Bolt-face signature   Marking embossed on a cartridge primer (or base of cartridge) by a breechblock or bolt.

Botanical evidence   Various plant structures, such as roots, stems, branches, leaves, fruits, or flowers, that may be used to determine time and season of death as well as possible prior location of remains.

Breechblock   Component of breech-loading firearm that rests against the base of the cartridge.

Brentamine fast blue   Chromogen used with a substrate such as alpha-naphthyl phosphate to detect acid phosphatase.

Brightness   Intensity.

Broach   Rifling tool consisting of a series of circular cutting tools mounted on a long rod. The rifling is cut in one pass of the broach through the gun barrel.

Bubble ring (vacuole)   Ring produced when blood containing air bubbles dries and retains the bubble configuration as a dry outline.

Bullet wipe   Soot, lubricant, or other material wiped from the surface of a bullet onto skin or other surfaces penetrated by the bullet.

Byte   Unit of computer storage capable of holding a single character.

CACHE   A method for increasing performance by keeping frequently used data in more accessible and faster storage.

Cadaver dogs   Dogs trained to recognize the scent or presence of decomposing remains.

Caliber   Diameter of a circle tangent with the tops of the lands of the rifling. Caliber is a nominal measurement; actual bore diameter may be different from the designated caliber.

Callus (callosity)   Overgrowth of woven bone surrounding area of injury.

Canids   Members of the dog family.

Cannabinoids   A term applied to marijuana and parts of the plant *Cannabis sativa* in which tetrahydrocannabinol (THC) is the active agent.

Carbon monoxide (CO) A highly toxic gas that is formed as a product of combustion.

Carcinogen A substance that produces cancer.

Castoff pattern Bloodstain pattern created when blood is released or thrown from a blood-bearing object in motion.

Catagen stage The intermediate or transition phase of hair growth.

Catalyst A substance which enhances the rate of a chemical reaction without actually taking part in the reaction itself. The catalyst is not changed as a result of the reaction.

Cathode A negative electrode.

Cation A positively charged ion which will migrate toward the cathode in an electric field.

Cause determination Developing an explanation of the circumstances and conditions that bring together a fuel, an ignition source, and an oxidizer to produce a fire.

Cause of death Disease or injury that initiates the lethal train of events leading to death, for example, coronary heart disease or a gunshot wound of the heart.

Cementum The outer covering of the root of the tooth.

Center fire cartridge Firearm cartridge in which the primer compound is contained in a centrally positioned primer cap.

Centromere Chromosomal region to which spindle fibers attach during cell division. It is composed of highly repetitive DNA sequences.

Certification Official recognition of professional development.

Chain of custody The documented process the evidence goes through from the point of gathering to the final presentation in court; intended to assure that there has been no tampering or altering the evidence.

Chemical color tests Chemical reactions producing colors when compounds or classes of compounds are brought into contact with various chemical reagents.

Chemical ionization A type of mass spectrometry in which a molecule reacts under relatively low energy with a reagent gas rather than fragmenting extensively.

Chemiluminescence The process by which light is emitted as a product of a chemical reaction.

Choke Constriction in the muzzle of a shotgun intended to concentrate the shot pattern.

Christmas tree stain Chemical that differentially stains sperm and other cells when mounted on microscope slides. Nuclear fast red and picroindigocarmine dyes are applied consecutively to the mount.

Chromatograms The display of separation data carried on with an instrument, usually as a chart indicating elution or retention time verses detector response for the materials separated. Qualitative and quantitative information can be obtained from a chromatogram.

Chromogen A compound which, when oxidized, displays a characteristic color.

Chromosome A long DNA molecule that also contains RNA and protein.

Circumstantial evidence Evidence requiring the trier of fact to infer certain events, for example, linking a defendant to a crime scene (and ultimately to the crime) via DNA, hair, fiber, glass, footprint, fingerprint, or ballistics evidence.

Clandestine drug laboratory Illicit location that manufactures controlled substances.

Class characteristic A feature of an item that is unique to a group of items or information in a nonindividual context about some aspect of a crime scene, e.g., a shoeprint was left by a male who wore a size 12 shoe or a hair came from a Caucasian female. Evidence that belongs to a class and is not considered unique.

Clot Gelatinous mass formed by a complex mechanism involving red blood cells, fibrinogen, platelets, and other clotting factors. Over time, the clot retracts, resulting in a clear separation of the mass from the more fluid, yellowish serum remaining at the periphery of the stain (see *serum stain*).

CMOS Complimentary metal oxide semiconductor; an integrated circuit design known for its low power consumption.

Cocaine A controlled substance derived from the erythroxylin cocoa plant that creates an excitatory condition (stimulation), state of wakefulness, and euphoria.

CODIS Combined DNA index system. A DNA database system.

Codominance A state where both forms of paired alleles at the same locus are expressed.

Coleoptera Major order of insects (beetles) important in forensic entomology and taphonomy.

Commissure Corner of the mouth.

Common origin The establishment by measurements or analyses that significant differences between two items cannot be shown; the conclusion is that they share the same source.

Comparison microscope Two microscopes linked by an optical bridge so that the observer can simultaneously view two independent images in one field each from a separate objective.

Comparison sample An uncontaminated or unmodified example to be used for comparison with a sample that may have been modified, altered, damaged, or contaminated.

Compound binocular microscope A microscope employing two eyepieces that magnify an image formed by a single common objective for high magnification in the range of 25 to 1200×.

Computer crime A criminal act where the computer is the victim, such as the introduction of a virus.

Computer-related crime A criminal act where the computer is involved, such as fraud.

Concentric cracks Those fractures that appear to circle around the point of impact.

Conchoidal lines Edge characteristics of glass fractures. They are stress marks shaped like arches that are perpendicular to one glass surface and curved nearly parallel to the opposite surface. The perpendicular surface faces the side where the crack originated.

Conditional evidence Evidence created by an action or event at a crime scene.

Congenital anomaly An abnormality, such as a spinal column defect, present at birth.

Connecting stroke Joining the ending stroke of one letter to the beginning stroke of another letter.

Contact wound A skin injury produced by a weapon in contact with or a fraction of an inch from the skin when discharged.

Contraband In forensic toxicology and drug testing facilities, this refers to suspected controlled substances.

Contrast Tonal ranges within an image.

Control A test performed in parallel with an experimental procedure and designed to yield predictable results that confirm the reliability of the experimental results.

Controlled substance Any substance, usually a drug, whose possession or use is regulated by law.

Contusion Bruise; leakage of blood from damaged blood vessels into tissues.

Contusion ring Bruising at the edges of a gunshot wound caused by penetration of the skin by a bullet.

Copybook Instruction manual for learning penmanship.

Coroner A court official in medieval England whose duties included investigating sudden and unexpected deaths and deaths from injury; in the United States, an elected official with death investigation duties.

Cortex The main body of hair containing protein fibrils, pigment, cortical fusi, and ovoid bodies.

Cortical bone (cortex) External layer of bone, characteristically dense and having a relatively smooth surface, as contrasted with inner, spongy bone.

Cortical fusi Microscopic air pockets or vacuoles within the cortex of hair.

Creatinine Component of urine that reacts with picric acid to form creatinine picrate, a detectable color product.

Crime scene Location of a crime, such as the room where a rape–homicide occurred or intersection where a truck was highjacked. The crime scene contains physical evidence used by forensic scientists to generate class or individual linking statements that constitute material evidence. Multiple crime scenes are possible, for example, a murder site and a subsequent burial site.

Crime scene assessment The process of analyzing cases to determine the "profile" of an unknown offender.

Crime scene documentation Permanent recording of the crime scene conditions and physical evidence present.

Crime scene management Teamwork approach of investigators and crime scene personnel that successfully resolves a case.

Crime scene reconstruction Analysis and reconstruction of a crime scene that logically links a detailed series of scientific *explanations* to provide an understanding of the sequence of events. Each explanation is developed, linked, and evaluated by applying *scientific method* to available data. This process involves proposing, testing, and evaluating explanatory connections among the physical evidence related to the events. The purpose of the analysis is to find the *best explanation* of related events.

Criminal investigative analysis Use of investigative techniques including indirect personality assessment, equivocal death analysis, investigative suggestion, trial strategy, characteristics and traits (profile) of unidentified.

Criminalistics Application of physical sciences to criminal investigation.

Cut or cut wound Incised wound that penetrates less than the maximal surface dimension.

Cut rifling Rifling created by hook cutting, scrape cutting, or broaching.

Cuticle The outermost layer of hair formed by overlapping scales.

Cyanide (CN) A highly toxic chemical especially in the form of gas (hydrogen cyanide).

Cyanoacrylate Important fuming method for the visualization of latent fingerprints; also called "Super Glue fuming."

Cyber forensics The extraction of evidence that particular digital data passed over some medium between two points in a network.

Cytosine (C) One of the four nucleotide bases in DNA. The others are adenine, guanine, and thymine.

Daubert test A standard for determining the reliability of scientific expert testimony in court currently adopted by many jurisdictions. Five factors are utilized to assess the scientific theory or technique: testing of theory, use of standards and controls, peer review, error rate, and acceptability in the relevant scientific community.

Dedicated dimensional standard The labeled ruler that is used in all analyses and photographs for a given bite mark case.

Deductive reasoning Using nonspecific details to infer a specific fact.

Defendant The suspect or accused in a criminal case or the person who is alleged to have caused the injury to the plaintiff.

Denaturation Loss of the natural configuration of a molecule through heat, chemical treatment, or pH change.

Dental stone Gypsum product, similar to plaster of Paris. Its hardness and durability make it a superior product for casting footwear or tire impressions.

Dentin The inner calcified skeleton of the tooth.

Dentine (dentin) Main substance of a tooth, consisting of a layer of tissue surrounding the pulp. The dentine is surrounded by the enamel of the tooth crown and the cementum of the tooth root.

Dentition The complement of teeth of an individual.

Deoxyribonucleic acid (DNA) Double-stranded helix molecule carrying genetic information; composed of four deoxyribonucleotides: adenine (A), thymine (T), cytosine (C), and guanine (G).

Depth of field (DF) Total distance, height, above and below the point of focus that also appears clearly focused. The DF decreases with increase of NA and ×.

Depressant Drug that reduces excitability and calms a person.

Destructive testing Testing that alters an original document, for example, barely noticeable hypodermic needle holes in an ink line (for TLC) or partial destruction of a small portion for paper testing.

Diaphysis Shaft of a developing long bone prior to union with the ends (epiphyses). The adjective form is *diaphyseal*.

Digital A computer device that processes data in digital form.

Digital evidence Contents of data packets traveling on a network.

Diluent Inert substance used to increase the mass of a controlled substance; exerts no physiological effect; is used to give the illusion that more controlled substance is present than actually is present.

Dinucleotide repeat Most common microsatellite repeat; two nucleotides are repeated in tandem.

Dipteria Order of insects that contains the true flies such as blowflies, houseflies, horseflies, craneflies, mosquitos, and midges.

Direct evidence Information that establishes a fact directly, without the need for further inference, for example, an eyewitness' testimony that the defendant fired the fatal shot.

Directionality Parameter that indicates the direction the blood was traveling when it impacted the target surface. Directionality of flight can usually be established from the geometric shape of the bloodstain.

Directionality angle Angle between the long axis of a bloodstain and a predetermined line on the plane of the target surface that represents zero degrees.

Direction of travel Trajectory or flight directionality of a blood drop that can be established by its angle of impact and directionality.

Disarticulated joint A joint that is no longer held together by soft tissue.

Discovery motion A formal legal request designed to allow the defense access to evidence in the possession of the prosecutor prior to trial.

Discriminant function Statistical method by which one subpopulation is quantitatively separated from another on the basis of specific numerical characteristics.

Disguised writing Alteration of handwriting for the purpose of concealment.

Disorganized criminal Commits crime impulsively with little or no planning. The perpetrator's lack of control over his victim and himself is apparent.

Distal Farthest from the center or point of attachment. In hair morphology, the tip is distal from the root.

Distant wound Firearm wound that lacks stippling, smoke, or soot. It generally indicates a distance of 1 meter or more from the skin to the gun muzzle at the time of discharge.

DMORTs Disaster Mortuary Operational Re-sponse Teams that process mass fatalities for the Federal Emergency Management Agency (FEMA).

DNA Advisory Board (DAB) A national board that provides guidelines and standards for DNA testing laboratories.

DNA band Visual image on an autoradiogram or agar gel showing a restriction fragment length polymorphism allele or DNA fragment.

DNA fingerprinting Series of techniques using the base sequence of an individual's DNA for identification purposes.

DNA fingerprint Multilocus genotype or phenotype that is sufficiently rare to be considered unique.

DNA probe Short segment of DNA marked with radioactive or chemical components and used to detect complementary fragments of DNA.

DNA polymerase Enzyme capable of linking deoxyribonucleotide subunits to make a DNA molecule. The original single-stranded DNA is called a template.

DNA polymorphism Two or more different DNA sequences at a particular gene locus or site in a population. ABO blood groups and hair and eye colors are polymorphisms. DNA polymorphisms and blood groups require special techniques to identify differences. Hair and eye colors are easier to identify.

DNA sequence Order of deoxyribonucleotide subunits (A, T, G, and C) in a DNA molecule, e.g., ATAAGGC.

dNTP Specific nucleotide used in the polymerase chain reaction process.

Document A written item employing marks and symbols to convey a message.

Dominant trait A trait that is expressed even if present only in one of the two alleles. If a gene has only one dominant allele (in which case, the one donated by the other parent is recessive), the physical manifestation of the trait will reflect the dominant allele.

Dot The smallest component of printer output. Resolution is expressed in dots per inch or pixels per inch.

DOT number A number that appears on every tire. It shows where and when the tire was made. The U.S. Department of Transportation has required such numbers since 1972.

Double helix Structure of the DNA molecule consisting of linked nucleotide subunits forming long strands joined by complementary bases.

Dpi (dots per inch) Measure of resolution of printer output. The more dots per inch, the higher the resolution. A 600 dpi resolution has 600 horizontal dots and 600 vertical dots per inch or 360,000 (600 × 600) dots.

Drawback effect Blood in the barrel of a firearm that has been drawn backward into the muzzle.

Drip pattern Bloodstain pattern resulting from blood dripping into blood.

Drop down The material dropped or the spreading fire by dropping burning materials onto unburned combustible materials and igniting them.

Drug Nonfood substance intended to affect the structure or function of the body.

DRUGFIRE A database used for acquiring, storing, and analyzing images of bullets and cartridge casings.

Druggist's fold Primary or inner container used to hold evidence.

Dry origin impression Impression that contains no significant moisture from itself or its substrate when made.

Dual tire assembly Wheel assembly with two tires mounted next to one another on each side of an axle.

Duquenois–Levine test Chemical color test used to confirm the presence of cannabinoids in plant material.

Dynamic loads Loads or forces that change and usually produce motion.

Ejector projection Device in an automatic or semiautomatic firearm that wrests the expended cartridge from the extractor and ejects it from the firearm; in a revolver, it ejects cartridges from the chambers in the cylinder.

Elastic collision Collision between two bodies in which the deformation of each body is directly proportional to the applied force (Hooke's law).

Electrochemical etching Rifling method in which the grooves of the rifling are produced by an electrochemical process.

Electrophoresis A technique for the separation of charged molecules by migration on a support medium under the influence of an electric potential.

Electrostatic charge Charge of negative ions, similar to the static charge emitted by newly dried socks.

Electrostatic lifting   Using a high voltage device to electrostatically transfer a dry-origin dust or residue impression to a black film, thus improving the contrast.

Embossing   A raised design or relief design usually found in custom printing.

Enamel   The outer covering of the crown of the tooth.

Encrypted evidence   Hidden data on a disk that is accessible with the use of a password or key.

Ending stroke   Terminal stroke of a letter.

Enhancement   Rendering an impression more visible through physical, photographic, chemical, or digital methods.

Enhancer sequence   Eukaryotic DNA sequence that increases the level of transcription; may be upstream or downstream from the promoter.

Entomology   The scientific study of insects.

Epiphyseal union   Joining of one bone growth center or epiphysis at either end of a long bone to another, commonly the diaphysis or shaft of a long bone.

Epiphysis   Bone growth centers at the ends of long bones, on the margins of some flat bones, and as projecting areas (processes or tubercles) connected to the main parts of bones by cartilage, and eventually fusing to them. The adjective form is epiphyseal.

Equifinality   In taphonomy, different agents may produce modifications to bone that cannot be differentiated from one another.

Equivocal death   Manner of death (homicide, suicide, accident) remains undetermined after a complete investigation.

ESDA   Electrostatic detection apparatus; instrument that recovers indented writing.

Ethanol   Ethyl alcohol or beverage alcohol found in beer, wine, and liquors.

Eukaryotic   A biological superkingdom containing organisms whose cells have true nuclei. Cell division occurs by mitosis and meiosis. Other kingdoms within this superkingdom are Fungi, Animalia (including humans), and Plantae.

Evidence   Information gathered by the prosecution and defense that satisfies the requirements of state or federal rules of evidence.

Examination quality photographs   Photographs taken by a camera held directly over evidence, such as a shoe or tire impression, that will be useful during a detailed examination of that evidence.

Exemplar   Example or representative item usually in undamaged or less damaged condition to which a damaged item can be compared.

Exon   DNA base sequence that encodes amino acid sequences for protein.

Expert witness   The specialist who is recognized for his/her expertise and is asked by the trier of fact to evaluate the facts in a case and render an opinion.

Expirated or exhaled blood   Blood propelled from the nose, mouth, or a wound as a result of air pressure and/or air flow.

Explosion   Sudden conversion of potential energy (often chemical) to kinetic energy accompanied by physical destruction of the container or structure via a high pressure wave front.

Exsanguination   Death after a significant amount (usually half or more) of blood is lost. Bleeding to death.

Extraction   Separation of the compound of interest from the rest of the sample.

Extractor   Component of a firearm that pulls an expended cartridge from the firing chamber.

Extruded powder   Smokeless powder manufactured by extrusion of nitrocellulose dough through a perforated steel plate. A sharp knife rotating against the plate cuts specified lengths of extruded dough.

Fatigue, material   A material becomes "tired" due to repeated applications of dynamic loads. The material fractures or fails at a strength level significantly less than it would fracture or fail if only static loads were applied.

Fibers   A common class of microscopic evidence. They are classified as animal, vegetable, or mineral, or natural, manufactured, or synthetic.

Filler Material added to paper during manufacture.

Firearm Heat engine that converts the chemical energy of a propellant into kinetic energy of a projectile.

Fire load Amount of material that can burn. The average fire load of a building is usually stated in British thermal units (BTUs) per square foot to enable the comparison of the fire sustaining potential of one building to that of another.

Fire patterns Marks left by fire, smoke, and soot on structures and devices. Several characteristic patterns help identify the relationship and orientation of the fire to the structure: horizontal patterns, plumes, V-shaped patterns, and saddle burns.

First instar First-stage fly larvae that cannot penetrate skin and must subsist on liquid protein.

First responder First person arriving at a crime scene; usually a law enforcement officer or other emergency personnel.

Fish system An image database for handwriting comparison developed in Germany.

Flanking region Region just adjacent to a region of interest, a gene, a repeat, or any other sequence.

Flight path Path of the blood drop as it moves through space from the impact site to the target.

Flintlock A weapon in which ignition of the propellant is accomplished by a piece of flint striking sparks from a piece of steel called the *frizzen*. The sparks fall into a pan containing fine black powder.

Flow pattern Change in the shape and direction of a wet bloodstain due to the influence of gravity or movement.

Fluorescence The property of producing light when acted upon by radiant energy.

Font A complete character set of type in one size and face.

Footwear Apparel worn on the foot, such as tennis shoes, boots, sandals, etc.

Footwear database Computerized compilation of shoe sole designs for the purpose of associating a crime scene impression with a manufacturer or to link crime scene impressions from one scene to others.

Forcing cone Flaring at the breech end of the barrel of a revolver. It serves to guide the bullet into the rifling.

Forensic anthropology Application of anthropology theory and methods, primarily human skeletal biology, taphonomy, and archaeological methods, to solve medicolegal problems.

Forensic archaeology Application of archaeological methods to recover human remains and interpret their spatial associations.

Forensic entomology The application of entomology to legal cases and insects associated with a dead body.

Forensic evidence Information generated by the prosecution or defense that satisfies the requirements of state or federal rules of evidence.

Forensic nursing The application of forensic science combined with the biological and psychological education of the registered nurse in the scientific investigation, evidence collection and preservation, analysis, prevention and treatment of trauma and death-related medical issues.

Forensic odontology or forensic dentistry Forensic dentistry; the application of the arts and sciences of dentistry to the legal system.

Forensic pathology The specialty of medicine and subspecialty of pathology dealing with investigating the causes of sudden and unexpected deaths.

Forensic psychology Application of psychological findings to the law, legal systems, and legal processes.

Forensic psychotherapy Application of pyschological knowledge to the treatment of mentally disordered persons (sometimes as the result of a court order) who commit violent or destructive acts against others.

Forensic taphonomy   Study of postmortem processes and their relationships to environmental contexts.

Forensic toxicology   The examination of all aspects of toxicology (the study of drugs and poisons that may have legal implications).

Forward spatter   Blood that travels in the same direction as the source of energy or force; often associated with exit gunshot wounds.

Foundation   A required showing to the trial court prior to the admissibility of certain evidence, such as bullet matching, that the party offering it has sufficient knowledge to be able to truthfully testify. One example is a ballistics examiner's rendition of reliable credentials.

Fragile X syndrome   A complex inherited syndrome of mental retardation usually seen in males and associated with a tendency for the X chromosome to break in culture at a trinucleotide repeat site.

Fragmented disk   Occurs when there are insufficient contiguous clusters to hold a large file so that it is broken up and spread around the disk taking advantage of available clusters.

Freehand simulation   Attempt to copy or draw a signature without the use of mechanical aids.

Friction ridge skin   Skin on the soles of the feet, palms of the hands, and fingers in humans and some primates that forms ridges and valleys. Friction ridge skin forms classifiable patterns on the end joints of the fingers.

Frye standard   A standard applied in some jurisdictions to the admissibility of scientific theory and method in court based upon the acceptance of the theory and method by the scientific community.

Gait measurement   Measurement of stride.

Gamete   Cell produced in the gonads of a male or female that contains half the DNA and half as many chromosomes as normal body or somatic cells. Sperm cells are produced by the testes and eggs are produced by the ovaries.

Ganser's syndrome   A syndrome often observed in prisoners. Individuals routinely give inaccurate answers to simple questions. Some experts regard this behavior as a form of malingering; others believe that it reflects a distinct syndrome that has existed more than 100 years.

Gas chromatography   Gas flowing through a coated tube separates compounds by their size, weight, and chemical reactivity with the coating of the tube or column.

Gas operation   An operating principle of automatic and semiautomatic firearms. A small amount of propellant gas is vented into a piston that pushes a rod to the rear, opening the breech, and extracting the expended cartridge. See *blowback* and *recoil operation*.

Gauge   Method of designating the diameter of a shotgun barrel. It equals the number of round lead balls of the diameter of the interior of the shotgun barrel required to weigh 1 pound.

GC-MS   Acronym for gas chromatography coupled with mass spectrometry.

Gel   Support medium, made from resin (acrylamide) or polysaccharide (agarose) that holds DNA molecules in place during the separation phase of electrophoresis.

Gene   DNA passed on to descendants. It functions by specifying the compositions of proteins. The gene is an ordered sequence of nucleotide base pairs that has a describable characteristic such as function (hemoglobin), size (15.2 kilobase pairs), or location (gene on the short arm of the seventh chromosome).

Gene frequency   Relative number or percentage of a specific allele, band, or DNA fragment in a population.

Genetic linkage   Quality that causes some traits to be inherited together, indicating that the traits are located close to each other on the same chromosome.

Genome The totality of an individual's DNA or genetic material; it is contained in every cell.

Genomic DNA DNA contained in a complete set of DNA molecules or chromosomes; the amount inherited from one parent in a sperm or an egg.

Genotype The configuration of genes at a specific locus.

Gigabyte 1,073,741,824 bytes.

Glycogen Storage form of starch in animal cells. It is produced by vaginal epithelia and is the basis for a presumptive test for these cells.

Glycoprotein A macromolecule composed of protein and carbohydrate.

Grabbers Mechanical "fingers" in a copy machine or printer that draw the paper through the machine.

Graphology Study of handwriting and how it relates to personality.

Gravimetric quantitation Using the ratio of preextraction and postextraction weights to determine concentration.

Gray ring *See* bullet wipe.

Green tire Unfinished tire that has all its components but has not yet been molded. It has no tread or sidewall design.

Grooves Void areas that run around and across a tire between the design elements. Recessed areas of rifling.

Guanine (G) One of the four nucleotide bases in DNA. The others are adenine, cytosine, and thymine.

Habit Repeated handwriting characteristic.

Hacker One who willfully penetrates or attempts to penetrate a computer system without authorization.

Hacking Unauthorized access or attempts to access computer assets.

Hallucinogen Psychoactive drug that induces hallucinations or alters sensory experiences.

Hammer forging Rifling method in which a barrel blank is hammered down over a mandrel. This method is used to make polygonal rifling.

Handgun Firearm designed to be fired with one hand.

Haploid cell Cell containing only one set of chromosomes, usually gametic cells.

Haploidy The ability of the human egg and sperm to carry only half the amount of DNA found in body cells.

Haplotype Set of closely linked genetic markers on one chromosome that tend to be inherited together.

Hardy–Weinberg equilibrium In a randomly breeding population, the number of individuals exhibiting a recessive trait is half the number of individuals who do not express the trait and carry only one of the recessive alleles. The assumptions used to derive this equilibrium are (1) no mutation or migration, (2) random mating, (3) no selection, and (4) infinite population size. The assumptions do not need to be met to use the Hardy–Weinberg to estimate genotype frequencies from allele frequencies.

Hashish The resin from marijuana that has been isolated from the plant material.

Hematoma A tumor of blood caused by leakage from damaged blood vessels; it contains enough blood to form a blood-filled space.

Heme The nonprotein portion of hemoglobin and a number of proteins in the body. Possess an iron protoporphyrin structure.

Hemoglobin gene Human hemoglobin is a tetramer composed of two alpha chains and two beta chains in adults. Other genes and pseudogenes are transcribed during fetal development. The alpha cluster is on chromosome 16 and the beta cluster is on chromosome 11.

Height ratio Distance from the base of a letter to the top relative to another letter.

Henry system Classification of 10-fingerprint cards so that they could be stored in large files. The system has been rendered largely obsolete by AFIS, but was widely used in the United States and the United Kingdom.

Heteroplasmy Genetic heterogeneity within populations of mitochondria in an individual.

Heterozygosity   Two alleles at one locus.

High risk crime   Crime committed at time or place that posed a great threat of discovery to the offender. This category of crime is normally attributed to the "disorganized" criminal. The use of alcohol or drugs will greatly enhance the risk potential of the offender by lowering inhibitions.

High risk victim   Person who, because of occupation, sexual history, lifestyle, or other circumstances, is highly vulnerable to violent crime.

HITS   Homicide investigation and tracking system. A database at the Washington State Attorney's Office for murder and sexual assault information.

Homology   Biological structures having common or shared ancestries.

Homozygote   Individual whose genetic makeup (genotype) is composed of identical alleles or bands at a particular gene (e.g., AA or aa).

Hook cutter   Rifling tool with a raised cutting edge used to cut one groove of rifling at a time.

Hybridization   DNA molecules are composed of two complementary halves that serve as templates for each other. Hybridization occurs when these halves separate and a half of different origin connects with one of the separated halves to form a hybrid molecule.

Hypervariable   Some segments of DNA molecules are identical or almost identical in all individuals while others show variability. A hypervariable is a DNA segment that is highly variable and differs in most individuals.

Hypervariable region   Locus with many alleles, especially those whose variation is due to variable numbers of tandem repeats.

IAI   International Association for Identification, the main professional organization for latent print examiners.

ICP–MS   Inductively coupled plasma–mass spectroscopy. A modern technique for metal analysis that utilizes radio frequency energy for the detection and quantitation of metals.

Identification evidence   Evidence that provides positive identification of the source.

Ignition source   Location of a flame, arc, spark, or chemical reaction that provides sufficient heat energy in the presence of a fuel and an oxidizer to initiate combustion.

Ilium   Upper blade-like portion of the adult pelvis. The adjective form is *iliac*.

Image resolution   Number of pixels displayed per unit of printed length in an image, usually measured in pixels per inch (ppi).

Immunoassays   Tests utilizing antibodies that react with a drug or substance that recognizes the antibody.

Immunoglobin gene   Immunoglobin is a collective term for all the antibodies produced by an organism. Immunoglobins are produced by complex heavy and light chain genes that undergo rearrangement during B cell development to produce an enormous variety of antibodies.

Impact site   Point on a bloody object or body that receives a blow. *Impact site* is used interchangeably with *point of origin*. An impact site may be an area on the surface of a target that is struck by blood in motion.

Impact spatter   Bloodstain pattern created when blood receives a blow or force resulting in the random dispersion of smaller drops.

Inattentional blinding   Phenomenon of not being able to see something that is within one's direct perceptual field because one is focused on something else.

Incendiary fire   Fire intentionally caused by human activity.

Incised wound   Injury produced by a sharp instrument and characterized by lack of surface abrasion and absence of bridging vessels, nerves, and smooth margins.

INDELS   Single-base insertion or deletion, also called SNP (single nucleotide polymorphism); can be more than one base, e.g., a two-base pair insertion or deletion.

Indented writing Writing impressed into the surface of a page of paper from pressure exerted upon the writing instrument when used on a previous page.

Independent testing The repeating of a scientific test by a defense expert or independent laboratory performed to ensure confidence in the result rendered by the prosecution expert.

Indirect personality assessment Assessment of a known individual believed to be responsible for the commission of a violent crime. This technique is utilized in preparing for cross-examination during trial, during interviews and interrogations, in investigating equivocal deaths, and in other situations.

Individual characteristic Feature that is unique to a specific item. Information in a specific context about some aspect of a crime scene, e.g., a print was left by a shoe consistent in all respects to the defendant's shoe, a hair was consistent in all respects to a sample of hair from a Caucasian female, or handwriting features are attributable to a particular person.

Individualization Establishment of uniqueness of an item through examination and experimentation; showing that no other item is exactly like the one in question. The adjective form is *individualistic.*

Inference Conclusion or deduction of a fact from examination of other case facts, e.g., the DNA found at the crime scene allows us to infer that the defendant was at the crime scene.

Infrared Type of light energy greater than that of visible light but shorter than microwave energy.

Infrared reflectance Tendency of an ink specimen to lighten when exposed to infrared light.

Infrared spectroscopy Use of the absorption of infrared radiation to produce a chemical fingerprint of a substance.

Inkjet printer Type of printer that sprays ionized ink onto a sheet of paper.

Insurance fraud Act intended to deliberately deceive an insurance carrier into paying a claim for a loss or issuing a policy based on false evidence. This may include a claim for the loss of a structure due to an intentionally set fire or a claim for reimbursement for items of greater value than those present in a fire-damaged structure.

Intermediate range gunshot wound Firearm wound that shows stippling but no smoke; generally indicates a distance of a few millimeters to a meter from skin to gun muzzle at the time of discharge.

Intrusion management A functional or operational model for describing the information protection process.

Iodine fuming Nondestructive method of visualizing latent fingerprints based on the interaction of iodine vapors with lipids in the latent residue; usually used to develop fingerprints on items with high intrinsic value.

Ion pairing A method of separation or isolation of a compound based on the affinity of the hydrochloride for chlorinated solvents, such as chloroform, over an acidic environment with a high chloride content, such as HCl.

Isoelectric Focusing (IEF) A process pf separating molecular species where the molecules accumulate or focus at their respective isoelectric points.

Isoelectric Point (IEP) The pH at which a normally charged molecule has no net charge.

Isotope Chemical element that exists in alternate forms containing identical numbers of protons and different numbers of neutrons.

Isozyme Multiple molecular forms of an enzyme occurring in a single species.

Junk DNA The discovery that much of the DNA in every cell was repeated sequence DNA that cannot code for polypeptides and led to speculation that the repeat sequence DNA had no function.

Karyotype In humans, the 46 chromosomes (23 pairs of homologous chromosomes) constituting our complete chromosomal inheritance. They are usually made microscopically visible by staining at the stage of cell division known as metaphase.

Kilobyte   1024 bytes.

λ Wavelength of the illumination.

Laceration   Injury produced by blunt instruments; characterized by surface abrasion, bridging vessels and nerves with irregular margins.

Land   Raised area of rifling.

Laser   Source of high intensity monochromatic (single wavelength) light. Certain methods use lasers to develop latent fingerprints.

Laser printer   Type of printer that uses a laser beam to produce an image on a drum.

Latent print   Fingerprint that cannot be seen under normal ambient lighting. A latent print requires some type of enhancement to clarify ridge details sufficiently to allow comparison and identification.

LC-MS   Liquid chromatography–mass spectrometry is a technique that replaces a gas chromatograph with a liquid chromatograph.

LD50   The quantity of a substance that kills 50 percent of the population.

Lens   An optical component that may be composed of one or multiple elements.

Levels I, II, and III   Ashbaugh's terminology for fingerprints. General patterns, such as whorl, ulnar loop, etc., constitute Level 1. Minutiae in the ridges constitute Level II. Sizes, shapes, and orientations of pores and edges constitute Level III.

Light spectrum   Distribution of light energy viewed as a range of different colors, e.g., a rainbow.

Lightning   Large-scale, high tension natural electric discharge in the atmosphere.

LIMS   Laboratory information management systems. A system of databases for maintaining evidence control.

Line quality   Appearance of a written stroke determined by a combination of factors, such as speed, shading, pen position, and skill; ranges from smooth and legible to tremulous and awkward.

Lingual   Next to the tongue.

Linkage equilibrium   Mathematical expectation that the frequencies of alleles at one gene are independent and can predict the frequencies of a double event. Analogies with gender and race or suits of playing cards may explain this concept. If males represent 50% of a population and Irish males represent 12.5%, one can estimate the percentage of the population consisting of Irish males by multiplying 50% by 12.5%. If the result is not an accurate prediction, the population is in linkage disequilibrium.

Liquid accelerant   Combustible or flammable liquid used to accelerate ignition and spread of a fire.

Living forensics   The identification and collection of evidence derived from living patients.

Livor mortis   The postmortem reddish discoloration of the body due to the settling of red blood cells due to gravity.

Locard's exchange principle   According to Edmond Locard, when two objects contact each other, materials are transferred from one object to another; the basis for proving contact by the analysis of microscopic evidence.

Locus   Chromosomal position where a particular gene is found. The plural is loci.

Low risk victim   Person whose victimology (sexual habits, lifestyle, pastimes, etc.) does not suggest that he or she may become the victim of a violent attack. If an attack occurs, one can hypothesize that the offender specifically targeted the victim, took advantage of his or her vulnerability, or knew the victim.

LPI   Lines per inch.

Lysis, cell   Disruption of a cell membrane that releases the cell contents including the DNA. The verb form is *lyse*.

Machine gun   Bipod- or tripod-mounted automatic weapon whose ammunition is fed from a magazine or a belt.

Machine pistol   Type of submachine gun.

Macroscopic Visible without magnification.

Macroscopic crime scene Description based on the size of a crime scene; the overall or "big picture" crime scene.

Macroscopic examination Visual examination, generally performed with the unaided eye; used to identify class characteristics.

Maggot Fly larva prior to the maturation of the adult fly.

Magnification (x) Amount the subject is enlarged by a lens system, the objective, eyepiece, or other.

Malingering Conscious attempt to feign a physical or mental illness; also called *simulation*.

Mandible Lower jaw.

Mandibular condyle Small, rounded projection of the lower jaw that forms a moveable joint with the cranium.

Manner of death Death occurs in one of four manners: natural, if caused solely by disease; accidental, if it occurs without apparent intent; suicidal, if caused by the deceased; and homicidal, if someone other than the deceased caused it.

Marginal abrasion See contusion ring.

Marijuana The common name for the plant *Cannabis sativa*.

Mass spectrometry Technique based of the detection of vaporized molecules and their ionized (charged) fragments. The detection and display of the spectra are based on the mass-to-charge ratios of the ions. The method is specific for qualitative analysis and useful for quantitative analysis.

Mastoid process Cone-shaped projection of the temporal bone located behind and below the opening for the ear.

Maternal lineage Genetic component passed through the female lines of a family, e.g., mitochondrial DNA.

Maxilla Upper jaw.

Mechanism of death Biochemical and/or physiologic abnormality produced by the cause of death which is incompatible with life, e.g., ventricular fibrillation or exsanguination.

Medical examiner Government official, always a physician and often a forensic pathologist, charged with investigating sudden and unexpected deaths or deaths from injuries.

Medulla The lengthwise central canal of a hair shaft.

Megabyte 1,048,576 bytes.

Megahertz (MHz) One million cycles per second.

Megapixel One million pixels.

Melanin The pigment that imparts color to hair and skin.

Mens rea The Latin term for "guilty mind", mens rea refers to a person's awareness of the fact that his or her conduct is criminal; it is the mental element of the crime.

Metameric Two or more materials that appear the same color under one type of illumination and different under another. Spectral analyses can differentiate metameric pairs.

Metaphysis Wide areas at both ends of an immature long bone shaft or diaphysis; contain growth zones and are attached and eventually united to epiphyseal discs. The adjective form is *metaphyseal*.

Microanalysis Application of a microscope and microscopical techniques to the observation, collection, and analysis of microevidence.

Microcrystal tests A reaction between the compound of interest and chemical reagent that results in the formation of unique crystals that can be observed with the microscope.

Micrometry A device utilizing a scale calibrated with stage micrometer for measurement of the physical dimensions of material viewed with a microscope.

Microorganisms Germs, including bacteria, yeasts, and viruses.

Microsatellite Short tandem repeat or simple sequence length polymorphism composed of di-, tri-, tetra-, or penta-nucleotide repeats.

Microscopic crime scene Crime scene description based on the type of physical evidence present.

Microscopic examination Visual examination utilizing some type of magnification; used to identify individual characteristics.

Microspectrophotometry Instruments that generate transmission, reflection, or absorption spectra from various translucent and opaque samples. The principal types are visible and infrared.

Minisatellite Simple sequence tandem repeat polymorphism in which the core repeat unit is usually 10 to 50 nucleotides long; variable number of tandem repeats.

Minutiae Ending ridges, bifurcations, and dots in the ridge patterns of fingerprints; the quality and quantity of these features serve as the bases of comparison and latent print identification.

Misting Blood reduced to a fine spray as the result of the application of energy or force.

Modus operandi (MO) Method of operation of a criminal. The principle that a criminal is likely to use the same technique repeatedly and that any analysis and record of the technique used in every serious crime will provide a means of identification in a particular crime.

Monitor resolution Number of pixels or dots displayed per unit of length on the monitor, usually measured in dots per inch (dpi). Monitor resolution depends on the size of the monitor and its pixel setting. Most new monitors have resolution of about 96 dpi.

Morphology Scientific study of the forms and functions of living organisms; shape and size of an organism in relationship to its function. The adjective form is *morphological*.

MSDS Descriptive information about a particular chemical, providing information about its physical and chemical characteristics and how it should be handled and stored.

mtDNA Mitochondrial DNA; the DNA found in mitochondria; a circular duplex with a genetic code differing from the universal genetic code.

Multilevel approach to crime scene security Assigning various levels of restrictions to areas within and around a crime scene.

Multiplexers Surveillance cameras recorded on a videotape often combined with an alarm system.

Mummification Drying of soft tissues due to exposure to hot or cold, dry environments unfavorable to bacterial growth.

Mutation rate Number of mutation events per gene per unit of time (for example, per cell generation); the proportion of mutations per cell division in bacteria and single-celled organisms or the proportion of mutations per gamete in higher organisms.

Myotomy The cutting of muscle; in the forensic odontology chapter, used to describe a technique of cutting certain facial muscles to release postmortem rigor mortis.

Myotonic dystrophy Common form of muscular dystrophy affecting adults. Its gene has a trinucleotide repeat.

N Lowest Refractive Index (RI) Material between the subject and the objective.

NAA Neutron activation analysis. A technique for metal analysis utilizing the characteristics of emitted radiation for the detection and quantitation of metals.

Nanotechnology Technology involving devices of only a few nanometers in size.

Narcoanalysis Interview conducted while a subject is under the influence of an intravenously injected substance such as sodium amytal.

Narcotic Addictive substance that reduces pain, alters mood and behavior, and usually induces sleep or stupor.

Natural cause Event, such as lightning, flood, tornado, earthquake, etc., that is not under the control of humans.

Negligence The failure to treat the patient at the highest level of care and competency.

Nessler's reagent Solution of mercuric iodide in potassium iodide; used to detect ammonia when urea is catalyzed to ammonia and carbon dioxide by urease; used forensically to detect urine.

Neutral population genetic theory Hypothesis that most genetic variation in natural populations is not maintained by selection because most alleles have equal fitness.

NIBIN National Integrated Ballistic Information Network. A database used for acquiring, storing, and analyzing images of bullets and cartridge casings.

NIDA The National Institute on Drug Abuse.

Ninhydrin Common name for triketohydrindene, a chemical that reacts with amino acids to form a recognizable bluish-purple compound called Ruhemann's purple; widely used to visualize latent fingerprints, often requires posttreatment.

NIST National Institute of Standards and Technology. Federal agency responsible for setting, approving, and maintaining measurements and materials standards in the United States (formerly National Bureau of Standards).

Noise treatment Arrangement of design elements of various sizes around the circumference of a tire to reduce noise.

Noncoding region Segment of DNA that does not have a nucleotide sequence that can be transcribed because stop codons are present or the sequence does not make sense.

Noninvasive analysis Analysis, testing, or examination that does not produce permanent and irreversible alteration of the evidence.

Normal hand forgery Writing another person's signature without attempting to simulate or disguise; written in one's own handwriting.

Nonrequest standard Normal writing, done without attention to the writing process.

Nuclear fast red Biological stain used to differentially stain spermatozoa to aid in their identification. It stains their nuclear material a dark red.

Nucleotide Molecule consisting of a base, a pentose sugar, and a phosphoric acid group.

Nucleus Cellular organelle surrounded by a nuclear envelope; it encloses the chromosomes during the interphase.

Numerical aperture (NA) NA = N sin AA/2.

Oblique lighting Lighting cast across a page of writing at an angle almost parallel with the page.

Obliteration Erasure or destruction of original information on a document.

Occipital condyles Pair of small, slightly rounded projections of the occipital bone, at the base of the cranium near the spinal cord opening, that form a joint with the first cervical vertebra.

Occlusal surface Surface of a tooth, which during chewing, comes in contact with teeth from the opposing jaw.

Odontology Dentistry; the study of the biology and repair of the teeth. The adjective form is *odontological*.

Opacity Imperviousness to the passage of light.

Opiates A term for the class of narcotic drugs derived from the opium plant, including morphine and codeine. Heroin is produced from morphine.

Organic compounds Class of chemical compounds with carbon bases; all hydrocarbons are organic compounds.

Organized offender Exhibits a great deal of thought and planning. The offender maintains control over himself and the victim. Little or no material of evidentiary value is present. Organized crime is carried out in a sophisticated and methodical manner.

Origin determination Observing a fire scene, collecting and analyzing evidence, and conducting interviews with witnesses to determine where the fire began.

Original equipment (OE) tire Original tire installed on a new vehicle.

OSHA Occupational Safety and Health Act that defines universal precautions for all types of work environments.

Ossification   Formation of bone from cartilage or other fibrous tissue.

Ossification centers   The areas of bone where development and growth occur, gradually replacing cartilage.

Osteoarthritis   Deterioration in joint integrity connected with use-wear, exacerbated by inflammation and related to reduction in bone density.

Osteology   Study of skeletal biology. The adjective form is *osteological.*

Osteometry   Scientific measurement of a skeleton. The adjective form is *osteometric.*

Overkill   Administering more trauma than necessary to end a life; overkill indicates personalized anger and suggests the offender knew the victim.

Ovoid bodies   Microscopic structures occasionally observed in the cortex of hair.

p-Dimethylaminocinnamaldehyde (DMAC)   Indicator chemical used to detect ammonia when urea is catalyzed by urease; used forensically to detect urine.

Palmer method   Style of writing taught in schools.

Paper opacity   See opacity.

Parent drop   Drop of blood from which a wave castoff or satellite spatter originates.

Patching   Addition of a written stroke to improve a defect in a written line.

Patent   A term used by latent print examiners to indicate *visible*; the opposite of *latent.*

Pathology   Medical specialty dealing with the diagnosis of disease by examining tissues and fluids.

PCR   Polymerase chain reaction; copying complementary strands of target DNA in a series of cycles or rounds to produce a large number of copies of the original strands.

PCR primer   DNA replication requires a primer to add nucleotides to a growing chain. *In vitro* replication in which a specific set of oligonucleotides is used to flank a target gene using oligonucleotides known as PCR primers.

PD   Acronym for physical developer.

Pedigree   Graphic method for summarizing data on the inheritance of particular phenotypes. Generally, squares represent males, circles represent females, parents are joined by horizontal lines, and parents are joined to offspring by vertical lines. Offspring are attached to each other by horizontal lines.

Pen lift   Break in a written line.

Penmanship   Style of writing.

Pen pressure   Amount of force applied to a pen or pencil while writing.

Perimortem   At or around the time of death.

Periodic acid Schiff (PAS) test   Presumptive test for the presence of vaginal material; glycogenated cells are stained bright magenta.

Periodontal ligament   The specialized connective tissue ligament that holds the tooth in the alveolus.

Permanent dentition   The 32 adult teeth.

Personality test   Any of a number of psychological tests that evaluate personality, psychopathology, and mental functioning.

Peyote   The common name for the small Mexican cactus, *Lophophora williamsii,* that contains the hallucinogen mescaline.

pH   The negative logarithm of the hydrogen ion concentration in a solution. The degree of acidity present.

Phadebas™ reagent   Commercial chemical consisting of a dye cross-linked to an insoluble starch. Upon digestion of the starch by amylase, the dye is released into solution. The intensity of color relates to the level of amylase present.

Phencyclidine (PCP)   A drug originally developed as a surgical anesthetic that was discontinued due to adverse patient reactions. Abusers of this drug often experience severe psychiatric manifestations.

Phenol   Toxic chemical used in the extraction of DNA from a cell.

Phenotype   The expressed characteristic of the genotype. Can be tested for and identified.

Physical anthropology Study of the biology, variation, and evolution of the human species.

Picking Adherence of a drug to the face of the punch used to produce a tablet. Picking creates holes in the surfaces of pressed tablets, usually near letters such as A or R.

Picroindigocarmine (PIC) Dye used to differentially stain spermatozoa for ease of identification. It stains the tail and midpiece green and the anterior head light pink.

Pistol A handgun; sometimes used to describe semiautomatic handguns.

Pixel Small continuous tone spots that comprise a digital image. *Pixel* is a combination of *picture* and *element.*

PPI Pixels per inch.

Plaintiff The injured party in a civil legal action.

Plasma The fluid portion of blood.

Plastic Indentation fingerprint impressed into a soft receiving surface; a plastic print has distinct 3D character.

Plastic collision Collision between two bodies in which a significant portion of the deformation is permanent.

Plastic deformation Nonelastic change in shape; warping.

Platters Multiple rotating disks on a hard drive that are usually read and written on both sides.

Point of origin Precise point where an ignition source, fuel, and oxidizer unite to create a fire.

Point or area of convergence A point or area to which a bloodstain pattern can be projected on a 2D surface; determined by tracing the long axes of well-defined bloodstains within the pattern back to a common point or area.

Point or area of origin A 3D point or area from which the blood that produced a bloodstain originated; determined by projecting angles of impact of well-defined bloodstains back to an axis constructed through the point or area of convergence.

Polar coordinates Method of crime scene measurement; measurement of items of evidence based upon their distances and angles from a fixed position.

Polarized light microscope (PLM) A microscope equipped with two polarizing elements positioned in the optical path of the microscope.

Polymerase Enzyme that serves as a catalyst in the formation of DNA and RNA.

Polymorphism The occurrence in a population of two or more genetically determined alternative phenotypes with frequencies greater than could be accounted for by mutation or drift.

Postconviction hearing Hearing granted upon a motion raising evidentiary issues. Such hearings often occur a significant time after conviction and are typically unsuccessful.

Postoffense behavior Behavior of a suspect within hours, days, and weeks after a crime. Such behavior distinguishes the offender from the rest of the suspect population.

Postmortem After death.

Postmortem drug testing Examinations performed on blood, urine, and/or body tissues to determine if drugs were a contributing factor in a death.

PPE Personal protective equipment; items such as gloves, face or eye shields, face masks, protective suits, and shoe covers that help minimize exposure to hazardous products.

Precipitin reaction An antigen–antibody reaction in which the result is a precipitate, usually visible.

Precision Ability to achieve the same result; reproducibility.

Preoffense behavior Behavior of an offender just before committing a crime. Often, a precipatory stressor is the catalyst for the commission of a violent crime.

Presumptive test A chemical test which, by production of color or light, indicates the presence of a body fluid of forensic interest (blood, semen, etc.).

Primary crime scene Description of a crime scene based on the location of the original criminal activity; the original scene.

Primary dentition The 20 baby teeth.

Primer Small piece of single-stranded DNA used for replication.

Proficiency test Simulated forensic case.

Profile Description of the results of an investigative analysis of an unsolved crime of violence; may cover victimology, crime reconstruction, significant facts of the autopsy, characteristics and traits of the offender, postoffense behavior, and investigative suggestions.

Projected blood pattern Pattern created when blood is projected or released as the result of force.

Projective test Psychological test based on the notion that if an individual is shown an ambiguous stimulus and asked to respond, his responses will reveal aspects of his personality, including inner thoughts, wishes, conflicts, and feelings.

Prognosis The predicted outcome of a patient's condition.

Prostate-specific antigen (PSA, p30) A 30-kD protein originating in the prostate gland; used forensically to confirm the presence of azoospermic semen following a positive presumptive test result.

Proteinase Enzyme that degrades polypeptides by facilitating the breakage of peptide bonds between amino acids.

Proteinase K Enzyme commonly used to degrade proteins.

Proximal Nearest the center or point of attachment. In hair morphology, the root is the proximal end.

Pseudogene Inactive DNA sequence usually derived from an adjacent active sequence. The inactive sequence is activated by accumulation.

Pseudomalingering A phenomenon whereby a mentally ill individual feigns the mental illness he actually has. The behavior is considered a temporary ego-supportive device that allows the individual to feel he has control over his illness.

Psilocin Controlled hallucinogenic substance contained in the psilocybe mushroom.

Psilocybin Controlled hallucinogenic substance contained in the psilocybe mushroom.

Psychological testing Quantitative or quasiquantitative evaluation of personality, psychopathology, and mental functioning.

Pubic symphysis Immovable joint formed by fibrous tissue and cartilage where the pubic bones (left and right pelvis) meet in the middle front of the abdomen.

Pulp The neurovascular tissue in the center of a tooth.

Pupa (pl. pupae) The stage between the larvae and the adult in insects having a complete metamorphosis.

Pupal cases The covering from which adult flies emerge and leave the body.

Purine A nitrogen base; the purine bases in DNA and RNA are adenine and guanine.

Putative Suspected or alleged, as in parentage.

Pyridine Common chemical used to dissolve ink components for thin layer chromatography.

Pyrimidine Nitrogenous base of which thymine is found in DNA; uridine in RNA; and cytosine in both RNA and DNA.

Quad compressors Surveillance cameras recorded on a videotape in which the image is divided into four parts for four cameras.

Qualitative analysis The determination of the identity of a substance.

Quality assurance Guarantee of value.

Quantitative analysis The determination of the amount of a particular substance present in a material substance.

Questioned document Document whose authenticity or origin is suspect.

Racemic mixture Combination of the different types of stereoisomers of the same compound.

Radial cracks Those fractures that originate from the impact point and propagate away.

RAM Random Access Memory; the main volatile memory of the personal computer.

Ramus of the mandible The vertical portion of the lower jaw that communicates with the skull.

Recoil operation An operating principle of automatic and semiautomatic firearms. When the weapon is fired, the barrel and breechblock initially recoil together. After traveling a short distance, the barrel and breechblock unlock and the breechblock continues to travel to the rear, extracting and ejecting the expended cartridge. See *blowback* and *gas operation.*

Reconstruction evidence Evidence of the events leading to, occurring during, and occurring after a crime is committed.

Refractive index (RI) Velocity of light in a vacuum divided by the velocity of light in the medium of interest $c_{vac}/c_{med.}$

Relative retention time (RRT) Ratio of the retention time of the substance of interest divided by the retention time of an internal standard run on the same gas or liquid chromatographic system at the same time.

Repetitive DNA DNA consisting of copies of the same or nearly the same nucleotide sequence; DNA sequences that are present in many copies per chromosome set. Repetitive DNA sequences may be closely linked as in satellite DNA or VNTR loci or dispersed throughout the genome or parts of the genome like the *alu* family of repetitive elements.

Request standards Handwritten standards issued in the presence of an investigator or examiner.

Resolution Measurement in units per inch of the amount of detail in an image file: dpi = dots per inch; ppi = pixels per inch; lpi = lines per inch.

Resolving power (RP) Ability to distinguish fine differences in structure. $RP = 0.6 \lambda/NA$.

Restriction enzyme Endonuclease that will recognize a specific target sequence of nucleotides in DNA and break the DNA chain at that point.

Retention time Time required for a substance to travel from the injection port to the detector in a gas or liquid chromatographic system.

Retouching Going back over a written line to correct a defect or improve its appearance; synonymous with patching.

Retread tire Tire carcass to which new tread rubber is added to produce a reusable tire.

Revolver Handgun that holds cartridges in a rotating cylinder.

RFLP Restriction fragment length polymorphism. Variation in banding patterns of DNA fragments generated by restriction digests. Polymorphic minisatellite repeats flanked by restriction enzyme recognition sites.

Ribosomal RNA gene Gene cluster that codes for the structural RNAs found in the ribosome. About 2000 copies of the 5S RNA gene are clustered in tandem on chromosome 1; tandem arrays of the genes for 28S, 5.8S, and 18S rRNA genes are found in clusters of roughly 50 repeats on chromosomes 13, 14, 15, 21, and 22 in humans.

Richter scale Arbitrary logarithmic scale used to measure and compare ground motion caused by seismic activity.

Ricochet Alteration of bullet trajectory by collision with a liquid or solid surface.

Ricochet or secondary splash Deflection of large volumes of blood after impact with a target surface that stains a second surface; does not occur when small drops of blood strike a surface.

Rifle Any rifled firearm including 16-inch naval guns; a firearm having a rifled barrel and designed to be fired from the shoulder.

Rifling System of spiral lands and grooves cut into the interior of a gun barrel; imparts rotation to fired bullets, improving their accuracy.

Rigor mortis Stiffening of the body after death due to the chemical breakdown of actin–myosin and the depletion of glycogen from muscles. A time-dependent change that helps determine time of death.

Rimfire cartridge Firearm cartridge in which the primer compound is placed within the rolled rim of the casing. The firing pin strikes the rim of the cartridge.

Rosin Sizing material made from turpentine gum. It is added to paper during manufacture. See *sizing*.

Rule 702 A federal rule of evidence that permits an expert who is qualified by experience, training, or education to offer an expert opinion if the scientific, technical, or other specialized knowledge will assist the trier of fact to understand the evidence or to determine a fact in issue. This rule was amended in 2000.

Sacroiliac joint Joint formed by the sacrum and both side of the pelvis (iliac bones).

Sacrum Portion of the lower part of the spine below the lumbar vertebrae and above the coccyx that forms a joint with the left and right pelvis. The adjective form is *sacral*.

Saliva Oral secretion comprised of water, mucus, proteins, salts, and enzymes. Its primary functions are to moisten the mouth, lubricate chewed food, and aid digestion.

Sample An entire submitted exhibit or sub-sample of the exhibit.

Saponification The conversion of fatty tissues of the body to a soapy, waxy substance called adipocere or grave wax.

Satellite DNA Highly repetitive eukaryotic DNA mainly located around centromeres and found in other places in the genome. The buoyant density of satellite DNA is usually different from that of the other DNA of a cell. The repetitive DNA forms a satellite or off-the-bell-curve fraction in a density gradient because of the base compositions of the repetitive regions.

Satellite spatter Small droplets of blood projected around a drop of blood upon impact with a surface. A wave castoff is considered a form of satellite spatter.

Scallop pattern Bloodstain produced by a single drop characterized by a wave-like, scalloped edge.

Scanning electron microscope (SEM) A microscope that permits viewing of samples at much greater magnification and resolution than is possible by light microscopes. Magnification is possible in the range of 10 to 100,000×.

Scavenger Organism that feeds on dead organisms.

Schlieren optics Imaging system in which the transparent or translucent object to be examined is placed between two spherical mirrors. The illuminant is a point light source placed at the focal point of one of the mirrors. Parallel light rays from the mirror pass through the object to the second mirror, which projects the image onto a screen. A knife edge is placed at the focal point of the second mirror to block unrefracted light rays. Only light rays refracted by the object reach the screen. Schlieren optics can produce images of thickness, density, and refractive index differences.

Sciatic notch Indentation in the lower part of the hipbone (ischium), through which the sciatic nerve extends.

Scrape cutter Rifling tool having one or two raised cutting edges that cut the grooves of rifling.

Script system An image database for handwriting comparison developed in The Netherlands.

Second instar Second-stage fly larvae resulting from molting of first-star larvae that can penetrate skin by using proteolytic enzymes and rasping action of their mouthparts.

Secondary crime scene Crime scene location after the original or primary crime scene.

Sectors Groupings of data on a single disk track.

Selenium Nonmetallic element sensitive to light; used as a coating on drums of photocopying machines.

Semen Complex mixture of organic and inorganic substances produced in the postpubertal male genital tract.

Semiautomatic weapon Firearm that fires and reloads itself before firing another shot; a self-loading weapon.

Sequence of strokes The order in which ink lines that cross each other were written.

Sequential switchers Surveillance cameras recorded on a videotape in a sequential method.

Seriation Analytical technique whereby a group of specimens is placed in a graduated order or series according to a trait or group of traits.

Serum The fluid portion of the blood minus the clotting factors and formed elements. Contains antibodies and other serum proteins.

Serum stain Clear, yellowish stain with a shiny surface often appearing around a bloodstain after the blood retracts due to clotting. The separation is affected by temperature, humidity, substrate, and/or air movement.

Sexual psychopath laws Legislation intended to regulate the evaluation, treatment, and legal disposition of convicted sex offenders.

Sexual ritualism A series of acts committed by an offender that are unnecessary to the accomplishment of the crime. The offender repeats the acts over a series of crimes. The intent is to increase psychosexual gratification. Ritualism should not be confused with *modus operandi*.

Shading Contrast between the written upstroke and downstroke of a line.

Shadow graph Image of a transparent or translucent object produced on a screen or photographic paper when the object is illuminated by light from a point source. Minute variations in thickness are accentuated.

Sharpey's fibers Specialized fibers that connect both the tooth and the alveolar bone to the periodontal ligament.

Shored exit wound A bullet exit wound that has many characteristics of a distant entrance wound. It is caused by supporting or shoring the skin as the bullet exits.

Short tandem repeat

Shotgun Smoothbore firearm designed to be fired from the shoulder.

Sign of elongation Microscopic characteristic of anisotropic material. It is positive (+) when the vibration direction of light along the length of the particle has a higher refractive index.

Signature A killer's psychological calling card left at each crime scene across a spectrum of several murders. Characteristics that distinguish one murder from all others.

SINE Acronym for short interspersed nuclear element.

Singleton A single (simple) nucleotide polymorphism appearing in less than 5 percent of the population.

Sipe Small groove in a tire design element intended to provide better traction. Sipes vary in depth and are useful for documenting tire wear.

Size marker Variable number of tandem repeats in FLP or STR form.

Sizing Material added to paper to change its smoothness, finish, absorbency, and appearance.

Skeletonized bloodstain Bloodstain consisting only of an outer periphery after the central area is removed by wiping when the liquid was partially dried; can also be produced by flaking away of the center of a completely dried stain.

Skid formula The formula used by accident investigators to determine how fast a car was traveling before the driver applied his breaks and initiated skidding. It involves the weight and mass of the vehicle, acceleration of gravity, frictional coefficient between the tires and pavement, and the distance skidded.

Skid mark Mark on the surface of a fired bullet made when the edges at the beginning of the rifling scrape the bullet surface or when the nose of the bullet slides on the surface of the forcing cone of a revolver barrel.

Skill level Proficiency of handwriting; line quality is a good indication of skill level.

Slack space The hidden space on a disk where DOS attempts to write a file to clear its RAM memory.

Slant Angle of writing with respect to a baseline.

SLICE Spectra library for identification and classification engine. A database for the archiving of x-ray spectra.

Slippage Mark on the surface of a fired bullet made when the bullet slides along the tops of the lands of the rifling. Slippage marks appear when the rifling is worn or when a subcaliber bullet is fired.

Smear Large volume of blood, usually 0.5 milliliters or more, that is distorted so that further classification is impossible. A smear is similar to a smudge, but is produced by a larger volume of blood.

Smokeless powder Propellant composed of nitrocellulose (single-base powders) or nitrocellulose plus nitroglycerin (double-base powders). Smokeless powders contain additives that increase shelf life and enhance performance. They are made in a variety of shapes (rods, perforated rods, spheres, disks, perforated disks, and flakes).

Smudge Bloodstain so distorted that no further classification is possible.

SNP Acronym for single nucleotide polymorphism (pronounced *SNIP*). A one-nucleotide change or difference from one individual to another.

Sole or outsole Part of a shoe that touches the ground.

Somatic Referring to the body or its structure.

Souvenir Personal item belonging to the victim of a violent crime which is taken by the offender—for example, jewelry, clothing, a photograph, or driver's license. The item serves as a reminder of a pleasurable encounter and may be used for masturbatory fantasies. The offender who takes a souvenir is usually an inadequate person who is likely to keep it for a long time or give it to a significant other. Also **Trophy**.

Spatter Dispersion of small blood droplets due to the forceful projection of blood.

Specialized forensic assessment techniques Specialized psychological tests and structured interviews focusing directly on a specific legal standard, question, or issue.

Spermatozoan Male reproductive cell contained in semen.

Spines Pointed edge characteristics that radiate away from the center of a bloodstain. Formation depends on impact velocity and surface texture.

Splash Bloodstain pattern (0.1 milliliter or more) created by a low velocity impact on a surface.

Spoliation of evidence Intentional or negligent destruction or alteration of evidence.

Stab wound Incised wound that penetrates farther than the maximal surface dimension.

Stabilizer Additive to smokeless powder that reacts with acidic breakdown products of nitrocellulose and nitroglycerin. Diphenylamine and ethyl centralite are common stabilizers.

Stage micrometer A microscope slide with a scale usually divided into 10-micrometer or 0.001-inch units. It is used to calibrate the eyepiece scale of a microscope used for measuring.

Staged scene A crime scene in which someone (usually the offender) arranges the scene or commits certain acts to have the scene convey a motivation different from the original motive or mislead investigators.

Standard of care The medical or psychological treatment guideline; can be general or specific. It specifies appropriate treatment based on scientific evidence and collaboration between medical and/or psychological professionals involved in the treatment of a given condition.

Starch-gel electrophoresis A method that uses purified starch-gel as a support medium to hold proteins while they are separated in an electric field.

Static load Load or force that does not change; it creates no net motion. Static loads balance each other.

Stellate defect Star-like tearing of soft tissue seen in contact wounds of the head or sternum.

Stereo binocular microscope Two similar but separate optical microscopes for observation by both eyes simultaneously for low to medium magnification in the range of 4 to 40×.

Stereoisomers Compounds with identical structural formulas; they differ in the way their molecules are arranged.

Sternum Breastbone. The adjective form is *sternal*.

Sticking Adherence of a drug formulation to the walls of a die used to produce tablets. Sticking causes unreproducible striations on the edges of the tablets.

Stimulant Drug that produces a temporary increase of functional activity or efficiency.

Stippling Disposition of fragments of powder into the skin as the result of a gunshot wound of relatively close range; also called powder tattooing.

Striae Noninked grooves left behind in an ink line by imperfections in the ball or ball housing of a pen.

Striation Fine scratch used in the microscopic comparison of bullets and tool marks. Striations are made by minute imperfections inside a gun barrel or on the surface of a tool.

Stringency Variable condition, such as temperature or salt concentration, used in renaturation assays.

Structural isomers Compounds that contain the same numbers and types of atoms but differ in the order of arrangement of the atoms. The types include chain, positional, and functional isomers.

Submachine gun Automatic weapon that fires pistol cartridges.

Subpubic angle Angle formed where the left and right pubic bones meet in the middle front of the body.

Subrogation Substituting one creditor for another as when an insurance company seeks to recover the costs paid to an insured by a manufacturer or service installer if the manufacturer's product can be shown to have inherent design flaws or the service provider engaged in negligent behavior.

Super Glue™ 2-Methyl and ethyl esters of cyanoacrylate; manufactured as an adhesive, and incidentally found to be useful in latent fingerprint development. See *cyanoacrylate fuming*.

Surface tension The force that pulls the surface molecules toward the interior of a liquid that decreases the surface area and causes a liquid droplet resist penetration.

Swaging Rifling method in which a rifling button is forced down a drilled-out barrel blank. The button simultaneously expands the barrel to its final diameter and embosses the lands and grooves on the interior.

Swap files Files used by Windows® for various functions, including temporary memory.

SWG Scientific working group; a representative group of practitioners in a forensic science specialty assembled to formulate and periodically review consensus standards for the specialty, including training, education, quality assurance, and interpretation of results.

SWGFAST Scientific Working Group on Friction Ridge Analysis, Study, and Technology; formed in 1995 for fingerprint and other friction ridge skin comparison and identification.

Swipe Transfer of blood onto a surface not already contaminated with blood. One edge is usually feathered and may indicate the direction of travel.

Tandem array Duplicated, triplicated, etc. arrangement in which the duplicated DNA segment is adjacent to the original and in the same order.

Taphonomy From the Greek taphos (burial or grave) and nomos (laws), literally translated as the laws of burial. In forensics, refers to the postmortem fate of human remains.

Taq polymerase Polymerase that can function at very high temperatures, typically above 95 degrees Celsius; synthesized from the thermophilic bacterium, *Thermus aquaticus*.

Target Surface upon which blood is deposited.

Telogen stage The dormant or resting phase of hair growth. Hair in the telogen phase is shed naturally.

Terminal velocity Maximum speed to which a free-falling drop of blood can accelerate in air; about 25.1 feet per second.

Test impression Impression made by using a known shoe or tire as a standard; used in the examination of shoe and tire impressions.

Tetranucleotide repeat Repeat of four nucleotides; member of a common class of repeats called microsatellites; generally referred to as short tandem repeats.

Thin layer chromatography The use of a solvent that travels through a porous medium to separate compounds by their chemical reactivity with the solvent.

Third instar Third-stage fly larvae that frequently aggregate in large masses and are voracious feeders.

Threat assessment Analysis of written or verbal communications containing direct or implied threats to harm or injure individuals, industries, institutions, or government agencies. The communication is analyzed for content and stylistic characteristics. Analysis may include a profile of the unknown perpetrator and evaluation of the unknown subject's potential to carry out the threat.

Three-dimensional impression Impression that has length, width, and depth.

Thymine (T) One of the four nucleotide bases in DNA. The others are adenine, guanine, and cytosine.

Tire impression When a tire contacts a surface, it results in the transfer of the class characteristics of design and size and possibly of wear and individual characteristics.

Tire track The path a tire makes when it crosses a surface.

Tire tread Part of a tire that contacts the road surface and contains a design.

TLC Wet chemical test known as thin layer chromatography.

Toxicogenomics Field of science that deals with how genomes respond to toxins.

Trabecular bone Spongy bone tissue forming a lace-like matrix of connective tissue strands; found beneath cortical bones.

Trace analysis Qualitative or quantitative analysis of the minor or ultraminor components of a sample.

Trace evidence Historically, a term used to describe any evidence small in size, such as hairs, fibers, and soil samples, that would be analyzed utilizing microscopical techniques.

Traced forgery Fraudulent signature produced by following the outline of a genuine signature.

Track width (vehicle stance) Measurement from the center of one wheel to the center of the opposite wheel on the same axle.

Trailer Pattern of poured or solid material, for example, a linear pour of a liquid accelerant or distribution of twisted sheets of paper that will lead a fire from one location to another.

Transcription Construction of an RNA molecule from a DNA template.

Transfer evidence Evidence exchanged or transferred from one location or source to another.

Transfer pattern Contact bloodstain created when a wet, bloody surface contacts a second surface as the result of compression or lateral movement. A recognizable mirror image or a recognizable portion of the original surface may be transferred to the second.

Transient evidence Temporary or easily lost evidence at a crime scene.

Transition Nucleotide-pair substitution involving the replacement of a purine with another purine, or a pyrimidine with another pyrimidine, for example, GC with AT.

Transposon Short DNA sequence, usually shorter than 15 kb; it has the ability to jump around the genome and insert itself.

Transversion Nucleotide-pair substitution involving the replacement of a purine with a pyrimidine, or vice versa, for example, GC with TA.

Trash mark Mark left on a finished copy during photocopying; results from imperfections or dirt on the cover glass, cover sheet, drum, or camera lens of a photocopy machine.

Tread Design Guide Annual publication that shows thousands of tire designs.

Tread wear indicator (wear bar) Raised rubber bar 1/16 inch above the bases of the tire grooves; it must appear at least six times on a tire.

Tremor Wavy back-and-forth movement on a written line.

Triangulation Method of measurement of a crime scene and physical evidence; every item of evidence is measured from two fixed points.

Trigger pull Force required to pull the trigger of a firearm and cause it to discharge.

Trinucleotide repeat Three nucleotides repeated in tandem. The repeats are sometimes associated with disease loci and fall within the class of microsatellite repeats.

Trophy Personal item belonging to the victim of a violent crime which is taken by the offender—for example, jewelry, clothing, a photograph, or driver's license. The item serves as a reminder of a pleasurable encounter and may be used for masturbatory fantasies. The offender who takes a souvenir is usually an inadequate person who is likely to keep it for a long time or give it to a significant other. Also **Souvenir.**

Turning diameter Diameter of the circle a vehicle makes when its steering wheel is fully turned; the tightest turn a vehicle can make.

TWG Technical working group. A precursor of the SWG. It served the same purpose in a forensic specialty area. Some TWGs still operate.

TWGDAM Technical Working Group on DNA Analysis Methods that publishes guidelines for DNA quality assurance and proficiency testing.

Two-dimensional impression Impression that has length and width, but no significant depth.

Typewriting system Typewriting device consisting of a machine, ribbon, and font.

Ultraviolet Area of the light spectrum just past visible violet and before the x-ray region.

Ultraviolet spectroscopy Use of the absorption of ultraviolet radiation to classify a substance.

Unallocated space The hidden space on a disk taken up by the "real" file when erasing it.

Undetermined cause Destruction accompanying some fires may be too extensive to allow determination of the origin and/or cause with a high degree of confidence.

Urea Nitrogenous compound formed from the catalysis of amino acids; found in high levels in urine.

Urobilinogen Intermediate product in the metabolism of bilirubin. When combined with zinc acetate, it forms a compound that fluoresces in ultraviolet light; used for the identification of feces.

V-shaped pattern Fire pattern seen when a flame impinges on a vertical surface.

Variant Organism that is recognizably different from a so-called standard type in the species.

Ventricular fibrillation Uncoordinated nonpropulsive quivering of the heart often produced by myocardial infarction or heart attack; also produced by low voltage electrocution.

Vestibule The circular space formed by the meeting of the cheeks and the jaw.

Victimology Victim's history (personality characteristics, strengths and weaknesses, occupation, hobbies, lifestyle, and sexual history) that impacts the analysis of a crime; a behavioral study of the victim of a violent crime (usually homicide). The analyst examines reputation, lifestyle, habits, associates, and pastimes to form an opinion about the individual's risk of becoming the victim of a violent crime.

Video spectral comparison Comparison and differentiation of inks by analyzing the infrared reflecting and luminescing qualities inherent to the ink; most often accomplished using a device made by Foster & Freeman, Ltd.

Vitreous humor Ocular fluid (fluid within the eye) that is often utilized as a sample for testing in postmortem toxicology.

Void or shadow Absence of bloodstain in an otherwise continuous bloodstain pattern. The geometry of the void may suggest an outline of the object that intercepted the blood, for example, furniture, a shoe, or a person.

VNTR locus Chromosomal locus at which a particular repetitive sequence is present in different numbers in different individuals of a population; a simple sequence tandem repeat polymorphism in which the core repeat unit is usually 250 bases long.

Wad Cardboard, fiber, or plastic disk found in shotshells; may be placed between the powder and the shot or over the shot.

Walk-through Preliminary crime scene survey performed to orient the crime scene investigator to the scene and the physical evidence at the scene.

Watermark Translucent design impressed into paper during manufacture. The design becomes visible when the paper is subjected to transmitted light and helps date a document.

Wave castoff Small blood droplet originating from a parent drop of blood caused by wave-like action of the liquid when it strikes a surface at an angle smaller than 90 degrees.

Wear Effect of frictional forces on a tire or shoe; wear eventually changes the design.

Wet origin impression Footwear impression containing significant moisture from the shoe sole or substrate.

Wheelbase Measurement between the centers of the hubs of the front wheels to the centers of the hubs of the rear wheels; it is very difficult to measure from the tracks made by a vehicle.

Wheellock Weapon in which ignition of the propellant is accomplished by a rotating wheel striking sparks from a piece of iron pyrite.

Wild type Genotype or phenotype found in nature or in standard laboratory stock; can also be a phenotype first observed in the wild.

Wipe Bloodstain pattern created when an object moves through an existing bloodstain, removes blood from the existing stain, and alters its appearance.

Working distance (WD) Distance between the subject and the closest portion of the objective when focused. The WD decreases as the NA and RP increase.

Workplace drug testing Examinations performed on primarily blood and urine from employees or job applicants for drug content.

X-ray diffraction The targeting of a beam of monochromatic x-rays on a sample so that the radiation is scattered. This technique indicates how the atoms or molecules are arranged in a given crystal.

X-ray fluorescence Technique based on the emission of characteristic x-ray radiation when a sample is exposed to exciting radiation from more energetic x-rays. The spectra are displayed as intensity versus energy or wavelength.

Yaw angle Angle between the axis of a bullet and its trajectory

Yawing Deviation of a bullet from the longitudinal axis of its flight.

Zaner–Bloser Style of writing taught in schools.

Contributors

James E. Starrs (foreword) is a professor of law at George Washington University Law School in Washington, D.C. He also holds a joint appointment as a professor in the Department of Forensic Sciences of the Columbian School of Arts and Sciences at the same university. He is a Distinguished Fellow of the American Academy of Forensic Sciences and is a past chairman of the Jurisprudence Section of the American Academy of Forensic Sciences, having served two 3-year terms on the Academy's Board of Directors. In February 1996, he received the Distinguished Fellow medallion from the American Academy of Forensic Sciences. He served as a member of the editorial board of the *Journal of Forensic Sciences* for 15 years and is presently the coauthor of *Scientific Evidence in Civil and Criminal Cases*, 4th ed., Foundation Press, 1995, and the author of some 100 articles in law and science. Professor Starrs has been a senior coeditor of *Scientific Sleuthing Review* for the past 25 years, and was on the advisory board for the three-volume *Encyclopedia of Forensic Sciences*. He has lectured widely, both nationally and internationally, to legal and scientific audiences on various subjects in the forensic sciences. He is probably most well known for having directed a number of exhumations of historical figures in controversial historical matters, such as the assassination of Senator Huey P. Long, the death and cannibalism of the five victims of Alfred Packer, the CIA LSD-related death of Frank R. Olson, and the identification of the remains of Jesse W. James. He has investigated the death of famed American explorer, Meriwether Lewis, and the mysteries underlying the death of FBI Director J. Edgar Hoover. He was, in 1999, the director of excavations in Charles Town, West Virginia, which sought to locate and identify the remains of Samuel Washington, George Washington's brother. In 2000, he organized a reinvestigation into the Boston Strangler killings. Among his other accomplishments, Professor Starrs crafted a computerized simulation of the killings of Nicole Brown Simpson and Ron Goldman, as well as the killings of the Menendez parents in Southern California, both of which were nationally televised.

Gail S. Anderson, MPM, Ph.D., is an associate professor and associate director in the School of Criminology at Simon Fraser University. She is also an adjunct professor at St. Mary's University and the University of Manitoba. She is a board certified forensic entomologist, a fellow of the American Academy of Forensic Sciences, a member of the Entomological Society of Canada, International Association of Identification, Canadian Identification Society, Canadian Society of Forensic Sciences, Entomological Society of America, and was president of The Entomological Society of B.C. in 1994 and 2003. She is a forensic consultant to the RCMP and city police across Canada and a regular instructor to law enforcement. Dr. Anderson has been analyzing forensic entomology cases since 1988, and has testified as an expert witness in court many times. With the collaboration of colleagues and graduate students, she is presently developing a database of forensic insects across Canada so that forensic entomology can be used with confidence across the country. She has published numerous papers in academic journals and book chapters on forensic entomology. Dr. Anderson was a recipient of Canada's Top 40 under 40 Award in 1999, a YWCA Women of Distinction Award for Science and Technology in 1999, and the Simon Fraser University Alumni Association Outstanding Alumni Award for Academic Achievement in 1995. She was recently listed in *Time* magazine as one of the top 5 innovators in the world, this century, in the field of Criminal Justice. Dr. Anderson was presented with the Derome Award in 2001, the most prestigious award bestowed by the Canadian Society of Forensic Science for outstanding contributions to the field of forensic science.

Kenneth P. Baker, Ed.S., is currently president and CEO of The Academy Group, Inc., a forensic behavioral science firm in Manassas, Virginia. He retired as a supervisor from the U.S. Secret Service in 1990, where he served in protection and criminal investigation, was Chief of Behavioral Research and Training in the Intelligence Division, and coordinated behavioral research related to assassination and threat assessments of potentially violent subjects. He served in the Behavioral Science Unit (BSU) at the FBI Academy, and graduated from the FBI Police Fellowship Program where he studied Criminal Investigative Analysis and Criminal Profiling. He served as an instructor at both the FBI and U.S. Secret Service Academies, and now lectures regularly at colleges and universities. Baker has 35 years of experience studying violent crime and is an expert in threat analysis, assassination, violence prediction, violent crime analysis, homicide, mass murder, serial murder, rape, equivocal death, extortion, indirect personality assessment, and premises security.

William J. Bodziak, M.S., holds a B.A. in biology from East Carolina University and an M.S. in forensic science from George Washington University. He was appointed a special agent of the Federal Bureau of Investigation (FBI) in 1970 after which he served in an investigative capacity in Connecticut, Maryland, and Florida. From 1973 until 1997, he was assigned to the FBI Laboratory in Washington, D.C. where he conducted forensic examinations of questioned documents and footwear and tire tread impression evidence. Mr. Bodziak is a fellow and past chairman of the Questioned Document Section of the American Academy of Forensic Sciences, a member of the American Society of Questioned Document Examiners and a certified diplomate of the American Board of Forensic Document Examiners. In addition, he is a member of the International Association for Identification, where he is a certified footwear examiner and has served as the chairman of the Footwear and Tire Track Section. He has testified as an expert witness in both domestic and foreign courts, and has lectured and instructed on topics related to examinations of footwear and tire tread impression evidence to members of the FBI as well as other international law enforcement agencies. Bodziak is the author of the well-known text, *Footwear Impression Evidence* (CRC Press), now in its second edition, and has also authored articles in professional journals in the areas of questioned document examination and footwear and tire tread impression evidence. Mr. Bodziak currently operates Bodziak Forensics in Jacksonville, Florida.

Donnell R. Christian, Jr., is the forensic science development coordinator for the U.S. Department of Justice's International Criminal Investigative Training Assistance Program (ICITAP) in Washington, D.C. With ICITAP, Christian has established forensic science programs in the developing democracies of Bosnia, Haiti, Kazakstan, Kyrgyzstan, Georgia, and Armenia. He specialized in forensic chemistry, with emphasis in the clandestine manufacture of controlled substances (i.e., drugs and explosives) with the Arizona Department of Public Safety Crime Laboratory and served 3 years as president and chairman of the board of directors for the Southwestern Association of Forensic Scientists (SWAFS). Christian has published articles on the analysis and the clandestine manufacture of controlled substances and also has acted as a private forensic consultant lecturing on a variety of forensic science issues.

Catherine Dougherty, M.A., RN, is currently a practicing emergency department nurse in the Baylor health care system in Texas. Her nearly 30-year nursing career has been spent primarily in the Trauma/Emergency Department as well as an 8-year stint in tissue banking. Ms. Dougherty is a founding member of the International Association of Forensic Nurses, a SANE nurse trained under Dr. Ann Burgess at the University of Pennsylvania, and a fellow in the American Academy of Forensic Sciences. She is the author of a chapter on forensic nursing in the major reference book, *The Encyclopedia of Forensic Sciences*, and an online instructor for the University of California at Riverside in forensic nursing. She was a contributor to the IAFN's Standards of Practice and the editor of the IAFN's newsletter. Ms. Dougherty is on the Editorial Board of *Critical Care Nursing Quarterly* and *Forensic Nurse*. She acts as a legal nurse consultant on cases involving nursing care in the emergency department.

Janet Barber Duval, M.S.N., RN (colonel, retired, USAF Nurse Corps), received her undergraduate and graduate degrees from the University of Cincinnati College of Nursing and Health and Indiana University, respectively. She has four decades of nursing experience in trauma and critical care. Duval is the principal author of several major nursing textbooks and journal articles, and has been the Editor of *Critical Care Nursing Quarterly* for 28 years. She is a distinguished fellow, International Association of Forensic Nurses; fellow, American Academy of Forensic Sciences; and holds memberships in several other professional organizations. In addition to serving as clinical consultant for Hill-Rom Company, she is adjunct associate professor, Indiana University School of Nursing, and is the online course coordinator and instructor for the Forensic Nursing Certificate Program, University of California, Riverside. Formerly, she has served as a forensic nursing course developer and educator for the University of Texas School of Nursing (Austin) and the University of New Mexico.

John Joseph Fenton, Ph.D., received his doctorate in biochemistry from the University of Minnesota. He is currently a clinical chemist and toxicologist at Crozer–Keystone Health Systems in Media, Pennsylvania, as well as a professor of chemistry and toxicology at West Chester University, where he also acts as the coordinator of the graduate program in Clinical Chemistry. Dr. Fenton is a diplomate of the American Board of Clinical Chemistry, a member of the National Academy of Clinical Biochemistry, the Society of Forensic Toxicologists, and the International Academy of Clinical Toxicology and Therapeutic Drug Monitoring. He is qualified as an expert witness in the area of toxicology for the Delaware County Court and is the author of or contributor to many scientific publications.

Robert E. Gaensslen, Ph.D., received his doctorate in biochemistry from Cornell University. He has been professor and director of forensic science at the University of New Haven, visiting fellow at the National Institute of Law Enforcement and Criminal Justice, and associate professor at the John Jay College of Criminal Justice, City University of New York. Dr. Gaensslen has participated in the publication of numerous books and has published over 60 papers in the refereed scientific literature. He was editor of the *Journal of Forensic Sciences* from 1992 to 2000 and also has given over 120 presentations at scientific meetings and to general audiences. Dr. Gaensslen is a fellow of the Criminalistics Section of the American Academy of Forensic Sciences and has received the Paul L. Kirk Distinguished Criminalist and Mary Cowan Service awards from the section; he was made a Distinguished Fellow by the Academy in 2000. He is a life member of the Northeastern Association of Forensic Scientists. Dr. Gaensslen is currently professor, director of graduate studies, and head of the Forensic Science Program at the University of Illinois at Chicago.

Zeno Geradts, Ph.D., is known worldwide for maintaining the exhaustive and highly regarded Zeno's Forensic Web site. He received his Ph.D. in pattern recognition in forensic image databases from the University of Utrecht in The Netherlands, holding positions over the course of his career in industry, government, and academia, with his focus always in image processing. While working for the Forensic Science Laboratory of The Netherlands, Dr. Geradts has worked as project manager for digital storage projects for shoeprints (REBEZO) and toolmarks (TRAX). He is currently the research and development coordinator for the department of forensic computer science, at the University of Utrecht, The Netherlands.

R. Tom Glass, D.D.S., Ph.D., received his undergraduate and dental training at Emory University (1965) and his Ph.D. in pathology (1972), as well as pathology/oral pathology residency training (certificate, 1971) at the University of Chicago. He is a board certified oral and maxillofacial pathologist by the American Board of Oral and Maxillofacial Pathology. Dr. Glass is presently chairman, Department of Forensic Sciences; director, Forensic Sciences Graduate Program, and professor of Forensic Sciences, Pathology, and Dental Medicine at Oklahoma State University Center for Health Sciences in Tulsa. While he has been involved in over 3000 cases in his career, he is probably best known for directing the dental team that identified the 167 bodies in the resolution of the Murrah Federal Building bombing in Oklahoma City on April 19, 1995. He has published over 100 papers and abstracts and has authored four textbook chapters on various aspects of forensic science. He has received multiple awards for his teaching and work in forensic dentistry. He continues to practice clinical oral and maxillofacial pathology and forensic dentistry.

Andrew Greenfield graduated with a B.S. degree from the University of Wales in 1989 and went on to complete a M.Sc. in forensic science at the University of Strathclyde. In 1993 he joined the Forensic Science Service in Birmingham, England, before taking a position at the Centre of Forensic Sciences in Toronto, Canada, in 1996. He is currently an assistant section head with the biology section and has been qualified as an expert in the areas of body fluid identification and DNA analysis. Greenfield regularly lectures in these subjects to students, peace officers, and members of the legal community.

William D. Haglund, Ph.D., is a forensic anthropologist residing in Shoreline, Washington. He received his B.S. degree in biological science from the University of California, Irvine, and his Ph.D. in physical anthropology from the University of Washington, Seattle. He served as chief medical investigator of the King County Medical Examiner's Office in Seattle for 14 years. In December 1995, he became the United Nations' senior forensic advisor for the International Criminal Tribunals for Rwanda and the former Yugoslavia. He is presently director of the International Forensic Program for Physicians for Human Rights (PHR), a Boston-based, nongovernmental organization. His duties with the United Nations and PHR have included investigation of human rights abuses, mass graves and potential genocide, as well as identification of victims. He also conducts extensive international training in forensics. Dr. Haglund has organized and directed forensic missions in numerous countries, including Guatemala, Honduras, Rwanda, Somaliland, Georgia/Abkhazia, the former Yugoslavia, Cyprus, Sri Lanka, Indonesia, and East Timor. He has been an affiliate member of the board of directors of the National Association of Medical Examiners, is a past three-times president of the Washington State Coroner/Medical Examiner's Association, and is a fellow of the Physical Anthropology Section of the American Academy of Forensic Sciences.

Susan Herrero has been involved in homicide and capital cases as a lawyer and as an investigator in trial and postconviction cases across the United States. Since 1989, she has worked with defense counsel throughout the country in hundreds of DNA cases, assisting with challenges to DNA evidence. She is a graduate of City University of New York and Seattle University Law School. She maintains offices in Seattle and in New Orleans at the Center for Equal Justice—a nonprofit, postconviction law firm representing accused persons on death row. Herrero has presented numerous lectures at seminars and conferences on forensic DNA.

Thomas A. Johnson is cofounder and chairman of the board of directors of the California Sciences Institute, and also serves as a member of the board of directors of the SANS Technology Institute. Dr. Johnson is one of the founding partners of the Forensic Data Center, a company focused on computer forensics. He received his bachelor's and master's degrees from Michigan State University and his doctorate from the University of California–Berkeley. Dr. Johnson founded the Center for Cybercrime and Forensic Computer Investigation and the Forensic Computer Investigation Graduate program, and was responsible for developing the online program in information protection and security and the Graduate National Security program offered at two National Nuclear Security Administration laboratories in California and New Mexico. Currently, he serves as a member of the FBI Infraguard program, and also is a member of the Electronic Crime Task Force, New York Field Office, U.S. Secret Service. The U.S. attorney general appointed Dr. Johnson a member of the Information Technology Working Group, and he served as chair of the Task Force Group on Combating High Technology Crime for the National Institute of Justice. Dr. Johnson was also appointed an advisor to the Judicial Council of California on the Court Technology Task Force by the California Supreme Court.

Patrick Jones is currently the Forensic Lab director, forensic science coordinator, and a graduate faculty member at Purdue University. He teaches graduate courses in forensic lasers and alternate light sources, forensic digital imaging, and undergraduate crime scene investigation, forensic digital imaging, evidence collection techniques, and scene documentation. He is doing research with forensic lasers and alternate light sources. He is a deputy coroner and Certified Medicolegal Death Investigator with the White County Indiana Coroner's Office. He is a retired Cook County (Illinois) sheriff police investigator with 21 years of service. His service assignments included 15 years as a crime scene investigator (CSI). He has a bachelor of science degree from Pacific Western University and a master of science degree from LaSalle University. He is a graduate of a National Law Enforcement Academy (DEA), a member IAI, the International Association for Identification, and a member of the board of directors of the International Association for Identification, Indiana Division. He is also a member of the American Academy for Forensic Science, the International Association of Homicide Investigators, the Indiana Association of Homicide and Violent Crimes Investigator Association, the Indiana Physics Association, the Indiana Microscopy Society, and the Indiana Coroner's Association.

Robert D. Keppel, Ph.D., is currently an associate professor for criminal justice at Sam Houston State University in Huntsville, Texas. He is especially known for his development of the Homicide Investigation and Tracking System (HITS) database. During his career, Dr. Keppel has served as chief criminal investigator for the attorney general of the State of Washington and homicide investigator for King County Sheriff's Department, in which capacity he was the lead investigator in the Ted Bundy serial murder case. He has investigated or consulted on more than 2000 homicide cases throughout the U.S. He is the author of The Riverman: Ted Bundy and I Hunt the Green River Killer and Signature Killers, as well as professional papers on the identification of critical factors in solving murder cases, and on the analysis of motives and signatures in serial murders. Dr. Keppel received his Ph.D. in criminal justice from the University of Washington where he is currently on the faculty.

Terrence F. Kiely, J.D., LL.M., graduated in 1964 with a B.S. degree in humanities from Loyola University. He went on to receive his J.D. in 1967 from DePaul University College of Law and an LL.M in foreign and comparative law from New York University School of Law in 1970. Kiely has been a full-time member of the DePaul University College of Law faculty since 1972, teaching in the areas of torts, products liability, criminal law, evidence, and forensic evidence. He is the author of five books: Preparing Products Liability Cases (John Wiley & Sons, 1987), Using Litigation Databases (John Wiley & Sons, 1989), Modern Tort Litigation (John Wiley & Sons, 1990), Forensic Evidence: Science and the Criminal Law (CRC Press, 2001), and Science and Litigation: Products Liability in Theory and Practice (CRC Press, 2002). Kiely is currently a professor of law and the director of the DePaul University College of Law's Center for Law and Science in Chicago, Illinois.

Paul E. Kish is a consulting bloodstain pattern analyst in Corning, New York. He holds a B.S. degree in criminal justice and an M.S. degree in education from Elmira College (Elmira, New York). He has been consulted on cases throughout the United States as well as in Australia, Canada, Denmark, New Zealand, and Sweden and given expert testimony in many of these jurisdictions. Kish has taught numerous basic and advanced bloodstain interpretation in schools in the United States as well as in Canada, England, The Netherlands, and Sweden. He has lectured extensively throughout the United States at forensic and law-related conferences and seminars, and authored various articles on the topic of bloodstain patterns including a contributing chapter to the text, Scientific and Legal Applications of Bloodstain Pattern Interpretation (CRC Press, 1998). He is a coauthor of the text Principles of Bloodstain Pattern Analysis—Theory and Practice (CRC Press, 2005). He is a fellow member of the American Academy of Forensic Sciences and a member of the International Association of Bloodstain Pattern Analysts and the FBI's Scientific Working Group on Bloodstain Pattern Analysis (SWGSTAIN).

Thomas A. Kubic is currently an instructor in forensic chemical instrumentation, scientific and expert testimony, electron microscopy, and advanced trace evidence analysis at John Jay College of Criminal Justice, CUNY, as well as being the director of TAKA Instructional Agency, a New York State not-for-profit educational institution concentrating on training in microscopy. He holds a master's degree in chemistry from Long Island University and is currently a doctoral candidate in forensic science at CUNY. He has been awarded the Paul L. Kirk Award (1997) for contributions to criminalistics and forensic science by the Criminalistics Section of the American Academy of Forensic Sciences, as well as the Arthur Neiderhoffer Scholarship for significant contributions to the field of criminal justice by the Graduate School City University of New York. Kubic is a member of the FBI-sponsored SWGMAT committee on forensic glass analysis, is a technical expert with NIST's laboratory accreditation program (NVLAP), and has been the laboratory director of a New York State accredited environmental laboratory. He is the author of 20 scientific and technical articles as well as a number of textbook chapters, and has made over 50 presentations at technical conferences. He is a fellow of the American Academy of Forensic Sciences and holds fellow status as a certified criminalist. Kubic is a member of the New York Microscopical Society, the Microscopy Society of America, the Microbeam Analysis Society, and the American Chemical Society, among other professional organizations.

Marilyn T. Miller is a graduate of Florida Southern College with a bachelor's degree in chemistry. She earned a master's degree in forensic chemistry from the University of Pittsburgh and a doctorate in education from Johnson & Wales University. She is currently an assistant professor in the Forensic Science Program and the Wilder School of Government and Public Affairs at Virginia Commonwealth University in Richmond, Virginia. Dr. Miller is a full member of the Criminalistics Section of the American Academy of Forensic Science, the Southern Association of Forensic Scientists, and the American Chemical Society. As a postsecondary educator she teaches a wide variety of forensic science and crime scene investigation classes to both forensic science and criminal justice majors at the undergraduate and graduate levels. She is a fellow of the Henry Lee Institute of Forensic Science and the National

Crime Scene Training Center. She has presented and taught as part of hundreds of forensic seminars across the United States. For over 15 years Dr. Miller worked as a supervisor and forensic scientist for law enforcement agencies in North Carolina, Pennsylvania, and Florida. She has testified over 350 times in county, state, and federal courts of law as an expert witness in the field of forensic sciences and crime scene reconstruction. Dr. Miller designed, opened, and operated forensic laboratories on the west coast of Florida. She has participated in hundreds of crime scene investigations, both as an active investigator and recently as a consultant for both state and defense attorneys. In addition to expertise in forensic science, other areas of emphasis are scientific crime scene investigation and reconstruction and bloodstain pattern analysis. Dr. Miller relates that her biggest accomplishment has been to prepare objective, ethical, and eager forensic science and criminal justice students for the use of forensic science in criminal investigations.

Michael R. Napier is currently a vice president and violent crime consultant with The Academy Group, a forensic behavioral science firm in Manassas, Virginia. He served in the FBI for over 27 years where he was trained as a polygraph examiner at the Department of Defense Polygraph Institute and was affiliated with its Behavioral Science Unit, as well as its operational successor, the National Center for the Analysis of Violent Crime. Napier served as a crisis-hostage negotiator and was deployed to assist in the resolution of international negotiations as a member of the FBI's elite Critical Incident Negotiating Team. He has trained law enforcement and other professionals in domestic and international venues, and was a field office supervisor in Los Angeles and Oklahoma City. Upon retirement, Napier served as a supervisory special agent in the FBI's Critical Incident Response Group at the FBI Academy, Quantico, Virginia.

Linda R. Netzel received a B.S. in chemistry with emphasis on criminalistics from Metropolitan State College of Denver in 1991. She has 15 years of forensic experience and is currently the director of the Kansas City Police Crime Laboratory. Prior to becoming the director she was a criminalist in the DNA and trace evidence sections of the laboratory. Additionally, Netzel has extensive homicide crime scene experience and instructs crime scene investigators on crime scene reconstruction and physical evidence collection and preservation. She has also provided instruction to hundreds of medical personnel, detectives, and attorneys regarding aspects of physical evidence that pertain to their respective fields. Netzel is a member of the American Society of Crime Laboratory Directors, the American Academy of Forensic Sciences, and the Midwestern Association of Forensic Scientists, and she is a diplomat with the American Board of Criminalistics.

Randall K. Noon is a licensed, professional engineer with Noon Consulting, Hiawatha, Kansas, working both in the United States and Canada. He has been investigating various types of failures related to construction for more than 25 years. Noon has authored the following forensic engineering text books published by CRC Press: Introduction to Forensic Engineering, Engineering Analysis of Fires and Explosions, Engineering Analysis of Vehicular Accidents, and, most recently, Forensic Engineering Investigations.

Frank H. Norwitch received his B.A. in general studies from the University of Miami. He worked at the Miami-Dade Police Department Crime Laboratory for 27 years, 17 of which he spent as the senior document examiner. Dating from that time to the present, he has maintained the Norwitch Document Laboratory (formerly Associated Forensic Services), a private document laboratory in West Palm Beach, Florida, specializing in civil and, occasionally, criminal document cases. In addition to specialized questioned document education received at the Miami-Dade Crime Laboratory, he received questioned document training from the FBI, U.S. Secret Service, U.S. Postal Service Laboratory, Rochester Institute of Technology, and the Florida Department of Law Enforcement—agencies that he assisted on numerous cases. Norwitch has examined over 5000 cases involving over 100,000 documents and testified as an expert witness over 300 times before local, state, federal, and military courts of the United States, U.S. Virgin Islands, the Bahamas, and the Cayman Islands. He is a former faculty member of the National College of District Attorneys and a certified instructor of forensic sciences by the State of Florida. He has taught numerous law enforcement seminars and lectured at civic, municipal, and security organization meetings. He is the author of articles pertaining to questioned document examination.

Nicholas Petraco is a lecturer in forensic science at John Jay College of CUNY. He earned his B.S. in chemistry and a M.S. in forensic science from John Jay College of Criminal Justice. Petraco is retired from the New York City Police Crime Laboratory where he served as a trace evidence analyst and forensic microscopist for almost 20 years. He currently is a forensic consultant to the City of New York Police Department's Forensic Investigative Service as well as other various city, state, and federal agencies, and he has testified to his determinations numerous times in criminal and civil trials. He is the past chairman of the FBI-sponsored SWGMAT forensic hair analysis committee. His current research interests are in the application of light microscopy to the analysis of soils, fibers, hairs, and art objects. Petraco has authored or coauthored over 50 scientific articles in journals and textbook chapters. He is a certified teacher, a certified criminalist, and a fellow of both the American Academy of Forensic Sciences and the New York Microscopical Society.

David R. Redsicker has been with Peter Vallas Associates, Inc., since 1983 and is the corporate director of Investigations working in the Endicott, New York, office. He has 31 years of origin and cause experience and has acted as a court-qualified forensic expert in both civil and criminal cases. He is a member of the New York State Chapter of the International Association of Arson Investigators, the International Association of Bloodstain Pattern Analysts, and the New York State Academy of Fire Science Student/Faculty Association, and is a past chairman of the Police Advisory Committee, International Association of Arson Investigators. Redsicker is a published author in the field of forensic science, including fire and crime investigation and forensic photography.

Walter F. Rowe, Ph.D., graduated from Emory University in 1967 with a B.S. (honors) in chemistry and went on to receive his M.A. in chemistry from Harvard University in 1968. Further graduate study was interrupted by 2 years of military service in the U.S. Army Criminal Investigation Laboratories at Ft. Gordon, Georgia, and Frankfurt-am-Main, during which time he served as a forensic chemistry specialist, was trained as a forensic drug chemist and a forensic serologist, received credentials as a U.S. Army criminal investigator, and participated in the investigation of crimes, including the notorious Jeffrey McDonald Case at Ft. Bragg, North Carolina, in 1970. Dr. Rowe graduated with honors from the U.S. Army Military Police School's criminal investigator course. After completing his military service in 1971, he returned to Harvard, where in 1976 he received his Ph.D. degree in chemistry. In 1975, he joined the faculty of the Department of Forensic Sciences at George Washington University in Washington, D.C. where he is now full professor of forensic sciences. Dr. Rowe has published over 50 articles in peer-reviewed scientific journals, as well as four book chapters, a number of which are devoted to the field of firearms examinations. He is a fellow of the Criminalistics Section, American Academy of Forensic Sciences, a former member of the editorial board of the Journal of Forensic Sciences, a member of Committee E30 of the American Society for Testing and Materials, and a member and former chair of the Council of Forensic Educators.

Robert L. Sadoff, M.D., is currently clinical professor of psychiatry, director of the Center for Studies in Social–Legal Psychiatry, and director of the Forensic Psychiatry Clinic at the University of Pennsylvania in Philadelphia. He received his M.D. from the University of Minnesota in 1959 and took his residency in psychiatry at UCLA Neuropsychiatric Institute where he received an M.S. in psychiatry in 1963. Dr. Sadoff is board-certified in psychiatry, forensic psychiatry, legal medicine, and additional qualifications in forensic psychiatry. He is the author of over 90 articles in medical and legal journals; a contributor of over 30 chapters in edited volumes; and has authored, edited, or coauthored six books including Forensic Psychiatry: A Practical Guide for Lawyers and Psychiatrists and Psychiatric Malpractice: Cases and Comments for Clinicians. During the past 40 years, Dr. Sadoff has examined over 9000 individuals charged with crimes and has testified for both prosecution and defense in criminal cases throughout the U.S. He has served as a consultant to the Norristown State Hospital, the Trenton Psychiatric Hospital, the Harrisburg State Hospital, the Forensic Psychiatric Hospital of New Jersey, and the Philadelphia Prison System.

Louis B. Schlesinger, Ph.D., is professor of psychology at John Jay College of Criminal Justice, City University of New York. He is a diplomate in forensic psychology of the American Board of Professional Psychology and served as president of the New Jersey Psychological Association in 1989. He was a member of the Council of Representatives of the American Psychological Association from 1991 to 1994. Dr. Schlesinger was the 1990 recipient of the New Jersey Psychological Association's "Psychologist of the Year" award, as well as recipient of the American Psychological Association's Karl F. Heiser Presidential Award (1993). He has testified in numerous forensic cases and has published articles, chapters, and eight books on the topic of forensic psychology, homicide, and criminal psychopathology.

Howard Seiden received a B.A. in chemistry and a B.S. in biology from Florida International University, as well as a master's in biological sciences from Florida Atlantic University. He has been with the Broward County Sheriff's Office Crime Laboratory in Fort Lauderdale for 25 years, during which time he has worked with many law enforcement organizations, including local municipalities, the FBI, Department of Transportation, Internal Revenue Service, U.S. Customs Service, State of Florida Office of the Auditor General, and State of Florida Office of the Attorney General/Medicaid Fraud. Seiden entered the field of document examinations in 1991 and received his training at the Miami-Dade Police Crime Laboratory. He is certified by the American Board of Forensic Document Examiners and is a member of the Southeastern Association of Document Examiners and the American Society of Testing and Materials's Subcommittee on Questioned Documents. Seiden is published in the field of questioned documents and has presented numerous training seminars to investigators from various governmental agencies.

Monica M. Sloan is a forensic biologist with the Biology Section, Centre of Forensic Sciences, in Toronto, Canada. After graduating with a B.S. (honors) from Queen's University, she joined the Centre of Forensic Sciences in 1991. Since 1992, she has provided expert testimony in over 50 cases both in Ontario and internationally in the areas of body fluid analysis and DNA typing. Sloan lectures frequently on forensic topics to many of the center's client groups. She is a member of the Canadian Society of Forensic Science and has been its biology section chair since 1997.

Marcella H. Sorg, RN, Ph.D., D-ABFA, has served as the consulting forensic anthropologist for Maine since 1977 and New Hampshire since 1980. She has been an associated faculty member of the University of Maine's Department of Anthropology in Orono,

Maine, since 1977, the School of Nursing since 1997, and is currently research associate at the University's Margaret Chase Smith Center for Public Policy. Dr. Sorg received her Ph.D. in physical anthropology from Ohio State University in 1979, and was certified by the American Board of Forensic Anthropology in 1984. She subsequently served as its secretary and later its president. A fellow of the American Academy of Forensic Sciences, Dr. Sorg has authored and coedited many books and articles on forensic anthropology and taphonomy, including Bone Modification (1987), Advances in Forensic Taphonomy: Method, Theory, and Archaeological Perspectives (CRC Press, 2002), Forensic Taphonomy: The Postmortem Fate of Human Remains (CRC Press, 1997), and Cadaver Dog Handbook: Forensic Training and Tactics for the Recovery of Human Remains (CRC Press, 2000). Her current research focuses on taphonomic approaches to remains exposed in both marine and terrestrial environments and the estimation of postmortem interval.

Robert P. Spalding, M.S., served in the Federal Bureau of Investigation for 28 years, retiring in January 1999. He performed duties as an investigative special agent in Cleveland and Washington, D.C., before being assigned to the Laboratory Division Serology Unit in 1973. As a laboratory examiner in serology and bloodstain pattern analysis, he performed casework examinations and presented expert testimony in criminal matters in United States and foreign courts on nearly 200 occasions. For 14 years, Mr. Spalding taught forensic serology at the Forensic Science Research and Training Center of the FBI Academy, available for graduate credit through the College of Continuing Education at the University of Virginia. In 1993, he was assigned to the Evidence Response Team (ERT) Unit where he taught crime scene investigation and bloodstain pattern analysis to FBI field office personnel. Mr. Spalding received his bachelor's and master's of science degrees in biochemistry from the University of Maine in 1965 and 1968, respectively. He is currently a member of the FBI's Scientific Working Group on Bloodstain Pattern Analysis (SWGSTAIN), and several professional organizations, and serves as a regional vice president of the International Association of Bloodstain Pattern Analysts. He is the owner of Spalding Forensics, LLC, a consulting and training firm specializing in casework involving bloodstain patterns and crime scene reconstruction.

Mary K. Sullivan, MSN, RNC, CARN, has worked for the Veterans Health Administration (VHA) for 19 years with most of her clinical background in psychiatric/mental health nursing. She is currently practicing as an outpatient mental health nurse for the Carl T. Hayden VA Medical Center in Phoenix, AZ, while also managing her collateral duties as a clinical forensic nurse. At this time these include teaching and consulting with nursing colleagues at the local and national levels within VHA on forensic nursing and its application to the clinical setting. She is responsible for pioneering the forensic nursing initiative currently in progress in VHA as well as being detailed out on special assignments to the Office of Inspector General on investigations of suspicious deaths and adverse patient events in VA medical centers. Ms. Sullivan was awarded the Veterans Health Administration Secretary's Award for Nursing Excellence in 2003, the Creighton University School of Nursing Alumni Merit award for 2004, and the International Association of Forensic Nursing's 2004 Virginia A. Lynch Pioneer Award, forensic nursing's top honor. She is an active member of the America Academy of Forensic Sciences, the International Association of Forensic Nurses, Sigma Theta Tau, and the American Psychiatric Nurses Association.

T. Paulette Sutton is currently a consulting bloodstain pattern analyst and associate professor of clinical laboratory sciences at the University of Tennessee Health Science Center in Memphis. She holds a B.S. in medical technology from the University of Tennessee and a M.S. in operations management—engineering from the University of Arkansas. She is retired from the Shelby County Medical Examiner's Office (Memphis). Ms. Sutton has given expert testimony in criminal as well as civil cases, in several state and federal courts for both prosecution and defense. She has taught courses and lectured extensively on bloodstain pattern analysis, and has received the Lecturer of Merit and the Distinguished Faculty awards from the National College of District Attorneys. She has contributed to the texts Interpretation of Bloodstain Evidence at Crime Scenes, Second Edition (CRC Press, 1998); Scientific and Legal Applications of Bloodstain Pattern Interpretation (CRC Press, 1998); and Forensic Science: An Introduction to Scientific and Investigative Techniques (CRC Press, 2002), and has authored the didactic and laboratory exercise manual Bloodstain Pattern Analysis in Violent Crimes. She also coauthored the text Principles of Bloodstain Pattern Analysis—Theory and Practice (CRC Press 2005). Ms. Sutton is a member of the FBI's Scientific Working Group on Bloodstain Pattern Analysis (SWGSTAIN), the International Association for Identification, and the International Association of Bloodstain Pattern Analysts.

Ronald K. Wright, M.D., J.D., is a forensic pathologist in private practice. He has a B.S. in biology and chemistry from Southwest Missouri University, an M.D. from Saint Louis University School of Medicine, and a J.D. from the University of Miami School of Law. He has practiced as a forensic pathologist and toxicologist for 30 years, holding positions of deputy chief medical examiner of the State of Vermont and Miami-Dade County, Florida, and chief medical examiner of Broward County, Florida.

Index